Student Solutions Manual
Laurel Technical Services

Intermediate
Algebra for College Students
Fifth Edition

Allen R. Angel

PRENTICE HALL, Upper Saddle River, NJ 07458

Executive Editor: Karin Wagner
Supplement Editor: Kate Marks
Special Projects Manager: Barbara A. Murray
Production Editor: Wendy Rivers
Supplement Cover Manager: Paul Gourhan
Supplement Cover Designer: Liz Nemeth
Manufacturing Buyer: Alan Fischer

Printed in the United States of America

10 9 8 7 6 5 4 3 2 1

ISBN 0-13-040245-1

Prentice-Hall International (UK) Limited, London
Prentice-Hall of Australia Pty. Limited, Sydney
Prentice-Hall Canada, Inc., Toronto
Prentice-Hall Hispanoamericana, S.A., Mexico
Prentice-Hall of India Private Limited, New Delhi
Prentice-Hall (Singapore) Pte. Ltd.
Prentice-Hall of Japan, Inc., Tokyo
Editora Prentice-Hall do Brazil, Ltda., Rio de Janeiro

Table of Contents

Chapter 1

Exercise Set 1.1

1–9. Answers will vary.

11. Answers will vary.

13. Do all the homework and preview the new material to be covered in class.

15. **(1)** Carefully write down any formulas or ideas that you need to remember.

(2) Look over the entire exam quickly to get an idea of its length and to make sure that no pages are missing. You will need to pace yourself to make sure that you complete the entire exam. Be prepared to spend more time on problems worth more points.

(3) Read the test directions carefully.

(4) Read each problem carefully. Answer each question completely and make sure you have answered the specific question asked.

(5) Starting with number 1, work each question in order. If you come across a problem that you are not sure of, do not spend too much time on it. Continue working the problems that you understand. After completing all other questions, come back to finish those questions you are not sure of. Do not spend too much time on any one question.

(6) Attempt each problem. You may be able to earn at least partial credit.

(7) Work carefully and write clearly so that your instructor can read your work. Also, it is easy to make mistakes when your writing is unclear.

(8) Check your work and your answers if you have time.

(9) Do not be concerned if others finish the test before you. Do not be disturbed if you are the last to finish. Use all your extra time to check your work.

17. The more you put into the course, the more you will get out of it.

19. Answers will vary.

Exercise Set 1.2

1. A variable is a letter used to represent various numbers.

3. A set is a collection of objects.

5. The set of natural or counting numbers is an infinite set. One can count, 1, 2, 3, …, infinitely high. Also, there is no largest element.

7. The 5 inequality symbols are:
$>$, is greater than
$\geq$, is greater than or equal to
$<$, is less than
$\leq$, is less than or equal to
$\neq$, is not equal to

9. $\{5, 6, 7, 8\}$

11. Every integer is also a rational number because it can be written with a denominator of 1.

13. True; $2 = \dfrac{2}{1}$ is both a rational number and an integer.

15. False; 0 is a whole number but not a natural number.

17. False; $\dfrac{3}{2}$ is not an integer.

19. True; this is how the rational and irrational numbers are defined.

21. True; there are no integers between 1 and 2.

23. $6 < 8$

25. $0 < 4$

27. $-3 > -3.5$

29. $-5 < -3$

31. $-4.6 > -4.7$

33. $1.1 < 1.9$

35. $-952 > -955$

37. $-\dfrac{7}{8} > -\dfrac{8}{9}$

39. $A = \{6\}$

41. $C = \{6, 8\}$

43. $E = \{0, 1, 2, 3, 4, 5, 6\}$

45. $H = \{0, 5, 10, 15, \ldots\}$

47. $J = \{-5\}$

49. a. 4 is a natural number.

    b. 4 and 0 are whole numbers.

    c. -3, 4 and 0 are integers.

    d. -3, 4, $\dfrac{1}{2}$, $\dfrac{5}{9}$, 0, -1.23 and $\dfrac{99}{100}$ are rational numbers.

    e. $\sqrt{2}$ and $\sqrt{8}$ are irrational numbers.

    f. -3, 4, $\dfrac{1}{2}$, $\dfrac{5}{9}$, 0, $\sqrt{2}$, $\sqrt{8}$, -1.23 and $\dfrac{99}{100}$ are real numbers.

51. $A \cup B = \{5, 6, 7, 8\}$
$A \cap B = \{6, 7\}$

53. $A \cup B = \{-1, -2, -4, -5, -6\}$
$A \cap B = \{-2, -4\}$

55. $A \cup B = \{0, 1, 2, 3\}$
$A \cap B = \{\ \}$ or $\varnothing$

57. $A \cup B = \{0, 1, 2, 3, 4, 5, 6, 7, 8\}$
$A \cap B = \{\ \}$ or $\varnothing$

59. $A \cup B = \{0.1, 0.2, 0.3, 0.4, \ldots\}$
$A \cap B = \{0.2, 0.3\}$

61. The set of natural numbers.

63. The set of even natural numbers greater than or equal to 8 and less than or equal to 30.

65. The set of odd integers.

67. a. Set A is the set of all x such that x is a natural number less than 8.

    b. $A = \{1, 2, 3, 4, 5,, 6, 7\}$

69. $\{x | x < 3\}$

71. $\{x | x \geq 6\}$

73. $\left\{x | -4 < x \leq \dfrac{3}{7}\right\}$

75. $\{x | x > 2 \text{ and } x \in N\}$

77. $\left\{x | x < \dfrac{40}{9} \text{ and } x \in N\right\}$

79. $\{x|x > -4\}$

81. $\{x|x \geq -3 \text{ and } x \in I\}$ or $\{x|x > -4 \text{ and } x \in I\}$

83. $\{x|-2 < x \leq 4.6\}$

85. $\left\{x|x \leq \dfrac{37}{4}\right\}$

87. $\{x|-1 \leq x \leq 3 \text{ and } x \in I\}$

89. Yes; the set of natural numbers is a subset of the set of whole numbers

91. Yes; the set of integers is a subset of the set of rational numbers.

93. No; the set of rational numbers is not a subset of the set of irrational numbers.

95. Yes; the set of irrational numbers is a subset of the set of real numbers.

97. Answers may vary.
Possible answer: {0.1, 0.2, 0.3, 0.4, 05}

99. Answers may vary.
Possible answer: $A = \{3, 5, 7, 8, 9\}$, $B = \{5, 7\}$
Therefore, $A \cup B = \{3, 5, 7, 8, 9\}$ and
$A \cap B = \{5, 7\}$

101. a. The set of brands that are in either category is: {Levi's Lands' End, L.L. Bean, Nike, Gold Toe, London Fog, Reebok, Hanes, Hanes Her Way, Arizona}

b. The set of brands that are in both categories is: {Levi's, L.L. Bean, Reebok}

c. Part a) represents the union because it asks for the brands in either category.

d. Part b) represents the intersection because it asks for the brands that are common to both categories.

103. a. The set of goods and services that have a weight of 17% or greater is: {Housing, Food & drinks, Transportation}

b. The set of goods and services that have a weight of less than 6% is: {Apparel & upkeep, Entertainment}

105. a. $A = \{1, 3, 4, 5, 6, 7\}$

b. $B = \{2, 3, 4, 6, 8\}$

c. $A \cup B = \{1, 2, 3, 4, 5, 6, 7, 8\}$

d. $A \cap B = \{3, 4, 6\}$

107. a. $\{x|x > 1\}$ includes fractions and decimal numbers which the other set does not contain.

b. $\{2, 3, 4, 5, \dots\}$

c. No, since it is not possible to list all real numbers greater than 1 in roster form.

109. a. $\dfrac{1}{9} = 0.111\dots$ so $\dfrac{1}{9} = 0.\overline{1}$

$\dfrac{2}{9} = 0.222\dots$ so $\dfrac{2}{9} = 0.\overline{2}$

$\dfrac{3}{9} = 0.333\dots$ so $\dfrac{3}{9} = 0.\overline{3}$

b. $\dfrac{4}{9} = 0.444\dots$ so $\dfrac{4}{9} = 0.\overline{4}$

$\dfrac{5}{9} = 0.555\dots$ so $\dfrac{5}{9} = 0.\overline{5}$

$\dfrac{6}{9} = 0.666\dots$ so $\dfrac{6}{9}$ or $\dfrac{2}{3} = 0.\overline{6}$

c. Based on (a) and (b), we deduce that
$0.\overline{9} = \dfrac{9}{9} = 1$.

Exercise Set 1.3

1. Additive inverses or opposites are two numbers that sum to 0.

3. $|a| = \begin{cases} a & \text{if} \quad a \geq 0 \\ -a & \text{if} \quad a < 0 \end{cases}$

5. Since a and $-a$ are the same distance from 0 on a number line, $|a| = |-a|$, is true for all real numbers, $\mathbb{R}$.

7. Since $|a| = \begin{cases} a, & a \geq 0 \\ -a, & a \leq 0 \end{cases}$, then $|a| = -a$ only when $a \leq 0$.

9. Since $|5| = 5$ and $|-5| = 5$, the desired values for a are 5 and -5.

15. $-\dfrac{a}{b}$ or $\dfrac{-a}{b}$

17. **a.** $(ab)c = a(bc)$

19. In general, $a + (b \cdot c) \neq (a+b) \cdot (a+c)$. To see this, consider $2 + (3 \cdot 4)$ and $(2+3) \cdot (2+4)$.
The left side is $2 + (3 \cdot 4) = 2 + 12 = 14$ and the right side is $(2+3) \cdot (2+4) = 5 \cdot 6 = 30$.

21. $|3| = 3$

23. $|-6| = 6$

25. $\left|-\dfrac{3}{4}\right| = \dfrac{3}{4}$

27. $|0| = 0$

29. $-|-7| = -7$

31. $-\left|\dfrac{5}{9}\right| = -\dfrac{5}{9}$

33. $|6| = 6$
$|-6| = 6$
$|6| = |-6|$

35. $-4 = -4$
$|-4| = 4$
$-4 < |-4|$

37. $-4 = -4$
$-|4| = -4$
$-4 = -|4|$

39. $|-7| = 7$
$-|3| = -3$
$|-7| > -|3|$

41. $-|-10| = -10$
$|-5| = 5$
$-|-10| < |-5|$

43. $|19| = 19$
$|-25| = 25$
$|19| < |-25|$

45. $-1, 2, |3|, |-5|, 6$

47. $-32, -|4|, 4, |-7|, 15$

49. $-|2.9|, -2.4, -2.1, -2, |-2.8|$

51. $-2, \dfrac{1}{3}, \left|-\dfrac{1}{2}\right|, \left|\dfrac{3}{5}\right|, \left|-\dfrac{3}{4}\right|$

53. $4 + (-3) = 1$

55. $-18 + 7 = -11$

57. $-14 - (-11) = -14 + 11 = -3$

59. $-9.5 - (-3.72) = -9.5 + 3.72 = -5.78$

61. $-\dfrac{3}{5} - \left(-\dfrac{2}{9}\right) = -\dfrac{3}{5} + \dfrac{2}{9}$
$= -\dfrac{27}{45} + \dfrac{10}{45}$
$= \dfrac{-27 + 10}{45}$
$= \dfrac{-17}{45}$ or $-\dfrac{17}{45}$

63. $5 + (-0.43) - 6.97 = 4.57 - 6.97$
$$= -2.40 \text{ or } -2.4$$

65. $8.9 - |8.5| - |17.6| = 8.9 - 8.5 - 17.6$
$$= 0.4 - 17.6$$
$$= -17.2$$

67. $|5 - 12| - |3| = |-7| - |3|$
$$= 7 - 3$$
$$= 4$$

69. $-|-3| - |7| + \left(6 + |-2|\right) = -|-3| - |7| + (6 + 2)$
$$= -|-3| - |7| + 8$$
$$= -3 - 7 + 8$$
$$= -10 + 8$$
$$= -2$$

71. $\left(\dfrac{3}{5} + \dfrac{3}{4}\right) - \dfrac{1}{2} = \left(\dfrac{12}{20} + \dfrac{15}{20}\right) - \dfrac{1}{2}$
$$= \left(\dfrac{12 + 15}{20}\right) - \dfrac{1}{2}$$
$$= \dfrac{27}{20} - \dfrac{1}{2}$$
$$= \dfrac{27}{20} - \dfrac{10}{20}$$
$$= \dfrac{27 - 10}{20}$$
$$= \dfrac{17}{20}$$

73. $-4 \cdot 12 = -48$

75. $-4\left(-\dfrac{5}{16}\right) = \left(-\dfrac{4}{1}\right)\left(-\dfrac{5}{16}\right)$
$$= -\dfrac{4}{1} \cdot \dfrac{5}{16}$$
$$= \dfrac{(-4)(-5)}{(1)(16)}$$
$$= \dfrac{20}{16}$$
$$= \dfrac{5}{4}$$

77. $(-1)(-2)(-1)(2)(-3) = 2(-1)(2)(-3)$
$$= -2(2)(-3)$$
$$= -4(-3)$$
$$= 12$$

79. $(-1.1)(3.4)(8.3)(-7.6)$
$$= -3.74(8.3)(-7.6)$$
$$= -31.042(-7.6)$$
$$= 235.9192$$

81. $-80 \div (-8) = \dfrac{-80}{-8}$
$$= 10$$

83. $-\dfrac{5}{9} \div \dfrac{-5}{9} = \dfrac{-5}{9} \cdot \dfrac{9}{-5}$
$$= \dfrac{-45}{-45}$$
$$= 1$$

85. $\left|\dfrac{-4}{7}\right| \div \dfrac{1}{14} = \dfrac{4}{7} \div \dfrac{1}{14}$
$$= \dfrac{4}{7} \cdot \dfrac{14}{1}$$
$$= \dfrac{4 \cdot 14}{7 \cdot 1}$$
$$= \dfrac{56}{7}$$
$$= 8$$

87. $\left|-\dfrac{5}{6}\right| \div \left|\dfrac{-1}{2}\right| = \dfrac{5}{6} \div \dfrac{1}{2}$
$$= \dfrac{5}{6} \cdot \dfrac{2}{1}$$
$$= \dfrac{5 \cdot 2}{6 \cdot 1}$$
$$= \dfrac{10}{6}$$
$$= \dfrac{5}{3}$$

89. $4 - 7 = 4 + (-7)$
$$= -3$$

91. $-20 \div (-4) = \dfrac{-20}{-4}$
$$= 5$$

93. $4\left(-\dfrac{8}{5}\right)\left(\dfrac{5}{2}\right) = -\dfrac{32}{5}\left(\dfrac{5}{2}\right)$

$$= -\dfrac{160}{10}$$
$$= -16$$

95. $(-2.1)(-22.3)(-8.6) = 46.83(-8.6)$
$$= -402.738$$

97. $-16.4 - (-9.6) - 14.8 = -16.4 + 9.6 - 14.8$
$$= -6.8 - 14.8$$
$$= -21.6$$

99. $-|8| \cdot \left|\dfrac{-1}{2}\right| = -(8)\left(\dfrac{1}{2}\right)$

$$= -4$$

101. $\left|-\dfrac{9}{4}\right| \div \left|-\dfrac{4}{9}\right| = \dfrac{9}{4} \div \dfrac{4}{9}$

$$= \dfrac{9}{4} \cdot \dfrac{9}{4}$$
$$= \dfrac{9 \cdot 9}{4 \cdot 4}$$
$$= \dfrac{81}{16}$$

103. $5 - |-7| + 3 - |-2| = 5 - 7 + 3 - 2$
$$= -2 + 3 - 2$$
$$= 1 - 2$$
$$= -1$$

105. $\left(-\dfrac{3}{5} - \dfrac{4}{9}\right) - \left(-\dfrac{2}{3}\right) = \left(-\dfrac{3}{5} - \dfrac{4}{9}\right) + \dfrac{2}{3}$

$$= \left(-\dfrac{27}{45} - \dfrac{20}{45}\right) + \dfrac{2}{3}$$
$$= \left(\dfrac{-27 - 20}{45}\right) + \dfrac{2}{3}$$
$$= -\dfrac{47}{45} + \dfrac{2}{3}$$
$$= -\dfrac{47}{45} + \dfrac{30}{45}$$
$$= \dfrac{-47 + 30}{45}$$
$$= \dfrac{-17}{45} \text{ or } -\dfrac{17}{45}$$

107. $(25 - |32|)(-6 - 5) = (25 - 32)(-6 - 5)$
$$= (-7)(-11)$$
$$= 77$$

109. $x + y = y + x$; Commutative property of addition

111. $x \cdot 0 = 0$; Multiplicative property of zero

113. $(x + 3) + 6 = x + (3 + 6)$; Associative property of addition

115. $x = 1 \cdot x$; Identity property of multiplication

117. $5(xy) = (5x)y$; Associative property of multiplication

119. $4(x + y + 2) = 4x + 4y + 8$; Distributive property

121. $5 + 0 = 5$; dentity property of addition

123. $3 + (-3) = 0$; Inverse property of addition

125. $x \cdot \dfrac{1}{x} = 1$; Inverse property of multiplication

127. $-(-x) = x$; Double negative property

129. -4 is the additive inverse
$\dfrac{1}{4}$ is the multiplicative inverse

131. $\dfrac{2}{3}$ is the additive inverse

$-\dfrac{3}{2}$ is the multiplicative inverse

133. The change in temperature is
$45° - (-4°) = 45° + 4° = 49°\,\text{F}$

135. Final depth is $-358.9 + 210.7 = -148.2$ ft or 148.2 ft below the starting point.

137. a. $47,600 - 60,000 = -12,400$
The author will owe $12,400 to the publisher.

b. $87,500 - 60,000 = 27,500$
The author will receive $27,500 from the publisher.

139. $4(3000) = 12,000$
$12,000 - 10,125 = 1875$

   a. She will receive a refund because she paid $12,000 and she only owed $10,125.

   b. She will receive $1875 as a refund.

143. a. $\quad NM - CT = \$3300 - (-\$2099)$
$\qquad\qquad\quad = \$3300 + \$2099$
$\qquad\qquad\quad = \$5399$

   b. $\quad OK - MI = \$1116 - (-\$1367)$
$\qquad\qquad\quad = \$1116 + \$1367$
$\qquad\qquad\quad = \$2483$

   c. $\quad VA - NJ = \$2695 - (-\$1883)$
$\qquad\qquad\quad = \$2695 + \$1883$
$\qquad\qquad\quad = \$4578$

   d. $\quad TX - CA = \$61 - (-\$197)$
$\qquad\qquad\quad = \$61 + \$197$
$\qquad\qquad\quad = \$258$

145. $\quad 1 - 2 + 3 - 4 + 5 - 6 + \ldots + 99 - 100$
Group in pairs:
$= (1-2) + (3-4) + (5-6) + \ldots + (99-100)$
Simplify all 50 pairs
$= -1 + (-1) + (-1) + \ldots + (-1)$
$= -1(50)$
$= -50$

147. $\dfrac{(1)|-2|(-3)|4|(-5)}{|-1|(-2)|-3|(4)|-5|} = \dfrac{(1)\cdot(2)\cdot(-3)\cdot(4)\cdot(-5)}{(1)\cdot(-2)\cdot(3)\cdot(4)\cdot(5)}$
$\qquad\qquad\qquad\qquad = \dfrac{120}{-120}$
$\qquad\qquad\qquad\qquad = -1$

149. True; the set of real numbers contains the irrational numbers.

150. The set of natural numbers is $\{1, 2, 3, 4, \ldots\}$

151. a. $\quad 3, 4, -2,$ and 0 are integers.

b. $\quad 3, 4, -2, \dfrac{5}{6},$ and 0 are rational numbers.

c. $\quad \sqrt{3}$ is an irrational number.

d. $\quad 3, 4, -2, \dfrac{5}{6}, \sqrt{3},$ and 0 are real numbers.

152. a. $\quad A \cup B = \{1, 4, 7, 9, 12, 15\}$

b. $\quad A \cap B = \{4, 7\}$

153. $\{x | -4 < x \le 6\}$

Exercise Set 1.4

1. a. In the expression a^n, a is called the base.

   b. In the expression a^n, n is called the exponent.

3. a. In the expression $\sqrt[n]{a}$, n is called the index.

   b. In the expression $\sqrt[n]{a}$, a is called the radicand.

5. The principal square root of a positive number radicand is the positive number whose square equals the radicand.

7. A positive number raised to an odd power is a positive number.

11. b. $\{5 - [4 - (3-8)]\}^2 = \{5 - [4 - (-5)]\}^2$
$\qquad\qquad\qquad\quad = [5 - (4+5)]^2$
$\qquad\qquad\qquad\quad = [5 - 9]^2$
$\qquad\qquad\qquad\quad = (-4)^2$
$\qquad\qquad\qquad\quad = 16$

13. $4^2 = 4 \cdot 4 = 16$

15. $(-5)^2 = (-5)(-5) = 25$

17. $-4^2 = -(4 \cdot 4) = -16$

19. $-\left(\dfrac{3}{5}\right)^4 = -\left(\dfrac{3}{5}\right)\left(\dfrac{3}{5}\right)\left(\dfrac{3}{5}\right)\left(\dfrac{3}{5}\right) = -\dfrac{81}{625}$

21. $-\left(-\dfrac{2}{3}\right)^4 = -\left(-\dfrac{2}{3}\right)\left(-\dfrac{2}{3}\right)\left(-\dfrac{2}{3}\right)\left(-\dfrac{2}{3}\right) = -\dfrac{16}{81}$

23. $-\sqrt{36} = -(6) = -6$ since $6 \cdot 6 = 36$.

25. $\sqrt[3]{-125} = -5$ since $(-5)(-5)(-5) = -125$

27. $\sqrt[3]{0.001} = 0.1$ since $(0.1)(0.1)(0.1) = 0.001$

29. $(0.42)^5 \approx 0.013$

31. $\left(\dfrac{5}{9}\right)^5 \approx 0.053$

33. $-(2.35)^{7.4} \approx -557.060$

35. $\sqrt[3]{5} \approx 1.710$

37. $\sqrt[5]{1246.5} \approx 4.160$

39. $-\sqrt[3]{\dfrac{20}{53}} \approx -0.723$

41. a. x^2 becomes $3^2 = 3 \cdot 3 = 9$

    b. $-x^2$ becomes $-3^2 = -3 \cdot 3 = -9$

43. a. x^2 becomes $1^2 = 1 \cdot 1 = 1$

    b. $-x^2$ becomes $-1^2 = -1 \cdot 1 = -1$

45. a. x^2 becomes $(-1)^2 = (-1)(-1) = 1$

    b. $-x^2$ becomes
$-(-1)^2 = -(-1)(-1) = -1$

47. a. x^2 becomes $\left(\dfrac{1}{3}\right)^2 = \left(\dfrac{1}{3}\right)\left(\dfrac{1}{3}\right) = \dfrac{1}{9}$

    b. $-x^2$ becomes $-\left(\dfrac{1}{3}\right)^2 = -\left(\dfrac{1}{3}\right)\left(\dfrac{1}{3}\right) = -\dfrac{1}{9}$

49. a. x^3 becomes $3^3 = 3 \cdot 3 \cdot 3 = 27$

    b. $-x^3$ becomes $-3^3 = -(3 \cdot 3 \cdot 3) = -27$

51. a. x^3 becomes
$(-3)^3 = (-3)(-3)(-3) = -27$

    b. $-x^3$ becomes
$-(-3)^3 = -(-3)(-3)(-3) = 27$

53. a. x^3 becomes
$(-2)^3 = (-2)(-2)(-2) = -8$

    b. $-x^3$ becomes
$-(-2)^3 = -(-2)(-2)(-2) = 8$

55. a. x^3 becomes $\left(\dfrac{2}{3}\right)^3 = \left(\dfrac{2}{3}\right)\left(\dfrac{2}{3}\right)\left(\dfrac{2}{3}\right) = \dfrac{8}{27}$

    b. $-x^3$ becomes
$-\left(\dfrac{2}{3}\right)^3 = -\left(\dfrac{2}{3}\right)\left(\dfrac{2}{3}\right)\left(\dfrac{2}{3}\right) = -\dfrac{8}{27}$

57. $\begin{aligned} 5^2 + 3^2 - 2^2 &= 25 + 9 - 4 \\ &= 34 - 4 \\ &= 30 \end{aligned}$

59. $\begin{aligned} -2^2 - 2^3 + 1^{10} + (-2)^3 &= -4 - 8 + 1 + (-8) \\ &= -4 - 8 + 1 - 8 \\ &= -12 + 1 - 8 \\ &= -11 - 8 \\ &= -19 \end{aligned}$

61. $\begin{aligned} (0.2)^2 &- (1.7)^2 - (3.2)^2 \\ &= 0.04 - 2.89 - 10.24 \\ &= -2.85 - 10.24 \\ &= -13.09 \end{aligned}$

63. $\left(-\dfrac{1}{2}\right)^3 - \left(\dfrac{1}{3}\right)^2 - \left(-\dfrac{2}{3}\right)^2 = -\dfrac{1}{8} - \dfrac{1}{9} - \dfrac{4}{9}$

$$= -\dfrac{9}{72} - \dfrac{8}{72} - \dfrac{32}{72}$$

$$= -\dfrac{17}{72} - \dfrac{32}{72}$$

$$= -\dfrac{49}{72}$$

65. $6 + 4 \cdot 5 = 6 + 20$
$$= 26$$

67. $20 - 6 \div 3 - 4 = 20 - 2 - 4 = 18 - 4 = 14$

69. $6 \div 3 + 5 \cdot \dfrac{3}{4} = 2 + \dfrac{15}{4} = \dfrac{8}{4} + \dfrac{15}{4} = \dfrac{23}{4}$

71. $\dfrac{3}{4} \div \dfrac{5}{6} + \dfrac{1}{2} \cdot \dfrac{9}{4} = \dfrac{3}{4} \cdot \dfrac{6}{5} + \dfrac{1}{2} \cdot \dfrac{9}{4}$

$$= \dfrac{18}{20} + \dfrac{9}{8}$$

$$= \dfrac{36}{40} + \dfrac{45}{40}$$

$$= \dfrac{81}{40}$$

73. $-2 + 5[3 - (2 - 4)] = -2 + 5[3 - (-2)]$
$$= -2 + 5(3 + 2)$$
$$= -2 + 5(5)$$
$$= -2 + 25$$
$$= 23$$

75. $3\left[(4 + 6)^2 - \sqrt[3]{8}\right] = 3\left[(10)^2 - \sqrt[3]{8}\right]$
$$= 3(100 - 2)$$
$$= 3(98)$$
$$= 294$$

77. $\{[(12 - 15) - 3] - 2\}^2 = \{[(-3) - 3] - 2\}^2$
$$= [(-6) - 2]^2$$
$$= (-8)^2$$
$$= 64$$

79. $4\left[5(13 - 3) \div (25 \div 5)^2\right]^2 = 4\left[5(10) \div (5)^2\right]^2$
$$= 4[50 \div 25]^2$$
$$= 4[2]^2$$
$$= 4 \cdot 4$$
$$= 16$$

81. $\dfrac{4 - (2 + 3)^2 - 6}{4(3 - 2) - 3^2} = \dfrac{4 - (5)^2 - 6}{4(1) - 3^2}$

$$= \dfrac{4 - 25 - 6}{4(1) - 9}$$

$$= \dfrac{4 - 25 - 6}{4 - 9}$$

$$= \dfrac{-21 - 6}{-5}$$

$$= \dfrac{-27}{-5}$$

$$= \dfrac{27}{5}$$

83. $\dfrac{8 + 4 \div 2 \cdot 3 + 4}{5^2 - 3^2 \cdot 2 - 7} = \dfrac{8 + 4 \div 2 \cdot 3 + 4}{25 - 9 \cdot 2 - 7}$

$$= \dfrac{8 + 2 \cdot 3 + 4}{25 - 18 - 7}$$

$$= \dfrac{8 + 6 + 4}{25 - 18 - 7}$$

$$= \dfrac{14 + 4}{7 - 7}$$

$$= \dfrac{18}{0} \text{ which is undefined}$$

85. $\dfrac{8 - [4 - (3 - 1)^2]}{5 - (-3)^2 + 4 \div 2} = \dfrac{8 - [4 - 2^2]}{5 - (-3)^2 + 4 \div 2}$

$$= \dfrac{8 - (4 - 4)}{5 - 9 + 4 \div 2}$$

$$= \dfrac{8 - 0}{5 - 9 + 4 \div 2}$$

$$= \dfrac{8 - 0}{5 - 9 + 2}$$

$$= \dfrac{8}{-4 + 2}$$

$$= \dfrac{8}{-2}$$

$$= -4$$

87. $-2|-3| - \sqrt{36} \div |2| + 3^2 = -2(3) - 6 \div 2 + 3^2$
$$= -2(3) - 6 \div 2 + 9$$
$$= -6 - 3 + 9$$
$$= -9 + 9$$
$$= 0$$

89. $\dfrac{6 - |-4| - 4|6-3|}{5 - 6 \cdot 2 \div |-6|} = \dfrac{6 - |-4| - 4|3|}{5 - 6 \cdot 2 \div 6}$
$$= \dfrac{6 - 4 - 4 \cdot 3}{5 - 6 \cdot 2 \div 6}$$
$$= \dfrac{6 - 4 - 12}{5 - 12 \div 6}$$
$$= \dfrac{2 - 12}{5 - 2}$$
$$= \dfrac{-10}{3} \text{ or } -\dfrac{10}{3}$$

91. $\dfrac{2}{5}\left[\sqrt[3]{27} - |-9| + 4 - 3^2\right]^2$
$$= \dfrac{2}{5}[3 - (9) + 4 - 9]^2$$
$$= \dfrac{2}{5}(3 - 9 + 4 - 9)^2$$
$$= \dfrac{2}{5}(-6 + 4 - 9)^2$$
$$= \dfrac{2}{5}(-2 - 9)^2$$
$$= \dfrac{2}{5}(-11)^2$$
$$= \dfrac{2}{5}(121)$$
$$= \dfrac{242}{5}$$

93. $\dfrac{24 - 5 - 4^2}{|-8| + 4 - 2(3)} + \dfrac{4 - (-3)^2 + |4|}{3^2 - 4 \cdot 3 + |-7|}$
$$= \dfrac{24 - 5 - 16}{8 + 4 - 2(3)} + \dfrac{4 - (9) + 4}{9 - 4 \cdot 3 + 7}$$
$$= \dfrac{24 - 5 - 16}{8 + 4 - 6} + \dfrac{4 - 9 + 4}{9 - 12 + 7}$$
$$= \dfrac{19 - 16}{12 - 6} + \dfrac{-5 + 4}{-3 + 7}$$
$$= \dfrac{3}{6} + \dfrac{-1}{4}$$
$$= \dfrac{2}{4} + \dfrac{-1}{4}$$
$$= \dfrac{2 - 1}{4}$$
$$= \dfrac{1}{4}$$

95. Substitute 1 for x:
$$-3x^2 - 4 = -3(1)^2 - 4$$
$$= -3(1) - 4$$
$$= -3 - 4$$
$$= -7$$

97. Substitute 5 for x:
$$-3x^2 + 6x + 5 = -3(5)^2 + 6(5) + 5$$
$$= -3(25) + 6(5) + 5$$
$$= -75 + 30 + 5$$
$$= -45 + 5$$
$$= -40$$

99. Substitute $-\dfrac{5}{6}$ for x:
$$4(x+1)^2 - 6x = 4\left(-\dfrac{5}{6} + 1\right)^2 - 6\left(-\dfrac{5}{6}\right)$$
$$= 4\left(\dfrac{1}{6}\right)^2 - 6\left(-\dfrac{5}{6}\right)$$
$$= 4\left(\dfrac{1}{36}\right) - 6\left(-\dfrac{5}{6}\right)$$
$$= \dfrac{1}{9} + 5$$
$$= \dfrac{1}{9} + \dfrac{45}{9}$$
$$= \dfrac{46}{9}$$

101. Substitute 1 for x and -3 for y:
$$
\begin{aligned}
6x^2 + 3y^2 - 5 &= 6(1)^2 + 3(-3)^3 - 5 \\
&= 6(1) + 3(-27) - 5 \\
&= 6 + (-81) - 5 \\
&= -75 - 5 \\
&= -80
\end{aligned}
$$

103. Substitute 4 for a and -1 for b:
$$
\begin{aligned}
3(a+b)^2 &+ 4(a+b) - 6 \\
&= 3[4 + (-1)]^2 + 4[4 + (-1)] - 6 \\
&= 3(3)^2 + 4(3) - 6 \\
&= 3(9) + 4(3) - 6 \\
&= 27 + 12 - 6 \\
&= 39 - 6 \\
&= 33
\end{aligned}
$$

105. Substitute 4 for x:
$$
\begin{aligned}
-6 &- \{x - [2x - (x-3)]\} \\
&= -6 - \{4 - [2 \cdot 4 - (4-3)]\} \\
&= -6 - \{4 - [2 \cdot 4 - 1]\} \\
&= -6 - \{4 - [8 - 1]\} \\
&= -6 - [4 - (7)] \\
&= -6 - (4 - 7) \\
&= -6 - (-3) \\
&= -6 + 3 \\
&= -3
\end{aligned}
$$

107. Substitute 6 for a, -11 for b, and 3 for c:
$$
\begin{aligned}
&\frac{-b + \sqrt{b^2 - 4ac}}{2a} \\
&= \frac{-(-11) + \sqrt{(-11)^2 - 4(6)(3)}}{2(6)} \\
&= \frac{11 + \sqrt{121 - 72}}{12} \\
&= \frac{11 + \sqrt{49}}{12} \\
&= \frac{11 + 7}{12} \\
&= \frac{18}{12} \\
&= \frac{3}{2}
\end{aligned}
$$

109. The expression is $(3x+6)^2$.
Now substitute 3 for x:
$$
\begin{aligned}
(3x+6)^2 &= (3 \cdot 3 + 6)^2 \\
&= (9+6)^2 \\
&= 15^2 \\
&= 225
\end{aligned}
$$

111. The expression is $6(3x+6) - 9$.
Now substitute 3 for x:
$$
\begin{aligned}
6(3x+6) - 9 &= 6(3 \cdot 3 + 6) - 9 \\
&= 6(9+6) - 9 \\
&= 6(15) - 9 \\
&= 90 - 9 \\
&= 81
\end{aligned}
$$

113. The expression is $\left(\dfrac{x+3}{2y}\right)^2 - 3$.
Now substitute 5 for x and 2 for y:
$$
\begin{aligned}
\left(\frac{x+3}{2y}\right)^2 - 3 &= \left(\frac{5+3}{2 \cdot 2}\right)^2 - 3 \\
&= \left(\frac{5+3}{4}\right)^2 - 3 \\
&= \left(\frac{8}{4}\right)^2 - 3 \\
&= 2^2 - 3 \\
&= 4 - 3 \\
&= 1
\end{aligned}
$$

115. a. 1945 means substitute 5 for x:
$$
\begin{aligned}
CO_2 &= 0.073x^2 - 0.39x + 0.55 \\
&= 0.073(5)^2 - 0.39(5) + 0.55 \\
&= 0.073(25) - 0.39(5) + 0.55 \\
&= 1.825 - 1.95 + 0.55 \\
&= 0.425
\end{aligned}
$$
There were approximately 0.425 million metric tons of CO_2 produced in 1945.

b. 1995 means substitute 10 for x:

$$CO_2 = 0.073x^2 - 0.39x + 0.55$$
$$= 0.073(10)^2 - 0.39(10) + 0.55$$
$$= 0.073(100) - 0.39(10) + 0.55$$
$$= 7.3 - 3.9 + 0.55$$
$$= 3.4 + 0.55$$
$$= 3.95$$

There were approximately 3.95 million metric tons of CO_2 produced in 1995.

117. **a.** 10-year-olds means substitute 10 for x:
percent of children

$$= 0.23x^2 - 1.98x + 4.42$$
$$= 0.23(10)^2 - 1.98(10) + 4.42$$
$$= 0.23(100) - 1.98(10) + 4.42$$
$$= 23 - 19.8 + 4.42$$
$$= 7.62$$

The percent of all 10-year-olds who are latchkey kids is 7.62%.

b. 14-year-olds mean substitute 14 for x.
percent of children

$$= 0.23x^2 - 1.98x + 4.42$$
$$= 0.23(14)^2 - 1.98(14) + 4.42$$
$$= 0.23(196) - 1.98(14) + 4.42$$
$$= 45.08 - 27.72 + 4.42$$
$$= 21.78$$

The percent of all 14-year-olds who are latchkey kids is 21.78%.

119. **a.** 1991 means substitute 1 for x:
speed

$$= 65.3x^3 - 350x^2 + 731x - 444$$
$$= 65.3(1)^3 - 350(1)^2 + 73(1) - 444$$
$$= 65.3 - 350 + 731 - 444$$
$$= 2.3$$

The speed of the fastest computer in 1991 was 2.3 billion operations per second.

b. 1999 means substitute 5 for x:
speed

$$= 65.3x^3 - 350x^2 + 731x - 444$$
$$= 65.3(5)^3 - 350(5)^2 + 731(5) - 444$$
$$= 65.3(125) - 350(25) + 731(5) - 444$$
$$= 8162.5 - 8750 + 3655 - 444$$
$$= 2623.5$$

The speed of the fast computer in 1999 is 2623.5 billion operations per second.

121. **a.** 1989 means substitute 7 for x:

$$\text{number} = 0.42x^2 - 3.44x + 5.80$$
$$= 0.42(7)^2 - 3.44(7) + 5.80$$
$$= 0.42(49) - 3.44(7) + 5.80$$
$$= 20.58 - 24.08 + 5.80$$
$$= 2.30$$

In 1989, 2.30 million people used cellular phones.

b. 1996 means substitute 14 for x:

$$\text{number} = 0.42x^2 - 3.44x + 5.80$$
$$= 0.42(14)^2 - 3.44(14) + 5.80$$
$$= 0.42(196) - 3.44(14) + 5.80$$
$$= 82.32 - 48.16 + 5.80$$
$$= 39.96$$

In 1996, 39.96 million people used cellular phones.

122. **a.** $A \cap B = \{b, c, f\}$

b. $A \cup B = \{a, b, c, d, f, g, h\}$

123. $|a| = |-a|$ for all real numbers or $\mathbb{R}$.

124. Since $|a| = \begin{cases} a & \geq 0 \\ -a & a < 0 \end{cases}$ then $|a| = a$ for $a \geq 0$.

125. $|a| = 4$ for $a = 4$ or $a = -4$ since $|4| = 4$ and $|-4| = 4$.

126. $-|6|, -4, -|-2|, 0, |-5|$

127. Associative property of addition

Exercise Set 1.5

1. a. $a^m \cdot a^n = a^{m+n}$

3. a. $a^{-m} = \dfrac{1}{a^m}, \; a \ne 0$

5. a. $(a^m)^n = a^{mn}$

7. a. $\left(\dfrac{a}{b}\right)^m = \dfrac{a^m}{b^m}, \; b \ne 0$

9. $x^{-1} = 5, \; \dfrac{1}{x} = 5, \; \dfrac{1}{x} = \dfrac{5}{1}$
Cross multiply: $5x = 1$
Solve for x: $x = \dfrac{1}{5}$

11. a. The opposite of x is $-x$.
The reciprocal of x is $\dfrac{1}{x}$.

    b. x^{-1} or $\dfrac{1}{x}$

    c. $-x$

13. $3^2 \cdot 3 = 3^2 \cdot 3^1$
$= 3^{2+1}$
$= 3^3$
$= 27$

15. $3^4 \cdot 3^0 = 3^{4+0}$
$= 3^4$
$= 81$

17. $\dfrac{4^3}{4} = 4^{3-1}$
$= 4^2$
$= 16$

19. $\dfrac{4^2}{4^0} = 4^{2-0}$
$= 4^2$
$= 16$

21. $4^{-2} = \dfrac{1}{4^2}$
$= \dfrac{1}{16}$

23. $-4^{-2} = -\dfrac{1}{4^2}$
$= -\dfrac{1}{16}$

25. $5^{-3} = \dfrac{1}{5^3}$
$= \dfrac{1}{125}$

27. $-5^{-3} = -\dfrac{1}{5^3}$
$= -\dfrac{1}{125}$

29. $5y^{-3} = \dfrac{5}{y^3}$

31. $\dfrac{1}{x^{-4}} = \dfrac{1}{\frac{1}{x^4}} = \dfrac{1}{1} \cdot \dfrac{x^4}{1} = x^4$ or $\dfrac{1}{x^{-4}} = x^4$

using the Helpful Hint following Example 3

33. $\dfrac{2a}{b^{-3}} = \dfrac{2a}{\frac{1}{b^3}} = \dfrac{2a}{1} \cdot \dfrac{b^3}{1} = 2ab^3$ or $\dfrac{2a}{b^{-3}} = 2ab^3$
(using the Helpful Hint)

35. $\dfrac{5m^{-2}n^{-3}}{2} = \dfrac{5}{2m^2n^3}$

37. $\dfrac{5x^{-2}y^{-3}}{z^{-4}} = \dfrac{5z^4}{x^2y^3}$

39. $\dfrac{6^{-1}x^{-1}}{y} = \dfrac{1}{6^1x^1y}$
$= \dfrac{1}{6xy}$

41. $x^0 = 1$

43. $-2x^0 = -2 \cdot x^0$
$$= -2(1)$$
$$= -2$$

45. $-(a+b)^0 = -(1)$
$$= -1$$

47. $3x^0 + 4y^0 = 3(1) + 4(1)$
$$= 3 + 4$$
$$= 7$$

49. $5^3 \cdot 5^{-4} = 5^{3+(-4)}$
$$= 5^{-1}$$
$$= \frac{1}{5}$$

51. $x^6 \cdot x^{-2} = x^{6+(-2)}$
$$= x^4$$

53. $\dfrac{3^4}{3^2} = 3^{4-2}$
$$= 3^2$$
$$= 9$$

55. $\dfrac{7^{-5}}{7^{-3}} = 7^{-5-(-3)} = 7^{-5+3} = 7^{-2} = \dfrac{1}{7^2} = \dfrac{1}{49}$
or, better yet
$$\dfrac{7^{-5}}{7^{-3}} = \dfrac{1}{7^{-3-(-5)}} = \dfrac{1}{7^{-3+5}} = \dfrac{1}{7^2} = \dfrac{1}{49}$$

57. $\dfrac{x^{-2}}{x} = x^{-2-1} = x^{-3} = \dfrac{1}{x^3}$
or, better yet
$$\dfrac{x^{-2}}{x} = \dfrac{1}{x^{1-(-2)}} = \dfrac{1}{x^{1+2}} = \dfrac{1}{x^3}$$

59. $\dfrac{5w^{-2}}{w^{-7}} = 5w^{-2-(-7)}$
$$= 5w^{-2+7}$$
$$= 5w^5$$

61. $2x^{-4} \cdot 6x^{-3} = 2 \cdot 6 \cdot x^{-4} \cdot x^{-3}$
$$= 12x^{-4+(-3)}$$
$$= 12x^{-7}$$
$$= \dfrac{12}{x^7}$$

63. $\left(-3p^{-2}\right)\left(-p^3\right) = (-3)(-1)p^{-2} \cdot p^3$
$$= 3p^{-2+3}$$
$$= 3p^1$$
$$= 3p$$

65. $\left(5r^2 s^{-2}\right)\left(-2r^5 s^2\right) = 5(-2)r^2 \cdot r^5 \cdot s^{-2} \cdot s^2$
$$= -10r^{2+5} \cdot s^{-2+2}$$
$$= -10r^7 s^0$$
$$= -10r^7$$

67. $\left(2x^4 y^7\right)\left(4x^3 y^{-5}\right) = 2 \cdot 4 \cdot x^4 \cdot x^3 \cdot y^7 \cdot y^{-5}$
$$= 8x^{4+3} y^{7+(-5)}$$
$$= 8x^7 y^2$$

69. $\dfrac{27x^5 y^{-4}}{9x^3 y^2} = \left(\dfrac{27}{9}\right)\dfrac{x^{5-3}}{y^{2-(-4)}}$
$$= \dfrac{3x^2}{y^6}$$

71. $\dfrac{9xy^{-4} z^3}{-3x^{-2} yz} = \left(\dfrac{9}{-3}\right)\dfrac{x^{1-(-2)} z^{3-1}}{y^{1-(-4)}}$
$$= -\dfrac{3x^3 z^2}{y^5}$$

73. $\left(2^2\right)^3 = 2^{2 \cdot 3}$
$$= 2^6$$
$$= 64$$

75. $\left(2^3\right)^{-2} = 2^{3(-2)}$

$\qquad = 2^{-6}$

$\qquad = \dfrac{1}{2^6}$

$\qquad = \dfrac{1}{64}$

77. $\left(x^{-4}\right)^{-2} = x^{-4(-2)}$

$\qquad\qquad = x^8$

79. $(-x)^3 = (-x)(-x)(-x)$

$\qquad\quad = -x^3$

81. $\left(-2x^{-2}\right)^3 = (-2)^3\left(x^{-2}\right)^3$

$\qquad\qquad = -8x^{-2(3)}$

$\qquad\qquad = -8x^{-6}$

$\qquad\qquad = -\dfrac{8}{x^6}$

83. $\left(\dfrac{3}{5}\right)^2 = \dfrac{3^2}{5^2}$

$\qquad\quad = \dfrac{9}{25}$

85. $\left(\dfrac{2}{7}\right)^{-2} = \dfrac{1}{\left(\dfrac{2}{7}\right)^2}$

$\qquad\qquad = \dfrac{1}{\dfrac{4}{49}}$

$\qquad\qquad = \dfrac{49}{4}$

or

$\left(\dfrac{2}{7}\right)^{-2} = \left(\dfrac{7}{2}\right)^2 \leftarrow$ Negative Exponent Rule

$\qquad\qquad = \dfrac{49}{4}$

87. $\left(\dfrac{2x}{3}\right)^{-2}$

$= \left(\dfrac{3}{2x}\right)^2 \leftarrow$ Negative Exponent Rule

$= \dfrac{3^2}{2^2 x^2}$

$= \dfrac{9}{4x^2}$

89. $\left(4x^2 y^{-2}\right)^2 = 4^2\left(x^2\right)^2\left(y^{-2}\right)^2$

$\qquad\qquad = 16x^{2\cdot2}y^{-2(2)}$

$\qquad\qquad = 16x^4 y^{-4}$

$\qquad\qquad = \dfrac{16x^4}{y^4}$

91. $\left(2a^3 b\right)^{-3} = \dfrac{1}{\left(2a^3 b\right)^3}$

$\qquad\qquad = \dfrac{1}{2^3\left(a^3\right)^3 b^3}$

$\qquad\qquad = \dfrac{1}{8a^{3\cdot3}b^3}$

$\qquad\qquad = \dfrac{1}{8a^9 b^3}$

93. $\left(-4x^{-3}y^5\right)^{-3} = (-4)^{-3}\left(x^{-3}\right)^{-3}\left(y^5\right)^{-3}$

$\qquad\qquad = (-4)^{-3}x^{(-3)(-3)}y^{5(-3)}$

$\qquad\qquad = (-4)^{-3}x^9 y^{-15}$

$\qquad\qquad = \dfrac{x^9}{(-4)^3 y^{15}}$

$\qquad\qquad = \dfrac{x^9}{-64y^{15}}$

$\qquad\qquad = -\dfrac{x^9}{64y^{15}}$

95.
$$\left(\frac{6x}{y^2}\right)^2 = \frac{6^2 x^2}{\left(y^2\right)^2}$$
$$= \frac{36x^2}{y^{2\cdot2}}$$
$$= \frac{36x^2}{y^4}$$

97.
$$\left(\frac{2r^4 s^5}{r^2}\right)^3 = \left(2r^{4-2} s^5\right)^3$$
$$= \left(2r^2 s^5\right)^3$$
$$= 2^3\left(r^2\right)^3\left(s^5\right)^3$$
$$= 8r^{2\cdot3} s^{5\cdot3}$$
$$= 8r^6 s^{15}$$

99.
$$\left(\frac{4xy}{y^3}\right)^{-3} = \left(\frac{4x}{y^{3-1}}\right)^{-3}$$
$$= \left(\frac{4x}{y^2}\right)^{-3}$$
$$= \left(\frac{y^2}{4x}\right)^3$$
$$= \frac{\left(y^2\right)^3}{4^3 x^3}$$
$$= \frac{y^{2\cdot3}}{64x^3}$$
$$= \frac{y^6}{64x^3}$$

101.
$$\left(\frac{4x^{-2}y}{x^{-5}}\right)^3 = \left(4x^{-2-(-5)} y\right)^3$$
$$= \left(4x^3 y\right)^3$$
$$= 4^3\left(x^3\right)^3 y^3$$
$$= 64x^{3\cdot3} y^3$$
$$= 64x^9 y^3$$

103.
$$\left(\frac{6x^2 y}{3xz}\right)^{-3} = \left[\left(\frac{6}{3}\right)\frac{x^{2-1} y}{z}\right]^{-3}$$
$$= \left(\frac{2xy}{z}\right)^{-3}$$
$$= \left(\frac{z}{2xy}\right)^3$$
$$= \frac{z^3}{8x^3 y^3}$$

105.
$$\left(\frac{x^6 y^{-2}}{x^{-2} y^3}\right)^2 = \left(\frac{x^{6-(-2)}}{y^{3-(-2)}}\right)^2$$
$$= \left(\frac{x^8}{y^5}\right)^2$$
$$= \frac{\left(x^8\right)^2}{\left(y^5\right)^2}$$
$$= \frac{x^{16}}{y^{10}}$$

107.
$$\left(\frac{4x^{-1} y^{-2} z^3}{2xy^2 z^{-3}}\right)^{-2} = \left(\frac{2z^{3-(-3)}}{x^{1-(-1)} y^{2-(-2)}}\right)^{-2}$$
$$= \left(\frac{2z^6}{x^2 y^4}\right)^{-2}$$
$$= \left(\frac{x^2 y^4}{2z^6}\right)^2$$
$$= \frac{\left(x^2\right)^2\left(y^4\right)^2}{2^2\left(z^6\right)^2}$$
$$= \frac{x^4 y^8}{4z^{12}}$$

109.
$$\left(\frac{-a^3b^{-1}c^{-3}}{2ab^3c^{-4}}\right)^{-3} = \left(\frac{-a^{3-1}c^{-3-(-4)}}{2b^{3-(-1)}}\right)^{-3}$$
$$= \left(\frac{-a^2c^1}{2b^4}\right)^{-3}$$
$$= \left(\frac{2b^4}{-a^2c}\right)^3$$
$$= \frac{2^3b^{4\cdot3}}{-a^{2\cdot3}c^3}$$
$$= -\frac{8b^{12}}{a^6c^3}$$

111.
$$\frac{\left(3x^{-4}y^2\right)^3}{\left(2x^3y^5\right)^3} = \frac{3^3x^{-4\cdot3}y^{2\cdot3}}{2^3x^{3\cdot3}y^{5\cdot3}}$$
$$= \frac{3^3x^{-12}y^6}{2^3x^9y^{15}}$$
$$= \frac{27}{8x^{9-(-12)}y^{15-6}}$$
$$= \frac{27}{8x^{21}y^9}$$

113.
$$x^{4a} \cdot x^{3a+4} = x^{4a+3a+4}$$
$$= x^{7a+4}$$

115.
$$w^{5b-2} \cdot w^{2b+3} = w^{5b-2+2b+3}$$
$$= w^{7b+1}$$

117.
$$\frac{x^{2w+3}}{x^{w-4}} = x^{2w+3-(w-4)}$$
$$= x^{2w+3-w+4}$$
$$= x^{w+7}$$

119.
$$\left(x^{3p+5}\right)\left(x^{2p-3}\right) = x^{3p+5+2p-3}$$
$$= x^{5p+2}$$

121.
$$x^{-m}\left(x^{3m+2}\right) = x^{-m}x^{3m+2}$$
$$= x^{-m+3m+2}$$
$$= x^{2m+2}$$

123.
$$\frac{25m^{a+b}n^{b-a}}{5m^{a-b}n^{a+b}} = \left(\frac{25}{5}\right)\frac{m^{a+b-(a-b)}}{n^{a+b-(b-a)}}$$
$$= \frac{5m^{a+b-a+b}}{n^{a+b-b+a}}$$
$$= \frac{5m^{2b}}{n^{2a}}$$

125.   **a.** $x^4 > x^3$ when $x < 0$ or $x > 1$

      b. $x^4 < x^3$ when $0 < x < 1$

      c. $x^4 = x^3$ when $x = 0$ or $x = 1$

      d. x^4 is not greater than x^3 when $0 \le x \le 1$

127.   **a.** $(-1)^n = 1$ for any even number n because an even number of negative factors is positive.

      b. $(-1)^n = -1$ for any odd number n because an odd number of negative factors is negative.

129.   **a.** $\left(-\dfrac{2}{3}\right)^{-2} = \dfrac{1}{\left(-\frac{2}{3}\right)^2} = \dfrac{1}{\frac{4}{9}} = \dfrac{9}{4}$ or

$$\left(-\frac{2}{3}\right)^{-2} = \left(-\frac{3}{2}\right)^2 = \frac{9}{4} \text{ by the negative}$$
exponent rule

Also, $\left(\dfrac{2}{3}\right)^{-2} = \dfrac{1}{\left(\frac{2}{3}\right)^2} = \dfrac{1}{\frac{4}{9}} = \dfrac{9}{4}$ or

$$\left(\frac{2}{3}\right)^{-2} = \left(\frac{3}{2}\right)^2 = \frac{9}{4} \text{ by the negative}$$
exponent rule

Thus, $\left(-\dfrac{2}{3}\right)^{-2} = \left(\dfrac{2}{3}\right)^{-2}$. So, yes they are equal.

      b.   Yes, because $x^{-2} = \dfrac{1}{x^2}$ and

$$(-x)^{-2} = \frac{1}{(-x)^2} = \frac{1}{x^2}$$

131. Let a represent the unknown exponent,

$$\left(\frac{x^2 y^{-2}}{x^{-3} y^a}\right)^2 = \left(x^{2-(-3)} y^{-2-a}\right)^2$$

$$= \left(x^5 y^{-2-a}\right)^2$$

$$= x^{10} y^{2(-2-a)}$$

$$= x^{10} y^{-4-2a}$$

Thus,

$$-4 - 2a = 2$$
$$2a = -4 - 2$$
$$2a = -6$$
$$a = -3$$

133. Let a and b represent the unknown exponents.

$$\left(\frac{x^a y^5 z^{-2}}{x^4 y^b z}\right)^{-1} = \left(\frac{y^{5-b}}{x^{4-a} z^{1-(-2)}}\right)^{-1}$$

$$= \frac{x^{4-a} z^3}{y^{5-b}}$$

Thus,

$$4 - a = 5$$
$$a = -1$$

and

$$5 - b = 2$$
$$b = 3$$

135.

$$\left(\frac{x^{5/8}}{x^{1/4}}\right)^3 = \left(x^{5/8-1/4}\right)^3$$

$$= \left(x^{3/8}\right)^3$$

$$= x^{(3/8)\cdot 3}$$

$$= x^{9/8}$$

137.

$$\frac{x^{1/2} y^{-3/2}}{x^5 y^{5/3}} = \frac{1}{x^{5-1/2} y^{5/3-(-3/2)}}$$

$$= \frac{1}{x^{9/2} y^{19/6}}$$

140.   **a.** $A \cup B = \{1, 2, 3, 4, 5, 6, 8\}$

      b. $A \cap B = \varnothing$ or $\{\}$

141.

142.

$$6 + |12| \div |-3| - 4 \cdot 2^2 = 6 + 12 \div 3 - 4 \cdot 2^2$$

$$= 6 + 12 \div 3 - 4 \cdot 4$$

$$= 6 + 4 - 16$$

$$= 10 - 16$$

$$= -6$$

143. $\sqrt[3]{-125} = -5$ since $(-5)(-5)(-5) = -125$

Exercise Set 1.6

1. The form of a number in scientific notation is a number greater than or equal to 1 and less than 10 multiplied by a power of 10.

3. No; If n is positive, then 10^{-n} or $\dfrac{1}{10^n}$ is positive, so 1 times a positive number is positive.

5. $7300 = 7.3 \times 10^3$

7. $0.047 = 4.7 \times 10^{-2}$

9. $19,000 = 1.9 \times 10^4$

11. $0.00000186 = 1.86 \times 10^{-6}$

13. $5,780,000 = 5.78 \times 10^6$

15. $0.000101 = 1.01 \times 10^{-4}$

17. $6.4 \times 10^3 = 6400$

19. $2.13 \times 10^{-5} = 0.0000213$

21. $3.12 \times 10^{-1} = 0.312$

23. $9 \times 10^6 = 9,000,000$

25. $2.07 \times 10^5 = 207,000$

27. $1 \times 10^6 = 1,000,000$

29. $\left(5\times10^3\right)\left(3\times10^4\right)=(5\times3)\left(10^3\times10^4\right)$

$$=15\times10^7$$
$$=150,000,000$$

31. $\dfrac{8.4\times10^{-6}}{4\times10^{-4}}=\left(\dfrac{8.4}{4}\right)\times10^{-6-(-4)}$

$$=2.1\times10^{-2}$$
$$=0.021$$

33. $\dfrac{5.85\times10^4}{4.5\times10^{-3}}=\left(\dfrac{5.85}{4.5}\right)\times10^{4-(-3)}$

$$=1.3\times10^7$$
$$=13,000,000$$

35. $\left(8.2\times10^5\right)\left(1.3\times10^{-2}\right)$

$$=(8.2\times1.3)\left(10^5\times10^{-2}\right)$$
$$=10.66\times10^3$$
$$=10,660$$

37. $\dfrac{9.2\times10^5}{2.3\times10^4}=\left(\dfrac{9.2}{2.3}\right)\times10^{5-4}$

$$=4\times10^1$$
$$=40$$

39. $\left(9.1\times10^{-4}\right)\left(6.3\times10^{-4}\right)$

$$=(9.1\times6.3)\left(10^{-4}\times10^{-4}\right)$$
$$=57.33\times10^{-8}$$
$$=0.0000005733$$

41. $(0.003)(0.00015)$

$$=\left(3\times10^{-3}\right)\left(1.5\times10^{-4}\right)$$
$$=(3\times1.5)\left(10^{-3}\times10^{-4}\right)$$
$$=4.5\times10^{-7}$$

43. $\dfrac{1,400,000}{700}=\dfrac{1.4\times10^6}{7.0\times10^2}$

$$=\left(\dfrac{1.4}{7}\right)\times10^{6-2}$$
$$=0.2\times10^4$$
$$=2.0\times10^3$$

45. $\dfrac{0.0000426}{200}=\dfrac{4.26\times10^{-5}}{2.0\times10^2}$

$$=\left(\dfrac{4.26}{2.0}\right)\times10^{-5-2}$$
$$=2.13\times10^{-7}$$

47. $(47,000)(35,000,000)$

$$=\left(4.7\times10^4\right)\left(3.5\times10^7\right)$$
$$=(4.7\times3.5)\left(10^4\times10^7\right)$$
$$=16.45\times10^{11}$$
$$=1.645\times10^{12}$$

49. $\dfrac{672}{0.0021}=\dfrac{6.72\times10^2}{2.1\times10^{-3}}$

$$=\left(\dfrac{6.72}{2.1}\right)\times10^{2-(-3)}$$
$$=3.2\times10^5$$

51. $\dfrac{0.00153}{0.00051}=\dfrac{1.53\times10^{-3}}{5.1\times10^{-4}}$

$$=\left(\dfrac{1.53}{5.1}\right)\times10^{-3+4}$$
$$=0.3\times10^1$$
$$=3.0\times10^0$$

53. $\left(1.23\times10^4\right)\left(5.67\times10^8\right)$

$$=(1.23\times5.67)\left(10^4\times10^8\right)$$
$$=6.974\times10^{12}$$

55. $\left(7.23\times10^{-3}\right)\left(1.37\times10^5\right)$

$$=(7.23\times1.37)\left(10^{-3}\times10^5\right)$$
$$=9.905\times10^2$$

57. $\dfrac{4.36 \times 10^{-4}}{8.17 \times 10^{-7}} = \left(\dfrac{4.36}{8.17}\right) \times 10^{-4-(-7)}$

$\qquad = 0.5337 \times 10^3$

$\qquad = 5.337 \times 10^2$

59. $\left(3.70 \times 10^{37}\right)\left(4.15 \times 10^{-30}\right)$

$\qquad = (3.70 \times 4.15)\left(10^{37} \times 10^{-30}\right)$

$\qquad = 15.355 \times 10^7$

$\qquad = 1.536 \times 10^8$

61. $\left(7.71 \times 10^3\right)\left(9.14 \times 10^{-31}\right)$

$\qquad = (7.71 \times 9.14)\left(10^3 \times 10^{-31}\right)$

$\qquad = 70.469 \times 10^{-28}$

$\qquad = 7.047 \times 10^{-27}$

63. $\dfrac{1.50 \times 10^{35}}{4.5 \times 10^{-26}} = \left(\dfrac{1.50}{4.5}\right) \times 10^{35-(-26)}$

$\qquad = 0.3333 \times 10^{61}$

$\qquad = 3.333 \times 10^{60}$

65. a. $10 = 10^1$; Add 1 to the exponent.

b. $100 = 10^2$; Add 2 to the exponent.

c. 1 million $= 1,000,000 = 10^6$; Add 6 to the exponent.

d. $7.59 \times 10^7 \times 1$ million

$\qquad = 7.59 \times 10^7 \times 10^6$

$\qquad = 7.59 \times 10^{13}$

67. a. $5.25 \times 10^4 - 4.25 \times 10^4 = 1 \times 10^4$

It is off by 1×10^4 or 10,000.

b. $5.25 \times 10^5 - 5.25 \times 10^4$

$\qquad = 52.5 \times 10^4 - 5.25 \times 10^4$

$\qquad = 47.25 \times 10^4$

$\qquad = 4.725 \times 10^5$

It is off by 4.725×10^5 or 472,500.

c. The error in part b is more serious because 472,500 is greater than 10,000.

69. a. $\dfrac{5.85 \times 10^8}{365} = \dfrac{5.85 \times 10^8}{3.65 \times 10^2}$

$\qquad = \left(\dfrac{5.85}{3.65}\right) \times 10^{8-2}$

$\qquad \approx 1.6027397 \times 10^6$

$\qquad = 1,602,739.7$

It travels approximately 1,602,739.7 miles per day.

b. Earth's speed

$\qquad = \dfrac{1.6027397 \times 10^6}{24} \approx 6.6781 \times 10^4$

Bullet's speed

$\qquad = \dfrac{6.6781 \times 10^4}{8} = 8.3476 \times 10^3$

The bullet's speed is approximately 8347.6 mph.

71. $6.09 \times 10^9 - 2.74 \times 10^8$

$\qquad = 6.09 \times 10^9 - 0.274 \times 10^9$

$\qquad = 5.816 \times 10^9$

About 5.816×10^9 people will live outside of the United States.

73. a. $4.2 \times 10^9 \times 5\% = 4.2 \times 10^9 \times 0.05$

$\qquad = 4.2 \times 10^9 \times 5 \times 10^{-2}$

$\qquad = (4.2 \times 5)\left(10^9 \times 10^{-2}\right)$

$\qquad = 21 \times 10^7$

$\qquad = 2.1 \times 10^8$

About 2.1×10^8 pounds are recycled.

b. $4.2 \times 10^9 \times 95\%$

$\qquad = 4.2 \times 10^9 \times 0.95$

$\qquad = 4.2 \times 10^9 \times 9.5 \times 10^{-1}$

$\qquad = (4.2 \times 9.5)\left(10^9 \times 10^{-1}\right)$

$\qquad = 39.9 \times 10^8$

$\qquad = 3.99 \times 10^9$

About 3.99×10^9 pounds are not recycled.

75. a. $1.157 \times 10^{11} \times 0.259$

$= 1.157 \times 10^{11} \times 2.59 \times 10^{-1}$

$= (1.157 \times 2.59)(10^{11} \times 10^{-1})$

$= 2.997 \times 10^{10}$

$\approx 3.00 \times 10^{10}$

Approximately 3.00×10^{10} passengers enplaned at Chicago's airport.

b. $1.157 \times 10^{11} \times 0.127$

$= 1.157 \times 10^{11} \times 1.27 \times 10^{-1}$

$= (1.157 \times 1.27)(10^{11} \times 10^{-1})$

$\approx 1.4 \times 10^{10}$

Approximately 1.47×10^{10} passengers enplaned at Denver's airport.

c. $\dfrac{2.997 \times 10^{10}}{1.470 \times 10^{10}} = \left(\dfrac{2.997}{1.469}\right) \times 10^{10-10}$

$\approx 2.04 \times 10^{0}$

≈ 2.04

The number of passengers that enplaned in Chicago was approximately 2.04 times greater than the number enplaned in Denver.

77. a. 1210 million − 265 million = 945 million. About 945 million more people lived in China than in the United States.

b. $\dfrac{1210 \text{ million}}{5,772,000,000} = \dfrac{1.210 \times 10^{9}}{5.772 \times 10^{9}}$

$= \left(\dfrac{1.210}{5.772}\right) \times 10^{9-9}$

$\approx 0.2096 \times 10^{0}$

$= 20.96\%$

Approximately 20.96% of the worlds poulation lived in China.

c. $\dfrac{1210 \text{ million}}{3.70 \times 10^{6}}$

$= \dfrac{1.21 \times 10^{9}}{3.70 \times 10^{6}}$

$= \left(\dfrac{1.21}{3.70}\right) \times 10^{9-6}$

$\approx 0.327 \times 10^{3}$

$= 327$ people per square mile

The population density in China was approximately 327.0 people per square mile.

d. $\dfrac{265 \text{ million}}{3.62 \times 10^{6}}$

$= \dfrac{2.65 \times 10^{8}}{3.62 \times 10^{6}}$

$= \left(\dfrac{2.65}{3.62}\right) \times 10^{8-6}$

$\approx 0.732 \times 10^{2}$

$= 73.2$

The population density in the United States was approximately 73.2 people per square mile.

79. a. $\dfrac{1.86 \times 10^{5} \text{ miles}}{1 \text{ second}} \cdot \dfrac{60 \text{ seconds}}{1 \text{ minute}} \cdot \dfrac{60 \text{ minutes}}{1 \text{ hour}} \cdot \dfrac{24 \text{ hours}}{1 \text{ day}} \cdot \dfrac{365 \text{ days}}{\text{year}}$

$= 1.86 \times 10^{5} \times 6.0 \times 10^{1} \times 6.0 \times 10^{1} \times 2.4 \times 10^{1} \times 3.65 \times 10^{2}$

$= (1.86 \times 6 \times 6 \times 2.4 \times 3.65)(10^{5} \times 10^{1} \times 10^{1} \times 10^{1} \times 10^{2})$

$\approx 587 \times 10^{10}$

$\approx 5.87 \times 10^{12} \text{ miles}$

b. $\dfrac{93,000,000}{1.86 \times 10^5} = \dfrac{9.3 \times 10^7}{1.86 \times 10^5}$

$\qquad = \left(\dfrac{9.3}{1.86}\right) \times 10^{7-5}$

$\qquad = 5 \times 10^2$

$\qquad = 500$ seconds or $8\dfrac{1}{3}$ minutes

c. $\dfrac{6.25 \times 10^{16}}{0.5 \times 1.86 \times 10^5} \approx 6.72 \times 10^{11}$ seconds or 21,309 years

Review Exercises

1. {3, 4, 5, 6}

2. {0, 3, 6, 9, …}

3. Yes; the set of natural numbers is a subset of the set of whole numbers.

4. Yes; the set of rational numbers is a subset of the set of real numbers.

5. Yes; the set of irrational numbers is a subset of the set of real numbers.

6. No; the set of rational numbers is not a subset of the set of irrational numbers.

7. 4 and 6 are natural numbers.

8. 4, 6 and 0 are whole numbers.

9. −2, 4, 6 and 0 are integers.

10. −2, 4, 6, $\dfrac{1}{2}$, 0, $\dfrac{15}{27}$, $-\dfrac{1}{5}$, and 1.47 are rational numbers.

11. $\sqrt{7}$ and $\sqrt{3}$ are irrational numbers.

12. −2, 4, 6, $\dfrac{1}{2}$, $\sqrt{7}$, $\sqrt{3}$, 0, $\dfrac{15}{27}$, $-\dfrac{1}{5}$, and 1.47 are real numbers.

13. False; $\dfrac{0}{1} = 0$.

14. True; 0, −2, and 4 can be written with a denominator of 1.

15. True; division by 0 is undefined.

16. True; the set of real numbers contains both the rational and irrational numbers.

17. $A \cup B = \{1, 2, 3, 4, 5\}$
$A \cap B = \{2, 3, 4, 5\}$

18. $A \cup B = \{2, 3, 4, 5, 6, 7, 8, 9\}$
$A \cap B = \varnothing$ or { }

19. $A \cup B = \{1, 2, 3, 4, …\}$
$A \cap B = \{2, 4, 6, …\}$

20. $A \cup B = \{3, 4, 5, 6, 9, 10, 11, 12\}$
$A \cap B = \{9, 10\}$

21.

22.

23.

24.

25. $-8 < 0$

26. $-4 < -3.9$

27. $1.06 < 1.6$

28. $|-3| = 3$
$3 = 3$
$|-3| = 3$

29. $|-4| = 4$
$|-6| = 6$
$|-4| < |-6|$

30. $13 = 13$
$|-5| = 5$
$13 > |-5|$

31. $\left|-\dfrac{2}{3}\right| = \dfrac{2}{3} = \dfrac{10}{15}$
$\dfrac{3}{5} = \dfrac{9}{15}$
$\left|-\dfrac{2}{3}\right| > \dfrac{3}{5}$

32. $-|-2| = -2$
$-5 = -5$
$-|-2| > -5$

33. $-5, -2, 4, |7|$

34. $0, \dfrac{3}{5}, 2.3, |-3|$

35. $-2, 3, |-5|, |-7|$

36. $-4, -|3|, -2.1, -2$

37. $-4, -|-3|, 5, 6$

38. $-3, 0, |1.6|, |-2.3|$

39. Distributive property

40. Commutative property of multiplication

41. Associative property of addition

42. Identity property of addition

43. Associative property of multiplication

44. Double negative property

45. Multiplicative property of zero

46. Inverse property of addition

47. Inverse property of multiplication

48. Identity property of multiplication

49. $4 - 2^2 + \sqrt{81} \div 3 = 4 - 4 + 9 \div 3$
$= 4 - 4 + 3$
$= 0 + 3$
$= 3$

50. $-4 \div (-2) + 16 - \sqrt{49} = -4 \div (-2) + 16 - 7$
$= 2 + 16 - 7$
$= 18 - 7$
$= 11$

51. $(4 - 6) - (-3 + 5) + 12 = (-2) - (2) + 12$
$= -2 - 2 + 12$
$= -4 + 12$
$= 8$

52. $3|-2| - (4 - 3) + 2(-3)$
$= 3(2) - (4 - 3) + 2(-3)$
$= 3(2) - 1 + 2(-3)$
$= 6 - 1 - 6$
$= 5 - 6$
$= -1$

53. $(6 - 9) \div (9 - 6) + 1 = -3 \div 3 + 1$
$= \dfrac{-3}{3} + 1$
$= -1 + 1$
$= 0$

54. $|6 - 3| \div 3 + 4 \cdot 8 - 12 = |3| \div 3 + 4 \cdot 8 - 12$
$= 3 \div 3 + 4 \cdot 8 - 12$
$= 1 + 32 - 12$
$= 33 - 12$
$= 21$

55. $\sqrt[3]{27} \div 3 + |4 - 2| + 4^2 = 3 \div 3 + |2| + 16$
$$= 1 + 2 + 16$$
$$= 3 + 16$$
$$= 19$$

56. $3^2 - 6 \cdot 9 + 4 \div 2^2 - 3 = 9 - 6 \cdot 9 + 4 \div 4 - 3$
$$= 9 - 54 + 1 - 3$$
$$= -45 + 1 - 3$$
$$= -44 - 3$$
$$= -47$$

57. $4 - (2 - 9)^0 + 3^2 \div 1 + 3$
$$= 4 - (-7)^0 + 3^2 \div 1 + 3$$
$$= 4 - 1 + 9 \div 1 + 3$$
$$= 4 - 1 + 9 + 3$$
$$= 3 + 9 + 3$$
$$= 12 + 3$$
$$= 15$$

58. $4^2 - \left(2 - 3^2\right)^2 + 4^3 = 4^2 - (2 - 9)^2 + 4^3$
$$= 4^2 - (-7)^2 + 4^3$$
$$= 16 - (49) + 64$$
$$= -33 + 64$$
$$= 31$$

59. $-3^2 + 14 \div 2 \cdot 3 - 6 = -9 + 14 \div 2 \cdot 3 - 6$
$$= -9 + 7 \cdot 3 - 6$$
$$= -9 + 21 - 6$$
$$= 12 - 6$$
$$= 6$$

60. $\left\{\left[(9 \div 3)^2 - 1\right]^2 \div 8\right\}^3 = \left\{\left[(3)^2 - 1\right]^2 \div 8\right\}^3$
$$= \left[(9 - 1)^2 \div 8\right]^3$$
$$= \left[8^2 \div 8\right]^3$$
$$= (64 \div 8)^3$$
$$= 8^3$$
$$= 512$$

61. $\dfrac{8 - 4 \div 2 + 3 \cdot 2}{\sqrt{36} \div 2 - 3} = \dfrac{8 - 4 \div 2 + 3 \cdot 2}{6 \div 2 - 3}$
$$= \dfrac{8 - 2 + 6}{3 - 3}$$
$$= \dfrac{6 + 6}{0}$$
$$= \dfrac{12}{0} \text{ is undefined}$$

62. $\dfrac{-(4 - 6)^2 - 3(-2) + |-6|}{18 - 9 \div 3 \cdot 5}$
$$= \dfrac{-(-2)^2 - 3(-2) + |-6|}{18 - 9 \div 3 \cdot 5}$$
$$= \dfrac{-4 - 3(-2) + |-6|}{18 - 9 \div 3 \cdot 5}$$
$$= \dfrac{-4 - 3(-2) + 6}{18 - 9 \div 3 \cdot 5}$$
$$= \dfrac{-4 + 6 + 6}{18 - 3 \cdot 5}$$
$$= \dfrac{-4 + 6 + 6}{18 - 15}$$
$$= \dfrac{2 + 6}{3}$$
$$= \dfrac{8}{3}$$

63. Substitute 2 for x:
$2x^2 + 3x + 1 = 2(2)^2 + 3(2) + 1$
$$= 2(4) + 3(2) + 1$$
$$= 8 + 6 + 1$$
$$= 14 + 1$$
$$= 15$$

64. Substitute 1 for x and $-\dfrac{1}{3}$ for y:

$$4x^2 - 3y^2 + 5 = 4(1)^2 - 3\left(-\dfrac{1}{3}\right)^2 + 5$$

$$= 4(1) - 3\left(\dfrac{1}{9}\right) + 5$$

$$= 4 - \dfrac{1}{3} + 5$$

$$= \dfrac{12}{3} - \dfrac{1}{3} + \dfrac{15}{3}$$

$$= \dfrac{11}{3} + \dfrac{15}{3}$$

$$= \dfrac{26}{3}$$

65. a. 1976 means substitute 7 for x.
   dollars

$$= 50.86x^2 - 316.75x + 541.48$$

$$= 50.86(7)^2 - 316.75(7) + 541.48$$

$$= 50.86(49) - 316.75(7) + 541.48$$

$$= 2492.14 - 2217.25 + 541.48$$

$$= 816.37$$

In 1976, the amount spent was
$816.37 million.

   b. 1996 means substitute 12 for x.
   dollars

$$= 50.86x^2 - 316.75x + 541.48$$

$$= 50.86(12)^2 - 316.75(12) + 541.48$$

$$= 50.86(144) - 316.75(12) + 541.48$$

$$= 7323.84 - 3801 + 541.48$$

$$= 4064.32$$

In 1996, the amount spent was
$4064.32 million.

66. a. 1980 means substitute 4 for x.
   freight hauled

$$= 14.04x^2 + 1.96x + 712.05$$

$$= 14.04(4)^2 + 1.96(4) + 712.05$$

$$= 14.04(16) + 1.96(4) + 712.05$$

$$= 224.64 + 7.84 + 712.05$$

$$= 944.53$$

In 1980, the amount of freight hauled by
trains was 944.53 ton-miles.

   b. 1995 means substitute 7 for x.
   freight hauled

$$= 14.04x^2 + 1.96x + 712.05$$

$$= 14.04(7)^2 + 1.96(7) + 712.05$$

$$= 14.04(49) + 1.96(7) + 712.05$$

$$= 687.96 + 13.72 + 712.05$$

$$= 1413.73$$

In 1995, the amount of freight hauled by
trains was 1413.73 ton-miles.

67. $4^2 \cdot 4^1 = 4^{2+1} = 4^3 = 64$

68. $x^3 \cdot x^5 = x^{3+5}$

$$= x^8$$

69. $\dfrac{x^6}{x^2} = x^{6-2}$

$$= x^4$$

70. $\dfrac{y^{12}}{y^3} = y^{12-3}$

$$= y^9$$

71. $\dfrac{x^4}{x^{-3}} = x^{4-(-3)}$

$$= x^{4+3}$$

$$= x^7$$

72. $x^4 \cdot x^{-7} = x^{4+(-7)}$

$$= x^{-3}$$

$$= \dfrac{1}{x^3}$$

73. $3^{-2} \cdot 3^{-1} = 3^{-2+(-1)}$

$$= 3^{-3}$$

$$= \dfrac{1}{3^3}$$

$$= \dfrac{1}{27}$$

74. $3x^0 = 3(1)$

$$= 3$$

75. $\left(3n^2\right)^2 = 3^2\left(n^2\right)^2$

$\qquad = 9n^{2\cdot2}$

$\qquad = 9n^4$

76. $\left(\dfrac{2}{3}\right)^{-1} = \left(\dfrac{3}{2}\right)^1$

$\qquad = \dfrac{3}{2}$

77. $\left(\dfrac{3}{4}\right)^{-2} = \left(\dfrac{4}{3}\right)^2$

$\qquad = \dfrac{4^2}{3^2}$

$\qquad = \dfrac{16}{9}$

78. $\left(\dfrac{x}{y^2}\right)^{-1} = \left(\dfrac{y^2}{x}\right)^1$

$\qquad = \dfrac{y^2}{x}$

79. $\left(7x^2y^5\right)\left(-3xy^4\right) = 7(-3)x^2\cdot x\cdot y^5\cdot y^4$

$\qquad = -21x^{2+1}y^{5+4}$

$\qquad = -21x^3y^9$

80. $\left(4x^2y^{-3}\right)\left(2x^{-4}y^2\right) = 4\cdot2\cdot x^2\cdot x^{-4}\cdot y^{-3}\cdot y^2$

$\qquad = 8\cdot x^{2+(-4)}\cdot y^{-3+2}$

$\qquad = 8x^{-2}y^{-1}$

$\qquad = \dfrac{8}{x^2y}$

81. $\dfrac{6x^{-3}y^5}{2x^2y^{-2}} = \dfrac{6y^{5-(-2)}}{2x^{2-(-3)}}$

$\qquad = \left(\dfrac{6}{2}\right)\dfrac{y^7}{x^5}$

$\qquad = \dfrac{3y^7}{x^5}$

82. $\dfrac{12x^{-3}y^{-4}}{4x^{-2}y^5} = \left(\dfrac{12}{4}\right)\dfrac{1}{x^{-2-(-3)}y^{5-(-4)}}$

$\qquad = \dfrac{3}{xy^9}$

83. $\dfrac{a^2b^{-7}c^{-10}}{a^{-3}b^{-3}c^{-4}} = \dfrac{a^{2-(-3)}}{b^{-3-(-7)}c^{-4-(-10)}}$

$\qquad = \dfrac{a^5}{b^4c^6}$

84. $\dfrac{16p^4q^{-2}r^{-3}}{4p^2q^{-1}r^3} = \left(\dfrac{16}{4}\right)\dfrac{p^{4-2}}{q^{-1-(-2)}r^{3-(-3)}}$

$\qquad = \dfrac{4p^2}{qr^6}$

85. $\left(\dfrac{5a^2b}{a}\right)^3 = \left(5a^{2-1}b\right)^3$

$\qquad = (5ab)^3$

$\qquad = 5^3a^3b^3$

$\qquad = 125a^3b^3$

86. $\left(\dfrac{x^5y}{-3y^2}\right)^2 = \left(\dfrac{x^5}{-3y^{2-1}}\right)^2$

$\qquad = \left(\dfrac{x^5}{-3y}\right)^2$

$\qquad = \dfrac{\left(x^5\right)^2}{(-3)^2y^2}$

$\qquad = \dfrac{x^{5\cdot2}}{9y^2}$

$\qquad = \dfrac{x^{10}}{9y^2}$

87. $\left(\dfrac{x^2y}{x^{-1}y^{-3}}\right)^2 = \left(x^{2-(-1)}y^{1-(-3)}\right)^2$

$\qquad = \left(x^3y^4\right)^2$

$\qquad = \left(x^3\right)^2\left(y^4\right)^2$

$\qquad = x^{3\cdot2}y^{4\cdot2}$

$\qquad = x^6y^8$

88. $\left(\dfrac{-5x^{-2}y}{z^3}\right)^3 = \left(\dfrac{-5y}{x^2z^3}\right)^3$

$\qquad = \dfrac{(-5)^3y^3}{\left(x^2\right)^3\left(z^3\right)^3}$

$\qquad = \dfrac{-125y^3}{x^{2\cdot3}z^{3\cdot3}}$

$\qquad = -\dfrac{125y^3}{x^6z^9}$

89. $\left(\dfrac{6xy^3}{z^2}\right)^{-2} = \left(\dfrac{z^2}{6xy^3}\right)^2$

$\qquad = \dfrac{(z^2)^2}{6^2x^2(y^3)^2}$

$\qquad = \dfrac{z^{2\cdot2}}{36x^2y^{3\cdot2}}$

$\qquad = \dfrac{z^4}{36x^2y^6}$

90. $\left(\dfrac{9m^{-2}n}{3mn}\right)^{-3} = \left[\left(\dfrac{9}{3}\right)\dfrac{n^{1-1}}{m^{1-(-2)}}\right]^{-3}$

$\qquad = \left(\dfrac{3}{m^3}\right)^{-3}$

$\qquad = \left(\dfrac{m^3}{3}\right)^3$

$\qquad = \dfrac{(m^3)^3}{3^3}$

$\qquad = \dfrac{m^9}{27}$

91. $\left(-2x^{-3}y^2\right)^{-4} = (-2)^{-4}\left(x^{-3}\right)^{-4}\left(y^2\right)^{-4}$

$\qquad = (-2)^{-4}x^{(-3)(-4)}y^{2(-4)}$

$\qquad = (-2)^{-4}x^{12}y^{-8}$

$\qquad = \dfrac{x^{12}}{(-2)^4y^8}$

$\qquad = \dfrac{x^{12}}{16y^8}$

92. $\left(\dfrac{16x^4y^3z^{-2}}{-4x^5y^2z^3}\right)^3 = \left[\left(\dfrac{16}{-4}\right)\dfrac{y^{3-2}}{x^{5-4}z^{3-(-2)}}\right]^3$

$\qquad = \left(\dfrac{-4y}{xz^5}\right)^3$

$\qquad = \dfrac{(-4)^3y^3}{x^3\left(z^5\right)^3}$

$\qquad = -\dfrac{64y^3}{x^3z^{15}}$

93. $\left(\dfrac{2x^{-1}y^5z^4}{3x^4y^{-2}z^{-2}}\right)^{-2} = \left(\dfrac{2y^{5-(-2)}z^{4-(-2)}}{3x^{4-(-1)}}\right)^{-2}$

$\qquad = \left(\dfrac{2y^7z^6}{3x^5}\right)^{-2}$

$\qquad = \left(\dfrac{3x^5}{2y^7z^6}\right)^2$

$\qquad = \dfrac{3^2x^{5\cdot2}}{2^2y^{7\cdot2}z^{6\cdot2}}$

$\qquad = \dfrac{9x^{10}}{4y^{14}z^{12}}$

94. $\left(\dfrac{8x^{-2}y^{-2}z}{-x^4y^{-4}z^3}\right)^{-1} = \left(\dfrac{8y^{-2-(-4)}}{-x^{4-(-2)}z^{3-1}}\right)^{-1}$

$\qquad = \left(\dfrac{8y^2}{-x^6z^2}\right)^{-1}$

$\qquad = \left(\dfrac{-x^6z^2}{8y^2}\right)^{1}$

$\qquad = -\dfrac{x^6z^2}{8y^2}$

95. $0.0000742 = 7.42 \times 10^{-5}$

96. $260,000 = 2.6 \times 10^5$

97. $183,000 = 1.83 \times 10^5$

98. $0.000001 = 1.0 \times 10^{-6}$

99. $\left(25 \times 10^{-3}\right)\left(1.2 \times 10^6\right)$

$\quad = (25 \times 1.2)\left(10^{-3} \times 10^6\right)$

$\quad = 30 \times 10^{-3+6}$

$\quad = 30 \times 10^3$

$\quad = 30,000$

100. $\dfrac{18 \times 10^3}{9 \times 10^5} = \left(\dfrac{18}{9}\right) \times \left(\dfrac{10^3}{10^5}\right)$

$\qquad = 2 \times 10^{3-5}$

$\qquad = 2 \times 10^{-2}$

$\qquad = 0.02$

101. $\dfrac{4,000,000}{0.02} = \dfrac{4 \times 10^6}{2 \times 10^{-2}}$

$\qquad = \left(\dfrac{4}{2}\right) \times \left(\dfrac{10^6}{10^{-2}}\right)$

$\qquad = 2 \times 10^{6-(-2)}$

$\qquad = 2 \times 10^8$

$\qquad = 200,000,000$

102. $(0.004)(500,000) = \left(4 \times 10^{-3}\right)\left(5 \times 10^5\right)$

$\qquad = (4 \times 5)\left(10^{-3} \times 10^5\right)$

$\qquad = 20 \times 10^{-3+5}$

$\qquad = 20 \times 10^2$

$\qquad = 2000$

103. a. $\$1.212 \times 10^8 - \$9.800 \times 10^7$

$\qquad = \$12.12 \times 10^7 - \$9.800 \times 10^7$

$\qquad = (\$12.12 - \$9.800) \times 10^7$

$\qquad = \$2.32 \times 10^7$ or $\$23,200,000$

b. $\dfrac{\$1.212 \times 10^8}{\$9.800 \times 10^7} = \left(\dfrac{1.212}{9.8}\right) \times 10^{8-7}$

$\qquad\qquad \approx 0.1237 \times 10^1$

$\qquad\qquad = 1.237$

It was about 1.237 times larger.

104. a. $1.04 \times 10^{10} = 10,400,000,000$

b. $1.04 \times 10^{10} = 10.4 \times 10^9$

$\qquad\qquad = 10.4$ billion

c. $\dfrac{1.04 \times 10^{10}}{20} = \dfrac{1.04 \times 10^{10}}{2.0 \times 10^1}$

$\qquad\qquad = \left(\dfrac{1.04}{2.0}\right) \times 10^{10-1}$

$\qquad\qquad = 0.52 \times 10^9$

$\qquad\qquad = 5.2 \times 10^8$

It traveled an average of 5.2×10^8 km or 520,000,000 km per year.

d. $1.04 \times 10^{10} \times 0.6$

$\qquad = 1.04 \times 10^{10} \times 6 \times 10^{-1}$

$\qquad = (1.04 \times 6)\left(10^{10} \times 10^{-1}\right)$

$\qquad = 6.24 \times 10^9$

It traveled about 6.24×10^9 miles or 6,240,000,000 miles.

Practice Test

1. $A = \{6, 7, 8, 9, \ldots\}$

2. True; the set of real numbers contains the set of rational numbers.

3. True; this is how the set of real numbers is defined.

4. $-\dfrac{3}{5},\ 2,\ -4,\ 0,\ \dfrac{19}{12}, 2.57,$ and -1.92 are rational numbers.

5. $-\dfrac{3}{5},\ 2,\ -4,\ 0,\ \dfrac{19}{12},\ 2.57,\ \sqrt{8},\ \sqrt{2},$ and $-1.92,$ are real numbers.

6. $A \cup B = \{5,\ 7,\ 8,\ 9,\ 10,\ 11,\ 14\}$
$A \cap B = \{8,\ 10\}$

7. $A \cup B = \{1,\ 3,\ 5,\ 7,\ \ldots\}$
$A \cap B = \{3,\ 5,\ 7,\ 9,\ 11\}$

8.
$$
\begin{array}{c}
-2.3 \qquad\qquad\qquad 5.2 \\
\end{array}
$$

9.

10. $-|4|,\ -2,\ |3|,\ 6$

11. Associative property of addition

12. Commutative property of addition

13. $\{4 - [6 - 3(4 - 5)]\}^2 \div (-5)$
$= \{4 - [6 - 3(-1)]\}^2 \div (-5)$
$= [4 - (6 + 3)]^2 \div (-5)$
$= [4 - 9]^2 \div (-5)$
$= (-5)^2 \div (-5)$
$= 25 \div (-5)$
$= -5$

14. $5^2 + 16 \div 4 - 3 \cdot 2 = 25 + 16 \div 4 - 3 \cdot 2$
$\qquad\qquad\qquad\qquad = 25 + 4 - 6$
$\qquad\qquad\qquad\qquad = 29 - 6$
$\qquad\qquad\qquad\qquad = 23$

15. $\dfrac{-3|4 - 8| \div 2 + 4}{-\sqrt{36} + 18 \div 3^2 + 4} = \dfrac{-3|-4| \div 2 + 4}{-\sqrt{36} + 18 \div 3^2 + 4}$

$\qquad\qquad\qquad = \dfrac{-3(4) \div 2 + 4}{-\sqrt{36} + 18 \div 3^2 + 4}$

$\qquad\qquad\qquad = \dfrac{-3(4) \div 2 + 4}{-6 + 18 \div 9 + 4}$

$\qquad\qquad\qquad = \dfrac{-12 \div 2 + 4}{-6 + 2 + 4}$

$\qquad\qquad\qquad = \dfrac{-6 + 4}{-6 + 2 + 4}$

$\qquad\qquad\qquad = \dfrac{-2}{-4 + 4}$

$\qquad\qquad\qquad = \dfrac{-2}{0}$

which is undefined.

16. $\dfrac{-6^2 + 3(4 - |6|) \div 6}{4 - (-3) + 12 \div 4 \cdot 5} = \dfrac{-6^2 + 3(4 - 6) \div 6}{4 - (-3) + 12 \div 4 \cdot 5}$

$\qquad\qquad\qquad = \dfrac{-6^2 + 3(-2) \div 6}{4 - (-3) + 12 \div 4 \cdot 5}$

$\qquad\qquad\qquad = \dfrac{-36 + 3(-2) \div 6}{4 - (-3) + 12 \div 4 \cdot 5}$

$\qquad\qquad\qquad = \dfrac{-36 + (-6) \div 6}{4 + 3 + 3 \cdot 5}$

$\qquad\qquad\qquad = \dfrac{-36 + (-1)}{4 + 3 + 15}$

$\qquad\qquad\qquad = \dfrac{-37}{7 + 15}$

$\qquad\qquad\qquad = \dfrac{-37}{22}$

$\qquad\qquad\qquad = -\dfrac{37}{22}$

17. Substitute 2 for x and 3 for y:
$-x^2 + 2xy + y^2 = -(2)^2 + 2(2)(3) + (3)^2$
$\qquad\qquad\qquad = -4 + 2(2)(3) + 9$
$\qquad\qquad\qquad = -4 + 12 + 9$
$\qquad\qquad\qquad = 8 + 9$
$\qquad\qquad\qquad = 17$

18. a. 1960 means substitute 3 for x:
Minimum wage
$$= -0.043x^3 + 0.52x^2 - 1.18x + 1.01$$
$$= -0.043(3)^3 + 0.52(3)^2 - 1.18(3) + 1.01$$
$$= -0.043(27) + 0.52(9) - 1.18(3) + 1.01$$
$$= -1.161 + 4.68 - 3.54 + 1.01$$
$$= 0.989$$
The minimum wage in 1960 was approximately \$0.99.

b. 1990 means substitute 6 for x:
Minimum wage
$$= -0.043x^3 + 0.52x^2 - 1.18x + 1.01$$
$$= -0.043(6)^3 + 0.52(6)^2 - 1.18(6) + 1.01$$
$$= -0.43(216) + 0.52(36) - 1.18(6) + 1.01$$
$$= -9.288 + 18.72 - 7.08 + 1.01$$
$$= 3.362$$
The minimum wage in 1990 was approximately \$3.36.

19. $3^{-2} = \dfrac{1}{3^2} = \dfrac{1}{9}$

20. $\left(\dfrac{3x^{-2}}{y}\right)^2 = \left(\dfrac{3}{x^2 y}\right)^2$
$$= \dfrac{3^2}{x^{2\cdot2} y^2}$$
$$= \dfrac{9}{x^4 y^2}$$

21. $\dfrac{3x^2 y^3 z^2}{9x^5 y^{-2} z^{-5}} = \left(\dfrac{3}{9}\right) \dfrac{y^{3-(-2)} z^{2-(-5)}}{x^{5-2}}$
$$= \dfrac{y^5 z^7}{3x^3}$$

22. $\left(\dfrac{-3x^3 y^{-2}}{x^{-1} y^5}\right)^{-3} = \left(\dfrac{-3x^{3-(-1)}}{y^{5-(-2)}}\right)^{-3}$
$$= \left(\dfrac{-3x^4}{y^7}\right)^{-3}$$
$$= \left(\dfrac{y^7}{-3x^4}\right)^3$$
$$= \dfrac{y^{7\cdot3}}{(-3)^3 x^{4\cdot3}}$$
$$= -\dfrac{y^{21}}{27x^{12}}$$

23. $242{,}000{,}000 = 2.42 \times 10^8$

24. $\dfrac{3.12 \times 10^6}{1.2 \times 10^{-2}} = \left(\dfrac{3.12}{1.2}\right) \times 10^{6-(-2)}$
$$= 2.6 \times 10^8$$
$$= 260{,}000{,}000$$

25. a. $0.542 \times 1.02 \times 10^8$
$$= 5.42 \times 10^{-1} \times 1.02 \times 10^8$$
$$= (5.42 \times 1.02)\left(10^{-1} \times 10^8\right)$$
$$\approx 5.528 \times 10^7$$

b. $0.458 \times 1.02 \times 10^8$
$$= 4.58 \times 10^{-1} \times 1.02 \times 10^8$$
$$= (4.58 \times 1.02)\left(10^{-1} \times 10^8\right)$$
$$\approx 4.672 \times 10^7$$

c. $5.528 \times 10^7 - 4.672 \times 10^7$
$$= (5.528 - 4.672) \times 10^7$$
$$= 0.856 \times 10^7$$
$$= 8.56 \times 10^6$$

Chapter 2

Exercise Set 2.1

1. The terms of an expression are the parts added.

3. **a.** The coefficient of $\dfrac{x+y}{4}$ or $\dfrac{1}{4}(x+y)$ is $\dfrac{1}{4}$.

   b. The coefficient of $-(x+3)$ or $-1(x+3)$ is -1.

   c. The coefficient of $-\dfrac{3(x+2)}{5}$ or $-\dfrac{3}{5}(x+2)$ is $-\dfrac{3}{5}$.

5. **a.** Like terms have the same variables and exponents.

   b. No; $3x$ and $3x^2$ are not like terms because the exponent on x is different for each term.

7. $2x+3 = x+5$
   $2(4)+3 \overset{?}{=} 4+5$
   $8+3 \overset{?}{=} 4+5$
   $11 \ne 9$

   No, 4 is not a solution to the equation because substituting 4 for x results in a false equation.

9. The addition property of equality states that if $a = b$, then $a + c = b + c$.

11. **a.** An identity is an equation that is always true.

    b. The solution set of an identity is $\mathbb{R}$.

13. **b.**
$$5x+2x-5 = 3(x-7)$$
$$7x-5 = 3(x-7)$$
$$7x-5 = 3x-21$$
$$7x-3x-5 = 3x-3x-21$$
$$4x-5 = -21$$
$$4x-5+5 = -21+5$$
$$4x = -16$$
$$\frac{4x}{4} = \frac{-16}{4}$$
$$x = -4$$

15. Symmetric property

17. Transitive property

19. Reflexive property

21. Addition property

23. Multiplication property

25. Multiplication property

27. $4x$ is a first degree term since $4x$ can be written as $4x^1$.

29. $3xy$ is a second degree term since $3xy$ can be written as $3x^1y^1$ and the sum of the exponents is $1 + 1 = 2$.

31. The degree of 7 is zero since 7 can be written as $7x^0$.

33. $-5x$ is a first degree term since $-5x$ can be written as $-5x^1$.

35. $3x^4y^6z^3$ is a thirteenth degree term since the sum of the exponents is $4 + 6 + 3 = 13$.

37. $3x^5y^6z$ is a twelfth degree term since $3x^5y^6z$ can be written as $3x^5y^6z^1$ and the sum of the exponents is $5 + 6 + 1 = 12$.

39. $8x + 7 + 7x - 12 = 8x + 7x + 7 - 12$
Combine like terms
$= 15x - 5$

41. $5x^2 - 3x + 2x - 5$
Combine like terms
$= 5x^2 - x - 5$

43. $10.6c^2 - 2.3c + 5.9c - 1.9c^2$
$= 10.6c^2 - 1.9c^2 - 2.3c + 5.9c$
Combine like terms
$= 8.7c^2 + 3.6c$

45. $6b^2 + 6b + 3a$ cannot be further simplified
since all of the terms are "unlike".

47. $xy + 3xy + y^2 - 2$
Combine like terms
$= 4xy + y^2 - 2$

49. $8.2(x - 3.4) - 1.2(9.8x + 12.4)$
Distributive property
$= 8.2x - 27.88 - 11.76x - 14.88$
Combine like terms
$= 8.2x - 11.76x - 27.88 - 14.88$
$= -3.56x - 42.76$

51. $3\left(x + \dfrac{1}{2}\right) - \dfrac{1}{3}x + 5$
Distributive property
$= 3x + \dfrac{3}{2} - \dfrac{1}{3}x + 5$
$= 3x - \dfrac{1}{3}x + \dfrac{3}{2} + 5$
$= \dfrac{9}{3}x - \dfrac{1}{3}x + \dfrac{3}{2} + \dfrac{10}{2}$
Combine like terms
$= \dfrac{8}{3}x + \dfrac{13}{2}$

53. $4 - [6(3x + 2) - x] + 4$
Distributive property
$= 4 - [18x + 12 - x] + 4$
Combine like terms
$= 4 - [17x - 12] + 4$
Distributive property
$= 4 - 17x - 12 + 4$
$= 4 - 12 + 4 - 17x$
Combine like terms
$= -4 - 17x$
$= -17x - 4$

55. $4x - [3x - (5x - 4y)] + y$
Distributive property
$= 4x - [3x - 5x + 4y] + y$
Combine like terms
$= 4x - [-2x + 4y] + y$
Distributive property
$= 4x + 2x - 4y + y$
Combine like terms
$= 6x - 3y$

57. $5b - \{7[2(3b - 2) - (4b + 9)] - 2\}$
Distributive property inside [ ]
$= 5b - \{7[6b - 4 - 4b - 9] - 2\}$
$= 5b - \{7[6b - 4b - 4 - 9] - 2\}$
Combine like terms
$= 5b - [7(2b - 13) - 2]$
Distributive property
$= 5b - [14b - 91 - 2]$
Combine like terms
$= 5b - (14b - 93)$
Distributive property
$= 5b - 14b + 93$
Combine like terms
$= -9b + 93$

59. $-\{[2rs - 3(r + 2s)] - 2(2r^2 - s)\}$
Distributive property
$= -\{[2rs - 3r - 6s] - 4r^2 + 2s\}$
$= -\{2rs - 3r - 6s - 4r^2 + 2s\}$
$= -(2rs - 3r - 6s + 2s - 4r^2)$
Combine like terms
$= -(2rs - 3r - 4s - 4r^2)$
Distributive property
$= -2rs + 3r + 4s + 4r^2$
$= 4r^2 - 2rs + 3r + 4s$

61.
$$3x + 5 = 17$$
$$3x + 5 - 5 = 17 - 5$$
$$3x = 12$$
$$\frac{3x}{3} = \frac{12}{13}$$
$$x = 4$$

63.
$$5x - 9 = 3(x - 2)$$
$$5x - 9 = 3x - 6$$
$$5x - 9 - 3x = 3x - 6 - 3x$$
$$2x - 9 = -6$$
$$2x - 9 + 9 = -6 + 9$$
$$2x = 3$$
$$\frac{2x}{2} = \frac{3}{2}$$
$$x = \frac{3}{2}$$

65.
$$4x - 8 = -4(2x - 3) + 4$$
$$4x - 8 = -8x + 12 + 4$$
$$4x - 8 = -8x + 16$$
$$4x - 8 + 8x = -8x + 16 + 8x$$
$$12x - 8 = 16$$
$$12x - 8 + 8 = 16 + 8$$
$$12x = 24$$
$$\frac{12x}{12} = \frac{24}{12}$$
$$x = 2$$

67.
$$4(x - 3) = 2(x + 9)$$
$$4x - 12 = 2x + 18$$
$$4x - 12 - 2x = 2x + 18 - 2x$$
$$2x - 12 = 18$$
$$2x - 12 + 12 = 18 + 12$$
$$2x = 30$$
$$\frac{2x}{2} = \frac{30}{2}$$
$$x = 15$$

69.
$$-3(t - 5) = 2(t - 5)$$
$$-3t + 15 = 2t - 10$$
$$-3t + 15 - 2t = 2t - 10 - 2t$$
$$-5t + 15 = -10$$
$$-5t + 15 - 15 = -10 - 15$$
$$-5t = -25$$
$$\frac{-5t}{-5} = \frac{-25}{-5}$$
$$t = 5$$

71.
$$3x + 4(x - 2) = 4x - 5$$
$$3x + 4x - 8 = 4x - 5$$
$$7x - 8 = 4x - 5$$
$$7x - 8 - 4x = 4x - 5 - 4x$$
$$3x - 8 = -5$$
$$3x - 8 + 8 = -5 + 8$$
$$3x = 3$$
$$\frac{3x}{3} = \frac{3}{3}$$
$$x = 1$$

73.
$$2 - (x + 5) = 4x - 8$$
$$2 - x - 5 = 4x - 8$$
$$-x - 3 = 4x - 8$$
$$-x - 3 - 4x = 4x - 8 - 4x$$
$$-5x - 3 = -8$$
$$-5x - 3 + 3 = -8 + 3$$
$$-5x = -5$$
$$\frac{-5x}{-5} = \frac{-5}{-5}$$
$$x = 1$$

75.
$$3y + 2(2y - 1) = 2(y - 6)$$
$$3y + 4y - 2 = 2y - 12$$
$$7y - 2 = 2y - 12$$
$$7y - 2 - 2y = 2y - 12 - 2y$$
$$5y - 2 = -12$$
$$5y - 2 + 2 = -12 + 2$$
$$5y = -10$$
$$\frac{5y}{5} = \frac{-10}{5}$$
$$y = -2$$

77.
$$5 - 3(2x + 1) = 4x - 8$$
$$5 - 6x - 3 = 4x - 8$$
$$2 - 6x = 4x - 8$$
$$2 - 6x - 4x = 4x - 8 - 4x$$
$$2 - 10x = -8$$
$$2 - 10x - 2 = -8 - 2$$
$$-10x = -10$$
$$\frac{-10x}{-10} = \frac{-10}{-10}$$
$$x = 1$$

79.
$$6 - (n + 3) = 3n + 5 - 2n$$
$$6 - n - 3 = 3n + 5 - 2n$$
$$3 - n = n + 5$$
$$3 - n + n = n + 5 + n$$
$$3 = 2n + 5$$
$$3 - 5 = 2n + 5 - 5$$
$$-2 = 2n$$
$$\frac{-2}{2} = \frac{2n}{2}$$
$$-1 = n$$
$$n = -1$$

81.
$$4(2x - 2) - 3(x + 7) = -4$$
$$8x - 8 - 3x - 21 = -4$$
$$5x - 29 = -4$$
$$5x - 29 + 29 = -4 + 29$$
$$5x = 25$$
$$\frac{5x}{5} = \frac{25}{5}$$
$$x = 5$$

83.
$$-4(3 - 4x) - 2(x - 1) = 12x$$
$$-12 + 16x - 2x + 2 = 12x$$
$$14x - 10 = 12x$$
$$14x - 10 - 14x = 12x - 14x$$
$$-10 = -2x$$
$$\frac{-10}{-2} = \frac{-2x}{-2}$$
$$5 = x$$
$$x = 5$$

85.
$$-(x - 8) = 6x - 4x - 2(x - 3)$$
$$-x + 8 = 6x - 4x - 2x + 6$$
$$-x + 8 = 6$$
$$-x + 8 - 8 = 6 - 8$$
$$-x = -2$$
$$\frac{-x}{-1} = \frac{-2}{-1}$$
$$x = 2$$

87.
$$5(x - 2) - 14x = -3x - (5 - 4x)$$
$$5x - 10 - 14x = -3x - 5 + 4x$$
$$-9x - 10 = x - 5$$
$$-9x - 10 + 9x = x - 5 + 9x$$
$$-10 = 10x - 5$$
$$-10 + 5 = 10x - 5 + 5$$
$$-5 = 10x$$
$$\frac{-5}{10} = \frac{10x}{10}$$
$$-\frac{1}{2} = x$$
$$x = -\frac{1}{2}$$

89.
$$2[3x - (4x - 6)] = 5(x - 6)$$
$$2(3x - 4x + 6) = 5x - 30$$
$$6x - 8x + 12 = 5x - 30$$
$$-2x + 12 = 5x - 30$$
$$-2x + 12 + 2x = 5x - 30 + 2x$$
$$12 = 7x - 30$$
$$12 + 30 = 7x - 30 + 30$$
$$42 = 7x$$
$$\frac{42}{7} = \frac{7x}{7}$$
$$6 = x$$
$$x = 6$$

91.
$$6 - \{4[x - (3x - 4) - x] + 4\} = 2(x + 3)$$
$$6 - [4(x - 3x + 4 - x) + 4] = 2(x + 3)$$
$$6 - [4(-3x + 4) + 4] = 2(x + 3)$$
$$6 - (-12x + 16 + 4) = 2(x + 3)$$
$$6 - (-12x + 20) = 2x + 6$$
$$6 + 12x - 20 = 2x + 6$$
$$-14 + 12x = 2x + 6$$
$$-14 + 12x - 2x = 2x + 6 - 2x$$
$$-14 + 10x = 6$$
$$-14 + 10x + 14 = 6 + 14$$
$$10x = 20$$
$$\frac{10x}{10} = \frac{20}{10}$$
$$x = 2$$

93.
$$-(3 - x) = 5 - \{6x - [2x - (3x - (5x - 8))]\}$$
$$-3 + x = 5 - [6x - (2x - (3x - 5x + 8))]$$
$$-3 + x = 5 - [6x - (2x - (-2x + 8))]$$
$$-3 + x = 5 - [6x - (2x + 2x - 8)]$$
$$-3 + x = 5 - [6x - (4x - 8)]$$
$$-3 + x = 5 - (6x - 4x + 8)$$
$$-3 + x = 5 - (2x + 8)$$
$$-3 + x = 5 - 2x - 8$$
$$-3 + x = -3 - 2x$$
$$-3 + x + 2x = -3 - 2x + 2x$$
$$-3 + 3x = -3$$
$$-3 + 3x + 3 = -3 + 3$$
$$3x = 0$$
$$\frac{3x}{3} = \frac{0}{3}$$
$$x = 0$$

95.
$$\frac{x}{3} = -12$$
$$3\left(\frac{x}{3}\right) = 3(-12)$$
$$x = -36$$

97.
$$\frac{4x - 2}{3} = -6$$
$$3\left(\frac{4x - 2}{3}\right) = 3(-6)$$
$$4x - 2 = -18$$
$$4x - 2 + 2 = -18 + 2$$
$$4x = -16$$
$$\frac{4x}{4} = \frac{-16}{4}$$
$$x = -4$$

99.
$$\frac{1}{3}x + \frac{1}{2}x = 10$$
$$6\left(\frac{1}{3}x + \frac{1}{2}x\right) = 6(10)$$
$$2x + 3x = 60$$
$$5x = 60$$
$$\frac{5x}{5} = \frac{60}{5}$$
$$x = 12$$

101.
$$4 - \frac{3}{4}x = 7$$
$$4 - \frac{3}{4}x - 4 = 7 - 4$$
$$-\frac{3}{4}x = 3$$
$$-4\left(-\frac{3}{4}x\right) = -4(3)$$
$$3x = -12$$
$$\frac{3x}{3} = \frac{-12}{3}$$
$$x = -4$$

103.
$$\frac{1}{2} = \frac{4}{5}x - \frac{1}{4}$$
$$20\left(\frac{1}{2}\right) = 20\left(\frac{4}{5}x - \frac{1}{4}\right)$$
$$10 = 16x - 5$$
$$10 + 5 = 16x - 5 + 5$$
$$15 = 16x$$
$$\frac{15}{16} = \frac{16x}{16}$$
$$\frac{15}{16} = x$$

105.
$$0.3x = x - 2.7$$
$$0.3x - x = x - 2.7 - x$$
$$-0.7x = -2.7$$
$$\frac{-0.7x}{-0.7} = \frac{-2.7}{-0.7}$$
$$x \approx 3.86$$

107.
$$4.7x - 3.6(x - 1) = 4.9$$
$$4.7x - 3.6x + 3.6 = 4.9$$
$$1.1x + 3.6 = 4.9$$
$$1.1x + 3.6 - 3.6 = 4.9 - 3.6$$
$$1.1x = 1.3$$
$$\frac{1.1x}{1.1} = \frac{1.3}{1.1}$$
$$x \approx 1.18$$

109.
$$0.047(3000 - x) = -0.06(x + 900)$$
$$141 - 0.047x = -0.06x - 54$$
$$141 - 0.047x + 0.06x = -0.06x - 54 + 0.06x$$
$$141 + 0.013x = -54$$
$$141 + 0.013x - 141 = -54 - 141$$
$$0.013x = -195$$
$$\frac{0.013x}{0.013} = \frac{-195}{0.013}$$
$$x = -15,000.00$$

111.
$$0.6(500 - 2.4x) = 3.6(2x - 4000)$$
$$300 - 1.44x = 7.2x - 14,400$$
$$300 - 1.44x + 1.44x = 7.2x - 14,400 + 1.44x$$
$$300 = 8.64x - 14,400$$
$$300 + 14,400 = 8.64x - 14,400 + 14,400$$
$$14,700 = 8.64x$$
$$\frac{14,700}{8.64} = \frac{8.64x}{8.64}$$
$$1701.39 \approx x$$
$$x \approx 1701.39$$

113.
$$0.04(1000) + 0.2(x + 2000) = 10,000$$
$$40 + 0.2x + 400 = 10,000$$
$$0.2x + 440 = 10,000$$
$$0.2x + 440 - 440 = 10,000 - 440$$
$$0.2x = 9560$$
$$\frac{0.2x}{0.2} = \frac{9560}{0.2}$$
$$x = 47,800$$

115.
$$2(x - 3) + x = 3(x + 4)$$
$$2x - 6 + x = 3x + 12$$
$$3x - 6 = 3x + 12$$
$$3x - 6 - 3x = 3x + 12 - 3x$$
$$-6 = 12$$
The solution set is $\varnothing$.
The equation is a contradiction.

117.
$$4(2x - 3) + 5 = -6(x - 4) + 12x - 31$$
$$8x - 12 + 5 = -6x + 24 + 12x - 31$$
$$8x - 7 = 6x - 7$$
$$8x - 7 + 7 = 6x - 7 + 7$$
$$8x = 6x$$
$$8x - 6x = 6x - 6x$$
$$2x = 0$$
$$x = 0$$
The solution set is $\{0\}$.
The equation is conditional.

119.
$$-4(2 + 4x) + 6x = -(6x + 8) - 4x$$
$$-8 - 16x + 6x = -6x - 8 - 4x$$
$$-8 - 10x = -10x - 8$$
$$-8 - 10x + 10x = -10x - 8 + 10x$$
$$-8 = -8$$
The solution set is $\mathbb{R}$.
The equation is an identity.

121.
$$6(x - 1) = -3(2 - x) + 3x$$
$$6x - 6 = -6 + 3x + 3x$$
$$6x - 6 = -6 + 6x$$
$$6x - 6 - 6x = -6 + 6x - 6x$$
$$-6 = -6$$
The solution set is $\mathbb{R}$.
The equation is an identity.

123.
$$3(x + 4) + 2x = 6 - (x - 3) + 6x$$
$$3x + 12 + 2x = 6 - x + 3 + 6x$$
$$12 + 5x = 9 + 5x$$
$$12 + 5x - 5x = 9 + 5x - 5x$$
$$12 = 9$$
The solution set is $\varnothing$.
The equation is a contradiction.

125. a. For 1994, substitute 2 for x.
$$I = 5x + 23$$
$$I = 5(2) + 23$$
$$I = 10 + 23$$
$$I = 33$$
There were 33 incidents in 1994.

b. Substitute 48 for I and solve for x.
$$48 = 5x + 23$$
$$48 - 23 = 5x + 23 - 23$$
$$25 = 5x$$
$$\frac{25}{5} = \frac{5x}{5}$$
$$5 = x$$
$x = 5$ represents the year 1997
There were 48 incidents in 1997.

127. a. For 1995, substitute 4 for x.
$$S = 10x + 20$$
$$S = 10(4) + 20$$
$$S = 40 + 20$$
$$S = 60$$
In 1995 the sales were $60 billion.

b. Substitute 100 for S and solve for x.
$$100 = 10x + 20$$
$$100 - 20 = 10x + 20 - 20$$
$$80 = 10x$$
$$\frac{80}{10} = \frac{10x}{10}$$
$$8 = x$$
$x = 8$ represents the year 1999
Sales of variable annuities will reach $100 billion in 1999.

129. Answers may vary. Possible answer:
$$2x = 8$$
$$x + 3 = 7$$
$$x - 2 = 2$$
All three equations, when simplified are equivalent to $x = 4$.

131. Answers may vary. One posible answer is $x + 5 = x + 5$. Make sure that the expressions on either side of the equal sign are equivalent.

133. Answers may vary. One possible answer is $4 + 3x + 1 = x + 9$.

135. $-3(x + 2) + 5x + 12 = n$
Substitute 6 for x and solve for n.
$$-3(6 + 2) + 5(6) + 12 = n$$
$$-3(8) + 30 + 12 = n$$
$$-24 + 30 + 12 = n$$
$$18 = n$$

137.
$$*\triangle - \square = \odot$$
$$*\triangle - \square + \square = \odot + \square$$
$$*\triangle = \odot + \square$$
$$\frac{*\triangle}{*} = \frac{\odot + \square}{*}$$
$$\triangle = \frac{\odot + \square}{*}$$

139.
$$\triangle(\odot + \square) = \otimes$$
$$\frac{\triangle(\odot + \square)}{\odot + \square} = \frac{\otimes}{\odot + \square}$$
$$\triangle = \frac{\otimes}{\odot + \square}$$

141. b. The definition of absolute value is
$$|a| = \begin{cases} a \text{ if } a \geq 0 \\ -a \text{ if } a < 0 \end{cases}$$

142. a. $-3^2 = -(3 \cdot 3) = -9$

b. $(-3)^2 = (-3)(-3) = 9$

143. $\left(-\frac{3}{4}\right)^3 = \left(-\frac{3}{4}\right)\left(-\frac{3}{4}\right)\left(-\frac{3}{4}\right) = -\frac{27}{64}$

144. $\sqrt[3]{-64} = -4$ since $(-4)^3 = -64$

Exercise Set 2.2

1. A mathematical model is an expression or equation that represents a real-life situation.

3. 1. Understand
2. Translate
3. Carry out
4. Check
5. Answer

5. a.
$$10 = 2(4) + 2w$$
$$10 = 8 + 2w$$
$$10 - 8 = 8 + 2w - 8$$
$$2 = 2w$$
$$\frac{2}{2} = \frac{2w}{2}$$
$$1 = w$$
$$w = 1$$

b.
$$P = 2l + 2w$$
$$P - 2l = 2l + 2w - 2l$$
$$P - 2l = 2w$$
$$\frac{P - 2l}{2} = \frac{2w}{2}$$
$$\frac{P - 2l}{2} = w$$
$$w = \frac{P - 2l}{2}$$

c. No; the same procedure was used for each solution.

d. $w = \dfrac{P - 2l}{2}$

$w = \dfrac{10 - 2(4)}{2}$

$w = \dfrac{10 - 8}{2}$

$w = \dfrac{2}{2}$

$w = 1$

They are the same.
The formula and the equation are equivalent when $P = 10$ and $l = 4$.

7. $A = lw$

$= 7(10)$

$= 70$

9. $P = 2l + 2w$

$= 2(15) + 2(6)$

$= 30 + 12$

$= 42$

11. $A = \pi r^2$

$= \pi(8)^2$

$= \pi(64)$

≈ 201.06

13. $A = \dfrac{1}{2} h\left(b_1 + b_2\right)$

$= \dfrac{1}{2}(10)(20 + 30)$

$= \dfrac{1}{2}(10)(50)$

$= 5(50)$

$= 250$

15. $P_1 = \dfrac{T_1 P_2}{T_2}$

$= \dfrac{250(300)}{500}$

$= \dfrac{75,000}{500}$

$= 150$

17. $m = \dfrac{y_2 - y_1}{x_2 - x_1}$

$= \dfrac{4 - (-3)}{-2 - (-6)}$

$= \dfrac{4 + 3}{-2 + 6}$

$= \dfrac{7}{4}$

19. $d = \sqrt{(x_2 - x_1)^2 + (y_2 - y_1)^2}$

$= \sqrt{(5 - (-3))^2 + (-6 - 3)^2}$

$= \sqrt{(5 + 3)^2 + (-6 - 3)^2}$

$= \sqrt{8^2 + (-9)^2}$

$= \sqrt{64 + 81}$

$= \sqrt{145} \approx 12.04$

21. $x = \dfrac{-b + \sqrt{b^2 - 4ac}}{2a}$

$= \dfrac{-(-5) + \sqrt{(-5)^2 - 4(2)(-12)}}{2(2)}$

$= \dfrac{5 + \sqrt{25 + 96}}{4}$

$= \dfrac{5 + \sqrt{121}}{4}$

$= \dfrac{5 + 11}{4}$

$= \dfrac{16}{4}$

$= 4$

23. $A = p\left(1 + \dfrac{r}{n}\right)^{nt}$

$= 100\left(1 + \dfrac{0.06}{1}\right)^{1 \cdot 3}$

$= 100(1.06)^3$

$= 100(1.191016)$

≈ 119.10

25.
$$3x + y = 5$$
$$3x + y - 3x = 5 - 3x$$
$$y = 5 - 3x$$
$$\text{or}$$
$$y = -3x + 5$$

27.
$$2x - y = -5$$
$$2x - y - 2x = -2x - 5$$
$$-y = -2x - 5$$
$$\frac{-y}{-1} = \frac{-2x - 5}{-1}$$
$$y = 2x + 5$$

29.
$$5x - 3y = -4$$
$$5x - 3y - 5x = -5x - 4$$
$$-3y = -5x - 4$$
$$\frac{-3y}{-3} = \frac{-5x - 4}{-3}$$
$$y = \frac{-1(5x + 4)}{-1 \cdot 3}$$
$$y = \frac{5x + 4}{3}$$

31.
$$\frac{1}{2}x + 2y = 6$$
$$2\left(\frac{1}{2}x + 2y\right) = 2 \cdot 6$$
$$x + 4y = 12$$
$$x + 4y - x = -x + 12$$
$$4y = -x + 12$$
$$\frac{4y}{4} = \frac{-x + 12}{4}$$
$$y = \frac{-x + 12}{4}$$

33.
$$3(x - 2) + 3y = 6x$$
$$3x - 6 + 3y = 6x$$
$$3x - 6 + 3y - 3x = 6x - 3x$$
$$-6 + 3y = 3x$$
$$-6 + 3y + 6 = 3x + 6$$
$$3y = 3x + 6$$
$$\frac{3y}{3} = \frac{3x + 6}{3}$$
$$y = x + 2$$

35.
$$3x - 5 = 2(3y + 6)$$
$$3x - 5 = 6y + 12$$
$$3x - 5 - 12 = 6y + 12 - 12$$
$$3x - 17 = 6y$$
$$\frac{3x - 17}{6} = \frac{6y}{6}$$
$$\frac{3x - 17}{6} = y$$
$$\text{or}$$
$$y = \frac{3x - 17}{6}$$

37.
$$d = rt$$
$$\frac{d}{r} = \frac{rt}{r}$$
$$\frac{d}{r} = t \text{ or } t = \frac{d}{r}$$

39.
$$A = lw$$
$$\frac{A}{w} = \frac{lw}{w}$$
$$\frac{A}{w} = l \text{ or } l = \frac{A}{w}$$

41.
$$P = 2l + 2w$$
$$P - 2l = 2l + 2w - 2l$$
$$P - 2l = 2w$$
$$\frac{P - 2l}{2} = \frac{2w}{2}$$
$$\frac{P - 2l}{2} = w \text{ or } w = \frac{P - 2l}{2}$$

43.
$$V = lwh$$
$$\frac{V}{lw} = \frac{lwh}{lw}$$
$$\frac{V}{lw} = h \text{ or } h = \frac{V}{lw}$$

45.
$$A = \frac{1}{2}bh$$
$$2A = 2 \cdot \frac{1}{2}bh$$
$$2A = bh$$
$$\frac{2A}{h} = \frac{bh}{h}$$
$$\frac{2A}{h} = b \text{ or } b = \frac{2A}{h}$$

47.
$$Ax + By = C$$
$$Ax + By - Ax = C - Ax$$
$$By = C - Ax$$
$$\frac{By}{B} = \frac{C - Ax}{B}$$
$$y = \frac{C - Ax}{B}$$

49.
$$y = mx + b$$
$$y - b = mx + b - b$$
$$y - b = mx$$
$$\frac{y - b}{x} = \frac{mx}{x}$$
$$\frac{y - b}{x} = m \text{ or } m = \frac{y - b}{x}$$

51.
$$y - y_1 = m(x - x_1)$$
$$\frac{y - y_1}{x - x_1} = \frac{m(x - x_1)}{x - x_1}$$
$$\frac{y - y_1}{x - x_1} = m \text{ or } m = \frac{y - y_1}{x - x_1}$$

53.
$$z = \frac{x - \mu}{\sigma}$$
$$\sigma z = \sigma\left(\frac{x - \mu}{\sigma}\right)$$
$$\sigma z = x - \mu$$
$$\sigma z - x = x - \mu - x$$
$$\sigma z - x = -\mu \quad \text{or} \quad \mu = x - z\sigma$$
$$x - \sigma z = \mu$$

55.
$$P_1 = \frac{T_1 P_2}{T_2}$$
$$T_2 P_1 = T_2\left(\frac{T_1 P_2}{T_2}\right)$$
$$T_2 P_1 = T_1 P_2$$
$$\frac{T_2 P_1}{P_1} = \frac{T_1 P_2}{P_1}$$
$$T_2 = \frac{T_1 P_2}{P_1}$$

57.
$$A = \frac{1}{2}h(b_1 + b_2)$$
$$2A = 2\left[\frac{1}{2}h(b_1 + b_2)\right]$$
$$2A = h(b_1 + b_2)$$
$$\frac{2A}{b_1 + b_2} = \frac{h(b_1 + b_2)}{b_1 + b_2}$$
$$\frac{2A}{b_1 + b_2} = h \text{ or } h = \frac{2A}{b_1 + b_2}$$

59.
$$S = \frac{n}{2}(f + l)$$
$$2S = 2\left[\frac{n}{2}(f + l)\right]$$
$$2S = n(f + l)$$
$$\frac{2S}{f + l} = \frac{n(f + l)}{f + l}$$
$$\frac{2S}{f + l} = n \text{ or } n = \frac{2S}{f + l}$$

61.
$$C = \frac{5}{9}(F - 32)$$
$$\frac{9}{5}C = \frac{9}{5} \cdot \frac{5}{9}(F - 32)$$
$$\frac{9}{5}C = F - 32$$
$$\frac{9}{5}C + 32 = F - 32 + 32$$
$$\frac{9}{5}C + 32 = F \text{ or } F = \frac{9}{5}C + 32$$

63.
$$F = \frac{km_1 m_2}{d^2}$$
$$Fd^2 = d^2\left(\frac{km_1 m_2}{d^2}\right)$$
$$Fd^2 = km_1 m_2$$
$$\frac{Fd^2}{km_2} = \frac{km_1 m_2}{km_2}$$
$$\frac{Fd^2}{km_2} = m_1 \text{ or } m_1 = \frac{Fd^2}{km_2}$$

65. a. If a ship is traveling k knots per hour, it travels $6076k$ feet per hour. If the speed is also m miles per hour, it is $5280m$ feet per hour. Thus, $5280m = 6076k$.

$$5280m = 6076k$$
$$m = \frac{6076k}{5280}$$
$$m \approx 1.15k$$

b. The quotient of 6076 and 5280 is about 1.15.

c. $m = 1.15(20.5)$
$m \approx 23.58$
It was traveling about 23.58 miles per hour.

67. $i = prt$
$ = 600(0.05)(3)$
$ = 90$
Paul must pay \$90 in simple interest.

69. $i = prt$
$1700 = 10,000(0.0425)t$
$1700 = 425t$
$\dfrac{1700}{425} = \dfrac{425t}{425}$
$ 4 = t$
The length of the loan was 4 years.

71. Let l = length, and w = width
$P = 2l + 2w$
$38 = 2(11) + 2w$
$38 = 22 + 2w$
$16 = 2w$
$8 = w$
The width is 8 feet.

73. Area of blue
= Area of big circle – Area of small circle
$= \pi r_1^2 - \pi r_2^2$
$= \pi(25)^2 - \pi(15)^2$
$= 625\pi - 225\pi$
$= 400\pi$
≈ 1256.64 square feet

75. a. $V = \pi r^2 h$
$ = \pi(4.5)^2(10.5)$
$ \approx 667.98$ cubic inches

b. $\dfrac{667.98}{231} \approx 2.89$ gallons

c. 2.89 gallons requires 2.89 ounces

77. $A = p\left(1 + \dfrac{r}{n}\right)^{nt}$
$ = 10,000\left(1 + \dfrac{0.06}{4}\right)^{4 \cdot 2}$
$ = 10,000(1.015)^8$
$ \approx 11,264.93$
Beth will have \$11,264.93 in her account.

79. $T_f = T_a(1 - F)$
$0.0425 = T_a(1 - 0.28)$
$0.0425 = T_a(0.72)$
$\dfrac{0.0425}{0.72} = T_a$
$ T_a \approx 0.059$

Since 5.9% is greater than 5%, the $4\frac{1}{4}\%$ tax-free investment is the better investment.

81. $m = -0.875x + 190$

a. $m = -0.875(50) + 190$
$ = -43.75 + 190$
$ = 146.25$
The maximum rate is 146.25 beats per minute.

b. $160 = -0.875x + 190$
$160 - 190 = -0.875x + 190 - 190$
$-30 = -0.875x$
$34.29 \approx x$
This person is about 34.29 years old.

83. a. $d = \dfrac{\frac{g}{3}}{u}$

b. $d = \dfrac{\frac{8600}{3}}{40}$

$d \approx \dfrac{2866.67}{40}$

$d \approx 71.67$

It should be drained and cleaned every 71.67 days.

85. a. $\text{BMI} = \dfrac{w}{h^2}$

b. $\text{BMI} = \dfrac{w(705)}{h^2}$

c. Answers will vary.

87. $-(5-8)^2 + |5-8| - 4^2 = -(-3)^2 + |-3| - 4^2$

$\qquad = -9 + 3 - 16$

$\qquad = -6 - 16$

$\qquad = -22$

88. $\dfrac{4 - 6 \div 3 + 5^2 - 6 \cdot 4}{5 - |6 \div (-2)|}$

$= \dfrac{4 - 6 \div 3 + 5^2 - 6 \cdot 4}{5 - |-3|}$

$= \dfrac{4 - 6 \div 3 + 25 - 6 \cdot 4}{5 - 3}$

$= \dfrac{4 - 2 + 25 - 24}{5 - 3}$

$= \dfrac{3}{2}$

89. Substitute 2 for x and 3 for y.

$6x^2 - 3xy + y^2 = 6(2)^2 - 3(2)(3) + (3)^2$

$\qquad = 6(4) - 3(2)(3) + 9$

$\qquad = 24 - 18 + 9$

$\qquad = 6 + 9$

$\qquad = 15$

90. $\qquad \dfrac{1}{3}x + 4 = \dfrac{2}{5}(x - 3)$

$15\left(\dfrac{1}{3}x\right) + 15(4) = 15\left[\dfrac{2}{5}(x - 3)\right]$

$\qquad 5x + 60 = 6(x - 3)$

$\qquad 5x + 60 = 6x - 18$

$\quad 5x - 5x + 60 = 6x - 5x - 18$

$\qquad 60 = x - 18$

$\qquad 60 + 18 = x - 18 + 18$

$\qquad 78 = x$

$\qquad x = 78$

Exercise Set 2.3

1. $\qquad B = 4A$

$\qquad A + B = 180$

$\qquad A + 4A = 180$

$\qquad 5A = 180$

$\qquad A = 36$

$\qquad B = 4A = 4(36) = 144$

The measure of angle A is 36° and B is 144°.

3. Let $x =$ smallest angle, then

$\qquad x + 20 =$ second angle

$\qquad 2x =$ third angle

$\qquad x + x + 20 + 2x = 180$

$\qquad 4x + 20 = 180$

$\qquad 4x = 160$

$\qquad x = 40$

$\qquad x + 20 = 40 + 20 = 60$

$\qquad 2x = 2(40) = 80$

The measure of the angles are 40°, 60°, and 80°.

5. Let $p =$ regular price of suit, then

$0.25p =$ amount of reduction

$\qquad p - 0.25p = 187.50$

$\qquad 0.75p = 187.50$

$\qquad p = 250$

The regular price of the suit is $250.

7. Let $x =$ number of weeks.

$\qquad 12.50x = 940$

$\qquad \dfrac{12.50x}{12.50} = \dfrac{940}{12.50}$

$\qquad x = 75.2$

It will take 75.2 weeks for the two costs to be the same.

9. Let p = maximum price, then
$0.073p$ = sales tax
$$p + 0.073p = 22,600$$
$$1.073p = 22,600$$
$$p \approx 21,062.44$$
The maximum price that Shane can pay for a new car is $21,062.44

11. Let n = number of miles.
$$0.20n + 35 = 80$$
$$0.20n = 45$$
$$\frac{0.20n}{0.20} = \frac{45}{0.20}$$
$$n = 225 \text{ miles}$$
Tanya can drive 255 miles.

13. Let t = number of trips, then
$2t$ = cost for one trip without pass
$$0.50t + 20 = 2t$$
$$20 = 1.5t$$
$$13.3 \approx t$$
The Oses would have to go more than 13 times for the cost of the monthly pass to be worthwhile.

15. Let t = tons of water displaced by an aircraft carrier, then $4.45t$ = tons of water displaced by the *Jahre Viking*
$$t + 4.45t = 622,608$$
$$5.45t = 622,608$$
$$t \approx 114,240$$
An aircraft carrier displaces 114,240 tons of water.
$4.45t = 4.45(114,240) = 508,368$
The *Jahre Viking* displaces 508,368 tons of water.

17. Let g = amount contributed to global equities fund, then $2g - 250$ = amount contributed to stock fund
$$g + 2g - 250 = 5000$$
$$3g - 250 = 5000$$
$$3g = 5250$$
$$g = 1750$$
$$2g - 250 = 2(1750) - 250 = 3250$$
Rich contributes $1750 to the global equities fund and $3250 to the stock fund.

19. Let x = number of grasses. Then $2x - 5$ = number of weeds and $2x + 2$ is the number of trees.
$$x + (2x - 5) + (2x + 2) = 57$$
$$5x - 3 = 57$$
$$5x = 60$$
$$\frac{5x}{5} = \frac{60}{5}$$
$$x = 12$$
There are 12 grasses,
$2(12) - 5 = 24 - 5 = 19$ weeds, and
$2(12) + 2 = 24 + 2 = 26$ trees.

21. Let x = number of years.
$$460(0.10)x = 260$$
$$46x = 260$$
$$\frac{46x}{46} = \frac{260}{46}$$
$$x \approx 5.65$$
It would take 5.65 years for the security device to pay for itself.

23. Let x = price of lunch
$$x + 0.07x + 0.15x = 15.75$$
$$1.22x = 15.75$$
$$x \approx 12.91$$
The maximum price of the lunch she can order is $12.91.

25. Let m = median number of hours worked in 1973 then $0.251m$ = increase in the median number of hours worked.
$$m + 0.251m = 50.8$$
$$1.251m = 50.8$$
$$m \approx 40.6$$
The median number of hours worked per week in 1973 was 40.6 hours.

27. Let d = number shipped by Dell, then $d + 428,000$ = number shipped by IBM,
$d + 428,000 + 812,000$ = number shipped by Packard Bell-NEC
$2d - 115,000$ = number shipped by Compaq.
$d + (d + 428,000) + (d + 428,000 + 812,000) + (2d - 115,000) = 10,508,000$

$$5d + 1,553,000 = 10,508,000$$
$$d = 1,797,000$$
$$d + 428,000 = 2,219,000$$
$$d + 428,000 + 812,000 = 3,031,000$$
$$2d - 115,000 = 3,467,000$$

The top 4 manufacturers shipped the following numbers of computers.
Dell: 1,791,000
IBM: 2,219,000
Packard Bell-NEC: 3,031,000
Compaq: 3,467,000

29. a. Let x = number of months for the total payments to be the same.
$$563.50x = 538.30x + 0.02(70,000) + 200$$
$$563.50x = 538.30x + 1400 + 200$$
$$563.50x = 538.30x + 1600$$
$$25.2x = 1600$$
$$\frac{25.2x}{25.2} = \frac{1600}{25.2}$$
$$x \approx 63.49$$
It takes about 63.49 months (5.29 years) for the total payments to be the same.

b. If they plan to keep the house for 30 years, the lower total cost would be with the mortgage from First National.

31. a. Let x = number of months (or monthly payments) necessary for the accumulated payments under the original mortgage plan to equal the accumulated payments and closing cost under the other plan.
$$510x = 420.50x + 2500$$
$$89.50x = 2500$$
$$\frac{89.50x}{89.50} = \frac{2500}{89.50}$$
$$x \approx 28$$
In about 28 months or 2.33 years, he would have paid the same amount under either plan.

b. Yes, any time after 2.33 years makes the refinancing worth it.

33. Let x = length of one side of the square. Since there are 7 sides, the total perimeter is $7x$.
$$7x = 91$$
$$\frac{7x}{7} = \frac{91}{7}$$
$$x = 13$$
The dimensions of each square will be 13 meters by 13 meters.

35. Let w = the width of each rectangle. Then $w + 1$ is the length. The fencing runs along 6 widths and 4 lengths.
$$6w + 4(w + 1) = 114$$
$$6w + 4w + 4 = 114$$
$$10w = 110$$
$$w = 11$$
$$w + 1 = 12$$
Each rectangle has width 11 meters and length 12 meters.

37. Let p = the original price of the calculator.

$$\begin{aligned} p - 0.10p - 5 &= 49 \\ 0.90p - 5 &= 49 \\ 0.90p &= 54 \\ \frac{0.90p}{0.90} &= \frac{54}{0.90} \\ p &= 60 \end{aligned}$$

The original price of the calculator was $60.

39. Let x = number of paintings to be sold. The break-even point occurs when sales equals expenses.

$$\begin{aligned} 50x &= 810 + 0.10(50x) \\ 50x &= 810 + 5x \\ 45x &= 810 \\ \frac{45x}{45} &= \frac{810}{45} \\ x &= 18 \end{aligned}$$

The break-even point occurs when 18 paintings are sold.

41. Let x = clothing allowance from Social Security.

$$\begin{aligned} x + (375 - x) + 0.07(375 - x) &= 375 + 17.50 \\ 375 + 0.07(375 - x) &= 375 + 17.50 \\ 0.07(375 - x) &= 17.50 \\ 26.25 - 0.07x &= 17.50 \\ -0.07x &= -8.75 \\ \frac{-0.07x}{-0.07} &= \frac{-8.75}{-0.07} \\ x &= 125 \end{aligned}$$

Thus, Stan's clothing allowance is $125.

43. Let x = bill before tax. then

$\frac{5}{8}x$ = amount paid by the Newton family

and $\frac{3}{8}x + 0.15x$ is the amount paid by the Lee family. The equation is

$$\begin{aligned} \frac{5}{8}x + \frac{3}{8}x + 0.15x &= 184.60 \\ 1.15x &= 184.60 \\ \frac{1.15x}{1.15} &= \frac{184.60}{1.15} \\ x &\approx 160.52 \end{aligned}$$

The amount paid by the Newton family is $\frac{5}{8}(160.52) = \$100.33$ and the amount paid by the Lee family is $\$184.60 - \$100.33 = \$84.27.$

45. a. Let x be the score for the final exam.

$$\begin{aligned} \frac{70 + 83 + 97 + 84 + 74 + x + x}{7} &= 80 \\ \frac{408 + 2x}{7} &= 80 \\ 7\left(\frac{408 + 2x}{7}\right) &= 7(80) \\ 408 + 2x &= 560 \\ 2x &= 152 \\ \frac{2x}{x} &= \frac{152}{2} \\ x &= 76 \end{aligned}$$

Philip needs to score 76 points on the final exam to have an average score of 80 points.

b. Again, let x be the score for the final exam.

$$\frac{70 + 83 + 97 + 84 + 74 + 2x}{7} = 90$$

$$\frac{408 + 2x}{7} = 90$$

$$7\left(\frac{408 + 2x}{7}\right) = 7(90)$$

$$408 + 2x = 630$$

$$2x = 222$$

$$\frac{2x}{x} = \frac{222}{2}$$

$$x = 111$$

No, in order to have a final average of 90, Philip will need a score of 111 on the final exam. Since this is impossible, he must settle for a final grade lower than 90.

47. Answers will vary.

49. Let x = number of miles driven.
$3(28) + 0.15x + 0.04[3(28) + 0.15x] = 121.68$
Original Charge   4% Sales Tax
$84 + 0.15x + 0.04(84 + 0.15x) = 121.68$
$84 + 0.15x + 3.36 + 0.006x = 121.68$
$87.36 + 0.156x = 121.68$
$0.156x = 34.32$
$$\frac{0.156x}{0.156} = \frac{34.32}{0.156}$$
$$x = 220$$
Denise drove a total of 220 miles during the three days.

51. $2 + \left|-\dfrac{3}{5}\right| = 2 + \dfrac{3}{5} = \dfrac{10}{5} + \dfrac{3}{5} = \dfrac{13}{5}$

52. $-6.4 - (-3.7) = -6.4 + 3.7 = -2.7$

53. $\left|-\dfrac{5}{8}\right| \div |-2| = \dfrac{5}{8} \div 2 = \dfrac{5}{8} \cdot \dfrac{1}{2} = \dfrac{5}{16}$

54. $5 - |-3| - |7| = 5 - 3 - 7 = 2 - 7 = -5$

55. $\left(2x^4 y^{-6}\right)^{-3} = 2^{-3}\left(x^4\right)^{-3}\left(y^{-6}\right)^{-3}$
$$= \dfrac{1}{8} x^{-12} y^{18}$$
$$= \dfrac{y^{18}}{8x^{12}}$$

Exercise Set 2.4

1.

	Rate	Time	Distance
Don	5	1.2	5(1.2)
Judy	4.5	1.2	4.5(1.2)

distance $= 5(1.2) + 4.5(1.2)$
$= 6 + 5.4$
$= 11.4$
The distance around the lake is 11.4 miles

3. Let t = time in hours

Balloon	Rate	Time	Distance
1	16	t	$16t$
2	14	t	$14t$

distance apart = balloon 1 distance – balloon 2 distance
$4 = 16t - 14t$
$4 = 2t$
$2 = t$
It will take 2 hours for the balloons to be 4 miles apart.

5. Let r = rate of freight train

	Rate	**Time**	**Distance**
Passenger train	$r + 18$	3	$3(r + 18)$
Freight train	r	$3 + 1.2 = 4.2$	$4.2r$

$3(r + 18) = 4.2r$
$3r + 54 = 4.2r$
$\quad\quad 54 = 1.2r$
$\quad\quad 45 = r$
$\quad\quad 63 = r + 18$

The passenger train travels at 63 mph and the freight train travels at 45 mph.

7. a. Let r = rate for the jogger

	Rate	**Time**	**Distance**
Jogger	r	2	$2r$
Cyclist	$4r$	2	$2(4r)$

cyclist's distance – jogger's distance = miles apart
$2(4r) - 2r = 18$
$\quad 8r - 2r = 18$
$\quad\quad\quad 6r = 18$
$\quad\quad\quad\quad r = 3$
$4r = 12$
Wayne rides at 12 mph.

b. distance = $8r = 8(3) = 24$
Wayne rode 24 miles.

c. $\$1.50(24) = \$36$
Bob Johnson pledged $36.

9. a. Let t = time to reach bottom of canyon

	Rate	**Time**	**Distance**
Trip down	3.4	t	$3.4t$
Trip Up	1.2	$12 - t$	$1.2(12 - t)$

distance down = distance up
$\quad\quad 3.4t = 1.2(12 - t)$
$\quad\quad 3.4t = 14.4 - 1.2t$
$\quad\quad 4.6t = 14.4$
$\quad\quad\quad\, t \approx 3.13$

It took her about 3.13 hours to reach the bottom of the canyon.

  b.
$$\begin{aligned}
\text{total distance} &= 2(\text{distance down})\\
&= 2(3.4 \cdot 3.13)\\
&= 2(10.642)\\
&= 21.284
\end{aligned}$$
The total distance traveled is about 21.3 miles.

11. Let t = time of operation for smaller machine

	Rate	Time	Amount
Smaller machine	400	t	$400t$
Larger machine	600	$t + 2$	$600(t + 2)$

$$\begin{aligned}
400t + 600(t + 2) &= 15{,}000\\
400t + 600t + 1200 &= 15{,}000\\
1000t &= 13{,}800\\
t &= 13.8
\end{aligned}$$
The smaller machine operated for 13.8 hours.

13. Let p = amount invested at 4%.

Principal	Rate	Time	Interest
p	0.04	1	$0.04p$
$11{,}000 - p$	0.05	1	$0.05(11{,}000 - p)$

The total interest is $530.
$$\begin{aligned}
0.04p + 0.05(11{,}000 - p) &= 530\\
0.04p + 550 - 0.05p &= 530\\
-0.01p &= -20\\
p &= 2000
\end{aligned}$$
$11000 - p = 9000$
Pat invested $2000 at 4% and $9000 at 5%.

15. a. Let m = the number of shares of microsoft

Stock	Price	Shares	Cost
Microsoft	108.75	m	$108.75m$
Hilton	27.25	$4m$	$27.25(4m)$
Total		$5m$	$108.75m + 27.25(4m)$

The total cost should be $10,000.
$$\begin{aligned}
108.75m + 27.25(4m) &= 10{,}000\\
108.75m + 109m &= 10{,}000\\
217.75m &= 10{,}000\\
m &\approx 45.92
\end{aligned}$$
Since only whole shares of stock can be purchased, Bob should buy 45 shares of Microsoft and $4(45) = 180$ shares of Hilton.

b. $108.75(45) + 27.25(180) = 4893.75 + 4905 = 9798.75$
$10,000 - 9798.75 = 201.25$
Bob had $201.25 left.

17. Let x = pounds of Kona coffee

Item	Cost	Pounds	Total
Kona	6.20	x	$6.20x$
Amaretto	5.80	18	$5.80(18)$
Mixture	6.10	$18 + x$	$6.10(18 + x)$

$6.20x + 5.80(18) = 6.10(18 + x)$
$6.2x + 104.4 = 109.8 + 6.1x$
$0.1x = 5.4$
$x = 54$

She should mix 54 pounds of Kona coffee with the amaretto coffee.

19. Let x = ounces of 12% solution

Solution	Strength	Ounces	Acid
Mail	12%	x	$0.12x$
Store	5%	40	$0.05(40)$
Mixture	8%	$x + 40$	$0.08(x + 40)$

$0.12x + 0.05(40) = 0.08(x + 40)$
$0.12x + 2 = 0.08x + 3.2$
$0.04x = 1.2$
$x = 30$

She should mix 30 ounces of the 12% vinegar

21. Let c = ounces of champagne to add

Item	Strength	Ounces	Amount of Alcohol
Champagne	0.115	c	$0.115c$
Juice-ale	0.0	32	$0(32)$
Mixture	0.05	$c + 32$	$0.05(c + 32)$

$0.115c + 0(32) = 0.05(c + 32)$
$0.115c = 0.05c + 1.6$
$0.065c = 1.6$
$c \approx 24.6$

Joaquin should add about 24.6 ounces of champagne to the mixture.

23. Let x = germination rate of the higher-quality seed.

Type	Germination Rate	Quantity	Amount
Low	0.76	16	0.76(16)
High	x	12	$12x$
Mixture	0.82	28	0.82(28)

$0.76(16) + 12x = 0.82(28)$
$12.16 + 12x = 22.96$
$\qquad 12x = 10.8$
$\qquad x = \dfrac{10.8}{12} = 0.90 \text{ or } 90\%$

The germination rate for the higher-quality seed is 90%.

25. Let x = Mr. Juenger's portion of the income tax deduction.

Person	Income	Deduction	Taxable
Mrs.	32,450	$6400 - x$	$32,450 - (6400 - x)$
Mr.	28,200	x	$28,200 - x$

The smallest amount of tax occurs when the taxable income is the same.
$28,200 - x = 32,450 - (6400 - x)$
$28,200 - x = 26,050 + x$
$\qquad 2150 = 2x$
$\qquad \dfrac{2150}{2} = x \text{ or } x = 1075$

Thus, Mr. Juenger's tax deduction is $1075 and Mrs. Juenger's tax deduction is $6400 - $1075 = $5325.

27. Let r = Bob's speed

	Rate	Time	Distance
Bob	r	4	$4r$
Julie	$r + 10$	4	$4(r + 10)$

Bob's distance + Julie's distance = Total distance apart
$4r + 4(r + 10) = 480$
$\quad 4r + 4r + 40 = 480$
$\qquad\qquad 8r = 440$
$\qquad\qquad r = 55$
$r + 10 = 65$
Bob is traveling at 55 mph and Julie at 65 mph.

29. Let x = amount invested at 6%.

Principal	Rate	Time	Interest
x	0.06	1	$0.06x$
$8000 - x$	0.10	1	$0.10(8000 - x)$

The interest is the same for both accounts.

$0.06x = 0.10(8000 - x)$
$0.06x = 800 - 0.10x$
$0.16x = 800$
$x = \dfrac{800}{0.16} = 5000$

Thus, $5000 was invested at 6% and the remaining amount of $8000 − $5000 = $3000 was invested at 10%.

31. Let t = amount of time needed to fly over land.

Trip	Rate	Time	Distance
Land	500	t	$500t$
Ocean	550	$2t$	$550(2t)$

The total distance is 5200 miles.

$500t + 550(2t) = 5200$
$500t + 1100t = 5200$
$1600t = 5200$
$t = \dfrac{5200}{1600} = 3.25$

The time over land was 3.25 hours and the time over the ocean was $2(3.25) = 6.50$ hours for a total of 9.75 hours.

33. a. Let t = time before the jets meet.

	Rate	Time	Distance
Jet	800	t	$800t$
Refueling plane	520	$t + 2$	$520(t + 2)$

The distances traveled are equal.

$800t = 520(t + 2)$
$800t = 520t + 1040$
$280t = 1040$
$t \approx 3.7143$

The two planes will meet in approximately 3.71 hours.

b. $800t = 800(3.7143) = 2971.44$
The refueling will take place approximately 2971.4 miles from the base.

35. Let x = number of hours worked at $6.00 per hour.

Job	Rate	No. of Hours	Total Pay
1	$6.00	x	$6.00x$
2	$6.50	$18 - x$	$6.50(18 - x)$

The pay for both jobs is $114,.

$$6.00x + 6.50(18 - x) = 114$$
$$6.00x + 117 - 6.50x = 114$$
$$-0.50x + 117 = 114$$
$$-0.50x = -3$$
$$x = \frac{-3}{-0.50} = 6 \text{ hours}$$

Kelli worked 6 hours at $6.00 per hour and $18 - 6 = 12$ hours at $6.50 per hour.

37. Let r = speed during the first hour.

Trip	Rate	Time	Distance
1st hour	r	1	$1(r)$
2nd hour	$r - 16$	1	$1(r - 16)$

The total distance is 100 miles.

$$1(r) + 1(r - 16) = 100$$
$$r + r - 16 = 100$$
$$2r - 16 = 100$$
$$2r = 116$$
$$r = \frac{116}{2} = 58$$

The rate for the first hour of the trip was 58 mph.

39. Let x = amount of 80% solution needed.

Solution	Strength of Solution	No. of Ounces	Amount of Alcohol
80%	0.80	x	$0.80x$
Water	0	$128 - x$	$0(128 - x)$
6%	0.06	128	$0.06(128)$

$$0.80x + 0(128 - x) = 0.06(128)$$
$$0.80x = 7.68$$
$$x = \frac{7.68}{0.80} = 9.6$$

Herb should combine 9.6 ounces of the 80% solution with $128 - 9.6 = 118.4$ ounces of water to produce the desired solution.

41. Let x = amount of 1.5% butterfat milk needed.

Type	Strength	No. of Quarts	Amount of Butterfat
5%	0.05	400	$0.05(400)$
1.5%	0.015	x	$0.015x$
2%	0.02	$400 + x$	$0.02(400 + x)$

$$0.05(400) + 0.015x = 0.02(400 + x)$$
$$20 + 0.015x = 8 + 0.02x$$
$$20 = 8 + 0.005x$$
$$12 = 0.005x$$
$$\frac{12}{0.005} = x$$
$$2400 = x$$

Sundance Dairy should combine 2400 quarts of 1.5% butterfat milk with 400 quarts of 5% butterfat milk to produce 2800 quarts of 2% butterfat milk.

43. Let t = time each machine works after the new machine is turned on.

Type	Rate	Time	No. of Cartons
Old	50	t	$50t + 200$
New	70	t	$70t$

$$70t = 50t + 200$$
$$20t = 200$$
$$t = \frac{200}{20} = 10 \text{ minutes}$$

It takes the new machine 10 minutes to produce the same total number of cartons as the old machine.

45. Let D = Total distance apart the actors should be when they start

Actor	Rate	Time	Distance
Bruce	6	5	$6(5)$
Robert	6	5	$6(5)$

Bruce's distance + Robert's distance = Total distance − 20

$$6(5) + 6(5) = D - 20$$
$$30 + 30 = D - 20$$
$$80 = D$$

The actors should start 80 feet apart.

47. Answers will vary.

49. It is possible to determine the times for the 2nd and 3rd parts of the trip.

2nd Part: $t = \dfrac{d}{r} = \dfrac{31}{90} \approx 0.344$ hour

3rd Part: $t = \dfrac{d}{r} = \dfrac{68}{45} \approx 1.511$ hours

The time for the first part (Paris to Calais) is $3.000 - 0.344 - 1.511 = 1.145$ hours. The distance is
(130 mph)(1.145 hours) ≈ 149 miles

51. Let x be the amount of 20% solution which must be drained. Then, $16 - x$ is the amount remaining.

$$0.20(16 - x) + 1.00x = 0.50(16)$$
$$3.2 - 0.20x + 1.00x = 8$$
$$3.2 + 0.80x = 8$$
$$0.80x = 4.8$$
$$x = \frac{4.8}{0.80} = 6$$

Thus, 6 quarts must be drained before adding the same amount of antifreeze.

52.
$$0.6x + 0.22 = 0.4(x - 2.3)$$
$$0.6x + 0.22 = 0.4x - 0.92$$
$$0.6x - 0.4x + 0.22 = 0.4x - 0.4x - 0.92$$
$$0.2x + 0.22 = -0.92$$
$$0.2x + 0.22 - 0.22 = -0.92 - 0.22$$
$$0.2x = -1.14$$
$$\frac{0.2x}{0.2} = \frac{-1.14}{0.2}$$
$$x = -5.7$$

53.
$$\frac{2}{9}x + 3 = x + \frac{1}{5}$$
$$45\left(\frac{2}{9}x\right) + 45(3) = 45(x) + 45\left(\frac{1}{5}\right)$$
$$10x + 135 = 45x + 9$$
$$10x - 10x + 135 = 45x - 10x + 9$$
$$135 = 35x + 9$$
$$135 - 9 = 35x + 9 - 9$$
$$126 = 35x$$
$$\frac{126}{35} = \frac{35x}{35}$$
$$\frac{18}{5} = x$$

54.
$$\frac{3}{5}(x - 2) = \frac{2}{7}(2x + 3y)$$
$$35\left[\frac{3}{5}(x - 2)\right] = 35\left[\frac{2}{7}(2x + 3y)\right]$$
$$21(x - 2) = 10(2x + 3y)$$
$$21x - 42 = 20x + 30y$$
$$21x - 20x - 42 = 20x - 20x + 30y$$
$$x - 42 = 30y$$
$$\frac{x - 42}{30} = \frac{30y}{30}$$
$$\frac{x - 42}{30} = y \text{ or } y = \frac{x - 42}{30}$$

55. Let x be the distance driven in one day.
$$30 + 0.14x = 16 + 0.24x$$
$$30 = 16 + 0.10x$$
$$14 = 0.10x$$
$$\frac{14}{0.10} = x$$
$$x = 140 \text{ miles}$$

The costs are the same when 140 miles are driven per day.

Exercise Set 2.5

1. $x \leq 7$ includes 7 and $x < 7$ does not include 7.

3. a. Use open circles when the endpoints are not included.

   b. Use closed circles when the endpoints are included.

  c. Answers may vary. One possible answer is $x < 4$.

  d. Answers may vary. One possible answer is $x \geq 4$.

5. No real number is both greater than 4 and less than 2.

7. a.

-4

  b. $(-\infty, -4)$

  c. $\{x | x < -4\}$

9. a.

5.2

  b. $[5.2, \infty)$

  c. $\{x | x \geq 5.2\}$

11. a.

$2 \qquad \frac{12}{5}$

  b. $\left[2, \dfrac{12}{5}\right)$

  c. $\left\{x \middle| 2 \leq x < \dfrac{12}{5}\right\}$

13. a.

$-7 \qquad -4$

  b. $(-7, -4]$

  c. $\{x | -7 < x \leq -4\}$

15. $x + 2 < 8$
$\qquad x < 6$

6

17. $3 - x < -4$
$\qquad -x < -7$
Reverse the inequality
$$\frac{-x}{-1} > \frac{-7}{-1}$$
$$x > 7$$

7

19. $\qquad 4.7x - 5.48 \geq 11.44$
$4.7x - 5.48 + 5.48 \geq 11.44 + 5.48$
$\qquad\qquad 4.7x \geq 16.92$
$$\frac{4.7x}{4.7} \geq \frac{16.92}{4.7}$$
$\qquad\qquad x \geq 3.6$

3.6

21. $4(x - 2) \leq 4x - 8$
$\qquad 4x - 8 \leq 4x - 8$
$\qquad\qquad -8 \leq -8$
Since this is a true statement, the solution is the entire real number line.

0

23. $4b - 6 \geq 3(b + 3) + 2b$
$\qquad 4b - 6 \geq 3b + 9 + 2b$
$\qquad 4b - 6 \geq 5b + 9$
$\qquad\qquad -6 \geq b + 9$
$\qquad\qquad\quad b \leq -15$

-15

25. $\qquad \dfrac{y}{3} + \dfrac{2}{5} \leq 4$
$$15\left(\frac{y}{3}\right) + 15\left(\frac{2}{5}\right) \leq 15(4)$$
$\qquad\qquad 5y + 6 \leq 60$
$\qquad\qquad\quad 5y \leq 54$
$$\frac{5y}{5} \leq \frac{54}{5}$$
$\qquad\qquad\quad y \leq \frac{54}{5}$

$\frac{54}{5}$

27. $\dfrac{c+4}{2}+9>c+2$

$\dfrac{c+4}{2}>c-7$

$2\left(\dfrac{c+4}{2}\right)>2(c)-2(7)$

$c+4>2c-14$

$4>c-14$

$18>c$

$(-\infty,\ 18)$

29. $4+\dfrac{4x}{3}<6$

$\dfrac{4x}{3}<2$

$3\left(\dfrac{4x}{3}\right)<3(2)$

$4x<6$

$\dfrac{4x}{4}<\dfrac{6}{4}$

$x<\dfrac{3}{2}$

$\left(-\infty,\ \dfrac{3}{2}\right)$

31. $\dfrac{5-6y}{3}\le1-2y$

$3\left(\dfrac{5-6y}{3}\right)\le3(1)-3(2y)$

$5-6y\le3-6y$

$5\le3$

This is a false statement which means there is no solution: $\varnothing$.

33. $-3x+1<3\big[(x+2)-2x\big]-1$

$-3x+1<3[-x+2]-1$

$-3x+1<-3x+6-1$

$-3x+1<-3x+5$

$1<5$

This is a true statement, so the solution set is $(-\infty,\infty)$.

35. $1<x+3<9$

$1-3<x+3-3<9-3$

$-2<x<6$

$(-2,6)$

37. $-3<5x\le8$

$\dfrac{-3}{5}<\dfrac{5x}{5}\le\dfrac{8}{5}$

$\dfrac{-3}{5}<x\le\dfrac{8}{5}$

$\left(-\dfrac{3}{5},\ \dfrac{8}{5}\right]$

39. $4\le2x-4<7$

$4+4\le2x-4+4<7+4$

$8\le2x<11$

$\dfrac{8}{2}\le\dfrac{2x}{2}<\dfrac{11}{2}$

$4\le x<\dfrac{11}{2}$

$\left[\dfrac{4}{2},\ \dfrac{11}{2}\right)$

41. $4.3<3.2x-2.1\le16.46$

$4.3+2.1<3.2x-2.1+2.1\le16.46+2.1$

$6.4<3.2x\le18.56$

$\dfrac{6.4}{3.2}<\dfrac{3.2x}{3.2}\le\dfrac{18.56}{3.2}$

$2<x\le5.8$

$(2,5.8]$

43. $4<\dfrac{4x-2}{2}\le12$

$2(4)<2\left(\dfrac{4x-2}{2}\right)\le2(12)$

$8<4x-2\le24$

$8+2<4x-2+2\le24+2$

$10<4x\le26$

$\dfrac{10}{4}<\dfrac{4x}{4}\le\dfrac{26}{4}$

$\dfrac{5}{2}<x\le\dfrac{13}{2}$

$\left\{x\,\middle|\,\dfrac{5}{2}<x\le\dfrac{13}{2}\right\}$

45. $6 \leq -3(2x - 4) < 12$

$6 \leq -6x + 12 < 12$

$6 - 12 \leq -6x + 12 - 12 < 12 - 12$

$-6 \leq -6x < 0$

Divide by -6 and reverse inequalities

$\dfrac{-6}{-6} \geq \dfrac{-6x}{-6} > \dfrac{0}{-6}$

$1 \geq x > 0$

$0 < x \leq 1$

$\{x \mid 0 < x \leq 1\}$

47. $0 < \dfrac{2(x - 3)}{5} \leq 12$

$5(0) < 5\left[\dfrac{2(x - 3)}{5}\right] \leq 5(12)$

$0 < 2(x - 3) \leq 60$

$\dfrac{0}{2} < \dfrac{2(x - 3)}{2} \leq \dfrac{60}{2}$

$0 < x - 3 \leq 30$

$0 + 3 < x - 3 + 3 \leq 30 + 3$

$3 < x \leq 33$

$\{x \mid 3 < x \leq 33\}$

49. $x < 4$

$x > 0$

$x < 4$ and $x > 0$ which is $0 < x < 4$ or

$\{x \mid 0 < x < 4\}$

51. $x < 2$

$x > 4$

$x < 2$ and $x > 4$

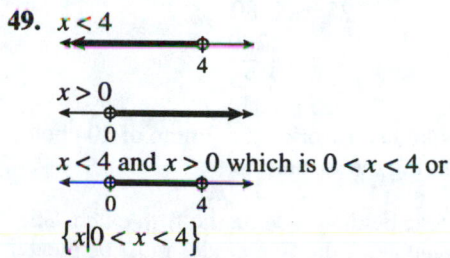

There is no overlap so the solution is the empty set, $\varnothing$.

53. $x + 1 < 3$ and $x + 1 > -4$

    $x < 2$ and $x > -5$

$x > -5$

$x < 2$

$x < -2$ and $x > -5$ which is $-5 < x < 2$ or

$\{x \mid -5 < x < 2\}$

55. $3x - 6 \leq 4$ or $2x - 3 < 5$

    $3x \leq 10$ or $2x < 8$

      $x \geq \dfrac{10}{3}$ or $x < 4$

$x \leq \dfrac{10}{3}$

$x < 4$

$x \leq \dfrac{10}{3}$ or $x < 4$ which is $x < 4$

$(-\infty, 4)$

57. $4x + 5 \geq 5$ and $3x - 4 \leq 2$

    $4x \geq 0$ and $3x \leq 6$

     $x \geq 0$ and $x \leq 2$

$x \geq 0$

$x \leq 2$

$x \geq 0$ and $x \leq 2$ which is $0 \leq x \leq 2$

$[0, 2]$

59.

$$4 - x < -2 \qquad 3x - 1 < -1$$
$$-x < -6 \qquad\quad 3x < 0$$
$$x > 6 \qquad\qquad x < 0$$

$x > 6$

$$6$$

$x < 0$

$$0$$

$x > 6$ or $x < 0$

$$0 \qquad 6$$

$$(-\infty,\ 0) \cup (6, \infty)$$

61. a. $l + w + d \le 61$

b. $l = 29, w = 21\dfrac{1}{2}$

$$l + w + d \le 61$$
$$29 + 21\frac{1}{2} + d \le 61$$
$$50\frac{1}{2} + d \le 61$$
$$d \le 10\frac{1}{2}$$

The maximum depth is $10\dfrac{1}{2}$ inches.

63. Let x be the maximum number of boxes.
$$70x \le 800$$
$$x \le \frac{800}{70}$$
$$x \le 11.43$$
The maximum number of boxes is 11.

65. Let x be the maximum number of minutes after the first three.
$$4.25 + 0.45x \le 9.50$$
$$0.45x \le 5.25$$
$$x \le \frac{5.25}{0.45}$$
$$x \le 11.6667$$
Thus, the maximum length of time a customer can talk is 11 minutes plus the first 3 minutes for a total of 14 minutes.

67. To make a profit, the cost must be less than the revenue: cost < revenue.
$$10,025 + 1.09x < 6.42x$$
$$10,025 < 5.33x$$
$$\frac{10,025}{5.33} < x$$
$$1880.86 < x$$
She needs to sell a minimum of 1881 books to make a profit.

69. Let x be the number of pieces of bulk mail.
$$85 + 0.256x < 0.33x$$
$$85 < 0.074x$$
$$\frac{85}{0.074} < x$$
$$1148.6 < x$$
The minimum number of pieces of bulk mail is 1149.

71. Let x be the maximum number of hours. The inequality is
$$6.25x + 8(90) \le 2000$$
$$6.25x + 720 \le 2000$$
$$6.25x \le 1280$$
$$x \le \frac{1280}{6.25}$$
$$x \le 204.8$$
Nikita can work a maximum of 204 hours during the summer.

73. Let x be the grade on the fifth exam. The average of the five grades must be greater than or equal to 90.
$$\frac{90 + 87 + 96 + 79 + x}{5} \ge 90$$
$$\frac{352 + x}{5} \ge 90$$
$$5\left(\frac{352 + x}{5}\right) \ge 5(90)$$
$$352 + x \ge 450$$
$$x \ge 98$$
Ray must make a 98 or higher on the fifth exam to earn a final grade of A in the course.

75. Let x be the score on the fifth exam.

$$80 \leq \frac{87 + 92 + 70 + 75 + x}{5} < 90$$

$$80 \leq \frac{324 + x}{5} < 90$$

$$5(80) \leq 5\left(\frac{324 + x}{5}\right) < 5(90)$$

$$400 \leq 324 + x < 450$$

$$76 \leq x < 126$$

To receive a final grade of B, Ms. Mahoney must score 76 or higher on the fifth exam. That is, the score must be
$76 \leq x \leq 100$ (maximum grade is 100).

77. Let x be the value of the third reading.

$$7.2 < \frac{7.48 + 7.15 + x}{3} < 7.8$$

$$7.2 < \frac{14.63 + x}{3} < 7.8$$

$$3(7.2) < 3\left(\frac{14.63 + x}{3}\right) < 3(7.8)$$

$$21.6 < 14.63 + x < 23.4$$

$$6.97 < x < 8.77$$

Any value between 6.97 and 8.77 would result in a normal pH reading.

79. a. crime index $\geq 600{,}000$ and welfare $\leq 1{,}000{,}000$
The years are 1990, 1991, and 1992. These are the years where the line representing the crime rate is on or above 600,000 and the line representing the number of people on welfare is on or below 1,000,000.

b. crime index $\leq 600{,}000$ or welfare $\geq 900{,}000$
The years are 1991 to 1996. These are the years where the line representing the number of people on welfare is on or above 900,000.

81. No, $-1 > -2$ but $(-1)^2 < (-2)^2$

83. Answers may vary.

85. First find the average of 82, 90, 74, 76, and 68.

$$\frac{82 + 90 + 74 + 76 + 68}{5} = \frac{390}{5} = 78$$

This represents $\frac{2}{3}$ of the final grade.

Let x be the score from the final exam. Since this represents $\frac{1}{3}$ of the final grade, the inequality is $80 \leq \frac{2}{3}(78) + \frac{1}{3}x < 90$

$$3(80) < 3\left[\frac{2}{3}(78) + \frac{1}{3}x\right] < 3(90)$$

$$240 \leq 2(78) + x < 270$$

$$240 \leq 156 + x < 270$$

$$84 \leq x < 114$$

Russell must score at least 84 points on the final exam to have a final grade of B. The range is $84 \leq x \leq 100$.

87. a. Answers may vary. One possible answer is: Write $x < 2x + 3 < 2x + 5$ as $x < 2x + 3$ and $2x + 3 < 2x + 5$

b. Solve each of the inequalities.

$x < 2x + 3$	and	$2x + 3 < 2x + 5$
$-x < 3$		$3 < 5$
$x > -3$		All real numbers

The final answer is $x > -3$ or $(-3, \infty)$.

89. a. $A \cup B = \{1, 2, 3, 4, 5, 6, 8, 9\}$

b. $A \cap B = \{1, 8\}$

90. a. 4 is a counting number.

b. 4 and 0 are whole numbers.

c. -3, 4, $\frac{5}{2}$, 0 and $-\frac{29}{80}$ are rational numbers.

d. -3, 4, $\frac{5}{2}$, $\sqrt{7}$, 0 and $-\frac{29}{80}$ are real numbers.

91. Associative property of addition.

92. Commutative property of addition

93.
$$R = L + (V - D)r$$
$$R = L + Vr - Dr$$
$$R - L + Dr = Vr$$
$$\frac{R - L + Dr}{r} = V \text{ or } V = \frac{R - L + Dr}{r}$$

Exercise Set 2.6

1. $|x| = a, \ a > 0$
Set $x = a$ or $x = -a$.

3. The solution to $|x| = -3$ is $\emptyset$. There is no real number whose absolute value is negative.

5. $|x| < a, \ a > 0$
Write $-a < x < a$.

7. $|x| > a, \ a > 0$
Write $x < -a$ or $x > a$.

9. a. $|ax + b| < c$
The solution is $m < x < n$ or
$\qquad m \qquad\quad n$

b. $|ax + b| > c$
The solution is $x < m$ or $x > n$ or
$\qquad m \qquad\quad n$

11. a. $|ax + b| = k, \ a \neq 0$
If $k \leq 0$, there are no solutions.

b. $|ax + b| = k, \ a \neq 0$
If $k = 0$, there is one solution.

c. $|ax + b| = k, \ a \neq 0$
If $k > 0$, there are two solutions.

13. a. $|x| = 4$, $x = -4$ or $x = 4$, C

b. $|x| < 4$, $(-4, 4)$, A

c. $|x| > 4$, $(-\infty, 4) \cup (4, \infty)$, D

d. $|x| \geq 4$, $(-\infty, 4] \cup [4, \infty)$, B

e. $|x| \leq 4$, $[-4, 4]$, E

15. $|x| = 5$
$x = 5$ or $x = -5$
The solution set is $\{-5, 5\}$.

17. $|x| = 12$
$x = 12$ or $x = -12$
The solution set is $\{-12, 12\}$.

19. $|x| = -2$
There is no solution since the right side is a negative number and the absolute value can never be equal to a negative number. The solution set $\emptyset$.

21. $|x + 5| = 7$
$x + 5 = 7 \qquad\qquad x + 5 = -7$
$\quad x = 2 \quad$ or $\qquad x = -12$
The solution set is $\{-12, 2\}$.

23. $|2.4 + 0.4x| = 4$
$2.4 + 0.4x = 4 \quad$ or $\quad 2.4 + 0.4x = -4$
$\qquad 0.4x = 1.6 \qquad\qquad 0.4x = -6.4$
$\qquad\quad x = \dfrac{1.6}{0.4} \qquad\qquad\quad x = \dfrac{-6.4}{0.4}$
$\qquad\quad x = 4 \qquad\qquad\qquad\quad x = -16$
The solution set is $\{-16, 4\}$.

25. $|5 - 3x| = \dfrac{1}{2}$

$$5 - 3x = \dfrac{1}{2} \qquad \text{or} \qquad 5 - 3x = -\dfrac{1}{2}$$

$$-3x = \dfrac{1}{2} - 5 \qquad\qquad -3x = -\dfrac{1}{2} - 5$$

$$-3x = -\dfrac{9}{2} \qquad\qquad -3x = -\dfrac{11}{2}$$

$$-\dfrac{1}{3}(3x) = -\dfrac{1}{3}\left(-\dfrac{9}{2}\right) \qquad -\dfrac{1}{3}(-3x) = -\dfrac{1}{3}\left(-\dfrac{11}{2}\right)$$

$$x = \dfrac{3}{2} \qquad\qquad x = \dfrac{11}{6}$$

The solution set is $\left\{\dfrac{3}{2}, \dfrac{11}{6}\right\}$.

27. $\left|\dfrac{x-3}{4}\right| = 5$

$$\dfrac{x-3}{4} = 5 \qquad \text{or} \qquad \dfrac{x-3}{4} = -5$$

$$4\left(\dfrac{x-3}{4}\right) = 4(5) \qquad 4\left(\dfrac{x-3}{4}\right) = 4(-5)$$

$$x - 3 = 20 \qquad\qquad x - 3 = -20$$

$$x = 23 \qquad\qquad x = -17$$

The solution set is $\{-17, 23\}$.

29. $\left|\dfrac{x-3}{4}\right| + 4 = 4$

$$\left|\dfrac{x-3}{4}\right| = 0$$

$$\dfrac{x-3}{4} = 0$$

$$4\left(\dfrac{x-3}{4}\right) = 4(0)$$

$$x - 3 = 0$$

$$x = 3$$

The solution set is $\{3\}$.

31. $|y| \le 5$

$$-5 \le y \le 5$$

The solution set is $\{y | -5 \le y \le 5\}$.

33. $|x - 7| \le 9$

$$-9 \le x - 7 \le 9$$

$$-9 + 7 \le x - 7 + 7 \le 9 + 7$$

$$-2 \le x \le 16$$

The solution set is $\{x | -2 \le x \le 16\}$.

35. $|3z - 5| \le 5$

$$-5 < 3z - 5 \le 5$$

$$-5 + 5 \le 3z - 5 + 5 \le 5 + 5$$

$$0 \le 3z \le 10$$

$$\dfrac{0}{3} \le \dfrac{3z}{3} \le \dfrac{10}{3}$$

$$0 \le z \le \dfrac{10}{3}$$

The solution set is $\left\{z \middle| 0 \le z \le \dfrac{10}{3}\right\}$.

37. $|2x+3|-5 \leq 10$

$|2x+3| \leq 15$

$-15 \leq 2x+3 \leq 15$

$-15-3 \leq 2x+3-3 \leq 15-3$

$-18 \leq 2x \leq 12$

$\dfrac{-18}{2} \leq \dfrac{2x}{2} \leq \dfrac{12}{2}$

$-9 \leq x \leq 6$

The solution set is $\{x|-9 \leq x \leq 6\}$

39. $|x-0.4| \leq 2.3$

$-2.3 \leq x-0.4 \leq 2.3$

$-2.3+0.4 \leq x-0.4+0.4 \leq 2.3+0.4$

$-1.9 \leq x \leq 2.7$

The solution set is $\{x|-1.9 \leq x \leq 2.7\}$.

41. $|2x-6|+5 \leq 2$

$|2x-6| \leq -3$

There is no solution since the right side is negative whereas the left side is non-negative; zero or a positive number is never less than a negative number. The solution set is $\varnothing$.

43. $\left|5-\dfrac{3x}{4}\right| < 8$

$-8 < 5-\dfrac{3x}{4} < 8$

$4(-8) < 4\left(5-\dfrac{3x}{4}\right) < 4(8)$

$-32 < 20-3x < 32$

$-32-20 < 20-20-3x < 32-20$

$-52 < -3x < 12$

$\dfrac{-52}{-3} > \dfrac{-3x}{-3} > \dfrac{12}{-3}$

$\dfrac{52}{3} > x > -4$

$-4 < x < \dfrac{52}{3}$

The solution set is $\left\{x \middle| -4 < x < \dfrac{52}{3}\right\}$.

45. $|4x-1| \leq 0$

$0 \leq 4x-1 \leq 0$

$4x-1=0$

$4x=1$

$x=\dfrac{1}{4}$

The solution is $\left\{\dfrac{1}{4}\right\}$.

47. $|x| > 3$

$x < -3$ or $x > 3$

The solution set is $\{x|x < -3 \text{ or } x > 3\}$.

49. $|x+4| > 5$

$x+4 < -5$ or $x+4 > 5$

$x < -9$ $x > 1$

The solution set is $\{x|x < -9 \text{ or } x > 1\}$.

51. $|4-3y| \geq 8$

$4-3y \leq -8$ or $4-3y \geq 8$

$-3y \leq -12$ $-3y \geq 4$

$y \geq \dfrac{-12}{-3}$ $y \leq \dfrac{4}{-3}$

$y \geq 4$ $y \leq -\dfrac{4}{3}$

The solution set is $\left\{y \middle| y \leq -\dfrac{4}{3} \text{ or } y \geq 4\right\}$.

53. $\left|\dfrac{5-3w}{4}\right| \geq 10$

$\dfrac{5-3w}{4} \leq -10$ or $\dfrac{5-3w}{4} \geq 10$

$4\left(\dfrac{5-3w}{4}\right) \leq 4(-10)$ $4\left(\dfrac{5-3w}{4}\right) \geq 4(10)$

$5-3w \leq -40$ $5-3w \geq 40$

$-3w \leq -45$ $-3w \geq 35$

$w \geq \dfrac{-45}{-3}$ $w \leq \dfrac{35}{-3}$

$w \geq 15$ $w \leq -\dfrac{35}{3}$

The solution set is $\left\{w \middle| w \leq -\dfrac{35}{3} \text{ or } w \geq 15\right\}$.

55. $|0.1x - 0.4| + 0.4 > 0.6$

$\quad |0.1x - 0.4| > 0.2$

$0.1x - 0.4 < -0.2 \quad$ or $\quad 0.1x - 0.4 > 0.2$

$\quad 0.1x < 0.2 \qquad\qquad 0.1x > 0.6$

$\qquad x < \dfrac{0.2}{0.1} \qquad\qquad x > \dfrac{0.6}{0.1}$

$\qquad x < 2 \qquad\qquad\qquad x > 6$

The solution set is $\{x | x < 2 \text{ or } x > 6\}$.

57. $\left| \dfrac{x}{2} + 4 \right| \geq 5$

$\dfrac{x}{2} + 4 \leq -5 \qquad$ or $\qquad \dfrac{x}{2} + 4 \geq 5$

$2\left(\dfrac{x}{2} + 4\right) \leq 2(-5) \qquad 2\left(\dfrac{x}{2} + 4\right) \geq 2(5)$

$\quad x + 8 \leq -10 \qquad\qquad x + 8 \geq 10$

$\qquad x \leq -18 \qquad\qquad\quad x \geq 2$

The solution set is $\{x | x \leq -18 \text{ or } x \geq 2\}$.

59. $|3x + 5| + 2 \geq 2$

$\quad |3x + 5| \geq 0$

$3x + 5 \leq 0 \qquad\qquad 3x + 5 \geq 0$

$\quad 3x \leq -5 \quad$ or $\quad 3x \geq -5$

$\qquad x \leq -\dfrac{5}{3} \qquad\qquad x \geq -\dfrac{5}{3}$

The solution set is $\left\{ x \middle| x \leq -\dfrac{5}{3} \text{ or } x \geq -\dfrac{5}{3} \right\}$

which is the set of real numbers or $\mathbb{R}$.
Another way of looking at this is to observe
that the absolute value of a number is always
greater than or equal to 0. Thus, the solution
is the set of real numbers or $\mathbb{R}$.

61. $|4 - 2x| > 0$

$4 - 2x < 0 \qquad$ or $\quad 4 - 2x > 0$

$\quad -2x < -4 \qquad\qquad -2x > -4$

$\qquad x > \dfrac{-4}{-2} \qquad\qquad x < \dfrac{-4}{-2}$

$\qquad x > 2 \qquad\qquad\qquad x < 2$

The solution set is $\{x | x < 2 \text{ or } x > 2\}$.

63. $|2x + 1| = |4x - 9|$

$2x + 1 = -(4x - 9) \quad$ or $\quad 2x + 1 = 4x - 9$

$2x + 1 = -4x + 9 \qquad\qquad 1 = 2x - 9$

$\quad 6x + 1 = 9 \qquad\qquad\qquad 10 = 2x$

$\qquad 6x = 8 \qquad\qquad\qquad \dfrac{10}{2} = x$

$\qquad x = \dfrac{8}{6} \qquad\qquad\qquad 5 = x$

$\qquad x = \dfrac{4}{3}$

The solution set is $\left\{ \dfrac{4}{3}, 5 \right\}$.

65. $|6x| = |3x - 9|$

$6x = -(3x - 9) \quad$ or $\quad 6x = 3x - 9$

$6x = -3x + 9 \qquad\qquad 3x = -9$

$\quad 9x = 9 \qquad\qquad\qquad x = \dfrac{-9}{3}$

$\qquad x = \dfrac{9}{9} \qquad\qquad\qquad x = -3$

$\qquad x = 1$

The solution set is $\{-3, 1\}$.

67. $\left|\dfrac{3}{4}x - 2\right| = \left|\dfrac{1}{2}x + 5\right|$

$\dfrac{3}{4}x - 2 = -\left(\dfrac{1}{2}x + 5\right)$ or $\quad \dfrac{3}{4}x - 2 = \dfrac{1}{2}x + 5$

$\dfrac{3}{4}x - 2 = -\dfrac{1}{2}x - 5 \qquad\qquad \dfrac{3}{4}x = \dfrac{1}{2}x + 7$

$\dfrac{3}{4}x + \dfrac{1}{2}x - 2 = -5 \qquad\qquad \dfrac{3}{4}x - \dfrac{1}{2}x = 7$

$\dfrac{5}{4}x - 2 = -5 \qquad\qquad\qquad \dfrac{1}{4}x = 7$

$\dfrac{5}{4}x = -3 \qquad\qquad\qquad 4\left(\dfrac{1}{4}x\right) = 4(7)$

$\dfrac{4}{5}\left(\dfrac{5}{4}x\right) = \dfrac{4}{5}(-3) \qquad\qquad x = 28$

$x = -\dfrac{12}{5}$

The solution set is $\left\{-\dfrac{12}{5},\ 28\right\}$.

69. $\left|\dfrac{1}{2}x + \dfrac{3}{5}\right| = \left|\dfrac{1}{2}x - 1\right|$

$\dfrac{1}{2}x + \dfrac{3}{5} = -\left(\dfrac{1}{2}x - 1\right)$ or $\dfrac{1}{2}x + \dfrac{3}{5} = \dfrac{1}{2}x - 1$

$\dfrac{1}{2}x + \dfrac{3}{5} = -\dfrac{1}{2}x + 1 \qquad \dfrac{3}{5} = -1$ False

$x + \dfrac{3}{5} = 1$

$x = \dfrac{2}{5}$

The solution set is $\left\{\dfrac{2}{5}\right\}$.

71. $|w| = 7$
$w = 7$ or $w = -7$
The solution set is $\{-7, 7\}$.

73. $|x - 3| < 5$
$-5 < x - 3 < 5$
$-2 < x < 8$
The solution set is $\{x | -2 < x < 8\}$.

75. $|x + 5| > 9$
$x + 5 < -9 \quad$ or $\quad x + 5 > 9$
$x < -14 \qquad\qquad x > 4$
The solution set is $\{x | x < -14 \text{ or } x > 4\}$.

77. $|4x + 2| = 9$
$4x + 2 = 9 \quad$ or $\quad 4x + 2 = -9$
$4x = 7 \qquad\qquad 4x = -11$
$x = \dfrac{7}{4} \qquad\qquad x = -\dfrac{11}{4}$
The solution set is $\left\{-\dfrac{11}{4},\ \dfrac{7}{4}\right\}$.

79. $|5x + 2| > 0$
$5 + 2x < 0 \quad$ or $\quad 5 + 2x > 0$
$2x < -5 \qquad\qquad 2x > -5$
$x < -\dfrac{5}{2} \qquad\qquad x > -\dfrac{5}{2}$
The solution set is $\left\{x \middle| x < -\dfrac{5}{2} \text{ or } x > -\dfrac{5}{2}\right\}$.

81. $|4 + 3x| \le 9$
$-9 \le 4 + 3x \le 9$
$-13 \le 3x \le 5$
$-\dfrac{13}{3} \le x \le \dfrac{5}{3}$
The solution set is $\left\{x \middle| -\dfrac{13}{3} \le x \le \dfrac{5}{3}\right\}$.

83. $|3x - 5| + 4 = 2$

$\qquad |3x - 5| = -2$

Since the right side is negative and the left side is non-negative, there is no solution since the absolute value can never equal a negative number. The solution set is $\varnothing$.

85. $\left|\dfrac{w + 4}{3}\right| - 1 < 3$

$\qquad \left|\dfrac{w + 4}{3}\right| < 4$

$\qquad -4 < \dfrac{w + 4}{3} < 4$

$\qquad 3(-4) < 3\left(\dfrac{w + 4}{3}\right) < 3(4)$

$\qquad -12 < w + 4 < 12$

$\qquad -16 < w < 8$

The solution set is $\{w | -16 < w < 8\}$.

87. $\left|\dfrac{3x - 2}{4}\right| + 5 \geq 5$

$\qquad \left|\dfrac{3x - 2}{4}\right| \geq 0$

Since the absolute value of a number is always greater than or equal to zero, the solution is the set of all real numbers or $\mathbb{R}$.

89. $|2x - 8| = \left|\dfrac{1}{2}x + 3\right|$

$2x - 8 = -\left(\dfrac{1}{2}x + 3\right)$ or $\quad 2x - 8 = \dfrac{1}{2}x + 3$

$2x - 8 = -\dfrac{1}{2}x - 3 \qquad\qquad \dfrac{3}{2}x - 8 = 3$

$\dfrac{5}{2}x - 8 = -3 \qquad\qquad\qquad \dfrac{3}{2}x = 11$

$\qquad \dfrac{5}{2}x = 5 \qquad\qquad \dfrac{2}{3}\left(\dfrac{3}{2}x\right) = \dfrac{2}{3}(11)$

$\dfrac{2}{5}\left(\dfrac{5}{2}x\right) = \dfrac{2}{5}(5) \qquad\qquad x = \dfrac{22}{3}$

$\qquad x = 2$

The solution set is $\left\{2, \dfrac{22}{3}\right\}$.

91. $|2 - 3x| = \left|4 - \dfrac{5}{3}x\right|$

$2 - 3x = -\left(4 - \dfrac{5}{3}x\right)$ or $\quad 2 - 3x = 4 - \dfrac{5}{3}x$

$2 - 3x = -4 + \dfrac{5}{3}x \qquad\qquad -3x = 2 - \dfrac{5}{3}x$

$-3x = -6 + \dfrac{5}{3}x \qquad\qquad -\dfrac{4}{3}x = 2$

$-\dfrac{14}{3}x = -6 \qquad -\dfrac{3}{4}\left(-\dfrac{4}{3}x\right) = -\dfrac{3}{2}(2)$

$\left(-\dfrac{3}{14}\right)\left(-\dfrac{14}{3}\right)x = \left(-\dfrac{3}{14}\right)(-6) \qquad x = -\dfrac{3}{2}$

$\qquad x = \dfrac{9}{7}$

The solution set is $\left\{-\dfrac{3}{2}, \dfrac{9}{7}\right\}$.

93. a. $\left| d - 4 \right| \le \dfrac{1}{2}$

$-\dfrac{1}{2} \le d - 4 \le \dfrac{1}{2}$

$-\dfrac{1}{2} + 4 \le d - 4 + 4 \le \dfrac{1}{2} + 4$

$\dfrac{7}{2} \le d \le \dfrac{9}{2}$

The solution is $\left[\dfrac{7}{2},\ \dfrac{9}{2} \right]$ or $[3.5, 4.5]$.

b. The spring will oscillate between 3.5 feet and 4.5 feet, inclusive.

95. $\{-5, 5\}$ is the solution set of $|x| = 5$.

97. $\left\{ x \mid x \le -5 \text{ or } x \ge 5 \right\}$ is the solution set of $|x| \ge 5$.

99. $|ax + b| \le 0$

$0 \le ax + b \le 0$

which is the same as

$ax + b = 0$

$ax = -b$

$x = -\dfrac{b}{a}$

101. a. Set $ax + b = -c$ or $ax + b = c$ and solve each equation for x.

b. $ax + b = -c$ or $ax + b = c$

$\quad ax = -c - b \qquad\qquad ax = c - b$

$\quad x = \dfrac{-c - b}{a} \qquad\qquad x = \dfrac{c - b}{a}$

The solution set is

$\left\{ x \,\middle|\, x = \dfrac{-c - b}{a} \text{ or } x = \dfrac{c - b}{a} \right\}$.

103. a. Write $ax + b < -c$ or $ax + b > c$ and solve each inequality for x.

b. $ax + b < -c$ or $ax + b > c$

$\quad ax < -c - b \qquad\qquad ax > c - b$

$\quad x < \dfrac{-c - b}{a} \qquad\qquad x > \dfrac{c - b}{a}$

The solution set is

$\left\{ x \,\middle|\, x < \dfrac{-c - b}{a} \text{ or } x > \dfrac{c - b}{a} \right\}$.

105. $|x - 3| = |3 - x|$

$x - 3 = -(3 - x)$ or $x - 3 = 3 - x$

$x - 3 = -3 + x \qquad\qquad 2x - 3 = 3$

$0 = 0 \qquad\qquad\qquad\quad 2x = 6$

$\text{True} \qquad\qquad\qquad\quad x = 3$

Since the first statement is always true all real values work. The solution set is $\mathbb{R}$.

107. $|x| = x$

By definition $|x| = \begin{cases} x, & x \ge 0 \\ -x, & x < 0 \end{cases}$

Thus, $|x| = x$ when $x \ge 0$

The solution set is $\{x \mid x \ge 0\}$.

109. $|x + 1| = 2x - 1$

$x + 1 = -(2x - 1)$ or $x + 1 = 2x - 1$

$x + 1 = -2x + 1 \qquad\qquad 1 = x - 1$

$3x + 1 = 1 \qquad\qquad\qquad 2 = x$

$3x = 0$

$x = 0$

Checking both possible solutions, only $x = 2$ checks. The solution set is $\{2\}$.

111. $|x - 2| = -(x - 2)$

By the definition,

$|x - 2| = \begin{cases} x - 2, & x - 2 \ge 0 \\ -(x - 2), & x - 2 \le 0 \end{cases}$ or

$\qquad\quad \begin{cases} x - 2, & x \ge 2 \\ -(x - 2), & x \le 2 \end{cases}$

Thus, $|x - 2| = -(x - 2)$ for $x \le 2$.

The solution set is $\{x \mid x \le 2\}$.

113. $x + |-x| = 6$

For $x \geq 0$: $x + |-x| = 6$
$$x + x = 6$$
$$2x = 6$$
$$x = 3$$
For $x < 0$: $x + |-x| = 6$
$$x - x = 6$$
$$0 = 6 \text{ False}$$
The solution set is $\{3\}$.

115. $x - |x| = 6$

For $x \geq 0$: $x - |x| = 6$
$$x - x = 6$$
$$0 = 6 \text{ False}$$
For $x < 0$: $x - |x| = 6$
$$x - (-x) = 6$$
$$x + x = 6$$
$$2x = 6$$
$$x = 3 \text{ Contradicts } x < 0$$
There are no values of x, so the solution set is $\varnothing$.

117.
$$\frac{1}{3} + \frac{1}{4} \div \frac{2}{5}\left(\frac{1}{3}\right)^2 = \frac{1}{3} + \frac{1}{4} \div \frac{2}{5} \cdot \frac{1}{9}$$
$$= \frac{1}{3} + \frac{1}{4} \cdot \frac{5}{2} \cdot \frac{1}{9}$$
$$= \frac{1}{3} + \frac{5}{72}$$
$$= \frac{1}{3} \cdot \frac{24}{24} + \frac{5}{72}$$
$$= \frac{24}{72} + \frac{5}{72}$$
$$= \frac{29}{72}$$

118. Substitute 1 for x and 3 for y.
$$4(x + 3y) - 5xy = 4(1 + 3 \cdot 3) - 5(1)(3)$$
$$= 4(1 + 9) - 5(1)(3)$$
$$= 4(10) - 5(1)(3)$$
$$= 40 - 15$$
$$= 25$$

119. Let x be the time needed to swim across the lake. Then $1.5 - x$ is the time needed to make the return trip.

	Rate	Time	Distance
First Trip	2	x	$2x$
Return Trip	1.6	$1.5 - x$	$1.6(1.5 - x)$

The distances are the same.
$$2x = 1.6(1.5 - x)$$
$$2x = 2.4 - 1.6x$$
$$3.6x = 2.4$$
$$x = \frac{2.4}{3.6} = \frac{2}{3}$$
The total distance across the lake is
$$2x = 2\left(\frac{2}{3}\right) = \frac{4}{3} \text{ or } 1.33 \text{ miles.}$$

120. $3(x - 2) - 4(x - 3) > 2$
$$3x - 6 - 4x + 12 > 2$$
$$-x + 6 > 2$$
$$-x > -4$$
$$\frac{-x}{-1} < \frac{-4}{-1}$$
$$x < 4$$
The solution set is $\{x | x < 4\}$.

Review Exercises

1. $15x^4y^6$ is a tenth degree term since the sum of the exponents is $4 + 6 = 10$.

2. $6x$ is a first degree term since $6x$ can be written as $6x^1$ and the only exponent is 1.

3. $-4xyz^5$ is a seventh degree term since $-4xyz^5$ can be written as $-4x^1y^1z^5$ and the sum of the exponents is $1 + 1 + 5 = 7$.

4. $x^2 + 3x + 6$ cannot be simplified since there are no like terms.

5. $x^2 + 2xy + 6x^2 - 4 = x^2 + 6x^2 + 2xy - 4$
$$= 7x^2 + 2xy - 4$$

6. $3(x + 4) - 3x - 4$
$= 3x + 12 - 3x - 4$
$= 3x - 3x + 12 - 4$
$= 0x + 8$
$= 8$

7. $2[-(x - y) + 3x] - 5y + 6$
$= 2[-x + y + 3x] - 5y + 6$
$= 2[2x + y] - 5y + 6$
$= 4x + 2y - 5y + 6$
$= 4x - 3y + 6$

8. $\dfrac{x - 4}{5} = 9 - x$

$5\left(\dfrac{x - 4}{5}\right) = 5(9) - 5(x)$

$x - 4 = 45 - 5x$
$x + 5x - 4 = 45 - 5x + 5x$
$6x - 4 = 45$
$6x - 4 + 4 = 45 + 4$
$6x = 49$
$\dfrac{6x}{6} = \dfrac{49}{6}$
$x = \dfrac{49}{6}$

9. $3(x + 2) - 6 = 4(x - 5)$
$3x + 6 - 6 = 4x - 20$
$3x + 0 = 4x - 20$
$3x = 4x - 20$
$3x - 4x = 4x - 4x - 20$
$-x = -20$
$\dfrac{-1x}{-1} = \dfrac{-20}{-1}$
$x = 20$

10. $3 + \dfrac{x}{2} = \dfrac{5}{6}$

$6(3) + 6\left(\dfrac{x}{2}\right) = 6\left(\dfrac{5}{6}\right)$

$18 + 3x = 5$
$18 - 18 + 3x = 5 - 18$
$3x = -13$
$\dfrac{3x}{3} = \dfrac{-13}{3}$
$x = -\dfrac{13}{3}$

11. $-6 - 2x = \dfrac{1}{2}(4x + 12) - 12$
$-6 - 2x = 2x + 6 - 12$
$6 - 12 = -6$
$-6 - 2x + 2x = 2x + 2x - 6$
$-6 = 4x - 6$
$-6 + 6 = 4x - 6 + 6$
$0 = 4x$
$\dfrac{0}{4} = \dfrac{4x}{4}$
$0 = x$

12. $2\left(\dfrac{x}{2} - 4\right) = 3\left(x + \dfrac{1}{3}\right)$
$x - 8 = 3x + 1$
$x - 8 + 8 = 3x + 1 + 8$
$x = 3x + 9$
$x - 3x = 3x - 3x + 9$
$-2x = 9$
$\dfrac{-2x}{-2} = \dfrac{9}{-2}$
$x = -\dfrac{9}{2}$

13. $3x - 4 = 6x + 4 - 3x$
$3x - 4 = 3x + 4$
$3x - 3x - 4 = 3x - 3x + 4$
$-4 = 4$
This is a false statement which means there is no solution.

14.
$$2(x-6) = 5 - \{2x - [4(x-3) - 5]\}$$
$$2x - 12 = 5 - \{2x - [4x - 12 - 5]\}$$
$$2x - 12 = 5 - \{2x - [4x - 17]\}$$
$$2x - 12 = 5 - \{2x - 4x + 17\}$$
$$2x - 12 = 5 - \{-2x + 17\}$$
$$2x - 12 = 5 + 2x - 17$$
$$2x - 12 = 2x - 12$$
$$2x - 12 - 2x = 2x - 12 - 2x$$
$$12 = 12$$

Since this is a true statement, the solution set is all real numbers, or $\mathbb{R}$.

15. $P = \dfrac{nRT}{V} = \dfrac{10(100)(4)}{20} = \dfrac{4000}{20} = 200$

16.
$$x = \dfrac{-b + \sqrt{b^2 - 4ac}}{2a}$$
$$= \dfrac{-10 + \sqrt{(10)^2 - 4(8)(-3)}}{2(8)}$$
$$= \dfrac{-10 + \sqrt{100 + 96}}{16}$$
$$= \dfrac{-10 + \sqrt{196}}{16}$$
$$= \dfrac{-10 + 14}{16}$$
$$= \dfrac{4}{16}$$
$$= \dfrac{1}{4}$$

17. $h = \dfrac{1}{2}at^2 + v_0 t + h_0$
$$= \dfrac{1}{2}(-32)(2)^2 + 60(2) + 120$$
$$= \dfrac{1}{2}(-32)(4) + 60(2) + 120$$
$$= -16(4) + 120 + 120$$
$$= -64 + 120 + 120$$
$$= 176$$

18. $z = \dfrac{\bar{x} - \mu}{\dfrac{\sigma}{\sqrt{n}}}$
$$= \dfrac{60 - 64}{\dfrac{5}{\sqrt{25}}}$$
$$= \dfrac{60 - 64}{\dfrac{5}{5}}$$
$$= \dfrac{60 - 64}{1}$$
$$= -4$$

19. $A = lw$
$$\dfrac{A}{w} = \dfrac{lw}{w}$$
$$\dfrac{A}{w} = l \text{ or } l = \dfrac{A}{w}$$

20. $A = \pi r^2 h$
$$\dfrac{A}{\pi r^2} = \dfrac{\pi r^2}{\pi r^2}$$
$$\dfrac{A}{\pi r^2} = h \text{ or } h = \dfrac{A}{\pi r^2}$$

21.
$$P = 2l + 2w$$
$$P - 2l = 2l - 2l + 2w$$
$$P = 2l = 2w$$
$$\dfrac{P - 2l}{2} = \dfrac{2w}{2}$$
$$\dfrac{P - 2l}{2} = w \text{ or } w = \dfrac{P - 2l}{2}$$

22.
$$d = rt$$
$$\dfrac{d}{t} = \dfrac{rt}{t}$$
$$\dfrac{d}{t} = r \text{ or } r = \dfrac{d}{t}$$

23.
$$y = mx + b$$
$$y - b = mx + b - b$$
$$y - b = mx$$
$$\dfrac{y - b}{x} = \dfrac{mx}{x}$$
$$\dfrac{y - b}{x} = m \text{ or } m = \dfrac{y - b}{x}$$

24.
$$2x - 3y = 5$$
$$2x - 2x - 3y = -2x + 5$$
$$-3y = -2x + 5$$
$$\frac{-3y}{-3} = \frac{-2x + 5}{-3}$$
$$y = \frac{-2x + 5}{-3} \text{ or } y = \frac{2x - 5}{3}$$

25.
$$P_1 V_1 = P_2 V_2$$
$$\frac{P_1 V_1}{P_2} = \frac{P_2 V_2}{P_2}$$
$$\frac{P_1 V_1}{P_2} = V_2 \text{ or } V_2 = \frac{P_1 V_1}{P_2}$$

26.
$$S = \frac{3a + b}{2}$$
$$2(S) = 2\left(\frac{3a + b}{2}\right)$$
$$2S = 3a + b$$
$$2S - b = 3a + b - b$$
$$2S - b = 3a$$
$$\frac{2S - b}{3} = \frac{3a}{3}$$
$$\frac{2S - b}{3} = a \text{ or } a = \frac{2S - b}{3}$$

27.
$$K = 2(d + l)$$
$$K = 2d + 2l$$
$$K - 2d = 2d - 2d + 2l$$
$$K - 2d = 2l$$
$$\frac{K - 2d}{2} = \frac{2l}{2}$$
$$\frac{D - 2d}{2} = l \text{ or } l = \frac{K - 2d}{2}$$

28. Let x be the original price.
$$x - 0.60x = 20$$
$$0.40x = 20$$
$$\frac{0.40x}{0.40} = \frac{20}{0.40}$$
$$x = 50$$
The original price was $50.

29. Let x be the number of years for the population to reach 5800.
$$4750 + 350x = 5800$$
$$350x = 1050$$
$$x = \frac{1050}{350}$$
$$x = 3$$
It will take 3 years for the population to grow from 4750 people to 5800 people.

30. Let x be the amount of sales.
$$300 + 0.06x = 650$$
$$0.06x = 350$$
$$\frac{0.06x}{0.06} = \frac{350}{0.06}$$
$$x = 5833.33$$
Dawn's sales must be $5833.33.

31. Let x be the number of round trips. The bus fare for a round trip is $2(\$1.65) = \$3.30$.
$$3.30x = 27.50$$
$$x = \frac{27.50}{3.30}$$
$$x \approx 8.33$$
The bus pass is worthwhile if 9 or more round trips are made.

32. Let x be the regular price.
$$x - 0.40x - 20 = 120$$
$$0.60x - 20 = 120$$
$$0.60x = 140$$
$$\frac{0.60x}{0.60} = \frac{140}{0.60}$$
$$x \approx 233.33$$
The regular price was $233.33.

33. Let r be the hourly inspection rate.
$$\text{Rate} \times \text{Time} = \text{Amount}$$
$$r \times 8 = 245$$
$$8r = 245$$
$$r = \frac{245}{8} \approx 30.6$$
Tanya can inspect about 30.6 rolls of film each hour.

34. Let x be the amount invested at 8%. Then $10,00 - x$ is the amount invested at 5%.

Account	Principal	Rate	Time	Interest
8%	x	0.08	1	$0.08x$
5%	$10,000 - x$	0.05	1	$0.05(10,000 - x)$

$$0.08x + 0.05(10,000 - x) = 680$$
$$0.08x + 500 - 0.05x = 680$$
$$500 + 0.03x = 680$$
$$0.03x = 180$$
$$x = \frac{180}{0.03} = 6000$$

Thus, the Sampsons invested \$6000 at 8% and $10,000 - 6000 = \$4000$ at 5%.

35. Let t be the amount of time needed.

Type	Rate	Time	Distance
One Train	60	t	$60t$
Other Train	90	t	$90t$

The total distance is 400 miles.
$$60t + 90t = 400$$
$$150t = 400$$
$$t = \frac{400}{150} = \frac{8}{3} = 2\frac{2}{3}$$

In $2\frac{2}{3}$ hours, the trains are 400 miles apart.

36. a. Let x be the speed of Shuttle 1. Then $x + 300$ is the speed of Shuttle 2.

Type	Rate	Time	Distance
Shuttle 1	x	5.5	$5.5x$
Shuttle 2	$x + 300$	5.0	$5.0(x + 300)$

The distances are the same.
$$5.5x = 5.0(x + 300)$$
$$5.5x = 5.0x + 1500$$
$$0.5x = 1500$$
$$x = \frac{1500}{0.5} = 3000$$

The speed of Shuttle 1 is 3000 mph.

b. The distance is $5.5(3000) = 16,500$ miles.

71

37. Let x be the amount of $6.00 coffee needed. Then $40 - x$ is the amount of $6.80 coffee needed.

Item	Cost per Pound	No. of Pounds	Total Value
$6.00 Coffee	$6.00	x	$6.00x$
$6.80 Coffee	$6.80	$40 - x$	$6.80(40 - x)$
Mixture	$6.50	40	$6.50(40)$

$$6.00x + 6.80(40 - x) = 6.50(40)$$
$$6.00x + 272 - 6.80x = 260$$
$$-0.80x + 272 = 260$$
$$-0.80x = -12$$
$$x = \frac{-12}{-0.80} = 15$$

Mr. Tomlins needs to combine 15 pounds of $6.00 coffee with $40 - 15 = 25$ pounds of $6.80 coffee to produce the mixture.

38. Let x be the original price of the blouse.
$$x - 0.12x = 22$$
$$0.88x = 22$$
$$x = \frac{22}{0.88} = 25$$

The original price of the blouse was $25.

39. Let x be the time spent jogging. Then $4 - x$ is the time spent walking.

Trip	Rate	Time	Distance
Jogging	7.2	x	$7.2x$
Walking	2.4	$4 - x$	$2.4(4 - x)$

a. The distances are the same.
$$7.2x = 2.4(4 - x)$$
$$7.2x = 9.6 - 2.4x$$
$$9.6x = 9.6$$
$$x = \frac{9.6}{9.6} = 1$$

Thus, Nicole jogged for 1 hour and walked for $4 - 1 = 3$ hours.

b. The distance one-way is $7.2(1) = 7.2$ miles. The total distance is twice this value or $2(7.2) = 14.4$ miles.

40. Let x be the measure of the smallest angle. The measure of the other two angles are $x + 25$ and $2x - 5$.
$$x + (x + 25) + (2x - 5) = 180$$
$$4x + 20 = 180$$
$$4x = 160$$
$$x = \frac{160}{4} = 40$$

The measures of the angles are $40°$, $40 + 25 = 65°$, and $2(40) - 5 = 80 - 5 = 75°$.

41. Let x be the flow rate of the smaller hose.

Type	Rate	Time	Amount (No. of Gallons)
Smaller	r	3	$3r$
Larger	$1.5r$	5	$5(1.5r)$

The total number of gallons of water is 3150 gallons.

$$3r + 5(1.5r) = 3150$$
$$3r + 7.5r = 3150$$
$$10.5r = 3150$$
$$r = \frac{3150}{10.5} = 300$$

The flow rate for the smaller hose is 300 gallons per hour and the flow rate for the larger hose is $1.5(300)$ = 450 gallons per hour.

42. Let x and $x + 1$ be the two consecutive integers.

$$x + (x + 1) = 49$$
$$2x + 1 = 49$$
$$2x = 48$$
$$x = \frac{48}{2} = 24$$

The two numbers are 24 and 24 + 1 = 25.

43. Let x be the amount of 20% solution.

Solution	Strength of Solution	No. of Ounces	Amount
20%	0.20	x	$0.20x$
6%	0.06	10	$0.06(10)$
Mixture	0.12	$x + 10$	$0.12(x + 10)$

$$0.20x + 0.06(10) = 0.12(x + 10)$$
$$0.20x + 0.6 = 0.12x + 1.2$$
$$0.08x + 0.6 = 1.2$$
$$0.08x = 0.6$$
$$x = \frac{0.6}{0.08} = 7.5$$

The clothier must combine 7.5 ounces of the 20% solution with 10 ounces of the 6% solution to obtain the
12% solution.

44. Let x be the amount invested at 10%. Then $12{,}000 - x$ is the amount invested at 6%.

Account	Principal	Rate	Time	Interest
10%	x	0.10	1	$0.10x$
6%	$12{,}000 - x$	0.06	1	$0.06(12{,}000 - x)$

$0.10x = 0.06(12{,}000 - x)$
$0.10x = 720 - 0.06x$
$0.16x = 720$
$x = \dfrac{720}{0.16} = 4500$

Thus, Ken invested \$4500 at 10% and $12000 - 4500 = \$7500$ at 6%.

45. Let x be the number of visits. The cost of the first plan = cost of second plan gives the equation
$40 + 1(x) = 25 + 4(x)$

$40 + x = 25 + 4x$
$15 + x = 4x$
$15 = 3x$
$\dfrac{15}{3} = x$ or $x = 5$

Erick needs to make more than 5 visits for the first plan to be advantageous.

46. Let x be the speed of the faster train. Then $x - 10$ is the speed of the slower train.

Train	Rate	Time	Distance
Faster	x	3	$3x$
Slower	$x - 10$	3	$3(x - 10)$

$3x + 3(x - 10) = 510$
$3x + 3x - 30 = 510$
$6x - 30 = 510$
$6x = 540$
$x = \dfrac{540}{6} = 90$

The speed of the faster train is 90 mph.

47. $x - 3 \geq 4$
$x \geq 7$

48. $2 - x \leq 5$
$-x \leq 3$
$x \geq -3$

49. $2x + 4 > 9$
$2x > 5$
$x > \dfrac{5}{2}$

50. $16 \le 4x - 5$

$21 \le 4x$

$\dfrac{21}{4} \le x$

$\dfrac{21}{4}$

51. $\dfrac{4x+3}{5} > -3$

$5\left(\dfrac{4x+3}{5}\right) > 5(-3)$

$4x + 3 > -15$

$4x > -18$

$x > \dfrac{-18}{4}$

$x > -\dfrac{9}{2}$

$-\dfrac{9}{2}$

52. $2(x-3) > 3x + 4$

$2x - 6 > 3x + 4$

$2x - 10 > 3x$

$-10 > x$

-10

53. $-4(x-2) \le 6x + 4$

$-4x + 8 \le 6x + 4$

$8 \le 10x + 4$

$4 \le 10x$

$\dfrac{4}{10} \le x$

$\dfrac{2}{5} \le x$

$\dfrac{2}{5}$

54. $\dfrac{x}{4} \ge 5 - 2x$

$4\left(\dfrac{x}{4}\right) \ge 4(5) - 4(2x)$

$x \ge 20 - 8x$

$9x \ge 20$

$x \ge \dfrac{20}{9}$

$\dfrac{20}{9}$

55. Let x be the maximum number of 80-pound boxes. Since the maximum load is 1525 pounds, the total weight of the passengers and boxes must be less than or equal to 1525 pounds.

$468 + 80x \le 1525$

$80x \le 1057$

$x \le \dfrac{1057}{80}$

$x \le 13.2125$

The maximum number of boxes the plane can hold is 13.

56. Let x be the number of additional minutes (beyond 3 minutes) of the phone call.

$4.50 + 0.95x \le 8.65$

$0.95x \le 4.15$

$x \le \dfrac{4.15}{0.95}$

$x \le 4.4$

The customer can talk for 3 minutes plus an additional 4 minutes for a total of 7 minutes.

57. Let x be the number of weeks (after the first week) needed to lose 27 pounds.

$3 + 1.5x \ge 27$

$1.5x \ge 24$

$x \ge \dfrac{24}{1.5}$

$x \ge 16$

The number of weeks is 16 plus the initial week for a total of 17 weeks.

58. $1 < x - 4 < 7$

$1 + 4 < x - 4 + 4 < 7 + 4$

$5 < x < 11$

$(5, 11)$

59. $2 \le x + 5 < 8$

$2 - 5 < x + 5 - 5 < 8 - 5$

$-3 \le x < 3$

$[-3, 3)$

60. $3 < 2x - 4 < 8$

$3 + 4 < 2x - 4 + 4 < 8 + 4$

$7 < 2x < 12$

$\dfrac{7}{2} < \dfrac{2x}{2} < \dfrac{12}{2}$

$\dfrac{7}{2} < x < 6$

$\left(\dfrac{7}{2}, 6 \right)$

61. $-12 < 6 - 3x < -2$

$-12 - 6 < 6 - 6 - 3x < -2 - 6$

$18 < -3x < -8$

$\dfrac{-18}{-3} > \dfrac{-3x}{-3} > \dfrac{-8}{-3}$

$6 > x > \dfrac{8}{3}$

$\dfrac{8}{3} < x < 6$

$\left(\dfrac{8}{3}, 6 \right)$

62. $-1 \le \dfrac{2x - 3}{4} < 5$

$4(-1) \le 4 \left(\dfrac{2x - 3}{4} \right) < 4(5)$

$-4 \le 2x - 3 < 20$

$-4 + 3 \le 2x - 3 + 3 < 20 + 3$

$-1 \le 2x < 23$

$\dfrac{-1}{2} \le \dfrac{2x}{2} < \dfrac{23}{2}$

$-\dfrac{1}{2} \le x < \dfrac{23}{2}$

$\left[-\dfrac{1}{2}, \dfrac{23}{2} \right)$

63. $-8 < \dfrac{4 - 2x}{3} < 0$

$3(-8) < 3 \left(\dfrac{4 - 2x}{3} \right) < 3(0)$

$-24 < 4 - 2x < 0$

$-24 - 4 < 4 - 4 - 2x < 0 - 4$

$-28 < -2x < -4$

$\dfrac{-28}{-2} > \dfrac{-2x}{-2} > \dfrac{-4}{-2}$

$14 > x > 2$

$2 < x < 14$

$(2, 14)$

64. Let x be the grade from the 5th exam. The inequality is

$80 \le \dfrac{94 + 73 + 72 + 80 + x}{5} < 90$

$80 \le \dfrac{319 + x}{5} < 90$

$5(80) \le 5 \left(\dfrac{319 + x}{5} \right) < 5(90)$

$400 \le 319 + x < 450$

$400 - 319 \le 319 + x < 450 - 319$

$81 \le x < 131$

(must use 100 here since it is not possible to score 131)

Thus, Jekeila needs to score 81 or higher on the 5th exam to receive a B.

$\{x | 81 \le x \le 100\}$

65. $x < 3$ and $2x - 4 > -10$

$x < 3$ and $2x > -6$

$x < 3$ and $x > -3$

$x < 3$

$x > -3$

$x < 3$ and $x > -3$ which is $-3 < x < 3$

$\{x | -3 < x < 3\}$

66. $2x - 1 > 5$ or $3x - 2 \leq 7$

      $2x > 6$ or $3x \leq 9$

        $x > 3$ or $x \leq 3$

$x > 3$

$x \leq 3$

$x > 3$ or $x \leq 3$

which is the entire real number line or $\mathbb{R}$.

67. $4x - 5 < 11$ and $-3x - 4 \geq 8$

     $4x < 16$ and $-3x \geq 12$

      $x < 4$ and $x \leq -4$

$x \leq 4$

$x \leq -4$

$x \leq -4$ and $x < 4$ which is $x \leq -4$

$\{x | x \leq -4\}$

68. $\dfrac{5x - 3}{2} > 7$ or $\dfrac{2x - 1}{3} \leq -3$

  $5x - 3 > 14$ or $2x - 1 \leq -9$

     $5x > 17$ or $2x \leq -8$

      $x > \dfrac{17}{5}$ or $x \leq -4$

$x \leq -4$

$x > \dfrac{17}{5}$

$x \leq -4$ or $x > \dfrac{17}{5}$

$\left\{ x \left| x \leq -4 \text{ or } x > \dfrac{17}{5} \right. \right\}$

69. $|x| = 4$

   $x = 4$ or $x = -4$

   The solution set is $\{-4, 4\}$.

70. $|x| < 3$

   $-3 < x < 3$

   The solution set is $\{x | -3 < x < 3\}$.

71. $|x| \geq 4$

   $x \leq -4$ or $x \geq 4$

   The solution set is $\{x | x \leq -4 \text{ or } x \geq 4\}$.

72. $|x - 4| = 9$

   $x - 4 = 9$ or $x - 4 = -9$

     $x = 13$ $x = -5$

   The solution set is $\{-5, 13\}$.

73. $|x - 2| \geq 5$

   $x - 2 \leq -5$ or $x - 2 \geq 5$

     $x \leq -3$ $x \geq 7$

   The solution set is $\{x | x \leq -3 \text{ or } x \geq 7\}$.

74. $|4 - 2x| = 5$

   $4 - 2x = 5$ or $4 - 2x = -5$

    $-2x = 1$ $-2x = -9$

     $x = \dfrac{1}{-2}$ $x = \dfrac{-9}{-2}$

     $x = -\dfrac{1}{2}$ $x = \dfrac{9}{2}$

   The solution set is $\left\{ -\dfrac{1}{2}, \dfrac{9}{2} \right\}$.

75. $|3 - 2x| < 7$

   $-7 < 3 - 2x < 7$

   $-10 < -2x < 4$

   $\dfrac{-10}{-2} > \dfrac{-2x}{-2} > \dfrac{4}{-2}$

   $5 > x > -2$

   $-2 < x < 5$

   The solution set is $\{x | -2 < x < 5\}$.

76. $\left|\dfrac{2x-3}{5}\right| = 1$

$\dfrac{2x-3}{5} = 1$ or $\dfrac{2x-3}{5} = -1$

$2x - 3 = 5$ $2x - 3 = -5$

$\qquad 2x = 8$ $2x = -2$

$\qquad\quad x = 4$ $x = -1$

The solution set is $\{-1, 4\}$.

77. $\left|\dfrac{x-4}{3}\right| < 6$

$-6 < \dfrac{x-4}{3} < 6$

$3(-6) < 3\left(\dfrac{x-4}{3}\right) < 3(6)$

$-18 < x - 4 < 18$

$-14 < x < 22$

The solution set is $\{x | -14 < x < 22\}$.

78. $|3x - 4| = |x + 3|$

$3x - 4 = -(x + 3)$ or $3x - 4 = x + 3$

$3x - 4 = -x - 3$ $2x - 4 = 3$

$4x - 4 = -3$ $2x = 7$

$4x = 1$ $x = \dfrac{7}{2}$

$x = \dfrac{1}{4}$

The solution set is $\left\{\dfrac{1}{4}, \dfrac{7}{2}\right\}$.

79. $|2x - 3| + 4 \geq -10$

$\quad |2x - 3| \geq -14$

Since the right side is negative and the left side is non-negative, the solution is the entire real number line since the absolute value of a number is always greater than a negative number. The solution set is all real numbers or $\mathbb{R}$.

80. $|x + 6| < -1$

Since the right side is negative and the left side is non-negative, there is no solution since the absolute value of a number can never be less than a negative number. $\varnothing$.

81. $3 < 2x - 5 \leq 9$

$3 + 5 < 2x - 5 + 5 \leq 9 + 5$

$8 < 2x \leq 14$

$\dfrac{8}{2} < \dfrac{2x}{2} \leq \dfrac{14}{2}$

$4 < x \leq 7$

The solution is $(4, 7]$.

82. $-6 \leq \dfrac{3 - 2x}{4} < 5$

$4(-6) \leq 4\left(\dfrac{3 - 2x}{4}\right) < 4(5)$

$-24 \leq 3 - 2x < 20$

$-27 \leq -2x < 17$

$\dfrac{-27}{-2} \geq \dfrac{-2x}{-2} > \dfrac{17}{-2}$

$\dfrac{27}{2} \geq x > -\dfrac{17}{2}$

$-\dfrac{17}{2} < x \leq \dfrac{27}{2}$

The solution is $\left(-\dfrac{17}{2}, \dfrac{27}{2}\right]$.

83. $x \leq 4$ and $4x - 6 \geq -14$

$x \leq 4$ $4x \geq -8$

$x \leq 4$ $x \geq -2$

$-2 \leq x \leq 4$

The solution is $[-2, 4]$.

84. $x - 3 \leq 4$ or $2x - 5 > 9$

$x - 3 + 3 \leq 4 + 3$ $2x - 5 + 5 > 9 + 5$

$x \leq 7$ $2x > 14$

$x > 7$

The solution is $(-\infty, \infty)$.

85. $-10 < 3(x - 4) \le 12$

$-10 < 3x - 12 \le 12$

$-10 + 12 < 3x - 12 + 12 \le 12 + 12$

$2 < 3x \le 24$

$\dfrac{2}{3} < x \le 8$

The solution is $\left(\dfrac{2}{3},\ 8 \right]$.

Practice Test

1. $-6xy^2z^3$ is a sixth degree term since $-6xy^2z^3$ can be written as $-6x^1y^2z^3$ and the sum of the exponents is $1 + 2 + 3 = 6$.

2. $2p - 3q + 2pq - 6p(q - 3) - 4p$
$= 2p - 3q + 2pq - 6pq + 18p - 4p$
$= (2p + 18p - 4p) - 3q + (2pq - 6pq)$
$= 16p - 3q - 4pq$

3. $4x - \{3 - [2(x - 2) - 5x]\}$
$= 4x - [3 - (2x - 4 - 5x)]$
$= 4x - [3 - (-3x - 4)]$
$= 4x - (3 + 3x + 4)$
$= 4x - (7 + 3x)$
$= 4x - 7 - 3x$
$= x - 7$

4. $\quad 3(x - 2) = 4(4 - x) + 5$
$3x - 6 = 16 - 4x + 5$
$3x - 6 = 21 - 4x$
$3x + 4x - 6 = 21 - 4x + 4x$
$7x - 6 = 21$
$7x - 6 + 6 = 21 + 6$
$7x = 27$
$\dfrac{7x}{7} = \dfrac{27}{7}$
$x = \dfrac{27}{7}$

5. $\quad \dfrac{3}{5} - \dfrac{x}{2} = 4$
$10\left(\dfrac{3}{5}\right) - 10\left(\dfrac{x}{2}\right) = 10(4)$
$6 - 5x = 40$
$6 - 6 - 5x = 40 - 6$
$-5x = 34$
$\dfrac{-5x}{-5} = \dfrac{34}{-5}$
$x = -\dfrac{34}{5}$

6. $\quad -3(x + 3) = 3\{[4 - (2x - 3)] - 4x\}$
$-3x - 9 = 3[(4 - 2x + 3) - 4x]$
$-3x - 9 = 3[(7 - 2x) - 4x]$
$-3x - 9 = 3(7 - 6x)$
$-3x - 9 = 21 - 18x$
$-3x - 9 + 18x = 21 - 18x + 18x$
$15x - 9 = 21$
$15x - 9 + 9 = 21 + 9$
$15x = 30$
$\dfrac{15x}{15} = \dfrac{30}{15}$
$x = 2$

7. $7x - 6(2x - 4) = 3 - (5x - 6)$
$7x - 12x + 24 = 3 - 5x + 6$
$-5x + 24 = -5x + 9$
$-5x + 24 + 5x = -5x + 9 + 5x$
$24 = 9$

This is a false statement which means there is no solution. $\varnothing$

8. $\quad -\dfrac{1}{2}(4x - 6) = \dfrac{1}{3}(3 - 6x) + 2$
$-2x + 3 = 1 - 2x + 2$
$-2x + 3 = -2x + 3$
$-2x + 3 + 2x = -2x + 3 + 2x$
$3 = 3$

This is always true which means the solution is any real number or $\mathbb{R}$.

9. $S_n = \dfrac{a_1(1 - r^n)}{1 - r}$

$S_3 = \dfrac{3\left[1 - \left(\dfrac{1}{3}\right)^3\right]}{1 - \dfrac{1}{3}} = \dfrac{3\left[1 - \dfrac{1}{27}\right]}{1 - \dfrac{1}{3}} = \dfrac{3\left(\dfrac{26}{27}\right)}{\dfrac{2}{3}}$

$\qquad = \dfrac{\dfrac{26}{9}}{\dfrac{2}{3}} = \dfrac{26}{9} \cdot \dfrac{3}{2} = \dfrac{13}{3}$

10. $\quad c = \dfrac{a - 3b}{2}$

$\quad 2(c) = 2\left(\dfrac{a - 3b}{2}\right)$

$\qquad 2c = a - 3b$

$\quad 2c - a = a - a - 3b$

$\quad 2c - a = -3b$

$\quad \dfrac{2c - a}{-3} = \dfrac{-3b}{-3}$

$\quad \dfrac{2c - a}{-3} = b \text{ or } b = \dfrac{a - 2c}{3}$

11. $\qquad A = \dfrac{1}{2}h(b_1 + b_2)$

$\qquad 2(A) = 2\left[\dfrac{1}{2}h(b_1 + b_2)\right]$

$\qquad 2A = h(b_1 + b_2)$

$\qquad 2A = hb_1 + hb_2$

$\qquad 2A - hb_1 = hb_1 - hb_1 + hb_2$

$\qquad 2A - hb_1 = hb_2$

$\qquad \dfrac{2A - hb_1}{h} = \dfrac{hb_2}{h}$

$\qquad \dfrac{2A - hb_1}{h} = b_2 \text{ or } b_2 = \dfrac{2A - hb_1}{h}$

12. Let x be the cost of the clubs before tax, then $0.07x$ is the tax.

$\quad x + 0.07x = 668.75$

$\qquad 1.07x = 668.75$

$\qquad x = \dfrac{668.75}{1.07}$

$\qquad x = 625$

The cost of the clubs before tax is \$625.

13. Let x be the number of miles.

$\quad 35 + 0.15x = 65$

$\qquad 0.15x = 30$

$\qquad x = \dfrac{30}{0.15}$

$\qquad x = 200$

Valerie can drive 200 miles.

14. Let x be the distance between the two joggers.

Person	Rate	Time	Distance
Homer	4	1.25	4(1.25)
Frances	5.25	1.25	(5.25)(1.25)

The total distance is

$x = 4(1.25) + (5.25)(1.25)$

$x = 5 + 6.5625$

$x = 11.5625$

In $1\dfrac{1}{4}$ hours, the joggers will be approximately 11.56 miles apart.

15. Let x be the amount of 12% solution.

Solution	Strength of Solution	No. of Liters	Amount of Salt
12%	0.12	x	$0.12x$
25%	0.25	10	$0.25(10)$
20%	0.20	$x + 10$	$0.20(x + 10)$

$$0.12x + 0.25(10) = 0.20(x + 10)$$
$$0.12x + 2.50 = 0.20x + 2.00$$
$$0.12x + 0.50 = 0.20x$$
$$0.50 = 0.08x$$
$$\frac{0.50}{0.08} = x$$
$$6.25 = x$$

Combine 6.25 liters of the 12% solution with 10 liters of the 25% solution to obtain the mixture.

16. Let x be the amount invested at 8%. Then $12,000 - x$ is the amount invested at 7%.

Account	Principal	Rate	Interest
8%	x	0.08	$0.08x$
7%	$12,000 - x$	0.07	$0.07(12,000 - x)$

The total interest is $910.
$$0.08x + 0.07(12,000 - x) = 910$$
$$0.08x + 840 - 0.07x = 910$$
$$0.01x + 840 = 910$$
$$0.01x = 70$$
$$x = \frac{70}{0.01}$$
$$x = 7000$$

Thus, $7000 was invested at 8% and the remaining amount of $12000 - 7000 = \$5000$ was invested at 7%.

17. $4(x-2) < 3(x-2) - 5$

$4x - 8 < 3x - 6 - 5$

$4x - 8 < 3x - 11$

$x - 8 < -11$

$x < -3$

18. $\dfrac{6-2x}{5} \ge -12$

$5\left(\dfrac{6-2x}{5}\right) \ge 5(-12)$

$6 - 2x \ge -60$

$-2x \ge -66$

$\dfrac{-2x}{-2} \le \dfrac{-66}{-2}$

$x \le 33$

19. $x - 3 \le 4$ and $2x - 4 > 5$

$x - 3 + 3 \le 4 + 3 \qquad 2x - 4 + 4 > 5 + 4$

$x \le 7 \qquad\qquad 2x > 9$

$x > \dfrac{9}{2}$

The solution is $\left(\dfrac{9}{2}, \ 7\right]$.

20. $-4 < \dfrac{x+4}{2} < 8$

$2(-4) < 2\left(\dfrac{x+4}{2}\right) < 2(8)$

$-8 < x + 4 < 16$

$-8 - 4 < x + 4 - 4 < 16 - 4$

$-12 < x < 12$

The solution is $(-12, 12)$.

21. $|x - 4| = 5$

$x - 4 = 5$ or $x - 4 = -5$

$x = 9 \qquad\quad x = -1$

The solution set is $\{-1, 9\}$.

22. $|2x - 3| = \left|\dfrac{1}{2}x - 10\right|$

$2x - 3 = -\left(\dfrac{1}{2}x - 10\right)$ or $2x - 3 = \dfrac{1}{2}x - 10$

$2x - 3 = -\dfrac{1}{2}x + 10 \qquad \dfrac{3}{2}x - 3 = -10$

$\dfrac{5}{2}x - 3 = 10 \qquad\qquad \dfrac{3}{2}x = -7$

$\dfrac{5}{2}x = 13 \qquad\qquad \dfrac{2}{3}\left(\dfrac{3}{2}x\right) = \dfrac{2}{3}(-7)$

$\dfrac{2}{5}\left(\dfrac{5}{2}x\right) = \dfrac{2}{5}(13) \qquad\qquad x = -\dfrac{14}{3}$

$x = \dfrac{26}{5}$

The solution set is $\left\{-\dfrac{14}{3}, \ \dfrac{26}{5}\right\}$.

23. $|3x - 2| = 0$

$3x - 2 = 0$

$3x = 2$

$x = \dfrac{2}{3}$

The solution set is $\left\{\dfrac{2}{3}\right\}$.

24. $|2x - 3| + 1 > 6$

$|2x - 3| > 5$

$2x - 3 < -5$ or $2x - 3 > 5$

$2x < -2 \qquad\qquad 2x > 8$

$x < -1 \qquad\qquad x > 4$

The solution set is $\{x | x < -1 \text{ or } x > 4\}$.

25. $\left|\dfrac{2x-3}{4}\right| \le \dfrac{1}{2}$

$-\dfrac{1}{2} \le \dfrac{2x-3}{4} \le \dfrac{1}{2}$

$4\left(-\dfrac{1}{2}\right) \le 4\left(\dfrac{2x-3}{4}\right) \le 4\left(\dfrac{1}{2}\right)$

$-2 \le 2x-3 \le 2$

$1 \le 2x \le 5$

$\dfrac{1}{2} \le x \le \dfrac{5}{2}$

The solution set is $\left\{x \middle| \dfrac{1}{2} \le x \le \dfrac{5}{2}\right\}$.

Cumulative Review Test

1. a. $A \cup B = \{1,2,3,4,5,6,7,9,10,12\}$

   b. $A \cap B = \{4,6,9,12\}$

2. a. Commutative property of addition

   b. Associative property of multiplication

   c. Distributive property

3. $-|-3| < |-5|$

$-|-3| = -3$

$-|-5| = 5$

4. $4-|-3|-(6+|-3|)^2 = 4-|-3|-(6+3)^2$

$= 4-|-3|-(9)^2$

$= 4-|-3|-81$

$= 4-3-81$

$= 1-81$

$= -80$

5. $-4^2+(-6)^2 \div (2^3-2)^2$

$= -4^2+(-6)^2 \div (8-2)^2$

$= -4^2+(-6)^2 \div (6)^2$

$= -16+36 \div 36$

$= -16+1$

$= -15$

6. Substitute -3 for x and -2 for y.

$x^3 - xy + y^2 = (-3)^3 - (-3)(-2) + (-2)^2$

$= -27 - (-3)(-2) + 4$

$= -27 - (6) + 4$

$= -33 + 4$

$= -29$

7. $\dfrac{8-\sqrt[3]{27}\cdot 3 \div 9}{|-5|-[5-(12\div 4)]^2} = \dfrac{8-\sqrt[3]{27}\cdot 3 \div 9}{|-5|-[5-3]^2}$

$= \dfrac{8-\sqrt[3]{27}\cdot 3 \div 9}{|-5|-2^2}$

$= \dfrac{8-3\cdot 3 \div 9}{5-4}$

$= \dfrac{8-9 \div 9}{5-4}$

$= \dfrac{8-1}{5-4}$

$= \dfrac{7}{1}$

$= 7$

8. $(2x^4y^3)^{-2} = \left(\dfrac{1}{2x^4y^3}\right)^2$

$= \dfrac{1^2}{2^2 x^{4\cdot 2} y^{3\cdot 2}}$

$= \dfrac{1}{4x^8 y^6}$

9. $\left(\dfrac{3m^2 n^{-4}}{m^{-3} n^2}\right)^2 = \left(\dfrac{3m^{2-(-3)}}{n^{2-(-4)}}\right)^2$

$= \left(\dfrac{3m^5}{n^6}\right)^2$

$= \dfrac{3^2 m^{5\cdot 2}}{n^{6\cdot 2}}$

$= \dfrac{9m^{10}}{n^{12}}$

10. $r^{3m-2} \cdot r^{2m-6} = r^{3m-2+2m-6}$

$= r^{5m-8}$

11. $40,600,000 = 4.06 \times 10^7$

12. $1.12 \times 10^{10} - 9.25 \times 10^9$
$= 11.2 \times 10^9 - 9.25 \times 10^9$
$= (11.2 - 9.25) \times 10^9$
$= 1.95 \times 10^9$ or $1,950,000,000$
The sales of carbonated beverages were about $\$1.95 \times 10^9$ or $\$1,950,000,000$ more than milk.

13. $3x - 4 = -2(x - 3) - 9$
$3x - 4 = -2x + 6 - 9$
$3x - 4 = -2x - 3$
$5x - 4 = -3$
$5x = 1$
$x = \dfrac{1}{5}$

14. $1.2(x - 3) = 2.4x - 4.98$
$1.2x - 3.6 = 2.4x - 4.98$
$1.2x = 2.4x - 1.38$
$-1.2x = -1.38$
$x = \dfrac{-1.38}{-1.2}$
$x = 1.15$

15. $\dfrac{x}{4} - 5 = 3x - \dfrac{1}{3}$
$12\left(\dfrac{x}{4}\right) - 12(5) = 12(3x) - 12\left(\dfrac{1}{3}\right)$
$3x - 60 = 36x - 4$
$3x = 36x + 56$
$-33x = 56$
$x = \dfrac{56}{-33}$
$x = -\dfrac{56}{33}$

16. $\dfrac{\frac{1}{4}x + 2}{3} = \dfrac{x - 4}{4}$
$12\left(\dfrac{\frac{1}{4}x + 2}{3}\right) = 12\left(\dfrac{x - 4}{4}\right)$
$4\left(\dfrac{1}{4}x + 2\right) = 3(x - 4)$
$x + 8 = 3x - 12$
$x + 20 = 3x$
$20 = 2x$
$10 = x$

17. A conditional equation is true only under specific conditions. An identity is true for all values of the variable. An inconsistent equation is never true. Answers may vary. One possible answer is: $3x + 4 = 13$ is a conditional linear equation.
$3(x + 7) = 2(x + 10) + x + 1$ is an identity.
$3x + 4 = 3x + 8$ is a contradiction.

18. $x = \dfrac{-b + \sqrt{b^2 - 4ac}}{2a}$
$= \dfrac{-(-8) + \sqrt{(-8)^2 - 4(3)(-3)}}{2(3)}$
$= \dfrac{-(-8) + \sqrt{64 + 36}}{6}$
$= \dfrac{-(-8) + \sqrt{100}}{6}$
$= \dfrac{8 + 10}{6}$
$= \dfrac{18}{6}$
$= 3$

19. $A = p + prt$
$A - p = p - p + prt$
$A - p = prt$
$\dfrac{A - p}{pr} = \dfrac{prt}{pr}$
$\dfrac{A - p}{pr} = t$ or $t = \dfrac{A - p}{pr}$

20. $-4 < \dfrac{5x-2}{3} < 2$

$3(-4) < 3\left(\dfrac{5x-2}{3}\right) < 3(2)$

$-12 < 5x - 2 < 6$

$-12 + 2 < 5x - 2 + 2 < 6 + 2$

$-10 < 5x < 8$

$\dfrac{-10}{5} < \dfrac{5x}{5} < \dfrac{8}{5}$

$-2 < x < \dfrac{8}{5}$

a.

b. $\left\{x \middle| -2 < x < \dfrac{8}{5}\right\}$

c. $\left(-2, \dfrac{8}{5}\right)$

21. $|4z + 8| = 12$

$4z + 8 = 12 \quad$ or $\quad 4z + 8 = -12$

$4z = 4 \qquad\qquad\qquad 4z = -20$

$z = 1 \qquad\qquad\qquad\quad z = -5$

Solution is $\{-5, 1\}$

22. $|2x - 4| - 6 \geq 18$

$|2x - 4| \geq 24$

$2x - 4 \leq -24 \quad$ or $\quad 2x - 4 \geq 24$

$2x \leq -20 \qquad\qquad\quad 2x \geq 28$

$x \leq -10 \qquad\qquad\quad x \geq 14$

The solution set is $\{x \mid x \leq -10 \text{ or } x \geq 14\}$.

23. Let x be the original price.

$x - 0.20x = 1800$

$0.80x = 1800$

$x = \dfrac{1800}{0.80}$

$x = 2250$

The original price was $2250.

24. Let x be the speed of the car traveling south. Then $x + 10$ is the speed of the car traveling north.

Car	Rate	Time	Distance
South	x	3	$3x$
North	$x + 10$	3	$3(x + 10)$

The total distance is 270 miles.

$3x + 3(x + 10) = 270$

$3x + 3x + 30 = 270$

$6x + 30 = 270$

$6x = 240$

$x = \dfrac{240}{6}$

$x = 40$

The speed of the car traveling south is 40 mph and the speed of the car traveling north is $40 + 10 = 50$ mph.

25. Let x be the amount of the 20% saltwater solution. Then $2 - x$ is the amount of the 50% saltwater solution.

Solution	Strength	No. of Liters	Amount of Salt
20%	0.20	x	$0.20x$
50%	0.50	$2 - x$	$0.50(2 - x)$
30%	0.30	2	$0.30(2)$

$$0.20x + 0.50(2 - x) = 0.30(2)$$
$$0.20x + 1.00 - 0.50x = 0.60$$
$$-0.30 + 1.00 = 0.60$$
$$-0.30x = -0.40$$
$$x = \frac{-0.40}{-0.30} = \frac{4}{3} \text{ or } 1\frac{1}{3} \text{ liters}$$

Ms. Kane should combine $1\frac{1}{3}$ liters of the 20% solution with $2 - 1\frac{1}{3} = \frac{2}{3}$ liter of the 50% solution to obtain the desired solution.

Chapter 3

Exercise Set 3.1

1. a. The graph of any linear equation looks like a straight line.

b. Two points are needed to graph a linear equation.
Two points uniquely determine a straight line.

3. When graphing $y = \dfrac{1}{x}$, we cannot substitute 0 for x.

$\dfrac{1}{0}$ is undefined.

5. $A(3, 1)$, $B(-6, 0)$, $C(2, -4)$, $D(-2, -4)$, $E(0, 3)$, $F(-8, 1)$, $G\left(\dfrac{3}{2}, -1\right)$

7.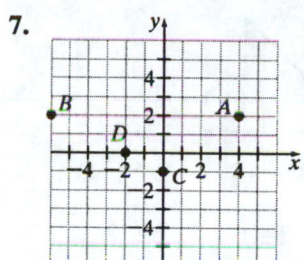

9. I

11. IV

13. II

15. III

17. $y = 6x + 9$
$21 \stackrel{?}{=} 6(2) + 9$
$21 \stackrel{?}{=} 12 + 9$
$21 = 21$ true
Yes, $(2, 21)$ is a solution to $y = 6x + 9$.

19. $y = |x| - 2$
$1 \stackrel{?}{=} |-3| - 2$
$1 \stackrel{?}{=} 3 - 2$
$1 = 1$ true
Yes, $(-3, 1)$ is a solution to $y = |x| - 2$.

21. $s = 2r^2 - r - 5$
$5 \stackrel{?}{=} 2(-2)^2 - (-2) - 5$
$5 \stackrel{?}{=} 8 + 2 - 5$
$5 = 5$ true
Yes, $(-2, 5)$ is a solution to $s = 2r^2 - r - 5$.

23. $2x^2 + y = 8$
$2(2)^2 + (0) \stackrel{?}{=} 8$
$8 + 0 \stackrel{?}{=} 8$
$8 = 8$ true

Yes, $(2, 0)$ is a solution to $2x^2 + y = 8$.

25. $2x^2 + 4x - y = 0$
$2\left(\dfrac{1}{2}\right)^2 + 4\left(\dfrac{1}{2}\right) - \left(\dfrac{3}{2}\right) \stackrel{?}{=} 0$
$2\left(\dfrac{1}{4}\right) + 4\left(\dfrac{1}{2}\right) - \left(\dfrac{3}{2}\right) \stackrel{?}{=} 0$
$\dfrac{1}{2} + 2 - \dfrac{3}{2} \stackrel{?}{=} 0$
$1 = 0$ false

No, $\left(\dfrac{1}{2}, \dfrac{3}{2}\right)$ is not a solution to

$2x^2 + 4x - y = 0$.

27.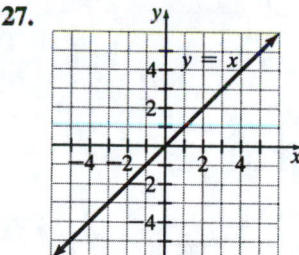

29.

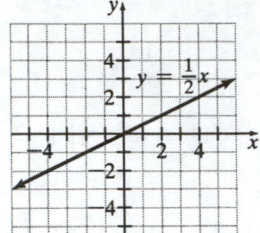

31.

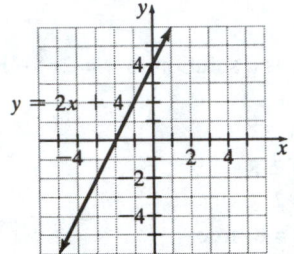

33. $y = -3x - 5$

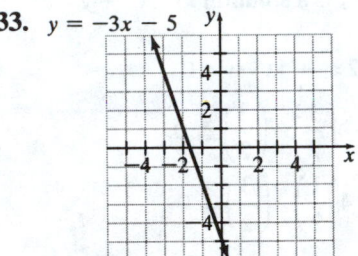

35.

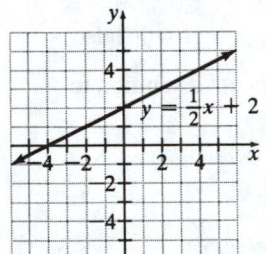

37.

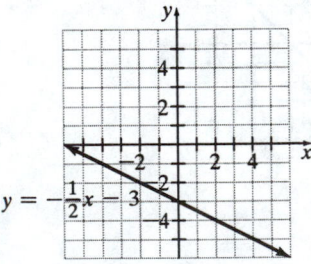

39.

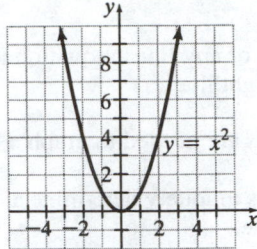

41.

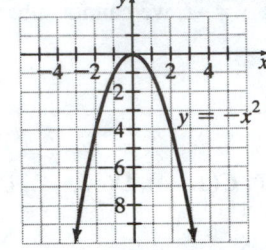

43.

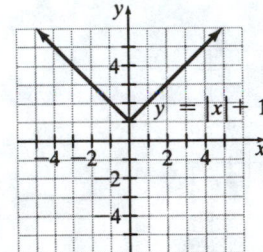

45.

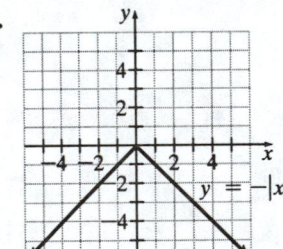

47.

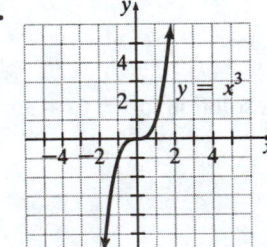

49.

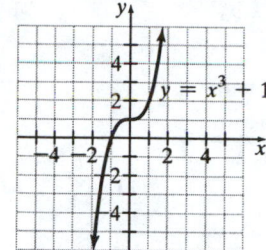

51.

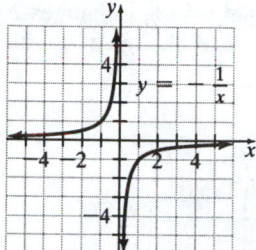

53.

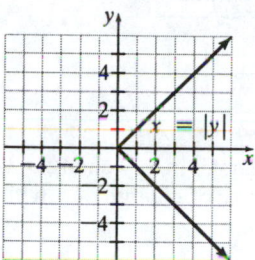

55.

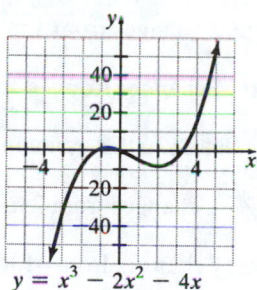

$y = x^3 - 2x^2 - 4x$

57.

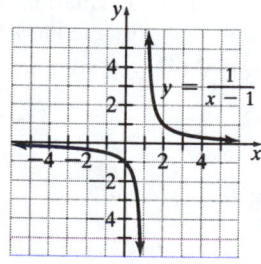

59.

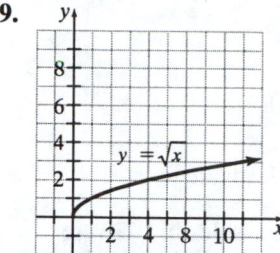

61.

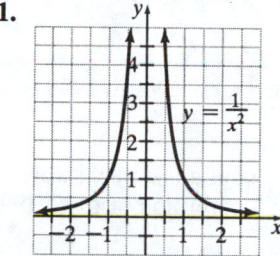

63. $y = \dfrac{x}{x^2 - 6}$

$$-\frac{2}{23} \stackrel{?}{=} \frac{\frac{1}{2}}{\left(\frac{1}{2}\right)^2 - 6}$$

$$-\frac{2}{23} \stackrel{?}{=} \frac{\frac{1}{2}}{-\frac{23}{4}}$$

$$-\frac{2}{23} = -\frac{2}{23} \quad \text{true}$$

Yes, $\left(\dfrac{1}{2},\ -\dfrac{2}{23}\right)$ is on the graph of the

equation $y = \dfrac{x}{x^2 - 6}$.

65. a.

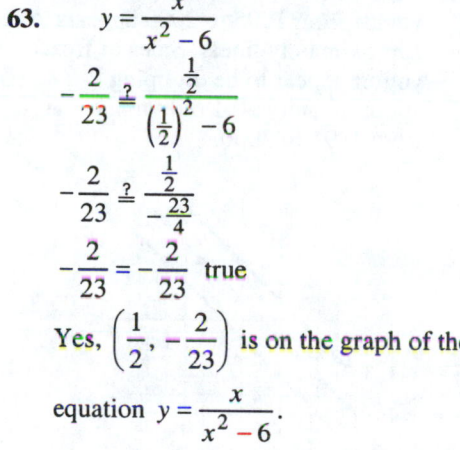

b. area $= \dfrac{1}{2}bh$

$= \dfrac{1}{2}(4)(4)$

$= 8$

The area is 8 square units.

67. a. The estimated sales in 1999 are $1.2 billion, $3.0 billion, and $3.6 billion, respectively.

b. $1.2 + 3.0 + 3.6 = 7.8$
The estimated total sales in 1999 is $7.8 billion.

c. Sales of low/nonfat ice cream were greater than $2.5 billion in 1998, 1999, and 2000.

d. Yes, the decrease in the sales of frozen yogurt from 1995 to 2000 appears to be approximately linear. Sales of frozen yogurt appear to be dropping approximately $0.1 billion per year from 1995 to 2000.

69.

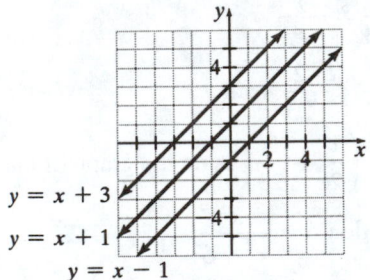

a. Each graph crosses the y-axis at the point corresponding to the constant term in the graph's equation.

b. Yes, all the equations seem to have the same slant or slope.

71.

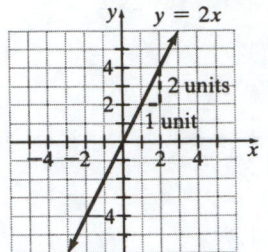

For each unit change in x, y changes 2 units. Therefore, the rate of change of y with respect to x is 2.

73.

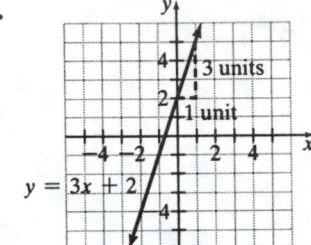

For each unit change in x, y changes 3 units. Therefore, the rate of change of y with respect to x is 3.

75. Starting at $(1, -4)$:
For a unit change, x changes from 1 to $1 + 1 = 2$.
At the same time, y changes from -4 to $-4 + 3 = -1$.
So, $(2, -1)$ is a solution to the equation.
Starting at $(2, -1)$:
For a unit change, x changes from 2 to $2 + 1 = 3$.
At the same time, y changes from -1 to $-1 + 3 = 2$.
So, $(3, 2)$ is a solution to the equation.
Answers may vary. One possible answer is the points $(2, -1)$ and $(3, 2)$.

77. c

79. a

81. b

83. a

85.

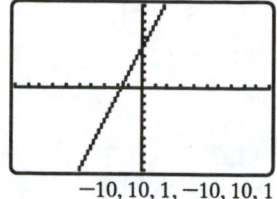

$-10, 10, 1, -10, 10, 1$

X	Y1
0	5
1	8
2	11
3	14
4	17
5	20
6	23

Y1◻3X+5

87.

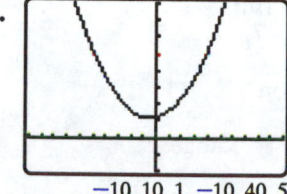

$-10, 10, 1, -10, 40, 5$

X	Y1
0	6
1	8
2	12
3	18
4	26
5	36
6	48

Y1◻X²+X+6

89.

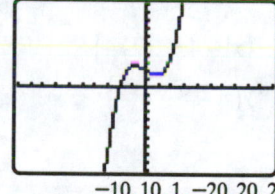

$-10, 10, 1, -20, 20, 2$

X	Y1
0	4
1	3
2	8
3	25
4	60
5	119
6	208

Y1◻X^3-2X+4

91.

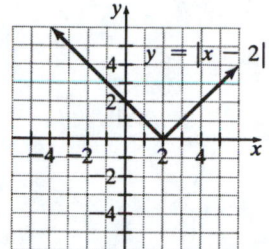

95. $\dfrac{-b + \sqrt{b^2 - 4ac}}{2a}$

$= \dfrac{-(7) + \sqrt{(7)^2 - 4(2)(-15)}}{2(2)}$

$= \dfrac{-7 + \sqrt{169}}{4}$

$= \dfrac{-7 + 13}{4}$

$= \dfrac{3}{2}$

96. Let x be the number of miles driven. The cost for renting from Hertz is $y_1 = 30 + 0.14x$.

The cost for renting from National Automobile Rental Agency is $y_2 = 16 + 0.24x$. The costs are equal when

$$y_1 = y_2$$
$$30 + 0.14x = 16 + 0.24x$$
$$14 = 0.10x$$
$$x = \frac{14}{0.10}$$
$$x = 140$$

You would have to drive 140 miles to make the costs equal.

97. $\qquad -4 \le \dfrac{4 - 3x}{2} < 5$

$$2(-4) \le 4 - 3x < 2(5)$$
$$-8 \le 4 - 3x < 10$$
$$-8 - 4 \le -3x < 10 - 4$$
$$-12 \le -3x < 6$$
$$\frac{-12}{-3} \ge x > \frac{6}{-3}$$
$$4 \ge x > -2$$
$$-2 < x \le 4$$
$$\{x | -2 < x \le 4\}$$

98. $|3x + 2| > 5$

$3x + 2 < -5$ or $3x + 2 > 5$

$3x < -7$ $\qquad$ $3x > 3$

$x < -\dfrac{7}{3}$ $\qquad$ $x > 1$

$\left\{ x \middle| x < -\dfrac{7}{3} \text{ or } x > 1 \right\}$

Exercise Set 3.2

1. A relation is any set of ordered pairs.

3. No, all relations are not functions. A relation can have two ordered pairs with the same first element but a function cannot.

5. If each vertical line drawn through any part of the graph intersects the graph in at most one point, the graph represents a function.

7. The range is the set of values for the dependent variable.

9. Domain: $\mathbb{R}$ or $(-\infty, \infty)$
There are no restrictions on values of x that can be used.
Range: $\mathbb{R}$ or $(-\infty, \infty)$
All values of y are represented in the function.

11. If y depends on x, then y is the dependent variable.

13. $f(x)$ is read "f of x."

15. a. Yes, the relation is a function.

 b. Domain: $\{3, 5, 10\}$, Range: $\{6, 10, 20\}$

17. a. Yes, the relation is a function.

 b. Domain: $\{$Ron, Jayne, Cecilia$\}$,
Range:$\{18, 19\}$

19. a. No, the relation is not a function.

 b. Domain: $\{1990, 1996, 1999\}$,
Range: $\{20, 32, 33\}$

21. a. A function

 b. Domain: $\{1, 2, 3, 4, 5\}$,
Range: $\{1, 2, 3, 4, 5\}$

23. a. A function

 b. Domain: $\{1, 2, 3, 4, 5, 7\}$,
Range: $\{-1, 0, 2, 4, 5\}$

25. a. Not a function

 b. Domain: $\{1, 2, 3, 5\}$,
Range: $\{-4, -1, 0, 1, 2\}$

27. a. Not a function

 b. Domain: $\{0, 1, 2\}$
Range: $\{-7, -1, 2, 3\}$

29. a. A function

 b. Domain: $\mathbb{R}$, Range: $\mathbb{R}$

 c. $x = 2$

31. a. Not a function

 b. Domain: $\{-2\}$, Range: $\mathbb{R}$

 c. $x = -2$

33. a. Not a function

 b. Domain: $\left\{ x \middle| -4 \le x \le 4 \right\}$
Range: $\left\{ y \middle| -2 \le y \le 2 \right\}$

 c. $x = 0$

35. a. Not a function

 b. Domain: $\mathbb{R}$, Range: $\mathbb{R}$

 c. $x = 2$

37. a. A function

 b. Domain: $\{1, 2, 3\}$, Range: $\{1\}$

 c. No values of x

39. a. A function

    b. Domain: $\{x|{-20} \le x \le 10\}$,

         Range: $\{y|{-2} \le y \le 2\}$

    c. $x = -17.5$ or $x = -7.5$ or $x = 2.5$

41. a. $f(5) = 3(5) + 6 = 15 + 6 = 21$

    b. $f(-2) = 3(-2) + 6 = -6 + 6 = 0$

43. a. $h(0) = (0)^2 - (0) - 6 = -6$

    b. $h(-1) = (-1)^2 - (-1) - 6 = 1 + 1 - 6 = -4$

45. a. $g(2) = (2)^3 + 2(2)^2 - 4 = 8 + 8 - 4 = 12$

    b. $g(-3) = (-3)^3 + 2(-3)^2 - 4$
$$= -27 + 18 - 4$$
$$= -13$$

47. a. $f(-5) = |(-5) + 3| = |-2| = 2$

    b. $f(-12.6) = |(-12.6) + 3| = |-9.6| = 9.6$

49. a. $f(3) = \sqrt{(3) + 1} = \sqrt{4} = 2$

    b. $f(24) = \sqrt{(24) + 1} = \sqrt{25} = 5$

51. a. $f(2) = \dfrac{(2)^2 - 4}{(2) + 2} = \dfrac{0}{4} = 0$

    b. $f(-3) = \dfrac{(-3)^2 - 4}{(-3) + 2} = \dfrac{5}{-1} = -5$

53. a. $d(4) = 60(4) = 240$
In 4 hours, the car travels 240 miles.

    b. $d(12) = 60(12) = 720$
In 12 hours, the car tavels 720 miles.

55. a. $C(r) = 2\pi r$

    b. $C(9) = 2\pi(9) = 18\pi \approx 56.5$
The circumference is about 56.5 feet.

57. a. $F(C) = \dfrac{9}{5}C + 32$

    b. $F(20) = \dfrac{9}{5}(20) + 32 = 68$
The Fahrenheit temperature is 68°F.

59. a. $T(3) = -0.03(3)^2 + 1.5(3) + 14$
$$= -0.27 + 4.5 + 14$$
$$= 18.23$$
The temperature is 18.23°C.

    b. $T(12) = -0.03(12)^2 + 1.5(12) + 14$
$$= -4.32 + 18 + 14$$
$$= 27.68$$
The temperature is 27.68°C.

61. a. $T(4) = -0.02(4)^2 - 0.34(4) + 80$
$$= -0.32 - 1.36 + 80$$
$$= 78.32$$
The temperature is 78.32°.

    b. $T(12) = -0.02(12)^2 - 0.34(12) + 80$
$$= -2.88 - 4.08 + 80$$
$$= 73.04$$
The temperature is 73.04°.

63. a. $T(6) = \dfrac{1}{3}(6)^3 + \dfrac{1}{2}(6)^2 + \dfrac{1}{6}(6)$
$$= 72 + 18 + 1$$
$$= 91$$
91 oranges

    b. $T(8) = \dfrac{1}{3}(8)^3 + \dfrac{1}{2}(8)^2 + \dfrac{1}{6}(8)$
$$= \dfrac{512}{3} + 32 + \dfrac{4}{3}$$
204 oranges

65. Answers may vary. One possible interpretation: The man walks on level ground, about 30 feet above sea level, for 5 minutes. For the next 5 minutes he walks uphill to 45 feet above sea level. For 5 minutes he walks on level ground then walks quickly downhill for 3 minutes to an elevation of 20 feet above sea level. For 7 minutes he walks on level ground. Then he walks quickly uphill.

67. Answers may vary. One possible interpretation: A woman drives in stop-and-go traffic for 5 minutes. Then she drives on the highway for 15 minutes, gets off onto a country road for a few minutes, stops for a couple of minutes, and returns to stop-and-go traffic.

69. a. Yes, the graph represents a function. Each
x-value has only one y-value.

b. The graph shows the number of active U.S. military personnel in all branches from 1975 to 1997.

c. In 1993, there were approximately 1,700,000 personnel.

d. From 1987 to 1997, the number of personnel decreased from 2,174,217 to 1,425,471—a decrease of 748,746. The percent decrease is $\dfrac{748,746}{2,174,217} \approx 0.34$, or 34%.

71. a. Yes, both graphs represent functions. Each x-value has only one y-value on each graph.

b. No, the graph is not a straight line.

c. Yes, the portion of the graph from 1988 to 1996 is fairly close to a straight line.

d. If $f(t) = 24$ million, then $t = 1994$.

e. If $g(t) = 70$, then $t = 1991$.

73. a.

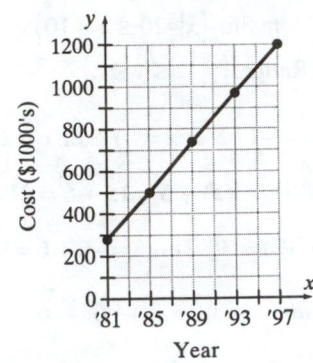

b. Yes, the points lie approximately on a straight line.

c. The cost of a 30-second commercial in 1995 was about 1100, or about $1,100,000.

75. a.

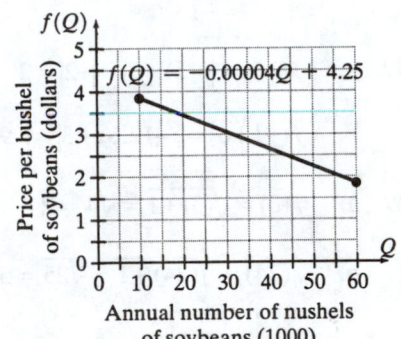

b. $f(40,000) = -0.00004(40,000) + 4.25$
$= -1.6 + 4.25$
$= 2.65$
The cost of a bushel of soybeans if 40,000 bushels are produced is approximately $2.65 per bushel.

78. $3x - 2 = \dfrac{1}{3}(3x - 3)$

$3x - 2 = \dfrac{1}{3}(3x) - \dfrac{1}{3}(3)$

$3x - 2 = x - 1$

$3x - x = -1 + 2$

$2x = 1$

$x = \dfrac{1}{2}$

79.
$$E = a_1p_1 + a_2p_2 + a_3p_3$$
$$E - a_1p_1 - a_3p_3 = a_2p_2$$
$$p_2 = \frac{E - a_1p_1 - a_3p_3}{a_2}$$

80. $\dfrac{3}{5}(x-3) > \dfrac{1}{4}(3-x)$

$20 \cdot \dfrac{3}{5}(x-3) > 20 \cdot \dfrac{1}{4}(3-x)$

$12(x-3) > 5(3-x)$

$12x - 36 > 15 - 5x$

$12x + 5x > 15 + 36$

$17x > 51$

$x > 3$

a.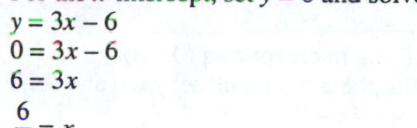
　　 3

b. $(3, \infty)$

c. $\{x \mid x > 3\}$

81. $\left|\dfrac{x-4}{3}\right| + 2 = 4$

$\left|\dfrac{x-4}{3}\right| = 2$

$\dfrac{x-4}{3} = -2$　or　$\dfrac{x-4}{3} = 2$

$x - 4 = -6$　　　　$x - 4 = 6$

$x = -2$　　　　　$x = 10$

The solution set is $\{-2, 10\}$.

Exercise Set 3.3

1. The standard form of a linear equation is $ax + by = c$, where a, b, and c are real numbers, and a and b are not both 0.

3. To find the x-intercept, set $y = 0$ and solve for x. To find the y-intercept, set $x = 0$ and solve for y.

5. The graph of $y = b$, for any real number b, will be a horizontal line.

7. The graph of $x = a$, for any real number a, will be a vertical line.

9. To solve an equation in one variable, graph both sides of the equation. The solution is the x-coordinate of the intersection.

11. $\quad y = 3x - 2$
　　 $3x - y = 2$

13. $\quad 2x = 3y - 4$
　　 $2x - 3y = -4$

15. $3(x-2) = 4(y-5)$
　　 $3x - 6 = 4y - 20$
　　 $3x - 4y = -14$

17. $y = 3x - 6$
For the y-intercept, set $x = 0$ and solve for y:
$y = 3x - 6$
$y = 3(0) - 6$
$y = 0 - 6$
$y = -6$
The y-intercept is at $(0, -6)$.
For the x-intercept, set $y = 0$ and solve for x:
$y = 3x - 6$
$0 = 3x - 6$
$6 = 3x$
$\dfrac{6}{3} = x$
$2 = x$
The x-intercept is at $(2, 0)$.

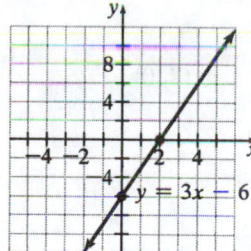

19. $y = 2x + 3$
For the y-intercept, set $x = 0$ and solve for y:
$y = 2x + 3$
$y = 2(0) + 3$
$y = 0 + 3$
$y = 3$
The y-intercept is at $(0, 3)$.
For the x-intercept, set $y = 0$ and solve for x:

$$y = 2x + 3$$
$$0 = 2x + 3$$
$$-3 = 2x$$
$$-\frac{3}{2} = x$$

The x-intercept is at $\left(-\frac{3}{2},\ 0\right)$.

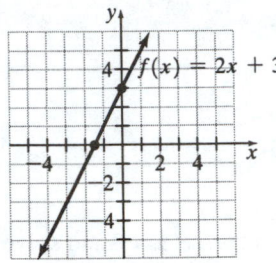

21. $y = 4x - 8$
For the y-intercept, set $x = 0$ and solve for y:
$$y = 4x - 8$$
$$y = 4(0) - 8$$
$$y = 0 - 8$$
$$y = -8$$
The y-intercept is at $(0, -8)$.
For the x-intercept, set $y = 0$ and solve for y:
$$y = 4x - 8$$
$$0 = 4x - 8$$
$$8 = 4x$$
$$\frac{8}{4} = x$$
$$2 = x$$
The x-intercept is at $(2, 0)$.

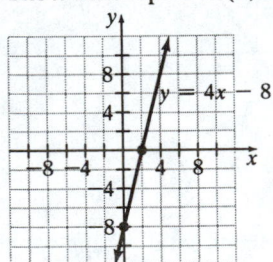

23. $4x = 3y - 9$
For the y-intercept, set $x = 0$ and solve for y:
$$4x = 3y - 9$$
$$4(0) = 3y - 9$$
$$0 = 3y - 9$$
$$9 = 3y$$
$$\frac{9}{3} = y \text{ or } 3 = y$$
The y-intercept is at $(0, 3)$.
For the x-intercept, set $y = 0$ and solve for x:
$$4x = 3y - 9$$
$$4x = 3(0) - 9$$
$$4x = 0 - 9$$
$$4x = -9$$
$$x = -\frac{9}{4}$$
The x-intercept is at $\left(-\frac{9}{4},\ 0\right)$.

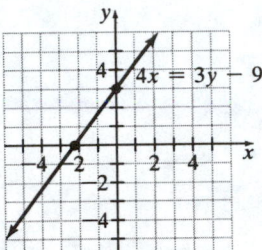

25. $30x + 25y = 50$
For the y-intercept, set $x = 0$ and solve for y:
$$30x + 25y = 50$$
$$30(0) + 25y = 50$$
$$0 + 25y = 50$$
$$25y = 50$$
$$y = \frac{50}{25} = 2$$
The y-intercept is at $(0, 2)$.
For the x-intercept, set $y = 0$ and solve for x:
$$30x + 25y = 50$$
$$30x + 25(0) = 50$$
$$30x + 0 = 50$$
$$30x = 50$$
$$x = \frac{50}{30} = \frac{5}{3}$$

The x-intercept is at $\left(\dfrac{5}{3},\ 0\right)$.

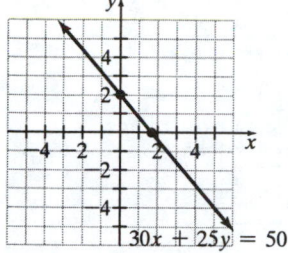

27. $0.25x + 0.50y = 1.00$
For the y-intercept, set $x = 0$ and solve for y:
$0.25x + 0.50y = 1.00$
$0.25(0) + 0.50y = 1.00$
$0 + 0.50y = 1.00$
$0.50y = 1.00$
$$y = \dfrac{1.00}{0.50} = 2$$
The y-intercept is at $(0, 2)$.
For the x-intercept, set $y = 0$ and solve for x:
$0.25x + 0.50y = 1.00$
$0.25x + 0.50(0) = 1.00$
$0.25x + 0 = 1.00$
$0.25x = 1.00$
$$x = \dfrac{1.00}{0.25} = 4$$
The x-intercept is $(4, 0)$.

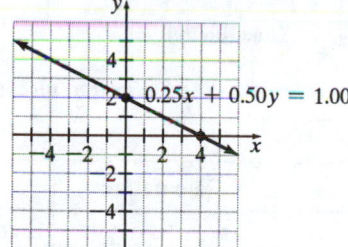

29. $120x - 360y = 720$
For the y-intercept, set $x = 0$ and solve for y:
$120x - 360y = 720$
$120(0) - 360y = 720$
$0 - 360y = 720$
$-360y = 720$
$$y = \dfrac{720}{-360} = -2$$
The y-intercept is at $(0, -2)$.
For the x-intercept, set $y = 0$ and solve for x:

$120x - 360y = 720$
$120x - 360(0) = 720$
$120x - 0 = 720$
$120x = 720$
$$x = \dfrac{720}{120} = 6$$
The x-intercept is 6 and the point is $(6, 0)$.

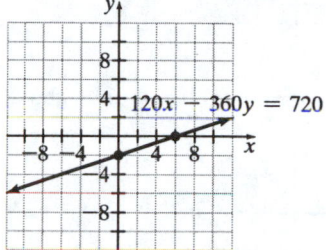

31. Multiply each term by the least common multiple, 12.
$$\dfrac{1}{3}x + \dfrac{1}{4}y = 12$$
$$12\left(\dfrac{1}{3}x\right) + 12\left(\dfrac{1}{4}y\right) = 12(12)$$
$$4x + 3y = 144$$
For the y-intercept, set $x = 0$ and solve for y:
$4x + 3y = 144$
$4(0) + 3y = 144$
$0 + 3y = 144$
$3y = 144$
$$y = \dfrac{144}{3} = 48$$
The y-intercept is at $(0, 48)$.
For the x-intercept, set $y = 0$ and solve for x:
$4x + 3y = 144$
$4x + 3(0) = 144$
$4x + 0 = 144$
$4x = 144$
$$x = \dfrac{144}{4} = 36$$
The x-intercept is at $(36, 0)$.

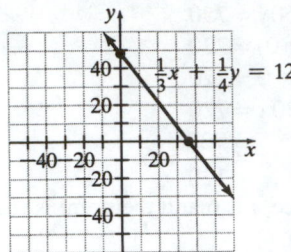

33. $y = 4$

This is a horizontal line 4 units above the *x*-axis.

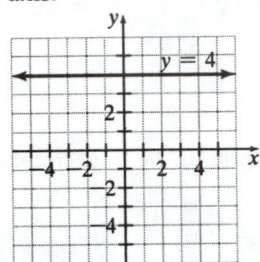

35. $x = -3$

This is a vertical line 3 units to the left of the *y*-axis.

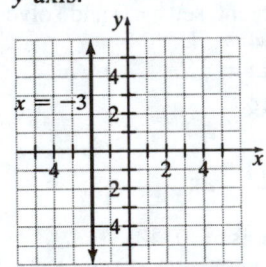

37. $f(x) = -3$

This is a horizontal line 3 units below the *x*-axis.

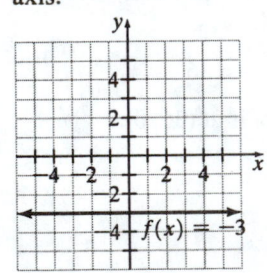

39. $g(x) = 0$

This is a horizontal line corresponding to the *x*-axis.

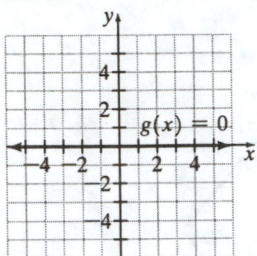

41. The equation is $d = 50t$. To graph, plot a few points.

t	Calculation	d
0	50(0)	0
1	50(1)	50
4	50(4)	200

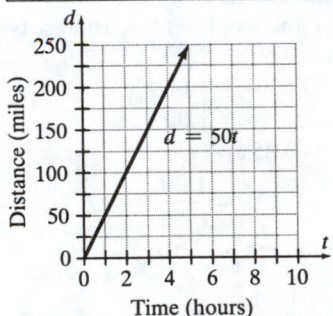

43. a. $p = 60x - 80{,}000$. To graph, plot a few points.

x	Calculation	p
0	60(0) – 80,000	–80,000
2500	60(2500) – 80,000	70,000
5000	60(5000) – 80,000	220,000

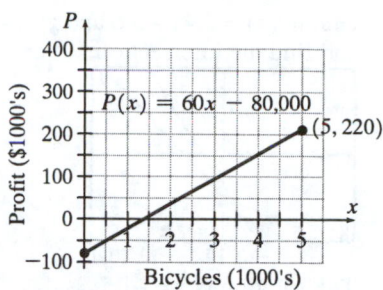

b. To break even, the profit would be zero. That is, set $p = 0$ and solve for x:

$0 = 60x - 80{,}000$

$-60x = -80{,}000$

$x = \dfrac{-80{,}000}{-60} \approx 1333$

The company must sell about 1,300 bicycles to break even.

c. To earn a profit of $150,000, set $p = 150{,}000$ and solve for x:

$150{,}000 = 60x - 80{,}000$

$230{,}000 = 60x$

$\dfrac{230{,}000}{60} = x$ or $x = 3833$

The company must sell about 3,800 bicycles to make a $150,000 profit.

45. a. $s(x) = 300 + 0.10x$

b. To graph, plot a few points.

x	Calculation	s
0	$300 + (0.10(0)$	300
2000	$300 + 0.10(2000)$	500
5000	$300 + 0.10(5000)$	800

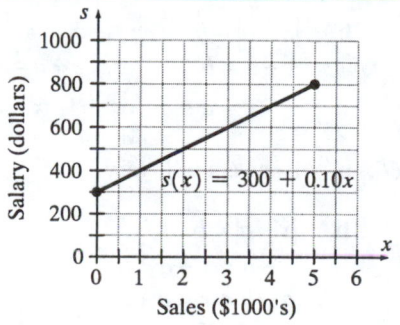

c. For weekly sales of $4000,

$s(4000) = 300 + 0.10(4000)$

$= 300 + 400$

$= 700$

Her salary is $700.

d. For a salary of $600, set $s = 600$ and solve for x.

$600 = 300 + 0.10x$

$300 = 0.10x$

$\dfrac{300}{0.10} = x$ or $x = 3000$

Her weekly sales are $3000.

47. a. There is only one y-value for each x-value.

b. The independent variable is length. The dependent variable is weight.

c. Yes, the graph of weight versus length is approximately linear.

d. The weight of the average girl who is 85 centimeters long is 11.5 kilograms.

e. The average length of a girl with a weight of 7 kilograms is 65 centimeters.

f. For a girl 95 centimeters long, the weights 12.0–15.5 kilograms are considered normal.

g. As the lengths increase, the normal range of weights increases. Yes, this is expected: as the girl grows, it is reasonable that her weight would increase with her length.

49. The x- and y-intercepts of a graph will be the same when the graph goes through the origin.

51. Answers may vary. One possible answer is, $f(x) = 4$ is a function whose graph has no x-intercept but has a y-intercept of $(0, 4)$.

53.

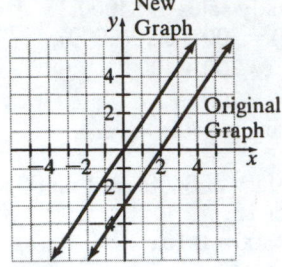

The x- and y-intercepts will both be 0.

55. a.

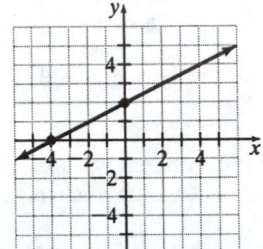

b. vertical change $= 2 - 0 = 2$

c. horizontal change $= 0 - (-4) = 4$

d. $\dfrac{\text{vertical change}}{\text{horizontal change}} = \dfrac{2}{4} = \dfrac{1}{2}$
   The ratio represents the slope of the line.

57. Graph $f(x) = 3(x + 2)$ and $g(x) = 6x + 12$, and find the intersection.

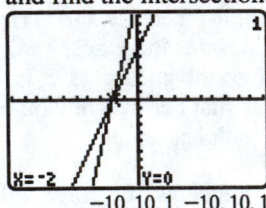

$-10, 10, 1, -10, 10, 1$

The solution is $x = -2$.

59. Graph $f(x) = 2.4x - 3.6$ and $g(x) = 1.6x - 4.8$, and find the intersection.

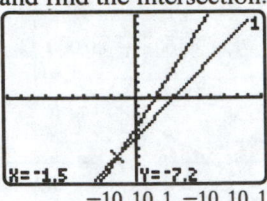

$-10, 10, 1, -10, 10, 1$

The solution is $x = -1.5$.

61.

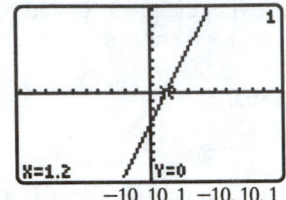

$-10, 10, 1, -10, 10, 1$

The x-intercept is $(1.2, 0)$.
The y-intercept is $(0, -3.6)$.

63. To use the graphing calculator, we must rewrite the equation in the form $y = f(x)$.
$$-4x - 3.2y = 8$$
$$-3.2y = 4x + 8$$
$$y = -\frac{1}{3.2}(4x + 8)$$

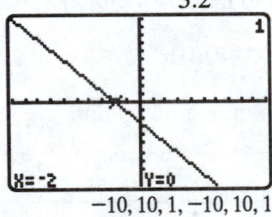

$-10, 10, 1, -10, 10, 1$

The x-intercept is $(-2, 0)$.
The y-intercept is $(0, -2.5)$.

65. a. Answers will vary.

   b. $|x - a| = b$
   $x - a = b$ or $x - a = -b$
        $x = a + b$ or $x = a - b$

66. a. Answers will vary.

   b. $|x - a| < b$
   $-b < x - a < b$
   $a - b < x < a + b$

67. a. Answers will vary.

    b. $|x - a| > b$

$$x - a < -b \quad \text{or} \quad x - a > b$$
$$x < a - b \quad \text{or} \quad x > a + b$$

68. $|x - 4| = |2x - 2|$

$$x - 4 = 2x - 2 \quad \text{or} \quad x - 4 = -(2x - 2)$$
$$2 - 4 = 2x - x \qquad\qquad x - 4 = -2x + 2$$
$$-2 = x \qquad\qquad\qquad -4 - 2 = -2x - x$$
$$-2 = x \qquad\qquad\qquad\qquad -6 = -3x$$
$$-2 = x \qquad\qquad\qquad\qquad\qquad 2 = x$$

The solution set is $\{-2, 2\}$.

Exercise Set 3.4

1. Select two points on the line. Then find $\dfrac{\Delta y}{\Delta x}$, the ratio of the vertical change (or rise) to the horizontal change (or run) between the two points.

3. The line falls going from left to right.

5. The horizontal change on a vertical line is zero, and we cannot divide by zero. So the slope is undefined.

7. To get the slope-intercept form from the standard form, solve for y in terms of x.

9. a. If a graph is translated up 3 units, it is lifted or moved up 3 units.

    b. If the y-intercept is $(0, -4)$ and the graph is translated up 5 units, the new y-intercept will be at $y = -4 + 5 = 1$. The new y-intercept is $(0, 1)$.

11. When the slope is given as a rate of change it means the change in y for a unit change in x.

13. $m = \dfrac{-4 - 2}{2 - 5} = \dfrac{-6}{-3} = 2$

15. $m = \dfrac{4 - 2}{1 - 5} = \dfrac{2}{-4} = -\dfrac{1}{2}$

17. $m = \dfrac{3 - 4}{0 - (-1)} = \dfrac{-1}{1} = -1$

19. $m = \dfrac{-1 - 2}{3 - 3} = \dfrac{-3}{0}$, undefined

21. $m = \dfrac{4 - 4}{-1 - (-3)} = \dfrac{0}{2} = 0$

23. $m = \dfrac{-3 - 3}{7 - (-2)} = \dfrac{-6}{9} = -\dfrac{2}{3}$

25. $\dfrac{4 - a}{3 - 2} = 1$

$$\dfrac{4 - a}{1} = 1$$
$$4 - a = 1$$
$$-a = -3$$
$$a = 3$$

27. $\dfrac{-4 - b}{2 - 5} = 2$

$$\dfrac{-4 - b}{-3} = 2$$
$$-3\left(\dfrac{-4 - b}{-3}\right) = -3 \cdot 2$$
$$-4 - b = -6$$
$$-b = -2$$
$$b = 2$$

29. $\dfrac{-4 - 2}{3 - x} = 2$

$$\dfrac{-6}{3 - x} = 2$$
$$(3 - x)\left(\dfrac{-6}{3 - x}\right) = (3 - x) \cdot 2$$
$$-6 = 6 - 2x$$
$$-12 = -2x$$
$$\dfrac{-12}{-2} = x$$
$$x = 6$$

31.

$$\frac{3-5}{x-3} = \frac{2}{3}$$

$$\frac{-2}{x-3} = \frac{2}{3}$$

$$3(x-3)\left(\frac{-2}{x-3}\right) = 3(x-3)\left(\frac{2}{3}\right)$$

$$3(-2) = (x-3)\cdot 2$$

$$-6 = 2x - 6$$

$$0 = 2x$$

$$x = 0$$

33. The slope is negative and y decreases 6 units when x increases 2 units. Thus,

$m = -\dfrac{6}{2} = -3$. The line crosses the y-axis at 0 so $b = 0$. Hence, $m = -3$ and $b = 0$ and the equation of the line is $y = -3x + 0$ or $y = -3x$.

35. The slope is undefined since the change in x is 0. The equation of this vertical line is $x = -2$.

37. The slope is negative and y decreases 1 unit when x increases 3 units. Thus, $m = -\dfrac{1}{3}$. The line crosses the y-axis at 2 so $b = 2$. Hence, $m = -\dfrac{1}{3}$ and $b = 2$ and the equation of the line is $y = -\dfrac{1}{3}x + 2$.

39. The slope is negative and y decreases 15 units when x increases 10 units. Thus, $m = -\dfrac{15}{10} = -\dfrac{3}{2}$. The line crosses the y-axis at 15 so $b = 15$. Hence, $m = -\dfrac{3}{2}$ and $b = 15$ and the equation of the line is $y = -\dfrac{3}{2}x + 15$.

41. The equation $y = -x + 2$ is given in slope-intercept form. The slope is -1 and the y-intercept is $(0, 2)$.

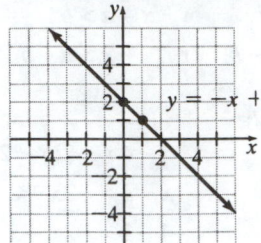

43. $20x - 30y = 60$

$$-30y = -20x + 60$$

$$y = \frac{-20x + 60}{-30}$$

$$y = \frac{2}{3}x - 2$$

The slope is $\dfrac{2}{3}$ and the y-intercept is $(0, -2)$.

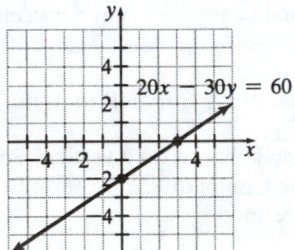

45. $-50x + 20y = 40$

$$20y = 50x + 40$$

$$y = \frac{50x + 40}{20}$$

$$y = \frac{5}{2}x + 2$$

The slope is $\dfrac{5}{2}$ and the y-intercept is $(0, 2)$.

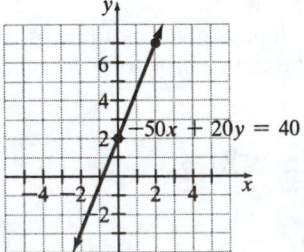

47. $f(x) = 3x - 5$
The slope is 3 and the y-intercept is $(0, -5)$.

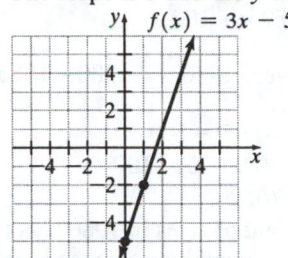

49. $g(x) = -\dfrac{2}{3}x + 3$

The slope is $-\dfrac{2}{3}$ and the y-intercept is $(0, 3)$.

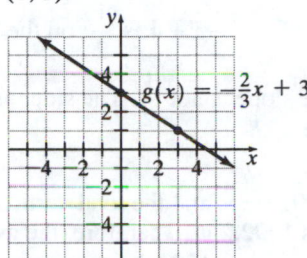

51. a. 2

    b. 4

    c. 1

    d. 3

53. If the slopes are the same and the y-intercepts are different, the lines are parallel.

55. Begin with $y = mx + b$. If $m = \dfrac{4}{3}$ and $(6, 3)$ is a point on the graph, then
$$y = \frac{4}{3}x + b$$
$$3 = \frac{4}{3}(6) + b$$
$$3 = 8 + b$$
$$-5 = b$$
The y-intercept is $(0, -5)$.

57. a. The slope is 3 and the y-intercept is $(0, 1)$, so the equation is $y = 3x + 1$.

    b. The slope is 3 and the y-intercept is $(0, -5)$, so the equation is $y = 3x - 5$.

59. a. The slope of the translated graph is 2.

    b. Using the y-intercept $b = 1$ is translated up 3 units, the y-intercept of the translated graph is at $y = 1 + 3 = 4$. The new y-intercept is $(0, 4)$.

    c. Using $m = 2$ and $b = 4$, the equation of the translated graph is $y = 2x + 4$.

61. First, rewrite the equation in the slope-intercept form by solving for y in terms of x.
$$3x - 2y = 6$$
$$-2y = -3x + 6$$
$$y = \frac{-3x + 6}{-2}$$
$$y = \frac{3}{2}x - 3$$
Thus, $m = \dfrac{3}{2}$ and $b = -3$. If the graph is translated down 4 units, then the y-intercept of the translated graph is at $y = -3 - 4 = -7$. Therefore, the equation of the translated graph is $y = \dfrac{3}{2}x - 7$.

63. $m = \dfrac{2 - 4}{-4 - 6} = \dfrac{-2}{-10} = \dfrac{1}{5}$

Thus, for a unit change in x, y changes $\dfrac{1}{5}$ or 0.2 unit.

65. a, b.

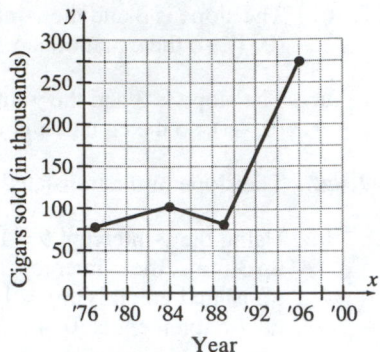

c. From 1977 to 1984,
$$m = \frac{102 - 78}{1984 - 1977} = \frac{24}{7} \approx 3.43$$
From 1984 to 1989,
$$m = \frac{81 - 102}{1989 - 1984} = \frac{-21}{5} = -4.2$$
From 1989 to 1996,
$$m = \frac{274 - 81}{1996 - 1989} = \frac{193}{7} \approx 27.57$$

d. The greatest average rate of change occurred during the period 1989 to 1996, because the largest slope corresponds to these years.

67. a. If x is the number of years after age 20, two points on the graph are (0, 200) and (50, 150). The slope is
$$m = \frac{150 - 200}{50 - 0} = \frac{-50}{50} = -1 \text{ and the}$$
y-intercept is (0, 200), so $b = 200$. Thus, the equation for the line is
$h(x) = -1 \cdot x + 200$, or $h(x) = -x + 200$.

b. For a 34-year-old man,
$x = 34 - 20 = 14$. Therefore,
$h(x) = -x + 200$
$h(14) = -14 + 200 = 186$
186 beats per minute is the maximum recommended heart rate.

69. a. If t is the number of years after 1955, two points on the graph are (0, 100) and (42, 60). The slope is
$$m = \frac{60 - 100}{42 - 0} = \frac{-40}{42} \approx -0.95 \text{ and the}$$

y-intercept is $b = 100$. Thus, the equation for the line is
$s(t) = -0.95t + 100$.

b. For the year 1980, $t = 1980 - 1955 = 25$. Therefore
$s(t) = -0.95t + 100$
$s(25) = -0.95(25) + 100$
$s(25) = 76.25$
The percent of sales in 1980 was about 76%. The graph indicates the level was about 76%.

71. The y-intercept of $y = 3x + 6$ is 6; on the screen, the y-intercept is not 6. The y-intercept is wrong.

73. The slope of $y = \frac{1}{2}x + 4$ is $\frac{1}{2}$; on the screen, the slope is not $\frac{1}{2}$. The slope is wrong.

75. There are 91 steps and the total vertical distance is 1292.2 in. Therefore, the average height of a step is $\frac{1292.2}{91} = 14.2$ inches.
If the slope is 2.21875 and the average height, or "rise", is 14.2 inches., the average width, or "run" is found as follows:
$$\text{slope} = \frac{\text{rise}}{\text{run}}$$
$$m = \frac{\text{height}}{\text{width}}$$
$$2.21875 = \frac{14.2}{\text{width}}$$
$$\text{width} = \frac{14.2}{2.21875} = 6.4$$
The average width is 6.4 inches.

78.
$$\frac{-6^2 - 16 \div 2 \div |-4|}{5 - 3 \cdot 2 - 4 \div 2^2} = \frac{-36 - 16 \div 2 \div 4}{5 - 6 - 4 \div 4}$$
$$= \frac{-36 - 8 \div 4}{5 - 6 - 1}$$
$$= \frac{-36 - 2}{-1 - 1}$$
$$= \frac{-38}{-2}$$
$$= 19$$

79. Multiply both sides by LCM, 60.
$$\frac{3}{4}x + \frac{1}{5} = \frac{2}{3}(x - 2)$$
$$60\left(\frac{3}{4}x + \frac{1}{5}\right) = 60 \cdot \frac{2}{3}(x - 2)$$
$$15 \cdot 3x + 12 = 20 \cdot 2(x - 2)$$
$$45x + 12 = 40x - 80$$
$$45x - 40x = -12 - 80$$
$$5x = -92$$
$$x = -\frac{92}{5}$$

80.
$$2.6x - (-1.4x + 3.4) = 6.2$$
$$2.6x + 1.4x - 3.4 = 6.2$$
$$2.6x + 1.4x = 6.2 + 3.4$$
$$4.0x = 9.6$$
$$x = \frac{9.6}{4.0}$$
$$x = 2.4$$

81. Let r be the rate of the second, slower train, in miles per hour. Then the first train travels at $r + 15$ miles per hour. The first train travels for a total of 6 hours at which time it is $6(r + 15)$ miles from Chicago. The second train travels for 3 hours at which time it is $3r$ miles from Chicago. If they are 270 miles apart, we have
$$6(r + 15) = 3r + 270$$
$$6r + 90 = 3r + 270$$
$$6r - 3r = 270 - 90$$
$$3r = 180$$
$$r = 60$$
The first train travels at $r + 15 = 60 + 15$, or 75 miles per hour, and the second train travels at $r = 60$, or 60 miles per hour.

82. a.
$$|2x + 1| > 3$$
$$2x + 1 < -3 \quad \text{or} \quad 2x + 1 > 3$$
$$2x < -4 \qquad\qquad 2x > 2$$
$$x < \frac{-4}{2} \qquad\qquad x > \frac{2}{2}$$
$$x < -2 \qquad\qquad x > 1$$
Solution: $x < -2$ or $x > 1$

    b.
$$|2x + 1| < 3$$
$$-3 < 2x + 1 < 3$$
$$-4 < 2x < 2$$
$$\frac{-4}{2} < x < \frac{2}{2}$$
$$-2 < x < 1$$

Exercise Set 3.5

1. The point-slope form of linear equation is $y - y_1 = m(x - x_1)$ where m is the slope of the line and $(x_1, \; y_1)$ is a point on the line.

3. Two lines are perpendicular if their slopes are negative reciprocals, or if one line is horizontal and the other is vertical.

5.
$$y - y_1 = m(x - x_1)$$
$$y - 4 = 3(x - 2)$$
$$y - 4 = 3x - 6$$
$$y = 3x - 2$$

7. $m = \dfrac{2 - 3}{5 - 6} = \dfrac{-1}{-1} = 1$
Use $m = 1$ and $(x_1, y_1) = (6, 3)$.
$$y - y_1 = m(x - x_1)$$
$$y - 3 = 1(x - 6)$$
$$y - 3 = x - 6$$
$$y = x - 3$$

9.
$$y - y_1 = m(x - x_1)$$
$$y - (-5) = \frac{1}{2}(x - (-1))$$
$$y + 5 = \frac{1}{2}(x + 1)$$
$$y + 5 = \frac{1}{2}x + \frac{1}{2}$$
$$y = \frac{1}{2}x - \frac{9}{2}$$

11. $m = \dfrac{-6-6}{4-(-4)} = \dfrac{-12}{8} = -\dfrac{3}{2}$

Use $m = -\dfrac{3}{2}$ and $(x_1, \ y_1) = (-4, \ 6)$.

$y - y_1 = m(x - x_1)$

$y - 6 = -\dfrac{3}{2}(x - (-4))$

$y - 6 = -\dfrac{3}{2}(x + 4)$

$y - 6 = -\dfrac{3}{2}x - 6$

$y = -\dfrac{3}{2}x$

13. $m_1 = \dfrac{8-4}{2-0} = \dfrac{4}{2} = 2$

$m_2 = \dfrac{5-(-1)}{3-1} = \dfrac{6}{3} = 2$

Since their slopes are equal, l_2 and l_2 are parallel.

15. $m_1 = \dfrac{-2-2}{-1-3} = \dfrac{-4}{-4} = 1$

$m_2 = \dfrac{-1-0}{3-2} = \dfrac{-1}{1} = -1$

Since the product of their slopes is -1, l_1 and l_2 are perpendicular.

17. $m_1 = \dfrac{2-3}{4-(-1)} = \dfrac{-1}{5} = -\dfrac{1}{5}$

$m_2 = \dfrac{2-(-3)}{4-1} = \dfrac{5}{3}$

Since their slopes are different and since the product of their slopes is not -1, l, and l_2 are neither parallel nor perpendicular.

19. $m_1 = \dfrac{4-(-2)}{2-2} = \dfrac{6}{0}$, undefined

$m_2 = \dfrac{4-4}{5-3} = \dfrac{0}{2} = 0$

l_1 is vertical and l_2 is horizontal. l_1 and l_2 are perpendicular.

21. $y = 3x - 5$, so $m_1 = 3$

$y = 3x + 1$, so $m_2 = 3$

Since their slopes are equal, the lines are parallel.

23. $\begin{aligned} 4x + 2y &= 8 \\ 2y &= -4x + 8 \\ y &= -2x + 4 \\ m_1 &= -2 \end{aligned}$ $\qquad \begin{aligned} 8x &= 4 - 4y \\ 4y &= -8x + 4 \\ y &= -2x + 1 \\ m_2 &= -2 \end{aligned}$

Since their slopes are equal, the lines are parallel.

25. $\begin{aligned} 4x + 2y &= 6 \\ 2y &= -4x + 6 \\ y &= -2x + 3 \\ m_1 &= -2 \end{aligned}$ $\qquad \begin{aligned} -x + 4y &= 4 \\ 4y &= x + 4 \\ y &= \tfrac{1}{4}x + 1 \\ \\ m_2 &= \tfrac{1}{4} \end{aligned}$

Since their slopes are different and since the product of their slopes is not -1, the lines are neither parallel nor perpendicular.

27. $\begin{aligned} y &= \tfrac{1}{2}x - 6 \\ \\ m_1 &= \tfrac{1}{2} \end{aligned}$ $\qquad \begin{aligned} -3y &= 6x + 9 \\ y &= -2x - 3 \\ m_2 &= -2 \end{aligned}$

Since the product of their slopes is -1, the lines are perpendicular.

29. $\begin{aligned} y &= 2x - 6 \\ m_1 &= 2 \end{aligned}$ $\qquad \begin{aligned} x &= -2y - 4 \\ 2y &= -x - 4 \\ y &= -\tfrac{1}{2}x - 2 \\ \\ m_2 &= -\tfrac{1}{2} \end{aligned}$

Since the product of their slopes is -1, the lines are perpendicular.

31. $\begin{aligned} x - 3y &= -9 \\ -3y &= -x - 9 \\ y &= \tfrac{1}{3}x + 3 \\ \\ m_1 &= \tfrac{1}{3} \end{aligned}$ $\qquad \begin{aligned} y &= 3x + 6 \\ m_2 &= 3 \end{aligned}$

Since their slopes are different and since the product of their slopes is not -1, the lines are neither parallel nor perpendicular.

33. The slope of the given line, $y = 2x + 4$, is $m_1 = 2$. So $m_2 = 2$. Now use the point-slope form with $m = 2$ and $(x_1, y_1) = (2, 5)$ to obtain the slope-intercept form.

$$y - y_1 = m(x - x_1)$$
$$y - 5 = 2(x - 2)$$
$$y - 5 = 2x - 4$$
$$y = 2x + 1$$

35. Find the slope of the given line.

$$-3x = 2y + 6$$
$$2y = -3x - 6$$
$$y = -\frac{3}{2}x - 3$$

$m_1 = -\frac{3}{2}$, so $m_2 = -\frac{3}{2}$. Now use the point-slope form with $m = -\frac{3}{2}$ and $(x_1, y_1) = \left(\frac{1}{5}, -\frac{2}{3}\right)$ to obtain the slope-intercept form.

$$y - y_1 = m(x - x_1)$$
$$y - \left(-\frac{2}{3}\right) = -\frac{3}{2}\left(x - \frac{1}{5}\right)$$
$$y + \frac{2}{3} = -\frac{3}{2}x + \frac{3}{10}$$
$$y = -\frac{3}{2}x + \frac{3}{10} - \frac{2}{3}$$
$$y = -\frac{3}{2}x - \frac{11}{30}$$

37. Find the slope of the given line.

$$\frac{1}{2}x = y - 6$$
$$y = \frac{1}{2}x + 6$$

$m_1 = \frac{1}{2}$, so $m_2 = -2$. Now use the point-slope form with $m = -2$ and $\left(-\frac{2}{3}, -4\right)$ to obtain the function notation.

$$y - y_1 = m(x - x_1)$$
$$y - (-4) = -2\left(x - \left(-\frac{2}{3}\right)\right)$$
$$y + 4 = -2\left(x + \frac{2}{3}\right)$$
$$y + 4 = -2x - \frac{4}{3}$$
$$y = -2x - \frac{4}{3} - 4$$
$$y = -2x - \frac{16}{3}$$
$$f(x) = -2x - \frac{16}{3}$$

39. Find the slope of the line with the given intercepts.

$$m_1 = \frac{3 - 0}{0 - 1} = \frac{3}{-1} = -3$$

So $m_2 = -3$. Now use the point-slope form with $m = -3$ and $(x_1, y_1) = (2, 5)$ to obtain the function notation.

$$y - y_1 = m(x - x_1)$$
$$y - 5 = -3(x - 2)$$
$$y - 5 = -3x + 6$$
$$y = -3x + 11$$
$$f(x) = -3x + 11$$

41. Find the slope of the line with the given intercepts.

$$m_1 = \frac{-3 - 0}{0 - 2} = \frac{3}{2}$$

So m_2 is the negative reciprocal, or

$$m_2 = -\frac{1}{m_1} = -\frac{1}{\frac{3}{2}} = -\frac{2}{3}.$$

Now use the point-slope form with $m = -\frac{2}{3}$ and $(x_1, y_1) = (6, 2)$ and obtain the slope-intercept form.

$$y - y_1 = m(x - x_1)$$
$$y - 2 = -\frac{2}{3}(x - 6)$$
$$y - 2 = -\frac{2}{3}x + 4$$
$$y = -\frac{2}{3}x + 6$$

43. a. To find the function $n(t)$, where t is the number of years since 1985, we use the points $(0, 63.1)$ and $(11, 78.7)$ to find the slope. (For the year 1996, $t = 1996 - 1985 = 11$.)
$$m = \frac{78.7 - 63.1}{11 - 0} = \frac{15.6}{11} \approx 1.42$$
Use the point-slope form with $m = 1.42$ and $(t_1, n_1) = (0, 63.1)$.
$$n - n_1 = m(t - t_1)$$
$$n - 63.1 = 1.42(t - 0)$$
$$n - 63.1 = 1.42t$$
$$n = 1.42t + 63.1$$
$$n(t) = 1.42t + 63.1$$

b. For the year 2020, $t = 2020 - 1985 = 35$.
$$n(t) = 1.42t + 63.1$$
$$n(35) = 1.42(35) + 63.1$$
$$= 49.7 + 63.1$$
$$= 112.8$$
About 112.8 thousand trademarks will be registered in 2020.

45. a. To find the function, use the points $(2.5, 210)$ and $(6, 370)$ to determine the slope.
$$m = \frac{370 - 210}{6 - 2.5} = \frac{160}{3.5} \approx 45.7$$
Now use the point-slope form with $m = 45.7$ and $(s_1, C_1) = (2.5, 210)$
$$C - C_1 = m(s - s_1)$$
$$C - 210 = 45.7(s - 2.5)$$
$$C - 210 = 45.7s - 114.25$$
$$C = 45.7s + 95.75$$
$$C(s) = 45.7s + 95.8$$

b. For a speed of 5 miles per hour:
$$C(s) = 45.7s + 95.8$$
$$C(5) = 45.7(5) + 95.8$$
$$= 228.5 + 95.8$$
$$= 324.3$$
The average person will burn about 324.3 calories.

47. a. For the year 1995, $t = 1995 - 1966 = 29$. Use the points $(0, 22.8)$ and $(29, 26.9)$ to determine the slope.
$$m = \frac{26.9 - 22.8}{29 - 0} = \frac{4.1}{29} \approx 0.14$$

Now use the point-slope form with $s = 0.14$ and $(t_1, m_1) = (0, 22.8)$.
$$m - m_1 = s(t - t_1)$$
$$m - 22.8 = 0.14(t - 0)$$
$$m - 22.8 = 0.14t$$
$$m = 0.14t + 22.8$$
$$m(t) = 0.14t + 22.8$$

b. For the year 2005, $t = 2005 - 1966 = 39$
$$m(t) = 0.14t + 22.8$$
$$m(39) = 0.14(39) + 22.8$$
$$= 5.46 + 22.8$$
$$= 28.26$$
The median age for males' first marriage in 2005 will be about 28.3 years.

49. a. For the year 1997, $t = 1997 - 1975 = 22$. Use the points $(0, 2.8)$ and $(22, 2.1)$ to determine the slope.
$$m = \frac{2.1 - 2.8}{22 - 0} = \frac{-0.7}{22} \approx -0.032$$
Now use the point-slope form with $m = -0.032$ and $(t_1, n_1) = (0, 2.8)$
$$n - n_1 = m(t - t_1)$$
$$n - 2.8 = -0.032(t - 0)$$
$$n - 2.8 = -0.032t$$
$$n = -0.032t + 2.8$$
$$n(t) = -0.032t + 2.8$$

b. For the year 2050, $t = 2050 - 1975 = 75$.
$$n(t) = -0.032t + 2.8$$
$$n(75) = -0.032(75) + 2.8$$
$$= -2.4 + 2.8$$
$$= 0.4$$
The number of farms in 2050 will be about 0.4 million.

51. a, b.

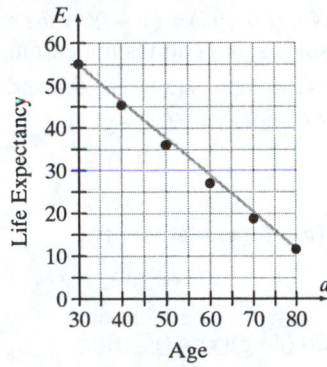

c. Use the points (30, 55.1) and (80, 11.6) to determine the slope of the line.
$$m = \frac{11.6 - 55.1}{80 - 30} = \frac{-43.5}{50} = -0.87$$
Now use the point-slope form with $m = -0.87$ and $(a_1, E_1) = (30, 55.1)$
$$E - E_1 = m(a - a_1)$$
$$E - 55.1 = -0.87(a - 30)$$
$$E - 55.1 = -0.87a + 26.1$$
$$E = -0.87a + 81.2$$
$$E(a) = -0.87a + 81.2$$

d. For a person aged 65:
$$E(a) = -0.87a + 81.2$$
$$E(65) = -0.87(65) + 81.2$$
$$= -56.55 + 81.2$$
$$= 24.65$$
Gretchen's life expectancy is about 24.65 years.

53. a. For the year 1995, $t = 1995 - 1967 = 28$. Use the points (0, 40) and (28, 140) to determine the slope of the line.
$$m = \frac{140 - 40}{28 - 0} = \frac{100}{28} \approx 3.57$$
Now use the point-slope form with $m = 3.57$ and $(t_1, V_1) = (0, 40)$.
$$V - V_1 = m(t - t_1)$$
$$V - 40 = 3.57(t - 0)$$
$$V - 40 = 3.57t$$
$$V = 3.57t + 40$$
$$V(t) = 3.57(t) + 40$$

b. For the year 2000, $t = 2000 - 1967 = 33$.
$$V(t) = 3.57(t) + 40$$
$$V(33) = 3.57(33) + 40$$
$$= 117.81 + 40$$
$$= 157.81$$
In the year 2000, about 157.81 million viewers will watch the Superbowl.

55. a. Fot the year 1996, $t = 1996 - 1984 = 12$. Use (0, 69) and (12, 63) to determine the slope of the line.
$$m = \frac{63 - 69}{12 - 0} = \frac{-6}{12} = -0.5$$
Now use the point-slope form with $m = -0.5$ and $(t_1, N_1) = (0, 69)$.
$$N - N_1 = m(t - t_1)$$
$$N - 69 = -0.5(t - 0)$$
$$N - 69 = -0.5t$$
$$N = -0.5t + 69$$
$$N(t) = -0.5t + 69$$

b. Use (0, 43) and (12, 65) to determine the slope of the line.
$$m = \frac{65 - 43}{12 - 0} = \frac{22}{12} \approx 1.83$$
Now use the point-slope form with $m = 1.83$ and $(t_1, H_1) = (0, 43)$.
$$H - H_1 = m(t - t_1)$$
$$H - 43 = 1.83(t - 0)$$
$$H - 43 = 1.83t$$
$$H = 1.83t + 43$$
$$H(t) = 183t + 43$$

57. a. With 1992 as the reference year, t is the number of years since 1992 so that, for the year 1997, $t = 1997 - 1992 = 5$. Use the points (0, 23.2) and (5, 20.3) to find the slope of the line.
$$m = \frac{20.3 - 23.2}{5 - 0} = \frac{-2.9}{5} = -0.58$$
Now use the point-slope form with $m = -0.58$ and $(t_1, n_1) = (0, 23.2)$
$$n - n_1 = m(t - t_1)$$
$$n - 23.2 = -0.58(t - 0)$$
$$n - 23.2 = -0.58t$$
$$n = -0.58t + 23.2$$
$$n(t) = -0.58t + 23.2$$
For the year 2010, $t = 2010 - 1992 = 18$.

$$n(t) = -0.58t + 23.2$$
$$n(18) = -0.58(18) + 23.2$$
$$= -10.44 + 23.2$$
$$= 12.76$$
There will be about 12,760 independent drug stores in 2010.

b. Yes, the function changed. Answers will vary.

c. No, the estimate for independent drug stores did not change. Answers will vary.

59. $4 - \dfrac{1}{2}x > 2x + 3$

$$-\dfrac{1}{2}x - 2x > 3 - 4$$

$$-\dfrac{5}{2}x > -1$$

$$\dfrac{-\frac{5}{2}x}{-\frac{5}{2}} < \dfrac{-1}{-\frac{5}{2}}$$

$$x < \dfrac{2}{5}$$

The solution is $\left(-\infty, \dfrac{2}{5}\right)$.

60. When dividing or multiplying both sides of an inequality by a negative number, reverse the direction of the inequality.

61. a. A relation is any set of ordered pairs.

b. A function is a correspondence where each member of the domain corresponds to exactly one member in the range.

c. Answers will vary.

62. D:{3, 4, 5, 6}, R:{−2, −1, 2, 3}

Exercise Set 3.6

1. Yes, $f(x) + g(x) = (f + g)(x)$ for all values of x. This is how addition of functions is defined.

3. $f(x)/g(x) = (f / g)(x)$ provided $g(x) \neq 0$. This is because division by zero is undefined.

5. No, $(f - g)(x) \neq (g - f)(x)$ for all values of x since subtraction is not commutative. For example, if $f(x) = x^2 + 1$ and $g(x) = x$; then
$$(f - g)(x) = f(x) - g(x)$$
$$= (x^2 + 1) - (x)$$
$$= x^2 - x + 1$$
$$(g - f)(x) = g(x) - f(x)$$
$$= (x) - (x^2 + 1)$$
$$= -x^2 + x - 1$$
So $(f - g)(x) \neq (g - f)(x)$.

7. a. $(f + g)(-2) = f(-2) + g(-2) = -3 + 5 = 2$

b. $(f - g)(-2) = f(-2) - g(-2) = -3 - 5 = -8$

c. $(f \cdot g)(-2) = f(-2) \cdot g(-2)$
$$= (-3) \cdot (5)$$
$$= -15$$

d. $(f / g)(-2) = f(-2)/g(-2) = \dfrac{-3}{5} = -\dfrac{3}{5}$

9. a. $(f + g)(x) = f(x) + g(x)$
$$= (x + 4) + (x^2 - 2x)$$
$$= x^2 - x + 4$$

b. $(f + g)(a) = a^2 - a + 4$

c. $(f + g)(2) = (2)^2 - (2) + 4$
$$= 4 - 2 + 4$$
$$= 6$$

11. a. $(f + g)(x) = f(x) + g(x)$
$$= (2x^2 - 3x + 5) + (x^3 - x^2)$$
$$= x^3 + x^2 - 3x + 5$$

b. $(f + g)(a) = a^3 + a^2 - 3a + 5$

c. $(f + g)(2) = (2)^3 + (2)^2 - 3(2) + 5$
$$= 8 + 4 - 6 + 5$$
$$= 11$$

13. a. $(f+g)(x)$
$$= f(x) + g(x)$$
$$= (4x^3 - 3x^2 - x) + (3x^2 + 4)$$
$$= 4x^3 - x + 4$$

b. $(f+g)(a) = 4a^3 - a + 4$

c. $(f+g)(2) = 4(2)^3 - (2) + 4$
$$= 32 - 2 + 4$$
$$= 34$$

15. $f(5) = 5^2 - 4 = 21$
$g(5) = -5(5) + 3 = -22$
$f(5) + g(5) = 21 + (-22) = -1$

17. $f(-2) = (-2)^2 - 4 = 0$
$g(-2) = -5(-2) + 3 = 13$
$f(-2) - g(-2) = 0 - 13 = -13$

19. $f(3) = 3^2 - 4 = 5$
$g(3) = -5(3) + 3 = -12$
$f(3) \cdot g(3) = 5(-12) = -60$

21. $f(4) = 4^2 - 4 = 12$
$g(4) = -5(4) + 3 = -17$
$$f(4) / g(4) = 12 / (-17) = -\frac{12}{17}$$

23. $f(-3) = (-3)^2 - 4 = 5$
$g(-3) = -5(-3) + 3 = 18$
$g(-3) - f(-3) = 18 - 5 = 13$

25. $f(0) = 0^2 - 4 = -4$
$g(0) = -5(0) + 3 = 3$
$$g(0) / f(0) = 3 / -4 = -\frac{3}{4}$$

27. $(f+g)(x) = f(x) + g(x)$
$$= (2x^2 - x) + (x - 6)$$
$$= 2x^2 - 6$$

29. $(f+g)(4) = 2(4)^2 - 6$
$$= 32 - 6$$
$$= 26$$

31. $(f-g)(6) = f(6) - g(6)$
$$= (2 \cdot 6^2 - 6) - (6 - 6)$$
$$= 66 - 0$$
$$= 66$$

33. $(f \cdot g)(-3) = f(-3) \cdot g(-3)$
$f(-3) = 2(-3)^2 - (-3) = 18 + 3 = 21$
$g(-3) = (-3) - 6 = -9$
$f(-3) \cdot g(-3) = 21 \cdot (-9) = -189$

35. $(f/g)(-1) = f(-1)/g(-1)$
$f(-1) = 2(-1)^2 - (-1) = 3$
$g(-1) = (-1) - 6 = -7$
$$f(-1) / g(-1) = 3 / (-7) = -\frac{3}{7}$$

37. $(g/f)(5) = g(5)/f(5)$
$f(5) = 2(5)^2 - 5 = 45$
$g(5) = 5 - 6 = -1$
$$g(5) / f(5) = (-1) / 45 = -\frac{1}{45}$$

39. $(g-f)(x) = g(x) - f(x)$
$$= (x - 6) - (2x^2 - x)$$
$$= -2x^2 + 2x - 6$$

41. $(f+g)(2) = f(2) + g(2) = 4 + (-1) = 3$

43. $(f \cdot g)(2) = f(2) \cdot g(2) = 4 \cdot (-1) = -4$

45. $(f-g)(-2) = f(-2) - g(-2) = 0 - 3 = -3$

47. $(g/f)(-2) = g(-2)/f(-2) = 3/0$, undefined

49. $(f+g)(3) = f(3) + g(3) = 1 + 3 = 4$

51. $(g-f)(2) = g(2) - f(2) = 2 - (-1) = 3$

53. $(f/g)(4) = f(4) /g(4) = 3/1 = -3$

55. $(g \cdot f)(0) = g(0) \cdot f(0) = (-2) \cdot 3 = -6$

57. a. The total number of GM vehicles sold in 1996 was approximately 4.9 million.

b. The number of GM cars sold in 1986 was approximately 4.7 million.

c. The number of GM cars sold in 1996 was approximately 2.9 million.

d. The number of GM trucks sold in 1986 was approximately 6.3 − 4.7 = 1.6 million.

e. The number of GM trucks sold in 1996 was approximately 4.9 − 2.9 = 2.0 million.

f. Answers will vary.

59. a. The number of PC's shipped in the United States in 1998 was approximately 32 million.

b. The number of PC's shipped worldwide in 1998 was approximately 100 million.

c. The graph of $(W − U)(t)$ will represent the number of PC's shipped outside of the U.S.

d. $(W − U)(1998) = W(1998) − U(1998)$
$= 100 − 32$
$= 68$ million

61. If $(f + g)(a) = 0$, then, $f(a)$ and $g(a)$ must either be opposites or both be equal to 0.

63. If $(f − g)(a) = 0$, then $f(a) = g(a)$.

65. If $(f/g)(a) < 0$, then $f(a)$ and $g(a)$ must have opposite signs.

67.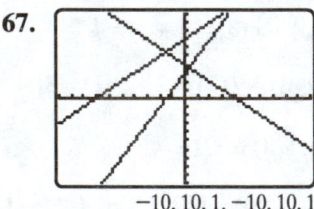
$$-10, 10, 1, -10, 10, 1$$

69.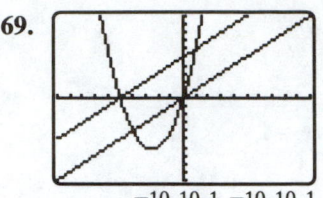
$$-10, 10, 1, -10, 10, 1$$

72. $A = \dfrac{1}{2}bh$

$2 \cdot A = 2 \cdot \dfrac{1}{2}bh$

$2A = bh$

$\dfrac{2A}{b} = \dfrac{bh}{b}$

$\dfrac{2A}{b} = h$ or $h = \dfrac{2A}{b}$

73. Let the pre-tax cost of the washing machine be x. $x + 0.06x = 477$
$$1.06x = 477$$
$$x = \frac{477}{1.06}$$
$$x = 450$$
The pre-tax cost of the washing machine was $450.

74.

x	y
−3	1
−2	0
−1	−1
0	−2
1	−1
2	0
3	1

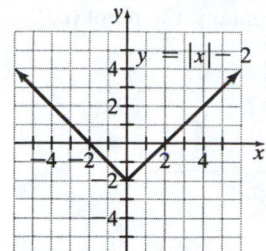

75. Set $y = 0$ to find the x-intercept.
$$3x - 4(0) = 12$$
$$3x = 12$$
$$x = 4$$
The x-intercept is $(4, 0)$. Set $x = 0$ to find the y-intercept.
$$3(0) - 4y = 12$$
$$-4y = 12$$
$$y = -3$$
The y-intercept is $(0, -3)$.

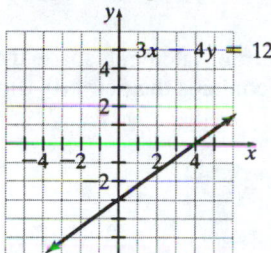

Exercise Set 3.7

1. The inequalities $\geq$ and $\leq$ include the corresponding equation; the points on the line satisfy the equation.

3. $(0, 0)$ cannot be used as a test point if the line passes through the origin.

5. $x > 1$
Graph the line $x = 1$ (vertical line) using a dashed line. For the check point, select $(0, 0)$:
$x > 1$
$0 > 1 \leftarrow$ Substitute 0 for x
Since this is a false statement, shade the region which does not contain $(0, 0)$.

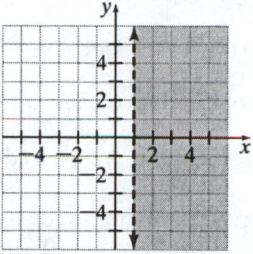

7. $y < -2$
Graph the line $y = -2$ (horizontal line) using a dashed line. For the check point, select $(0, 0)$.
$y < -2$
$0 < -2 \leftarrow$ Substitute 0 for y.
Since this is a false statement, shade the region which does not contain $(0, 0)$.

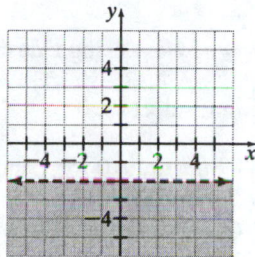

9. $y \geq 2x$
Graph the line $y = 2x$ using a solid line. For the check point, select $(0, 2)$.
$y \geq 2x$
$2 \geq 2(0) \leftarrow$ Substitute 0 for x, 2 for y
$2 \geq 0$
Since this is a true statement, shade the region which contains the point $(0, 2)$.

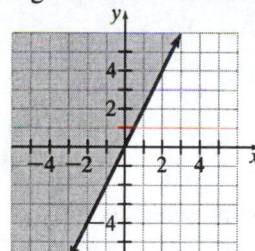

11. $y < 2x + 1$
Graph the line $y = 2x + 1$ using a dashed line.
For the check point, select $(0, 0)$.
$y < 2x + 1$
$0 < 2(0) + 1 \leftarrow$ Substitute 0 for x and y
$0 < 1$
Since this is a true statement, shade the region which contains the point $(0, 0)$.

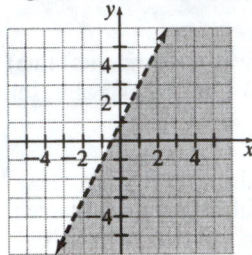

13. $y < -3x + 4$
Graph the line $y = -3x + 4$ using a dashed line.
For the check point, select $(0, 0)$.
$y < -3x + 4$
$0 < -3(0) + 4 \leftarrow$ Substitute 0 for x and y
$0 < 4$
Since this is a true statement, shade the region which contains the point $(0, 0)$.

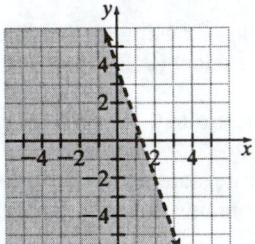

15. $y \geq \dfrac{1}{2}x - 3$

Graph the line $y = \dfrac{1}{2}x - 3$ using a solid line.

For the check point, select $(0, 0)$.

$y \geq \dfrac{1}{2}x - 3$

$0 \geq \dfrac{1}{2}(0) - 3 \leftarrow$ Substitute 0 for x and y

$0 \geq -3$
Since this is a true statement, shade the

region which contains the point $(0, 0)$.

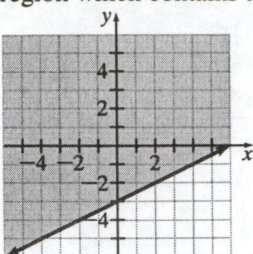

17. $y \leq \dfrac{1}{3}x + 6$

Graph the line $y = \dfrac{1}{3}x + 6$ using a solid line.

For the check point, select $(0, 0)$.

$y \leq \dfrac{1}{3}x + 6$

$0 \leq \dfrac{1}{3}(0) + 6 \leftarrow$ Substitute 0 for x and y

$0 \leq 6$
Since this is a true statement, shade the region which contains the point $(0, 0)$.

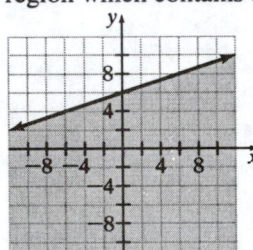

19. $y \leq -3x + 5$
Graph the line $y = -3x + 5$ using a solid line.
For the check point, select $(0, 0)$.
$y \leq -3x + 5$
$0 \leq -3(0) + 5 \leftarrow$ Substitute 0 for x and y
$0 \leq 5$
Since this is a true statement, shade the region which contains the point $(0, 0)$.

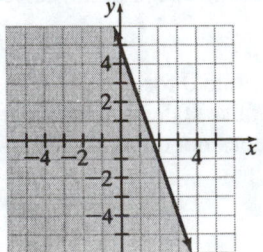

21. $2x + y < 4$
Graph the line $2x + y = 4$ using a dashed line.
For the check point, select $(0, 0)$.
$2x + y < 4$
$2(0) + 0 < 4$ ← Substitute 0 for x and y
$0 < 4$
Since this is a true statement, shade the region which contains the point $(0, 0)$.

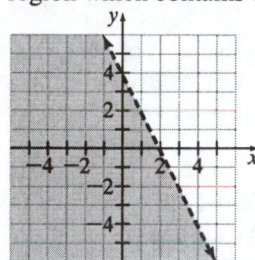

23. $2x \leq 5y + 10$
Graph the line $2x = 5y + 10$ using a solid line. For the check point, select $(0, 0)$.
$2x \leq 5y + 10$
$2(0) \leq 5(0) + 10$ ← Substitute 0 for x and y
$0 \leq 10$
Since this is a true statement, shade the region which contains the point $(0, 0)$.

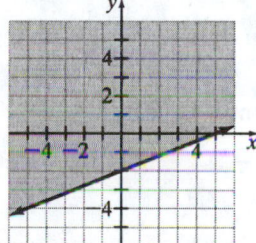

25. a, b.

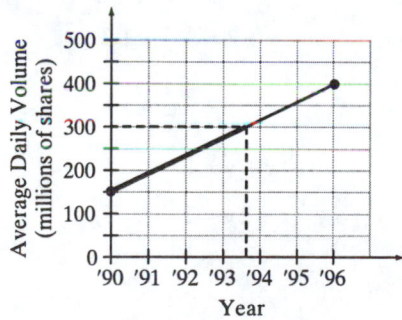

c. The number of shares first exceeded 300 million shares in 1993.

27. a, b.

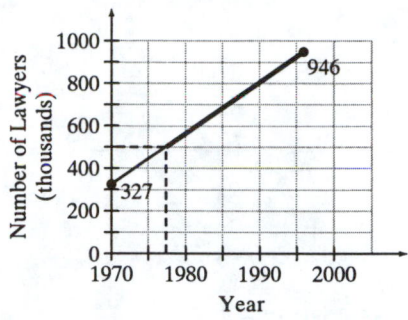

c. The number of lawyers exceeded 500 thousand for the first time in 1978.

29. a.

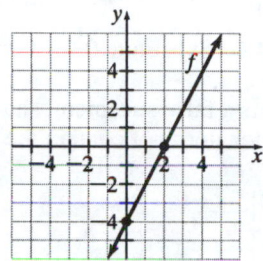

b.

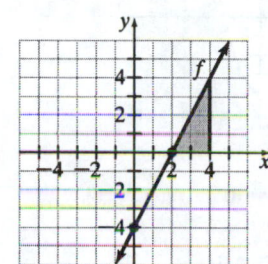

31. $y < |x|$
Graph the equation $y = |x|$ using a dashed line. For the check point, select $(0, 2)$.
$y < |x|$
$2 < (0)$ ← Substitute 0 for x and 2 for y
$2 < 0$
Since this is a false statement, shade the region which does not contain the point $(0, 2)$.

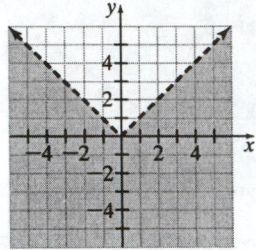

33. $y < x^2 - 4$

Graph the equation $y = x^2 - 4$ using a dashed line. For the check point, select $(0, 0)$.

$y < x^2 - 4$

$0 < 0^2 - 4 \leftarrow$ Substitute 0 for x and y
$0 < -4$

Since this is a false statement, shade the region which does not contain the point $(0, 0)$.

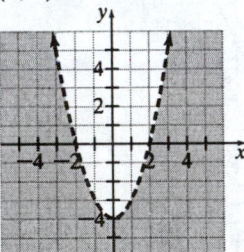

34. $4 - \dfrac{5x}{3} = -6$

$-\dfrac{5x}{3} = -10$

$3\left(-\dfrac{5x}{3}\right) = 3(-10)$

$-5x = -30$

$x = \dfrac{-30}{-5}$

$x = 6$

35. $C = \bar{x} + Z\dfrac{\sigma}{\sqrt{n}}$

$C = 80 + 1.96\dfrac{3}{\sqrt{25}}$

$C = 80 + 1.96\left(\dfrac{3}{5}\right)$

$C = 80 + 1.176$

$C = 81.176$

36. Let x be the original cost of the CD. The first week, the price was reduced by 10%, and the second week, the price was reduced an additional $2.

$x - 0.10x - 2.00 = 12.15$

$0.90x - 2.00 = 12.15$

$0.90x = 14.15$

$x = \dfrac{14.15}{0.90}$

$x \approx 15.72$

The original cost of the CD was $15.72.

37. $2x - y = 4$

$-y = -2x + 4$

$y = 2x - 4$

So $m_1 = 2$. The slope of a line perpendicular

to this one is $m_2 = -\dfrac{1}{m_1} = -\dfrac{1}{2}$. Use the

point-slope form with $m = -\dfrac{1}{2}$ and

$(x_1, \ y_1) = (6, \ -2)$.

$y - y_1 = m(x - x_1)$

$y - (-2) = -\dfrac{1}{2}(x - 6)$

$y + 2 = -\dfrac{1}{2}x + 3$

$y = -\dfrac{1}{2}x + 1$

$\dfrac{1}{2}x + y = 1$

$x + 2y = 2$

Review Exercises

1.

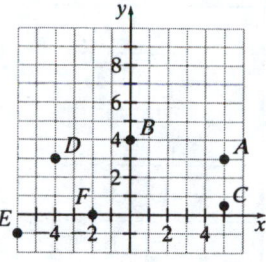

2. $y = 4x$

x	y
-1	$y = 4(-1) = -4$
0	$y = 4(0) = 0$
1	$y = 4(1) = 4$

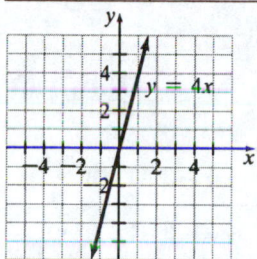

3. $y = -3x + 4$

x	y
0	$y = -3(0) + 4 = 4$
1	$y = -3(1) + 4 = 1$
2	$y = -3(2) + 4 = -2$

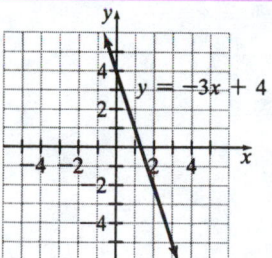

4. $y = \dfrac{3}{2}x - 3$

x	y
0	$y = \frac{3}{2}(0) - 3 = -3$
2	$y = \frac{3}{2}(2) - 3 = 0$
4	$y = \frac{3}{2}(4) - 3 = 3$

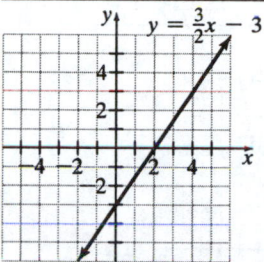

5. $y = -\dfrac{1}{2}x + 2$

x	y
-2	$y = -\frac{1}{2}(-2) + 2 = 3$
0	$y = -\frac{1}{2}(0) + 2 = 2$
2	$y = -\frac{1}{2}(2) + 2 = 1$

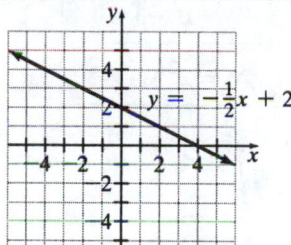

6. $y = x^2$

x	y
–3	$y = (-3)^2 = 9$
–1	$y = (-1) =^2 -1$
0	$y = 0^2 = 0$
2	$y = 2^2 = 4$

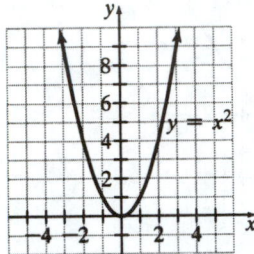

7. $y = x^2 - 1$

x	y
–3	$y = (-3)^2 - 1 = 8$
–1	$y = (-1)^2 - 1 = 0$
0	$y = 0^2 - 1 = 0$
1	$y = 1^2 - 1 = 0$
2	$y = 2^2 - 1 = 3$

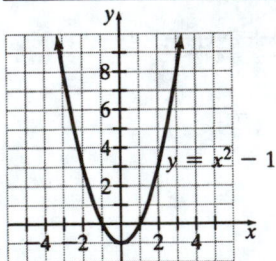

8. $y = |x|$

x	y
–4	$y = |-4| = 4$
–1	$y = |-1| = 1$
0	$y = |0| = 0$
2	$y = |2| = 2$

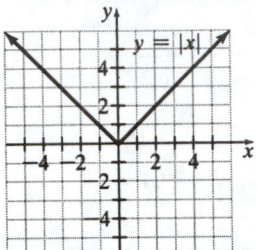

9. $y = |x| - 1$

x	y
–4	$y = |-4| - 1 = 3$
–1	$y = |-1| - 1 = 0$
0	$y = |0| - 1 = -1$
2	$y = |2| - 1 = 1$

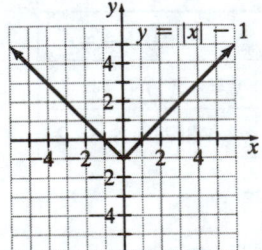

10. $y = x^3$

x	y
-2	$y = (-2)^3 = -8$
-1	$y = (-1)^3 = -1$
0	$y = 0^3 = 0$
1	$y = 1^3 = 1$
2	$y = 2^3 = 8$

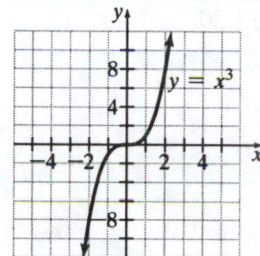

11. $y = x^3 + 4$

x	y
-2	$y = (-2)^3 + 4 = -4$
-1	$y = (-1)^3 + 4 = 3$
0	$y = 0^3 + 4 = 4$
1	$y = 1^3 + 4 = 5$

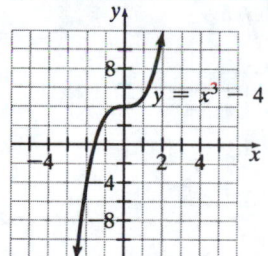

12. A function is a correspondence where each member of the domain corresponds to exactly one member of the range.

13. No, every relation is not a function. $\{(4, 2), (4, -2)\}$ is a relation but not a function.
Yes, every function is a relation because it is a set of ordered pairs.

14. Yes, each member of the domain corresponds to exactly one member of the range.

15. No, the domain element 4 corresponds to more than one member of the range (2 and 0).

16. a. No, the relation is not a function.

    b. Domain: $\{x | -1 \le x \le 1\}$
         Range: $\{y | -1 \le y \le 1\}$

17. a. No, the relation is not a function.

    b. Domain: $\{x | -2 \le x \le 2\}$
         Range: $\{y | -1 \le y \le 1\}$

18. a. Yes, the relation is a function.

    b. Domain: $\mathbb{R}$
         Range: $\{y | y \le 0\}$

19. a. Yes, the relation is a function.

    b. Domain: $\mathbb{R}$
         Range: $\mathbb{R}$

20. $f(x) = x^2 + 2x - 5$

    a. $f(3) = 3^2 + 2(3) - 5 = 9 + 6 - 5 = 10$

    b. $f(a) = a^2 + 2a - 5$

21. $g(t) = t^3 - 5t + 2$

    a. $g(-3) = (-3)^3 - 5(-3) + 2$
             $= -27 + 15 + 2$
             $= -10$

b. $g(4) = 4^3 - 5(4) + 2$
$= 64 - 20 + 2$
$= 46$

22. Answers will vary.

23. $N(x) = 40x - 0.2x^2$

    a. $N(20) = 40(20) - 0.2(20)^2$
$= 800 - 80$
$= 720$
720 baskets of apples are produced by 20 trees.

    b. $N(50) = 40(50) - 0.2(50)^2$
$= 2000 - 500$
$= 1500$
1500 baskets of apples are produced by 50 trees.

24. $h(t) = -16t^2 + 100$

    a. $h(1) = -16(1)^2 + 100 = 84$
After 1 second, the height of the ball is 84 feet.

    b. $h(2) = -16(2)^2 + 100 = 36$
After 2 seconds, the height of the ball is 36 feet.

25. $y = \dfrac{1}{2}x - 4$
To find the x-intercept, set $y = 0$.
$0 = \dfrac{1}{2}x - 4$
$4 = \dfrac{1}{2}x$
$8 = x$
The x-intercept is (8, 0).
To find the y-intercept , set $x = 0$.
$y = \dfrac{1}{2}(0) - 4$
$y = -4$

The y-intercept is (0, –4).

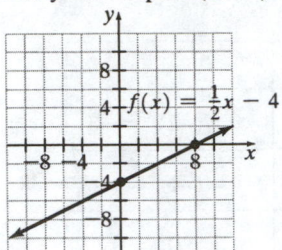

26. $\dfrac{2}{3}x = \dfrac{1}{4}y + 20$
To find the x-intercept, set $y = 0$.
$\dfrac{2}{3}x = \dfrac{1}{4}(0) + 20$
$\dfrac{2}{3}x = 20$
$\dfrac{3}{2}\left(\dfrac{2}{3}x\right) = \dfrac{3}{2}(20)$
$x = 30$
The x-intercept is (30, 0).
To find the y-intercept, set $x = 0$.
$\dfrac{2}{3}(0) = \dfrac{1}{4}y + 20$
$0 = \dfrac{1}{4}y + 20$
$-20 = \dfrac{1}{4}y$
$4(-20) = 4\left(\dfrac{1}{4}y\right)$
$-80 = y$
The y-intercept is (0 –80).

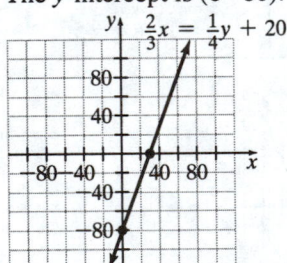

27. $f(x) = 4$ is a horizontal line 4 units above the x-axis.

x	y
–2	4
0	4
2	4

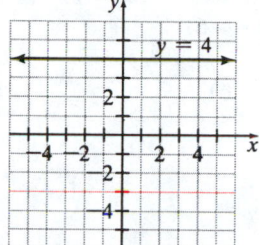

28. $x = -2$ is a vertical line 2 units to the left of the y-axis.

x	y
–2	2
–2	0
–2	–2

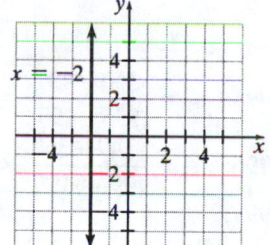

29. a. $p = 0.1x - 5000$

x	y
0	$p = 0.1(0) - 5000 = -5000$
50,000	$p = 0.1(50,000) - 5000 = 0$
100,000	$p = 0.1(100,000) - 5000 = 5000$

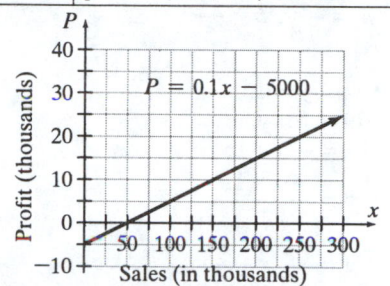

b. Approximately 50,000 bagels are sold when the comany breaks even.

c. Approximately 250,000 bagels are sold for $20,000 profit.

30. The principle is $12,000 and the time is one year. Use the decimal form of the interest rate.

$I = 12,000r$

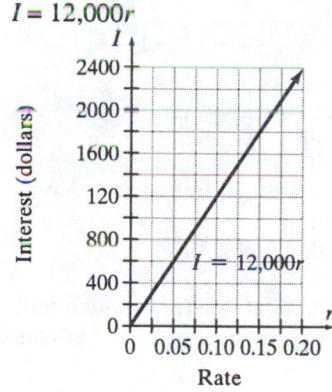

31. $y = -x + 5$ or $y = -1x + 5$
$m = -1, b = 5$

32. $f(x) = -4x + \dfrac{1}{2}$

$m = -4, \ b = \dfrac{1}{2}$

33. $3x + 5y = 12$
Solve for y.
$$5y = -3x + 12$$
$$y = \frac{-3x + 12}{5}$$
$$y = -\frac{3}{5}x + \frac{12}{5}$$
$$m = -\frac{3}{5}, \ b = \frac{12}{5}$$

34. $9x + 7y = 15$
Solve for y.
$$7y = -9x + 15$$
$$y = \frac{-9x + 15}{7}$$
$$y = -\frac{9}{7}x + \frac{15}{7}$$
$$m = -\frac{9}{7}, \ b = \frac{15}{7}$$

35. $x = -2$ is a vertical line so m is undefined and there is no y-intercept.

36. $y = 6$ is a horizontal line so that $m = 0$ and $b = 6$.

37. $m = \dfrac{6 - (-1)}{4 - 5} = \dfrac{6 + 1}{4 - 5} = \dfrac{7}{-1} = -7$

38. $m = \dfrac{3 - 1}{-2 - 4} = \dfrac{2}{-6} = -\dfrac{1}{3}$

39. This is a horizontal line ($m = 0$) having a y-intercept of 3.
The equation is $y = 3$.

40. This is a vertical line (m is undefined) having an x-intercept of 2. The equation is $x = 2$.

41. The slope is negative and y changes 2 units when x changes 4 units. Thus,
$$m = -\frac{2}{4} = -\frac{1}{2}. \text{ Hence, } m = -\frac{1}{2} \text{ and } b = 2.$$
$$y = -\frac{1}{2}x + 2$$

42. Find the slope.
$$m = \frac{8 - 5}{-2 - 3} = \frac{3}{-5} = -\frac{3}{5} \text{ or } -0.6.$$
For a one unit change in x, y changes $-\dfrac{3}{5}$ or -0.6 unit.

43. Use the point-slope form.
$$y - y_1 = m(x - x_1)$$
$$y - (-8) = \frac{4}{3}(x - (-6))$$
$$y + 8 = \frac{4}{3}(x + 6)$$
$$y + 8 = \frac{4}{3}x + 8$$
$$y = \frac{4}{3}x$$
The y-intercept is $(0, 0)$.

44. a.

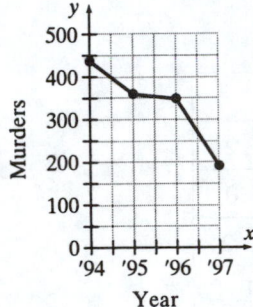

b. 1994 to 1995: $m_1 = \dfrac{360 - 437}{1995 - 1994} = -77$

1995 to 1996: $m_2 = \dfrac{349 - 360}{1996 - 1995} = -11$

1996 to 1997:
$$m_3 = \frac{192 - 349}{1997 - 1996} = -157$$

45. First, find the slope.
$$m = \frac{98.2 - 35.6}{2070 - 1980} = \frac{62.6}{90} \approx 0.7$$
Let t be the number of years since 1980. Use the point-slope form with $m = 0.7$ and $(t_1, n_1) = (0, 35.6)$.

$$n - m_1 = m(t - t_1)$$
$$n - 35.6 = 0.7(t - 0)$$
$$n - 35.6 = 0.7t$$
$$n = 0.7t + 35.6$$
$$n(t) = 0.7t + 35.6$$

46. Write each equation in slope-intercept form by solving for y.

$$y = 3x - 6 \qquad 6y = 18x + 6$$
$$y = \frac{18x + 6}{6}$$
$$y = 3x + 1$$

Since $m = 3$ for both lines, the lines are parallel.

47. Write each equation in slope-intercept form by solving for y.

$$2x - 3y = 9$$
$$-3y = -2x + 9$$
$$y = \frac{-2x + 9}{-3}$$
$$y = \frac{2}{3}x - 3$$
$$-3x - 2y = 6$$
$$-2y = 3x + 6$$
$$y = \frac{3x + 6}{-2}$$
$$y = -\frac{3}{2}x - 3$$

Since the slopes are $\frac{2}{3}$ and $-\frac{3}{2}$ which are negative reciprocals, the lines are perpendicular.

48. Write each equation in slope-intercept form by solving for y.

$$4x - 2y = 10 \qquad -2x + 4y = -8$$
$$-2y = -4x + 10 \qquad 4y = 2x - 8$$
$$y = \frac{-4x + 10}{-2} \qquad y = \frac{2x - 8}{4}$$
$$y = 2x - 5 \qquad y = \frac{1}{2}x - 2$$

Since the slopes are 2 and $\frac{1}{2}$ which are neither equal nor negative reciprocals, the lines are neither parallel nor perpendicular.

49. Use the point-slope form with $m = -\frac{2}{3}$ and $(x_1,\ y_1) = (3,\ 2)$.

$$y - 2 = -\frac{2}{3}(x - 3)$$
$$y - 2 = -\frac{2}{3}x + 2$$
$$y = -\frac{2}{3}x + 4$$

50. First, find the slope: $m = \frac{3 - 1}{4 - 2} = \frac{2}{2} = 1$.

Now, use the point-slope form with $m = 1$ and $(x_1,\ y_1) = (4,\ 3)$.

$$y - 3 = 1(x - 4)$$
$$y - 3 = x - 4$$
$$y = x - 1$$

51. The slope of the line $y = 3x - 4$ is 3. Since the new line is parallel to this line, its slope is also 3. Use the point-slope form with $m = 3$ and $(x_1,\ y_1) = (-6,\ 2)$.

$$y - 2 = 3[x - (-6)]$$
$$y - 2 = 3(x + 6)$$
$$y - 2 = 3x + 18$$
$$y = 3x + 20$$

52. To find the slope of the line $2x - 5y = 6$, solve for y.

$$-5y = -2x + 6$$
$$y = \frac{-2x + 6}{-5}$$
$$y = \frac{2}{5}x - \frac{6}{5}$$

The slope of this line is $\frac{2}{5}$, and since the new line is parallel to this line, its slope is also $\frac{2}{5}$. Use the point-slope form with $m = \frac{2}{5}$ and $(x_1,\ y_1) = (4, -2)$.

$$y - (-2) = \frac{2}{5}(x - 4)$$
$$y + 2 = \frac{2}{5}x - \frac{8}{5}$$
$$y = \frac{2}{5}x - \frac{18}{5}$$

53. The slope of the line $y = \dfrac{3}{5}x + 5$ is $\dfrac{3}{5}$. Since the new line is perpendicular to this line, its slope is $-\dfrac{5}{3}$. Use the point-slope form with $m = -\dfrac{5}{3}$ and $(x_1, y_1) = (-3, 1)$.

$$y - 1 = -\frac{5}{3}[x - (-3)]$$
$$y - 1 = -\frac{5}{3}(x + 3)$$
$$y - 1 = -\frac{5}{3}x - 5$$
$$y = -\frac{5}{3}x - 4$$

54. To find the slope of the line $4x - 2y = 8$, solve for y.
$$-2y = -4x + 8$$
$$y = \frac{-4x + 8}{-2}$$
$$y = 2x - 4$$
The slope of this line is 2. Since the new line is perpendicular to this line, its slope is $-\dfrac{1}{2}$.

Use the point-slope form with $m = -\dfrac{1}{2}$ and $(x_1, y_1) = (4, 2)$.

$$y - 2 = -\frac{1}{2}(x - 4)$$
$$y - 2 = -\frac{1}{2}x + 2$$
$$y = -\frac{1}{2}x + 4$$

55. $m_1 = \dfrac{3 - (-3)}{4 - 0} = \dfrac{3 + 3}{4 - 0} = \dfrac{6}{4} = \dfrac{3}{2}$

$m_2 = \dfrac{-1 - (-2)}{1 - 2} = \dfrac{-1 + 2}{1 - 2} = \dfrac{1}{-1} = -1$

Since the slopes are neither the same nor negative reciprocals, the lines are neither parallel nor perpendicular.

56. $m_1 = \dfrac{2 - 3}{3 - 2} = \dfrac{-1}{1} = -1$

$m_2 = \dfrac{1 - 4}{4 - 1} = \dfrac{-3}{3} = -1$

Since the slopes are the same, the lines are parallel.

57. $m_1 = \dfrac{0 - 3}{4 - 1} = \dfrac{-3}{3} = -1$

$m_2 = \dfrac{2 - 3}{5 - 6} = \dfrac{-1}{-1} = 1$

Since the slopes are negative reciprocals, the lines are perpendicular.

58. $m_1 = \dfrac{5 - 3}{-3 - 2} = \dfrac{2}{-5} = -\dfrac{2}{5}$

$m_2 = \dfrac{-2 - 2}{-4 - (-1)} = \dfrac{-2 - 2}{-4 + 1} = \dfrac{-4}{-3} = \dfrac{4}{3}$

Since the slopes are neither the same nor negative reciprocals, the lines are neither parallel nor perpendicular.

59. a. First, find the slope. For the year 1989, the point is $(0, 206)$. For the year 1994, $t = 1994 - 1989 = 5$, and the point is $(5, 11)$.
$$m = \frac{11 - 206}{5 - 0} = \frac{-195}{5} = -39$$
Use the point-slope form with $m = -39$ and $(t_1, b_1) = (0, 206)$
$$b - b_1 = m(t - t_1)$$
$$b - 206 = -39(t - 0)$$
$$b - 206 = -39t$$
$$b = -39t + 206$$
$$b(t) = -39t + 206$$

b. For the year 1990, $t = 1990 - 1989 = 1$.
$$b(t) = -39t + 206$$
$$b(1) = -39(1) + 206$$
$$= -39 + 206$$
$$= 167$$
There were approximately 167 bank failures in 1990.

60. a. In 1996, $t = 1996 - 1986 = 10$. The two points are $(0, 522)$ and $(10, 1138)$.

$$m = \frac{1138 - 522}{10 - 0} = \frac{616}{10} = 61.6$$

Use the point-slope form with $m = 61.6$ and $(t_1, p_1) = (0, 522)$.

$$p - p_1 = m(t - t_1)$$
$$p - 522 = 61.6(t - 0)$$
$$p - 522 = 61.6t$$
$$p = 61.6t + 522$$
$$p(t) = 61.6t + 522$$

b. For the year 1993, $t = 1993 - 1986 = 7$.
$$p(t) = 61.6t + 522$$
$$p(7) = 61.6(7) + 522$$
$$= 431.2 + 522$$
$$= 953.2$$

There were about 953,200 prisoners in 1993.

c. Yes, the graph shows approximately 950,000 prisoners in 1993.

61. $(f + g)(x) = f(x) + g(x)$
$$= (x^2 - 3x + 4) + (2x - 5)$$
$$= x^2 - x - 1$$

62. $(f + g)(3) = f(3) + g(3)$
$$= (3)^2 - 3 - 1$$
$$= 9 - 3 - 1$$
$$= 5$$

63. $(g - f)(x) = (2x - 5) - (x^2 - 3x + 4)$
$$= -x^2 + 5x - 9$$

64. $(g - f)(-1) = g(-1) - f(-1)$
$$= -(-1)^2 + 5(-1) - 9$$
$$= -1 - 5 - 9$$
$$= -15$$

65. $(f \cdot g)(-1) = f(-1) \cdot g(-1)$
$$= ((-1)^2 - 3(-1) + 4) \cdot (2(-1) - 5)$$
$$= 8(-7)$$
$$= -56$$

66. $(f \cdot g)(5) = f(5) \cdot g(5)$
$$= (5^2 - 3(5) + 4) \cdot (2(5) - 5)$$
$$= 14(5)$$
$$= 70$$

67. $(f / g)(1) = f(1) / g(1)$
$$= (1^2 - 3(1) + 4) / (2(1) - 5)$$
$$= 2 / -3$$
$$= -\frac{2}{3}$$

68. $(f / g)(2) = f(2) / g(2)$
$$= (2^2 - 3(2) + 4) / (2(2) - 5)$$
$$= 2(-1)$$
$$= -2$$

69. a. The number of male deaths from Alzheimer's disease in 1994 was about 6000.

b. The number of female deaths from Alzheimer's disease in 1994 was about 13,000.

c. The total number of deaths from Alzheimer's disease in 1994 was about 19,000.

70. a. $d(1995)$ was about 720.

b. $l(1995)$ was about 900.

c. $(d + l)(1995)$ was about 1620.

71. $y \geq -3$
Graph the line $y = -3$ using a solid line. For the check point, select $(0, 0)$.
$y \geq -3$
$0 \geq -3$ ← Substitute 0 for y
Since this is a true statement, shade the region which contains $(0, 0)$.

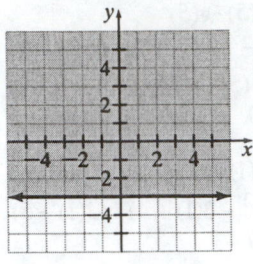

72. $x < 4$

Graph the line $x = 4$ using a dashed line. For the check point, select $(0, 0)$.

$x < 4$

$0 < 4$ ← Substitute 0 for x

Since this is a true statement, shade the region which contains $(0, 0)$.

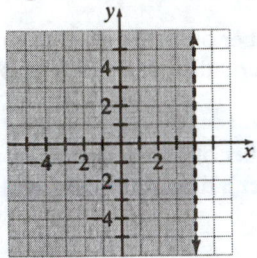

73. $y \le 4x - 3$

Graph the line $y = 4x - 3$ using a solid line. For the check point, select $(0, 0)$.

$y \le 4x - 3$

$0 \le 4(0) - 3$ ← Substitute 0 for x and y

$0 \le -3$

Since this is a false statement, shade the region which does not contain $(0, 0)$.

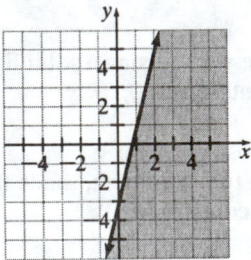

74. $y < \dfrac{1}{3}x - 2$

Graph the line $y = \dfrac{1}{3}x - 2$ using a dashed line. For the check point, select $(0, 0)$.

$y < \dfrac{1}{3}x - 2$

$0 < \dfrac{1}{3}(0) - 2$ ← Substitute 0 for x and y

$0 < -2$

Since this is a false statement, shade the region which does not contain $(0, 0)$.

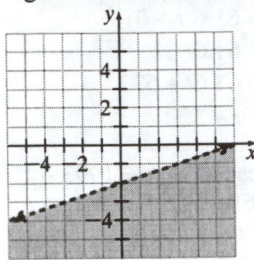

Practice Test

1. $y = 4x - 2$

x	y
-1	$y = 4(-1) - 2 = -6$
0	$y = 4(0) - 2 = -2$
1	$y = 4(1) - 2 = 2$

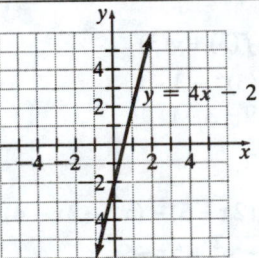

2. $y = x^2$

x	y
-3	$y = (-3)^2 = 9$
-1	$y = (-1)^2 = 1$
0	$y = 0^2 = 0$
2	$y = 2^2 = 4$

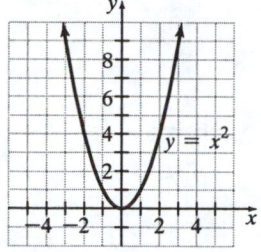

3. $y = x^3 - 1$

x	y
-2	$y = (-2)^3 - 1 = -9$
-1	$y = (-1)^3 - 1 = -2$
0	$y = 0^3 - 1 = -1$
1	$y = 1^3 - 1 = 0$
2	$y = 2^3 - 1 = 7$

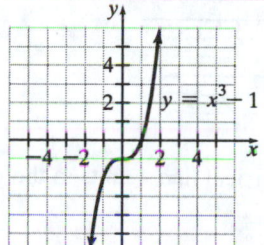

4. $y = |x|$

x	y
-3	$y = |-3| = 3$
0	$y = |0| = 0$
4	$y = |4| = 4$

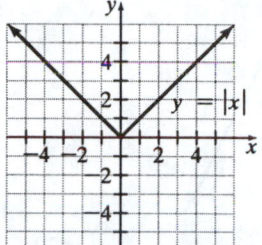

5. A function is a correspondence where each member in the domain corresponds with exactly one member in the range.

6. Yes, because each member in the domain corresponds with exactly one member in the range.

7. Yes, it is a function.
Domain: $\mathbb{R}$
Range: $\{y | y \le 4\}$

8. No, it is not a function.
Domain: $\{x | -3 \le x \le 3\}$
Range: $\{y | -2 \le y \le 2\}$

9. $f(x) = 3x^2 - 6x + 2$
$f(-2) = 3(-2)^2 - 6(-2) + 2$
$\qquad = 12 + 12 + 2$
$\qquad = 26$

10. $100y + 200x = 400$
To find the x-intercept, set $y = 0$.
$100(0) + 200x = 400$
$\qquad\qquad 200x = 400$
$\qquad\qquad\quad x = \dfrac{400}{200}$
$\qquad\qquad\quad x = 2$

The x-intercept is (2, 0).
To find the y-intercept, set $x = 0$.

$$100y + 200(0) = 400$$
$$100y = 400$$
$$y = \frac{400}{100}$$
$$y = 4$$

The y-intercept is (0, 4).

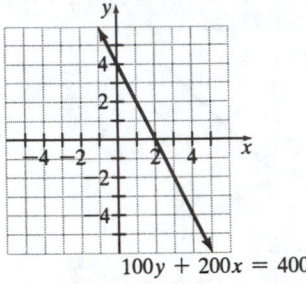

$$100y + 200x = 400$$

11. $\frac{1}{2}y = -\frac{1}{3}x + 4$

To find the x-intercept, set $y = 0$.

$$\frac{1}{2}(0) = -\frac{1}{3}x + 4$$
$$\frac{1}{3}x = 4$$
$$3\left(\frac{1}{3}x\right) = 3(4)$$
$$x = 12$$

The x-intercept is (12, 0).
To find the y-intercept, set $x = 0$.

$$\frac{1}{2}y = -\frac{1}{3}x + 4$$
$$\frac{1}{2}y = -\frac{1}{3}(0) + 4$$
$$\frac{1}{2}y = 4$$
$$2\left(\frac{1}{2}y\right) = 2(4)$$
$$y = 8$$

The y-intercept is (0, 8).

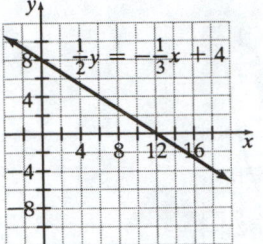

12. $f(x) = -3$ is a horizontal line 3 units below the
x-axis.

x	y
−2	−3
0	−3
2	−3

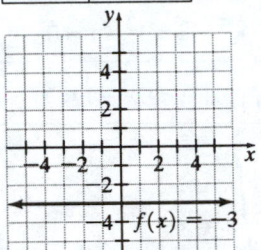

13. $x = 4$ is a vertical line 4 units to the right of the
y-axis.

x	y
4	−2
4	0
4	2

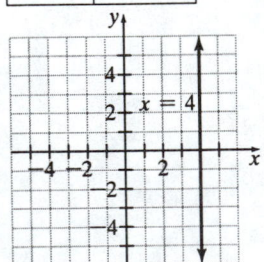

14. a. $p(x) = 10.2x - 50,000$

b. The company breaks even when $p(x) = 0$.

$$10.2x - 50,000 = 0$$
$$10.2x = 50,000$$
$$x = \frac{50,000}{10.2} = 4900$$

The company breaks even when it sells 4900 books.

c.
$$10.2x - 50,000 = 100,000$$
$$10.2x = 150,000$$
$$x = \frac{150,000}{10.2} \approx 14,700$$

The company needs to sell about 14,700 books to break even.

15. To determine the slope and y-intercept, solve for y.

$$4x - 3y = 9$$
$$-3y = -4x + 9$$
$$y = \frac{-4x + 9}{-3}$$
$$y = \frac{4}{3}x - 3$$
$$m = \frac{4}{3}, \quad b = -3$$

16. $m = \dfrac{-1 - 2}{4 - (-6)} = \dfrac{-3}{10} = -\dfrac{3}{10}$

17. For the year 2050, $t = 2050 - 2000 = 50$. The points are $(0, 274.634)$ and $(50, 393.931)$.

$$m = \frac{393.931 - 274.634}{50 - 0}$$
$$= \frac{119.297}{50}$$
$$= 2.38594$$
$$\approx 2.386$$

Use the point-slope form with $m = 2.386$ and $(t_1, \ p_1) = (0, 274.634)$.

$$p - p_1 = m(t - t_1)$$
$$p - 274.634 = 2.386(t - 0)$$
$$p - 274.634 = 2.386t$$
$$p = 2.386t + 274.634$$
$$p(t) = 2.386t + 274.634$$

18. Write each equation in slope-intercept form by solving for y.

$$3x - 6y = -4 \qquad\qquad -6x - 3y = 10$$
$$-6y = -3x - 4 \qquad\qquad -3y = 6x + 10$$
$$y = \frac{-3x - 4}{-6} \qquad\qquad y = \frac{6x + 10}{-3}$$
$$y = \frac{1}{2}x + \frac{2}{3} \qquad\qquad y = -2x - \frac{10}{3}$$

Since the slopes are $\dfrac{1}{2}$ and -2, which are negative reciprocals, the lines are perpendicular.

19. To find the slope of the line $3x - 2y = 6$, solve for y.

$$3x - 2y = 6$$
$$-2y = -3x + 6$$
$$y = \frac{-3x + 6}{-2}$$
$$y = \frac{3}{2}x - 3$$

The slope of this line is $\dfrac{3}{2}$. Since the new line is perpendicular, its slope is $-\dfrac{2}{3}$. Use the point-slope form with $m = -\dfrac{2}{3}$ and $(x_1, \ y_1) = (3, -4)$.

$$y - y_1 = m(x - x_1)$$
$$y - (-4) = -\frac{2}{3}(x - 3)$$
$$y + 4 = -\frac{2}{3}x + 2$$
$$y = -\frac{2}{3}x - 2$$

20. a. For the year 1994, $t = 1994 - 1985 = 9$.
The points are (0, 6851) and (9, 47,761).

$$m = \frac{47{,}761 - 6851}{9 - 0}$$
$$= \frac{40{,}910}{9}$$
$$\approx 4545.6$$

Use the point-slope form with
$m = 4545.6$ and $(t_1, n_1) = (0, 6851)$.
$$n - n_1 = m(t - t_1)$$
$$n - 6851 = 4545.6(t - 0)$$
$$n - 6851 = 4545.6t$$
$$n = 4545.6t + 6851$$
$$n(t) = 4545.6t + 6581$$

b. For the year 1990, $t = 1990 - 1985 = 5$.
$$n(t) = 4545.6t + 6851$$
$$n(5) = 4545.6(5) + 6851$$
$$= 22{,}728 + 6851$$
$$= 29{,}579$$

There were about 29,579 AIDS deaths
in 1990.

21. $(f + g)(3) = f(3) + g(3)$
$$= (2(3)^2 - 3) + (3 - 5)$$
$$= 15 - 2 = 13$$

22. $(f / g)(-1) = f(-1) / g(-1)$
$$= (2(-1)^2 - (-1)) / ((-1) - 5)$$
$$= 3 / (-6) = -\frac{1}{2}$$

23. $f(a) = 2a^2 - a$

24. a. The total number of tons of paper to be
used in 2010 will be about 44 million
tons.

b. The number of tons of paper to be used
by businesses in 2010 will be about
18 million tons.

c. The number of tons of paper to be used
for reference, print media, and
household use in 2010 will be about
44 − 18 = 26, or 26 million tons.

25. Graph the line $y = 3x - 2$ using a dashed
line. For the check point, select (0, 0).
$$y < 3x - 2$$
$$0 < 3 \cdot 0 - 2 \leftarrow \text{Substitute 0 for } x \text{ and } y$$
$$0 < -2$$
Since the statement is false, shade the region
that does not contain (0, 0).

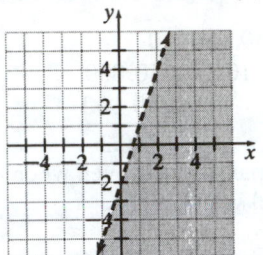

Cumulative Review Test

1. a. $A \cap B = \{2, \ 4, \ 6\}$

b. $A \cup B = \{1, \ 2, \ 3, \ 4, \ 5, \ 6, \ 8\}$

2. a. None of the numbers are natural
numbers.

b. $-6, \ -4, \ -\sqrt{2}, \ 0, \ \dfrac{1}{3}, \ \sqrt{3}, \ 4.67,$

and $\dfrac{37}{2}$ are real numbers.

3. $2 - \{3[6 - 4(6^2 \div 4)]\}$
$$= 2 - \{3[6 - 4(36 \div 4)]\}$$
$$= 2 - \{3[6 - 4(9)]\}$$
$$= 2 - \{3[6 - 36]\}$$
$$= 2 - \{3[-30]\}$$
$$= 2 - \{-90\}$$
$$= 2 + 90$$
$$= 92$$

4. $\left(\dfrac{4x^2}{y^{-3}}\right)^2 = (4x^2y^3)^2$

$\qquad\qquad = 4^2(x^2)^2(y^3)^2$

$\qquad\qquad = 16x^4y^6$

5. $\left(\dfrac{2x^4y^{-2}}{4xy^3}\right)^3 = \left(\dfrac{x^3}{2y^5}\right)^3 = \dfrac{(x^3)^3}{2^3(y^5)^3} = \dfrac{x^9}{8y^{15}}$

6. a.   Air: 40.6% of $0.406 \times 1.23 \times 10^9$

$\qquad = 4.06 \times 10^{-1} \times 1.23 \times 10^9$

$\qquad = 4.9938 \times 10^8$ or $499,380,000$

    b.   Bus: 29.2% of $0.292 \times 1.23 \times 10^9$

$\qquad = 2.92 \times 10^{-1} \times 1.23 \times 10^9$

$\qquad = 3.5916 \times 10^8$ or $359,160,000$

    c. $(4.9938 \times 10^8) - (3.5916 \times 10^8)$

$\qquad = (4.9938 - 3.5916) \times 10^8$

$\qquad = 1.4022 \times 10^8$ or $140,220,000$

7. $4(x-3) - 2 = 4[x - (-3 + x)]$

$\quad 4x - 12 - 2 = 4[x + 3 - x]$

$\qquad 4x - 14 = 4[3]$

$\qquad 4x - 14 = 12$

$\qquad\quad 4x = 26$

$\qquad\qquad x = \dfrac{26}{4}$

$\qquad\qquad x = \dfrac{13}{2}$

8. $\qquad \dfrac{4}{5} - \dfrac{x}{3} = 10$

$\quad 15\left(\dfrac{4}{5} - \dfrac{x}{3}\right) = 15(10)$

$\qquad 12 - 5x = 150$

$\qquad\quad -5x = 138$

$\qquad\qquad x = -\dfrac{138}{5}$

9. $5x - \{4 - [2(x-4)] - 5\}$

$= 5x - \{4 - [2x - 8] - 5\}$

$= 5x - \{4 - 2x + 8 - 5\}$

$= 5x - \{-2x + 7\}$

$= 5x + 2x - 7$

$= 7x - 7$

10. $\qquad A = \dfrac{1}{2}h(b_1 + b_2)$

$\qquad 2A = h(b_1 + b_2)$

$\qquad \dfrac{2A}{h} = b_1 + b_2$

$\quad \dfrac{2A}{h} - b_2 = b_1$

$\qquad\quad b_1 = \dfrac{2A}{h} - b_2$

11. Let x be amount of 10% salt water solution.

Solution	Strength	Amount	Salt
10%	0.10	x	$0.10x$
6%	0.06	8	$0.06(8)$
Mixture	0.09	$x + 8$	$0.09(x + 8)$

$0.10x + 0.06(8) = 0.09(x + 8)$

$0.10x + 0.48 = 0.09x + 0.72$

$0.10x - 0.09x = 0.72 - 0.48$

$0.01x = 0.24$

$x = \dfrac{0.24}{0.01}$

$x = 24$

24 liters of the 10% salt water solution must be used.

12. $3(x - 4) < 6(2x + 3)$

$3x - 12 < 12x + 18$

$3x - 12x < 18 + 12$

$-9x < 30$

$\dfrac{-9x}{-9} > \dfrac{30}{-9}$

$x > -\dfrac{10}{3}$

13. $-4 < 3x - 7 < 8$

$-4 + 7 < 3x - 7 + 7 < 8 + 7$

$3 < 3x < 15$

$\dfrac{3}{3} < \dfrac{3x}{3} < \dfrac{15}{3}$

$1 < x < 5$

14. $|2x - 3| > 4$

$2x - 3 < -4 \quad$ or $\quad 2x - 3 > 4$

$\quad 2x < -1 \qquad\qquad 2x > 7$

$\quad x < -\dfrac{1}{2} \qquad\qquad x > \dfrac{7}{2}$

The solution set is $\left\{ x \middle| x < -\dfrac{1}{2} \text{ or } x > \dfrac{7}{2} \right\}$.

15. $|2x - 4| = \left| \dfrac{1}{2}x - 2 \right|$

$2x - 4 = \dfrac{1}{2}x - 2 \quad$ or $\quad 2x - 4 = -\left(\dfrac{1}{2}x - 2 \right)$

$2x - \dfrac{1}{2}x = -2 + 4 \qquad\qquad 2x - 4 = -\dfrac{1}{2}x + 2$

$\dfrac{3}{2}x = 2 \qquad\qquad\qquad 2x + \dfrac{1}{2}x = 2 + 4$

$\dfrac{2}{3}\left(\dfrac{3}{2}x \right) = \dfrac{2}{3}(2) \qquad\qquad \dfrac{5}{2}x = 6$

$x = \dfrac{4}{3} \qquad\qquad\qquad \dfrac{2}{5}\left(\dfrac{5}{2}x \right) = \dfrac{2}{5}(6)$

$\qquad\qquad\qquad\qquad\qquad x = \dfrac{12}{5}$

The solution set is $\left\{ \dfrac{4}{3}, \dfrac{12}{5} \right\}$.

16. Find the x-intercept by setting $y = 0$.

$2x + 4y = 10$

$2x + 4(0) = 10$

$2x + 0 = 10$

$2x = 10$

$x = \dfrac{10}{2}$

$x = 5$

The x-intercept is $(5, 0)$.

Find the y-intercept by setting $x = 0$.

$2x + 4y = 10$

$2(0) + 4y = 10$

$4y = 10$

$y = \dfrac{10}{4}$

$y = \dfrac{5}{2}$

The y-intercept is $\left(0, \dfrac{5}{2} \right)$.

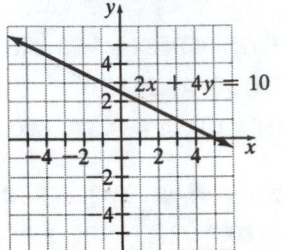

17. Write each equation in slope-intercept form by solving for y.

$2x - 5y = 6 \qquad\qquad 5x - 2y = 9$

$-5y = -2x + 6 \qquad\qquad -2y = -5x + 9$

$y = \dfrac{-2x + 6}{-5} \qquad\qquad y = \dfrac{-5x + 9}{-2}$

$y = \dfrac{2}{5}x - \dfrac{6}{5} \qquad\qquad y = \dfrac{5}{2}x - \dfrac{9}{2}$

Since the slopes are $\dfrac{2}{5}$ and $\dfrac{5}{2}$ which are niether equal nor negative reciprocals, the lines are neither parallel nor perpendicular.

18. a. The graph is not a function.

b. Domain: $\{ x \mid x \le 2 \}$; Range: $\mathbb{R}$

19. $(f + g)(x) = f(x) + g(x)$

$= (x^2 + 3x - 2) + (4x - 6)$

$= x^2 + 7x - 8$

20. $(f \cdot g)(4) = f(4) \cdot g(4)$

$= (4^2 + 3 \cdot 4 - 2) \cdot (4 \cdot 4 - 6)$

$= (16 + 12 - 2) \cdot (16 - 6)$

$= 26 \cdot 10$

$= 260$

Chapter 4

1. The solution to a system of linear equations is the point(s) that satisfy all equations in the system.

3. A consistent system of equations has a solution.

5. An inconsistent system of equations is a system of equations that has no solutions.

7. Compare the slopes and y-intercepts of the equations. If the slopes are different, the system is consistent. If the slopes and y-intercepts are the same, the system is dependent. If the slopes are the same and the y-intercepts are different, the system is inconsistent.

9. You will get a false statement, like $6 = 0$.

11. $y = 2x + 4$ and $y = 2x - 1$

    a. $(0, 4)$ does not satisfy the second equation since the left side is 4, whereas the right side is $2(0) - 1 = 0 - 1 = -1$.

    b. $(3, 10)$ does not satisfy the second equation since the left side is 10, whereas the right side is $2(3) - 1 = 6 - 1 = 5$.

13. $0.5s = -0.5r + 2$ and $2s = -2r + 8$
    Observe that both equations are identical since if the first equation is multiplied by 4 the result is
    $4(0.5s) = 4(-0.5r) + 4(2)$
    $2s = -2r + 8$
    which is precisely the second equation. Thus, we need to check each ordered pair in only the second equation. If the ordered pair satisfies the second equation, it automatically satisfies the first equation.

    a. $(2, 5)$ does not satisfy the second equation since the left side is $2(5) = 10$, whereas the right side is $-2(2) + 8 = -4 + 8 = 4$.

    b. $(1, 3)$ satisfies the second equation. The left side is $2(3) = 6$ and the right side is $-2(1) + 8 = -2 + 8 = 6$. Thus, it satisfies both equations.

15. $x + 2y - z = -5$
    $2x - y + 2z = 8$
    $3x + 3y + 4z = 5$

    a. $(1, 3, -2)$ does not satisfy the first equation since the left side is $1 + 2(3) - (-2) = 1 + 6 + 2 = 9$, whereas the right side is -5.

    b. $(1, -2, 2)$ satisfies all three equations. For the first equation, the left side is $1 + 2(-2) - (2) = 1 - 4 - 2 = -5$ and the right side is -5. For the second equation, the left side is $2(1) - (-2) + 2(2) = 2 + 2 + 4 = 8$ and the right side is 8. Finally, for the third equation, the left side is $3(1) + 3(-2) + 4(2) = 3 - 6 + 8 = 5$ and the right side is 5.

17. Write each equation in slope-intercept form.
    $3y = -x + 6 \qquad\qquad x - 2y = 1$
    $y = -\dfrac{1}{3}x + 2 \qquad\quad -2y = -x + 1$
    $\qquad\qquad\qquad\qquad\quad y = \dfrac{1}{2}x - \dfrac{1}{2}$

    Since the slope of the first line is $-\dfrac{1}{3}$ and the slope of the second line is $\dfrac{1}{2}$, the slopes are different so that the lines intersect to produce one solution. This is a consistent system.

19. Write each equation in slope-intercept form.
    $y = \dfrac{1}{3}x + 4 \qquad\qquad 3y = x + 12$
    $\qquad\qquad\qquad\qquad\quad y = \dfrac{1}{3}x + 4$

    Since both equations are identical, the line is the same for both of them to produce an infinite number of solutions. This is a dependent system.

21. Write each equation in slope-intercept form.

$$3x - 3y = 9 \qquad\qquad 2x - 2y = -4$$
$$-3y = -3x + 9 \qquad\quad -2y = -2x - 4$$
$$y = x - 3 \qquad\qquad\quad y = x + 2$$

Since the slope of each line is 1, but the y-intercepts are different ($b = -3$ for the first equation, $b = 2$ for the second equation), the two lines are parallel and produce no solution. This is an inconsistent system.

23. Write each equation in slope-intercept form.

$$y = \frac{3}{2}x + \frac{1}{2} \qquad\qquad 3x - 2y = -\frac{1}{2}$$
$$-2y = -3x - \frac{1}{2}$$
$$y = \frac{3}{2}x + \frac{1}{4}$$

Since the slope of each line is $\frac{3}{2}$, but the

y-intercepts are different ($b = \frac{1}{2}$ for the first

equation, $b = \frac{1}{4}$ for the second equation) the

two lines are parallel and produce no solution. This is an inconsistent system.

25. Graph the equations $y = x + 5$ and $y = -x + 3$.

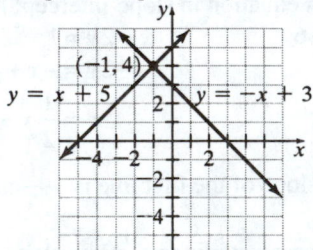

The lines intersect and the point of intersection is (–1, 4). This is a consistent system.

27. Graph the equations $y = 4x - 1$ and $2y = 8x + 6$.

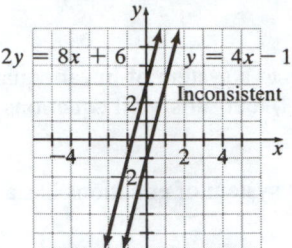

The lines are parallel. The system is inconsistent and there is no solution.

29. Graph the equations $2x + 3y = 6$ and $4x = -6y + 12$.

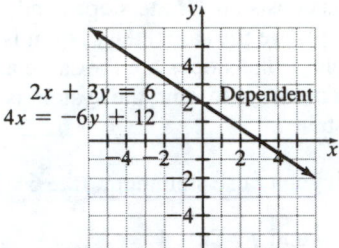

The equations produce the same line. The system is dependent and there are an infinite number of solutions.

31. Graph the equations $x + 3y = 4$ and $x = 1$.

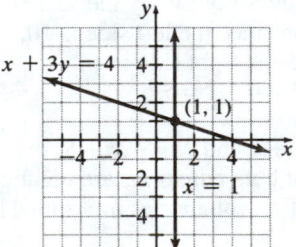

The lines intersect and the point of intersection is (1, 1). This is a consistent system.

33. Graph the equations $y = -5x + 5$ and $y = 2x - 2$.

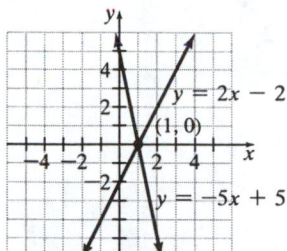

The lines intersect and the point of intersection is $(1, 0)$. This is a consistent system.

35. Graph the equations $2x - y = -4$ and $2y = 4x - 6$.

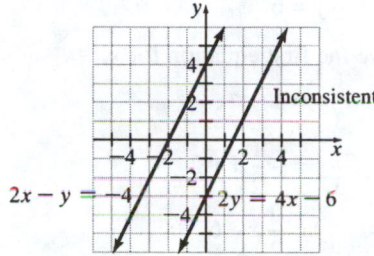

The lines are parallel and do not intersect. The system is inconsistent and there is no solution.

37. $x + 2y = 5$
$$x = 2y + 1$$
Substitute $2y + 1$ for x in the first equation.
$$(2y + 1) + 2y = 5$$
$$4y + 1 = 5$$
$$4y = 4$$
$$y = 1$$
Now, substitute 1 for y in the second equation.
$$x = 2y + 1$$
$$= 2(1) + 1$$
$$= 2 + 1$$
$$= 3$$
The solution is $(3, 1)$.

39. $x + y = 10$
$$x = y$$
Substitute x for y in first equation.
$$x + x = 10$$
$$2x = 10$$
$$x = 5$$
Now, substitute 5 for x in the second equation.
$$x = y$$
$$5 = y$$
The solution is $(5, 5)$.

41. $2r + s = 4$
$$2r + s + 6 = 0$$
Solve the second equation for s.
$$2r + s + 6 = 0$$
$$s = -2r - 6$$
Substitute $-2r - 6$ for s in the first equation.
$$2r + (-2r - 6) = 4$$
$$-6 = 4$$
This is a false statement which means there is no solution. This is an inconsistent system.

43. $x = \dfrac{1}{2}$
$$x + \frac{1}{3}y + 6 = 0$$

Substitute $\dfrac{1}{2}$ for x in the second equation.
$$\frac{1}{2} + \frac{1}{3}y + 6 = 0$$
$$\frac{1}{3}y + \frac{13}{2} = 0$$
$$\frac{1}{3}y = -\frac{13}{2}$$
$$3\left(\frac{1}{3}y\right) = 3\left(-\frac{13}{2}\right)$$
$$y = -\frac{39}{2}$$
The solution is $\left(\dfrac{1}{2}, -\dfrac{39}{2}\right)$.

45. $a - \dfrac{1}{2}b = 2$

$b = 2a - 4$

Substitute $2a - 4$ for b in the first equation.

$a - \dfrac{1}{2}(2a - 4) = 2$

$a - a + 2 = 2$

$2 = 2$

Since this is an identity, there are an infinite number of solutions. This is a dependent system.

47. $x = y + 4$

$3x + 7y = -18$

Substitute $y + 4$ for x in the second equation.

$3(y + 4) + 7y = -18$

$3y + 12 + 7y = -18$

$10y + 12 = -18$

$10y = -30$

$y = \dfrac{-30}{10} = -3$

Now, substitute -3 for y in the first equation.

$x = y + 4$

$x = -3 + 4$

$x = 1$

The solution is $(1, -3)$.

49. $5x - 4y = -7$

$x - \dfrac{3}{5}y = -2$

First, solve the second equation for x.

$x - \dfrac{3}{5}y = -2$

$x = \dfrac{3}{5}y - 2$

Now substitute $\dfrac{3}{5}y - 2$ for x in the first equation.

$5\left(\dfrac{3}{5}y - 2\right) - 4y = -7$

$3y - 10 - 4y = -7$

$-10 - y = -7$

$-y = 3$

$y = -3$

Finally, substitute -3 for y in the equation

$x = \dfrac{3}{5}y - 2$.

$x = \dfrac{3}{5}(-3) - 2$

$x = -\dfrac{9}{5} - 2$

$x = -\dfrac{9}{5} - \dfrac{10}{5}$

$x = -\dfrac{19}{5}$

The solution is $\left(-\dfrac{19}{5}, -3\right)$.

51. $\dfrac{1}{2}x - \dfrac{1}{3}y = 2$

$\dfrac{1}{4}x + \dfrac{2}{3}y = 6$

Solve the first equation for y.

$\dfrac{1}{2}x - \dfrac{1}{3}y = 2$

$-\dfrac{1}{3}y = -\dfrac{1}{2}x + 2$

$y = \dfrac{3}{2}x - 6$

Substitute $\dfrac{3}{2}x - 6$ for y in the second equation.

$\dfrac{1}{4}x + \dfrac{2}{3}\left(\dfrac{3}{2}x - 6\right) = 6$

$\dfrac{1}{4}x + x - 4 = 6$

$\dfrac{5}{4}x - 4 = 6$

$\dfrac{5}{4}x = 10$

$x = \dfrac{4}{5}(10) = 8$

Substitute 8 for x in the equation

$y = \dfrac{3}{2}x - 6$.

$y = \dfrac{3}{2}(8) - 6 = 12 - 6 = 6$

The solution is $(8, 6)$.

53.

$$x + y = 0$$
$$x - y = 4$$

Add: $\quad 2x \quad = 4$
$$x \quad = 2$$

Substitute 2 for x in the first equation.
$$x + y = 0$$
$$2 + y = 0$$
$$y = -2$$

The solution is $(2, -2)$.

55.

$$3x + 2y = 15$$
$$x - 2y = -7$$

Add: $\quad 4x \quad\quad = 8$
$$x = 2$$

Substitute 2 for x in the first equation.
$$3x + 2y = 15$$
$$3(2) + 2y = 15$$
$$6 + 2y = 15$$
$$2y = 9$$
$$y = \frac{9}{2}$$

The solution is $\left(2, \frac{9}{2}\right)$.

57.

$$3p + q = 6$$
$$-6p - 2q = 10$$

To eliminate q, multiply the first equation by 2 and then add.
$$2[3p + q = 6]$$
$$-6p - 2q = 10$$
gives
$$6p + 2q = 12$$
$$-6p - 2q = 10$$

Add: $\quad\quad 0 = 22$

Since $0 = 22$ is a false statement, the system has no solution. It is an inconsistent system.

59.

$$2x + y = 6$$
$$3x - 2y = 16$$

To eliminate y, multiply the first equation by 2 and then add.
$$2[2x + y = 6]$$
$$3x - 2y = 16$$
gives

$$4x + 2y = 12$$
$$3x - 2y = 16$$

Add: $\quad 7x \quad\quad = 28$
$$x = 4$$

Substitute 4 for x in the first equation.
$$2x + y = 6$$
$$2(4) + y = 6$$
$$8 + y = 6$$
$$y = -2$$

The solution is $(4, -2)$.

61. $\quad 2a - 5b = 13$
$$5a + 3b = 17$$

To eliminate b, multiply the first equation by 3 and the second equation by 5 and then add.
$$3[2a - 5b = 13]$$
$$5[5a + 3b = 17]$$
gives
$$6a - 15b = 39$$
$$25a + 15b = 85$$

Add: $\quad 31a \quad\quad = 124$
$$a = 4$$

Substitute 4 for a in the second equation.
$$5a + 3b = 17$$
$$5(4) + 3a = 17$$
$$20 + 3b = 17$$
$$3b = -3$$
$$b = -1$$

The solution is $(4, -1)$.

63. $\quad 3y = 2x + 4$
$$3y = 2x + 4$$

Since the equations are the same, there are an infinite number of solutions and the system is dependent.

65.

$$2x - y = 8$$
$$3x + y = 6$$

Add: $\quad 5x \quad\quad = 14$
$$x = \frac{14}{5}$$

Substitute $\frac{14}{5}$ for x in the second equation.

$$3x + y = 6$$

$$3\left(\frac{14}{5}\right) + y = 6$$

$$\frac{42}{5} + y = 6$$

$$y = 6 - \frac{42}{5}$$

$$y = \frac{30}{5} - \frac{42}{5}$$

$$y = -\frac{12}{5}$$

The solution is $\left(\frac{14}{5}, -\frac{12}{5}\right)$.

67. $3x - 4y = 5$

$\qquad 2x = 5y - 3$

Write the system in standard form.

$3x - 4y = 5$

$2x - 5y = -3$

To eliminate x, multiply the first equation by -2 and the second equation by 3 and then add.

$-2[3x - 4y = 5]$

$\;\;3[2x - 5y = -3]$

gives

$$\begin{array}{r} -6x + 8y = -10 \\ 6x - 15y = -9 \\ \hline \end{array}$$

Add: $\;\;\; -7y = -19$

$$y = \frac{-19}{-7}$$

$$= \frac{19}{7}$$

Substitute $\frac{19}{7}$ for y in the second equation.

$2x = 5y - 3$

$2x = 5\left(\frac{19}{7}\right) - 3$

$2x = \frac{95}{7} - 3$

$2x = \frac{74}{4}$

$x = \frac{1}{2}\left(\frac{74}{7}\right) = \frac{37}{7}$

The solution is $\left(\frac{37}{7}, \frac{19}{7}\right)$.

69. $0.2x + 0.5y = 1.6$

$\;\;\; -0.3x + 0.4y = -0.1$

To eliminate x, multiply the first equation by 3 and the second equation by 2 and then add.

$3[0.2x + 0.5y = 1.6]$

$2[-0.3x + 0.4y = -0.1]$

gives

$$\begin{array}{r} 0.6x + 1.5y = 4.8 \\ -0.6x + 0.8y = -0.2 \\ \hline \end{array}$$

Add: $\qquad 2.3y = 4.6$

$$y = \frac{4.6}{2.3} = 2$$

Now, substitute 2 for y in the first equation.

$0.2x + 0.5y = 1.6$

$0.2x + 0.5(2) = 1.6$

$0.2x + 1 = 1.6$

$0.2x = 0.6$

$$x = \frac{0.6}{0.2} = 3$$

The solution is $(3, 2)$.

71. $2.1m - 0.6n = 8.4$

$\;\;\; -1.5m - 0.3n = -6.0$

To eliminate n, multiply the second equation by -2 and then add.

$2.1m - 0.6n = 8.4$

$-2[-1.5x - 0.3y = -6.0]$

gives

$$\begin{array}{r} 2.1m - 0.6n = 8.4 \\ 3.0m + 0.6n = 12.0 \\ \hline \end{array}$$

Add: $5.1m \qquad\;\; = 20.40$

$$m = \frac{20.40}{5.1} = 4$$

Substitute 4 for m into the second equation.

$-1.5m - 0.3n = -6.0$

$-1.5(4) - 0.3n = -6.0$

$-6.0 - 0.3n = -6.0$

$-0.3n = 0$

$n = 0$

The solution is $(4, 0)$.

73. $\dfrac{1}{2}x - \dfrac{1}{3}y = 1$

$\;\;\; \dfrac{1}{4}x - \dfrac{1}{9}y = \dfrac{2}{3}$

To clear fractions, multiply the first equation by 6 and the second equation by 36:

$$6\left[\frac{1}{2}x - \frac{1}{3}y = 1\right]$$

$$36\left[\frac{1}{4}x - \frac{1}{9}y = \frac{2}{3}\right]$$

gives

$$3x - 2y = 6$$
$$9x - 4y = 24$$

To eliminate y, multiply the first equation by -2 and then add.

$$-2[3x - 2y = 6]$$
$$9x - 4y = 24$$

gives

$$-6x + 4y = -12$$
$$\underline{9x - 4y = \ 24}$$

Add: $3x \qquad = 12$

$$x = 4$$

Substitute 4 for x in the equation $3x - 2y = 6$.

$$3(4) - 2y = 6$$
$$12 - 2y = 6$$
$$6 - 2y = 0$$
$$6 = 2y$$
$$3 = y$$

The solution is (4, 3).

75. $\frac{1}{5}x + \frac{1}{2}y = 4$

$$\frac{2}{3}x - y = \frac{8}{3}$$

To clear fractions and to eliminate x, multiply the first equation by 10 and the second equation by -3 and then add.

$$10\left[\frac{1}{5}x + \frac{1}{2}y = 4\right]$$

$$-3\left[\frac{2}{3}x - y = \frac{8}{3}\right]$$

gives

$$2x + 5y = 40$$
$$\underline{-2x + 3y = -8}$$

Add: $8y = 32$

$$y = 4$$

Substitute 4 for y in the equation $2x + 5y = 40$.

$$2x + 5(4) = 40$$
$$2x + 20 = 40$$
$$2x = 20$$
$$x = 10$$

The solution is (10, 4).

77. Answers will vary. The system should involve a variable that has a coefficient of 1.

79. $N(t) = -11.2t + 80$
$M(t) = 18.7t + 2$

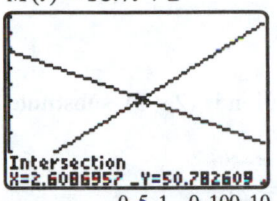

Intersection
X=2.6086957 _Y=50.782609

0, 5, 1,   0, 100, 10

The solution is approximately (2.61, 50.78). Therefore, the market shares will be equal 2.61 years after 1995 or in 1997.

81. Multiply the first equation by -2 and notice that the x- and y-terms have the same coefficients but the constant terms are different.

83. **a.** If a system has more than one solution, then it must have an infinite number of solutions. All of these solutions must lie on the same line.

b. Slope is $m = \dfrac{6-4}{-2-2} = \dfrac{2}{-4} = -\dfrac{1}{2}$. Use

the point-slope form with $m = -\dfrac{1}{2}$ and

$(x_1, y_1) = (2, 4)$.

$$y - y_1 = m(x - x_1)$$

$$y - 4 = -\frac{1}{2}(x - 2)$$

$$y - 4 = -\frac{1}{2}x + 1$$

$$y = -\frac{1}{2} + 5$$

The y-intercept is (0, 5).

c. Yes, the graph of a non-vertical line is a function.

85. Answers may vary. One example is
$$x + y = 1$$
$$2x + 2y = 2.$$

87. a. Answers may vary. One example is
$$x + y = 7$$
$$x - y = -3.$$

b. Choose coefficients for x and y, then use the given coordinates to find the constants.

89. $Ax + 4y = -8$
$3x - By = 21$
Since the solution is $(2, -3)$, substitute 2 for x and -3 for y.
$A(2) + 4(-3) = -8$
$3(2) - B(-3) = 21$
or
$2A - 12 = -8$
$6 + 3B = 21$
This is a system of two equations in the two unknowns A and B. To solve, solve each equation for the unknown variable. In the first equation, solve for A.
$2A - 12 = -8$
$\quad\quad 2A = 4$
$\quad\quad\quad A = \dfrac{4}{2} = 2$
In the second equation, solve for B.
$6 + 3B = 21$
$\quad\quad 3B = 15$
$\quad\quad\quad B = \dfrac{15}{3} = 5$
Thus, $A = 2$ and $B = 5$.

91. $f(x) = mx + b$
Substitute $(2, 6)$ and $(-1, -6)$ into the equation to get a system.
$6 = 2m + b$
$-6 = -m + b$
Multiply the second equation by -1 and add.
$\quad 6 = 2m + b$
$\quad 6 = m - b$
$\overline{12 = 3m}$
$\quad 4 = m$
Substitute 4 for m in the first equation and solve for b.

$6 = 2(4) + b$
$-2 = b$
Thus, $m = 4$ and $b = -2$.

93. The system is dependent or one graph is not in the viewing window.

95. $\dfrac{x+2}{2} - \dfrac{y+4}{3} = 4$
$\dfrac{x+y}{2} = \dfrac{1}{2} + \dfrac{x-y}{3}$

Start by writing each equation in standard form after clearing fractions.
$$6\left(\dfrac{x+2}{2}\right) - 6\left(\dfrac{y+4}{3}\right) = 6(4)$$
$$6\left(\dfrac{x+y}{2}\right) = 6\left(\dfrac{1}{2}\right) + 6\left(\dfrac{x-y}{3}\right)$$

$3(x+2) - 2(y+4) = 24$
$3(x+y) = 3 + 2(x-y)$

$3x + 6 - 2y - 8 = 24$
$3x + 3y = 3 + 2x - 2y$

$3x - 2y = 26 \quad (1)$
$x + 5y = 3 \quad\quad (2)$
To eliminate x, multiply equation (2) by -3 and then add to equation (1).
$\quad 3x - 2y = 26$
$-3[x + 5y = 3]$
gives
$\quad\quad\quad 3x - 2y = 26$
$\quad\quad\quad \underline{-3x - 15y = -9}$
Add:$\quad\quad -17y = 17$
$$y = \dfrac{17}{-17} = -1$$
Now, substitute -1 for y in equation (2).
$x + 5(-1) = 3$
$\quad x - 5 = 3$
$\quad\quad x = 8$
The solution is $(8, -1)$.

97. Rewrite the system using the hint.

$$3 \cdot \frac{1}{a} + 4 \cdot \frac{1}{b} = -1$$

$$\frac{1}{a} + 6 \cdot \frac{1}{b} = 2$$

Now let $x = \frac{1}{a}$ and $y = \frac{1}{b}$.

$3x + 4y = -1$ (1)

$x + 6y = 2$ (2)

Multiply equation (2) by -3 and add.

$3x + 4y = -1$

$-3[x + 6y = 2]$

gives

$$3x + 4y = -1$$
$$\underline{-3x - 18y = -6}$$

Add: $-14y = -7$

$$y = \frac{-7}{-14} = \frac{1}{2}$$

Substitute $\frac{1}{2}$ for y in equation (2).

$$x + 6\left(\frac{1}{2}\right) = 2$$

$$x + 3 = 2$$

$$x = -1$$

Now find the values of a and b.

$$x = \frac{1}{a}$$

$$-1 = \frac{1}{a}$$

$$a = -1$$

$$y = \frac{1}{b}$$

$$\frac{1}{2} = \frac{1}{b}$$

$$b = 2$$

The solution is $(-1, 2)$.

99. $4ax + 3y = 19$

$-ax + y = 4$

Solve the second equation for y.

$-ax + y = 4$

$y = ax + 4$

Substitute $ax + 4$ for y in the first equation.

$4ax + 3y = 19$

$4ax + 3(ax + 4) = 19$

$4ax + 3ax + 12 = 19$

$7ax + 12 = 19$

$7ax = 7$

$$x = \frac{7}{7a} = \frac{1}{a}$$

Now substitute $\frac{1}{a}$ for x in the equation

$y = ax + 4$.

$$y = a\left(\frac{1}{a}\right) + 4$$

$$y = 1 + 4$$

$$y = 5$$

The solution is $\left(\frac{1}{a}, 5\right)$.

102. Rational numbers can be expressed as quotients of two integers. Irrational numbers cannot.

103. a. Yes, the set of real numbers includes the set of rational numbers.

    b. Yes, the set of real numbers includes the set of irrational numbers.

104. $|x - 4| = |4 - x|$

$x - 4 = 4 - x$ or $x - 4 = -(4 - x)$

$2x - 4 = 4$ $x - 4 = -4 + x$

$2x = 8$ $-4 = -4$

$x = 4$

This means that the solution is all real numbers or $\mathbb{R}$.

105. $A = p\left(1 + \dfrac{r}{n}\right)^t = 500\left(1 + \dfrac{0.08}{2}\right)^1$

    $= 500(1.04) = 520$

106. No, the points $(-3, 4)$ and $(-3, 2)$ have the same first coordinate but different second coordinates.

Exercise Set 4.2

1. The graph will be a plane.

3.
$$x = 3$$
$$x + 2y = 7$$
$$-3x - y + 4z = 9$$

Substitute 3 for x in the second equation.
$$x + 2y = 7$$
$$3 + 2y = 7$$
$$2y = 4$$
$$y = 2$$

Substitute 3 for x and 2 for y in the third equation.
$$-3x - y + 4z = 9$$
$$-3(3) - 2 + 4z = 9$$
$$-9 - 2 + 4z = 9$$
$$-11 + 4z = 9$$
$$4z = 20$$
$$z = 5$$

The solution is (3, 2, 5).

5.
$$5x - 6z = -17$$
$$3x - 4y + 5z = -1$$
$$2z = -6$$

Solve the third equation for z.
$$2z = -6$$
$$z = -3$$

Substitute –3 for z in the first equation.
$$5x - 6z = -17$$
$$5x - 6(-3) = -17$$
$$5x + 18 = -17$$
$$5x = -35$$
$$x = -7$$

Substitute –7 for x and –3 for z in the second equation.
$$3x - 4y + 5z = -1$$
$$3(-7) - 4y + 5(-3) = -1$$
$$-21 - 4y - 15 = -1$$
$$-4y - 36 = -1$$
$$-4y = 35$$
$$y = \frac{35}{-4} = -\frac{35}{4}$$

The solution is $\left(-7, -\dfrac{35}{4}, -3\right)$.

7.
$$x + 2y = 6$$
$$3y = 9$$
$$x + 2z = 12$$

Solve the second equation for y.

$$3y = 9$$
$$y = 3$$

Substitute 3 for y in the first equation.
$$x + 2y = 6$$
$$x + 2(3) = 6$$
$$x + 6 = 6$$
$$x = 0$$

Substitute 0 for x in the third equation.
$$x + 2z = 12$$
$$0 + 2z = 12$$
$$2z = 12$$
$$z = 6$$

The solution is (0, 3, 6).

9.
$$x + y + z = 3 \quad (1)$$
$$x - z = -5 \quad (2)$$
$$2x - y + 2z = 0 \quad (3)$$

To eliminate y between equation (1) and (3), simply add.
$$x + y + z = 3$$
$$\underline{2x - y + 2z = 0}$$
Add: $3x + 3z = 3 \quad (4)$

Equations (2) and (4) are two equations in two unknowns. To eliminate z, multiply equation (4) by $\dfrac{1}{3}$ and add to equation (2).

$$x - z = -5$$
$$\frac{1}{3}[3x + 3z = 3]$$
gives
$$x - z = -5$$
$$\underline{x + z = 1}$$
Add: $2x = -4$
$$x = -2$$

Substitute –2 for x in equation (2).
$$x - z = -5$$
$$-2 - z = -5$$
$$-z = -3$$
$$z = 3$$

Substitute –2 for x and 3 for z in equation (1).
$$x + y + z = 3$$
$$-2 + y + 3 = 3$$
$$y + 1 = 3$$
$$y = 2$$

The solution is (–2, 2, 3).

11. $x - 2z = -5$ (1)
$-y + 3z = 3$ (2)
$-2x + z = 4$ (3)

To eliminate x between equations (1) and (3), multiply equation (1) by 2 and then add.

$2[x - 2z = -5]$
$-2x + z = 4$

gives

$2x - 4z = -10$
$\underline{-2x + z = 4}$
Add: $-3z = -6$
$z = 2$

Substitute 2 for z in equation (1).

$x - 2z = -5$
$x - 2(2) = -5$
$x - 4 = -5$
$x = -1$

Substitute 2 for z in equation (2).

$-y + 3z = 3$
$-y + 3(2) = 3$
$-y + 6 = 3$
$-y = -3$
$y = 3$

The solution is $(-1, 3, 2)$.

13. $3p + 2q = 11$ (1)
$4q - r = 6$ (2)
$2p + 2r = 2$ (3)

To eliminate r between equations (2) and (3), multiply equation (2) by 2 and add to equation (3).

$2[4q - r = 6]$
$2p + 2r = 2$

gives

$8q - 2r = 12$
$\underline{2p + 2r = 2}$
Add: $2p + 8q = 14$ (4)

Equations (1) and (4) are two equations in two unknowns. To eliminate q, multiply equation (1) by -4 and add to equation (4).

$4[3p + 2q = 11]$
$2p + 8q = 14$

gives

$-12p - 8q = -44$
$\underline{2p + 8q = 14}$
Add: $-10p = -30$
$p = 3$

Substitute 3 for p in equation (1).

$3p + 2q = 11$
$3(3) + 2q = 11$
$9 + 2q = 11$
$2q = 2$
$q = 1$

Substitute 3 for p in equation (3).

$2p + 2r = 2$
$2(3) + 2r = 2$
$6 + 2r = 2$
$2r = -4$
$r = -2$

The solution is $(3, 1, -2)$.

15. $p + q + 4 = 4$ (1)
$p - 2q - r = 1$ (2)
$2p - q - 2r = -1$ (3)

To eliminate q between equations (1) and (3), simply add.

$p + q + r = 4$
$\underline{2p - q - 2r = -1}$
Add: $3p - r = 3$ (4)

To eliminate q between equations (1) and (2), multiply equation (1) by 2 and then add.

$2[p + q + r = 4]$
$p - 2q - r = 1$

gives

$2p + 2q + 2r = 8$
$\underline{p - 2q - r = 1}$
Add: $3p + r = 9$ (5)

Equations (4) and (5) are two equations in two unknowns.

$3p - r = 3$
$3p + r = 9$

To eliminate r, simply add these two equations.

$3p - r = 3$
$\underline{3p + r = 9}$
Add: $6p = 12$
$p = 2$

Substitute 2 for p in equation (5).

$3p + r = 9$
$3(2) + r = 9$
$6 + r = 9$
$r = 3$

Substitute 2 for p and 3 for r in equation (1).

$$p + q + r = 4$$
$$2 + q + 3 = 4$$
$$q + 5 = 4$$
$$q = -1$$

The solution is $(2, -1, 3)$.

17. $2x - 2y + 3z = 5$ (1)
$\quad 2x + y - 2z = -1$ (2)
$\quad 4x - y - 3z = 0$ (3)

To eliminate y between equations (2) and (3), simply add.

$$2x + y - 2z = -1$$
$$\underline{4x - y - 3z = 0}$$

Add: $\quad 6x - 5z = -1$ (4)

To eliminate y between equations (1) and (2), multiply equation (2) by 2 and then add.

$$2x - 2y + 3z = 5$$
$$2[2x + y - 2z = -1]$$

gives

$$2x - 2y + 3z = 5$$
$$\underline{4x + 2y - 4z = -2}$$

Add: $\quad 6x - z = 3$ (5)

Equations (4) and (5) are two equations in two unknowns.

$$6x - 5z = -1$$
$$6x - z = 3$$

To eliminate x, multiply equation (5) by -1 and then add.

$$6x - 5z = -1$$
$$-1[6x - z = 3]$$

gives

$$6x - 5z = -1$$
$$\underline{-6x + z = -3}$$

Add: $\quad -4z = -4$
$$z = 1$$

Substitute 1 for z in equation (5).

$$6x - z = 3$$
$$6x - 1 = 3$$
$$6x = 4$$
$$x = \frac{4}{6} = \frac{2}{3}$$

Substitute $\frac{2}{3}$ for x and 1 for z in equation (2).

$$2x + y - 2z = -1$$
$$2\left(\frac{2}{3}\right) + y - 2(1) = -1$$
$$\frac{4}{3} + y - 2 = -1$$
$$y - \frac{2}{3} = -1$$
$$y = -1 + \frac{2}{3} = -\frac{1}{3}$$

The solution is $\left(\frac{2}{3}, -\frac{1}{3}, 1\right)$.

19. $2r + 2s - 3t = 12$ (1)
$\quad -2r + s + t = -11$ (2)
$\quad 3r + 4s + 2t = 4$ (3)

To eliminate s between equations (1) and (2), multiply equation (2) by -2 and then add.

$$2r + 2s - 3t = 12$$
$$2[-2r + s + t = -11]$$

gives

$$2r + 2s - 3t = 12$$
$$\underline{4r - 2s - 2t = 22}$$
$$6r - 5t = 34$ (4)$$

To eliminate s between equations (2) and (3), multiply equation (2) by -4 and then add.

$$-4[-2r + s + t = -11]$$
$$3r + 4s + 2t = 4$$

gives

$$8r - 4s - 4t = 44$$
$$\underline{3r + 4s + 2t = 4}$$
$$11r - 2t = 48$ (5)$$

Equations (4) and (5) are two equations in two unknowns.

$$6r - 5t = 34$$
$$11r - 2t = 48$$

To eliminate t, multiply equation (4) by -2 and equation (5) by 5, then add.

$$-2[6r - 5t = 34]$$
$$5[11r - 2t = 48]$$

gives

144

$$-12r + 10t = -68$$
$$\underline{55r - 10t = 240}$$
Add: $43r = 172$
$$r = 4$$

Substitute 4 for r in equation (5).
$$11r - 2t = 48$$
$$11(4) - 2t = 48$$
$$44 - 2t = 48$$
$$-2t = 4$$
$$t = -2$$

Substitute 4 for r and -2 for t in equation (2).
$$-2r + s + t = -11$$
$$-2(4) + s + (-2) = -11$$
$$-8 + s - 2 = -11$$
$$-10 + s = -11$$
$$s = -1$$

The solution is $(4, -1, -2)$.

21. $2a + 2b - c = 2$ (1)
$3a + 4b + c = -4$ (2)
$5a - 2b - 3c = 5$ (3)

To eliminate c between equations (1) and (2), simply add.
$$2a + 2b - c = 2$$
$$\underline{3a + 4b + c = -4}$$
Add: $5a + 6b = -2$ (4)

To eliminate c between equations (2) and (3), multiply equation (2) by 3 and then add.
$$3[3a + 4b + c = -4]$$
$$5a - 2b - 3c = 5$$
gives
$$9a + 12b + 3c = -12$$
$$\underline{5a - 2b - 3c = 5}$$
Add: $14a + 10b = -7$ (5)

Equations (4) and (5) are two equations in two unknowns.
$$5a + 6b = -2$$
$$14a + 10b = -7$$

To eliminate b, multiply equation (4) by -5 and multiply equation (5) by 3 and then add.
$$-5[5a + 6b = -2]$$
$$3[14a + 10b = -7]$$
gives

$$-25a - 30b = 10$$
$$\underline{42a + 30b = -21}$$
Add: $17a = -11$
$$a = -\frac{11}{17}$$

Substitute $-\dfrac{11}{17}$ for a in equation (4).
$$5a + 6b = -2$$
$$5\left(-\frac{11}{17}\right) + 6b = -2$$
$$-\frac{55}{17} + 6b = -2$$
$$6b = -2 + \frac{55}{17}$$
$$b = \frac{1}{6} \cdot \frac{21}{17} = \frac{7}{34}$$

Substitute $-\dfrac{11}{17}$ for a and $\dfrac{7}{34}$ for b in equation (2).
$$3a + 4b + c = -4$$
$$3\left(-\frac{11}{17}\right) + 4\left(\frac{7}{34}\right) + c = -4$$
$$-\frac{33}{17} + \frac{14}{17} + c = -4$$
$$-\frac{19}{17} + c = -4$$
$$c = -4 + \frac{19}{17}$$
$$c = -\frac{49}{17}$$

The solution is $\left(-\dfrac{11}{17}, \dfrac{7}{34}, -\dfrac{49}{17}\right)$.

23. $-x + 3y + z = 0$ (1)
$-2x + 4y - z = 0$ (2)
$3x - y + 2z = 0$ (3)

To eliminate z between equations (1) and (2), simply add.
$$-x + 3y + z = 0$$
$$\underline{-2x + 4y - z = 0}$$
Add: $-3x + 7y = 0$ (4)

To eliminate z between equations (2) and (3), multiply equation (2) by 2 and then add.
$$2[-2x + 4y - z] = 0$$
$$3x - y + 2z = 0$$

gives

$$-4x + 8y - 2z = 0$$
$$3x - y + 2z = 0$$

Add: $-x + 7y = 0$ (5)

Equations (4) and (5) are two equations in two unknowns.

$$-3x + 7y = 0$$
$$-x + 7y = 0$$

To eliminate y, multiply equation (4) by -1 and then add.

$$-1[-3x + 7y = 0]$$
$$-x + 7y = 0$$

gives

$$3x - 7y = 0$$
$$-x + 7y = 0$$

Add: $2x = 0$

$$x = 0$$

Substitute 0 for x in equation (5).

$$-x + 7y = 0$$
$$-0 + 7y = 0$$
$$7y = 0$$
$$y = 0$$

Finally, substitute 0 for x and 0 for y into equation (1).

$$-x + 3y + z = 0$$
$$-0 + 3(0) + z = 0$$
$$0 + z = 0$$
$$z = 0$$

The solution is (0, 0, 0).

25.
$$\frac{2}{3}x + y - \frac{1}{3}z = \frac{1}{3} \quad (1)$$

$$\frac{1}{2}x + y + z = \frac{5}{2} \quad (2)$$

$$\frac{1}{4}x - \frac{1}{4}y + \frac{1}{4}z = \frac{3}{2} \quad (3)$$

To clear fractions, multiply equation (1) by 3, equation (2) by 2, and equation (3) by 4. The resulting system is:

$$3\left(\frac{2}{3}x + y - \frac{1}{3}z = \frac{1}{3}\right)$$

$$2\left(\frac{1}{2}x + y + z = \frac{5}{2}\right)$$

$$4\left(\frac{1}{4}x - \frac{1}{4}y + \frac{1}{4}z = \frac{3}{2}\right)$$

gives

$$4\left(\frac{1}{4}x - \frac{1}{4}y + \frac{1}{4}z = \frac{3}{2}\right)$$

$$2x + 3y - z = 1 \quad (4)$$
$$x + 2y + 2z = 5 \quad (5)$$
$$x - y + z = 6 \quad (6)$$

To eliminate z between equations (4) and (6), simply add.

$$2x + 3y - z = 1$$
$$x - y + z = 6$$

Add: $3x + 2y = 7$ (7)

To eliminate z between equations (4) and (5), multiply equation (4) by 2 and then add.

$$2[2x + 3y - z = 1]$$
$$x + 2y + 2z = 5$$

gives

$$4x + 6y - 2z = 2$$
$$x + 2y + 2z = 5$$

Add: $5x + 8y = 7$ (8)

Equations (7) and (8) are two equations in two unknowns.

$$3x + 2y = 7$$
$$5x + 8y = 7$$

To eliminate y, multiply equation (7) by -4 and then add.

$$-4[3x + 2y = 7]$$
$$5x + 8y = 7$$

gives

$$-12x - 8y = -28$$
$$5x + 8y = 7$$

Add: $-7x = -21$

$$x = \frac{-21}{-7} = 3$$

Substitute 3 for x in equation (7).

$$3x + 2y = 7$$
$$3(3) + 2y = 7$$
$$9 + 2y = 7$$
$$2y = -2$$
$$y = -1$$

Substitute 3 for x and -1 for y in equation (6).

$$x - y + z = 6$$
$$3 - (-1) + z = 6$$
$$3 + 1 + z = 6$$
$$4 + z = 6$$
$$z = 2$$

The solution is (3, -1, 2).

27. Multiply each equation by 10.
$$10(0.2x + 0.3y + 0.3z = 1.1)$$
$$10(0.4x - 0.2y + 0.1z = 0.4)$$
$$10(-0.1x - 0.1y + 0.3z = 0.4)$$
gives
$$2x + 3y + 3z = 11 \quad (1)$$
$$4x - 2y + z = 4 \quad (2)$$
$$-x - y + 3z = 4 \quad (3)$$
To eliminate multiply equation (2) by –3 and then add.
$$2x + 3y + 3z = 11$$
$$3[4x - 2y + z = 4]$$
gives
$$\begin{aligned} 2x + 3y + 3z &= 11 \\ -12x + 6y - 3z &= -12 \end{aligned}$$
Add: $-10x + 9y \quad = -1 \quad (4)$
To eliminate z between equations (1) and (3) multiply equation (1) by –1 and then add.
$$-1[2x + 3y + 3z = 11]$$
$$-x - y + 3z = 4$$
gives
$$\begin{aligned} -2x - 3y - 3z &= -11 \\ -x - y + 3z &= 4 \end{aligned}$$
Add: $-3x - 4y \quad = -7 \quad (5)$
Equations (4) and (5) are two equations in two unknowns.
$$-10x + 9y = -1$$
$$-3x - 4y = -7$$
To eliminate y, multiply equation (4) by –3 and equation (5) by 10.
$$-3[-10 + 9y = -1]$$
$$10[-3x - 4y = -7]$$
gives
$$\begin{aligned} 30x - 27y &= 3 \\ -30x - 40y &= -70 \end{aligned}$$
Add: $\quad -67y = -67$
$$y = 1$$
Substitute 1 for y in equation (4).
$$-10x + 9y = -1$$
$$-10x + 9(1) = -1$$
$$-10x = -10$$
$$x = 1$$
Substitute 1 for x and 1 for y in equation (1).

$$2x + 3y + 3z = 11$$
$$2(1) + 3(1) + 3z = 11$$
$$5 + 3z = 11$$
$$3z = 6$$
$$z = 2$$
The solution is (1, 1, 2).

29. $2x + y + 2z = 1 \quad (1)$
$\quad x - 2y - z = 0 \quad (2)$
$\quad 3x - y + z = 2 \quad (3)$
To eliminate z between equations (2) and (3), simply add.
$$\begin{aligned} x - 2y - z &= 0 \\ 3x - y + z &= 2 \end{aligned}$$
Add: $4x - 3y \quad = 2 \quad (4)$
To eliminate z between equations (1) and (2), multiply equation (2) by 2 and then add.
$$2x + y + 2z = 1$$
$$2[x - 2y - z = 0]$$
gives
$$\begin{aligned} 2x + y + 2z &= 1 \\ 2x - 4y - 2z &= 0 \end{aligned}$$
Add: $4x - 3y \quad = 1 \quad (5)$
Equations (4) and (5) are two equations in two unknowns.
$$4x - 3y = 2$$
$$4x - 3y = 1$$
To eliminate x, multiply equation (4) by –1 and then add.
$$-1[4x - 3y = 2]$$
$$4x - 3y = 1$$
gives
$$\begin{aligned} -4x + 3y &= -2 \\ 4x - 3y &= 1 \end{aligned}$$
Add: $\quad 0 = -1 \quad$ False
Since this is a false statement, there is no solution and the system is inconsistent.

31. $3x - 4y + z = 4 \quad (1)$
$\quad x + 2y + z = 4 \quad (2)$
$\quad -6x + 8y - 2z = -8 \quad (3)$
To eliminate x between equations (1) and (3), multiply equation (1) by 2 and then add.
$$2[3x - 4y + z = 4]$$
$$-6x + 8y - 2z = -8$$
gives

$$6x - 8y + 2z = 8$$
$$-6x + 8y - 2z = -8$$
Add: $\qquad\qquad 0 = 0$

Since $0 = 0$ is a true statement, the system is dependent.

33. $\quad x + 3y + 2z = 6 \quad (1)$
$\qquad x - 2y - z = 8 \quad (2)$
$\quad -3x - 9y - 6z = -4 \quad (3)$

To eliminate x between equations (1) and (3), multiply equation (1) by 3 and then add.
$\quad 3[x + 3y + 2z = 6]$
$\quad -3x - 9y - 6z = -4$
gives

$$3x + 9y + 6z = 18$$
$$-3x - 9y - 6z = -4$$
Add: $\qquad\qquad 0 = 14$

Since $0 = 14$ is a false statement, the system is inconsistent.

35. No point is common to all three planes. Therefore, the system is inconsistent.

37. A straight line is common to all three planes. Therefore, there are an infinite number of points common to all three planes and the system is dependent.

39. a. Yes, if two or more of the planes are parallel, there will be no solution.

b. Yes, three planes may intersect at a single point.

c. No, the possibilities are no solution, one solution, or infinitely many solutions.

41. $Ax + By + Cz = -2$

Substitute $(1, 2, -1)$, $(1, 1, -3)$, and $(2, 3, -2)$ into the equation forming three equations in the three unknowns A, B, and C.
$\quad A(1) + B(2) + C(-1) = -2$
$\quad A(1) + B(1) + C(-3) = -2$
$\quad A(2) + B(3) + C(-2) = -2$
gives
$\quad A + 2B - C = -2 \quad (1)$
$\quad A + B - 3C = -2 \quad (2)$
$\quad 2A + 3B - 2C = -2 \quad (3)$

To eliminate A between equations (1) and (2), multiply equation (2) by -1 and then add.
$\quad A + 2B - C = -2$
$\quad -1[A + B - 3C = -2]$
gives

$$A + 2B - C = -2$$
$$-A - B + 3C = 2$$
Add: $\qquad B + 2C = 0 \quad (4)$

To eliminate A between equations (1) and (3), multiply equation (1) by -2 and add.
$\quad -2[A + 2B - C = -2]$
$\quad 2A + 3B - 2C = -2$
gives

$$-2A - 4B + 2C = 4$$
$$2A + 3B - 2C = -2$$
Add: $\qquad -B \qquad = 2$
$\qquad\qquad\qquad B = -2$

Substitute -2 for B in equation (4).
$\quad B + 2C = 0$
$\quad -2 + 2C = 0$
$\qquad 2C = 2$
$\qquad C = 1$

Substitute -2 for B and 1 for C in equation (2).
$\quad A + B - 3C = -2$
$\quad A + (-2) - 3(1) = -2$
$\qquad A - 5 = -2$
$\qquad A = 3$

Therefore, $A = 3$, $B = -2$, $C = 1$. and the equation is $3x - 2y + z = -2$.

43. One example is
$\quad x + y + z = 10$
$\quad x + 2y + z = 11$
$\quad x + y + 2z = 16$

Choose coefficients for x, y, and z, then use the given coordinates to find the constants.

45. a. $y = ax^2 + bx + c$
For the point $(1, -1)$,
let $y = -1$ and $x = 1$.
$\quad -1 = a(1)^2 + b(1) + c$
$\quad -1 = a + b + c \qquad (1)$
For the point $(-1, -5)$,
let $y = -5$ and $x = -1$.
$\quad -5 = a(-1)^2 + b(-1) + c$
$\quad -5 = a - b + c \qquad (2)$

For the point $(3, 11)$,
let $y = 11$ and $x = 3$.
$11 = a(3)^2 + b(3) + c$
$11 = 9a + 3b + c$ (3)
Equations (1), (2), and (3) give us a
system of three equations.
$$a + b + c = -1 \quad (1)$$
$$a - b + c = -5 \quad (2)$$
$$9a + 3b + c = 11 \quad (3)$$
To eliminate a and c between
equations (1) and (2) multiply
equation (2) by -1 and then add.
$$a + b + c = -1$$
$$-1[a - b + c = -5]$$
gives
$$a + b + c = -1$$
$$\underline{-a + b - c = 5}$$
Add: $2b = 4$
$$b = 2$$
Substitute 2 for b in equations (1)
and (3).
Equation (1) becomes
$a + b + c = -1$
$a + 2 + c = -1$
$a + c = -3$ (4)
Equation (3) becomes
$9a + 3b + c = 11$
$9a + 3(2) + c = 11$
$9a + c = 5$ (5)
Equations (4) and (5) are two
equations in two unknowns. To
eliminate c, multiply equation (4)
by -1 and then add.
$-1[a + c = -3]$
$9a + c = 5$
gives
$$-a - c = 3$$
$$\underline{9a + c = 5}$$
Add: $8a = 8$
$$a = 1$$
Finally, substitute 1 for a in equation
(4).
$a + c = -3$
$1 + c = -3$
$c = -4$
Thus, $a = 1$, $b = 2$, and $c = -4$.

b.   The quadratic equation is
$y = x^2 + 2x - 4$.
This is the equation determined by the
values found in part a.

47. $3a + 2b - c = 0$ (1)
$2a + 2c + d = 5$ (2)
$a + 2b - d = -2$ (3)
$2a - b + c + d = 2$ (4)
To eliminate d between equations (2)
and (3), simply add.
$$2a + 2c + d = 5$$
$$\underline{a + 2b - d = -2}$$
Add: $3a + 2b + 2c = 3$ (5)
Using equations (1) and (5), we can
eliminate a and b and then solve for c.
To do this, multiply equation (1) by -1
and then add.
$-1[3a + 2b - c = 0]$
$3a + 2b + 2c = 3$
gives
$$-3a - 2b + c = 0$$
$$\underline{3a + 2b + 2c = 3}$$
Add: $3c = 3$
$$c = 1$$
To eliminate d from equations (3) and
(4), simply add.
$$a + 2b - d = -2$$
$$\underline{2a - b + c + d = 2}$$
Add: $3a + b + c = 0$ (6)
Since $c = 1$, equation (6) becomes
$3a + b + 1 = 0$
$3a + b = -1$ (7)
Substitute 1 for c in equation (1).
$3a + 2b - c = 0$
$3a + 2b - 1 = 0$
$3a + 2b = 1$ (8)
Using equations (7) and (8), we now
have two equations system in two
unknowns.
$3a + b = -1$
$3a + 2b = 1$
To eliminate a, multiply equation (7) by
-1 and then add.
$-1[3a + b = -1]$
$3a + 2b = 1$
gives

$$-3a - b = 1$$
$$\underline{3a + 2b = 1}$$
Add: $b = 2$

Now, substitute 2 for b in equation (7).

$$3a + b = -1$$
$$3a + 2 = -1$$
$$3a = -3$$
$$a = -1$$

Finally, substitute -1 for a and 1 for c in equation (2).

$$2a + 2c + d = 5$$
$$2(-1) + 2(1) + d = 5$$
$$-2 + 2 + d = 5$$
$$d = 5$$

The solution is $(-1, 2, 1, 5)$.

49. Let x be the time for Cameron.

Then, $x - \dfrac{1}{6}$ is the time for Phillipa.

	rate	time	distance
Phillipa	5	$t - \dfrac{1}{6}$	$5\left(t - \dfrac{1}{6}\right)$
Cameron	3	t	$3t$

a. The distances traveled are the same.

$$5\left(t - \frac{1}{6}\right) = 3t$$
$$5t - \frac{5}{6} = 3t$$
$$5t = 3t + \frac{5}{6}$$
$$2t = \frac{5}{6}$$
$$t = \frac{1}{2}\left(\frac{5}{6}\right) = \frac{5}{12} \text{ hr}$$

or $\dfrac{5}{12}(60) = 25$ min

b. The distance is

$$3t = 3\left(\frac{5}{12}\right) = \frac{15}{12} = 1\frac{1}{4}$$

or 1.25 miles.

50. $\left| 4 - \dfrac{2}{3}x \right| > 5$

$$4 - \frac{2}{3}x > 5 \quad \text{or} \quad 4 - \frac{2}{3}x < -5$$
$$-\frac{2}{3}x > 1 \qquad\qquad -\frac{2}{3}x < -9$$
$$x < -\frac{3}{2} \qquad\qquad x > \frac{27}{2}$$

The solution is $\left\{ x \middle| x < -\dfrac{3}{2} \text{ or } x > \dfrac{27}{2} \right\}$.

51. $\left| \dfrac{3x - 4}{2} \right| - 1 < 5$

$$\left| \frac{3x - 4}{2} \right| < 6$$
$$-6 < \frac{3x - 4}{2} < 6$$
$$2(-6) < 2\left(\frac{3x - 4}{2} \right) < 2(6)$$
$$-12 < 3x - 4 < 12$$
$$-12 + 4 < 3x - 4 + 4 < 12 + 4$$
$$-8 < 3x < 16$$
$$-\frac{8}{3} < \frac{3x}{3} < \frac{16}{3}$$
$$-\frac{8}{3} < x < \frac{16}{3}$$

The solution is $\left\{ x \middle| -\dfrac{8}{3} < x < \dfrac{16}{3} \right\}$.

52. $\left| 2x - \dfrac{1}{2} \right| = -5$

There is no solution since the right side is a negative number and the left side is non-negative and it is not possible for a non-negative quantity to be equal to a negative number. The solution is $\varnothing$.

Exercise Set 4.3

1. a. Let $g =$ gross, in millions of dollars
Tommy Lee Jones' income: $7 + 0.05g$
Steven Spielberg's income: $0.20g$
$$7 + 0.05g = 0.20g$$
$$7 = 0.15g$$
$$46.7 \approx g$$
The movie has to gross about $46.7 million.

b. $7 + 0.05g = 7 + 0.05(600)$
$$= 7 + 30$$
$$= 37$$
Jones received \$37 million.

c. $0.20g = 0.20(600)$
$$= 120$$
Spielberg received \$120 million.

3. Let F = grams of fat in fries
    H = grams of fat in hamburger
    $F = 3H + 4$
$$F - H = 46$$
Substitute $3H + 4$ for F in the second equation.
$$F - H = 46$$
$$3H + 4 - H = 46$$
$$2H = 42$$
$$H = 21$$
Substitute 21 for H in the first equation.
$$F = 3H + 4$$
$$F = 3(21) + 4$$
$$F = 63 + 4$$
$$F = 67$$
The hamburger has 21 grams of fat and the fries have 67 grams of fat.

5. Let B = FCI for the Braves
    E = FCI for the Expos
    $B = 2E - 26.68$
$$B - E = 53.74$$
Substitute $2E - 26.68$ for B in the second equation.
$$B - E = 53.74$$
$$2E - 26.68 - E = 53.74$$
$$E = 80.42$$
Substitute 80.42 for E in the first equation.
$$B = 2E - 26.68$$
$$B = 2(80.42) - 26.68$$
$$B = 134.16$$
The FCI for the Expos is \$80.42 and the FCI for the Braves is \$134.16.

7. Let A and B be the measures of the two angles.
$$A + B = 180$$
$$A = 3B - 28$$
Substitute $3B - 28$ for A in the first equation.

$$A + B = 180$$
$$3B - 28 + B = 180$$
$$4B - 28 = 180$$
$$4B = 208$$
$$B = 52$$
Now substitute 52 for B in the second equation.
$$A = 3B - 28$$
$$A = 3(52) - 28$$
$$A = 128$$
The two angles measure 52° and 128°.

9. Let x be the weekly salary and y be the commission rate.
$$x + 4000y = 660$$
$$x + 6000y = 740$$
Multiply the first equation by –1 and then add.
$$-1[x + 4000y = 660]$$
$$x + 6000y = 740$$
gives
$$\begin{array}{r} -x - 4000y = -660 \\ x + 6000y = 740 \\ \hline \end{array}$$
Add: $2000y = 80$
$$y = \frac{80}{2000} = 0.04$$
Substitute 0.04 for y in the first equation.
$$x + 4000y = 660$$
$$x + 4000(0.04) = 660$$
$$x + 160 = 660$$
$$x = 500 \text{ dollars}$$
Her weekly salary is \$500 and the commission rate is 4%.

11. Let f = fixed charge
    a = additional charge per flier
$$f + 1000a = 550$$
$$f + 2000a = 800$$
Multiply the first equation by –1 and add to the second equation.
$$-1[f + 1000a = 550]$$
$$f + 2000a = 800$$
gives
$$\begin{array}{r} -f - 1000a = -550 \\ f + 2000a = 800 \\ \hline \end{array}$$
Add: $1000a = 250$
$$a = 0.25$$

Substitute 0.25 for a in the first equation.
$$f + 1000a = 550$$
$$f + 1000(0.25) = 550$$
$$f + 250 = 550$$
$$f = 300$$
The fixed charge is \$300 and the charge for each flier is \$0.25.

13. Let x be the amount of 20% solution and y be the amount of 4% solution.
$$x + y = 10$$
$$0.20x + 0.04y = 0.10(10)$$
Solve the first equation for x.
$$x = 10 - y$$
Substitute $10 - y$ for x in the second equation.
$$0.20x + 0.04y = 0.10(10)$$
$$0.20(10 - y) + 0.04y = 1$$
$$2 - 0.20y + 0.04y = 1$$
$$-0.16y = -1$$
$$y = 6.25$$
Substitute 6.25 for y in the first equation.
$$x + y = 10$$
$$x + 6.25 = 10$$
$$x = 3.75$$
Dave Visser should mix 3.75 gallons of the 20% solution with 6.25 gallons of the 4% solution.

15. Let x = gallons of concentrate (18% solution) and y = gallons of water (0% solution).
$$x + y = 200$$
$$0.18x + 0y = 0.009(200)$$
Solve the second equation for x.
$$0.18x + 0y = 0.009(200)$$
$$0.18x = 1.8$$
$$x = 10$$
Substitute 10 for x in the first equation.
$$x + y = 200$$
$$10 + y = 200$$
$$y = 190$$
The mixture should contain 10 gallons of concentrate and 190 gallons of water.

17. Let R = orders of regular wings
J = orders of jumbo wings
$$5.99R + 8.99J = 1024.66$$
$$R + J = 134$$
Solve the second equation for R.
$$R = 134 - J$$
Substitute $134 - J$ for R in the first equation.
$$5.99R + 8.99J = 1024.66$$
$$5.99(134 - J) + 8.99J = 1024.66$$
$$802.66 - 5.99J + 8.99J = 1024.66$$
$$3J = 222$$
$$J = 74$$
Substitute 74 for J in the second equation
$$R + J = 134$$
$$R + 74 = 134$$
$$R = 60$$
Jimmy Stephen's Wing House sold 60 regular orders and 74 jumbo orders.

19. Let x be the amount of the 5% butterfat milk and y be the amount of the skim milk (0% butterfat).
$$x + y = 100$$
$$0.05x + 0y = 0.035(100)$$
Solve the second equation for x.
$$0.05x + 0y = 0.035(100)$$
$$0.05x = 3.5$$
$$x = \frac{3.5}{0.05} = 70$$
Now substitute 70 for x in the first equation.
$$x + y = 100$$
$$70 + y = 100$$
$$y = 30$$
Laura needs to mix 70 gallons of the 5% butterfat milk with 30 gallons of skim milk to produce 100 gallons of 3.5% butterfat milk.

21. Let x be the amount of heavy cream and y be the amount of half-and-half.
$$x + y = 16$$
$$0.36x + 0.105y = 0.2(16)$$
Solve the first equation for x.
$$x + y = 16$$
$$x = 16 - y$$
Substitute $16 - y$ for x in the second equation.

$0.36x + 0.105y = 0.2(16)$
$0.36(16 - y) + 0.105y = 0.2(16)$
$5.76 - 0.36y + 0.105y = 3.2$
$-0.255y + 5.76 = 3.2$
$-0.255y = -2.56$
$y = \dfrac{-2.56}{-0.255} \approx 10.04$

Now substitute 10.04 for y in the first equation.
$x + y = 16$
$x + 10.04 \approx 16$
$x \approx 5.96$ ounces

Steve needs to mix about 5.96 ounces of heavy cream with about 10.04 ounces of half-and-half to produce 16 ounces of 20% butterfat.

23. Let x be the amount of apple juice and y be the amount of raspberry juice.
$x + y = 8$
$8.3x + 9.3y = 8(8.7)$
Solve the first equation for y.
$x + y = 8$
$y = 8 - x$
Substitute $8 - x$ for y in the second equation.
$8.3x + 9.3y = 8(8.7)$
$8.3x + 9.3(8 - x) = 8(8.7)$
$8.3x + 74.4 - 9.3x = 69.6$
$74.4 - 1.0x = 69.6$
$-x = -4.8$
$x = 4.8$
Substitute 4.8 for x in the equation $y = 8 - x$.
$y = 8 - x$
$y = 8 - 4.8$
$y = 3.2$

The company should mix 4.8 ounces of apple juice with 3.2 ounces of raspberry juice.

25. Pull toward $= -\dfrac{1}{5}d + 70$

Pull away $= -\dfrac{4}{3}d + 230$

a. Since the pull values are equal, substitute $-\dfrac{1}{5}d + 70$ for the pull in the second equation.
$-\dfrac{1}{5}d + 70 = -\dfrac{4}{3}d + 230$
$-\dfrac{1}{5}d = -\dfrac{4}{3}d + 160$
$15\left(-\dfrac{1}{5}d\right) = 15\left(-\dfrac{4}{3}d\right) + 15(160)$
$-3d = -20d + 2400$
$17d = 2400$
$d = \dfrac{2400}{17} \approx 141.2$ cm

b. Pull toward $= -\dfrac{1}{5}d + 70$
$= -\dfrac{1}{5}(100) + 70$
$= -20 + 70$
$= 50$
Pull away $= -\dfrac{4}{3}d + 230$
$= -\dfrac{4}{3}(100) + 230$
$= -\dfrac{400}{3} + \dfrac{690}{3}$
$= \dfrac{290}{3}$
$= 96\dfrac{2}{3}$

The rat will pull away.

27. Let s = number of small cones
l = number of large cones
The system is
$s + l = 260$
$1s + 1.5l = 299$
Solve the first equation for s.
$s = 260 - l$
Substitute $260 - l$ for s in the second equation.
$s + 1.5l = 299$
$260 - l + 1.5l = 299$
$0.5l = 39$
$l = 78$
Substitute 78 for l in the first equation.

$$s + l = 260$$
$$s + 78 = 260$$
$$s = 182$$

Rita Pendegrass sold 182 small cones and 78 large cones.

29. Let x be the time the slower machine was working and y be the time the faster machine was working.
$$y = x + 3$$
$$75x + 120y = 1335$$
Substitute $x + 3$ for y in the second equation.
$$75x + 120y = 1335$$
$$75x + 120(x + 3) = 1335$$
$$75x + 120x + 360 = 1335$$
$$195x + 360 = 1335$$
$$195x = 975$$
$$x = 5$$
Substitute 5 for x in the first equation.
$$y = x + 3$$
$$y = 5 + 3$$
$$y = 8$$
The slower machine was working for 5 minutes and the faster machine was working for
8 minutes.

31. Let x be the amount of pure antifreeze and y be the amount of 18% antifreeze.
$$x + y = 16 \qquad \text{or} \qquad x + y = 16$$
$$1x + 0.18y = 0.2(16) \qquad x + 0.18y = 3.2$$
To solve, multiply the second equation by −1 and then add.
$$x + y = 16$$
$$-1[x + 0.18y = 3.2]$$
gives
$$x + y = 16$$
$$-x - 0.18y = -3.2$$
Add: $\quad 0.82y = 12.8$
$$y = \frac{12.8}{0.82} \approx 15.61$$
Now substitute 15.61 for y in the first equation.
$$x + y = 16$$
$$x + 15.61 \approx 16$$
$$x \approx 0.39$$
0.39 liters of pure antifreeze must be added to 15.61 liters of the mixture.

33. Let x be the number of Model A chairs and y be the number of Model B chairs.

Model	Assemble	Paint
A	1	0.5
B	3.2	0.4

$$1x + 3.2y = 46.4$$
$$0.5x + 0.4y = 8.8$$
To solve, multiply the second equation by −2 and then add.
$$1x + 3.2y = 46.4$$
$$-2[0.5x + 0.4y = 8.8]$$
gives
$$x + 3.2y = 46.4$$
$$-x - 0.8y = -17.6$$
Add: $\quad 2.4y = 28.8$
$$y = \frac{28.8}{2.4} = 12$$
Substitute 12 for y in the first equation.
$$x + 3.2y = 46.4$$
$$x + 3.2(12) = 46.4$$
$$x + 38.4 = 46.4$$
$$x = 8$$
The company should make 8 Model A chairs and 12 Model B.

35. Let x = speed of Melissa's car and y = speed of Tom's car
$$x = y + 15$$
$$\frac{150}{x} = \frac{120}{y}$$
Substitute $y + 15$ for x in the second equation.
$$\frac{150}{y + 15} = \frac{120}{y}$$
$$150y = 120y + 1800$$
$$30y = 1800$$
$$y = 60$$
Substitute 60 for y in the first equation.
$$x = y + 15$$
$$x = 60 + 15$$
$$x = 75$$
Tom traveled at 60 mph and Melissa traveled at 75 mph.

37. $I(t) = -1.2t + 27.2$
$C(t) = 0.24t + 17.1$
$-1.2t + 27.2 = 0.24t + 17.1$
$27.2 - 17.1 = 0.24t + 1.2t$
$10.1 = 1.44t$
$7.0 \approx t$
They will be equal approximately 7 years after 1992 or in 1999.

39. a. Let c = cost
and m = minutes
Plan 1: $c = 0.15m$
Plan 2: $c = 0.10m + 4.95$

b.

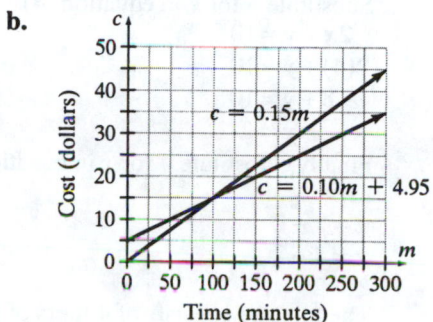

c. The cost is the same at about 100 minutes.

d. $0.15m = 0.10m + 4.95$
$0.05m = 4.95$
$m = \dfrac{4.95}{0.05} = 99$
The actual point at which the costs are equal is 99 minutes. The graph cannot be read with enough accuracy to distinguish between 99 and 100 minutes, which are very close.

41. a. Let x = number of chief petty officers
y = number of officers
z = number of enlisted men
$x + y + z = 141$
$x = y + 4$
$z = 8y - 3$

b. Substitute $y + 4$ for x and $8y - 3$ for z in the first equation.
$x + y + z = 141$
$y + 4 + y + 8y - 3 = 141$
$10y = 140$
$y = 14$
Substitute 14 for y in the second and third equations.
$x = y + 4$ $z = 8y - 3$
$x = 14 + 4$ $z = 8(14) - 3$
$x = 18$ $z = 109$
There are 14 officers, 18 chief petty officers, and 109 enlisted men.

43. a. Let x = number of land mines in Iraq
y = number of land mines in Angola
z = number of land mines in Iran
$x + y + z = 41$
$z = 3x - 14$
$y = 2x - 5$

b. Substitute $3x - 14$ for z and $2x - 5$ for y in the first equation.
$x + y + z = 41$
$x + 2x - 5 + 3x - 14 = 41$
$6x = 60$
$x = 10$
Substitute 10 for x in the second and third equations.
$z = 3x - 14$ $y = 2x - 5$
$z = 3(10) - 14$ $y = 2(10) - 5$
$z = 16$ $y = 15$
The number of land mines in the countries is
Iraq: 10 million
Angola: 15 million
Iran: 16 million.

45. a. Let x, y, and z be the measures of the three angles.
$x + y + z = 180$
$x = \dfrac{2}{3}y$
$z = 3y - 30$

b. Substitute $\dfrac{2}{3}y$ for x and $3y - 30$ for z in the first equation.

$$x + y + z = 180$$
$$\frac{2}{3}y + y + 3y - 30 = 180$$
$$\frac{14}{3}y - 30 = 180$$
$$\frac{14}{3}y = 210$$
$$\frac{3}{14}\left(\frac{14}{3}y\right) = \frac{3}{14}(210)$$
$$y = 45$$

Substitute 45 for y in the second equation.

$$x = \frac{2}{3}y$$
$$x = \frac{2}{3}(45)$$
$$x = 30$$

Substitute 45 for y in the third equation.

$$z = 3y - 30$$
$$z = 3(45) - 30$$
$$z = 135 - 30$$
$$z = 105$$

The three angles are 30°, 45°, and 105°.

47. a. Let x be the amount of the 10% solution, y be the amount of the 12% solution, and z be the amount of the 20% solution.

$$x + y + z = 8 \qquad (1)$$
$$0.10x + 0.12y + 0.20z = (0.13)8 \quad (2)$$
$$z = x - 2 \qquad (3)$$

b. Substitute $x - 2$ for z in equation (1).

$$x + y + z = 8$$
$$x + y + (x - 2) = 8$$
$$2x + y - 2 = 8$$
$$2x + y = 10 \qquad (4)$$

Substitute $x - 2$ for z in equation (2).

$$0.10x + 0.12y + 0.20z = (0.13)8$$
$$0.10x + 0.12y + 0.20(x - 2) = (0.13)8$$
$$0.10x + 0.12y + 0.20x - 0.40 = 1.04$$
$$0.30x + 0.12y = 1.44 \qquad (5)$$

Equations (4) and (5) are a system of two equations in two unknowns.

$$2x + y = 10$$
$$0.30x + 0.12y = 1.44$$

To solve, multiply equation (5) by 100 and equation (4) by -12 and then add.

$$-12[2x + y = 10]$$
$$100[0.30x + 0.12y = 1.44]$$

gives

$$-24x - 12y = -120$$
$$\underline{30x + 12y = 144}$$

Add: $6x \qquad\quad = 24$

$$x = 4$$

Substitute 4 for x in equation (4).

$$2x + y = 10$$
$$2(4) + y = 10$$
$$8 + y = 10$$
$$y = 2$$

Finally, substitute 4 for x in equation (3).

$$z = x - 2$$
$$z = 4 - 2$$
$$z = 2$$

The mixture consists of 4 liters of the 10% solution, 2 liters of the 12% solution, and 2 liters of the 20% solution.

49. a. Let x be the number of children's chairs, y be the number of standard chairs, and z be the number of executive chairs.

$$5x + 4y + 7z = 154 \quad (1)$$
$$3x + 2y + 5 = 94 \quad (2)$$
$$2x + 2y + 4z = 76 \quad (3)$$

b. To eliminate y between equations (1) and (2), multiply equation (2) by -2 and add.

$$5x + 4y + 7z = 154$$
$$-2[3x + 2y + 5z = 94]$$

gives

$$5x + 4y + 7z = 154$$
$$\underline{-6x - 4y - 10z = -188}$$

Add: $-x \qquad - 3z = -34 \quad (4)$

To eliminate y between equations (2) and (3), multiply equation (3) by -1 and add.

$$3x + 2y + 5z = 94$$
$$-1[2x + 2y + 4z = 76]$$

gives

$$3x + 2y + 5z = 94$$
$$-2x - 2y - 4z = -76$$

Add: $x \quad\quad + z = 18$ (5)

Equations (4) and (5) are a system of two equations in two unknowns. To eliminate x, simply add.

$$-x - 3z = -34$$
$$x + z = 18$$

Add: $-2z = -16$

$$z = \frac{-16}{-2} = 8$$

Substitute 8 for z in equation (5).

$$x + z = 18$$
$$x + 8 = 18$$
$$x = 10$$

Substitute 10 for x and 8 for z in equation (3).

$$2x + 2y + 4z = 76$$
$$2(10) + 2y + 4(8) = 76$$
$$20 + 2y + 32 = 76$$
$$2y + 52 = 76$$
$$2y = 24$$
$$y = 12$$

The Donaldson Furniture Company should produce 10 children's chairs, 12 standard chairs, and 8 executive chairs.

51. $I_A + I_B + I_C = 0$ (1)
$$-8I_B + 10I_C = 0 \quad (2)$$
$$4I_A - 8I_B = 6 \quad (3)$$

To eliminate I_A between equations (1) and (3), multiply equation (1) by -4 and add.

$$-4[I_A + I_B + C = 0]$$
$$4I_A - 8I_B = 6$$

gives

$$-4I_A - 4I_B - 4I_C = 0$$
$$4I_A - 8I_B = 6$$

Add: $-12I_B - 4I_C = 6$

or $-6I_B - 2I_C = 3$ (4)

Equations (4) and (2) are a system of two equations in two unknowns.

$$-8I_B + 10I_C = 0$$
$$-6I_B - 2I_C = 3$$

Multiply equation (4) by 5 and add this result to equation (2).

$$-8I_B + 10I_C = 0$$
$$5[-6I_B - 2I_C = 3]$$

gives

$$-8I_B + 10I_C = 0$$
$$-30I_B - 10I_C = 15$$

Add: $-38I_B = 15$

$$I_B = \frac{15}{-38} = -\frac{15}{38}$$

Substitute $-\dfrac{15}{38}$ for I_B in equation (2).

$$-8I_B + 10I_C = 0$$
$$-8\left(-\frac{15}{38}\right) + 10I_C = 0$$
$$\frac{120}{38} + 10I_C = 0$$
$$10I_C = -\frac{120}{38}$$
$$\frac{1}{10}(10I_C) = \frac{1}{10}\left(-\frac{120}{38}\right)$$
$$I_C = -\frac{12}{38} = -\frac{6}{19}$$

Finally, substitute $-\dfrac{15}{38}$ for I_B in equation (3).

$$4I_A - 8I_B = 6$$

$$4I_A - 8\left(-\frac{15}{38}\right) = 6$$

$$4I_A + \frac{120}{38} = 6$$

$$4I_A = 6 - \frac{120}{38}$$

$$4I_A = 6 - \frac{60}{19}$$

$$4I_A = \frac{114}{19} - \frac{60}{19}$$

$$4I_A = \frac{54}{19}$$

$$\frac{1}{4}(4I_A) = \frac{1}{4}\left(\frac{54}{19}\right)$$

$$I_A = \frac{27}{38}$$

The current in branch A is $\dfrac{27}{38}$, the current in

branch B is $-\dfrac{15}{38}$ and the current in branch

C is $-\dfrac{6}{19}$.

54. Substitute -2 for x and 5 for y.

$$\frac{1}{2}x + \frac{2}{5}xy + \frac{1}{8}y = \frac{1}{2}(-2) + \frac{2}{5}(-2)(5) + \frac{1}{8}(5)$$

$$= -1 - 4 + \frac{5}{8}$$

$$= -5 + \frac{5}{8}$$

$$= -\frac{40}{8} + \frac{5}{8}$$

$$= -\frac{35}{8}$$

55. $4 - 2[(x-5) + 2x] = -(x+6)$

$$4 - 2(x - 5 + 2x) = -x - 6$$
$$4 - 2(3x - 5) = -x - 6$$
$$4 - 6x + 10 = -x - 6$$
$$-6x + 14 = -x - 6$$
$$-6x + x = -6 - 14$$
$$-5x = -20$$
$$x = 4$$

56. Use the vertical line test. If a vertical line cannot be drawn to intersect the graph in more than one point, the graph is a function.

57. The slope is
$$m = \frac{-4 - (-8)}{6 - 2} = \frac{-4 + 8}{6 - 2} = \frac{4}{4} = 1$$
Use the point-slope form with $m = 1$ and $(x_1, y_1) = (6, -4)$.
$$y - y_1 = m(x - x_1)$$
$$y - (-4) = 1(x - 6)$$
$$y + 4 = x - 6$$
$$y = x - 10$$

Exercise Set 4.4

1. A square matrix has the same number of rows and columns.

3. The next step is to change the -1 in the second row to 1 by multiplying the second row of numbers by -1.

5. a. A row of numbers contains all 0's.

  b. All the numbers in a row on the left side of the augmented matrix are 0's, but the number on the right side is not 0.

7. $\begin{bmatrix} -8 & 4 & | & 10 \\ 3 & 5 & | & -1 \end{bmatrix} \begin{bmatrix} 1 & -\frac{1}{2} & | & -\frac{5}{4} \\ 3 & 5 & | & -1 \end{bmatrix} -\frac{1}{8}R$

9. $\begin{bmatrix} 4 & 0 & 3 & | & 8 \\ 5 & -7 & 2 & | & 14 \\ -1 & 3 & 5 & | & 12 \end{bmatrix}$

$\begin{bmatrix} 4 & 0 & 3 & | & 8 \\ -\frac{5}{7} & 1 & -\frac{2}{7} & | & -2 \\ -1 & 3 & 5 & | & 12 \end{bmatrix} -\frac{1}{7}R_2$

11. $\begin{bmatrix} 1 & 3 & | & 12 \\ -3 & 8 & | & -6 \end{bmatrix} \begin{bmatrix} 1 & 3 & | & 12 \\ 0 & 17 & | & 30 \end{bmatrix} 3R_1 + R_2$

13. $\begin{bmatrix} 1 & 0 & 8 & | & \frac{1}{4} \\ 5 & 2 & 2 & | & -2 \\ 6 & -3 & 1 & | & 0 \end{bmatrix}$

$\begin{bmatrix} 1 & 0 & 8 & | & \frac{1}{4} \\ 0 & 2 & -38 & | & -\frac{13}{4} \\ 6 & -3 & 1 & | & 0 \end{bmatrix} -5R_1 + R_2$

15. $x + 3y = 3$

$-x + y = -3$

$\begin{bmatrix} 1 & 3 & | & 3 \\ -1 & 1 & | & -3 \end{bmatrix}$

$\begin{bmatrix} 1 & 3 & | & 3 \\ 0 & 4 & | & 0 \end{bmatrix} R_1 + R_2$

$\begin{bmatrix} 1 & 3 & | & 3 \\ 0 & 1 & | & 0 \end{bmatrix} \frac{1}{4}R_2$

The system is

$x + 3y = 3$

$y = 0$

Substitute 0 for y in the first equation.

$x + 3y = 3$

$x + 3(0) = 3$

$x + 0 = 3$

$x = 3$

The solution is (3, 0).

17. $x - 4y = -1$

$3x - 2y = 7$

$\begin{bmatrix} 1 & -4 & | & -1 \\ 3 & -2 & | & 7 \end{bmatrix}$

$\begin{bmatrix} 1 & -4 & | & -1 \\ 0 & 10 & | & 10 \end{bmatrix} -3R_1 + R_2$

$\begin{bmatrix} 1 & -4 & | & -1 \\ 0 & 1 & | & 1 \end{bmatrix} \frac{1}{10}R_2$

The system is

$x - 4y = -1$

$y = 1$

Substitute 1 for y in the first equation.

$x - 4y = -1$

$x - 4(1) = -1$

$x = 3$

The solution is (3, 1).

19. $-3a + 6b = 3$

$4a - 2b = -1$

$\begin{bmatrix} -3 & 6 & | & 3 \\ 4 & -2 & | & -1 \end{bmatrix}$

$\begin{bmatrix} 1 & -2 & | & -1 \\ 4 & -2 & | & -1 \end{bmatrix} -\frac{1}{3}R_1$

$\begin{bmatrix} 1 & -2 & | & -1 \\ 0 & 6 & | & 3 \end{bmatrix} -4R_1 + R_2$

$\begin{bmatrix} 1 & 2 & | & 1 \\ 0 & 1 & | & \frac{1}{2} \end{bmatrix} \frac{1}{6}R_2$

The system is

$a + 2b = 1$

$b = \frac{1}{2}$

Substitute $\frac{1}{2}$ for b in the first equation.

$a + 2b = 1$

$a + 2\left(\frac{1}{2}\right) = 1$

$a + 1 = 1$

$a = 0$

The solution is $\left(0, \frac{1}{2}\right)$.

21. $2x - 5y = -6$

$-4x + 10y = 12$

$\begin{bmatrix} 2 & -5 & | & -6 \\ -4 & 10 & | & 12 \end{bmatrix}$

$\begin{bmatrix} 1 & -\frac{5}{2} & | & -3 \\ -4 & 10 & | & 12 \end{bmatrix} \frac{1}{2}R_1$

$\begin{bmatrix} 1 & -\frac{5}{2} & | & -3 \\ 0 & 0 & | & 0 \end{bmatrix} 4R_1 + R_2$

Since the last row contains all 0's, this is a dependent system of equations.

23. $12x + 10y = -14$

$4x - 3y = -11$

$\begin{bmatrix} 12 & 10 & | & -14 \\ 4 & -3 & | & -11 \end{bmatrix}$

$\begin{bmatrix} 1 & \frac{5}{6} & | & -\frac{7}{6} \\ 4 & -3 & | & -11 \end{bmatrix} \frac{1}{12}R_1$

$$\begin{bmatrix} 1 & \frac{5}{6} & \bigm| & -\frac{7}{6} \\ 0 & -\frac{19}{3} & \bigm| & -\frac{19}{3} \end{bmatrix} -4R_1 + R_2$$

$$\begin{bmatrix} 1 & \frac{5}{6} & \bigm| & -\frac{7}{6} \\ 0 & 1 & \bigm| & 1 \end{bmatrix} -\frac{3}{19}R_2$$

The system is

$$x + \frac{5}{6}y = -\frac{7}{6}$$
$$y = 1$$

Substitute 1 for y in the first equation.

$$x + \frac{5}{6}y = -\frac{7}{6}$$
$$x + \frac{5}{6}(1) = -\frac{7}{6}$$
$$x + \frac{5}{6} = -\frac{7}{6}$$
$$x = -\frac{12}{6}$$
$$x = -2$$

The solution is $(-2, 1)$.

25. $-3x + 6y = 5$
$\quad\quad 2x - 4y = 8$

$$\begin{bmatrix} -3 & 6 & \bigm| & 5 \\ 2 & -4 & \bigm| & 8 \end{bmatrix}$$

$$\begin{bmatrix} 1 & -2 & \bigm| & -\frac{5}{3} \\ 2 & -4 & \bigm| & 8 \end{bmatrix} -\frac{1}{3}R_1$$

$$\begin{bmatrix} 1 & -2 & \bigm| & -\frac{5}{3} \\ 0 & 0 & \bigm| & \frac{34}{3} \end{bmatrix} -2R_1 + R_2$$

Since the last row contains zeros on the left and a nonzero number on the right, this is an inconsistent system and there is no solution.

27. $\quad 9x - 8y = 4$
$\quad\quad -3x + 4y = -1$

$$\begin{bmatrix} 9 & -8 & \bigm| & 4 \\ -3 & 4 & \bigm| & -1 \end{bmatrix}$$

$$\begin{bmatrix} 1 & -\frac{8}{9} & \bigm| & \frac{4}{9} \\ -3 & 4 & \bigm| & -1 \end{bmatrix} \frac{1}{9}R_1$$

$$\begin{bmatrix} 1 & -\frac{8}{9} & \bigm| & \frac{4}{9} \\ 0 & \frac{4}{3} & \bigm| & \frac{1}{3} \end{bmatrix} 3R_1 + R_2$$

$$\begin{bmatrix} 1 & -\frac{8}{9} & \bigm| & \frac{4}{9} \\ 0 & 1 & \bigm| & \frac{1}{4} \end{bmatrix} \frac{3}{4}R_2$$

The system is

$$x - \frac{8}{9}y = \frac{4}{9}$$
$$y = \frac{1}{4}$$

Substitute $\frac{1}{4}$ for y in the first equation.

$$x - \frac{8}{9}y = \frac{4}{9}$$
$$x - \frac{8}{9}\left(\frac{1}{4}\right) = \frac{4}{9}$$
$$x - \frac{2}{9} = \frac{4}{9}$$
$$x = \frac{4}{9} + \frac{2}{9}$$
$$x = \frac{6}{9}$$
$$x = \frac{2}{3}$$

The solution is $\left(\frac{2}{3}, \frac{1}{4}\right)$.

29. $10m = 8n + 15$
$\quad\quad 16n = -15m - 2$
Write the system in standard form.
$\quad 10m - 8n = 15$
$\quad 15m + 16n = -2$

$$\begin{bmatrix} 10 & -8 & \bigm| & 15 \\ 15 & 16 & \bigm| & -2 \end{bmatrix}$$

$$\begin{bmatrix} 1 & -\frac{4}{5} & \bigm| & \frac{3}{2} \\ 15 & 16 & \bigm| & -2 \end{bmatrix} \frac{1}{10}R_1$$

$$\begin{bmatrix} 1 & -\frac{4}{5} & \bigm| & \frac{3}{2} \\ 0 & 28 & \bigm| & -\frac{49}{2} \end{bmatrix} -15R_1 + R_2$$

$$\begin{bmatrix} 1 & -\frac{4}{5} & \bigm| & \frac{3}{2} \\ 0 & 1 & \bigm| & -\frac{7}{8} \end{bmatrix} \frac{1}{28}R_2$$

The system is

$$m - \frac{4}{5}n = \frac{3}{2}$$
$$n = -\frac{7}{8}$$

Substitute $-\dfrac{7}{8}$ for n in the first equation.

$$m - \frac{4}{5}n = \frac{3}{2}$$

$$m - \frac{4}{5}\left(-\frac{7}{8}\right) = \frac{3}{2}$$

$$m + \frac{7}{10} = \frac{3}{2}$$

$$m = \frac{3}{2} - \frac{7}{10}$$

$$m = \frac{15}{10} - \frac{7}{10}$$

$$m = \frac{8}{10}$$

$$m = \frac{4}{5}$$

The solution is $\left(\dfrac{4}{5},\ -\dfrac{7}{8}\right)$.

31.
$$x + y - 3z = -1$$
$$2x + y - z = 3$$
$$-x + 2y - z = -3$$

$$\begin{bmatrix} 1 & 1 & -3 & | & -1 \\ 2 & 1 & -1 & | & 3 \\ -1 & 2 & -1 & | & -3 \end{bmatrix}$$

$$\begin{bmatrix} 1 & 1 & -3 & | & -1 \\ 0 & -1 & 5 & | & 5 \\ -1 & 2 & -1 & | & -3 \end{bmatrix} -2R_1 + R_2$$

$$\begin{bmatrix} 1 & 1 & -3 & | & -1 \\ 0 & -1 & 5 & | & 5 \\ 0 & 3 & -4 & | & -4 \end{bmatrix} R_1 + R_3$$

$$\begin{bmatrix} 1 & 1 & -3 & | & -1 \\ 0 & 1 & -5 & | & -5 \\ 0 & 3 & -4 & | & -4 \end{bmatrix} -1R_2$$

$$\begin{bmatrix} 1 & 1 & -3 & | & -1 \\ 0 & 1 & -5 & | & -5 \\ 0 & 0 & 11 & | & 11 \end{bmatrix} -3R_2 + R_3$$

$$\begin{bmatrix} 1 & 1 & -3 & | & -1 \\ 0 & 1 & -5 & | & -5 \\ 0 & 0 & 1 & | & 1 \end{bmatrix} \frac{1}{11}R_3$$

The system is

$$x + y - 3z = -1$$
$$y - 5z = -5$$
$$z = 1$$

Substitute 1 for z in the second equation.

$$y - 5z = -5$$
$$y - 5(1) = -5$$
$$y - 5 = -5$$
$$y = 0$$

Substitute 0 for y and 1 for z in the first equation.

$$x + y - 3z = -1$$
$$x + 0 - 3(1) = -1$$
$$x - 3 = -1$$
$$x = 2$$

The solution is $(2, 0, 1)$.

33.
$$x + 2y = 5$$
$$y - z = -1$$
$$2x - 3z = 0$$

Write the system in standard form.

$$x + 2y + 0z = 5$$
$$0x + y - z = -1$$
$$2x + 0y - 3z = 0$$

$$\begin{bmatrix} 1 & 2 & 0 & | & 5 \\ 0 & 1 & -1 & | & -1 \\ 2 & 0 & -3 & | & 0 \end{bmatrix}$$

$$\begin{bmatrix} 1 & 2 & 0 & | & 5 \\ 0 & 1 & -1 & | & -1 \\ 0 & -4 & -3 & | & -10 \end{bmatrix} -2R_1 + R_3$$

$$\begin{bmatrix} 1 & 2 & 0 & | & 5 \\ 0 & 1 & -1 & | & -1 \\ 0 & 0 & -7 & | & -14 \end{bmatrix} 4R_2 + R_3$$

$$\begin{bmatrix} 1 & 2 & 0 & | & 5 \\ 0 & 1 & -1 & | & -1 \\ 0 & 0 & 1 & | & 2 \end{bmatrix} -\frac{1}{7}R_3$$

The system is
$$x + 2y = 5$$
$$y - z = -1$$
$$z = 2$$

Substitute 2 for z in the second equation.

$$y - z = -1$$
$$y - 2 = -1$$
$$y = 1$$

Substitute 1 for y in the first equation.

$x + 2y = 5$
$x + 2(1) = 5$
$x + 2 = 5$
$x = 3$
The solution is (3, 1, 2).

35. $2x + 12y + 4z = 0$
$-4x + 9y - z = 11$
$-x - 12y + 3z = -2$

$$\begin{bmatrix} 2 & 12 & 4 & | & 0 \\ -4 & 9 & -1 & | & 11 \\ -1 & -12 & 3 & | & -2 \end{bmatrix}$$

$$\begin{bmatrix} 1 & 6 & 2 & | & 0 \\ -4 & 9 & -1 & | & 11 \\ -1 & -12 & 3 & | & -2 \end{bmatrix} \tfrac{1}{2}R_1$$

$$\begin{bmatrix} 1 & 6 & 2 & | & 0 \\ 0 & 33 & 7 & | & 11 \\ -1 & -12 & 3 & | & -2 \end{bmatrix} 4R_1 + R_2$$

$$\begin{bmatrix} 1 & 6 & 2 & | & 0 \\ 0 & 33 & 7 & | & 11 \\ 0 & -6 & 5 & | & -2 \end{bmatrix} R_1 + R_3$$

$$\begin{bmatrix} 1 & 6 & 2 & | & 0 \\ 0 & 1 & \tfrac{7}{33} & | & \tfrac{1}{3} \\ 0 & -6 & 5 & | & -2 \end{bmatrix} \tfrac{1}{33}R_2$$

$$\begin{bmatrix} 1 & 6 & 2 & | & 0 \\ 0 & 1 & \tfrac{7}{33} & | & \tfrac{1}{3} \\ 0 & 0 & \tfrac{69}{11} & | & 0 \end{bmatrix} 6R_2 + R_3$$

$$\begin{bmatrix} 1 & 6 & 2 & | & 0 \\ 0 & 1 & \tfrac{7}{33} & | & \tfrac{1}{3} \\ 0 & 0 & 1 & | & 0 \end{bmatrix} \tfrac{11}{69}R_3$$

The system is
$x + 6y + 2z = 0$
$y + \dfrac{7}{33}z = \dfrac{1}{3}$
$z = 0$
Substitute 0 for z in the second equation.

$y + \dfrac{7}{33}z = \dfrac{1}{3}$
$y + \dfrac{7}{33}(0) = \dfrac{1}{3}$
$y + 0 = \dfrac{1}{3}$
$y = \dfrac{1}{3}$

Substitute $\dfrac{1}{3}$ for y and 0 for z in the first equation.
$x + 6y + 2z = 0$
$x + 6\left(\dfrac{1}{3}\right) + 2(0) = 0$
$x + 2 = 0$
$x = -2$
The solution is $\left(-2, \dfrac{1}{3}, 0\right)$.

37. $6x - 2y + 8z = 26$
$-3x + y - 4z = -13$
$x + 2y + 3z = 14$

$$\begin{bmatrix} 6 & -2 & 8 & | & 26 \\ -3 & 1 & -4 & | & -13 \\ 1 & 2 & 3 & | & 14 \end{bmatrix}$$

$$\begin{bmatrix} 1 & -\tfrac{1}{3} & \tfrac{4}{3} & | & \tfrac{13}{3} \\ -3 & 1 & -4 & | & -13 \\ 1 & 2 & 3 & | & 14 \end{bmatrix} \tfrac{1}{6}R_1$$

$$\begin{bmatrix} 1 & -\tfrac{1}{3} & \tfrac{4}{3} & | & \tfrac{13}{3} \\ 0 & 0 & 0 & | & 0 \\ 1 & 2 & 3 & | & 14 \end{bmatrix}$$

Since there is a row of all zeros, the system is dependent.

39. $4p - q + r = 4$
$-6p + 3q - 2r = -5$
$2p + 5q - r = 7$

$$\begin{bmatrix} 4 & -1 & 1 & | & 4 \\ -6 & 3 & -2 & | & -5 \\ 2 & 5 & -1 & | & 7 \end{bmatrix}$$

$$\left[\begin{array}{ccc|c} 1 & -\frac{1}{4} & \frac{1}{4} & 1 \\ -6 & 3 & -2 & -5 \\ 2 & 5 & -1 & 7 \end{array}\right] \frac{1}{4}R_1$$

$$\left[\begin{array}{ccc|c} 1 & -\frac{1}{4} & \frac{1}{4} & 1 \\ 0 & \frac{3}{2} & -\frac{1}{2} & 1 \\ 2 & 5 & -1 & 7 \end{array}\right] 6R_1+R_2$$

$$\left[\begin{array}{ccc|c} 1 & -\frac{1}{4} & \frac{1}{4} & 1 \\ 0 & \frac{3}{2} & -\frac{1}{2} & 1 \\ 0 & \frac{11}{2} & -\frac{3}{2} & 5 \end{array}\right] -2R_1+R_3$$

$$\left[\begin{array}{ccc|c} 1 & -\frac{1}{4} & \frac{1}{4} & 1 \\ 0 & 1 & -\frac{1}{3} & \frac{2}{3} \\ 0 & \frac{11}{2} & -\frac{3}{2} & 5 \end{array}\right] \frac{2}{3}R_2$$

$$\left[\begin{array}{ccc|c} 1 & -\frac{1}{4} & \frac{1}{4} & 1 \\ 0 & 1 & -\frac{1}{3} & \frac{2}{3} \\ 0 & 0 & \frac{1}{3} & \frac{4}{3} \end{array}\right] -\frac{11}{2}R_2+R_3$$

$$\left[\begin{array}{ccc|c} 1 & -\frac{1}{4} & \frac{1}{4} & 1 \\ 0 & 1 & -\frac{1}{3} & \frac{2}{3} \\ 0 & 0 & 1 & 4 \end{array}\right]$$

The system is
$$x-\frac{1}{4}y+\frac{1}{4}z=1$$
$$y-\frac{1}{3}z=\frac{2}{3}$$
$$z=4$$

Substitute 4 for z in the second equation.
$$y-\frac{1}{3}z=\frac{2}{3}$$
$$y-\frac{1}{3}(4)=\frac{2}{3}$$
$$y-\frac{4}{3}=\frac{2}{3}$$
$$y=\frac{6}{3}$$
$$y=2$$

Substitute 2 for y and 4 for z in the first equation.

$$x-\frac{1}{4}y+\frac{1}{4}z=1$$
$$x-\frac{1}{4}(2)+\frac{1}{4}(4)=1$$
$$x-\frac{1}{2}+1=1$$
$$x+\frac{1}{2}=1$$
$$x=\frac{1}{2}$$

The solution is $\left(\frac{1}{2},\,2,\,4\right)$.

41. $2x-4y+3z=-12$
$3x-y+2z=-3$
$-4x+8y-6z=10$

$$\left[\begin{array}{ccc|c} 2 & -4 & 3 & -12 \\ 3 & -1 & 2 & -3 \\ -4 & 8 & -6 & 10 \end{array}\right]$$

$$\left[\begin{array}{ccc|c} 1 & -2 & \frac{3}{2} & -6 \\ 3 & -1 & 2 & -3 \\ -4 & 8 & -6 & 10 \end{array}\right] \frac{1}{2}R_1$$

$$\left[\begin{array}{ccc|c} 1 & -2 & \frac{3}{2} & -6 \\ 0 & 5 & -\frac{5}{2} & 15 \\ -4 & 8 & -6 & 10 \end{array}\right] -3R_1+R_2$$

$$\left[\begin{array}{ccc|c} 1 & -2 & \frac{3}{2} & -6 \\ 0 & 5 & -\frac{5}{2} & 15 \\ 0 & 0 & 0 & -14 \end{array}\right] 4R_1+R_3$$

Since the last row contains zeros on the left and a nonzero number on the right, the system is inconsistent and there is no solution.

43. $5x - 3y + 4z = 22$
  $-x - 15y + 10z = -15$
  $-3x + 9y - 12z = -6$

$$\begin{bmatrix} 5 & -3 & 4 & | & 22 \\ -1 & -15 & 10 & | & -15 \\ -3 & 9 & -12 & | & -6 \end{bmatrix}$$

$$\begin{bmatrix} 1 & -\frac{3}{5} & \frac{4}{5} & | & \frac{22}{5} \\ -1 & -15 & 10 & | & -15 \\ -3 & 9 & -12 & | & -6 \end{bmatrix} \frac{1}{5}R_1$$

$$\begin{bmatrix} 1 & -\frac{3}{5} & \frac{4}{5} & | & \frac{22}{5} \\ 0 & -\frac{78}{5} & \frac{54}{5} & | & -\frac{53}{5} \\ -3 & 9 & -12 & | & -6 \end{bmatrix} R_1 + R_2$$

$$\begin{bmatrix} 1 & -\frac{3}{5} & \frac{4}{5} & | & \frac{22}{5} \\ 0 & -\frac{78}{5} & \frac{54}{5} & | & -\frac{53}{5} \\ 0 & \frac{36}{5} & -\frac{48}{5} & | & \frac{36}{5} \end{bmatrix} 3R_1 + R_3$$

$$\begin{bmatrix} 1 & -\frac{3}{5} & \frac{4}{5} & | & \frac{22}{5} \\ 0 & 1 & -\frac{9}{13} & | & \frac{53}{78} \\ 0 & \frac{36}{5} & -\frac{48}{5} & | & \frac{36}{5} \end{bmatrix} -\frac{5}{78}R_2$$

$$\begin{bmatrix} 1 & -\frac{3}{5} & \frac{4}{5} & | & \frac{22}{5} \\ 0 & 1 & -\frac{9}{13} & | & \frac{53}{78} \\ 0 & 0 & -\frac{60}{13} & | & \frac{30}{13} \end{bmatrix} -\frac{36}{5}R_2 + R_3$$

$$\begin{bmatrix} 1 & -\frac{3}{5} & \frac{4}{5} & | & \frac{22}{5} \\ 0 & 1 & -\frac{9}{13} & | & \frac{53}{78} \\ 0 & 0 & 1 & | & -\frac{1}{2} \end{bmatrix}$$

The system is
$$x - \frac{3}{5}y + \frac{4}{5}z = \frac{22}{5}$$
$$y - \frac{9}{13}z = \frac{53}{78}$$
$$z = -\frac{1}{2}$$

Substitute $-\frac{1}{2}$ for z in the second equation.

$$y - \frac{9}{13}z = \frac{53}{78}$$
$$y - \frac{9}{13}\left(-\frac{1}{2}\right) = \frac{53}{78}$$
$$y + \frac{9}{26} = \frac{53}{78}$$
$$y = \frac{53}{78} - \frac{27}{78}$$
$$y = \frac{26}{78}$$
$$y = \frac{1}{3}$$

Substitute $\frac{1}{3}$ for y and $-\frac{1}{2}$ for z in the first equation.

$$x - \frac{3}{5}y + \frac{4}{5}z = \frac{22}{5}$$
$$x - \frac{3}{5}\left(\frac{1}{3}\right) + \frac{4}{5}\left(-\frac{1}{2}\right) = \frac{22}{5}$$
$$x - \frac{1}{5} - \frac{2}{5} = \frac{22}{5}$$
$$x - \frac{3}{5} = \frac{22}{5}$$
$$x = \frac{25}{5}$$
$$x = 5$$

The solution is $\left(5, \frac{1}{3}, -\frac{1}{2}\right)$.

45. No, this is the same as switching the order of the equations.

47. Let x = smallest angle
  y = remaining angle
  z = largest angle
  $z = x + 55$
  $z = y + 20$
  $x + y + z = 180$

Write the system in standard form:
  $x - z = -55$
  $y - z = -20$
  $x + y + z = 180$

$$\begin{bmatrix} 1 & 0 & -1 & | & -55 \\ 0 & 1 & -1 & | & -20 \\ 1 & 1 & 1 & | & 180 \end{bmatrix}$$

$$\begin{bmatrix} 1 & 0 & -1 & | & -55 \\ 0 & 1 & -1 & | & -20 \\ 0 & 1 & 2 & | & 235 \end{bmatrix} -1R_1 + R_3$$

$$\begin{bmatrix} 1 & 0 & -1 & | & -55 \\ 0 & 1 & -1 & | & -20 \\ 0 & 0 & 3 & | & 255 \end{bmatrix} -1R_2 + R_3$$

$$\begin{bmatrix} 1 & 0 & -1 & | & -55 \\ 0 & 1 & -1 & | & -20 \\ 0 & 0 & 1 & | & 85 \end{bmatrix} \tfrac{1}{3}R_3$$

The system is
$$x - z = -55$$
$$y - z = -20$$
$$z = 85$$
Substitute 85 for z in the second equation.
$$y - z = -20$$
$$y - 85 = -20$$
$$y = 65$$
Substitute 85 for z in the first equation.
$$x - z = -55$$
$$x - 85 = -55$$
$$x = 30$$
The angles are 30°, 65°, and 85°.

49. Let x = amount Chiquita controls,
　　y = amount Dole controls,
　　z = amount Del Monte controls
$$x = z + 12$$
$$y = 2z - 3$$
$$x + y + z = 65$$
Write the system in standard form.
$$x - z = 12$$
$$y - 2z = -3$$
$$x + y + z = 65$$

$$\begin{bmatrix} 1 & 0 & -1 & | & 12 \\ 0 & 1 & -2 & | & -3 \\ 1 & 1 & 1 & | & 65 \end{bmatrix}$$

$$\begin{bmatrix} 1 & 0 & -1 & | & 12 \\ 0 & 1 & -2 & | & -3 \\ 0 & 1 & 2 & | & 53 \end{bmatrix} -1R_1 + R_3$$

$$\begin{bmatrix} 1 & 0 & -1 & | & 12 \\ 0 & 1 & -2 & | & -3 \\ 0 & 0 & 4 & | & 56 \end{bmatrix} -1R_2 + R_3$$

$$\begin{bmatrix} 1 & 0 & -1 & | & 12 \\ 0 & 1 & -2 & | & -3 \\ 0 & 0 & 1 & | & 14 \end{bmatrix} \tfrac{1}{4}R_3$$

The system is
$$x - z = 12$$
$$y - 2z = -3$$
$$z = 14$$
Substitute 14 for z in the second equation.
$$y - 2z = -3$$
$$y - 2(14) = -3$$
$$y - 28 = -3$$
$$y = 25$$
Substitute 14 for z in the first equation.
$$x - z = 12$$
$$x - 14 = 12$$
$$x = 26$$
Thus, Del Monte controls 14% of the bananas, Dole controls 25% and Chiquita controls 26%, with the remaining 100% – 65% = 35% being controlled by "other."

51. a.　$A \cup B = \{1, 2, 3, 4, 5, 6, 9, 10\}$

　　b.　$A \cap B = \{4, 6\}$

52. a.　
　　　　$-2 \qquad 4$

　　b.　$\{x | -2 < x \le 4\}$

　　c.　$(-2, 4]$

53. A graph is the set of points whose coordinates satisfy an equation.

54.　$f(x) = -2x^2 + 4x - 6$
$$f(-5) = -2(-5)^2 + 4(-5) - 6$$
$$= -50 - 20 - 6$$
$$= -76$$

Exercise Set 4.5

1. Answers will vary.

3. If $D = 0$ and D_x, D_y, and D_z also equal 0, the system is dependent.

5. $\begin{vmatrix} 5 & 3 \\ -1 & 4 \end{vmatrix} = 5(4) - (-1)(3)$

$= 20 - (-3)$

$= 20 + 3$

$= 23$

7. $\begin{vmatrix} \frac{1}{2} & 3 \\ 2 & -4 \end{vmatrix} = \frac{1}{2}(-4) - (2)(3) = -2 - 6 = -8$

9. $\begin{vmatrix} 3 & 2 & 0 \\ 0 & 5 & 3 \\ -1 & 4 & 2 \end{vmatrix} = 3\begin{vmatrix} 5 & 3 \\ 4 & 2 \end{vmatrix} - 0\begin{vmatrix} 2 & 0 \\ 4 & 2 \end{vmatrix} + (-1)\begin{vmatrix} 2 & 0 \\ 5 & 3 \end{vmatrix}$

$= 3(10 - 12) - 0(4 - 0) - 1(6 - 0)$

$= 3(-2) - 0(4) - 1(6)$

$= -6 - 0 - 6$

$= -12$

11. $\begin{vmatrix} 2 & 3 & 1 \\ 1 & -3 & -6 \\ -4 & 5 & 9 \end{vmatrix}$

$= 2\begin{vmatrix} -3 & -6 \\ 5 & 9 \end{vmatrix} - 1\begin{vmatrix} 3 & 1 \\ 5 & 9 \end{vmatrix} + (-4)\begin{vmatrix} 3 & 1 \\ -3 & -6 \end{vmatrix}$

$= 2[-27 - (-30)] - 1(27 - 5) - 4[-18 - (-3)]$

$= 2(3) - 1(22) - 4(-15)$

$= 6 - 22 + 60$

$= 44$

13. $x + 2y = 5$

$x - 2y = 1$

To solve, first calculate D, D_x, and D_y.

$D = \begin{vmatrix} 1 & 2 \\ 1 & -2 \end{vmatrix} = (1)(-2) - (1)(2) = -2 - 2 = -4$

$D_x = \begin{vmatrix} 5 & 2 \\ 1 & -2 \end{vmatrix}$

$= (5)(-2) - (1)(2)$

$= -10 - 2$

$= -12$

$D_y = \begin{vmatrix} 1 & 5 \\ 1 & 1 \end{vmatrix} = (1)(1) - (1)(5) = 1 - 5 = -4$

$x = \dfrac{D_x}{D} = \dfrac{-12}{-4} = 3$ and $y = \dfrac{D_y}{D} = \dfrac{-4}{-4} = 1$

The solution is (3, 1).

15. $x - 2y = -1$

$x + 3y = 9$

To solve, first calculate D, D_x, and D_y.

$D = \begin{vmatrix} 1 & -2 \\ 1 & 3 \end{vmatrix} = (1)(3) - (1)(-2) = 3 + 2 = 5$

$D_x = \begin{vmatrix} -1 & -2 \\ 9 & 3 \end{vmatrix}$

$= (-1)(3) - (9)(-2)$

$= -3 + 18$

$= 15$

$D_y = \begin{vmatrix} 1 & -1 \\ 1 & 9 \end{vmatrix} = (1)(9) - (1)(-1) = 9 + 1 = 10$

$x = \dfrac{D_x}{D} = \dfrac{15}{5} = 3$ and $y = \dfrac{D_y}{D} = \dfrac{10}{5} = 2$

The solution is (3, 2).

17. $3a + 4b = 8$

$2a - 3b = 9$

To solve, first calculate D, D_a, and D_b.

$D = \begin{vmatrix} 3 & 4 \\ 2 & -3 \end{vmatrix}$

$= (3)(-3) - (2)(4)$

$= -9 - 8$

$= -17$

$D_a = \begin{vmatrix} 8 & 4 \\ 9 & -3 \end{vmatrix}$

$= (8)(-3) - (9)(4)$

$= -24 - 36$

$= -60$

$D_b = \begin{vmatrix} 3 & 8 \\ 2 & 9 \end{vmatrix} = (3)(9) - (2)(8) = 27 - 16 = 11$

$a = \dfrac{D_a}{D} = \dfrac{-60}{-17} = \dfrac{60}{17}$ and

$b = \dfrac{D_b}{D} = \dfrac{11}{-17} = -\dfrac{11}{17}$

The solution is $\left(\dfrac{60}{17}, -\dfrac{11}{17} \right)$.

19. $\quad 2x = y + 5$

$6x + 2y = -5$

Rewrite the system in standard form:

$2x - y = 5$

$6x + 2y = -5$

Now calculate D, D_x, and D_y.

$$D = \begin{vmatrix} 2 & -1 \\ 6 & 2 \end{vmatrix} = (2)(2) - (6)(-1) = 4 + 6 = 10$$

$$D_x = \begin{vmatrix} 5 & -1 \\ -5 & 2 \end{vmatrix}$$
$$= (5)(2) - (-5)(-1)$$
$$= 10 - 5$$
$$= 5$$

$$D_y = \begin{vmatrix} 2 & 5 \\ 6 & -5 \end{vmatrix}$$
$$= (2)(-5) - (6)(5)$$
$$= -10 - 30$$
$$= -40$$

$$x = \frac{D_x}{D} = \frac{5}{10} = \frac{1}{2} \text{ and } y = \frac{D_y}{D} = \frac{-40}{10} = -4$$

The solution is $\left(\frac{1}{2}, -4 \right)$.

21. $3r = -4s - 6$
$3s = -5r + 1$
Rewrite the system in standard form.
$3r + 4s = -6$
$5r + 3s = 1$
Now calculate D, D_r, and D_s.

$$D = \begin{vmatrix} 3 & 4 \\ 5 & 3 \end{vmatrix} = (3)(3) - (5)(4) = 9 - 20 = -11$$

$$D_r = \begin{vmatrix} -6 & 4 \\ 1 & 3 \end{vmatrix}$$
$$= (-6)(3) - (1)(4)$$
$$= -18 - 4$$
$$= -22$$

$$D_s = \begin{vmatrix} 3 & -6 \\ 5 & 1 \end{vmatrix}$$
$$= (3)(1) - (5)(-6)$$
$$= 3 + 30$$
$$= 33$$

$$r = \frac{D_r}{D} = \frac{-22}{-11} = 2 \text{ and } s = \frac{D_s}{D} = \frac{33}{-11} = -3$$

The solution is $(2, -3)$.

23. $6.3x - 4.5y = -9.9$
$-9.1x + 3.2y = -2.2$
Here, you can work with decimals in the determinants. If you do not want to use decimals, then you need to multiply each equation by 10 to clear the decimals. First, calculate D, D_x, and D_y.

$$D = \begin{vmatrix} 6.3 & -4.5 \\ -9.1 & 3.2 \end{vmatrix}$$
$$= (6.3)(3.2) - (-9.1)(-4.5)$$
$$= 20.16 - 40.95$$
$$= -20.79$$

$$D_x = \begin{vmatrix} -9.9 & -4.5 \\ -2.2 & 3.2 \end{vmatrix}$$
$$= (-9.9)(3.2) - (-2.2)(-4.5)$$
$$= -31.68 - 9.90$$
$$= -41.58$$

$$D_y = \begin{vmatrix} 6.3 & -9.9 \\ -9.1 & -2.2 \end{vmatrix}$$
$$= (6.3)(-2.2) - (-9.1)(-9.9)$$
$$= -13.86 - 90.09$$
$$= -103.95$$

$$x = \frac{D_x}{D} = \frac{-41.58}{-20.79} = 2 \text{ and}$$

$$y = \frac{D_y}{D} = \frac{-103.95}{-20.79} = 5$$

The solution is $(2, 5)$.

25. $x + y - z = -2$
$x + 0y + z = 5$
$2x - y + 2z = 11$
To solve, first calculate D, D_x, D_y, and D_z.

$$D = \begin{vmatrix} 1 & 1 & -1 \\ 1 & 0 & 1 \\ 2 & -1 & 2 \end{vmatrix}$$

$$= 1\begin{vmatrix} 0 & 1 \\ -1 & 2 \end{vmatrix} - 1\begin{vmatrix} 1 & -1 \\ -1 & 2 \end{vmatrix} + 2\begin{vmatrix} 1 & -1 \\ 0 & 1 \end{vmatrix}$$
$$= 1(0 + 1) - 1(2 - 1) + 2(1 - 0)$$
$$= 1(1) - 1(1) + 2(1)$$
$$= 1 - 1 + 2$$
$$= 2$$

$$D_x = \begin{vmatrix} -2 & 1 & -1 \\ 5 & 0 & 1 \\ 11 & -1 & 2 \end{vmatrix}$$

$$= -2\begin{vmatrix} 0 & 1 \\ -1 & 2 \end{vmatrix} - 5\begin{vmatrix} 1 & -1 \\ -1 & 2 \end{vmatrix} + 11\begin{vmatrix} 1 & -1 \\ 0 & 1 \end{vmatrix}$$

$$= -2(0+1) - 5(2-1) + 11(1-0)$$

$$= -2(1) - 5(1) + 11(1)$$

$$= -2 - 5 + 11$$

$$= 4$$

$$D_y = \begin{vmatrix} 1 & -2 & -1 \\ 1 & 5 & 1 \\ 2 & 11 & 2 \end{vmatrix}$$

$$= 1\begin{vmatrix} 5 & 1 \\ 11 & 2 \end{vmatrix} - 1\begin{vmatrix} -2 & -1 \\ 11 & 2 \end{vmatrix} + 2\begin{vmatrix} -2 & -1 \\ 5 & 1 \end{vmatrix}$$

$$= 1(10-11) - 1(-4+11) + 2(-2+5)$$

$$= 1(-1) - 1(7) + 2(3)$$

$$= -1 - 7 + 6$$

$$= -2$$

$$D_z = \begin{vmatrix} 1 & 1 & -2 \\ 1 & 0 & 5 \\ 2 & -1 & 11 \end{vmatrix}$$

$$= 1\begin{vmatrix} 0 & 5 \\ -1 & 11 \end{vmatrix} - 1\begin{vmatrix} 1 & -2 \\ -1 & 11 \end{vmatrix} + 2\begin{vmatrix} 1 & -2 \\ 0 & 5 \end{vmatrix}$$

$$= 1(0+5) - 1(11-2) + 2(5-0)$$

$$= 1(5) - 1(9) + 2(5)$$

$$= 5 - 9 + 10$$

$$= 6$$

$$x = \frac{D_x}{D} = \frac{4}{2} = 2, \ y = \frac{D_y}{D} = \frac{-2}{2} = -1, \text{ and}$$

$$z = \frac{D_z}{D} = \frac{6}{2} = 3$$

The solution is $(2, -1, 3)$

27. $\quad -x + y + 0z = 1$
$\quad\quad\ \ 0x + y - z = 2$
$\quad\quad\ \ x + 0y + z = -2$

To solve, first calculate D, D_x, D_y, and D_z.

$$D = \begin{vmatrix} -1 & 1 & 0 \\ 0 & 1 & -1 \\ 1 & 0 & 1 \end{vmatrix}$$

$$= (-1)\begin{vmatrix} 1 & -1 \\ 0 & 1 \end{vmatrix} - 0\begin{vmatrix} 1 & 0 \\ 0 & 1 \end{vmatrix} + 1\begin{vmatrix} 1 & 0 \\ 1 & -1 \end{vmatrix}$$

$$= (-1)(1-0) - 0(1-0) + 1(-1-0)$$

$$= (-1)(1) - 0(1) + 1(-1)$$

$$= -1 - 0 - 1$$

$$= -2$$

$$D_x = \begin{vmatrix} 1 & 1 & 0 \\ 2 & 1 & -1 \\ -2 & 0 & 1 \end{vmatrix}$$

$$= 1\begin{vmatrix} 1 & -1 \\ 0 & 1 \end{vmatrix} - 2\begin{vmatrix} 1 & 0 \\ 0 & 1 \end{vmatrix} + (-2)\begin{vmatrix} 1 & 0 \\ 1 & -1 \end{vmatrix}$$

$$= 1(1-0) - 2(1-0) - 2(-1-0)$$

$$= 1(1) - 2(1) - 2(-1)$$

$$= 1 - 2 + 2$$

$$= 1$$

$$D_y = \begin{vmatrix} -1 & 1 & 0 \\ 0 & 2 & -1 \\ 1 & -2 & 1 \end{vmatrix}$$

$$= (-1)\begin{vmatrix} 2 & -1 \\ -2 & 1 \end{vmatrix} - 0\begin{vmatrix} 1 & 0 \\ -2 & 1 \end{vmatrix} + 1\begin{vmatrix} 1 & 0 \\ 2 & -1 \end{vmatrix}$$

$$= (-1)(2-2) - 0(1-0) + 1(-1-0)$$

$$= (-1)(0) - 0(1) + 1(-1)$$

$$= 0 - 0 - 1$$

$$= -1$$

$$D_z = \begin{vmatrix} -1 & 1 & 1 \\ 0 & 1 & 2 \\ 1 & 0 & -2 \end{vmatrix}$$

$$= -1\begin{vmatrix} 1 & 2 \\ 0 & -2 \end{vmatrix} - 0\begin{vmatrix} 1 & 1 \\ 0 & -2 \end{vmatrix} + 1\begin{vmatrix} 1 & 1 \\ 1 & 2 \end{vmatrix}$$

$$= -1(-2-0) - 0(-2-0) + 1(2-1)$$

$$= -1(-2) - 0(-2) + 1(1)$$

$$= 2 + 0 + 1$$

$$= 3$$

$$x = \frac{D_x}{D} = \frac{1}{-2} = -\frac{1}{2}, \ y = \frac{D_y}{D} = \frac{-1}{-2} = \frac{1}{2}, \text{ and}$$

$$z = \frac{D_z}{D} = \frac{3}{-2} = -\frac{3}{2}$$

The solution is $\left(-\frac{1}{2}, \ \frac{1}{2}, \ -\frac{3}{2} \right)$.

29. $2x + 2y + 2z = 0$
$-x - 3y + 7z = 15$
$3x + y + 4z = 21$

To solve, first calculate D, D_x, D_y, and D_z

$$D = \begin{vmatrix} 2 & 2 & 2 \\ -1 & -3 & 7 \\ 3 & 1 & 4 \end{vmatrix}$$

$$= 2\begin{vmatrix} -3 & 7 \\ 1 & 4 \end{vmatrix} - (-1)\begin{vmatrix} 2 & 2 \\ 1 & 4 \end{vmatrix} + 3\begin{vmatrix} 2 & 2 \\ -3 & 7 \end{vmatrix}$$

$$= 2(-12 - 7) + 1(8 - 2) + 3(14 + 6)$$
$$= 2(-19) + 1(6) + 3(20)$$
$$= -38 + 6 + 60$$
$$= 28$$

$$D_x = \begin{vmatrix} 0 & 2 & 2 \\ 15 & -3 & 7 \\ 21 & 1 & 4 \end{vmatrix}$$

$$= 0\begin{vmatrix} -3 & 7 \\ 1 & 4 \end{vmatrix} - 15\begin{vmatrix} 2 & 2 \\ 1 & 4 \end{vmatrix} + 21\begin{vmatrix} 2 & 2 \\ -3 & 7 \end{vmatrix}$$

$$= 0(-12 - 7) - 15(8 - 2) + 21(14 + 6)$$
$$= 0(-19) - 15(6) + 21(20)$$
$$= 0 - 90 + 420$$
$$= 330$$

$$D_y = \begin{vmatrix} 2 & 0 & 2 \\ -1 & 15 & 7 \\ 3 & 21 & 4 \end{vmatrix}$$

$$= 2\begin{vmatrix} 15 & 7 \\ 21 & 4 \end{vmatrix} - (-1)\begin{vmatrix} 0 & 2 \\ 21 & 4 \end{vmatrix} + 3\begin{vmatrix} 0 & 2 \\ 15 & 7 \end{vmatrix}$$

$$= 2(60 - 147) + 1(0 - 42) + 3(0 - 30)$$
$$= 2(-87) + 1(-42) + 3(-30)$$
$$= -174 - 42 - 90$$
$$= -306$$

$$D_z = \begin{vmatrix} 2 & 2 & 0 \\ -1 & -3 & 15 \\ 3 & 1 & 21 \end{vmatrix}$$

$$= 2\begin{vmatrix} -3 & 15 \\ 1 & 21 \end{vmatrix} - (-1)\begin{vmatrix} 2 & 0 \\ 1 & 21 \end{vmatrix} + 3\begin{vmatrix} 2 & 0 \\ -3 & 15 \end{vmatrix}$$

$$= 2(-63 - 15) + 1(42 - 0) + 3(30 - 0)$$
$$= 2(-78) + 1(42) + 3(30)$$
$$= -156 + 42 + 90$$
$$= -24$$

$$x = \frac{D_x}{D} = \frac{330}{28} = \frac{165}{14},$$

$$y = \frac{D_y}{D} = \frac{-306}{28} = -\frac{153}{14}, \text{ and}$$

$$z = \frac{D_z}{D} = \frac{-24}{28} = -\frac{6}{7}$$

The solution is $\left(\frac{165}{14}, \ -\frac{153}{14}, \ -\frac{6}{7} \right)$.

31. $a - b + 2c = 3$
$a - b + c = 1$
$2a + b + 2c = 2$

To solve, first calculate D, D_a, D_b, and D_c.

$$D = \begin{vmatrix} 1 & -1 & 2 \\ 1 & -1 & 1 \\ 2 & 1 & 2 \end{vmatrix}$$

$$= 1\begin{vmatrix} -1 & 1 \\ 1 & 2 \end{vmatrix} - 1\begin{vmatrix} -1 & 2 \\ 1 & 2 \end{vmatrix} + 2\begin{vmatrix} -1 & 2 \\ -1 & 1 \end{vmatrix}$$

$$= 1(-2 - 1) - 1(-2 - 2) + 2(-1 + 2)$$
$$= 1(-3) - 1(-4) + 2(1)$$
$$= -3 + 4 + 2$$
$$= 3$$

$$D_a = \begin{vmatrix} 3 & -1 & 2 \\ 1 & -1 & 1 \\ 2 & 1 & 2 \end{vmatrix}$$

$$= 3\begin{vmatrix} -1 & 1 \\ 1 & 2 \end{vmatrix} - 1\begin{vmatrix} -1 & 2 \\ 1 & 2 \end{vmatrix} + 2\begin{vmatrix} -1 & 2 \\ -1 & 1 \end{vmatrix}$$

$$= 3(-2 - 1) - 1(-2 - 2) + 2(-1 + 2)$$
$$= 3(-3) - 1(-4) + 2(1)$$
$$= -9 + 4 + 2$$
$$= -3$$

$$D_b = \begin{vmatrix} 1 & 3 & 2 \\ 1 & 1 & 1 \\ 2 & 2 & 2 \end{vmatrix}$$

$$= 1\begin{vmatrix} 1 & 1 \\ 2 & 2 \end{vmatrix} - 1\begin{vmatrix} 3 & 2 \\ 2 & 2 \end{vmatrix} + 2\begin{vmatrix} 3 & 2 \\ 1 & 1 \end{vmatrix}$$

$$= 1(2 - 2) - 1(6 - 4) + 2(3 - 2)$$
$$= 1(0) - 1(2) + 2(1)$$
$$= 0 - 2 + 2$$
$$= 0$$

$$D_c = \begin{vmatrix} 1 & -1 & 3 \\ 1 & -1 & 1 \\ 2 & 1 & 2 \end{vmatrix}$$

$$= 1\begin{vmatrix} -1 & 1 \\ 1 & 2 \end{vmatrix} - 1\begin{vmatrix} -1 & 3 \\ 1 & 2 \end{vmatrix} + 2\begin{vmatrix} -1 & 3 \\ -1 & 1 \end{vmatrix}$$

$$= 1(-2-1) - 1(-2-3) + 2(-1+3)$$

$$= 1(-3) - 1(-5) + 2(2)$$

$$= -3 + 5 + 4$$

$$= 6$$

$$a = \frac{D_a}{D} = \frac{-3}{3} = -1, \; b = \frac{D_b}{D} = \frac{0}{3} = 0, \text{ and}$$

$$c = \frac{D_c}{D} = \frac{6}{3} = 2$$

The solution is $(-1, 0, 2)$.

33. $a + 2b + c = 1$
$a - b + c = 1$
$2a + b + 2c = 2$
To solve, first calculate D, D_a, D_b, and D_c.

$$D = \begin{vmatrix} 1 & 2 & 1 \\ 1 & -1 & 1 \\ 2 & 1 & 2 \end{vmatrix}$$

$$= 1\begin{vmatrix} -1 & 1 \\ 1 & 2 \end{vmatrix} - 1\begin{vmatrix} 2 & 1 \\ 1 & 2 \end{vmatrix} + 2\begin{vmatrix} 2 & 1 \\ -1 & 1 \end{vmatrix}$$

$$= 1(-2-1) - 1(4-1) + 2(2+1)$$

$$= 1(-3) - 1(3) + 2(3)$$

$$= -3 - 3 + 6$$

$$= 0$$

$$D_a = \begin{vmatrix} 1 & 2 & 1 \\ 1 & -1 & 1 \\ 2 & 1 & 2 \end{vmatrix}$$

$$= 1\begin{vmatrix} -1 & 1 \\ 1 & 2 \end{vmatrix} - 1\begin{vmatrix} 2 & 1 \\ 1 & 2 \end{vmatrix} + 2\begin{vmatrix} 2 & 1 \\ -1 & 1 \end{vmatrix}$$

$$= 1(-2-1) - 1(4-1) + 2(2+1)$$

$$= 1(-3) - 1(3) + 2(3)$$

$$= -3 - 3 + 6$$

$$= 0$$

$$D_b = \begin{vmatrix} 1 & 1 & 1 \\ 1 & 1 & 1 \\ 2 & 2 & 2 \end{vmatrix}$$

$$= 1\begin{vmatrix} 1 & 1 \\ 2 & 2 \end{vmatrix} - 1\begin{vmatrix} 1 & 1 \\ 2 & 2 \end{vmatrix} + 2\begin{vmatrix} 1 & 1 \\ 1 & 1 \end{vmatrix}$$

$$= 1(2-2) - 1(2-2) + 2(1-1)$$

$$= 1(0) - 1(0) + 2(0)$$

$$= 0 - 0 + 0$$

$$= 0$$

$$D_c = \begin{vmatrix} 1 & 2 & 1 \\ 1 & -1 & 1 \\ 2 & 1 & 2 \end{vmatrix}$$

$$= 1\begin{vmatrix} -1 & 1 \\ 1 & 2 \end{vmatrix} - 1\begin{vmatrix} 2 & 1 \\ 1 & 2 \end{vmatrix} + 2\begin{vmatrix} 2 & 1 \\ -1 & 1 \end{vmatrix}$$

$$= 1(-2-1) - 1(4-1) + 2(2+1)$$

$$= 1(-3) - 1(3) + 2(3)$$

$$= -3 - 3 + 6$$

$$= 0$$

Since $D = 0$, $D_a = 0$, $D_b = 0$, and $D_c = 0$, there are an infinite number of solutions to the system and it is a dependent system.

35. $1.1x + 2.3y - 4.0z = -9.2$
$-2.3x + 0y + 4.6z = 6.9$
$0x - 8.2y - 7.5z = -6.8$

Here, you can work with decimals in the determinants. If you do not want to use decimals, then you need to multiply each equation by 10 to clear the decimals. To solve, first calculate D, D_x, D_y, and D_z.

$$D = \begin{vmatrix} 1.1 & 2.3 & -4.0 \\ -2.3 & 0 & 4.6 \\ 0 & -8.2 & -7.5 \end{vmatrix}$$

$$= 1.1 \begin{vmatrix} 0 & 4.6 \\ -8.2 & -7.5 \end{vmatrix} - (-2.3) \begin{vmatrix} 2.3 & -4.0 \\ -8.2 & -7.5 \end{vmatrix} + 0 \begin{vmatrix} 2.3 & -4.0 \\ 0 & 4.6 \end{vmatrix}$$

$$= 1.1(0 + 37.72) + 2.3(-17.25 - 32.8) + 0(10.58 - 0)$$

$$= 1.1(37.72) + 2.3(-50.05) + 0(10.58)$$

$$= 41.492 - 115.115 + 0$$

$$= -73.623$$

$$D_x = \begin{vmatrix} -9.2 & 2.3 & -4.0 \\ 6.9 & 0 & 4.6 \\ -6.8 & -8.2 & -7.5 \end{vmatrix}$$

$$= -9.2 \begin{vmatrix} 0 & 4.6 \\ -8.2 & -7.5 \end{vmatrix} - 6.9 \begin{vmatrix} 2.3 & -4.0 \\ -8.2 & -7.5 \end{vmatrix} + (-6.8) \begin{vmatrix} 2.3 & -4.0 \\ 0 & 4.6 \end{vmatrix}$$

$$= -9.2(0 + 37.72) - 6.9(-17.25 - 32.8) - 6.8(10.58 - 0)$$

$$= -9.2(37.72) - 6.9(-50.05) - 6.8(10.58)$$

$$= -347.024 + 345.345 - 71.944$$

$$= -73.623$$

$$D_y = \begin{vmatrix} 1.1 & -9.2 & -4.0 \\ -2.3 & 6.9 & 4.6 \\ 0 & -6.8 & -7.5 \end{vmatrix}$$

$$= 1.1 \begin{vmatrix} 6.9 & 4.6 \\ -6.8 & -7.5 \end{vmatrix} - (-2.3) \begin{vmatrix} -9.2 & -4.0 \\ -6.8 & -7.5 \end{vmatrix} + 0 \begin{vmatrix} -9.2 & -4.0 \\ 6.9 & 4.6 \end{vmatrix}$$

$$= 1.1(-51.75 + 31.28) + 2.3(69 - 27.2) + 0(-42.32 + 27.6)$$

$$= 1.1(-20.47) + 2.3(41.8) + 0(-14.72)$$

$$= -22.517 + 96.14 + 0$$

$$= 73.623$$

$$D_z = \begin{vmatrix} 1.1 & 2.3 & -9.2 \\ -2.3 & 0 & 6.9 \\ 0 & -8.2 & -6.8 \end{vmatrix}$$

$$= 1.1 \begin{vmatrix} 0 & 6.9 \\ -8.2 & -6.8 \end{vmatrix} - (-2.3) \begin{vmatrix} 2.3 & -9.2 \\ -8.2 & -6.8 \end{vmatrix} + 0 \begin{vmatrix} 2.3 & -9.2 \\ 0 & 6.9 \end{vmatrix}$$

$$= 1.1(0 + 56.58) + 2.3(-15.64 - 75.44) + 0(15.87 - 0)$$

$$= 1.1(56.58) + 2.3(-91.08) + 0(15.87)$$

$$= 62.238 - 209.484 + 0$$

$$= 147.246$$

$x = \dfrac{D_x}{D} = \dfrac{-73.623}{-73.623} = 1$, $y = \dfrac{D_y}{D} = \dfrac{73.623}{-73.623} = -1$, and $z = \dfrac{D_z}{D} = \dfrac{-147.246}{-73.623} = 2$

The solution is $(1, -1, 2)$.

37.
$$x + y + z = 1$$
$$2x + 2y + 2z = 2$$
$$3x + 3y + 3z = 3$$
To solve, first calculate D, D_x, D_y, and D_z.

$$D = \begin{vmatrix} 1 & 1 & 1 \\ 2 & 2 & 2 \\ 3 & 3 & 3 \end{vmatrix}$$

$$= 1\begin{vmatrix} 2 & 2 \\ 3 & 3 \end{vmatrix} - 2\begin{vmatrix} 1 & 1 \\ 3 & 3 \end{vmatrix} + 3\begin{vmatrix} 1 & 1 \\ 2 & 2 \end{vmatrix}$$
$$= 1(6-6) - 2(3-3) + 3(2-2)$$
$$= 1(0) - 2(0) + 3(0)$$
$$= 0 - 0 + 0$$
$$= 0$$

$$D_x = \begin{vmatrix} 1 & 1 & 1 \\ 2 & 2 & 2 \\ 3 & 3 & 3 \end{vmatrix}$$

This is identical to the determinant computed above. Thus, $D_x = 0$.
Similarly,

$$D_y = \begin{vmatrix} 1 & 1 & 1 \\ 2 & 2 & 2 \\ 3 & 3 & 3 \end{vmatrix} = 0$$

$$D_z = \begin{vmatrix} 1 & 1 & 1 \\ 2 & 2 & 2 \\ 3 & 3 & 3 \end{vmatrix} = 0$$

Since $D = 0$, $D_x = 0$, $D_y = 0$, and $D_z = 0$, the system is dependent so there are an infinite number of solutions.

39.
$$4x - 3y + 8z = 12$$
$$2x - \frac{3}{2}y + 4z = 11$$
$$x - 5z = -10$$
To clear the system of fractions, multiply the second equation by 2.
$$4x - 3y + 8z = 12$$
$$4x - 3y + 8z = 22$$
$$x + 0y - 5z = -10$$
To solve, first calculate D, D_x, D_y, and D_z.

$$D = \begin{vmatrix} 4 & -3 & 8 \\ 4 & -3 & 8 \\ 1 & 0 & -5 \end{vmatrix}$$

$$= 4\begin{vmatrix} -3 & 8 \\ 0 & -5 \end{vmatrix} - 4\begin{vmatrix} -3 & 8 \\ 0 & -5 \end{vmatrix} + 1\begin{vmatrix} -3 & 8 \\ -3 & 8 \end{vmatrix}$$
$$= 4(15 - 0) - 4(15 - 0) + 1(-24 + 24)$$
$$= 4(15) - 4(15) + 1(0)$$
$$= 60 - 60 + 0$$
$$= 0$$

$$D_x = \begin{vmatrix} 12 & -3 & 8 \\ 22 & -3 & 8 \\ -10 & 0 & -5 \end{vmatrix}$$

$$= 12\begin{vmatrix} -3 & 8 \\ 0 & -5 \end{vmatrix} - 22\begin{vmatrix} -3 & 8 \\ 0 & -5 \end{vmatrix} + (-10)\begin{vmatrix} -3 & 8 \\ -3 & 8 \end{vmatrix}$$
$$= 12(15 - 0) - 22(15 - 0) - 10(-24 + 24)$$
$$= 12(15) - 22(15) - 10(0)$$
$$= 180 - 330 - 0$$
$$= -150$$
Since $D = 0$ and $D_x = -150 \neq 0$, there is no solution to the system and it is an inconsistent system.

41.
$$0.2x - 0.1y - 0.3z = -0.1$$
$$0.2x - 0.1y + 0.1z = -0.9$$
$$0.1x + 0.2y - 0.4z = 1.7$$
To clear decimals multiply each equation by 10.
$$2x - y - 3x = -1$$
$$2x - y + z = -9$$
$$x + 2y - 4z = 17$$
To solve, first calculate D, D_x, D_y, and D_z.

$$D = \begin{vmatrix} 2 & -1 & -3 \\ 2 & -1 & 1 \\ 1 & 2 & -4 \end{vmatrix}$$

$$= 2\begin{vmatrix} -1 & 1 \\ 2 & -4 \end{vmatrix} - 2\begin{vmatrix} -1 & -3 \\ 2 & -4 \end{vmatrix} + 1\begin{vmatrix} -1 & -3 \\ -1 & 1 \end{vmatrix}$$
$$= 2(4 - 2) - 2(4 + 6) + 1(-1 - 3)$$
$$= 2(2) - 2(10) + 1(-4)$$
$$= 4 - 20 - 4$$
$$= -20$$

$$D_x = \begin{vmatrix} -1 & -1 & -3 \\ -9 & -1 & 1 \\ 17 & 2 & -4 \end{vmatrix}$$

$$= -1 \begin{vmatrix} -1 & 1 \\ 2 & -4 \end{vmatrix} - (-9) \begin{vmatrix} -1 & -3 \\ 2 & -4 \end{vmatrix} + 17 \begin{vmatrix} -1 & -3 \\ -1 & 1 \end{vmatrix}$$

$$= -1(4-2) + 9(4+6) + 17(-1-3)$$

$$= -1(2) + 9(10) + 17(-4)$$

$$= -2 + 90 - 68$$

$$= 20$$

$$D_y = \begin{vmatrix} 2 & -1 & -3 \\ 2 & -9 & 1 \\ 1 & 17 & -4 \end{vmatrix}$$

$$= 2 \begin{vmatrix} -9 & 1 \\ 17 & -4 \end{vmatrix} - 2 \begin{vmatrix} -1 & -3 \\ 17 & -4 \end{vmatrix} + 1 \begin{vmatrix} -1 & -3 \\ -9 & 1 \end{vmatrix}$$

$$= 2(36-17) - 2(4+51) + 1(-1-27)$$

$$= 2(19) - 2(55) + 1(-28)$$

$$= 38 - 110 - 28$$

$$= -100$$

$$D_z = \begin{vmatrix} 2 & -1 & -1 \\ 2 & -1 & -9 \\ 1 & 2 & 17 \end{vmatrix}$$

$$= 2 \begin{vmatrix} -1 & -9 \\ 2 & 17 \end{vmatrix} - 2 \begin{vmatrix} -1 & -1 \\ 2 & 17 \end{vmatrix} + 1 \begin{vmatrix} -1 & -1 \\ -1 & -9 \end{vmatrix}$$

$$= 2(-17+18) - 2(-17+2) + 1(9-1)$$

$$= 2(1) - 2(-15) + 1(8)$$

$$= 2 + 30 + 8$$

$$= 40$$

$$x = \frac{D_x}{D} = \frac{20}{-20} = -1, \ y = \frac{D_y}{D} = \frac{-100}{-20} = 5, \text{ and } z = \frac{D_z}{D} = \frac{40}{-20} = -2$$

The solution is $(-1, 5, -2)$.

43. $\begin{vmatrix} a_1 & b_1 \\ a_2 & b_2 \end{vmatrix} = a_1 b_2 - a_2 b_1$

$\begin{vmatrix} b_1 & a_1 \\ b_2 & a_2 \end{vmatrix} = b_1 a_2 - b_2 a_1 = a_2 b_1 - a_1 b_2$

The second result is the negative of the first result. Thus, the second determinant has the opposite sign.

45. 0

47. Yes, the determinant will become the opposite of the original value.

49. Yes, the determinant will become the opposite of the original value.

51. $\begin{vmatrix} 4 & 6 \\ -2 & y \end{vmatrix} = 32$

$(4)(y) - (-2)(6) = 32$

$4y + 12 = 32$

$4y = 20$

$y = \dfrac{20}{4} = 5$

53. $\begin{vmatrix} 3 & x & -2 \\ 0 & 5 & -6 \\ -1 & 4 & -7 \end{vmatrix} = -31$

$3\begin{vmatrix} 5 & -6 \\ 4 & -7 \end{vmatrix} - 0\begin{vmatrix} x & -2 \\ 4 & -7 \end{vmatrix} + (-1)\begin{vmatrix} x & -2 \\ 5 & -6 \end{vmatrix} = -31$

$3(-35 + 24) - 0(-7x + 8) - 1(-6x + 10) = -31$

$3(-11) - 0(-7x + 8) - 1(-6x + 10) = -31$

$-33 - 0 + 6x - 10 = -31$

$6x - 43 = -31$

$6x = 12$

$x = \dfrac{12}{6}$

$= 2$

55. a. To eliminate y, multiply the first equation by b_2 and the second equation by $-b_1$ and then add.

$b_2[a_1x + b_1y = c_1]$

$-b_1[a_2x + b_2y = c_2]$

gives

$a_1b_2x + b_1b_2y = c_1b_2$

$\underline{-a_2b_1x - b_1b_2y = -c_2b_1}$

Add: $(a_1b_2 - a_2b_1)x = c_1b_2 - c_2b_1$

$x = \dfrac{c_1b_2 - c_2b_1}{a_1b_2 - a_2b_1}$

b. To eliminate x, multiply the first equation by $-a_2$ and the second equation by a_1 and then add.

$-a_2[a_1x + b_1y = c_1]$

$a_1[a_2x + b_2y = c_2]$

gives

$-a_1a_2x - a_2b_1y = -a_2c_1$

$\underline{a_1a_2x + a_1b_2y = a_1c_2}$

Add: $(a_1b_2 - a_2b_1)y = a_1c_2 - a_2c_1$

$y = \dfrac{a_1c_2 - a_2c_1}{a_1b_2 - a_2b_1}$

56. $3(x-2) < \dfrac{4}{5}(x-4)$

$5[3(x-2)] < 5\left[\dfrac{4}{5}(x-4)\right]$

$15(x-2) < 4(x-4)$

$15x - 30 < 4x - 16$

$11x - 30 < -16$

$11x < 14$

$x < \dfrac{14}{11}$

$\left(-\infty, \dfrac{14}{11}\right)$

57. $3x + 4y = 8$
Solve for y.
$4y = -3x + 8$

$y = -\dfrac{3}{4}x + 2$

x	y
-4	$y = -\dfrac{3}{4}(-4) + 2 = 3 + 2 = 5$
0	$y = -\dfrac{3}{4}(0) + 2 = 0 + 2 = 2$
4	$y = -\dfrac{3}{4}(4) + 2 = -3 + 2 = -1$

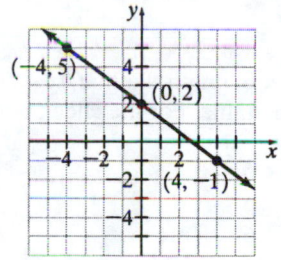

58. $3x + 4y = 8$
For the x-intercept, let $y = 0$.
$3x + 4y = 8$
$3x + 4(0) = 8$
$3x + 0 = 8$
$3x = 8$

$x = \dfrac{8}{3} = 2\dfrac{2}{3}$

For the y-intercept, let $x = 0$.

$3x + 4y = 8$
$3(0) + 4y = 8$
$0 + 4y = 8$
$4y = 8$
$y = 2$

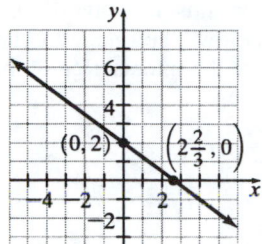

59. $3x + 4y = 8$
Solve for y.
$4y = -3x + 8$

$y = -\dfrac{3}{4}x + 2$

The slope is $-\dfrac{3}{4}$ and the y-intercept is 2.

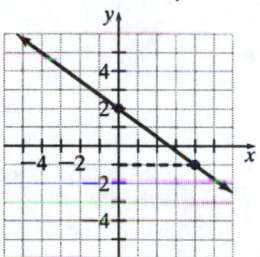

Exercise Set 4.6

1. Answers will vary.

3. $x - y > 2$
   $y < -2x + 3$
For $x - y > 2$, graph the line $x - y = 2$ using a dashed line. For the check point, select $(0, 0)$:
$x - y > 2$
$0 - 0 > 2$
$\quad 0 > 2 \qquad$ False
Since this is a false statement, shade the region which does not contain the point $(0, 0)$. This is the region "below" the line. For $y < -2x + 3$, graph the line $y = -2x + 3$ using a dashed line. For the check point,

select (0, 0):

$y < -2x + 3$

$0 < -2(0) + 3$

$0 < 3$ True

Since this is a true statement, shade the region which contains the point (0, 0). This is the region "below" the line. To obtain the final region, take the intersection of the above two regions.

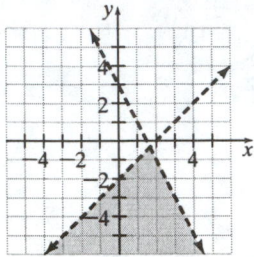

5. $y \leq x - 4$

$y < -2x + 4$

For $y \leq x - 4$, graph the line $y = x - 4$ using a solid line. For the check point, select (0, 0):

$y \leq x - 4$

$0 \leq 0 - 4$

$0 \leq -4$ False

Since this is a false statement, shade the region which does not contain the point (0, 0). This is the region "below" the line.

For $y < -2x + 4$, graph the line $y = -2x + 4$ using a dashed line. For the check point, select (0, 0):

$y < -2x + 4$

$0 < -2(0) + 4$

$0 < 4$ True

Since this is a true statement, shade the region which contains the point (0, 0). This is the region "below" the line. To obtain the final region, take the intersection of the above two regions.

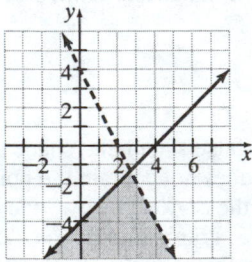

7. $y < x$

$y \geq 3x + 2$

For $y < x$, graph the line $y = x$ using a dashed line. For the check point, select (4, 0):

$y < x$

$0 < 4$ True

Since this is a true statement, shade the region which contains the point (4, 0). This is the region "below" the line.

For $y \geq 3x + 2$, graph the line $y = 3x + 2$ using a solid line. For the check point, select (0, 0):

$y \geq 3x + 2$

$0 \geq 3(0) + 2$

$0 \geq 2$ False

Since this is a false statement, shade the region which does not contain the point (0, 0). This is the region "above" the line. To obtain the final region, take the intersection of the above two regions.

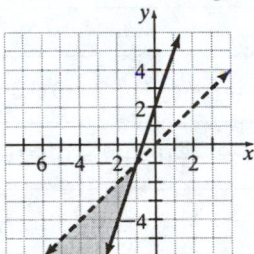

9. $4x - 2y < 6$

$y \leq -x + 4$

For $4x - 2y < 6$, graph the line $4x - 2y = 6$ using a dashed line. For the check point, select (0, 0):

$4x - 2y < 6$

$4(0) - 2(0) < 6$

$0 < 6$ True

Since this is a true statement, shade the region which contains the point (0, 0). This is the region "above" the line.

For $y \leq -x + 4$, graph the line $y = -x + 4$ using a solid line. For the check point, select (0, 0):

$y \leq -x + 4$

$0 \leq -0 + 4$

$0 \leq 4$ True

Since this is a true statement, shade the region which contains the point (0, 0). This is the region "below" the line.

To obtain the final region, take the
intersection of the above two regions.

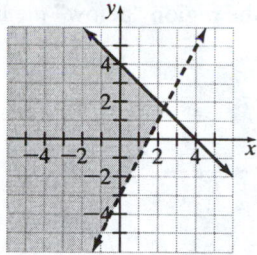

11. $-4x + 5y < 20$
$\quad\quad x \geq -3$
For $-4x + 5y < 20$, graph the line
$-4x + 5y = 20$ using a dashed line. For the
check point, select $(0, 0)$:
$\quad\quad -4x + 5y < 20$
$\quad -4(0) + 5(0) < 20$
$\quad\quad\quad 0 < 20 \quad$ True
Since this is a true statement, shade the
region which contains the point $(0, 0)$. This
is the region "below" the line.
For $x \geq -3$, the graph is the line $x = -3$ along
with the region to the right of $x = -3$. To
obtain the final region, take the intersection
of the above two regions.

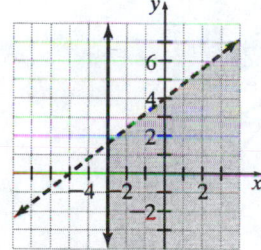

13. $x \leq 4$
$\quad\quad y \geq -2$
For $x \leq 4$, the graph is the line $x = 4$ along
with the region to the left of $x = 4$. For
$y \geq -2$, the graph is the line $y = -2$ along
with the region above the line $y = -2$. To
obtain the final region, take the intersection
of the above two regions.

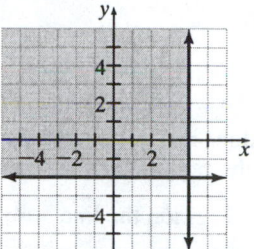

15. $5x + 2y > 10$
$\quad\quad 3x - y > 3$
For $5x + 2y > 10$, graph the line $5x + 2y = 10$
using a dashed line. For the check point,
select $(0, 0)$:
$\quad\quad 5x + 2y > 10$
$\quad 5(0) + 2(0) > 10$
$\quad\quad\quad 0 > 10 \quad$ False
Since this is a false statement, shade the
region which does not contain the point
$(0, 0)$. This is the region "above" the line.
For $3x - y > 3$, graph the line $3x - y = 3$
using a dashed line. For the check point,
select $(0, 0)$:
$\quad\quad 3x - y > 3$
$\quad 3(0) - 0 > 3$
$\quad\quad\quad 0 > 3 \quad\quad$ False
Since this is a false statement, shade the
region which does not contain the point
$(0, 0)$. This is the region "below" the line. To
obtain the final region, take the intersection
of the above two regions.

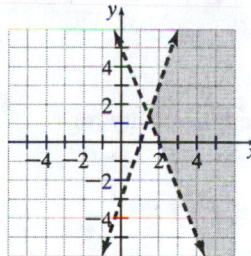

17. $-2x > y + 4$
$\quad\quad -x < \dfrac{1}{2}y - 1$
For $-2x > y + 4$, graph the line $-2x = y + 4$
using a dashed line. For the check point,
select $(0, 0)$:

$$-2x > y + 4$$
$$-2(0) > 0 + 4$$
$$0 > 4 \qquad \text{False}$$

Since this is a false statement, shade the region which does not contain the point $(0, 0)$. This is the region "below" the line.

For $-x < \dfrac{1}{2}y - 1$, graph the line

$-x = \dfrac{1}{2}y - 1$ using a dashed line. For the

check point, select $(0, 0)$:

$$-x < \frac{1}{2}y - 1$$
$$-0 < \frac{1}{2}(0) - 1$$
$$0 < -1 \qquad \text{False}$$

Since this is a false statement, shade the region which does not contain the point $(0, 0)$. This is the region "above" the line. To obtain the final region take the intersection of the above two regions. Since the regions do not overlap, the final result is the empty set which means there is no solution.

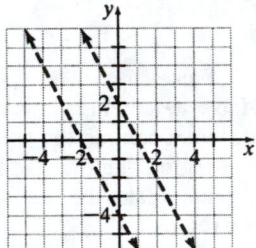

19. $\quad y < 3x - 4$
$\quad 6x \ge 2y + 8$

Solve the second inequality for y.

$$6x \ge 2y + 8$$
$$6x - 2y > 8$$
$$-2y \ge -6x + 8$$
$$y \le 3x - 4$$

The second inequality is now identical to the first except that the second inequality includes the line.

For $y < 3x - 4$, graph the line $y = 3x - 4$ using a dashed line. For the check point, select $(0, 0)$:

$$y < 3x - 4$$
$$0 < 3(0) - 4$$
$$0 < -4 \qquad \text{False}$$

Since this is a false statement, shade the region which does not contain the point $(0, 0)$. This is the region "below" the line.

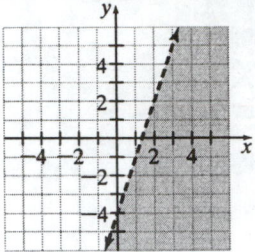

21. $\quad x \ge 0$
$\quad y \ge 0$
$\quad 5x + 4y \le 20$
$\quad x + 2y \le 6$

The first two inequalities indicate that the region must be in the first quadrant. For $5x + 4y \le 20$, the graph is the line $5x + 4y = 20$ along with the region below this line. For $x + 2y < 6$, the graph is the line $x + 2y = 6$ along with the region below this line. To obtain the final region, take the intersection of these regions.

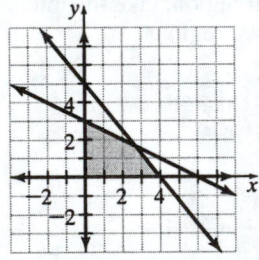

23. $\quad x \ge 0$
$\quad y \ge 0$
$\quad x + y \le 6$
$\quad 7x + 4y \le 28$

The first two inequalities indicate that the region must be in the first quadrant. For $x + y \le 6$, the graph is the line $x + y = 6$ along with the region below this line. For $7x + 4y \le 28$, the graph is the line $7x + 4y = 28$ along with the region below the line. To obtain the final region, take the intersection of these regions.

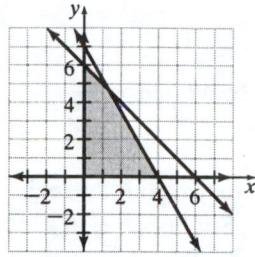

25. $x \geq 0$
$y \geq 0$
$7x + 4y \leq 24$
$2x + 5y \leq 20$

The first two inequalities indicate that the region must be in the first quadrant. For $7x + 4y \leq 24$, the graph is the line $7x + 4y = 24$ along with the region below this line. For $2x + 5y \leq 20$, the graph is the line $2x + 5y = 20$ along with the region below the line. To obtain the final region, take the intersection of these regions. The final answer is

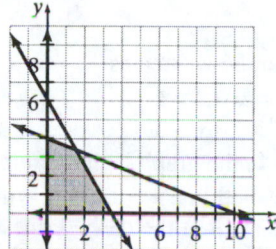

27. $x \geq 0$
$y \geq 0$
$x \leq 4$
$x + y \leq 6$
$x + 2y \leq 8$

The first two inequalities indicate that the region must be in the first quadrant. The third inequality indicates that the region must be on or to the left of the line $x = 4$. For $x + y \leq 6$, the graph is the line $x + y = 6$ along with the region below this line. For $x + 2y \leq 8$, the graph is the line $x + 2y = 8$ along with the region below the line. To obtain the final region, take the intersection of these regions.

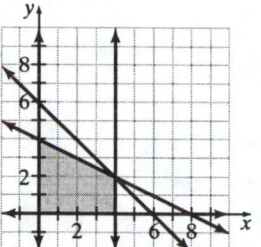

29. $x \geq 0$
$y \geq 0$
$x \leq 15$
$30x + 25y \leq 750$
$10x + 40y \leq 800$

The first two inequalities indicate that the region must be in the first quadrant. The third inequality indicates that the region must be on or to the left of the line $x = 15$. For $30x + 25y \leq 750$ the graph is the line $30x + 25y = 750$ along with the region below this line. For $10x + 40y \leq 800$, the graph is the line $10x + 40y = 800$ along with the region below the line. To obtain the final region, take the intersection of these regions.

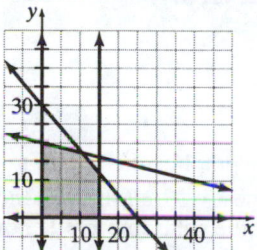

31. $|y| > 2$
$y \leq x + 3$

For $|y| > 2$, the graph is the region above the dashed line $y = 2$ along with the region below the dashed line $y = -2$. For $y \leq x + 3$, the graph is the region below the solid line $y = x + 3$. To obtain the final region, take the intersection of these regions.

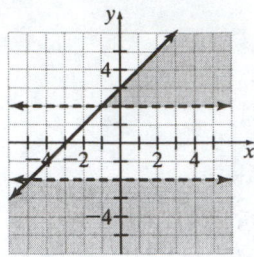

33. $|y| < 4$

$y \geq -2x + 2$

For $|y| < 4$, the graph is the region between the dashed lines $y = -4$ and $y = 4$. For $y \geq -2x + 2$, the graph is the region above the solid line $y = -2x + 2$. To obtain the final region, take the intersection of these regions.

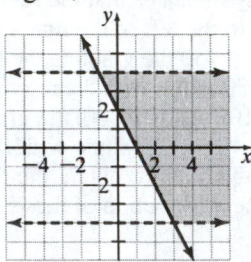

35. $|x| \geq 1$

$|y| \geq 2$

For $|x| \geq 1$, the graph is the region to the left of the solid line $x = -1$ along with the region to the right of the solid line $x = 1$. For $|y| \geq 2$, the graph is the region above the solid line $y = 2$ along with the region below the solid line $y = -2$. To obtain the final region, take the intersection of these regions.

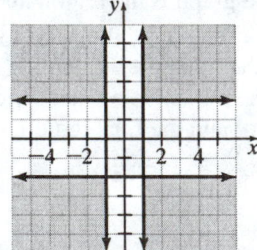

37. $|x + 2| < 3$

$|y| > 4$

$|x + 2| < 3$ can be written as

$-3 < x + 2 < 3$

$-5 < x < 1$

For $|x + 2| < 3$, the graph is the region between the dashed lines $x = -5$ and $x = 1$. For $|y| > 4$, the graph is the region above the dashed line $y = 4$ along with the region below the dashed line $y = -4$. To obtain the final region, take the intersection of these regions.

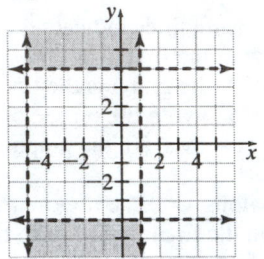

39. $|x - 2| > 1$

$y > -2$

$|x - 2| > 1$ can be written as

$x - 2 < -1$ or $x - 2 > 1$

$x < 1$ $x > 3$

For $|x - 2| > 1$, the graph is the region to the left of the dashed line $x = 1$ along with the region to the right of the dashed line $x = 3$. For $y > -2$, the graph is the region above the dashed line $y = -2$. To obtain the final region, take the intersection of these regions.

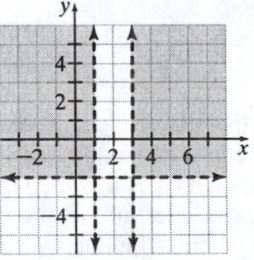

41. If the boundary lines are parallel, there may be no solution. For example, the system $y < x$ and $y > x + 1$ has no solution.

43. There are no solutions. Opposite sides of the same line are being shaded, but not the line itself.

45. There are an infinite number of solutions. Both inequalities include the line $5x - 2y = 3$.

47. There are an infinite number of solutions. The lines are parallel but the same side of each line is being shaded.

49. $y < |x|$
$y < 4$
For $y < |x|$, graph the equation $y = |x|$ using a dashed line. For the check point, select $(0, 3)$.
$y < |x|$
$3 < |0|$
$3 < 0$ False
Since this is a false statement, shade the region which does not contain the point $(0, 3)$. This is the region below the graph of $y = |x|$.
For $y < 4$, the graph is the region below the dashed line $y = 4$. To obtain the final region, take the intersection of these regions.

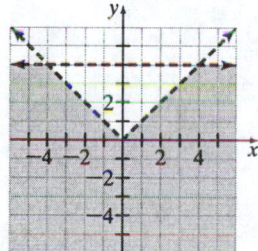

51. $f_1 d_1 + f_2 d_2 = f_3 d_3$
$f_1 d_1 - f_1 d_1 + f_2 d_2 = f_3 d_3 - f_1 d_1$
$f_2 d_2 = f_3 d_3 - f_1 d_1$
$\dfrac{f_2 d_2}{d_2} = \dfrac{f_3 d_3 - f_1 d_1}{d_2}$
$f_2 = \dfrac{f_3 d_3 - f_1 d_1}{d_2}$

52. Domain: $\{-1, 0, 4, 5\}$
Range: $(-5, -2, 2, 3\}$

53. Domain: $\mathbb{R}$
Range: $\mathbb{R}$

54. Domain: $\mathbb{R}$
Range: $\{y | y \geq -1\}$

Review Exercises

1. Write each equation in slope-intercept form.
$x + 2y = 8$ $3x + 6y = 12$
$2y = -x + 8$ $6y = -3x + 12$
$y = -\dfrac{1}{2}x + 4$ $y = -\dfrac{1}{2}x + 2$
Since the slope of each line is $-\dfrac{1}{2}$ but the y-intercepts are different ($b = 4$ for first equation, $b = 2$ for second equation), the two lines are parallel and produce no solution. This is an inconsistent system.

2. Write each equation in slope-intercept form.
$y = -3x - 6$ is already in this form.
$2x + 3y = 8$
$3y = -2x + 8$
$y = -\dfrac{2}{3}x + \dfrac{8}{3}$
Since the slope of the first line is -3 and the slope of the second line is $-\dfrac{2}{3}$, the slopes are different so that the lines intersect to produce one solution. This is a consistent system.

3. Write each equation in slope-intercept form.
$y = \dfrac{1}{2}x + 4$ is already in this form.
$x + 2y = 8$
$2y = -x + 8$
$y = -\dfrac{1}{2}x + 4$
Since the slope of the first line is $\dfrac{1}{2}$ and the slope of the second line is $-\dfrac{1}{2}$, the slopes are different so that the lines intersect to produce one solution. This is a consistent system.

4. Write each equation in slope-intercept form.

$$6x = 4y - 8 \qquad\qquad 4x = 6y + 8$$
$$6x + 8 = 4y \qquad\qquad 4x - 8 = 6y$$
$$\frac{6x + 8}{4} = y \qquad\qquad \frac{4x - 8}{6} = y$$
$$\frac{3}{2}x + 2 = y \qquad\qquad \frac{2}{3}x - \frac{4}{3} = y$$

Since the slope of the first line is $\frac{3}{2}$ and the slope of the second line is $\frac{2}{3}$, the slopes are different so that the lines intersect to produce one solution. This is a consistent system.

5. Graph the equations $y = x + 3$ and $y + 2x + 5$.

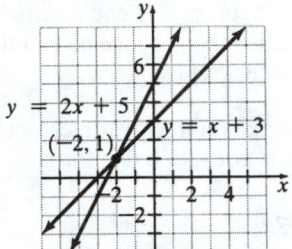

The lines intersect and the point of intersection is $(-2, 1)$.

6. Graph the equations $x = -2$ and $y = 3$.

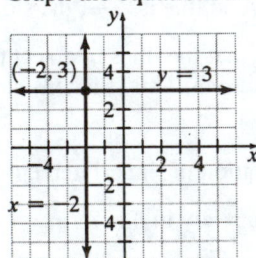

The lines intersect and the point of intersection is $(-2, 3)$.

7. Graph the equations $2x + 2y = 8$ and $2x - y = -4$.

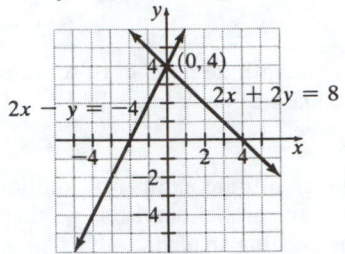

The lines intersect and the point of intersection is $(0, 4)$.

8. Graph the equations $2y = 2x - 6$ and $\frac{1}{2}x - \frac{1}{2}y = \frac{3}{2}$.

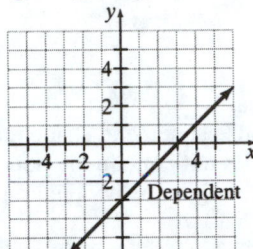

Both equations produce the same line. This is a dependent system.

9. $y = 2x + 1$
$y = 3x - 2$
Substitute $3x - 2$ for y in the first equation.
$$y = 2x + 1$$
$$3x - 2 = 2x + 1$$
$$x - 2 = 1$$
$$x = 3$$
Now, substitute 3 for x in the first equation.
$$y = 2x + 1$$
$$y = 2(3) + 1$$
$$y = 6 + 1$$
$$y = 7$$
The solution is $(3, 7)$.

10. $y = -x + 5$
$y = 2x - 1$
Substitute $2x - 1$ for y in the first equation.

$$y = -x + 5$$
$$2x - 1 = -x + 5$$
$$3x - 1 = 5$$
$$3x = 6$$
$$x = \frac{6}{3} = 2$$

Now, substitute 2 for x in the second equation.
$$y = 2x - 1$$
$$y = 2(2) - 1$$
$$y = 4 - 1$$
$$y = 3$$
The solution is (2, 3).

11. $a = 2b - 8$
$$2b - 5a = 0$$
Substitute $2b - 8$ for a in the second equation.
$$2b - 5a = 0$$
$$2b - 5(2b - 8) = 0$$
$$2b - 10b + 40 = 0$$
$$-8b + 40 = 0$$
$$-8b = -40$$
$$b = \frac{-40}{-8} = 5$$
Now, substitute 5 for b in the first equation.
$$a = 2b - 8$$
$$a = 2(5) - 8$$
$$a = 10 - 8$$
$$a = 2$$
The solution is (2, 5).

12. $3x + y = 17$
$$\frac{1}{2}x - \frac{3}{4}y = 1$$
First multiply the second equation by 4 to eliminate fractions.
$$3x + y = 17$$
$$4\left(\frac{1}{2}x - \frac{3}{4}y = 1\right)$$
gives
$3x + y = 17 \leftarrow$ first equation
$2x - 3y = 4 \leftarrow$ second equation
Now, solve the first equation for y.
$$3x + y = 17$$
$$y = -3x + 17$$
Substitute $-3x + 17$ for y in the second

equation.
$$2x - 3y = 4$$
$$2x - 3(-3x + 17) = 4$$
$$2x + 9x - 51 = 4$$
$$11x - 51 = 4$$
$$11x = 55$$
$$x = \frac{55}{11} = 5$$
Finally, substitute 5 for x in the equation $y = -3x + 17$.
$$y = -3x + 17$$
$$y = -3(5) + 17$$
$$y = -15 + 17$$
$$y = 2$$
The solution is (5, 2).

13. $x + y = 6$
           $\underline{x - y = 10}$
Add: $2x \quad\;\; = 16$
$$x = \frac{16}{2} = 8$$
Substitute 8 for x in the first equation.
$$x + y = 6$$
$$8 + y = 6$$
$$y = -2$$
The solution is (8, –2).

14. $x + 2y = -3$
          $\underline{2x - 2y = 6}$
Add: $3x \quad\;\; = 3$
$$x = \frac{3}{3} = 1$$
Substitute 1 for x in the first equation.
$$x + 2y = -3$$
$$1 + 2y = -3$$
$$2y = -4$$
$$y = \frac{-4}{2} = -2$$
The solution is (1, –2).

15. $2x + 3y = 4$
     $x + 2y = -6$
To eliminate x, multiply the second equation by –2 and then add.
$$2x + 3y = 4$$
$$-2[x + 2y = -6]$$
gives

$$2x + 3y = 4$$
$$-2x - 4y = 12$$

Add: $-y = 16$

$$y = -16$$

Substitute -16 for y in the second equation.

$$x + 2y = -6$$
$$x + 2(-16) = -6$$
$$x - 32 = -6$$
$$x = 26$$

The solution is $(26, -16)$.

16. $0.6x + 0.5y = 2$

$0.25x - 0.2y = 1.65$

To eliminate y, multiply the first equation by 2 and the second equation by 5 and then add.

$$2[0.6x + 0.5y = 2]$$
$$5[0.25x - 0.2y = 1.65]$$

gives

$$1.20x + 1.0y = 4$$
$$1.25x - 1.0y = 8.25$$

Add: $2.45x = 12.25$

$$x = \frac{12.25}{2.45} = 5$$

Substitute 5 for x in the first equation.

$$0.6x + 0.5y = 2$$
$$0.6(5) + 0.5y = 2$$
$$3 + 0.5y = 2$$
$$0.5y = -1$$
$$y = \frac{-1}{0.5} = -2$$

The solution is $(5, -2)$.

17. $4r - 3s = 8$

$2r + 5s = 8$

To eliminate r, multiply the second equation by -2 and then add.

$$4r - 3s = 8$$
$$-2[2r + 5s = 8]$$

gives

$$4r - 3s = 8$$
$$-4r - 10s = -16$$

Add: $-13s = -8$

$$s = \frac{-8}{-13} = \frac{8}{13}$$

Substitute $\frac{8}{13}$ for s in the first equation.

$$4r - 3s = 8$$
$$4r - 3\left(\frac{8}{13}\right) = 8$$
$$4r - \frac{24}{13} = 8$$
$$4r = 8 + \frac{24}{13}$$
$$4r = \frac{104}{13} + \frac{24}{13}$$
$$4r = \frac{128}{13}$$
$$r = \frac{128}{13} \cdot \frac{1}{4}$$
$$r = \frac{32}{13}$$

The solution is $\left(\frac{32}{15}, \frac{8}{13}\right)$.

18. $-2m + 3n = 15$

$3m + 3n = 10$

To eliminate n, multiply the second equation by -1 and then add.

$$-2m + 3n = 15$$
$$-1[3m + 3n = 10]$$

gives

$$-2m + 3n = 15$$
$$-3m - 3n = -10$$

Add: $-5m = 5$

$$m = \frac{5}{-5} = -1$$

Substitute -1 for m in the second equation.

$$3m + 3n = 10$$
$$3(-1) + 3n = 10$$
$$-3 + 3n = 10$$
$$3n = 13$$
$$n = \frac{13}{3}$$

The solution is $\left(-1, \frac{13}{3}\right)$.

19. $x + \frac{2}{5}y = \frac{9}{5}$

$x - \frac{3}{2}y = -2$

To clear fractions and to eliminate x, multiply the first equation by 10 and the

184

second equation by –10 and then add.

$$10\left(x+\frac{2}{5}y=\frac{9}{5}\right)$$

$$-10\left(x-\frac{3}{2}y=-2\right)$$

gives

$$\begin{aligned}10x+\ \ 4y&=18\\-10x+15y&=20\end{aligned}$$

Add: $\qquad 19y=38$

$$y=2$$

Now substitute 2 for y in the equation $10x+4y=18$.

$$10x+4y=18$$
$$10x+4(2)=18$$
$$10x+8=18$$
$$10x=10$$
$$x=1$$

The solution is $(1, 2)$.

20. $\quad 2x+2y=8$
$$y=4x-3$$

Write the system in standard form and divide the first equation by 2.

$$x+y=4$$
$$-4x+y=-3$$

To eliminate x, multiply the first equation by 4 and then add.

$$4[x+y=4]$$
$$-4x+y=-3$$

gives

$$\begin{aligned}4x+4y&=16\\-4x+\ \ y&=-3\end{aligned}$$

Add: $\qquad 5y=13$

$$y=\frac{13}{5}$$

Substitute $\frac{13}{5}$ for y in the first equation.

$$x+y=4$$
$$x+\frac{13}{5}=4$$
$$x=4-\frac{13}{5}$$
$$x=\frac{20}{5}-\frac{13}{5}$$
$$x=\frac{7}{5}$$

The solution is $\left(\frac{7}{5},\frac{13}{5}\right)$.

21. $\qquad\qquad y=-\frac{3}{4}x+\frac{5}{2}$

$$x+\frac{5}{4}y=\frac{7}{2}$$

Write the system in standard form.

$$\frac{3}{4}x+y=\frac{5}{2}$$

$$x+\frac{5}{4}y=\frac{7}{2}$$

To clear fractions and to eliminate x, multiply the first equation by 16 and the second equation by –12 and then add.

$$16\left[\frac{3}{4}x+y=\frac{5}{2}\right]$$

$$-12\left[x+\frac{5}{4}y=\frac{7}{2}\right]$$

gives

$$\begin{aligned}12x+16y&=\ \ 40\\-12x-15y&=-42\end{aligned}$$

Add: $\qquad\qquad y=\ -2$

Now, substitute –2 for y in the equation

$x+\frac{5}{4}y=\frac{7}{2}$ and then solve for x.

$$x + \frac{5}{4}y = \frac{7}{2}$$

$$x + \frac{5}{4}(-2) = \frac{7}{2}$$

$$x - \frac{5}{2} = \frac{7}{2}$$

$$x = \frac{5}{2} + \frac{7}{2}$$

$$x = \frac{12}{2}$$

$$x = 6$$

The solution is $(6, -2)$.

22. $2x - 5y = 12$

$$x - \frac{4}{3}y = -2$$

To clear fractions and to eliminate x, multiply the first equation by -3 and the second equation by 6 and then add.

$$-3[2x - 5y = 12]$$

$$6\left[x - \frac{4}{3}y = -2\right]$$

gives

$$-6x + 15y = -36$$
$$\underline{6x - \ 8y = -12}$$
Add: $7y = -48$

$$y = -\frac{48}{7}$$

Now substitute $-\dfrac{48}{7}$ for y in the first equation.

$$2x - 5y = 12$$

$$2x - 5\left(-\frac{48}{7}\right) = 12$$

$$2x + \frac{240}{7} = 12$$

$$2x = 12 - \frac{240}{7}$$

$$2x = \frac{84}{7} - \frac{240}{7}$$

$$2x = -\frac{156}{7}$$

$$x = \frac{1}{2}\left(-\frac{156}{7}\right) = -\frac{78}{7}$$

The solution is $\left(-\dfrac{78}{7}, -\dfrac{48}{7}\right)$.

23. $2x + y = 4$

$$x + \frac{1}{2}y = 2$$

To eliminate y, multiply the second equation by -2 and then add.

$$2x + y = 4$$

$$-2\left[x + \frac{1}{2}y = 2\right]$$

gives

$$2x + y = \ \ 4$$
$$\underline{-2x - y = -4}$$
Add: $0 = 0$ True

The system has an infinite number of solutions.

24. $2x = 4y + 5$

$2y = x - 6$

Write the system in standard form.

$2x - 4y = 5$

$-x + 2y = -6$

To eliminate x, multiply the second equation by 2 and then add.

$$2x - 4y = 5$$

$$2[-x + 2y = -6]$$

gives

$$2x - 4y = \ \ 5$$
$$\underline{-2x + 4y = -12}$$
Add: $0 = -7$ False

Since this a false statement, there is no solution to the system. The system is inconsistent.

25. $x + 2y = 12$

$$4x = 8$$

$$3x - 4y + 5z = 20$$

Solve the second equation for x.

$4x = 8$

$x = 2$

Substitute 2 for x in the first equation.

$x + 2y = 12$

$2 + 2y = 12$

$2y = 10$

$y = 5$

Substitute 2 for x and 5 for y in the third equation.

$$3x - 4y + 5z = 20$$
$$3(2) - 4(5) + 5z = 20$$
$$6 - 20 + 5z = 20$$
$$-14 + 5z = 20$$
$$5z = 34$$
$$z = \frac{34}{5}$$

The solution is $\left(2, 5, \dfrac{34}{5}\right)$.

26. $3x + 4y - 5z = 10$
　　　$4x + 2z = 16$
　　　　　$2z = -4$

Solve the third equation for z.
$$2z = -4$$
$$z = \frac{-4}{2} = -2$$

Substitute –2 for z in the second equation.
$$4x + 2z = 16$$
$$4x + 2(-2) = 16$$
$$4x - 4 = 16$$
$$4x = 20$$
$$x = \frac{20}{4} = 5$$

Substitute 5 for x and –2 for z in the first equation.
$$3x + 4y - 5z = 10$$
$$3(5) + 4y - 5(-2) = 10$$
$$15 + 4y + 10 = 10$$
$$4y + 25 = 10$$
$$4y = -15$$
$$y = -\frac{15}{4}$$

The solution is $\left(5, -\dfrac{15}{4}, -2\right)$.

27. $x + 5y + 5z = 6$　(1)
　　　$3x + 3y - z = 10$　(2)
　　　$x + 3y + 2z = 5$　(3)

To eliminate x between equations (1) and (2), multiply equation (1) by –3 and then add.
$$-3[x + 5y + 5z = 6]$$
$$3x + 3y - z = 10$$
gives

$$-3x - 15y - 15z = -18$$
$$\underline{3x + 3y - z = 10}$$
Add:　$-12y - 16z = -8$
or
$3y + 4z = 2$　(4)

To eliminate x between equations (2) and (3), multiply equation (3) by –3 and then add.
$$3x + 3y - z = 10$$
$$-3[x + 3y + 2z = 5]$$
gives

$$3x + 3y - z = 10$$
$$\underline{-3x - 9y - 6z = -15}$$
Add:　$-6y - 7z = -5$　(5)

Equations (4) and (5) are two equations in two unknowns.
$$3y + 4z = 2$$
$$-6y - 7z = -5$$

To eliminate y, multiply equation (4) by 2 and then add.
$$2[3y + 4z = 2]$$
$$-6y - 7z = -5$$
gives

$$6y + 8z = 4$$
$$\underline{-6y - 7z = -5}$$
Add:　　　$z = -1$

Substitute –1 for z in equation (4).
$$3y + 4z = 2$$
$$3y + 4(-1) = 2$$
$$3y - 4 = 2$$
$$3y = 6$$
$$y = 2$$

Finally, substitute 2 for y and –1 for z in equation (1).
$$x + 5y + 5z = 6$$
$$x + 5(2) + 5(-1) = 6$$
$$x + 10 - 5 = 6$$
$$x + 5 = 6$$
$$x = 1$$
The solution is $(1, 2, -1)$.

28. $-x - y - z = -6$ (1)
　　　$2x + 3y - z = 7$　(2)
　　　$-3x + y + z = -6$ (3)

To eliminate y and z between equations (1) and (3) simply add.

$$-x - y - z = -6$$
$$-3x + y + z = -6$$
Add: $-4x \quad\quad = -12$

$$x = \frac{-12}{-4} = 3$$

Substitute 3 for x in equations (2) and (3) to produce a system of two equations in the unknowns y and z.

Equation (2) becomes
$$2x + 3y - z = 7$$
$$2(3) + 3y - z = 7$$
$$6 + 3y - z = 7$$
$$3y - z = 1 \quad (4)$$

Equation (3) becomes
$$-3x + y + z = -6$$
$$-3(3) + y + z = -6$$
$$-9 + y + z = -6$$
$$y + z = 3 \quad (5)$$

To eliminate z, add equations (4) and (5).
$$3y - z = 1$$
$$y + z = 3$$
Add: $4y \quad = 4$

$$y = \frac{4}{4} = 1$$

Finally, substitute 1 for y in equation (5).
$$y + z = 3$$
$$1 + z = 3$$
$$z = 2$$

The solution is (3, 1, 2).

29. $3y - 2z = -4 \quad (1)$
$3x - 5z = -7 \quad (2)$
$2x + y = 6 \quad (3)$

To eliminate y between equations (1) and (3), multiply equation (3) by -3 and then add.
$$3y - 2z = -4$$
$$-3[2x + y = 6]$$
gives
$$3y - 2z = -4$$
$$-6x - 3y \quad = -18$$
Add: $-6x \quad -2z = -22$
or
$$-3x - z = -11 \quad\quad (4)$$

Equations (4) and (2) are two equations into two unknowns. To eliminate x, simply add.

$$3x - 5z = -7$$
$$-3x - z = -11$$
Add: $\quad -6z = -18$
$$z = 3$$

Substitute 3 for z in equation (2).
$$3x - 5z = -7$$
$$3x - 5(3) = -7$$
$$3x - 15 = -7$$
$$3x = 8$$
$$x = \frac{8}{3}$$

Substitute $\frac{8}{3}$ for x in equation (3).
$$2x + y = 6$$
$$2\left(\frac{8}{3}\right) + y = 6$$
$$\frac{16}{3} + y = 6$$

$$y = 6 - \frac{16}{3}$$
$$y = \frac{18}{3} - \frac{16}{3}$$
$$y = \frac{2}{3}$$

The solution is $\left(\frac{8}{3}, \frac{2}{3}, 3\right)$.

30. $3a + 2b - 5c = 19 \quad (1)$
$2a - 3b + 3c = -15 \quad (2)$
$5a - 4b - 2c = -2 \quad (3)$

To eliminate b between equations (1) and (2), multiply equation (1) by 3 and equation (2) by 2 and then add.
$$3[3a + 2b - 5c = 19]$$
$$2[2a - 3b + 3c = -15]$$
gives
$$9a + 6b - 15c = 57$$
$$4a - 6b + 6c = -30$$
Add: $13a \quad\quad -9c = 27 \quad (4)$

To eliminate b between equations (1) and (3), multiply equation (1) by 2 and then add.
$$2[3a + 2b - 5c = 19]$$
$$5a - 4b - 2c = -2$$
gives

$$6a + 4b - 10c = 38$$
$$\underline{5a - 4b - 2c = -2}$$
Add: $11a \quad\quad - 12c = 36 \quad$ (5)

Equations (4) and (5) are two equations into two unknowns.

$$13a - 9c = 27$$
$$11a - 12c = 36$$

To eliminate c, multiply equation (4) by 4 and equation (5) by –3 and then add.

$$4[13a - 9c = 27]$$
$$-3[11a - 12c = 36]$$

gives

$$52a - 36c = \quad 108$$
$$\underline{-33a + 36c = -108}$$
Add: $19a \quad\quad = \quad 0$
$$a = 0$$

Substitute 0 for a in equation (4).

$$13a - 9c = 27$$
$$13(0) - 9c = 27$$
$$-9c = 27$$
$$c = \frac{27}{-9} = -3$$

Finally, substitute 0 for a and –3 for c in equation (1).

$$3a + 2b - 5c = 19$$
$$3(0) + 2b - 5(-3) = 19$$
$$0 + 2a + 15 = 19$$
$$2a + 15 = 19$$
$$2a = 4$$
$$a = \frac{4}{2} = 2$$

The solution is (0, 2, –3).

31.
$$x - y + 3z = 1 \quad (1)$$
$$-x + 2y - 2z = 1 \quad (2)$$
$$x - 3y + z = 2 \quad (3)$$

To eliminate x between equations (1) and (2), simply add.

$$x - y + 3z = 1$$
$$\underline{-x + 2y - 2z = 1}$$
Add: $\quad\quad y \quad + z = 2 \quad\quad$ (4)

To eliminate x between equations (2) and (3), simply add.

$$-x + 2y - 2z = 1$$
$$\underline{x - 3y + \quad z = 2}$$
Add: $\quad -y - \quad z = 3 \quad\quad$ (5)

Equations (4) and (5) are two equations in two unknowns. To eliminate y and z simply add.

$$y + z = 2$$
$$\underline{-y - z = 3}$$
Add: $\quad\quad 0 = 5 \quad$ False

Since this is a false statement, there is no solution to the system. This is an inconsistent system.

32. $-2x + 2y - 3z = 6 \quad (1)$
$\quad\quad 4x - y + 2z = -2 \quad (2)$
$\quad\quad 2x + y - z = 4 \quad (3)$

To eliminate x between equations (1) and (2), multiply equation (1) by 2 and then add.
$2[-2x + 2y - 3z = 6]$
$\quad\quad 4x - y + 2z = -2$
gives

$$-4x + 4y - 6z = 12$$
$$\underline{4x - \quad y + 2z = -2}$$
Add: $\quad\quad 3y - 4z = 10 \quad$ (4)

To eliminate x between equations (1) and (3), simply add.

$$-2x + 2y - 3z = \quad 6$$
$$\underline{2x + \quad y - \quad z = \quad 4}$$
Add: $\quad\quad 3y - 4z = 10 \quad$ (5)

Equations (4) and (5) are two equations in two unknowns.
$$3y - 4z = 10$$
$$3y - 4z = 10$$

Since they are identical, there are an infinite number of solutions. This is a dependent system.

33. Let x be Bob Edward's brother's age and y be his niece's age.
$$x - y = 18$$
$$x = 4y$$

Substitute $4y$ for x in the first equation.
$$x - y = 18$$
$$4y - y = 18$$
$$3y = 18$$
$$y = \frac{18}{3} = 6$$

Now substitute 6 for y in the second equation.

$x = 4y$
$x = 5(6)$
$x = 24$
His brother is 24 years old and his niece is 6 years old.

34. Let x be the speed of the plane in still air and y be the speed of the wind.
$x + y = 600$
$x - y = 530$
To eliminate y, simply add.

$$\begin{array}{r} x + y = 600 \\ x - y = 530 \\ \hline \text{Add: } 2x \quad\;\; = 1130 \\ x \quad\;\; = 565 \end{array}$$

Substitute 565 for x in the first equation.
$x + y = 600$
$565 + y = 600$
$y = 35$
The speed of the plane in still air is 565 mph and the speed of the wind is 35 mph.

35. Let x be the amount of 20% acid solution and y be the amount of 50% acid solution.
$x + y = 6$
$0.2x + 0.5y = 0.4(6)$
To clear decimals, multiply the second equation by 10.
$x + y = 6$
$2x + 5y = 24$
Solve the first equation for y.
$x + y = 6$
$y = -x + 6$
Substitute $-x + 6$ for y in the second equation.
$2x + 5y = 24$
$2x + 5(-x + 6) = 24$
$2x - 5x + 30 = -24$
$-3x + 30 = 24$
$-3x = -6$
$x = \dfrac{-6}{-3} = 2$
Finally, substitute 3 for x in the equation $y = -x + 6$.
$y = -x + 6$
$y = -3 + 6$
$y = 4$

James should combine 2 liters of the 20% acid solution to 4 liters of the 50% acid solution.

36. Let x be the number of adult tickets and y be the number of children's tickets.
$x + y = 650$
$15x + 11y = 8790$
To solve, multiply the first equation by -11 and then add.
$-11[x + y = 650]$
$15x + 11y = 8790$
gives

$$\begin{array}{r} -11x - 11y = -7150 \\ 15x + 11y = \;\; 8790 \\ \hline \text{Add: } 4x \quad\quad = \;\; 1640 \\ x = \dfrac{1640}{4} = 410 \end{array}$$

Substitute 410 for x in the first equation.
$x + y = 650$
$410 + y = 650$
$y = 240$
Thus, 410 adult tickets and 240 children's tickets were sold.

37. Let $x =$ age at first time and $y =$ age at second time.
$y = 2x - 5$
$x + y = 118$
Substitute $2x - 5$ for y in the second equation.
$x + y = 118$
$x + 2x - 5 = 118$
$3x - 5 = 118$
$3x = 123$
$x = 41$
Substitute 41 for x in the first equation.
$y = 2x - 5$
$y = 2(41) - 5$
$y = 82 - 5$
$y = 77$
His ages were 41 years and 77 years.

38. Let x be the amount invested at 7%, y the amount invested at 5%, and z the amount invested at 3%.

$$x + y + z = 40,000 \quad (1)$$
$$y = x - 5000 \quad (2)$$
$$0.07x + 0.05y + 0.03z = 2300 \quad (3)$$

Substitute $x - 5000$ for y in equations (1) and (3). Equation (1) becomes

$$x + y + z = 40,000$$
$$x + x - 5000 + z = 40,000$$
$$2x + z = 45,000 \quad (4)$$

Equation (3) becomes

$$0.07x + 0.05y + 0.03z = 2300$$
$$0.07x + 0.05(x - 5000) + 0.03z = 2300$$
$$0.07x + 0.05x - 250 + 0.03z = 2300$$
$$0.12x + 0.03z = 2550 \quad (5)$$

Equation (4) and (5) are a system of two equations in two unknowns. Solve equation (4) for z.

$$2x + z = 45,000$$
$$z = -2x + 45,000$$

Substitute $-2x + 45,000$ for z in equation (5).

$$0.12x + 0.03z = 2550$$
$$0.12x + 0.03(-2x + 45,000) = 2550$$
$$0.12x - 0.06x + 1350 = 2550$$
$$0.06x = 1200$$
$$x = 20,000$$

Now substitute 20,000 for x in equation (2).

$$y = x - 5000$$
$$y = 20,000 - 5000 = 15,000$$

Finally, substitute 20,000 for x and 15,000 for y in equation (1).

$$x + y + z = 40,000$$
$$20,000 + 15,000 + z = 40,000$$
$$35,000 + z = 40,000$$
$$z = 5000$$

Thus, $20,000 was invested at 7%, $15,000 at 5%, and $5000 at 3%.

39. $\quad -4x + 9y = 7$
$\quad\quad 5x + 6y = -3$

$$\begin{bmatrix} -4 & 9 & | & 7 \\ 5 & 6 & | & -3 \end{bmatrix}$$

$$\begin{bmatrix} 1 & -\frac{9}{4} & | & -\frac{7}{4} \\ 5 & 6 & | & -3 \end{bmatrix} -\frac{1}{4}R_1$$

$$\begin{bmatrix} 1 & -\frac{9}{4} & | & -\frac{7}{4} \\ 0 & \frac{69}{4} & | & \frac{23}{4} \end{bmatrix} -5R_1 + R_2$$

$$\begin{bmatrix} 1 & -\frac{9}{4} & | & -\frac{7}{4} \\ 0 & 1 & | & \frac{1}{3} \end{bmatrix} \frac{4}{69}R_2$$

The system is

$$x - \frac{9}{4}y = -\frac{7}{4}$$
$$y = \frac{1}{3}$$

Substitute $\frac{1}{3}$ for y in the first equation.

$$x - \frac{9}{4}y = -\frac{7}{4}$$
$$x - \frac{9}{4}\left(\frac{1}{3}\right) = -\frac{7}{4}$$
$$x - \frac{3}{4} = -\frac{7}{4}$$
$$x = -\frac{7}{4} + \frac{3}{4} = -\frac{4}{4} = -1$$

The solution is $\left(-1, \frac{1}{3}\right)$.

40. $\quad 2x - 3y = 4$
$\quad\quad 2x = y - 2$

Write the system in standard form:
$$2x - 3y = 4$$
$$2x - y = -2$$

$$\begin{bmatrix} 2 & -3 & | & 4 \\ 2 & -1 & | & -2 \end{bmatrix}$$

$$\begin{bmatrix} 1 & -\frac{3}{2} & | & 2 \\ 2 & -1 & | & -2 \end{bmatrix} \frac{1}{2}R_1$$

$$\begin{bmatrix} 1 & -\frac{3}{2} & | & 2 \\ 0 & 2 & | & -6 \end{bmatrix} -2R_1 + R_2$$

$$\begin{bmatrix} 1 & -\frac{3}{2} & | & 2 \\ 0 & 1 & | & -3 \end{bmatrix} \frac{1}{2}R_2$$

The system is

$$x - \frac{3}{2}y = 2$$
$$y = -3$$

Substitute -3 for y in the first equation.

$$x - \frac{3}{2}y = 2$$

$$x - \frac{3}{2}(-3) = 2$$

$$x + \frac{9}{2} = 2$$

$$x = 2 - \frac{9}{2}$$

$$x = \frac{4}{2} - \frac{9}{2} = -\frac{5}{2}$$

The solution is $\left(-\frac{5}{2}, -3\right)$.

41. $y = 2x - 4$
$4x = 2y + 8$

Write the system in standard form.

$-2x + y = -4$
$4x - 2y = 8$

$$\begin{bmatrix} -2 & 1 & | & -4 \\ 4 & -2 & | & 8 \end{bmatrix}$$

$$\begin{bmatrix} 1 & -\frac{1}{2} & | & 2 \\ 4 & -2 & | & 8 \end{bmatrix} -\frac{1}{2}R_1$$

$$\begin{bmatrix} 1 & -\frac{1}{4} & | & 2 \\ 0 & 0 & | & 0 \end{bmatrix} -4R_1 + R_2$$

Since the last row is all zeros, the system is dependent.

42. $2x - y - z = 5$
$x + 2y + 3z = -2$
$3x - 2y + z = 2$

$$\begin{bmatrix} 2 & -1 & -1 & | & 5 \\ 1 & 2 & 3 & | & -2 \\ 3 & -2 & 1 & | & 2 \end{bmatrix}$$

$$\begin{bmatrix} 1 & -\frac{1}{2} & -\frac{1}{2} & | & \frac{5}{2} \\ 1 & 2 & 3 & | & -2 \\ 3 & -2 & 1 & | & 2 \end{bmatrix} \frac{1}{2}R_1$$

$$\begin{bmatrix} 1 & -\frac{1}{2} & -\frac{1}{2} & | & \frac{5}{2} \\ 0 & \frac{5}{2} & \frac{7}{2} & | & -\frac{9}{2} \\ 3 & -2 & 1 & | & 2 \end{bmatrix} -1R_1 + R_2$$

$$\begin{bmatrix} 1 & -\frac{1}{2} & -\frac{1}{2} & | & \frac{5}{2} \\ 0 & \frac{5}{2} & \frac{7}{2} & | & -\frac{9}{2} \\ 0 & -\frac{1}{2} & \frac{5}{2} & | & -\frac{11}{2} \end{bmatrix} -3R_1 + R_3$$

$$\begin{bmatrix} 1 & -\frac{1}{2} & -\frac{1}{2} & | & \frac{5}{2} \\ 0 & 1 & \frac{7}{5} & | & -\frac{9}{5} \\ 0 & -\frac{1}{2} & \frac{5}{2} & | & -\frac{11}{2} \end{bmatrix} \frac{2}{5}R_2$$

$$\begin{bmatrix} 1 & -\frac{1}{2} & -\frac{1}{2} & | & \frac{5}{2} \\ 0 & 1 & \frac{7}{5} & | & -\frac{9}{5} \\ 0 & 0 & \frac{16}{5} & | & -\frac{32}{5} \end{bmatrix} \frac{1}{2}R_2 + R_3$$

$$\begin{bmatrix} 1 & -\frac{1}{2} & -\frac{1}{2} & | & \frac{5}{2} \\ 0 & 1 & \frac{7}{5} & | & -\frac{9}{5} \\ 0 & 0 & 1 & | & -2 \end{bmatrix} \frac{5}{16}R_3$$

The system is

$$x - \frac{1}{2}y - \frac{1}{2}z = \frac{5}{2}$$

$$y + \frac{7}{5}z = -\frac{9}{5}$$

$$z = -2$$

Substitute -2 for z in the second equation.

$$y + \frac{7}{5}z = -\frac{9}{5}$$

$$y + \frac{7}{5}(-2) = -\frac{9}{5}$$

$$y - \frac{14}{5} = -\frac{9}{5}$$

$$y = \frac{5}{5} = 1$$

Substitute 1 for y and -2 for z in the first equation.

$$x - \frac{1}{2}y - \frac{1}{2}z = \frac{5}{2}$$

$$x - \frac{1}{2}(1) - \frac{1}{2}(-2) = \frac{5}{2}$$

$$x - \frac{1}{2} + 1 = \frac{5}{2}$$

$$x + \frac{1}{2} = \frac{5}{2}$$

$$x = \frac{4}{2} = 2$$

The solution is $(2, 1, -2)$.

43.
$$3a - b + c = 2$$
$$2a - 3b + 4c = 4$$
$$a + 2b - 3c = -6$$

$$\begin{bmatrix} 3 & -1 & 1 & | & 2 \\ 2 & -3 & 4 & | & 4 \\ 1 & 2 & -3 & | & -6 \end{bmatrix}$$

$$\begin{bmatrix} 1 & -\frac{1}{3} & \frac{1}{3} & | & \frac{2}{3} \\ 2 & -3 & 4 & | & 4 \\ 1 & 2 & -3 & | & -6 \end{bmatrix} \frac{1}{3}R_1$$

$$\begin{bmatrix} 1 & -\frac{1}{3} & \frac{1}{3} & | & \frac{2}{3} \\ 0 & -\frac{7}{3} & \frac{10}{3} & | & \frac{8}{3} \\ 1 & 2 & -3 & | & -6 \end{bmatrix} -2R_1 + R_2$$

$$\begin{bmatrix} 1 & -\frac{1}{3} & \frac{1}{3} & | & \frac{2}{3} \\ 0 & -\frac{7}{3} & \frac{10}{3} & | & \frac{8}{3} \\ 0 & \frac{7}{3} & -\frac{10}{3} & | & -\frac{20}{3} \end{bmatrix} -1R_1 + R_3$$

$$\begin{bmatrix} 1 & -\frac{1}{3} & \frac{1}{3} & | & \frac{2}{3} \\ 0 & 1 & -\frac{10}{7} & | & -\frac{8}{7} \\ 0 & \frac{7}{3} & -\frac{10}{3} & | & -\frac{20}{3} \end{bmatrix} -\frac{3}{7}R_2$$

$$\begin{bmatrix} 1 & -\frac{1}{3} & \frac{1}{3} & | & \frac{2}{3} \\ 0 & 1 & -\frac{10}{7} & | & -\frac{8}{7} \\ 0 & 0 & 0 & | & -4 \end{bmatrix}$$

Since the last row has all zeros on the left side and a nonzero number on the right side, the system is inconsistent.

44.
$$x + y + z = 3$$
$$3x + 2y = 1$$
$$y - 3z = -10$$

$$\begin{bmatrix} 1 & 1 & 1 & | & 3 \\ 3 & 2 & 0 & | & 1 \\ 0 & 1 & -3 & | & -10 \end{bmatrix}$$

$$\begin{bmatrix} 1 & 1 & 1 & | & 3 \\ 0 & -1 & -3 & | & -8 \\ 0 & 1 & -3 & | & -10 \end{bmatrix} -3R_1 + R_2$$

$$\begin{bmatrix} 1 & 1 & 1 & | & 3 \\ 0 & 1 & 3 & | & 8 \\ 0 & 1 & -3 & | & -10 \end{bmatrix} -1R_2$$

$$\begin{bmatrix} 1 & 1 & 1 & | & 3 \\ 0 & 1 & 3 & | & 8 \\ 0 & 0 & -6 & | & -18 \end{bmatrix} -1R_2 + R_3$$

$$\begin{bmatrix} 1 & 1 & 1 & | & 3 \\ 0 & 1 & 3 & | & 8 \\ 0 & 0 & 1 & | & 3 \end{bmatrix} -\frac{1}{6}R_3$$

The system is
$$x + y + z = 3$$
$$y + 3z = 8$$
$$z = 3$$
Substitute 3 for z in the second equation.
$$y + 3z = 8$$
$$y + 3(3) = 8$$
$$y + 9 = 8$$
$$y = -1$$
Substitute -1 for y and 3 for z in the first equation.
$$x + y + z = 3$$
$$x - 1 + 3 = 3$$
$$x + 2 = 3$$
$$x = 1$$
The solution is $(1, -1, 3)$.

45.
$$5x + 6y = 14$$
$$x - 3y = 7$$
To solve, first calculate D, D_x, and D_y.

$$D = \begin{vmatrix} 5 & 6 \\ 1 & -3 \end{vmatrix}$$
$$= (5)(-3) - (1)(6)$$
$$= -15 - 6$$
$$= -21$$

$$D_x = \begin{vmatrix} 14 & 6 \\ 7 & -3 \end{vmatrix}$$
$$= (14)(-3) - (7)(6)$$
$$= -42 - 42$$
$$= -84$$

$$D_y = \begin{vmatrix} 5 & 14 \\ 1 & 7 \end{vmatrix}$$
$$= (5)(7) - (1)(14)$$
$$= 35 - 14$$
$$= 21$$

$$x = \frac{D_x}{D} = \frac{-84}{-21} = 4 \text{ and } y = \frac{D_y}{D} = \frac{21}{-21} = -1$$
The solution is $(4, -1)$.

46. $3x + 5y = -2$
$5x + 3y = 2$
To solve, first calculate D, D_x, and D_y.

$D = \begin{vmatrix} 3 & 5 \\ 5 & 3 \end{vmatrix}$

$= (3)(3) - (5)(5)$
$= 9 - 25$
$= -16$

$D_x = \begin{vmatrix} -2 & 5 \\ 2 & 3 \end{vmatrix}$

$= (-2)(3) - (2)(5)$
$= -6 - 10$
$= -16$

$D_y = \begin{vmatrix} 3 & -2 \\ 5 & 2 \end{vmatrix}$

$= (3)(2) - (5)(-2)$
$= 6 + 10$
$= 16$

$x = \dfrac{D_x}{D} = \dfrac{-16}{-16} = 1$ and $y = \dfrac{D_y}{D} = \dfrac{16}{-16} = -1$
The solution is $(1, -1)$.

47. $4m + 3n = 2$
$7m - 2n = -11$
To solve, first calculate D, D_m, and D_n.

$D = \begin{vmatrix} 4 & 3 \\ 7 & -2 \end{vmatrix}$

$= (4)(-2) - (7)(3)$
$= -8 - 21$
$= -29$

$D_m = \begin{vmatrix} 2 & 3 \\ -11 & -2 \end{vmatrix}$

$= (2)(-2) - (-11)(3)$
$= -4 + 33$
$= 29$

$D_n = \begin{vmatrix} 4 & 2 \\ 7 & -11 \end{vmatrix}$

$= (4)(-11) - (7)(2)$
$= -44 - 14$
$= -58$

$m = \dfrac{D_m}{D} = \dfrac{29}{-29} = -1$ and

$n = \dfrac{D_n}{D} = \dfrac{-58}{-29} = 2$.
The solution is $(-1, 2)$.

48. $r + s + t = 8$
$r - s - t = 0$
$r + 2s + t = 9$
To solve, calculate D, D_r, D_s, and D_t.

$D = \begin{vmatrix} 1 & 1 & 1 \\ 1 & -1 & -1 \\ 1 & 2 & 1 \end{vmatrix}$

$= 1 \begin{vmatrix} -1 & -1 \\ 2 & 1 \end{vmatrix} - 1 \begin{vmatrix} 1 & 1 \\ 2 & 1 \end{vmatrix} + 1 \begin{vmatrix} 1 & 1 \\ -1 & -1 \end{vmatrix}$

$= 1(-1 + 2) - 1(1 - 2) + 1(-1 + 1)$
$= 1(1) - 1(-1) + 1(0)$
$= 1 + 1 + 0$
$= 2$

$D_r = \begin{vmatrix} 8 & 1 & 1 \\ 0 & -1 & -1 \\ 9 & 2 & 1 \end{vmatrix}$

$= 8 \begin{vmatrix} -1 & -1 \\ 2 & 1 \end{vmatrix} - 0 \begin{vmatrix} 1 & 1 \\ 2 & 1 \end{vmatrix} + 9 \begin{vmatrix} 1 & 1 \\ -1 & -1 \end{vmatrix}$

$= 8(-1 + 2) - 0(1 - 2) + 9(-1 + 1)$
$= 8(1) - 0(-1) + 9(0)$
$= 8 + 0 + 0$
$= 8$

$D_s = \begin{vmatrix} 1 & 8 & 1 \\ 1 & 0 & -1 \\ 1 & 9 & 1 \end{vmatrix}$

$= 1 \begin{vmatrix} 0 & -1 \\ 9 & 1 \end{vmatrix} - 1 \begin{vmatrix} 8 & 1 \\ 9 & 1 \end{vmatrix} + 1 \begin{vmatrix} 8 & 1 \\ 0 & -1 \end{vmatrix}$

$= 1(0 + 9) - 1(8 - 9) + 1(-8 - 0)$
$= 1(9) - 1(-1) + 1(-8)$
$= 9 + 1 - 8$
$= 2$

$$D_t = \begin{vmatrix} 1 & 1 & 8 \\ 1 & -1 & 0 \\ 1 & 2 & 9 \end{vmatrix}$$

$$= 1\begin{vmatrix} -1 & 0 \\ 2 & 9 \end{vmatrix} - 1\begin{vmatrix} 1 & 8 \\ 2 & 9 \end{vmatrix} + 1\begin{vmatrix} 1 & 8 \\ -1 & 0 \end{vmatrix}$$

$$= 1(-9-0) - 1(9-16) + 1(0+8)$$

$$= 1(-9) - 1(-7) + 1(8)$$

$$= -9 + 7 + 8$$

$$= 6$$

$$r = \frac{D_r}{D} = \frac{8}{2} = 4, \ s = \frac{D_s}{D} = \frac{2}{2} = 1, \text{ and}$$

$$t = \frac{D_t}{D} = \frac{6}{2} = 3$$

The solution is (4, 1, 3).

49. $x + 2y - 4z = 17$
$2x - y + z = -9$
$2x - y - 3z = -1$

To solve, calculate D, D_x, D_y, and D_z.

$$D = \begin{vmatrix} 1 & 2 & -4 \\ 2 & -1 & 1 \\ 2 & -1 & -3 \end{vmatrix}$$

$$= 1\begin{vmatrix} -1 & 1 \\ -1 & -3 \end{vmatrix} - 2\begin{vmatrix} 2 & -4 \\ -1 & -3 \end{vmatrix} + 2\begin{vmatrix} 2 & -4 \\ -1 & 1 \end{vmatrix}$$

$$= 1(3+1) - 2(-6-4) + 2(2-4)$$

$$= 1(4) - 2(-10) + 2(-2)$$

$$= 4 + 20 - 4$$

$$= 20$$

$$D_x = \begin{vmatrix} 17 & 2 & -4 \\ -9 & -1 & 1 \\ -1 & -1 & -3 \end{vmatrix}$$

$$= 17\begin{vmatrix} -1 & 1 \\ -1 & -3 \end{vmatrix} - (-9)\begin{vmatrix} 2 & -4 \\ -1 & -3 \end{vmatrix} - 1\begin{vmatrix} 2 & -4 \\ -1 & 1 \end{vmatrix}$$

$$= 17(3+1) + 9(-6-4) - 1(2-4)$$

$$= 17(4) + 9(-10) - 1(-2)$$

$$= 68 - 90 + 2$$

$$= -20$$

$$D_y = \begin{vmatrix} 1 & 17 & -4 \\ 2 & -9 & 1 \\ 2 & -1 & -3 \end{vmatrix}$$

$$= 1\begin{vmatrix} -9 & 1 \\ -1 & -3 \end{vmatrix} - 2\begin{vmatrix} 17 & -4 \\ -1 & -3 \end{vmatrix} + 2\begin{vmatrix} 17 & -4 \\ -9 & 1 \end{vmatrix}$$

$$= 1(27+1) - 2(-51-4) + 2(17-36)$$

$$= 1(28) - 2(-55) + 2(-19)$$

$$= 28 + 110 - 38$$

$$= 100$$

$$D_z = \begin{vmatrix} 1 & 2 & 17 \\ 2 & -1 & -9 \\ 2 & -1 & -1 \end{vmatrix}$$

$$= 1\begin{vmatrix} -1 & -9 \\ -1 & -1 \end{vmatrix} - 2\begin{vmatrix} 2 & 17 \\ -1 & -1 \end{vmatrix} + 2\begin{vmatrix} 2 & 17 \\ -1 & -9 \end{vmatrix}$$

$$= 1(1-9) - 2(-2+17) + 2(-18+17)$$

$$= 1(-8) - 2(15) + 2(-1)$$

$$= -8 - 30 - 2$$

$$= -40$$

$$x = \frac{D_x}{D} = \frac{-20}{20} = -1, \ y = \frac{D_y}{D} = \frac{100}{20} = 5,$$

and $z = \frac{D_z}{D} = \frac{-40}{20} = -2$

The solution is (−1, 5, −2).

50. $\quad y + 3z = 4$
$-x - y + 2z = 0$
$x + 2y + z = 1$

To solve, first calculate D, D_x, D_y, and D_z.

$$D = \begin{vmatrix} 0 & 1 & 3 \\ -1 & -1 & 2 \\ 1 & 2 & 1 \end{vmatrix}$$

$$= 0\begin{vmatrix} -1 & 2 \\ 2 & 1 \end{vmatrix} - (-1)\begin{vmatrix} 1 & 3 \\ 2 & 1 \end{vmatrix} + 1\begin{vmatrix} 1 & 3 \\ -1 & 2 \end{vmatrix}$$

$$= 0(-1-4) + 1(1-6) + 1(2+3)$$

$$= 0(-5) + 1(-5) + 1(5)$$

$$= 0 - 5 + 5$$

$$= 0$$

$$D_x = \begin{vmatrix} 4 & 1 & 3 \\ 0 & -1 & 2 \\ 1 & 2 & 1 \end{vmatrix}$$

$$= 4 \begin{vmatrix} -1 & 2 \\ 2 & 1 \end{vmatrix} - 0 \begin{vmatrix} 1 & 3 \\ 2 & 1 \end{vmatrix} + 1 \begin{vmatrix} 1 & 3 \\ -1 & 2 \end{vmatrix}$$

$$= -4(-1-4) - 0(1-6) + 1(2+3)$$

$$= 4(-5) - 0(-5) + 1(5)$$

$$= -20 + 0 + 5$$

$$= -15$$

Since $D = 0$ and $D_x = -15$, the system is inconsistent.

51. $-x + 3y > 6$
$2x - y \leq 2$

For $-x + 3y > 6$, graph the line $-x + 3y = 6$ using a dashed line. For the check point, select $(0, 0)$:

$-x + 3y > 6$
$-0 + 3(0) > 6$
$\qquad 0 > 6 \qquad$ False

Since this is a false statement, shade the region which does not contain the point $(0, 0)$.
For $2x - y \leq 2$, graph the line $2x - y = 2$ using a solid line. For the check point, select $(0, 0)$:

$2x - y \leq 2$
$2(0) - 0 \leq 2$
$\qquad 0 \leq 2 \qquad$ True

Since this is a true statement, shade the region which contains the point $(0, 0)$. To obtain the final region, take the intersection of the above two regions.

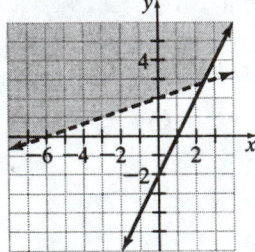

52. $5x - 2y \leq 10$
$3x + 2y > 6$

For $5x - 2y \leq 10$, graph the line $5x - 2y = 10$ using a solid line. For the check point, select $(0, 0)$:

$5x - 2y \leq 10$
$5(0) - 2(0) \leq 10$
$\qquad 0 \leq 10 \qquad$ True

Since this is a true statement, shade the region which contains the point $(0, 0)$.
For $3x + 2y > 6$, graph the line $3x + 2y = 6$ using a dashed line. For the check point, select $(0, 0)$:

$3x + 2y > 6$
$3(0) + 2(0) > 6$
$\qquad 0 > 6 \qquad$ False

Since this is a false statement, shade the region which does not contain the point $(0, 0)$. To obtain the final region, take the intersection of the above two regions.

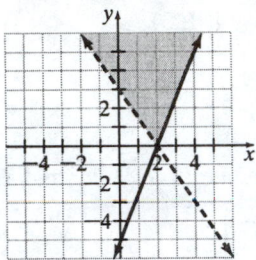

53. $y > 2x + 3$
$y < -x + 4$

For $y > 2x + 3$, graph the line $y = 2x + 3$ using a dashed line. For the check point, select $(0, 0)$:

$y > 2x + 3$
$0 > 2(0) + 3$
$0 > 3 \qquad$ False

Since this is a false statement, shade the region which does not contain the point $(0, 0)$.
For $y < -x + 4$, graph the line $y = -x + 4$ using a dashed line. For the check point, select $(0, 0)$:

$y < -x + 4$
$0 < -0 + 4$
$0 < 4 \qquad$ True

Since this is a true statement, shade the region which contains the point $(0, 0)$. To obtain the final region, take the intersection of the above two regions.

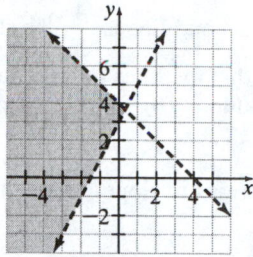

54. $x > -2y + 4$

$y < -\dfrac{1}{2}x - \dfrac{3}{2}$

For $x > -2y + 4$, graph the line $x = -2y + 4$ using a dashed line. For the check point, select $(0, 0)$:

$x > -2y + 4$

$0 > -2(0) + 4$

$0 > 4$ False

Since this is a false statement, shade the region which does not contain the point $(0, 0)$.

For $y < -\dfrac{1}{2}x - \dfrac{3}{2}$, graph the line

$y = -\dfrac{1}{2}x - \dfrac{3}{2}$ using a dashed line. For the

check point, select $(0, 0)$:

$y < -\dfrac{1}{2}x - \dfrac{3}{2}$

$0 < -\dfrac{1}{2}(0) - \dfrac{3}{2}$

$0 < -\dfrac{3}{2}$ False

Since this is a false statement, shade the region which does not contain the point $(0, 0)$. To obtain the final region, take the intersection of the above two regions. The regions do not overlap, so there are no solutions.

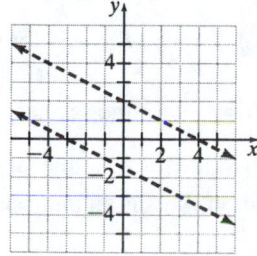

55. $x \geq 0$

$y \geq 0$

$x + y \leq 6$

$4x + y \leq 8$

The first two inequalities indicate that the solution must be in the first quadrant. For $x + y \leq 6$, the graph is the line $x + y = 6$ along with the region below this line. For $4x + y \leq 8$, the graph is the line $4x + y = 8$ along with the region below this line. To obtain the final region, take the intersection of these regions.

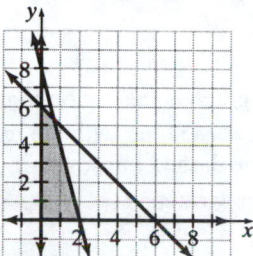

56. $x \geq 0$

$y \geq 0$

$2x + y \leq 6$

$4x + 5y \leq 20$

The first two inequalities indicate that the solution must be in the first quadrant. For $2x + y \leq 6$, graph the line $2x + y = 6$ along with the region below this line. For $4x + 5y \leq 20$, graph the line $4x + 5y = 20$ along with the region below this line. To obtain the final region, take the intersection of these regions.

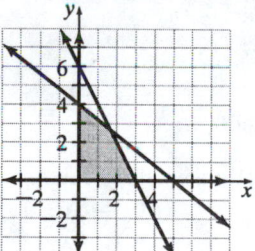

57. $|x| \leq 3$

$|y| > 2$

For $|x| \leq 3$, the graph is the region between the solid lines $x = -3$ and $x = 3$. For $|y| > 2$, the graph is the region above the dashed line

$y = 2$ along with the region below the dashed line $y = -2$. To obtain the final region, take the intersection of these regions.

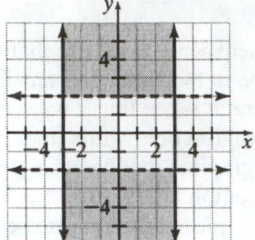

58. $|x| > 4$

$|y - 2| \leq 3$

For $|x| > 4$, the graph is the region to the left of dashed line $x = -4$ along with the region to the right of the dashed line $x = 4$.

$|y - 2| \leq 3$ can be written as

$-3 \leq y - 2 \leq 3$

$-1 \leq y \leq 5$

For $|y - 2| \leq 3$, the graph is the region between the solid lines $y = -1$ and $y = 5$. To obtain the final region, take the intersection of these regions.

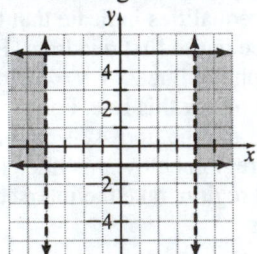

Practice Test

1. Answers will vary.

2. Write both equations in slope-intercept form.

$4x + 3y = -6 \qquad 6y = 8x + 4$

$\quad 3y = -4x - 6 \qquad y = \dfrac{8x + 4}{6}$

$\quad y = \dfrac{-4x - 6}{3} \qquad y = \dfrac{4}{3}x + \dfrac{2}{3}$

$\quad y = -\dfrac{4}{3}x - 2$

Since the slope of the first line is $-\dfrac{4}{3}$ and

the slope of the second line is $\dfrac{4}{3}$, the slopes are different so that the lines intersect to produce one solution. This is a consistent system.

3. Write both equations in slope-intercept form.

$5x + 3y = 9$

$\quad 3y = -5x + 9$

$\quad y = \dfrac{-5x + 9}{3}$

$\quad y = -\dfrac{5}{3}x + 3$

$\quad 2y = -\dfrac{10}{3}x + 6$

$\quad \dfrac{1}{2}(2y) = \dfrac{1}{2}\left(-\dfrac{10}{3}x + 6\right)$

$\quad y = -\dfrac{5}{3}x + 3$

Since the equations are identical, there is an infinite number of solutions and this is a dependent system.

4. Write both equations in slope-intercept form.

$5x - 4y = 6$

$\quad -4y = -5x + 6$

$\quad y = \dfrac{-5x + 6}{-4}$

$\quad y = \dfrac{5}{4}x - \dfrac{3}{2}$

$\quad -10x + 8y = -10$

$\quad 8y = 10x - 10$

$\quad y = \dfrac{10x - 10}{8}$

$\quad y = \dfrac{5}{4}x - \dfrac{5}{4}$

Since the slope of each line is $\dfrac{5}{4}$, but the

y-intercepts are different $\left(b = -\dfrac{3}{2}\right.$ for the

first equation, $b = -\dfrac{5}{4}$ for the second

equation$\Big)$, the two lines are parallel and produce no solution. This is an inconsistent system.

5. Graph the equations $y = 3x - 2$ and $y = -2x + 8$.

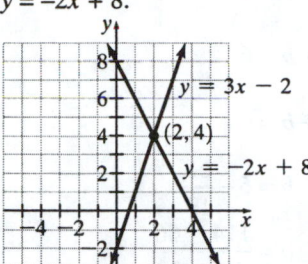

The lines intersect and the point of intersection is $(2, 4)$.

6. Graph the equations $y = -x + 6$ and $y = 2x + 3$.

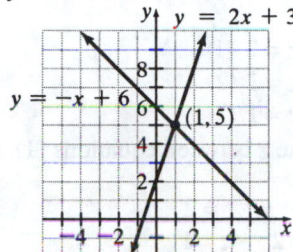

The lines intersect and the point of intersection is $(1, 5)$.

7. $y = 6x - 12$
$y = 4x - 8$
Substitute $6x - 12$ for y in the second equation.
$$y = 4x - 8$$
$$6x - 12 = 4x - 8$$
$$6x - 4x = 12 - 8$$
$$2x = 4$$
$$x = 2$$
Substitute 2 for x in the first equation.
$$y = 6x - 12$$
$$y = 6(2) - 12$$
$$y = 12 - 12$$
$$y = 0$$
The solution is $(2, 0)$.

8. $3x + y = 8$
$x - 2y = 5$
Solve the first equation for y.
$$3x + y = 8$$
$$y = -3x + 8$$
Substitute $-3x + 8$ for y in the second

equation.
$$x - 2y = 5$$
$$x - 2(-3x + 8) = 5$$
$$x + 6x - 16 = 5$$
$$7x = 21$$
$$x = 3$$
Substitute 3 for x in the equation
$y = -3x + 8$.
$$y = -3(3) + 8$$
$$y = -9 + 8$$
$$y = -1$$
The solution is $(3, -1)$.

9. $4x - 3y = 10$
$2x + y = 5$
To eliminate y, multiply the second equation by 3 and then add.
$$4x - 3y = 10$$
$$3[2x + y = 5]$$
gives
$$4x - 3y = 10$$
$$\underline{6x + 3y = 15}$$
Add: $10x = 25$
$$x = \frac{5}{2}$$
Substitute $\frac{5}{2}$ for x in the second equation.
$$2x + y = 5$$
$$2\left(\frac{5}{2}\right) + y = 5$$
$$5 + y = 5$$
$$y = 0$$
The solution is $\left(\frac{5}{2}, 0\right)$.

10. $ 0.3x = 0.2y + 0.4$
$-1.2x + 0.8y = -1.6$
Write the system in standard form.
$$0.3x - 0.2y = 0.4$$
$$-1.2x + 0.8y = -1.6$$
To eliminate x, multiply the first equation by 4 and then add.
$$4[0.3x - 0.2y = 0.4]$$
$$-1.2x + 0.8y = -1.6$$
gives

$$1.2x - 0.8y = 1.6$$
$$-1.2x + 0.8y = -1.6$$

Add: $0 = 0$ True

Since this is a true statement, there are an infinite number of solutions and this is a dependent system.

11. $\dfrac{3}{2}a + b = 6$

$a - \dfrac{5}{2}b = -4$

To clear fractions, multiply both equations by 2.

$2\left[\dfrac{3}{2}a + b = 6\right]$

$2\left[a - \dfrac{5}{2}b = -4\right]$

gives

$3a + 2b = 12$
$2a - 5b = -8$

Now, to eliminate b, multiply the first equation by 5 and the second equation by 2 and then add.

$5[3a + 2b = 12]$
$2[2a - 5b = -8]$

gives

$$15a + 10b = 60$$
$$\underline{4a - 10b = -16}$$

Add: $19a = 44$

$a = \dfrac{44}{19}$

Substitute $\dfrac{44}{19}$ for a in the first equation.

$\dfrac{3}{2}a + b = 6$

$\dfrac{3}{2}\left(\dfrac{44}{19}\right) + b = 6$

$\dfrac{66}{19} + b = 6$

$b = 6 - \dfrac{66}{19}$

$b = \dfrac{114}{19} - \dfrac{66}{19}$

$b = \dfrac{48}{19}$

The solution is $\left(\dfrac{44}{19}, \dfrac{48}{19}\right)$.

12. $x + y + z = 2$ (1)
$-2x - y + z = 1$ (2)
$x - 2y - z = 1$ (3)

To eliminate z between equations (1) and (3) simply add.

$$x + y + z = 2$$
$$\underline{x - 2y - z = 1}$$

Add: $2x - y = 3$ (4)

To eliminate z between equations (2) and (3) simply add.

$$-2x - y + z = 1$$
$$\underline{x - 2y - z = 1}$$

Add: $-x - 3y = 2$ (5)

Equations (4) and (5) are two equations in two unknowns.

$2x - y = 3$
$-x - 3y = 2$

To eliminate x, multiply equation (5) by 2 and then add.

$2x - y = 3$
$2[-x - 3y = 2]$

gives

$$2x - y = 3$$
$$\underline{-2x - 6y = 4}$$

Add: $-7y = 7$

$y = \dfrac{7}{-7} = -1$

Substitute -1 for y in equation (4).

$$2x - y = 3$$
$$2x - (-1) = 3$$
$$2x + 1 = 3$$
$$2x = 2$$
$$x = \frac{2}{2} = 1$$

Finally, substitute 1 for x and -1 for y in equation (1).

$$x + y + z = 2$$
$$1 - 1 + z = 2$$
$$0 + z = 2$$
$$z = 2$$

The solution is $(1, -1, 2)$.

13. $4x - 5y + 3z = 2$
$2x - y - 2z = 4$
$3x + 2y - z = -3$

The augmented matrix is $\begin{bmatrix} 4 & -5 & 3 & | & 2 \\ 2 & -1 & -2 & | & 4 \\ 3 & 2 & -1 & | & -3 \end{bmatrix}$

14. $\begin{bmatrix} 6 & -2 & 4 & | & 4 \\ 4 & 3 & 5 & | & 6 \\ 2 & -1 & 4 & | & -3 \end{bmatrix}$

$\begin{bmatrix} 6 & -2 & 4 & | & 4 \\ 0 & 5 & -3 & | & 12 \\ 2 & -1 & 4 & | & -3 \end{bmatrix} -2R_3 + R_2$

15. $x - 5y = -2$
$3x - y = 8$

$\begin{bmatrix} 1 & -5 & | & -2 \\ 3 & -1 & | & 8 \end{bmatrix}$

$\begin{bmatrix} 1 & -5 & | & -2 \\ 0 & 14 & | & 14 \end{bmatrix} -3R_1 + R_2$

$\begin{bmatrix} 1 & -5 & | & -2 \\ 0 & 1 & | & 1 \end{bmatrix} \frac{1}{14}R_2$

The system is
$$x - 5y = -2$$
$$y = 1$$

Substitute 1 for y in the first equation.

$$x - 5y = -2$$
$$x - 5(1) = -2$$
$$x = 3$$

The solution is $(3, 1)$.

16. $x - 2y + z = 7$
$-2x - y - z = -7$
$3x - 2y + 2z = 15$

$\begin{bmatrix} 1 & -2 & 1 & | & 7 \\ -2 & -1 & -1 & | & -7 \\ 3 & -2 & 2 & | & 15 \end{bmatrix}$

$\begin{bmatrix} 1 & -2 & 1 & | & 7 \\ 0 & -5 & 1 & | & 7 \\ 3 & -2 & 2 & | & 15 \end{bmatrix} 2R_1 + R_2$

$\begin{bmatrix} 1 & -2 & 1 & | & 7 \\ 0 & -5 & 1 & | & 7 \\ 0 & 4 & -1 & | & -6 \end{bmatrix} -3R_1 + R_3$

$\begin{bmatrix} 1 & -2 & 1 & | & 7 \\ 0 & 1 & -\frac{1}{5} & | & -\frac{7}{5} \\ 0 & 4 & -1 & | & -6 \end{bmatrix} -\frac{1}{5}R_2$

$\begin{bmatrix} 1 & -2 & 1 & | & 7 \\ 0 & 1 & -\frac{1}{5} & | & -\frac{7}{5} \\ 0 & 0 & -\frac{1}{5} & | & -\frac{2}{5} \end{bmatrix} -4R_2 + R_3$

$\begin{bmatrix} 1 & -2 & 1 & | & 7 \\ 0 & 1 & -\frac{1}{5} & | & -\frac{7}{5} \\ 0 & 0 & 1 & | & 2 \end{bmatrix} -5R_3$

The system is
$$x - 2y + z = 7$$
$$y - \frac{1}{5}z = -\frac{7}{5}$$
$$z = 2$$

Substitute 2 for z in the second equation.

$$y - \frac{1}{5}z = -\frac{7}{5}$$
$$y - \frac{1}{5}(2) = -\frac{7}{5}$$
$$y - \frac{2}{5} = -\frac{7}{5}$$
$$y = -1$$

Substitute -1 for y and 2 for z in the first equation.

$$x - 2y + z = 7$$
$$x - 2(-1) + 2 = 7$$
$$x + 2 + 2 = 7$$
$$x + 4 = 7$$
$$x = 3$$

The solution is $(3, -1, 2)$.

17. $\begin{vmatrix} 4 & 6 \\ -2 & 5 \end{vmatrix} = 4(5) - (-2)(6)$

$$= 20 - (-12)$$
$$= 20 + 12$$
$$= 32$$

18. $\begin{vmatrix} 8 & 2 & -1 \\ 3 & 0 & 5 \\ 6 & -3 & 4 \end{vmatrix} = 8\begin{vmatrix} 0 & 5 \\ -3 & 4 \end{vmatrix} - 3\begin{vmatrix} 2 & -1 \\ -3 & 4 \end{vmatrix} + 6\begin{vmatrix} 2 & -1 \\ 0 & 5 \end{vmatrix}$

$$= 8(0 + 15) - 3(8 - 3) + 6(10 - 0)$$
$$= 8(15) - 3(5) + 6(10)$$
$$= 120 - 15 + 60$$
$$= 165$$

19. $5x - 2y = -13$
$2x + y = 11$

To solve, first calculate D, D_x, and D_y.

$$D = \begin{vmatrix} 5 & -2 \\ 2 & 1 \end{vmatrix}$$
$$= (5)(1) - (2)(-2)$$
$$= 5 + 4$$
$$= 9$$

$$D_x = \begin{vmatrix} -13 & -2 \\ 11 & 1 \end{vmatrix}$$
$$= (-13)(1) - (11)(-2)$$
$$= -13 + 22$$
$$= 9$$

$$D_y = \begin{vmatrix} 5 & -13 \\ 2 & 11 \end{vmatrix}$$
$$= (5)(11) - (2)(-13)$$
$$= 55 + 26$$
$$= 81$$

$$x = \frac{D_x}{D} = \frac{9}{9} = 1 \text{ and } y = \frac{D_y}{D} = \frac{81}{9} = 9.$$

The solution is $(1, 9)$.

20. $2x - y - z = -3$
$3x - 2y - 2z = -5$
$-x + y + 2z = 4$

To solve, first calculate D, D_x, D_y, and D_z.

$$D = \begin{vmatrix} 2 & -1 & -1 \\ 3 & -2 & -2 \\ -1 & 1 & 2 \end{vmatrix}$$

$$= 2\begin{vmatrix} -2 & -2 \\ 1 & 2 \end{vmatrix} - 3\begin{vmatrix} -1 & -1 \\ 1 & 2 \end{vmatrix} - 1\begin{vmatrix} -1 & -1 \\ -2 & -2 \end{vmatrix}$$
$$= 2(-4 + 2) - 3(-2 + 1) - 1(2 - 2)$$
$$= 2(-2) - 3(-1) - 1(0)$$
$$= -4 + 3 - 0$$
$$= -1$$

$$D_x = \begin{vmatrix} -3 & -1 & -1 \\ -5 & -2 & -2 \\ 4 & 1 & 2 \end{vmatrix}$$

$$= -3\begin{vmatrix} -2 & -2 \\ 1 & 2 \end{vmatrix} - (-5)\begin{vmatrix} -1 & -1 \\ 1 & 2 \end{vmatrix} + 4\begin{vmatrix} -1 & -1 \\ -2 & -2 \end{vmatrix}$$
$$= -3(-4 + 2) + 5(-2 + 1) + 4(2 - 2)$$
$$= -3(-2) + 5(-1) + 4(0)$$
$$= 6 - 5 + 0$$
$$= 1$$

$$D_y = \begin{vmatrix} 2 & -3 & -1 \\ 3 & -5 & -2 \\ -1 & 4 & 2 \end{vmatrix}$$

$$= 2\begin{vmatrix} -5 & -2 \\ 4 & 2 \end{vmatrix} - 3\begin{vmatrix} -3 & -1 \\ 4 & 2 \end{vmatrix} - 1\begin{vmatrix} -3 & -1 \\ -5 & -2 \end{vmatrix}$$
$$= 2(-10 + 8) - 3(-6 + 4) - 1(6 - 5)$$
$$= 2(-2) - 3(-2) - 1(1)$$
$$= -4 + 6 - 1$$
$$= 1$$

$$D_z = \begin{vmatrix} 2 & -1 & -3 \\ 3 & -2 & -5 \\ -1 & 1 & 4 \end{vmatrix}$$

$$= 2\begin{vmatrix} -2 & -5 \\ 1 & 4 \end{vmatrix} - 3\begin{vmatrix} -1 & -3 \\ 1 & 4 \end{vmatrix} - 1\begin{vmatrix} -1 & -3 \\ -2 & -5 \end{vmatrix}$$
$$= 2(-8 + 5) - 3(-4 + 3) - 1(5 - 6)$$
$$= 2(-3) - 3(-1) - 1(-1)$$
$$= -6 + 3 + 1$$
$$= -2$$

$x = \dfrac{D_x}{D} = \dfrac{1}{-1} = -1, \; y = \dfrac{D_y}{D} = \dfrac{1}{-1} = -1,$

$z = \dfrac{D_z}{D} = \dfrac{-2}{-1} = 2$

The solution is $(-1, -1, 2)$.

21. Let x be the number of pounds of cashews and y be the number of pounds of peanuts.

$x + y = 20$

$7x + 5.5y = 20(6)$

Solve the first equation for y.

$x + y = 20$

$\quad y = -x + 20$

Substitute $-x + 20$ for y in the second equation.

$7x + 5.5y = 20(6)$

$7x + 5.5(-x + 20) = 20(6)$

$7x - 5.5x + 110 = 120$

$\quad 1.5x + 110 = 120$

$\quad\quad 1.5x = 10$

$x = \dfrac{10}{1.5} \cdot \dfrac{10}{10} = \dfrac{100}{15} = \dfrac{20}{3}$ or $6\dfrac{2}{3}$

Thus, Max should mix $6\dfrac{2}{3}$ lb of cashews

with $20 - 6\dfrac{2}{3} = 13\dfrac{1}{3}$ lb of peanuts to obtain the desired mixture.

22. Let x = amount of 6% solution

$\quad y$ = amount of 15% solution

$\quad\quad x + y = 10$

$0.06x + 0.15y = 0.09(10)$

The system can be written as

$\quad x + y = 10$

$6x + 15y = 90$

Solve the first equation for y.

$y = 10 - x$

Substitute $10 - x$ for y in the second equation.

$\quad 6x + 15y = 90$

$\quad 6x + 15(10 - x) = 90$

$\quad 6x + 150 - 15x = 90$

$\quad\quad -9x = -60$

$\quad\quad x = \dfrac{-60}{-9} = \dfrac{20}{3} = 6\dfrac{2}{3}$

Substitute $6\dfrac{2}{3}$ for x into $y = 10 - x$

$y = 10 - 6\dfrac{2}{3}$

$y = 3\dfrac{1}{3}$

She should mix $6\dfrac{2}{3}$ liters of 6% solution

and $3\dfrac{1}{3}$ liters of 15% solution.

23. Let x = smallest number

$\quad y$ = remaining number

$\quad z$ = largest number

$x + y + z = 25$

$\quad\quad z = 3x$

$\quad\quad y = 2x + 1$

Substitute $2x + 1$ for y and $3x$ for z in the first equation.

$\quad x + y + z = 25$

$x + 2x + 1 + 3x = 25$

$\quad\quad 6x = 24$

$\quad\quad x = 4$

Substitute 4 for x in the third equation.

$y = 2x + 1$

$y = 2(4) + 1$

$y = 9$

Substitute 4 for x in the second equation.

$z = 3x$

$z = 3(4)$

$z = 12$

The three numbers are 4, 9, and 12.

24. $\quad 3x + 2y < 9$

$\quad -2x + 5y \le 10$

For $3x + 2y < 9$, graph the line $3x + 2y = 9$ using a dashed line. For the check point, select $(0, 0)$.

$\quad 3x + 2y < 9$

$3(0) + 2(0) < 9$

$\quad\quad 0 < 9 \quad\quad$ True

Since this is a true statement, shade the region which contains the point $(0, 0)$. this is the region "below" the line.

For $-2x + 5y \le 10$, graph the line $-2x + 5y = 10$ using a solid line. For the check point, select $(0, 0)$.

$\quad -2x + 5y \le 10$

$-2(0) + 5(0) \le 10$

$\quad\quad 0 \le 10 \quad\quad$ True

Since this is a true statement, shade the region which contains the point (0, 0). this is the region "below" the line. To obtain the final region, take the intersection of the above two regions.

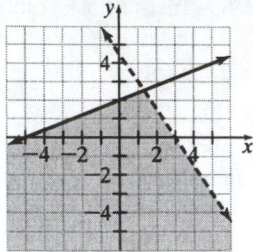

25. $|x| > 3$

    $|y| \le 1$

    For $|x| > 3$, the graph is the region to the left of the dashed line $x = -3$ along with the region to the right of the dashed line $x = 3$. For $|y| \le 1$, the graph is the region between the solid lines $y = -1$ and $y = 1$. To obtain the final region, take the intersection of these regions.

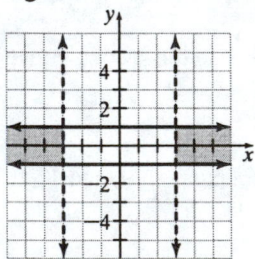

Cumulative Review Test

1. $24 \div 4[2 - (5 - 2)]^2 - 6$
   $= 24 \div 4[2 - 3]^2 - 6$
   $= 24 \div 4(-1)^2 - 6$
   $= 24 \div 4 \times 1 - 6$
   $= 6 \times 1 - 6$
   $= 6 - 6$
   $= 0$

2. **a.** 9 and 1 are natural numbers.

   b. $\dfrac{1}{2}$, -4, 9, 0, -4.63, and 1 are rational numbers.

   c. $\dfrac{1}{2}$, -4, 9, 0, $\sqrt{3}$, -4.63, and 1 are real numbers.

3. $-|-8|$, -1, $\dfrac{5}{8}$, $\dfrac{3}{4}$, $|-4|$, $|-10|$

4. $-[3 - 2(x - 4)] = 3(x - 6)$
   $-[3 - 2x + 8] = 3(x - 6)$
   $-(-2x + 11) = 3(x - 6)$
   $2x - 11 = 3x - 18$
   $-11 = x - 18$
   $7 = x$
   $x = 7$

5. $\dfrac{1}{3}x = \dfrac{3}{5}x + 4$
   $15\left(\dfrac{1}{3}x\right) = 15\left(\dfrac{3}{5}x\right) + 15(4)$
   $5x = 9x + 60$
   $-4x = 60$
   $x = \dfrac{60}{-4} = -15$

6. $|4x - 3| + 2 = 10$
   $|4x - 3| = 8$
   $4x - 3 = -8$ or $4x - 3 = 8$
   $4x = -5$ $4x = 11$
   $x = -\dfrac{5}{4}$ $x = \dfrac{11}{4}$

7. $R = 3(a + b)$
   $R = 3a + 3b$
   $R - 3a = 3a - 3a + 3b$
   $R - 3a = 3b$
   $\dfrac{R - 3a}{3} = \dfrac{3b}{3}$
   $\dfrac{R - 3a}{3} = b$ or $b = \dfrac{R - 3a}{3}$

8.

$$0 < \frac{3x-2}{4} \le 8$$

$$4(0) < 4\left(\frac{3x-2}{4}\right) \le 4(8)$$

$$0 < 3x - 2 \le 32$$

$$0 + 2 < 3x - 2 + 2 \le 32 + 2$$

$$2 < 3x \le 34$$

$$\frac{2}{3} < \frac{3x}{3} \le \frac{34}{3}$$

$$\frac{2}{3} < x \le \frac{34}{3}$$

The solution set is $\left\{ x \mid \frac{2}{3} < x \le \frac{34}{3} \right\}$.

9. $\left(\dfrac{3x^2 y^{-2}}{y^3}\right)^{-2} = \left(\dfrac{y^3}{3x^2 y^{-2}}\right)^{2}$

$$= \left(\frac{y^5}{3x^2}\right)^2$$

$$= \frac{y^{5 \cdot 2}}{3^2 x^{2 \cdot 2}}$$

$$= \frac{y^{10}}{9x^4}$$

10. $2y = 3x - 8$

$$y = \frac{3x - 8}{2}$$

$$y = \frac{3}{2}x - 4$$

x	y
0	$y = \frac{3}{2}(0) - 4 = 0 - 4 = -4$
2	$y = \frac{3}{2}(2) - 4 = 3 - 4 = -1$
4	$y = \frac{3}{2}(4) - 4 = 6 - 4 = 2$

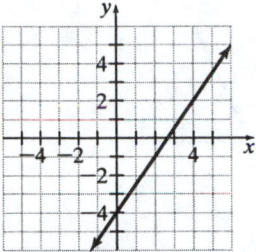

11. $2x - 3y = 8$

To find the slope, solve for y.

$$2x - 3y = 8$$

$$-3y = -2x + 8$$

$$y = \frac{-2x + 8}{-3}$$

$$y = \frac{2}{3}x - \frac{8}{3}$$

The slope of this line is $\frac{2}{3}$. Since the new line is parallel to this line, its slope is also $\frac{2}{3}$. Use the point-slope form with $m = \frac{2}{3}$ and $(x_1,\, y_1) = (2,\, 3)$.

$$y - y_1 = m(x - x_1)$$

$$y - 3 = \frac{2}{3}(x - 2)$$

$$y - 3 = \frac{2}{3}x - \frac{4}{3}$$

$$y = \frac{2}{3}x - \frac{4}{3} + 3$$

$$y = \frac{2}{3}x - \frac{4}{3} + \frac{9}{3}$$

$$y = \frac{2}{3}x + \frac{5}{3}$$

12. $6x - 3y < 12$

Graph the line $6x - 3y = 12$ using a dashed line. For the check point, select $(0, 0)$.

$$6x - 3y < 12$$

$$6(0) - 3(0) < 12$$

$$0 - 0 < 12$$

$$0 < 12 \qquad \text{True}$$

Since this is a true statement, shade the region containing the point $(0, 0)$.

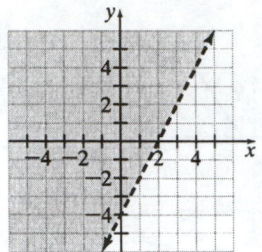

13. a. It is a function since it passes the vertical line test.

    b. It is a function since it passes the vertical line test.

    c. It is not a function since it fails the vertical line test.

14. a. $f(x) = x^2 - 3x + 4$
$$f(-2) = (-2)^2 - 3(-2) + 4$$
$$= 4 + 6 + 4$$
$$= 14$$

    b. $f(x) = x^2 - 3x + 4$
$$f(c) = c^2 - 3c + 4$$

15. $3x + y = 6$
$y = 2x + 1$
Substitute $2x + 1$ for y in the first equation.
$$3x + y = 6$$
$$3x + 2x + 1 = 6$$
$$5x + 1 = 6$$
$$5x = 5$$
$$x = \frac{5}{5} = 1$$
Substitute 1 for x in the second equation.
$$y = 2x + 1$$
$$y = 2(1) + 1$$
$$y = 2 + 1$$
$$y = 3$$
The solution is $(1, 3)$.

16. $5x + 4y = 10$
$3x + 5y = -7$
To eliminate y, multiply the first equation by 5 and the second equation by –4 and then add.

$5[5x + 4y = 10]$
$-4[3x + 5y = -7]$
gives
$$25x + 20y = 50$$
$$\underline{-12x - 20y = 28}$$
Add: $13x = 78$
$$x = \frac{78}{13} = 6$$
Substitute 6 for x in the first equation.
$$5x + 4y = 10$$
$$5(6) + 4y = 10$$
$$30 + 4y = 10$$
$$4y = -20$$
$$y = \frac{-20}{4} = -5$$
The solution is $(6, -5)$.

17. $x - 2y = 0$ (1)
$2x + z = 7$ (2)
$y - 2z = -5$ (3)
To eliminate z between equations (2) and (3), multiply equation (2) by 2 and then add.
$2[2x + z = 7]$
$y - 2z = -5$
gives
$$4x + 2z = 14$$
$$\underline{ y - 2z = -5}$$
Add: $4x + y = 9$ (4)
Equations (4) and (1) are two equations in two unknowns:
$$x - 2y = 0$$
$$4x + y = 9$$
To eliminate y, multiply equation (4) by 2 and then add.
$x - 2y = 0$
$2[4x + y = 9]$
gives
$$x - 2y = 0$$
$$\underline{8x + 2y = 18}$$
Add: $9x = 18$
$$x = \frac{18}{9} = 2$$
Substitute 2 for x in equation (4).

$$4x + y = 9$$
$$4(2) + y = 9$$
$$8 + y = 9$$
$$y = 1$$

Finally, substitute 2 for x in equation (2).
$$2x + z = 7$$
$$2(2) + z = 7$$
$$4 + z = 7$$
$$z = 3$$
The solution is $(2, 1, 3)$.

18. Let x be the measure of the smallest angle. Then $9x$ is the measure of the largest angle and $x + 70$ is the measure of the remaining angle. The sum of the measures of the three angles is $180°$.

$$x + (x + 70) + 9x = 180$$
$$11x + 70 = 180$$
$$11x = 110$$
$$x = \frac{110}{11} = 10$$
$$x + 70 = 10 + 70 = 80$$
$$9x = 9(10) = 90$$

The three angles are $10°$, $90°$, and $80°$.

19. Let t be the time for Judy to catch up to Dawn.

	rate	time	distance
Judy	6	t	$6t$
Dawn	4	$t + \frac{1}{2}$	$4\left(t + \frac{1}{2}\right)$

$$6t = 4\left(t + \frac{1}{2}\right)$$
$$6t = 4t + 2$$
$$2t = 2$$
$$t = \frac{2}{2} = 1$$

It takes 1 hour for Judy to catch up to Dawn.

20. Let x be the number of \$20 tickets sold. Then $1000 - x$ is the number of \$16 tickets sold.

$$20x + 16(1000 - x) = 18,400$$
$$20x + 16,000 - 16x = 18,400$$
$$4x + 16,000 = 18,400$$
$$4x = 2400$$
$$x = \frac{2400}{4} = 600$$

Thus, 600 \$20 tickets and $1000 - 600 = 400$ \$16 tickets were sold for the concert.

Chapter 5

Exercise Set 5.1

1. The terms are the parts that are added or subtracted.

3. A polynomial is a finite sum of terms in which all variables have whole number exponents and no variable appears in a denominator.

5. The leading coefficient is the coefficient of the leading term.

7. a. It is the same as that of the highest-degree term.

b. The degree of $-4x^4 + 6x^3y^4 + z^5$ is the same as the degree of the highest-degree term $\left(6x^3y^4\right)$. The degree is $3 + 4 = 7$.

9. a. A polynomial is linear if its degree is 0 or 1.

b. Answers will vary. One example is $x + 4$

11. a. A polynomial is cubic if it has degree 3 and is in one variable.

b. Answers will vary. One example is $x^3 + x - 4$

13. Answers will vary. One example is $x^5 + x + 1$

15. Since $5y$ has only one term, it is a monomial.

17. Since -10 has only term, it is a monomial.

19. The polynomial $8x^2 - 2x + 8y^2$ has three terms. It is a trinomial.

21. Since $-2x^2 + 5x^{-1}$ has a negative exponent, it is not a polynomial.

23. $-8 - 4x - x^2 = -x^2 - 4x - 8$; second degree

25. $6y^2 + 3xy + 10x^2 = 10x^2 + 3xy + 6y^2$; second degree

27. $-2x^4 + 5x^2 - 4$ is already in descending order; fourth degree

29. $x^4 + 3x^6 - 2x - 10 = 3x^6 + x^4 - 2x - 10$

a. $3x^6$: The degree of the polynomial is 6.

b. $3x^6$: The leading coefficient is 3.

31. $4x^2y^3 + 6xy^4 + 9xy^5$
$= 9xy^5 + 6xy^4 + 4x^2y^3$

a. $9xy^5$ or $9x^1y^5$: The degree of the polynomial is $1 + 5 = 6$.

b. $9xy^5$: The leading coefficient is 9.

33. $-\dfrac{1}{2}m^4n^5p^8 + \dfrac{3}{5}m^3p^6 - \dfrac{5}{9}n^4p^6q$

a. $-\dfrac{1}{2}m^4n^5p^8$: The degree of the polynomial is $4 + 5 + 8 = 17$.

b. $-\dfrac{1}{2}m^4n^5p^8$: The leading coefficient is $-\dfrac{1}{2}$.

35. $P(x) = x^2 - 6x + 4$
$P(2) = 2^2 - 6(2) + 4$
$= 4 - 12 + 4$
$= -4$

37. $P(x) = 2x^2 - 3x - 6$

$$P\left(\frac{1}{2}\right) = 2\left(\frac{1}{2}\right)^2 - 3\left(\frac{1}{2}\right) - 6$$
$$= \frac{1}{2} - \frac{3}{2} - 6$$
$$= -7$$

39. $P(x) = 0.2x^3 + 1.6x^2 - 2.3$

$$P(0.4) = 0.2(0.4)^3 + 1.6(0.4)^2 - 2.3$$
$$= 0.0128 + 0.256 - 2.3$$
$$= -2.0312$$

41. $\left(x^2 + 5x - 2\right) + (6x - 7) = x^2 + 5x - 2 + 6x - 7 \leftarrow$ Remove parentheses

$$= x^2 + 5x + 6x - 2 - 7 \leftarrow \text{Rearrange terms}$$
$$= x^2 + 11x - 9 \qquad\quad \leftarrow \text{Combine like terms}$$

43. $\left(x^2 - 6x + 3\right) - (2x + 5) = x^2 - 6x + 3 - 2x - 5 \leftarrow$ Remove parentheses

$$= x^2 - 6x - 2x + 3 - 5 \leftarrow \text{Rearrange terms}$$
$$= x^2 - 8x - 2 \qquad\quad \leftarrow \text{Combine like terms}$$

45. $\left(4y^2 + 6y - 3\right) - \left(2y^2 + 6\right) = 4y^2 + 6y - 3 - 2y^2 - 6 \leftarrow$ Remove parentheses

$$= 4y^2 - 2y^2 + 6y - 3 - 6 \leftarrow \text{Rearrange terms}$$
$$= 2y^2 + 6y - 9 \qquad\quad\;\; \leftarrow \text{Combine like terms}$$

47. $\left(-\frac{5}{9}a + 8\right) + \left(-\frac{2}{3}a^2 - \frac{1}{4}a - 5\right) = -\frac{5}{9}a + 8 - \frac{2}{3}a^2 - \frac{1}{4}a - 5 \qquad \leftarrow$ Remove parentheses

$$= -\frac{2}{3}a^2 - \frac{5}{9}a - \frac{1}{4}a + 8 - 5 \qquad \leftarrow \text{Rearrange terms}$$
$$= -\frac{2}{3}a^2 - \frac{20}{36}a - \frac{9}{36}a + 8 - 5 \leftarrow \text{Common denominator}$$
$$= -\frac{2}{3}a^2 - \frac{29}{36}a + 3 \qquad\qquad\;\; \leftarrow \text{Combine like terms}$$

49. $\left(1.4x^2 + 0.6x - 8.3\right) - \left(4.9x^2 + 3.7x + 19.2\right)$

$$= 1.4x^2 + 0.6x - 8.3 - 4.9x^2 - 3.7x - 19.2 \quad \leftarrow \text{Remove parentheses}$$
$$= 1.4x^2 - 4.9x^2 + 0.6x - 3.7x - 8.3 - 19.2 \leftarrow \text{Rearrange terms}$$
$$= -3.5x^2 - 3.1x - 27.5 \qquad\qquad\qquad\;\; \leftarrow \text{Combine like terms}$$

51. $\left(-\dfrac{1}{3}x^3 + \dfrac{1}{4}x^2y + 3xy^2\right) + \left(-x^3 - \dfrac{1}{2}x^2y + xy^2\right)$

$= -\dfrac{1}{3}x^3 + \dfrac{1}{4}x^2y + 3xy^2 - x^3 - \dfrac{1}{2}x^2y + xy^2$ ← Remove parentheses

$= -\dfrac{1}{3}x^3 - x^3 + \dfrac{1}{4}x^2y - \dfrac{1}{2}x^2y + 3xy^2 + xy^2$ ← Rearrange terms

$= -\dfrac{4}{3}x^3 - \dfrac{1}{4}x^2y + 4xy^2$ ← Combine like terms

53. $(3a - 6b + 5c) - (-2a + 4b - 8c) = 3a - 6b + 5c + 2a - 4b + 8c$ ← Remove parentheses

$= 3a + 2a - 6b - 4b + 5c + 8c$ ← Rearrange terms

$= 5a - 10b + 13c$ ← Combine like terms

55. $\left(3a^2b - 6ab + 5b^2\right) - \left(4ab - 6b^2 - 5a^2b\right)$

$= 3a^2b - 6ab + 5b^2 - 4ab + 6b^2 + 5a^2b$ ← Remove parentheses

$= 3a^2b + 5a^2b - 6ab - 4ab + 5b^2 + 6b^2$ ← Rearrange terms

$= 8a^2b - 10ab + 11b^2$ ← Combine like terms

57. $\left(3r^2 - 5t^2 + 2rt\right) + \left(-6rt + 2t^2 - r^2\right) = 3r^2 - 5t^2 + 2rt - 6rt + 2t^2 - r^2$ ← Remove parentheses

$= 3r^2 - r^2 + 2rt - 6rt - 5t^2 + 2t^2$ ← Rearrange terms

$= 2r^2 - 4rt - 3t^2$ ← Combine like terms

59. $6x^2 - 2x - \left[3x - \left(4x^2 - 6\right)\right] = 6x^2 - 2x - \left[3x - 4x^2 + 6\right]$ ← Remove parentheses

$= 6x^2 - 2x - 3x + 4x^2 - 6$ ← Remove brackets

$= 6x^2 + 4x^2 - 2x - 3x - 6$ ← Rearrange terms

$= 10x^2 - 5x - 6$ ← Combine like terms

61. $5w - 6w^2 - \left[\left(3w - 2w^2\right) - \left(4w + w^2\right)\right] = 5w - 6w^2 - \left[3w - 2w^2 - 4w - w^2\right]$ ← Remove parentheses

$= 5w - 6w^2 - \left[3w - 4w - 2w^2 - w^2\right]$ ← Rearrange terms

$= 5w - 6w^2 - \left(-w - 3w^2\right)$ ← Combine like terms

$= 5w - 6w^2 + w + 3w^2$ ← Remove parenthesess

$= -6w^2 + 3w^2 + 5w + w$ ← Rearrange terms

$= -3w^2 + 6w$ ← Combine like terms

63. $(3x + 5) - (4x - 6) = 3x + 5 - 4x + 6$ ← Remove parentheses

$= 3x - 4x + 5 + 6$ ← Rearrange terms

$= -x + 11$ ← Combine like terms

65. $\left(-2x^2 + 4x - 12\right) + \left(-x^2 - 2x\right) = -2x^2 + 4x - 12 - x^2 - 2x \leftarrow$ Remove parentheses

$$= -2x^2 - x^2 + 4x - 2x - 12 \leftarrow \text{Rearrange terms}$$
$$= -3x^2 + 2x - 12 \qquad\qquad \leftarrow \text{Combine like terms}$$

67. $\left(-4.2a^2 - 9.6a\right) - \left(0.2a^2 - 3.9a + 26.4\right)$

$$= -4.2a^2 - 9.6a - 0.2a^2 + 3.9a - 26.4$$
$$= -4.2a^2 - 0.2a^2 - 9.6a + 3.9a - 26.4$$
$$= -4.4a^2 - 5.7a - 26.4$$

69. $\left(-\dfrac{1}{2}x^2y + 6xy^2 + \dfrac{3}{5}\right) - \left(5x^2y + \dfrac{5}{9}\right)$

$$= -\frac{1}{2}x^2y + 6xy^2 + \frac{3}{5} - 5x^2y - \frac{5}{9}$$
$$= -\frac{1}{2}x^2y - 5x^2y + 6xy^2 + \frac{3}{5} - \frac{5}{9}$$
$$= -\frac{11}{2}x^2y + 6xy^2 + \frac{2}{45}$$

71. $\left(3x^{2r} - 2x^r + 6\right) + \left(2x^{2r} - 6x^r - 3\right)$

$$= 3x^{2r} - 2x^r + 6 + 2x^{2r} - 6x^r - 3$$
$$= 3x^{2r} + 2x^{2r} - 2x^r - 6x^r + 6 - 3$$
$$= 5x^{2r} - 8x^r + 3$$

73. $\left(x^{2s} - 6x^s + 4\right) - \left(2x^{2s} - 4x^s - 3\right)$

$$= x^{2s} - 6x^s + 4 - 2x^{2s} + 4x^s + 3$$
$$= x^{2s} - 2x^{2s} - 6x^s + 4x^s + 4 + 3$$
$$= -x^{2s} - 2x^s + 7$$

75. $\left(7b^{4n} - 3b^{2n} - 1\right) - \left(5b^{3n} - b^{2n}\right)$

$$= 7b^{4n} - 3b^{2n} - 1 - 5b^{3n} + b^{2n}$$
$$= 7b^{4n} - 5b^{3n} - 3b^{2n} + b^{2n} - 1$$
$$= 7b^{4n} - 5b^{3n} - 2b^{2n} - 1$$

77. No; answers will vary.

79. No; answers will vary.

81. $A(r) = \pi r^2$

$A(6) = \pi 6^2$

$A = \pi(36)$

$A \approx 113.10 \text{ in}^2$

83. $h = P(t) = -16t^2 + 1250$

$h = P(6) = -16(6)^2 + 1250 = 674$

The object is 674 feet from the ground.

85. a. $R(x) = 2x^2 - 60x + 4000$, $C(x) = 8050 - 420x$

$P(x) = R(x) - C(x)$

$P(x) = \left(2x^2 - 60x + 4000\right) - (8050 - 420x)$

$P(x) = 2x^2 - 60x + 4000 - 8050 + 420x$

$P(x) = 2x^2 + 360x - 4050$

b. $P(100) = 2(100)^2 + 360(100) - 4050$

$= 20,000 + 36,000 - 4050$

$= 51,950$

The profit is $51,950.

87. $y = x^2 + 3x - 4$ is graph c.

The coefficient of the leading term is positive, so the graph opens up. The *y*-intercept is –4.

89. $y = -x^3 + 2x - 6$ is graph c.

The coefficient of the leading term is negative, so that eliminates graphs (a) and (b).

91. a. $S(t) = 45.4t^2 - 26.31t + 261.9$

$t = 2000 - 1996 = 4$

$S(4) = 45.4(4)^2 - 26.31(4) + 261.9$

$= 726.4 - 105.24 + 261.9$

$= 883.06$ thousand or $883,060$

The number of digital cameras expected to be sold in 2000 is 883,060.

b. Yes

93. $N(t) = 0.28t^2 - 2.84t + 78.97$

a. $t = 2020 - 1995 = 25$

$N(25) = 0.28(25)^2 - 2.84(25) + 78.97$

$= 175 - 71 + 78.97$

$= 182.97$ thousand or $182,970$

The number of centenarians in 2020 is 182,970.

b. $t = 2045 - 1995 = 50$

$N(50) = 0.28(50)^2 - 2.84(50) + 78.97$

$= 700 - 142 + 78.97$

$= 636.97$ thousand or $636,970$

The number of centenarians in 2045 is 636,970.

95. a. $y_1 = x^3$

$y_2 = x^3 - 3x^2 - 3$

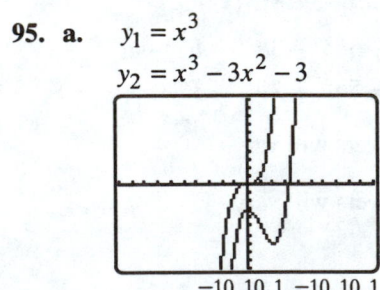

$-10, 10, 1, -10, 10, 1$

b. increase

c. Answers will vary.

d. decrease

e. Answers will vary.

97. $y = -x^4 + 3x^3 - 5$ is graph b.
The leading coefficient is negative, therefore eliminating graph (c). The y-intercept is –5.

101. $x - 4y = -16$
$2x + 3y = -10$
To eliminate x, multiply the first equation by –2 and then add to the second equation.
$-2(x - 4y = -16)$
$\quad 2x + 3y = -10$
gives
$$\begin{array}{r} -2x + 8y = 32 \\ 2x + 3y = -10 \\ \hline \text{Add:} \quad 11y = 22 \end{array}$$
$$y = \frac{22}{11} = 2$$
Now, substitute 2 for y into the first equation and then solve for x:
$x - 4y = -16$
$x - 4(2) = -16$
$x - 8 = -16$
$x = -8$
The solution is (–8, 2).

102. $\begin{bmatrix} 1 & -4 & | & -16 \\ 2 & 3 & | & -10 \end{bmatrix} \xrightarrow{-2R_1 + R_2} \begin{bmatrix} 1 & -4 & | & -16 \\ 0 & 11 & | & 22 \end{bmatrix}$

$\xrightarrow{\frac{1}{11}R_2} \begin{bmatrix} 1 & -4 & | & -16 \\ 0 & 1 & | & 2 \end{bmatrix}$

The system is
$x - 4y = -16$
$\quad\quad y = 2$
Substitute 2 for y into the first equation.
$x - 4y = -16$
$x - 4(2) = -16$
$x - 8 = -16$
$x = -8$
The solution is (–8, 2).

103. $x - 4y = -16$
$2x + 3y = -10$　First find D, D_x, and D_y.
$$D = \begin{vmatrix} 1 & -4 \\ 2 & 3 \end{vmatrix} = (1)(3) - (2)(-4) = 3 + 8 = 11$$
$$\begin{aligned} D_x &= \begin{vmatrix} -16 & -4 \\ -10 & 3 \end{vmatrix} \\ &= (-16)(3) - (-10)(-4) \\ &= -48 - 40 \\ &= -88 \end{aligned}$$
$$\begin{aligned} D_y &= \begin{vmatrix} 1 & -16 \\ 2 & -10 \end{vmatrix} \\ &= (1)(-10) - (2)(-16) \\ &= -10 + 32 \\ &= 22 \end{aligned}$$
Then, $x = \dfrac{D_x}{D} = \dfrac{-88}{11} = -8$ and
$y = \dfrac{D_y}{D} = \dfrac{22}{11} = 2$
The solution is (–8, 2).

104. Let x, y, z be the three numbers.
The system is
(1)　$x + y + z = 12$
(2)　$\quad\quad x + y = z$
(3)　$\quad\quad\quad z = 2y - 4$
To solve, substitute $x + y$ for z in the first equation and then solve for z:
$x + y + z = 12$
$\quad\quad z + z = 12$
$\quad\quad\quad 2z = 12$
$\quad\quad\quad\quad z = 6$
Substitute 6 for z in equation (3) and then solve for y:
$z = 2y - 4$
$6 = 2y - 4$
$10 = 2y$
$5 = y$
Finally, substitute 5 for y and 6 for z in equation (1) and then solve for x:
$x + y + z = 12$
$x + 5 + 6 = 12$
$\quad x + 11 = 12$
$\quad\quad\quad x = 1$
The three numbers are 1, 5, and 6.

Exercise Set 5.2

1. Answers will vary.

3. **a.** Answers will vary.

   b. $(4+x)(x^2-6x+3)$
   $$= 4(x^2-6x+3) + x(x^2-6x+3)$$
   $$= 4x^2 - 24x + 12 + x^3 - 6x^2 + 3x$$
   $$= x^3 - 2x^2 - 21x + 12$$

5. **a.** Answers will vary.

   b. Answers will vary. One example is $(x+4)(x-4)$

   c. Answers will vary.

   d. Answers will vary. One example is $x^2 - 16$

7. Yes, answers may vary.

9. $(4xy)(6xy^4) = 4 \cdot 6 \cdot x \cdot x \cdot y \cdot y^4$
   $$= 24x^{1+1}y^{1+4}$$
   $$= 24x^2 y^5$$

11. $\left(\dfrac{5}{9}x^2y^5\right)\left(\dfrac{1}{5}x^5y^3z^2\right) = \dfrac{5}{9} \cdot \dfrac{1}{5} \cdot x^2 \cdot x^5 \cdot y^5 \cdot y^3 \cdot z^2$
    $$= \dfrac{1}{9}x^{2+5}y^{5+3}z^2$$
    $$= \dfrac{1}{9}x^7 y^8 z^2$$

13. $-3x^2y(-2x^4y^2 + 3xy^3 + 4) = (-3x^2y)(-2x^4y^2) + (-3x^2y)(3xy^3) + (-3x^2y)(4)$
    $$= 6x^6y^3 - 9x^3y^4 - 12x^2y$$

15. $\dfrac{2}{3}yz(3x + 4y - 9y^2) = \left(\dfrac{2}{3}yz\right)(3x) + \left(\dfrac{2}{3}yz\right)(4y) + \left(\dfrac{2}{3}yz\right)(-9y^2)$
    $$= 2xyz + \dfrac{8}{3}y^2z - 6y^3z$$

17. $0.3a^5b^4(9.5a^6b - 4.6a^4b^3 + 1.2ab^5) = 0.3a^5b^4(9.5a^6b) + 0.3a^5b^4(-4.6a^4b^3) + 0.3a^5b^4(1.2ab^5)$
    $$= (0.3)(9.5)a^{5+6}b^{4+1} + 0.3(-4.6)a^{5+4}b^{4+3} + 0.3(1.2)a^{5+1}b^{4+5}$$
    $$= 2.85a^{11}b^5 - 1.38a^9b^7 + 0.36a^6b^9$$

19. $(4x-6)(3x-5) = (4x)(3x) + (4x)(-5) + (-6)(3x) + (-6)(-5)$
$$= 12x^2 - 20x - 18x + 30$$
$$= 12x^2 - 38x + 30$$

21. $(4-x)(3+2x^2) = (4)(3) + (4)(2x^2) + (-x)(3) + (-x)(2x^2)$
$$= 12 + 8x^2 - 3x - 2x^3$$
$$= -2x^3 + 8x^2 - 3x + 12 \leftarrow \text{ Rearrange terms}$$

23. $\left(\dfrac{2}{5}x - \dfrac{1}{5}z\right)\left(\dfrac{1}{3}x + z\right) = \left(\dfrac{2}{5}x\right)\left(\dfrac{1}{3}x\right) + \left(\dfrac{2}{5}x\right)(z) + \left(-\dfrac{1}{5}z\right)\left(\dfrac{1}{3}x\right) + \left(-\dfrac{1}{5}z\right)(z)$
$$= \dfrac{2}{15}x^2 + \dfrac{2}{5}xz - \dfrac{1}{15}xz - \dfrac{1}{5}z^2$$
$$= \dfrac{2}{15}x^2 + \dfrac{1}{3}xz - \dfrac{1}{5}z^2$$

25. $(2.3a - 1.4b)(5.6a + 4.2b) = 2.3a(5.6a) + 2.3a(4.2b) - 1.4b(5.6a) - 1.4b(4.2b)$
$$= 12.88a^2 + 9.66ab - 7.84ab - 5.88b^2$$
$$= 12.88a^2 + 1.82ab - 5.88b^2$$

27.
$$
\begin{array}{r}
x^2 - 3x + 2 \\
x - 4 \\
\hline
-4x^2 + 12x - 8 \\
x^3 - 3x^2 + 2x \\
\hline
x^3 - 7x^2 + 14x - 8
\end{array}
$$

$\leftarrow$ Multiply top expression by -4

$\leftarrow$ Multiply top expression by x

$\leftarrow$ Add like terms

29.
$$
\begin{array}{r}
4x^2 + 9x - 2 \\
x - 2 \\
\hline
-8x^2 - 18x + 4 \\
4x^3 + 9x^2 - 2x \\
\hline
4x^3 + x^2 - 20x + 4
\end{array}
$$

$\leftarrow$ Multiply top expression by -2

$\leftarrow$ Multiply top expression by x

$\leftarrow$ Add like terms

31.
$$
\begin{array}{r}
2a^2 - ab + 2b^2 \\
a - 3b \\
\hline
-6a^2b + 3ab^2 - 6b^3 \\
2a^3 - a^2b + 2ab^2 \\
\hline
2a^3 - 7a^2b + 5ab^2 - 6b^3
\end{array}
$$

33. $(3x-1)^3 = (3x-1)(3x-1)(3x-1)$

$$= \left[(3x)^2 - 2(3x)(1) + 1^2\right](3x-1)$$

$$= \left(9x^2 - 6x + 1\right)(3x-1)$$

To complete the solution, multiply vertically

$$
\begin{array}{r}
9x^2 - 6x + 1 \\
3x - 1 \\
\hline
-9x^2 + 6x - 1 \\
27x^3 - 18x^2 + 3x \\
\hline
27x^3 - 27x^2 + 9x - 1
\end{array}
$$

35.

$$
\begin{array}{r}
2a^2 - \ 6a + 3 \\
3a^2 - \ 5a - 2 \\
\hline
-4a^2 + 12a - 6 \\
-10a^3 + 30a^2 - 15a \\
6a^4 - 18a^3 + \ 9a^2 \\
\hline
6a^4 - 28a^3 + 35a^2 - 3a - 6
\end{array}
$$

$\leftarrow$ Multiply top expression by -6

$\leftarrow$ Multiply top expression by $-5a$

$\leftarrow$ Multiply top expression by $3a^2$

$\leftarrow$ Add like terms

37.

$$
\begin{array}{r}
5r^2 - rs + 2s^2 \\
2r^2 - s^2 \\
\hline
-5r^2s^2 + rs^3 - 2s^4 \\
10r^4 - 2r^3s + 4r^2s^2 \\
\hline
10r^4 - 2r^3s - r^2s^2 + rs^3 - 2s^4
\end{array}
$$

$\leftarrow$ Multiply top expression by $-s^2$

$\leftarrow$ Multiply top expression by $2r^2$

$\leftarrow$ Add like terms

39. $(x+2)(x+2) = (x+2)^2$

$$= (x)^2 + 2(x)(2) + (2)^2$$

$$= x^2 + 4x + 4$$

41. $(2x-3y)^2 = (2x)^2 - 2(2x)(3y) + (3y)^2$

$$= 4x^2 - 12xy + 9y^2$$

43. $\left(5m^2 + 2n\right)\left(5m^2 - 2n\right) = \left(5m^2\right)^2 - (2n)^2$

$$= 25m^4 - 4n^2$$

45. $\left[y + (4-2x)\right]^2 = (y)^2 + 2(y)(4-2x) + (4-2x)^2$

$$= y^2 + 8y - 4xy + 16 - 16x + 4x^2$$

47. $[4-(x-3y)]^2 = (4)^2 - 2(4)(x-3y) + (x-3y)^2$
$$= 16 - 8(x-3y) + (x-3y)^2$$
$$= 16 - 8x + 24y + x^2 - 6xy + 9y^2$$

49. $[a+(b+2)][a-(b+2)] = (a)^2 - (b+2)^2$
$$= a^2 - \left(b^2 + 4b + 4\right)$$
$$= a^2 - b^2 - 4b - 4$$

51. $(3y+4)(2y-3)$
$$= (3y)(2y) + (3y)(-3) + (4)(2y) + (4)(-3)$$
$$= 6y^2 - 9y + 8y - 12$$
$$= 6y^2 - y - 12$$

53. $\left(2x - \dfrac{3}{4}\right)\left(2x + \dfrac{3}{4}\right) = (2x)^2 - \left(\dfrac{3}{4}\right)^2$
$$= 4x^2 - \dfrac{9}{16}$$

55. $\dfrac{2}{3}x^2 y^4 \left(\dfrac{3}{5}xy^3 - \dfrac{1}{4}x^4 y + 2xy^3 z^5\right) = \left(\dfrac{2}{3}x^2 y^4\right)\left(\dfrac{3}{5}xy^3\right) + \left(\dfrac{2}{3}x^2 y^4\right)\left(-\dfrac{1}{4}x^4 y\right) + \left(\dfrac{2}{3}x^2 y^4\right)\left(2xy^3 z^5\right)$
$$= \dfrac{2}{5}x^3 y^7 - \dfrac{1}{6}x^6 y^5 + \dfrac{4}{3}x^3 y^7 z^5$$

57.
$$
\begin{array}{r}
2x^2 + 4x - 3 \\
x + 3 \\
\hline
6x^2 + 12x - 9 \\
2x^3 + 4x^2 - 3x \\
\hline
2x^3 + 10x^2 + 9x - 9
\end{array}
$$
     $6x^2 + 12x - 9$ ← Multiply top expression by 3

     $2x^3 + 4x^2 - 3x$ ← Multiply top expression by x

     $2x^3 + 10x^2 + 9x - 9$ ← Add like terms

59.
     $3x^2 + 4xy - 2y^2$

     $2x - 3y$

     $-9x^2 y - 12xy^2 + 6y^3$ ← Multiply top expression by $-3x$

     $6x^3 + 8x^2 y - 4xy^2$ ← Multiply top expression by $2x$

     $6x^3 - x^2 y - 16xy^2 + 6y^3$ ← Add like terms

61. $(x+3)^3 = (x+3)(x+3)(x+3)$

$\left[x^2+2(x)(3)+3^2\right](x+3) = \left(x^2+6x+9\right)(x+3)$

To complete the solution, multiply vertically:

$$x^2+6x+9$$
$$\underline{\qquad x+3}$$
$$3x^2+18x+27$$
$$\underline{x^3+6x^2\ +9x\qquad\quad}$$
$$x^3+9x^2+27x+27$$

63. $\left[w+(3x+4)\right]\left[w-(3x+4)\right]$

$= (w)^2 - (3x+4)^2$

$= w^2 - (9x^2+24x+16)$

$= w^2 - 9x^2 - 24x - 16$

65. $(a+b)(a-b)\left(a^2-b^2\right)$

$= \left(a^2-b^2\right)\left(a^2-b^2\right)$

$= \left(a^2\right)^2 - 2\left(a^2\right)\left(b^2\right) + \left(b^2\right)^2$

$= a^4 - 2a^2b^2 + b^4$

67. $(x-4)(6+x)(2x-8)$

$= \left(6x+x^2-24-4x\right)(2x-8)$

$= \left(x^2+2x-24\right)(2x-8)$

$= x^2(2x-8) + 2x(2x-8) - 24(2x-8)$

$= 2x^3 - 8x^2 + 4x^2 - 16x - 48x + 192$

$= 2x^3 - 4x^2 - 64x + 192$

69. a. $(f \cdot g)(x) = f(x) \cdot g(x)$

$= (x-5)(x+4)$

$= x^2 + 4x - 5x - 20$

$= x^2 - x - 20$

b. $(f \cdot g)(4) = 4^2 - 4 - 20$

$= 16 - 4 - 20$

$= -8$

71. a. $(f \cdot g)(x) = f(x) \cdot g(x)$

$= \left(3x^2+2\right)(4-x)$

$= 12x^2 - 3x^3 + 8 - 2x$

$= -3x^3 + 12x^2 - 2x + 8$

b. $(f \cdot g)(4) = -3(4)^3 + 12(4)^2 - 2(4) + 8$

$= -192 + 192 - 8 + 8$

$= 0$

73. a. $(f \cdot g)(x)$

$= f(x) \cdot g(x)$

$= \left(2x^2+6x-4\right)(5x+3)$

$= 2x^2(5x+3) + 6x(5x+3) - 4(5x+3)$

$= 10x^3 + 6x^2 + 30x^2 + 18x - 20x - 12$

$= 10x^3 + 36x^2 - 2x - 12$

b. $(f \cdot g)(4) = 10(4)^3 + 36(4)^2 - 2(4) - 12$

$= 640 + 576 - 8 - 12$

$= 1196$

75. a. The sum of the areas of the four pieces is $x^2 + 5x + 3x + 15 = x^2 + 8x + 15$

b. The area is length × width

$= (x+5)(x+3)$

$= x^2 + 5x + 3x + 15$

$= x^2 + 8x + 15$

77. The area is
length $\times$ width $= (6+x)(6-x)$.
$$= 6^2 - x^2$$
$$= 36 - x^2$$

79. The area is the sum of the area of the large square and the area of small square
$$= x^2 + y^2$$

81. a. The area of the larger rectangle is
$$(2x+3)(x+4) = 2x^2 + 8x + 3x + 12$$
$$= 2x^2 + 11x + 12$$
The area of the smaller rectangle is
$(2x)(x) = 2x^2$.
The area of the shaded portion is the area of the larger rectangle – area of the smaller
rectangle.
$$\text{shaded portion} = \left(2x^2 + 11x + 12\right) - 2x^2$$
$$= 11x + 12$$

b. Since the area of the shaded portions is 67 sq in., set $11x + 12$ equal to 67 and then solve:
$$11x + 12 = 67$$
$$11x = 55$$
$$x = \frac{55}{11} = 5$$
Now, the dimensions of the larger rectangle are $2(5)+3 = 10+3 = 13$ inches by
$5+4 = 9$ inches.
The area is $(13)(9) = 117$ square inches. Also, the dimensions of the smaller rectangle are $2(5) = 10$ by 5 in.
The area is $(10)(5) = 50$ sq in.

83. $x^2 - 36 = x^2 - 6^2 = (x-6)(x+6)$

85. $x^2 + 12x + 36 = x^2 + 2(6x) + 6^2$
$$= (x+6)(x+6)$$

87. $a(x-n)^3 = a(x-n)(x-n)(x-n)$

89. a. Answers will vary. One example is Observe that the length is $a + b$ and the width is $a + b$. Since the area is the product of the length and the width, this gives
$$\text{area} = (a+b)(a+b) = (a+b)^2$$

b. The sum of the area of the four pieces is
$$(a)(a) + (a)(b) + (b)(a) + (b)(b)$$
$$= a^2 + ab + ab + b^2$$
$$= a^2 + 2ab + b^2$$

c. $(a+b)^2 = a^2 + 2ab + b^2$

d. The are the same.

91. $A = P\left(1 + \dfrac{r}{n}\right)^{nt}$

a. $A = P\left(1 + \dfrac{r}{1}\right)^{1 \cdot t}$
$$A = P(1+r)^t$$

b. $A = 1000(1 + 0.06)^2$
$$= 1000(1.06)^2$$
$$= 1123.6$$
The amount is $1123.60.

93. $f(x) = x^2 + 3x + 4$
Then $f(a+b) = (a+b)^2 + 3(a+b) + 4$ ← Replace x by $a+b$
$$= a^2 + 2ab + b^2 + 3a + 3b + 4$$

95. $3x^t\left(5x^{2t-1}+4x^{3t}\right)=\left(3x^t\right)\left(5x^{2t-1}\right)+\left(3x^t\right)\left(4x^{3t}\right)$

$\qquad\qquad\qquad = 15x^{t+(2t-1)}+12x^{t+3t}$

$\qquad\qquad\qquad = 15x^{3t-1}+12x^{4t}$

97. $\left(6x^m-5\right)\left(2x^{2m}-3\right)=\left(6x^m\right)\left(2x^{2m}\right)+\left(6x^m\right)(-3)+(-5)\left(2x^{2m}\right)+(-5)(-3)$

$\qquad\qquad\qquad\qquad = 12x^{m+2m}-18x^m-10x^{2m}+15$

$\qquad\qquad\qquad\qquad = 12x^{3m}-18x^m-10x^{2m}+15$

99. $\left(y^{a-b}\right)^{a+b}=y^{(a-b)(a+b)}$

$\qquad\qquad\quad = y^{a^2-b^2}$

101. First, find

$(x-3y)^2=(x)^2-2(x)(3y)+(3y)^2$

$\qquad\qquad = x^2-6xy+9y^2$

Then,

$(x-3y)^4=\left[(x-3y)^2\right]^2=\left(x^2-6xy+9y^2\right)^2$

$$
\begin{array}{r}
x^2-6xy+9y^2 \\
x^2-6xy+9y^2 \\
\hline
9x^2y^2-54xy^3+81y^4 \\
-6x^3y+36x^2y^2-54xy^3 \\
x^4-6x^3y+\ 9x^2y^2 \\
\hline
x^4-12x^3y+54x^2y^2-108xy^3+81y^4
\end{array}
$$

103. a. Answers will vary. Possible answer.
Graph $y_1=\left(x^2+2x+3\right)(x+2)$ and
$y_2=x^3+4x^2+7x+6$.
If they coincide, then the multiplication
is correct.

b.

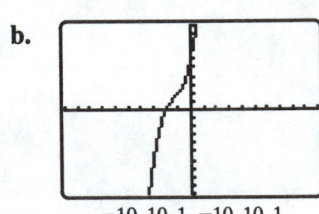

$-10, 10, 1, -10, 10, 1$

It is correct.

105. $\left[(y+1)-(x+2)\right]^2$

$= \left[y+1-x-2\right]^2$

$= \left[(y-x)-1\right]^2$

$= (y-x)^2-2(y-x)(1)+(1)^2$

$= y^2-2xy+x^2-2y+2x+1$

107. Every function is a relation, but, not every
relation is a function. A relation is any set of
ordered pairs. A function is a set of ordered
pairs, no two of which have the same first
coordinate and a different second coordinate.

108. a. $ax+by=c$

b. $y=mx+b$

c. $y-y_1=m(x-x_1)$

109. (1) $\quad -2x+3y+4z=17$

(2) $\quad -5x-3y+z=-1$

(3) $\quad -x-2y+3z=18$

To eliminate x between equations (1) and
(3), multiply equation (3) by -2 and then
add:

$\qquad -2x+3y+4z=17$

$-2(-x-2y+3z=18)$

gives

$\qquad\quad -2x+3y+4z=\ \ 17$

$\qquad\quad\ \underline{2x+4y-6z=-36}$

Add: $\qquad\quad 7y-2z=-19$ $\quad$ (4)

To eliminate x between equations (2) and
(3), multiply equation (3) by -5 and then
add:

$$-5x - 3y + z = -1$$
$$-5(-x - 2y + 3z = 18)$$
gives
$$-5x - 3y + z = -1$$
$$\underline{5x + 10y - 15z = -90}$$
Add: $\quad 7y - 14z = -91$ or $y - 2z = -13 \quad$ (5)

Equations (4) and (5) give us two equations in two unknowns. To solve, multiply equation (5) by -1 and then add:
$$7y - 2z = -19$$
$$-1(y - 2z = -13)$$
gives
$$7y - 2z = -19$$
$$\underline{-y + 2z = 13}$$
Add: $\quad 6y = -6$
$$y = -1$$
Substitute -1 for y into equation (5) and then solve for z:
$$y - 2z = -13$$
$$-1 - 2z = -13$$
$$-2z = -12$$
$$z = \frac{-12}{-2} = 6$$
Finally, substitute -1 for y and 6 for z into equation (1) and then solve for x:
$$-2x + 3y + 4z = 17$$
$$-2x + 3(-1) + 4(6) = 17$$
$$-2x - 3 + 24 = 17$$
$$-2x + 21 = 17$$
$$-2x = -4$$
$$x = \frac{-4}{-2} = 2$$
The solution is $(2, -1, 6)$.

110. Let x be the number of first class envelopes. Then $550 - x$ is the number of bulk postage envelopes.
The equation is
$$0.33x + 0.228(550 - x) = 140.70$$
$$0.33x + 125.4 - 0.228x = 140.70$$
$$0.102x + 125.4 = 140.70$$
$$0.102x = 15.3$$
$$x = \frac{15.3}{0.102} = 150$$
Thus, there are 150 first class envelopes and $550 - 150 = 400$ bulk postage envelopes.

Exercise Set 5.3

1. a. Answers will vary.

  b. $\dfrac{5x^4 - 6x^3 - 4x^2 - 12x + 7}{3x}$
$$= \frac{5x^4}{3x} - \frac{6x^3}{3x} - \frac{4x^2}{3x} - \frac{12x}{3x} + \frac{7}{3x}$$
$$= \frac{5}{3}x^3 - 2x^2 - \frac{4}{3}x - 4 + \frac{7}{3x}$$

3. Yes; answers will vary.

5. Place them in descending order of the variable.

7. a. Answers will vary.

  b.

$$\begin{array}{r|rrr}
5 & 1 & 3 & -4 \\
  &   & 5 & 40 \\
\hline
  & 1 & 8 & 36 \\
\end{array}$$ $\leftarrow$ Remainder

Thus, $\dfrac{x^2 + 3x - 4}{x - 5} = x + 8 + \dfrac{36}{x - 5}$

9. $\dfrac{6x + 8}{2} = \dfrac{6x}{2} + \dfrac{8}{2} = 3x + 4$

11. $\dfrac{4x^2+2x}{2x} = \dfrac{4x^2}{2x} + \dfrac{2x}{2x} = 2x+1$

13. $\dfrac{12x^2-4x-8}{4} = \dfrac{12x^2}{4} - \dfrac{4x}{4} - \dfrac{8}{4}$
$= 3x^2 - x - 2$

15. $\dfrac{4x^5 - 6x^4 + 12x^3 - 8x^2}{4x^2}$
$= \dfrac{4x^5}{4x^2} - \dfrac{6x^4}{4x^2} + \dfrac{12x^3}{4x^2} - \dfrac{8x^2}{4x^2}$
$= x^3 - \dfrac{3}{2}x^2 + 3x - 2$

17. $\dfrac{4x^2y^2 - 8xy^3 + 3y^4}{2y^2}$
$= \dfrac{4x^2y^2}{2y^2} - \dfrac{8xy^3}{2y^2} + \dfrac{3y^4}{2y^2}$
$= 2x^2 - 4xy + \dfrac{3}{2}y^2$

19. $\dfrac{6x^2y - 12x^3y^2 + 9y^3}{2xy^2}$
$= \dfrac{6x^2y}{2xy^2} - \dfrac{12x^3y^2}{2xy^2} + \dfrac{9y^3}{2xy^2}$
$= \dfrac{3x}{y} - 6x^2 + \dfrac{9y}{2x}$

21. $\dfrac{3xyz + 6xyz^2 - 9x^3y^5z^7}{6xy}$
$= \dfrac{3xyz}{6xy} + \dfrac{6xyz^2}{6xy} - \dfrac{9x^3y^5z^7}{6xy}$
$= \dfrac{z}{2} + z^2 - \dfrac{3}{2}x^2y^4z^7$

23. $x+1 \overline{)\,x^2 + 4x + 3}$ quotient $x+3$

$\underline{x^2 + x} \qquad \leftarrow x(x+1)$
$\quad\; 3x + 3$
$\quad\; \underline{3x + 3} \quad \leftarrow 3(x+1)$
$\qquad\quad 0 \quad \leftarrow$ Remainder

Thus, $\dfrac{x^2 + 4x + 3}{x+1} = x+3$

25. $x+5 \overline{)\,2x^2 + 13x + 15}$ quotient $2x+3$

$\underline{2x^2 + 10x} \qquad \leftarrow 2x(x+5)$
$\quad\; 3x + 15$
$\quad\; \underline{3x + 15} \leftarrow 3(x+5)$
$\qquad\quad 0 \leftarrow$ Remainder

Thus, $\dfrac{2x^2 + 13x + 15}{x+5} = 2x+3$

27. $a-2 \overline{)\,2a^2 + a - 9}$ quotient $2a+5$

$\underline{2a^2 - 4a} \qquad \leftarrow 2a(a-2)$
$\quad\; 5a - 9$
$\quad\; \underline{5a - 10} \quad \leftarrow 5(a-2)$
$\qquad\quad 1 \quad \leftarrow$ Remainder

Thus, $\dfrac{2a^2 + a - 9}{a-2} = 2a + 5 + \dfrac{1}{a-2}$

29. $2x-1 \overline{)\,6x^2 + x - 2}$ quotient $3x+2$

$\underline{6x^2 - 3x} \qquad \leftarrow 3x(2x-1)$
$\quad\; 4x - 2$
$\quad\; \underline{4x - 2} \leftarrow 2(2x-1)$
$\qquad\quad 0 \leftarrow$ Remainder

Thus, $\dfrac{6x^2 + x - 2}{2x-1} = 3x + 2$

31.
$$2r-3 \overline{)\begin{array}{l} 2r+3 \\ 4r^2+0r-9 \end{array}} \leftarrow \text{Write } 4r^2-9$$
$$\text{as } 4r^2+0r-9$$
$$\underline{4r^2-6r} \leftarrow 2r(2r-3)$$
$$6r-9$$
$$\underline{6r-9} \leftarrow 3(2r-3)$$
$$0 \leftarrow \text{Remainder}$$

Thus $\dfrac{4r^2-9}{2r-3}=2r+3$

33.
$$3x+2 \overline{)\begin{array}{l} 3x^2-3x+1 \\ 9x^3-3x^2-3x+4 \end{array}}$$
$$\underline{9x^3+6x^2} \leftarrow 3x^2(3x+2)$$
$$-9x^2-3x$$
$$\underline{-9x^2-6x} \leftarrow -3x(3x+2)$$
$$3x+4$$
$$\underline{3x+2} \leftarrow 1(3x+2)$$
$$2 \leftarrow \text{Remainder}$$

Thus,
$$\dfrac{9x^3-3x^2-3x+4}{3x+2}=3x^2-3x+1+\dfrac{2}{3x+2}$$

35.
$$x-1 \overline{)\begin{array}{l} -x^2-7x-5 \\ -x^3-6x^2+2x-3 \end{array}}$$
$$\underline{-x^3+x^2} \leftarrow -x^2(x-1)$$
$$-7x^2+2x$$
$$\underline{-7x^2+7x} \leftarrow -7x(x-1)$$
$$-5x-3$$
$$\underline{-5x+5} \leftarrow -5x(x-1)$$
$$-8 \leftarrow \text{Remainder}$$

Thus,
$$\dfrac{-x^3-6x^2+2x-3}{x-1}=-x^2-7x-5-\dfrac{8}{x-1}$$

37.
$$2y+3 \overline{)\begin{array}{l} 2y^2+3y-1 \\ 4y^3+12y^2+7y-3 \end{array}}$$
$$\underline{4y^3+6y^2} \leftarrow 2y^2(2y+3)$$
$$6y^2+7y$$
$$\underline{6y^2+9y} \leftarrow 3y(2y+3)$$
$$-2y-3$$
$$\underline{-2y-3} \leftarrow -1(2y+3)$$
$$0 \leftarrow \text{Remainder}$$

Thus, $\dfrac{4y^3+12y^2+7y-3}{2y+3}=2y^2+3y-1$

39.
$$2x-1 \overline{)\begin{array}{l} 2x^2+x-2 \\ 4x^3+0x^2-5x+0 \end{array}} \leftarrow \text{Write } 4x^3-5x \text{ as } 4x^3+0x^2-5x+0$$
$$\underline{4x^3-2x^2} \leftarrow 2x^2(2x-1)$$
$$2x^2-5x$$
$$\underline{2x^2-x} \leftarrow x(2x-1)$$
$$-4x+0$$
$$\underline{-4x+2} \leftarrow -2(2x-1)$$
$$-2 \leftarrow \text{Remainder}$$

Thus, $\dfrac{4x^2-5x}{2x-1}=2x^2+x-2-\dfrac{2}{2x-1}$

41.

$$x^2 + 0x - 2 \overline{\smash{\big)}\, 3x^5 + 0x^4 + 0x^3 + 4x^2 - 12x - 8} \quad \begin{array}{c} 3x^3 \qquad + 6x + 4 \end{array}$$

$$\underline{3x^5 + 0x^4 - 6x^3} \qquad \leftarrow 3x^3(x^2 + 0x - 2)$$

$$6x^3 + 4x^2 - 12x$$

$$\underline{6x^3 + 0x^2 - 12x} \qquad \leftarrow 6x(x^2 + 0x - 2)$$

$$4x^2 + 0x - 8$$

$$\underline{4x^2 + 0x - 8} \leftarrow 4(x^2 + 0x - 2)$$

$$0 \leftarrow \text{Remainder}$$

Thus, $\dfrac{3x^5 + 4x^2 - 12x - 8}{x^2 - 2} = 3x^3 + 6x + 4$

43.

$$3x^3 - 8x^2 + 0x - 5 \overline{\smash{\big)}\, 3x^4 + 4x^3 - 32x^2 - 5x - 20} \quad \begin{array}{c} x + 4 \end{array}$$

$$\underline{3x^4 - 8x^3 + 0x^2 - 5x} \qquad \leftarrow x\left(3x^3 - 8x^2 + 0x - 5\right)$$

$$12x^3 - 32x^2 + 0x - 20$$

$$\underline{12x^3 - 32x^2 + 0x - 20} \leftarrow 4\left(3x^3 - 8x^2 + 0x - 5\right)$$

$$0 \leftarrow \text{Remainder}$$

Thus, $\dfrac{3x^4 + 4x^3 - 32x^2 - 5x - 20}{3x^3 - 8x^2 - 5} = x + 4$

45.

$$x + 4 \overline{\smash{\big)}\, 3x^4 + 4x^3 - 32x^2 - 5x - 20} \quad \begin{array}{c} 3x^3 - 8x^2 \qquad - 5 \end{array}$$

$$\underline{3x^4 + 12x^3} \qquad\qquad \leftarrow 3x^3(x + 4)$$

$$-8x^3 - 32x^2$$

$$\underline{-8x^3 - 32x^2} \qquad\qquad \leftarrow -8x^2(x + 4)$$

$$-5x - 20$$

$$\underline{-5x - 20} \qquad \leftarrow -5(x + 4)$$

$$0 \qquad \leftarrow \text{Remainder}$$

Thus, $\dfrac{3x^4 + 4x^3 - 32x^2 - 5x - 20}{x + 4} = 3x^3 - 8x^2 - 5$

47.

$$\begin{array}{r|rrr} 2 & 1 & 1 & -6 \\ & & 2 & 6 \\ \hline & 1 & 3 & 0 \end{array}$$

Thus, $\dfrac{x^2 + x - 6}{x - 2} = x + 3$

49.

$$\begin{array}{r|rrr} 3 & 1 & 5 & -12 \\ & & 3 & 24 \\ \hline & 1 & 8 & 12 \end{array}$$

Thus, $\dfrac{x^2 + 5x - 12}{x - 3} = x + 8 + \dfrac{12}{x - 3}$

51.

$$
\begin{array}{r|rrr}
4 & 3 & -7 & -10 \\
  &   & 12 & 20 \\
\hline
  & 3 & 5 & 10
\end{array}
$$

Thus, $\dfrac{3x^2 - 7x - 10}{x - 4} = 3x + 5 + \dfrac{10}{x - 4}$

53.

$$
\begin{array}{r|rrrr}
1 & 4 & -3 & 2 & 0 \\
  &   & 4 & 1 & 3 \\
\hline
  & 4 & 1 & 3 & 3
\end{array}
$$

Thus, $\dfrac{4x^3 - 3x^2 + 2x}{x - 1} = 4x^2 + x + 3 + \dfrac{3}{x - 1}$

55.

$$
\begin{array}{r|rrrr}
-3 & 3 & 7 & -4 & 12 \\
   &   & -9 & 6 & -6 \\
\hline
   & 3 & -2 & 2 & 6
\end{array}
$$

Thus, $\dfrac{3x^3 + 7x^2 - 4x + 12}{x + 3} = 3x^2 - 2x + 2 + \dfrac{6}{x + 3}$

57.

$$
\begin{array}{r|rrrrr}
1 & 1 & 0 & 0 & 0 & -1 \\
  &   & 1 & 1 & 1 & 1 \\
\hline
  & 1 & 1 & 1 & 1 & 0
\end{array}
$$

Thus, $\dfrac{y^4 - 1}{y - 1} = y^3 + y^2 + y + 1$

59.

$$
\begin{array}{r|rrrrrr}
-1 & 1 & 1 & 0 & 0 & 0 & -10 \\
   &   & -1 & 0 & 0 & 0 & 0 \\
\hline
   & 1 & 0 & 0 & 0 & 0 & -10
\end{array}
$$
$\leftarrow$ Write $y^5 + y^4 - 10$ as $y^5 + y^4 + 0y^3 + 0y^2 + 0y - 10$

Thus, $\dfrac{y^5 + y^4 - 10}{y + 1} = y^4 - \dfrac{10}{y + 1}$

61.

$$
\begin{array}{r|rrrr}
\frac{1}{3} & 3 & 2 & -4 & 1 \\
            &   & 1 & 1 & -1 \\
\hline
            & 3 & 3 & -3 & 0
\end{array}
$$

Thus, $\dfrac{3x^3 + 2x^2 - 4x + 1}{x - \frac{1}{3}} = 3x^2 + 3x - 3$

63.

$$
\begin{array}{r|rrrrr}
\frac{1}{2} & 2 & -1 & 2 & -3 & 1 \\
            &   & 1 & 0 & 1 & -1 \\
\hline
            & 2 & 0 & 2 & -2 & 0
\end{array}
$$

Thus,

$\dfrac{2x^4 - x^3 + 2x^2 - 3x + 1}{x - \frac{1}{2}} = 2x^3 + 2x - 2$

65. To find the remainder, use synthetic division:

$$\begin{array}{r|rrr} 2 & 4 & -5 & 4 \\ & & 8 & 6 \\ \hline & 4 & 3 & 10 \end{array}$$

The remainder is 10.

67. To find the remainder, use synthetic division:

$$\begin{array}{r|rrrr} 2 & 1 & -2 & 4 & -8 \\ & & 2 & 0 & 8 \\ \hline & 1 & 0 & 4 & 0 \end{array}$$

The remainder is 0 which means that $x - 2$ is a factor.

69. To find the remainder, use synthetic division.

$$\begin{array}{r|rrrr} \frac{1}{2} & -2 & -6 & 2 & -4 \\ & & -1 & -\frac{7}{2} & -\frac{3}{4} \\ \hline & -2 & -7 & -\frac{3}{2} & -\frac{19}{4} \end{array} \text{ or } -4.75$$

The remainder is $-\dfrac{19}{4}$ or -4.75.

71. Width $= \dfrac{\text{area}}{\text{length}} = \dfrac{6x^2 - 8x - 8}{2x - 4}$

$$\begin{array}{r} 3x + 2 \\ 2x - 4 \overline{)6x^2 - 8x - 8} \end{array}$$

$$\underline{6x^2 - 12x} \quad \leftarrow 3x(2x - 4)$$
$$4x - 8$$
$$\underline{4x - 8} \quad \leftarrow 2(2x - 4)$$
$$0 \quad \leftarrow \text{Remainder}$$

Thus the width is $3x + 2$.

73. To compare the areas, compare the lengths and widths of the two figures (rectangles). Length: observe that $12x + 24$ is six times $2x + 4$

Width: observe that $\dfrac{1}{2}x + 4$ is $\dfrac{1}{2}$ of $x + 8$

The area of the larger rectangle is $\dfrac{1}{2}(6) = 3$ times the area of the smaller rectangle.

75. No; answers will vary.

77. Answers will vary. One example is If the remainder is 0, $x - a$ is a factor.

79. $\dfrac{P(x)}{x - 4} = x + 2$

$P(x) = (x - 4)(x + 2)$

$P(x) = x^2 - 2x - 8$

81. $\dfrac{P(x)}{x - 4} = x + 3 + \dfrac{4}{x - 4}$

$P(x) = (x - 4)\left(x + 3 + \dfrac{4}{x - 4}\right)$

$ = x(x - 4) + 3(x - 4) + \dfrac{4}{x - 4}(x - 4)$

$ = x^2 - 4x + 3x - 12 + 4$

$ = x^2 - x - 8$

83.

$$x-2y{\overline{\smash{\big)}\,2x^3-x^2y-7xy^2+2y^3}}$$

quotient: $2x^2+3xy-y^2$

$$\underline{2x^3-4x^2y} \qquad \leftarrow 2x^2(x-2y)$$
$$3x^2y-7xy^2$$
$$\underline{3x^2y-6xy^2} \qquad \leftarrow 3xy(x-2y)$$
$$-xy^2+2y^3$$
$$\underline{-xy^2+2y^3} \leftarrow -y^2(x-2y)$$
$$0 \quad \leftarrow \text{Remainder}$$

Thus, $\dfrac{2x^3-x^2y-7xy^2+2y^3}{x-2y}=2x^2+3xy-y^2$

85.

$$2x-3{\overline{\smash{\big)}\,2x^2+2x-2}}$$

quotient: $x+\dfrac{5}{2}$

$$\underline{2x^2-3x} \qquad \leftarrow x(2x-3)$$
$$5x-2$$
$$\underline{5x-\dfrac{15}{2}} \leftarrow \dfrac{5}{2}(2x-3)$$
$$\dfrac{11}{2} \quad \leftarrow \text{Remainder}$$

Thus, $\dfrac{2x^2+2x-2}{2x-3}=x+\dfrac{5}{2}+\dfrac{11}{2(2x-3)}$

87. Two of the dimensions are r and $2r+2$. This product is $r(2r+2)=2r^2+2r$. To find the third side, divide the volume by the above product. That is,

$$\text{the third side} = \frac{\text{volume}}{\text{product of two sides}}$$
$$= \frac{2r^3+4r^2+2r}{2r^2+2r}$$

$$2r^2+2r{\overline{\smash{\big)}\,2r^3+4r^2+2r}}$$

quotient: $r+1$

$$\underline{2r^3+2r^2} \qquad \leftarrow r\left(2r^2+2r\right)$$
$$2r^2+2r$$
$$\underline{2r^2+2r} \leftarrow 1\left(2r^2+2r\right)$$
$$0 \quad \leftarrow \text{Remainder}$$

Thus, the third side is $w=r+1$.

89. The polynomial is the product of $x - 3$ with $x^2 - 3x + 4$ plus the remainder of 2.

$$
\begin{array}{r}
x^2 - 3x + 4 \\
x - 3 \\
\hline
-3x^2 + 9x - 12 \\
x^3 - 3x^2 + 4x \\
\hline
x^3 - 6x^2 + 13x - 12
\end{array}
$$

Now, add on the remainder of 2 to get $x^3 - 6x^2 + 13x - 10$ for the polynomial.

91.
$$\frac{4x^{n+1} + 2x^n - 3x^{n-1} - x^{n-2}}{2x^n} = \frac{4x^{n+1}}{2x^n} + \frac{2x^n}{2x^n} - \frac{3x^{n-1}}{2x^n} - \frac{x^{n-2}}{2x^n}$$

$$= 2x^{n+1-n} + x^{n-n} - \frac{3}{2}x^{n-1-n} - \frac{1}{2}x^{n-2-n}$$

$$= 2x + x^0 - \frac{3}{2}x^{-1} - \frac{1}{2}x^{-2}$$

$$= 2x + 1 - \frac{3}{2x} - \frac{1}{2x^2}$$

93. Let $P(x) = x^{100} + x^{99} + \ldots + x + 1$. To determine if $x - 1$ is a factor, compute $P(1)$. If it is 0, then $x - 1$ is a factor. Now,

$P(1) = 1^{100} + 1^{99} + 1^{98} + 1^{97} + \ldots + 1^2 + 1^1 + 1$

$= 1 + 1 + 1 + 1 + \ldots + 1 + 1 + 1 = 101$

Note: Here you are adding up the number 1 a total of 101 times. Since $P(1) = 101$ which is not zero, then $x - 1$ is not a factor.

95. Let $P(x) = x^{99} + x^{98} + \ldots + x + 1$. To determine if $x + 1$ is a factor, compute $P(-1)$. If it is 0, then $x + 1$ is a factor. Now,

$P(-1) = (-1)^{99} + (-1)^{98} + (-1)^{97} + (-1)^{96} + \ldots + (-1)^3 + (-1)^2 + (-1) + 1$

$= -1 + 1 - 1 + 1 + \ldots - 1 + 1 - 1 + 1$

$= \quad 0 + 0 + \ldots + 0 \quad + 0 = 0$

Since $P(-1) = 0$, $x + 1$ is a factor.

97. a.

$$
\begin{array}{r|rrrr}
-\frac{5}{3} & 9 & 9 & 5 & 12 \\
& & -15 & 10 & -25 \\
\hline
& 9 & -6 & 15 & -13
\end{array}
$$

Now, divide $9x^2 - 6x + 15$ by 3 to get $3x^2 - 2x + 5$.

Thus,

$$\frac{9x^3 + 9x^2 + 5x + 12}{3x + 5}$$

$$= 3x^2 - 2x + 5 - \frac{13}{3x + 5}$$

b. Because we are expressing the remainder in terms of $3x + 5$ rather than $x + \frac{5}{3}$, the denominator of the remainder is altered rather than the numerator.

98. $\left(4x^{-2}y^3\right)^{-2} = 4^{-2}x^{-2(-2)}y^{3(-2)}$

$$= \frac{1}{16}x^4y^{-6}$$

$$= \frac{x^4}{16y^6}$$

99. $-1 < \dfrac{4(3x-2)}{3} \le 5$

$-3 < 4(3x-2) \le 15$ ← Multiply by 3

$-3 < 12x - 8 \le 15$ ← Distributive property

$5 < 12x \le 23$ ← Add 8

$\dfrac{5}{12} < x \le \dfrac{23}{12}$ ← Divide by 12

The graph is

$\begin{array}{c} \\ \xrightarrow{\hspace{4cm}} \\ \frac{5}{12} \qquad \frac{23}{12} \end{array}$

100. $f(x) = \dfrac{1}{2}x + \dfrac{3}{7}$

Then,

$$f\left(-\frac{2}{3}\right) = \frac{1}{2}\left(-\frac{2}{3}\right) + \frac{3}{7}$$

$$= -\frac{1}{3} + \frac{3}{7}$$

$$= -\frac{7}{21} + \frac{9}{21}$$

$$= \frac{-7+9}{21}$$

$$= \frac{2}{21}$$

101. $20x - 60y = 120$. To graph, find the intercepts

$x = 0$: $20(0) - 60y = 120$

$-60y = 120$

$y = \dfrac{120}{-60} = -2$

The y-intercept is -2

$y = 0$: $20x - 60(0) = 120$

$20x = 120$

$x = \dfrac{120}{20} = 6$

The x-intercept is 6.

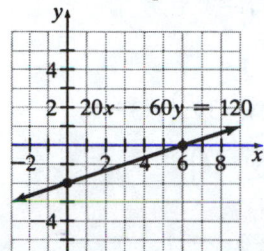

Exercise Set 5.4

1. Determine if all the terms contain a greatest common factor and, if so, factor it out.

3. a. Answers will vary.

    b. The greatest common factor is $2x^2y$.

    c. $2x^2y\left(3y^4 - x + 6x^7y^2\right)$

5. a. Answers will vary.

    b. $6x^3 - 2xy^3 + 3x^2y^2 - y^5$

$$= 2x\left(3x^2 - y^3\right) + y^2\left(3x^2 - y^3\right)$$

$$= \left(2x + y^2\right)\left(3x^2 - y^3\right)$$

7. $x^4y^6,\ x^3y^5,\ xy^6,\ x^2y^4$

The lowest power of x is 1. The lowest power of y is 4. Therefore the GCF is xy^4.

9. $8n + 8 = 8 \cdot n + 8 \cdot 1 = 8(n + 1)$

11. $6x^2 + 3x - 9 = 3 \cdot 2x^2 + 3 \cdot x + 3(-3)$
$\qquad = 3(2x^2 + x - 3)$

13. $8p^2 - 6p + 4 = 2 \cdot 4p^2 - 2 \cdot 3p + 2 \cdot 2$
$\qquad = 2(4p^2 - 3p + 2)$

15. $7x^5 - 9x^4 + 3x^3 = x^3 \cdot 7x^2 + x^3(-9x) + x^3 \cdot 3$
$\qquad = x^3(7x^2 - 9x + 3)$

17. $-24y^{15} + 9y^3 - 3y$
$\qquad = -3y \cdot 8y^{14} - 3y(-3y^2) - 3y \cdot 1$
$\qquad = -3y(8y^{14} - 3y^2 + 1)$

19. $3x^2 y + 6x^2 y^2 + 3xy$
$\qquad = 3xy \cdot x + 3xy \cdot 2xy + 3xy \cdot 1$
$\qquad = 3xy(x + 2xy + 1)$

21. $40a^5 b^4 c - 8a^4 b^2 c^2 + 4a^2 c$
$\qquad = 4a^2 c \cdot 10a^3 b^4 + 4a^2 c(-2a^2 b^2 c) + 4a^2 c(1)$
$\qquad = 4a^2 c(10a^3 b^4 - 2a^2 b^2 c + 1)$

23. $9p^4 q^5 r - 3p^2 q^2 r^2 + 6pq^5 r^3$
$\qquad = 3pq^2 r \cdot 3p^3 q^3 + 3pq^2 r(-pr) + 3pq^2 r(2q^3 r^2)$
$\qquad = 3pq^2 r(3p^3 q^3 - pr + 2q^3 r^2)$

25. $-52x^2 y^2 - 16xy^3 + 26z$
$\qquad = -2 \cdot 26x^2 y^2 - 2 \cdot 8xy^3 - 2(-13z)$
$\qquad = -2(26x^2 y^2 + 8xy^3 - 13z)$

27. $-6x + 2 = -2(3x) - 2(-1)$
$\qquad = -2(3x - 1)$

29. $-w^2 + 7w - 5 = -(w^2 - 7w + 5)$

31. $-3r^2 - 6r + 9 = -3(r^2 + 2r - 3)$

33. $-6r^4 s^3 + 4r^2 s^4 + 2rs^5$
$\qquad = -2(3r^4 s^3 - 2r^2 s^4 - rs^5)$
$\qquad = -2rs^3(3r^3 - 2rs - s^2)$

35. $-a^4 b^2 c + 5a^3 bc^2 + a^2 b$
$\qquad = -a^2 b(a^2 bc) - a^2 b(-5ac^2) - a^2 b(-1)$
$\qquad = -a^2 b(a^2 bc - 5ac^2 - 1)$

37. $x(a - 2) + 1(a - 2) = (x + 1)(a - 2)$

39. $4a(x - 1) - 3(x - 1) = (4a - 3)(x - 1)$

41. $(x - 2)(3x + 5) - (x - 2)(5x - 4)$
$\qquad = (x - 2)(3x + 5 - 5x + 4)$
$\qquad = (x - 2)(-2x + 9)$
$\qquad = -(x - 2)(2x - 9)$

43. $(2a + 4)(a - 3) - (2a + 4)(2a - 1)$
$\qquad = (2a + 4)(a - 3 - 2a + 1)$
$\qquad = (2a + 4)(-a - 2)$
$\qquad = -2(a + 2)(a + 2)$

45. $x^2 + 3x - 5x - 15 = x(x + 3) - 5(x + 3)$
$\qquad = (x - 5)(x + 3)$

47. $8n^2 - 4n - 20n + 10$
$\qquad = 2(4n^2 - 2n - 10n + 5)$
$\qquad = 2[2n(2n - 1) - 5(2n - 1)]$
$\qquad = 2(2n - 5)(2n - 1)$

49. $ax + ay + bx + by = a(x + y) + b(x + y)$
$\qquad = (a + b)(x + y)$

51. $x^3 - 3x^2 + 4x - 12 = x^2(x - 3) + 4(x - 3)$
$\qquad = (x^2 + 4)(x - 3)$

53. $10m^2 - 12mn - 25mn + 30n^2$
$\qquad = 2m(5m - 6n) - 5n(5m - 6n)$
$\qquad = (2m - 5n)(5m - 6n)$

55. $6x^3 + 18x^2 - 12x - 36$

$= 6\left(x^3 + 3x^2 - 2x - 6\right)$

$= 6\left[x^2(x+3) - 2(x+3)\right]$

$= 6\left(x^2 - 2\right)(x+3)$

57. $a^5 - a^4 + a^3 - a^2 = a^2\left(a^3 - a^2 + a - 1\right)$

$= a^2\left[a^2(a-1) + 1(a-1)\right]$

$= a^2\left(a^2 + 1\right)(a-1)$

59. a. $h(t) = -16t^2 + 128t$

$h(3) = -16(3)^2 + 128(3)$

$= -16(9) + 128(3)$

$= -144 + 384$

$= 240$

The height after 3 seconds is 240 feet.

b. $h(t) = -16t^2 + 128t$

$h(t) = -16t(t) - 16t(-8)$

$h(t) = -16t(t-8)$

c. $h(3) = -16(3)(3-8)$

$= -48(-5)$

$= 240$ feet

61. a. $A = \pi r^2 + 2rl$

$= \pi(20)^2 + 2(20)(40)$

$= 400\pi + 1600$

≈ 2856.64

The area is about 2856.64 ft^2.

b. $A = \pi r^2 + 2rl$

$A = r(\pi r) + r(2l)$

$A = r(\pi r + 2l)$

c. $A = 20(\pi \cdot 20 + 2 \cdot 40)$

$A \approx 2856.64$ ft^2

63. a. Sale price is

$(x + 0.06x) - 0.06(x + 0.06x)$

$= (1 - 0.06)(x + 0.06x)$

$= (0.94)(1.06x)$

b. Now, $(0.94)(1.06x) = 0.9964x$ upon multiplication. This represents 99.64% of the 1999 price which means that the new price is slightly lower than the price of the 1999 model.

65. a. Final price is

$(x + 0.15x) - 0.20(x + 0.15x)$

$= 0.80(x + 0.15x)$

b. $(x + 0.15x) - 0.20(x + 0.15x)$

$= (1 - 0.20)(x + 0.15x)$

$= 0.80(1.15x)$

$= 0.92x$

The sale price is 92% of the regular price.

67. $5a(3x - 2)^5 + 4(3x - 2)^4$

$= 5a(3x - 2) \cdot (3x - 2)^4 + 4 \cdot (3x - 2)^4$

$= (3x - 2)^4\left[5a(3x - 2) + 4\right]$

$= (3x - 2)^4(15ax - 10a + 4)$

69. $4x^2(x - 3)^3 - 6x(x - 3)^2 + 4(x - 3)$

$= 2(x - 3)\left[2x^2(x - 3)^2 - 3x(x - 3) + 2\right]$ $\leftarrow (2x - 3)$ is a common factor

$= 2(x - 3)\left[2x^2\left(x^2 - 6x + 9\right) - 3x(x - 3) + 2\right]$

$= 2(x - 3)\left(2x^4 - 12x^3 + 18x^2 - 3x^2 + 9x + 2\right)$

$= 2(x - 3)\left(2x^4 - 12x^3 + 15x^2 + 9x + 2\right)$

71. $ax^2 + 2ax - 3a + bx^2 + 2bx - 3b$

$= ax^2 + bx^2 + 2ax + 2bx - 3a - 3b$

$= x^2(a+b) + 2x(a+b) - 3(a+b)$

$= (x^2 + 2x - 3)(a+b)$

73. $x^{6m} - 2x^{4m} = x^{4m} \cdot x^{2m} - x^{4m} \cdot 2$

$= x^{4m}(x^{2m} - 2)$

75. $3x^{4m} - 2x^{3m} + x^{2m}$

$= x^{2m} \cdot 3x^{2m} - x^{2m} \cdot 2x^m + x^{2m} \cdot 1$

$= x^{2m}(3x^{2m} - 2x^m + 1)$

77. $a^r b^r + c^r b^r - a^r d^r - c^r d^r$

$= b^r(a^r + c^r) - d^r(a^r + c^r)$

$= (b^r - d^r)(a^r + c^r)$

79. a. $6x^3 - 3x^2 + 9x$

$= 3x(2x^2) + 3x(-x) + 3x(3)$

$= 3x(2x^2 - x + 3)$

Yes

b. 0

c. Answers will vary.

81. a. They should be the same graph.

b. $y_1 = 8x^3 - 16x^2 - 4x$

$y_2 = 4x(2x^2 - 4x - 1)$

$-10, 10, 1, -20, 10, 4$

c. Answers will vary.

d. The factoring is not correct.

83. $\dfrac{\left(\left|\frac{1}{2}\right| - \left|-\frac{1}{3}\right|\right)^2}{-\left|\frac{1}{3}\right| \cdot \left|-\frac{2}{5}\right|} = \dfrac{\left(\frac{1}{2} - \frac{1}{3}\right)^2}{-\left(\frac{1}{3}\right)\left(\frac{2}{5}\right)}$ ← Do absolute values first

$= \dfrac{\left(\frac{3}{6} - \frac{2}{6}\right)^2}{-\left(\frac{1}{3}\right)\left(\frac{2}{5}\right)}$ ← Simplify numerator

$= \dfrac{\left(\frac{1}{6}\right)^2}{-\left(\frac{1}{3}\right)\left(\frac{2}{5}\right)}$ ← Simplify numerator

$= \dfrac{\frac{1}{36}}{-\left(\frac{1}{3}\right)\left(\frac{2}{5}\right)}$ ← Simplify numerator

$= \dfrac{\frac{1}{36}}{-\frac{2}{15}}$ ← Multiply denominator

$= \dfrac{1}{36}\left(-\dfrac{15}{2}\right)$ ← Perform division

$= -\dfrac{15}{72} = -\dfrac{5}{24}$ ← Simplify/multiply

84. $-4 < \dfrac{6-3x}{2} \le 5$

$2(-4) < 2\left(\dfrac{6-3x}{2}\right) \le 2(5)$ ← Multiply all three parts by 2

$-8 < 6 - 3x \le 10$ ← Simplify

$-8 - 6 < 6 - 6 - 3x \le 10 - 6$ ← Subtract 6 from all three parts

$-14 < -3x \le 4$ ← Simplify

$\dfrac{-14}{-3} > \dfrac{-3x}{-3} \ge \dfrac{4}{-3}$ ← Divide all three parts by –3 and reverse both inequality symbols

$\dfrac{14}{3} > x \ge -\dfrac{4}{3}$ ← Simplify

$\left\{x \mid -\dfrac{4}{3} \le x < \dfrac{14}{3}\right\}$

85. Let x be the amount invested at 5%. Then
$10{,}000 - x$ was invested at 6% and the equation is
$$0.05x + 0.06(10{,}000 - x) = 560$$
$$0.05x + 600 - 0.06x = 560$$
$$600 - 0.01x = 560$$
$$-0.01x = -40$$
$$x = \dfrac{-40}{-0.01}$$
$$= 4{,}000 \text{ dollars}$$
Thus, $4,000 was invested at 5% and
$10{,}000 - 4{,}000 = \$6{,}000$ was invested at 6%.

86. Let x be the amount invested at 5% and y be the amount invested at 6%. The system is
$$x + y = 10{,}000$$
$$0.05x + 0.06y = 560$$
Solve the first equation for y:
$$x + y = 10{,}000$$
$$y = -x + 10{,}000$$
Substitute $-x + 10{,}000$ for y into the second equation and then solve for x
$$0.05x + 0.06y = 560$$
$$0.05x + 0.06(-x + 10{,}000) = 560$$
$$0.05x - 0.06x + 600 = 560$$
$$-0.01x + 600 = 560$$
$$-0.01x = -40$$
$$x = \dfrac{-40}{-0.01}$$
$$= 4{,}000 \text{ dollars}$$
Thus, $4,000 was invested at 5% and $6,000 was invested at 6%.

87. First, solve $3x - 4y = 12$ for y:

$$3x - 4y = 12$$
$$-4y = -3x + 12$$
$$y = \frac{-3x + 12}{-4} = \frac{3}{4}x - 3$$

The slope of this line is $\frac{3}{4}$. Since the new line is parallel to this line, its slope is also $\frac{3}{4}$. Using

$y = mx + b$ with $m = \frac{3}{4}$ and $b = -2$ (the y-intercept is –2), the equation becomes

$$y = mx + b$$
$$y = \frac{3}{4}x - 2$$
$$4y = 3x - 8 \quad \leftarrow \text{ Multiply by 4 to clear fractions}$$
$$3x - 4y = 8$$

Exercise Set 5.5

1. Factor out the greatest common factor if it is present.

3. a. Answers will vary.

b. $6x^2 - x - 12$
Observe that $6(-12) = -72$. The two numbers whose product is –72 and whose sum is –1 are –9 and 8, since $(-9)(8) = -72$ and $-9 + 8 = -1$. Then the middle term, $-x$, can be written as $-x = -9x + 8x$ and the factorization is
$$6x^2 - x - 12 = 6x^2 - 9x + 8x - 12$$
$$= 3(2x - 3) + 4(2x - 3)$$
$$= (3x + 4)(2x - 3)$$

5. Both are +.

7. One is +, one is –.

9. $x^2 + 7x + 6$
The two numbers whose product is 6 and whose sum is 7 are 1 and 6, since $(1)(6) = 6$ and $1 + 6 = 7$. Thus,
$$x^2 + 7x + 6 = (x + 1)(x + 6).$$

11. $y^2 - 12y + 11$
The two numbers whose product is 11 and whose sum is –12 are –11 and –1, since $(-11)(-1) = 11$ and $-11 + (-1) = -12$. Thus,
$$y^2 - 12y + 11 = (y - 11)(y - 1).$$

13. $x^2 - 16x + 64$
The two numbers whose product is 64 and whose sum is –16 are –8 and –8, since $(-8)(-8) = 64$ and $-8 + (-8) = -16$. Thus,
$$x^2 - 16x + 64 = (x - 8)(x - 8) \text{ or } (x - 8)^2.$$

15. $x^2 - 13x - 30$
The two numbers whose product is –30 and whose sum is –13 are 2 and –15, since $(2)(-15) = -30$ and $2 - 15 = -13$. Thus,
$$x^2 - 13x - 30 = (x + 2)(x - 15).$$

17. $x^2 - 4xy + 3y^2$
The two numbers whose product is 3 and whose sum is –4 are –1 and –3, since $(-1)(-3) = 3$ and $-1 + (-3) = -4$. Thus,
$$x^2 - 4xy + 3y^2 = (x - y)(x - 3y).$$

19. $z^2 - 7yz + 10y^2$
The two numbers whose product is 10 and whose sum is –7 are –5 and –2, since $(-5)(-2) = 10$ and $-5 + (-2) = -7$. Thus,
$$z^2 - 7y^2 + 10y^2 = (z - 2y)(z - 5y).$$

21. $4x^2 + 12x - 16$

$= 4\left(x^2 + 3x - 4\right) \leftarrow 4$ is a common factor

Now factor $x^2 + 3x - 4$. The two numbers whose product is –4 and whose sum is 3 are 4 and –1, since $(4)(-1) = -4$ and $4 + (-1) = 3$. Now,

$x^2 + 3x - 4 = (x + 4)(x - 1)$ and the answer is $4x^2 + 12x - 16 = 4(x + 4)(x - 1)$.

23. $x^3 - 3x^2 - 18x$

$= x\left(x^2 - 3x - 18\right) \leftarrow x$ is a common factor

Now, factor $x^2 - 3x - 18$. The two numbers whose product is –18 and whose sum is –3 are –6 and 3 since, $(-6)(3) = -18$ and $-6 + 3 = -3$. Thus,

$x^2 - 3x - 18 = (x - 6)(x + 3)$ and the answer is $x^3 - 3x^2 - 18x = x(x - 6)(x + 3)$.

25. $5p^2 - 8p + 3$

Observe that $(5)(3) = 15$. The two numbers whose product is 15 and whose sum is –8 are –5 and –3, since $(-5)(-3) = 15$ and $(-5) + (-3) = -8$. Now the middle term, $-8p$, can be written as $-5p - 3p$ and the factorization is

$5p^2 - 8p + 3 = 5p^2 - 5p - 3p + 3$
$= 5p(p - 1) - 3(p - 1)$
$= (5p - 3)(p - 1)$

27. $3x^2 - 3x - 6 = 3\left(x^2 - x - 2\right)$

Now factor $\left(x^2 - x - 2\right)$. The two numbers whose product is –2 and sum is –1 are –2 and 1, since $(-2)(-1) = -2$ and $-2 + 1 = -1$. Thus

$3x^2 - 3x - 6 = 3\left(x^2 - x - 2\right)$
$= 3(x - 2)(x + 1)$

29. $6x^2 - 13x - 63$

Observe that $6(-63) = -378$. The two numbers whose product is –378 and whose sum is –13 are –27 and 14, since $(-27)(14) = -378$ and $-27 + 14 = -13$. Now

the middle term $-13x$ can be written as $-27x + 14x$ and the factorization is

$6x^2 - 13x - 63 = 6x^2 - 27x + 14x - 63$
$= 3x(2x - 9) + 7(2x - 9)$
$= (3x + 7)(2x - 9)$

31. $30x^2 - 71x + 35$

Observe that $(30)(35) = 1050$. The two numbers whose product is 1050 and sum is –71 are –50 and –21, since $(-50)(-21) = 1050$ and $-50 - 21 = -71$. Now, the middle term can be written as $-71x = -50x - 21x$ and the factorization is

$30x^2 - 71x + 35 = 30x^2 - 50x - 21x + 35$
$= 10x(3x - 5) - 7(3x - 5)$
$= (10x - 7)(3x - 5)$

33. $32x^2 - 22x + 3$

Observe that $32(3) = 96$. The two numbers whose product is 96 and sum is –22 are –16 and –6, since $(-16)(-6) = 96$ and $-16 - 6 = -22$. Now, the middle term can be written as $-22x = -16x - 6x$ and the factorization is

$32x^2 - 22x + 3 = 32x^2 - 16x - 6x + 3$
$= 16x(2x - 1) - 3(2x - 1)$
$= (16x - 3)(2x - 1)$

35. $18w^2 + 18wz - 8z^2 = 2\left(9w^2 + 9wz - 4z^2\right)$

Now, factor $9w^2 + 9wz - 4z^2$. Observe that $(9)(-4) = -36$. The two numbers whose product is –36 and whose sum is 9 are 12 and –3, since $(12)(-3) = -36$ and $12 + (-3) = 9$. Then the middle term, $9wz$, can be written as $9wz = 12wz - 3wz$ and the factorization is

$9w^2 + 9wz - 4z^2$
$= 9w^2 + 12wz - 3wz - 4z^2$
$= 3w(3w + 4z) - z(3w + 4z)$
$= (3w - z)(3w + 4z)$

Thus,

$18w^2 + 18wz - 8z^2 = 2(3w - z)(3w + 4z)$.

37. $8x^2 + 30xy - 27y^2$

Observe that $8(-27) = -216$. The two numbers whose product is -216 and whose sum is 30 are 36 and -6, since $(36)(-6) = -216$ and $36 - 6 = 30$. The middle term can be written as $36xy - 6xy = 30xy$ and the factorization is

$8x^2 + 30xy - 27y^2$
$= 8x^2 + 36xy - 6xy - 27y^2$
$= 4x(2x + 9y) - 3(2x + 9y)$
$= (4x - 3y)(2x + 9y)$

39. $100b^2 - 90b + 20 = 10(10b^2 - 9b + 2)$

Now, factor $10b^2 - 9b + 2$. Observe that $(10)(2) = 20$. The two numbers whose product is 20 and whose sum is -9 are -4 and -5, since $(-4)(-5) = 20$ and $-4 + (-5) = -9$. Then the middle term, $-9b$, can be written as $-4b - 5b$ and the factorization is

$10b^2 - 9b + 2 = 10b^2 - 4b - 5b + 2$
$\qquad = 2b(5b - 2) - 1(5b - 2)$
$\qquad = (2b - 1)(5b - 2)$

Thus,
$100b^2 - 90b + 20 = 10(2b - 1)(5b - 2)$.

41. $a^3b^5 - a^2b^5 - 12ab^5$
$= ab^5(a^2 - a - 12) \leftarrow ab^5$ is a common factor

Now, factor $a^2 - a - 12$. The two numbers whose product is -12 and whose sum is -1 are -4 and 3, since $(-4)(3) = -12$ and $-4 + 3 = -1$

Thus,
$a^2 - ab - 12 = (a - 4)(a + 3)$
and the answer is
$a^3b^5 - a^2b^5 - 12ab^5 = ab^5(a - 4)(a + 3)$.

43. $3b^4c - 18b^3c^2 + 27b^2c^3$
$= 3b^2c(b^2 - 6bc + 9c^2)$

$[3b^2c$ is a common factor.]
Now factor $b^2 - 6bc + 9c^2$. The two numbers whose product is 9 and whose sum is -6 are -3 and -3, since $(-3)(-3) = 9$ and

$-3 + (-3) = -6$. Thus,
$b^2 - 6bc + 9c^2 = (b - 3c)(b - 3c)$
and the answer is
$3b^4c - 18b^3c^2 + 27b^2c^3$
$= 3b^2c(b - 3c)(b - 3c)$
$= 3b^2c(b - 3c)^2$

45. $8m^8n^3 + 4m^7n^4 - 24m^6n^5$
$= 4m^6n^3(2m^2 + mn - 6n^2) \leftarrow 4m^6n^3$ is a common factor. Now factor $2m^2 + mn - 6n^2$. Observe that $(2)(-6) = -12$. The two numbers whose product is -12 and whose sum is 1 are 4 and -3, since $(4)(-3) = -12$ and $4 + (-3) = 1$. Then the middle term, mn, can be written as $4mn - 3mn$ and the factorization is

$2m^2 + mn - 6n^2 = 2m^2 + 4mn - 3mn - 6n^2$
$\qquad = 2m(m + 2n) - 3n(m + 2n)$
$\qquad = (2m - 3n)(m + 2n)$

Thus,
$8m^8n^3 + 4m^7n^4 - 24m^6n^5$
$= 4m^6n^3(2m - 3n)(m + 2n)$.

47. $30x^2 - x - 20$

Observe that $(30)(-20) = -600$. The two numbers whose product is -600 and whose sum is -1 are -25 and 24, since $(-25)(24) = -600$ and $-25 + 24 = -1$. Then the middle term, $-x$, can be written as $-25x + 24x$ and the factorization is

$30x^2 - x - 20 = 30x^2 - 25x + 24x - 20$
$\qquad = 5x(6x - 5) + 4(6x - 5)$.
$\qquad = (5x + 4)(6x - 5)$

49. $8x^4y^4 + 24x^3y^4 - 32x^2y^4 = 8x^2y^4\left(x^2 + 3x - 4\right)$ $\left[8x^2y^4 \text{ is a common factor}\right]$

Now, factor $x^2 + 3x - 4$. The two numbers whose product is –4 and whose sum is 3 are 4 and –1, since $(4)(-1) = -4$ and $4 + (-1) = 3$. Thus, $8x^4y^4 + 2x^3y^4 - 32x^2y^4 = 8x^2y^4(x+4)(x-1)$.

51. $x^4 + x^2 - 6 = \left(x^2\right)^2 + x^2 - 6$

$\begin{aligned} &= y^2 + y - 6 &&\leftarrow \text{Replace } x^2 \text{ by } y \\ &= (y+3)(y-2) &&\leftarrow \text{Use 3 and } -2 \text{ since } (3)(-2) = -6,\ 3 + (-2) = 1 \\ &= \left(x^2 + 3\right)\left(x^2 - 2\right) &&\leftarrow \text{Replace } y \text{ by } x^2 \end{aligned}$

53. $x^4 + 5x^2 + 6 = \left(x^2\right)^2 + 5x^2 + 6$

$\begin{aligned} &= y^2 + 5y + 6 &&\leftarrow \text{Replace } x^2 \text{ by } y \\ &= (y+3)(y+2) &&\leftarrow \text{Use 3 and 2 since } (3)(2) = 6,\ 3 + 2 = 5 \\ &= \left(x^2 + 3\right)\left(x^2 + 2\right) &&\leftarrow \text{Replace } y \text{ by } x^2 \end{aligned}$

55. $6a^4 + 5a^2 - 25$

$\begin{aligned} &= 6\left(a^2\right)^2 + 5a^2 - 25 \\ &= 6w^2 + 5w - 25 &&\leftarrow \text{Replace } a^2 \text{ by } w \\ &= 6w^2 - 10w + 15w - 25 &&\leftarrow \text{Use } -10w + 15w \text{ for } 5w \text{ since } (-10)(15) = -150,\ -10 + 15 = 5 \\ &= 2w(3w - 5) + 5(3w - 5) &&\leftarrow \text{Factor by grouping} \\ &= (2w + 5)(3w - 5) \\ &= \left(2a^2 + 5\right)\left(3a^2 - 5\right) &&\leftarrow \text{Replace } w \text{ by } a^2 \end{aligned}$

57. $4(x+1)^2 + 8(x+1) + 3$

$\begin{aligned} &= 4y^2 + 8y + 3 &&\leftarrow \text{Replace } x+1 \text{ by } y \\ &= 4y^2 + 6y + 2y + 3 &&\leftarrow \text{Use } 6y + 2y \text{ for } 8y \text{ since } (6)(2) = 12,\ \text{and } 6 + 2 = 8 \\ &= 2y(2y + 3) + 1(2y + 3) \\ &= (2y + 1)(2y + 3) \\ &= \left[2(x+1) + 1\right]\left[2(x+1) + 3\right] &&\leftarrow \text{Replace } y \text{ by } x+1 \\ &= (2x + 2 + 1)(2x + 2 + 3) \\ &= (2x + 3)(2x + 5) \end{aligned}$

59. $6(a+2)^2 - 7(a+2) - 5$

$= 6y^2 - 7y - 5$ ← Replace $a+2$ by y

$= 6y^2 + 3y - 10y - 5$ ← Use $3y - 10y$ for $-7y$ since $(3)(-10) = -30$ and $3 + (-10) = -7$

$= 3y(2y+1) - 5(2y+1)$

$= (3y-5)(2y+1)$

$= [3(a+2) - 5][2(a+2) + 1]$ ← Replace y by $a+2$

$= (3a + 6 - 5)(2a + 4 + 1)$

$= (3a+1)(2a+5)$

61. $a^2 b^2 - 8ab + 15 = (ab)^2 - 8ab + 15$

$= x^2 - 8x + 15$ ← Replace ab by x

$= (x-5)(x-3)$ ← Use -5 and -3 since $(-5)(-3) = 15$, and $-5 + (-3) = -8$

$= (ab - 5)(ab - 3)$ ← Replace x by ab

63. $3x^2 y^2 - 2xy - 5 = 3(xy)^2 - 2xy - 5$

$= 3w^2 - 2w - 5$ ← Replace xy by w

$= 3w^2 + 3w - 5w - 5$ ← Use $3w - 5w$ for $-2w$ since $(3)(-5) = -15$ and $3 + (-15) = -2$

$= 3w(w+1) - 5(w+1)$

$= (3w - 5)(w+1)$

$= (3xy - 5)(xy + 1)$ ← Replace w by xy

65. $2a^2(5-a) - 7a(5-a) + 5(5-a)$

$= (5-a)(2a^2 - 7a + 5)$ ← $5-a$ is a common factor

$= (5-a)(2a^2 - 2a - 5a + 5)$ ← Use $-2a - 5a$ for $-7a$ since $(-2)(-5) = 10$ and $-2 + (-5) = -7$

$= (5-a)[2a(a-1) - 5(a-1)]$ ← Factor by grouping

$= (5-a)(2a-5)(a-1)$

67. $2x^2(x-3) + 7x(x-3) + 6(x-3)$

$= (x-3)(2x^2 + 7x + 6)$ ← $x-3$ is a common factor

$= (x-3)(2x^2 + 4x + 3x + 6)$ ← Use $4x + 3x$ for $7x$ since $(4)(3) = 12$ and $4 + 3 = 7$

$= (x-3)[2x(x+2) + 3(x+2)]$ ←Factor by grouping

$= (x-3)(2x+3)(x+2)$

69. $y^4 - 7y^2 - 30 = (y^2) - 7y^2 - 30$

$= b^2 - 7b - 30$ ← Replace y^2 by b

$= (b-10)(b+3)$ ← Use -10 and 3 since $(-10)(3) = -30$ and $-10 + 3 = -7$

$= (y^2 - 10)(y^2 + 3)$ ← Replace b by y^2

71. $x^2(x+3)+3x(x+3)+2(x+3)=(x+3)\left(x^2+3x+2\right)$ ← $x+3$ is a common factor

$\qquad\qquad\qquad\qquad\qquad\quad = (x+3)(x+2)(x+1)$ ← Use 2 and 1 since $(2)(1)=2$ and $2+1=3$

73. $5a^5b^2-8a^4b^3+3a^3b^4$

$\quad = a^3b^2\left(5a^2-8ab+3b^2\right)$ ← a^3b^2 is a common factor

$\quad = a^3b^2\left[5a^2-5ab-3ab+3b^2\right]$ ← Use $-5ab-3ab$ for $8ab$ since $(-5)(-3)=-15,\ -5+(-3)=-8$

$\quad = a^3b^2\left[5a(a-b)-3b(a-b)\right]$ ← Factor by grouping

$\quad = a^3b^2(5a-3b)(a-b)$

75. To find the polynomial multiply the factors.

$(2x+3y)(x-4y)=2x(x-4y)+3y(x-4y)$

$\qquad\qquad\qquad\quad = 2x^2-8xy+3xy-12y^2$

$\qquad\qquad\qquad\quad = 2x^2-5xy-12y^2$

77. To find the other factor, simply divide $x^2+3x-18$ by $x-3$.

$$\begin{array}{r} x+6 \\ x-3\overline{)x^2+3x-18} \end{array}$$

$\qquad\quad \underline{x^2-3x} \qquad\quad$ ← $x(x-3)$

$\qquad\qquad\quad 6x-18$

$\qquad\qquad\quad \underline{6x-18} \qquad$ ← $6(x-3)$

$\qquad\qquad\qquad\quad 0 \qquad$ ← Remainder

The other factor is $x+6$.

79. a. Answers will vary.

    b. $30x^2+23x-40$

$\qquad = 30x^2-25x+48x-40$

$\qquad = 5x(6x-5)+8(6x-5)$

$\qquad = (5x+8)(6x-5)$

$\qquad 49x^2-98x+13$

$\qquad = 49x^2-7x-91x+13$

$\qquad = 7x(7x-1)-13(7x-1)$

$\qquad = (7x-13)(7x-1)$

81. To factor $2x^2+bx-5$, the factors must be of the form

$\qquad (2x-5)(x+1)$ which gives $2x^2-3x-5$

or $(2x-1)(x+5)$ which gives $2x^2+9x-5$

or $(2x+5)(x-1)$ which gives $2x^2+3x-5$

or $(2x+1)(x-5)$ which gives $2x^2-9x-5$

Therefore $b=\pm 3,\ \pm 9$.

83. To factor x^2+bx+5, the factors must be of the form $(x+1)(x+5)$ which gives x^2+6x+5 or $(x-1)(x-5)$ which gives x^2-6x+5.

Therefore, $b=6$ or -6.

85. a. $b^2-4ac=(-8)^2-4(1)(15)$

$\qquad\qquad = 64-60$

$\qquad\qquad = 4$

a perfect square

    b. $x^2-8x+15=x^2-5x-3x+15$

$\qquad\qquad\qquad = x(x-5)-3(x-5)$

$\qquad\qquad\qquad = (x-3)(x-5)$

87. a. $b^2-4ac=(-4)^2-4(1)(6)$

$\qquad\qquad = 16-24$

$\qquad\qquad = -8$

not a perfect square

    b. Not factorable

89. Answers will vary. One example is $x^2 + 2x + 1$

91.
$$4a^{2n} - 4a^n - 15 = 4\left(a^n\right)^2 - 4a^n - 15$$
$$= 4y^2 - 4y - 15 \qquad\qquad \leftarrow \text{Replace } a^n \text{ by } y$$
$$= 4y^2 - 10y + 6y - 15$$
$$= 2y(2y - 5) + 3(2y - 5)$$
$$= (2y + 3)(2y - 5)$$
$$= \left(2a^n + 3\right)\left(2a^n - 5\right) \qquad \leftarrow \text{Replace } y \text{ by } a^n$$

93.
$$x^2(x + y)^2 - 7xy(x + y)^2 + 12y^2(x + y)^2 = (x + y)^2\left(x^2 - 7xy + 12y^2\right)$$
$$= (x + y)^2\left(x^2 - 3xy - 4xy + 12y^2\right)$$
$$= (x + y)^2\left[x(x - 3y) - 4y(x - 3y)\right]$$
$$= (x + y)^2(x - 4y)(x - 3y)$$

95.
$$x^{2n} + 3x^n - 10 = \left(x^n\right)^2 + 3\left(x^n\right) - 10$$
$$= y^2 + 3y - 10 \qquad \leftarrow \text{Replace } x^n \text{ with } y$$
$$= y^2 - 2y + 5y - 10$$
$$= y(y - 2) + 5(y - 2)$$
$$= (y + 5)(y - 2)$$
$$= \left(x^n + 5\right)\left(x^n - 2\right) \qquad \leftarrow \text{Replace } y \text{ with } x^n$$

97. a. Answers will vary.

b.

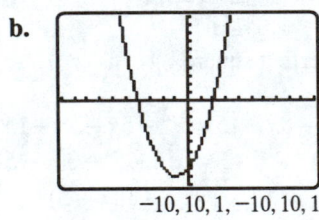

$$-10, 10, 1, -10, 10, 1$$

The graphs of $y_1 = x^2 + 2x - 8$ and $y_2 = (x + 4)(x - 2)$ are the same. The factoring is correct.

99.
$$\frac{36,000,000}{0.0004} = \frac{3.6 \times 10^7}{4 \times 10^{-4}}$$
$$= \frac{3.6}{4} \times 10^{7-(-4)}$$
$$= 0.9 \times 10^{11}$$
$$= 9.0 \times 10^{10}$$

100. The slope of a horizontal line is 0.
The change in y is 0.

101. The slope of a vertical line is undefined.

102.
$$2x - 3y = 6$$
$$-3y = -2x + 6$$
$$y = \frac{2}{3}x - 2$$

The slope of a line perpendicular to this line

is

$$-\frac{1}{\frac{2}{3}} = -\frac{3}{2}.$$

$$y - y_1 = m(x - x_1)$$

$$y - (-2) = -\frac{3}{2}(x - 5)$$

$$y + 2 = -\frac{3}{2}x + \frac{15}{2}$$

$$y = -\frac{3}{2}x + \frac{15}{2} - \frac{4}{2}$$

$$y = -\frac{3}{2}x + \frac{11}{2}$$

103. $2x^2y - 6xy^2 - \left(3x^2y + 2xy^2 - 6\right)$

$$= 2x^2y - 6xy^2 - 3x^2y - 2xy^2 + 6$$

$$= 2x^2y - 3x^2y - 6xy^2 - 2xy^2 + 6$$

$$= -x^2y - 8xy^2 + 6$$

Exercise Set 5.6

1. a. Answers will vary.

   b. $x^2 - 16 = x^2 - 4^2 = (x + 4)(x - 4)$

3. Answers will vary.

5. $a^3 + b^3 = (a + b)\left(a^2 - ab + b^2\right)$

7. $x^2 - 81 = (x)^2 - (9)^2 = (x + 9)(x - 9)$

9. $x^2 - 16 = (x)^2 - (4)^2 = (x + 4)(x - 4)$

11. $1 - 9a^2 = (1)^2 - (3a)^2 = (1 + 3a)(1 - 3a)$

13. $25 - 16y^4 = (5)^2 - \left(4y^2\right)^2 = \left(5 + 4y^2\right)\left(5 - 4y^2\right)$

15. $\dfrac{1}{16} - x^2 = \left(\dfrac{1}{4}\right)^2 - (x)^2 = \left(\dfrac{1}{4} + x\right)\left(\dfrac{1}{4} - x\right)$

17. $a^2b^2 - 49c^2 = (ab)^2 - (7c)^2 = (ab + 7c)(ab - 7c)$

19. $0.4x^2 - 0.9 = (0.2x)^2 - (0.3)^2$

$$= (0.2x + 0.3)(0.2x - 0.3)$$

21. $36 - (x - 6)^2 = (6)^2 - (x - 6)^2$

$$= [6 + (x - 6)][6 - (x - 6)]$$

$$= (6 + x - 6)(6 - x + 6)$$

$$= x(12 - x)$$

23. $a^2 - (3b + 2)^2 = [a + (3b + 2)][a - (3b + 2)]$

$$= (a + 3b + 2)(a - 3b - 2)$$

25. $x^2 + 10x + 25 = x^2 + 2(x)(5) + 5^2$

$$= (x + 5)^2$$

27. $4 + 4x + x^2 = 2^2 + 2(2)(x) + x^2$

$$= (2 + x)^2$$

29. $4x^2 - 20xy + 25y^2$

$$= (2x)^2 - 2(2x)(5y) + (5y)^2$$

$$= (2x - 5y)^2$$

31. $0.81x^2 - 0.36x + 0.04$

$$= (0.9x)^2 - 2(0.9)(0.2)x + (0.2)^2$$

$$= (0.9x - 0.2)^2$$

33. $w^4 + 16w^2 + 64 = \left(w^2\right)^2 + 2\left(w^2\right)(8) + 8^2$

$$= \left(w^2 + 8\right)^2$$

35. $(x + y)^2 + 2(x + y) + 1$

$$= (x + y)^2 + 2(x + y)(1) + 1^2$$

$$= [(x + y) + 1]^2$$

$$= (x + y + 1)^2$$

37. $a^4 - 2a^2b^2 + b^4$

$= \left(a^2\right)^2 - 2\left(a^2\right)\left(b^2\right) + \left(b^2\right)^2$

$= \left(a^2 - b^2\right)^2$

$= \left[(a+b)(a-b)\right]^2$

$= (a+b)^2(a-b)^2$

39. $x^2 + 6x + 9 - y^2 = (x+3)^2 - (y)^2$

$= \left[(x+3)+y\right]\left[(x+3)-y\right]$

$= (x+3+y)(x+3-y)$

41. $25 - \left(x^2 + 4x + 4\right) = (5)^2 - (x+2)^2$

$= \left[5+(x+2)\right]\left[5-(x+2)\right]$

$= (5+x+2)(5-x-2)$

$= (x+7)(-x+3)$

43. $9a^2 - 12ab + 4b^2 - 9$

$= (3a - 2b)^2 - (3)^2$

$= \left[(3a-2b)+3\right]\left[(3a-2b)-3\right]$

$= (3a-2b+3)(3a-2b-3)$

45. $(x+y)^2 - (x-y)^2$

$= \left[(x+y)+(x-y)\right]\left[(x+y)-(x-y)\right]$

$= \left[x+y+x-y\right]\left[x+y-x+y\right]$

$= (2x)(2y)$

$= 4xy$

47. $x^3 - 27 = (x)^3 - (3)^3$

$= (x-3)\left[x^2 + x(3) + 3^2\right]$

$= (x-3)\left(x^2 + 3x + 9\right)$

49. $x^3 + y^3 = (x)^3 + (y)^3$

$= (x+y)\left(x^2 - xy + y^2\right)$

51. $64 - a^3 = (4)^3 - (a)^3$

$= (4-a)\left[4^2 + 4(a) + a^2\right]$

$= (4-a)\left(16 + 4a + a^2\right)$

53. $27y^3 - 8x^3$

$= (3y)^3 - (2x)^3$

$= (3y - 2x)\left[(3y)^2 + (3y)(2x) + (2x)^2\right]$

$= (3y - 2x)\left(9y^2 + 6xy + 4x^2\right)$

55. $24x^3 - 81y^3$

$= 3\left(8x^3 - 27y^3\right)$

$= 3\left[(2x)^3 - (3y)^3\right]$

$= 3(2x - 3y)\left[(2x)^2 + (2x)(3y) + (3y)^2\right]$

$= 3(2x - 3y)\left(4x^2 + 6xy + 9y^2\right)$

57. $5x^3 - 625y^3 = 5\left(x^3 - 125y^3\right)$

$= 5\left[(x)^3 - (5y)^3\right]$

$= 5(x - 5y)\left[x^2 + x(5y) + (5y)^2\right]$

$= 5(x - 5y)\left(x^2 + 5xy + 25y^2\right)$

59. $(x+1)^3 + 1$

$= (x+1)^3 + (1)^3$

$= \left[(x+1)+1\right]\left[(x+1)^2 - (x+1)(1) + 1^2\right]$

$= (x+2)\left(x^2 + 2x + 1 - x - 1 + 1\right)$

$= (x+2)\left(x^2 + x + 1\right)$

61. $(x-y)^3 - 27$

$= (x-y)^3 - (3)^3$

$= \left[(x-y)-3\right]\left[(x-y)^2 + (x-y)(3) + 3^2\right]$

$= (x-y-3)\left(x^2 - 2xy + y^2 + 3x - 3y + 9\right)$

63. $b^3 - (b+3)^3$

$= \left[b - (b+3) \right]\left[b^2 + b(b+3) + (b+3)^2 \right]$

$= (b - b - 3)\left(b^2 + b^2 + 3b + b^2 + 6b + 9 \right)$

$= (-3)\left(3b^2 + 9b + 9 \right)$

$= (-3) \cdot 3\left(b^2 + 3b + 3 \right)$

$= -9\left(b^2 + 3b + 3 \right)$

65. $121y^4 - 49x^2 = \left(11y^2 \right)^2 - (7x)^2$

$\qquad\qquad\qquad = \left(11y^2 + 7x \right)\left(11y^2 - 7x \right)$

67. $16y^2 - 81x^2 = (4y)^2 - (9x)^2$

$\qquad\qquad\quad = (4y + 9x)(4y - 9x)$

69. $25x^4 - 81y^6 = \left(5x^2 \right)^2 - \left(9y^3 \right)^2$

$\qquad\qquad\quad = \left(5x^2 + 9y^3 \right)\left(5x^2 - 9y^3 \right)$

71. $x^3 - 64 = (x)^3 - (4)^3$

$\qquad\quad = (x - 4)\left[x^2 + x(4) + 4^2 \right]$

$\qquad\quad = (x - 4)\left(x^2 + 4x + 16 \right)$

73. $9x^2y^2 + 24xy + 16$

$= (3xy)^2 + 2(3xy)(4) + (4)^2$

$= (3xy + 4)^2$

75. $a^4 + 2a^2b^2 + b^4$

$= \left(a^2 \right)^2 + 2\left(a^2 \right)\left(b^2 \right) + \left(b^2 \right)^2$

$= \left(a^2 + b^2 \right)^2$

77. $x^2 - 2x + 1 - y^2 = (x - 1)^2 - (y)^2$

$\qquad\qquad\qquad = \left[(x - 1) + y \right]\left[(x - 1) - y \right]$

$\qquad\qquad\qquad = (x - 1 + y)(x - 1 - y)$

79. $(x + y)^3 + 1$

$= (x + y)^3 + (1)^3$

$= \left[(x + y) + 1 \right]\left[(x + y)^2 - (x + y)(1) + 1^2 \right]$

$= (x + y + 1)\left(x^2 + 2xy + y^2 - x - y + 1 \right)$

81. $(m + n)^2 - (2m - n)^2$

$= \left[(m + n) + (2m - n) \right]\left[(m + n) - (2m - n) \right]$

$= (3m)(-m + 2n)$

$= 3m(-m + 2n)$

83. Area of larger square is $(a)(a) = a^2$.

Area of smaller square is $(b)(b) = b^2$.

a. Area of shaded region is $a^2 - b^2$.

b. In factored form, the area is
$a^2 - b^2 = (a + b)(a - b)$.

85. Volume of larger solid is $(6a)(a)(a) = 6a^3$.
Volume of smaller solid is
$(6a)(b)(b) = 6ab^2$.

a. Volume of shaded region is
$6a^3 - 6ab^2$.

b. In factored form, the volume is
$6a^3 - 6ab^2 = 6a\left(a^2 - b^2 \right)$
$\qquad\qquad\qquad = 6a(a + b)(a - b)$.

87. Volume of larger sphere is $\dfrac{4}{3}\pi R^3$.

Volume of smaller sphere is $\dfrac{4}{3}\pi r^3$.

a. Volume of shaded region is
$\dfrac{4}{3}\pi R^3 - \dfrac{4}{3}\pi r^3$.

b. In factored form the volume is

$$\frac{4}{3}\pi R^3 - \frac{4}{3}\pi r^3$$

$$= \frac{4}{3}\pi\left(R^3 - r^3\right)$$

$$= \frac{4}{3}\pi(R-r)\left(R^2 + Rr + r^2\right).$$

89. Express $4x^2 + bx + 9$ as $(2x)^2 + bx + (3)^2$.
Now,

$$bx = 2(2x)(3) \quad \text{or} \quad bx = -2(2x)(3)$$

$$bx = 12 \qquad\qquad bx = -12x$$

$$b = 12 \qquad\qquad b = -12$$

91. Express c as a^2. Then $25x^2 + 20x + c$ becomes

$$25x^2 + 20x + a^2 \text{ or } (5x)^2 + 20x + (a)^2. \text{ But,}$$

$$20x = 2(5x)(a)$$

$$20x = 10xa$$

$$\frac{20x}{10x} = a$$

$$2 = a$$

Now, $a = 2$ and $c = a^2 = (2)^2 = 4$.

93. a. Find an expression whose square is $25x^2 - 30x + 9$.

b.
$$A(x) = \left[s(x)\right]^2$$

$$A(x) = 25x^2 - 30x + 9$$

$$= (5x - 3)^2$$

Therefore $s(x) = 5x - 3$

c. $s(2) = 5(2) - 3 = 10 - 3 = 7$

95. $x^2 + 64 = \left(x^4 + 16x^2 + 64\right) - 16x^2$

$$= \left(x^2 + 8\right)^2 - (4x)^2$$

$$\left(x^2 + 8 + 4x\right)\left(x^2 + 8 - 4x\right)$$

$$= \left(x^2 + 4x + 8\right)\left(x^2 - 4x + 8\right)$$

97. $P(x) = x^2$

$$P(a+h) - P(a) = (a+h)^2 - a^2$$

$$= a^2 + 2ah + h^2 - a^2$$

$$= 2ah + h^2$$

$$= h(2a + h)$$

99. a. The area is $4 \cdot 4 = 16$

b. The sum is $x^2 + 8x + 16$

c. $x^2 + 8x + 16 = (x + 4)^2$

101. $64x^{4a} - 9y^{6a}$

$$= \left(8x^{2a}\right)^2 - \left(3y^{3a}\right)^2$$

$$= \left(8x^{2a} + 3y^{3a}\right)\left(8x^{2a} - 3y^{3a}\right)$$

103. $a^{2n} - 16a^n + 64 = \left(a^n\right)^2 - 2\left(a^n\right)(8) + (8)^2$

$$= \left(a^n - 8\right)^2$$

105. $x^{3n} - 8 = \left(x^n\right)^3 - 2^3$

$$= \left(x^n - 2\right)\left[\left(x^n\right)^2 + \left(x^n\right)(2) + (2)^2\right]$$

$$= \left(x^n - 2\right)\left(x^{2n} + 2x^n + 4\right)$$

107. $y_1 = 2x^2 - 18$
$y_2 = 2(x+3)(x-3)$

$$-10, 10, 1, -30, 10, 2$$

The factoring is correct because the graphs are the same.

109. a. $x^6 - 1 = \left(x^3\right)^2 - 1^2 = \left(x^3 - 1\right)\left(x^3 + 1\right)$

b.
$$x^6 - 1 = \left(x^2\right)^3 - 1^3$$
$$= \left(x^2 - 1\right)\left[\left(x^2\right)^2 + \left(x^2\right)(1) + (1)^2\right]$$
$$= \left(x^2 - 1\right)\left(x^4 + x^2 + 1\right)$$

c.
$$x^6 - 1$$
$$= \left(x^3 - 1\right)\left(x^3 + 1\right)$$
$$= (x - 1)\left(x^2 + x + 1\right)(x + 1)\left(x^2 - x + 1\right)$$
$$= (x - 1)(x + 1)\left(x^2 + x + 1\right)\left(x^2 - x + 1\right)$$
$$= \left(x^2 - 1\right)\left(x^4 + x^2 + 1\right)$$
upon multiplication

111. a. $3, 6$

    b. $-2, \dfrac{5}{9}, -1.67, 0, 3, 6$

    c. $\sqrt{3}, -\sqrt{6}$

    d. $-2, \dfrac{5}{9}, -1.67, 0, \sqrt{3}, -\sqrt{6}, 3, 6$

112. $\{4, 5, 6, 7\}$

113. $\{4, 5, 6, 7, \ldots\}$

114. $z = \dfrac{x - \bar{x}}{\dfrac{s}{\sqrt{n}}}$

$$= \dfrac{15 - 13}{\dfrac{3}{\sqrt{9}}}$$

$$= \dfrac{2}{1} = 2$$

115. Let x be the width. Then $2x + 2$ is the length and use $P = 2l + 2w$ to obtain
$$22 = 2(2x + 2) + 2(x)$$
$$22 = 4x + 4 + 2x$$
$$22 = 6x + 4$$
$$18 = 6x$$
$$3 = x$$
Thus, the width is 3 ft and the length is $2(3) + 2 = 6 + 2 = 8$ ft.

Exercise Set 5.7

1. a. Factor out the GCF if it is present. If the remaining polynomial is the difference of two squares, difference of two cubes, or the sum of two cubes, use the following factoring formulas.
$$a^2 - b^2 = (a + b)(a - b)$$
$$a^3 - b^3 = (a - b)\left(a^2 + ab + b^2\right)$$
$$a^3 + b^3 = (a + b)\left(a^2 - ab + b^2\right)$$

    b. Factor out the GCF if it is present. If the remaining polynomial is a perfect square, use the special factoring formulas:
$$a^2 + 2ab + b^2 = (a + b)^2$$
$$a^2 - 2ab + b^2 = (a - b)^2$$

    c. Factor out the GCF if it is present. Factor the remaining polynomial using the method of grouping.

3. $4x^2 + 4x - 48 = 4(x^2 + x - 12)$ ← 4 is a common factor

$\qquad\qquad = 4(x+4)(x-3)$ ← Use 4 and -3 since $(4)(-3) = -12$, and $4 + (-3) = 1$

5. $10s^2 + 19s - 15 = 10s^2 + 25s - 6s - 15$ ← Use $19s = 25s - 6s$

$\qquad\qquad = 5s(2s+5) - 3(2s+5)$

$\qquad\qquad = (5s-3)(2s+5)$

7. $-8r^2 + 26r - 15 = -(8r^2 - 26r + 15)$ ← -1 is a common factor

$\qquad\qquad = -(8r^2 - 20r - 6r + 15)$ ← Use $-26r = -20r - 6r$

$\qquad\qquad = -[4r(2r-5) - 3(2r-5)]$

$\qquad\qquad = -(4r-3)(2r-5)$

9. $0.4x^2 - 0.036 = 0.4(x^2 - 0.09)$ ← 0.4 is a common factor

$\qquad\qquad = 0.4(x^2 - 0.3^2)$ ← Difference of two squares

$\qquad\qquad = 0.4(x+0.3)(x-0.3)$

11. $5x^5 - 45x = 5x(x^4 - 9)$ ← $5x$ is a common factor

$\qquad\qquad = 5x\left[(x^2)^2 - (3)^2\right]$ ← Difference of two squares

$\qquad\qquad = 5x(x^2+3)(x^2-3)$

13. $3x^6 - 3x^5 - 12x^5 + 12x^4 = 3x^4(x^2 - x - 4x + 4)$ ← $3x^4$ is a common factor

$\qquad\qquad = 3x^4[x(x-1) - 4(x-1)]$ ← Factor by grouping

$\qquad\qquad = 3x^4(x-4)(x-1)$

15. $5x^4y^2 + 20x^3y^2 - 15x^3y^2 - 60x^2y^2 = 5x^2y^2(x^2 + 4x - 3x - 12)$ ← $5x^2y^2$ is a common factor

$\qquad\qquad = 5x^2y^2[x(x+4) - 3(x+4)]$ ← Factor by grouping

$\qquad\qquad = 5x^2y^2(x-3)(x+4)$

17. $x^4 - x^2y^2 = x^2(x^2 - y^2)$ ← x^2 is a common factor

$\qquad\qquad = x^2(x+y)(x-y)$ ← Difference of two squares

19. $x^7y^2 - x^4y^2 = x^4y^2(x^3 - 1)$ ← x^4y^2 is a common factor

$\qquad\qquad = x^4y^2(x-1)(x^2+x+1)$ ← Difference of two cubes

21. $x^5 - 16x = x\left(x^4 - 16\right)$ $\leftarrow$ x is a common factor

$= x\left[\left(x^2\right)^2 - (4)^2\right]$ $\leftarrow$ Difference of two squares

$= x\left(x^2 + 4\right)\left(x^2 - 4\right)$ $\leftarrow$ Difference of two squares on $x^2 - 4$

$= x\left(x^2 + 4\right)(x + 2)(x - 2)$

23. $4x^6 + 32y^3 = 4\left(x^6 + 8y^3\right)$ $\leftarrow$ 4 is a common factor

$= 4\left[\left(x^2\right)^3 + (2y)^3\right]$ $\leftarrow$ Sum of two cubes

$= 4\left(x^2 + 2y\right)\left[\left(x^2\right)^2 - \left(x^2\right)(2y) + (2y)^2\right]$ $\leftarrow$ Simplify

$= 4\left(x^2 + 2y\right)\left(x^4 - 2x^2y + 4y^2\right)$

25. $2(a + b)^2 - 18 = 2\left[(a + b)^2 - 9\right]$ $\leftarrow$ 2 is a common factor

$= 2[(a + b) + 3][(a + b) - 3]$ $\leftarrow$ Difference of two squares

$= 2(a + b + 3)(a + b - 3)$ $\leftarrow$ Simplify

27. $6x^2 + 36xy + 54y^2 = 6\left(x^2 + 6xy + 9y^2\right)$ $\leftarrow$ 6 is a common factor

$= 6\left[x^2 + 2(x)(3y) + (3y)^2\right]$ $\leftarrow$ Perfect square trinomial

$= 6(x + 3y)^2$

29. $(x + 2)^2 - 4 = [(x + 2) + 2][(x + 2) - 2]$ $\leftarrow$ Difference of two squares

$= (x + 2 + 2)(x + 2 - 2)$ $\leftarrow$ Simplify

$= (x + 4)(x) \text{ or } x(x + 4)$

31. $(2a + b)(2a - 3b) - (2a + b)(a - b) = (2a + b)[(2a - 3b) - (a - b)]$ $\leftarrow$ $(2a + b)$ is a common factor

$= (2a + b)(2a - 3b - a + b)$ $\leftarrow$ Simplify

$= (2a + b)(a - 2b)$

33. $(y + 3)^2 + 4(y + 3) + 4 = (y + 3)^2 + 2(y + 3)(2) + (2)^2$ $\leftarrow$ Perfect square trinomial

$= [(y + 3) + 2]^2$

$= (y + 5)^2$

35. $45a^4 - 30a^3 + 5a^2 = 5a^2\left(9a^2 - 6a + 1\right)$ $\leftarrow$ $5a^2$ is a common factor

$= 5a^2\left[(3a)^2 - 2(3a)(1) + 1^2\right]$ $\leftarrow$ Perfect square trinomial

$= 5a^2(3a - 1)^2$

37. $x^3 + \dfrac{1}{27} = x^3 + \left(\dfrac{1}{3}\right)^3$ ← Sum of two cubes

$\qquad = \left(x + \dfrac{1}{3}\right)\left[x^2 - x\left(\dfrac{1}{3}\right) + \left(\dfrac{1}{3}\right)^2\right]$

$\qquad = \left(x + \dfrac{1}{3}\right)\left[x^2 - \dfrac{x}{3} + \dfrac{1}{9}\right]$

39. $3x^3 + 2x^2 - 27x - 18 = x^2(3x + 2) - 9(3x + 2)$ ← Factor by grouping

$\qquad\qquad\qquad\qquad = \left(x^2 - 9\right)(3x + 2)$ ← Difference of two squares

$\qquad\qquad\qquad\qquad = (x + 3)(x - 3)(3x + 2)$

41. $a^3 b - 16ab^3 = ab\left(a^2 - 16b^2\right)$ ← ab is a common factor

$\qquad\qquad\quad = ab\left[a^2 - (4b)^2\right]$ ← Difference of two squares

$\qquad\qquad\quad = ab(a + 4b)(a - 4b)$

43. $81 - (x^2 + 2xy + y^2) = 81 - (x + y)^2$ ← Perfect square trinomial

$\qquad\qquad\qquad\qquad = [9 + (x + y)][9 - (x + y)]$ ← Difference of two squares

$\qquad\qquad\qquad\qquad = (9 + x + y)(9 - x - y)$ ← Simplify

45. $24x^2 - 34x + 12 = 2\left(12x^2 - 17x + 6\right)$ ← 2 is a common factor

$\qquad\qquad\qquad = 2\left(12x^2 - 8x - 9x + 6\right)$ ← Use $-17x = -8x - 9x$

$\qquad\qquad\qquad = 2[4x(3x - 2) - 3(3x - 2)]$ ← Factor by grouping

$\qquad\qquad\qquad = 2(4x - 3)(3x - 2)$

47. $16x^2 - 34x - 15 = 16x^2 - 40x + 6x - 15$ ← Use $-34x = -40x + 6$

$\qquad\qquad\qquad = 8x(2x - 5) + 3(2x - 5)$ ← Factor by grouping

$\qquad\qquad\qquad = (8x + 3)(2x - 5)$

49. $x^4 - 81 = \left(x^2\right)^2 - (9)^2$ ← Difference of two squares

$\qquad\quad = \left(x^2 + 9\right)\left(x^2 - 9\right)$ ← Difference of two squares $(x^2 - 9)$

$\qquad\quad = \left(x^2 + 9\right)(x + 3)(x - 3)$

51. $5bc - 10cx - 6by + 12xy = 5c(b - 2x) - 6y(b - 2x)$ ← Factor by grouping

$\qquad\qquad\qquad\qquad = (5c - 6y)(b - 2x)$ ← $(b - 2x)$ is a common factor

53. $3x^4 - x^2 - 4 = 3(x^2)^2 - x^2 - 4$

$\qquad = 3(x^2)^2 - 4x^2 + 3x^2 - 4 \quad \leftarrow$ Use $-x^2 = -4x^2 + 3x^2$

$\qquad = x^2(3x^2 - 4) + 1(3x^2 - 4) \quad \leftarrow$ Factor by grouping

$\qquad = (x^2 + 1)(3x^2 - 4) \qquad\quad \leftarrow (3x^2 - 4)$ is a common factor

55. $y^2 - (x^2 - 8x + 16) = y^2 - (x-4)^2 \qquad\qquad \leftarrow$ Perfect square trinomial

$\qquad = [y + (x-4)][y - (x-4)] \quad \leftarrow$ Difference of two squares

$\qquad = (y + x - 4)(y - x + 4) \qquad \leftarrow$ Simplify

57. $2(y+4)^2 + 5(y+4) - 12 = 2x^2 + 5x - 12 \qquad\qquad \leftarrow$ Replace $y + 4$ with x

$\qquad = 2x^2 + 8x - 3x - 12 \qquad \leftarrow$ Use $5x = 8x - 3x$

$\qquad = 2x(x+4) - 3(x+4) \qquad \leftarrow$ Factor by grouping

$\qquad = (2x - 3)(x + 4) \qquad\qquad \leftarrow x + 4$ is a common factor

$\qquad = [2(y+4) - 3][(y+4) + 4] \quad \leftarrow$ Replace x with $y + 4$

$\qquad = (2y + 8 - 3)(y + 8) \qquad\quad \leftarrow$ Simplify

$\qquad = (2y + 5)(y + 8) \qquad\qquad \leftarrow$ Simplify

59. $a^2 + 12ab + 36b^2 - 16c^2 = (a + 6b)^2 - (4c)^2 \qquad\qquad \leftarrow$ Perfect square trinomial

$\qquad = [(a + 6b) + 4c][(a + 6b) - 4c] \quad \leftarrow$ Difference of two squares

$\qquad = (a + 6b + 4c)(a + 6b - 4c) \qquad \leftarrow$ Simplify

61. $6x^4 y + 15x^3 y - 9x^2 y = 3x^2 y(2x^2 + 5x - 3) \qquad\qquad \leftarrow 3x^2 y$ is a common factor

$\qquad = 3x^2 y(2x^2 - x + 6x - 3) \qquad \leftarrow$ Use $5x = -x + 6x$

$\qquad = 3x^2 y[x(2x - 1) + 3(2x - 1)] \quad \leftarrow$ Factor by grouping

$\qquad = 3x^2 y(x + 3)(2x - 1) \qquad\qquad \leftarrow 2x + 1$ is a common factor

63. $x^4 - 2x^2 y^2 + y^4 = (x^2)^2 - 2(x^2)(y^2) + (y^2)^2 \quad \leftarrow$ Perfect square trinomial

$\qquad = (x^2 - y^2)^2 \qquad\qquad \leftarrow$ Difference of two squares

$\qquad = [(x + y)(x - y)]^2$

$\qquad = (x + y)^2 (x - y)^2$

65. $a^2 + b^2$ is not factorable, (e)

67. $a^3 + b^3 = (a + b)(a^2 - ab + b^2)$, (a)

69. $a^3 - b^3 = (a - b)(a^2 + ab + b^2)$, (f)

71. $a^3 + b^3 = (a + b)(a^2 - ab + b^2)$, (c)

73. The area of larger rectangle is $a(a+b)$.
The area of center rectangle is $b(a+b)$.

   a. The area of shaded region is
$$a(a+b)-b(a+b)=a^2-b^2.$$

   b. In factored form, the area is
$$a(a+b)-b(a+b)=(a-b)(a+b)$$

75. **a.** The area of the shaded region is the sum of the areas of the three regions. This is
$$(a)(a)+(2b)(a)+(b)(b)=a^2+2ab+b^2$$

   b. In factored form, the area is
$$a^2+2ab+b^2=(a+b)^2.$$

77. The area of the left side is $b(a-b)$.
The area of the right side is $b(a-b)$.
The area of the front side is $a(a-b)$.
The area of the back side is $a(a-b)$.

   a. The surface area is
$$2a(a-b)+2b(a-b).$$

   b. In factored form, the area is
$$2a(a-b)+2b(a-b)=2(a+b)(a-b).$$

79. **a.** Answers will vary.

   b. Answers will vary.

81. **a.** $x^{-3}-2x^{-4}-3x^{-5}=x^{-5}\left(x^2-2x-3\right)$

   b. $x^{-5}\left(x^2-2x-3\right)=x^{-5}(x-3)(x+1)$

83. **a.** $5x^{1/2}+2x^{-1/2}-3x^{-3/2}$
$$=5x^{-3/2}x^2+2x^{-3/2}x^1-3x^{-3/2}$$
$$=x^{-3/2}\left(5x^2+2x-3\right)$$

   b. $x^{-3/2}(5x^2+2x-3)$
$$=x^{-3/2}(5x-3)(x+1)$$

84. $4(x-2)=3(x-4)-4$
$$4x-8=3x-12-4$$
$$4x-8=3x-16$$
$$x-8=-16$$
$$x=-8$$

85. $-5(x-2)+3=-5x-6$
$$-5x+10+3=-5x-6$$
$$-5x+13=-5x-6$$
$$13=-6$$
Since this is a FALSE statement, there is NO solution.

86. $4(x-3)<6(x-4)$
$$4x-12<6x-24$$
$$-12<2x-24$$
$$12<2x$$
$$6<x$$
Solution is $\left\{x\mid x>6\right\}$.

87. $|2x-3|>-4$
Observe that the left side is non-negative and the right side is negative. Since a non-negative value is always larger than a negative number, the solution is all real numbers or $\mathbb{R}$.

Exercise Set 5.8

1. The degree of a polynomial function is the same as the degree of the leading term.

3. $ax^2+bx+c=0$ is the standard form of a quadratic equation.

5. **a.** The zero factor property only holds when one side of the equation is 0.

   b. $(x+3)(x+4)=2$
$$x^2+7x+12=2$$
$$x^2+7x+10=0$$
$$(x+5)(x+2)=0$$
$$x+5=0 \quad \text{or} \quad x+2=0$$
$$x=-5 \qquad\qquad x=-2$$
The solutions are –2 and –5.

7. a. Answers will vary. Possible answer: Factor the polynomial, set the factors equal to 0, and then solve.

b.
$$-x - 20 = -12x^2$$
$$12x^2 - x - 20 = 0$$
$$(3x - 4)(4x + 5) = 0$$
$$3x - 4 = 0 \quad \text{or} \quad 4x + 5 = 0$$
$$3x = 4 \quad \text{or} \quad 4x = -5$$
$$x = \frac{4}{3} \quad \text{or} \quad x = -\frac{5}{4}$$
The solutions are $\frac{4}{3}, -\frac{5}{4}$.

9. a. The two shorter sides of a right triangle are called the legs.

b. The longest side of a right triangle is called the hypotenuse.

11. -8 and -2; answers will vary.

13. $x(x + 5) = 0$
$$x = 0 \quad \text{or} \quad x + 5 = 0$$
$$x = -5$$
The solutions are $0, -5$.

15. $5x(x + 9) = 0$
$$5x = 0 \quad \text{or} \quad x + 9 = 0$$
$$x = 0 \qquad\qquad x = -9$$
The solutions are $0, -9$.

17. $(2x + 5)(x - 3)(3x + 6) = 0$
$$2x + 5 = 0 \quad \text{or} \quad x - 3 = 0 \quad \text{or} \quad 3x + 6 = 0$$
$$2x = -5 \qquad\qquad x = 3 \qquad\qquad 3x = -6$$
$$x = -\frac{5}{2} \qquad\qquad\qquad\qquad x = -\frac{6}{3}$$
$$x = -2$$
The solutions are $-\frac{5}{2}, 3, -2$.

19. $4x - 12 = 0$
$$4(x - 3) = 0$$
$$x - 3 = 0$$
$$x = 3$$
The solution is 3.

21. $-x^2 + 12x = 0$
$$x(-x + 12) = 0$$
$$x = 0 \quad \text{or} \quad -x + 12 = 0$$
$$x = 12$$
The solutions are $0, 12$.

23. $9x^2 = -18x$
$$9x^2 + 18x = 0$$
$$9x(x + 2) = 0$$
$$9x = 0 \quad \text{or} \quad x + 2 = 0$$
$$x = 0 \qquad\qquad x = -2$$
The solutions are $0, -2$.

25. $x^2 + x - 12 = 0$
$$(x + 4)(x - 3) = 0$$
$$x + 4 = 0 \quad \text{or} \quad x - 3 = 0$$
$$x = -4 \qquad\qquad x = 3$$
The solutions are $-4, 3$.

27. $x(x - 12) = -20$
$$x^2 - 12x = -20$$
$$x^2 - 12x + 20 = 0$$
$$(x - 10)(x - 2) = 0$$
$$x - 10 = 0 \quad \text{or} \quad x - 2 = 0$$
$$x = 10 \qquad\qquad x = 2$$
The solutions are $10, 2$.

29. $-z^2 - 3z = -18$
$$-z^2 - 3z + 18 = 0$$
$$z^2 + 3z - 18 = 0$$
$$(z + 6)(z - 3) = 0$$
$$z + 6 = 0 \quad \text{or} \quad z - 3 = 0$$
$$z = -6 \qquad\qquad z = 3$$
The solutions are $-6, 3$.

31. $3x^2 - 6x - 72 = 0$
$$3(x^2 - 2x - 24) = 0$$
$$3(x - 6)(x + 4) = 0$$
$$x - 6 = 0 \quad \text{or} \quad x + 4 = 0$$
$$x = 6 \qquad\qquad x = -4$$
The solutions are $6, -4$.

33.
$$x^3 + 19x^2 = 42x$$
$$x^3 + 19x^2 - 42x = 0$$
$$x\left(x^2 + 19x - 42\right) = 0$$
$$x(x + 21)(x - 2) = 0$$
$$x = 0 \quad \text{or} \quad x + 21 = 0 \quad \text{or} \quad x - 2 = 0$$
$$x = -21 \qquad x = 2$$
The solutions are 0, –21, 2.

35.
$$2y^2 + 22y + 60 = 0$$
$$2\left(y^2 + 11y + 30\right) = 0$$
$$2(y + 6)(y + 5) = 0$$
$$y + 6 = 0 \quad \text{or} \quad y + 5 = 0$$
$$y = -6 \qquad y = -5$$
The solutions are –6, –5.

37.
$$-7x - 10 = -6x^2$$
$$6x^2 - 7x - 10 = 0$$
$$(x - 2)(6x + 5) = 0$$
$$x - 2 = 0 \quad \text{or} \quad 6x + 5 = 0$$
$$x = 2 \qquad 6x = -5$$
$$x = -\frac{5}{6}$$
The solutions are $2, -\frac{5}{6}$.

39.
$$-2y^2 + 24y - 22 = 0$$
$$-2\left(y^2 - 12y + 11\right) = 0$$
$$-2(y - 11)(y - 1) = 0$$
$$y - 11 = 0 \quad \text{or} \quad y - 1 = 0$$
$$y = 11 \qquad y = 1$$
The solutions are 11, 1.

41.
$$z^3 + 16z^2 = -64z$$
$$z^3 + 16z^2 + 64z = 0$$
$$z\left(z^2 + 16z + 64\right) = 0$$
$$z(z + 8)^2 = 0$$
$$z = 0 \quad \text{or} \quad z + 8 = 0$$
$$z = -8$$
The solutions are 0, –8.

43.
$$4x^3 + 4x^2 - 48x = 0$$
$$4x\left(x^2 + x - 12\right) = 0$$
$$4x(x + 4)(x - 3) = 0$$
$$4x = 0 \quad \text{or} \quad x + 4 = 0 \quad \text{or} \quad x - 3 = 0$$
$$x = 0 \qquad x = -4 \qquad x = 3$$
The solutions are 0, –4, 3.

45.
$$6x^2 = 16x$$
$$6x^2 - 16x = 0$$
$$2x(3x - 8) = 0$$
$$2x = 0 \quad \text{or} \quad 3x - 8 = 0$$
$$x = 0 \qquad 3x = 8$$
$$x = \frac{8}{3}$$
The solutions are $0, \frac{8}{3}$.

47.
$$25x^3 - 16x = 0$$
$$x\left(25x^2 - 16\right) = 0$$
$$x\left[(5x)^2 - 4^2\right] = 0$$
$$x(5x + 4)(5x - 4) = 0$$
$$x = 0 \quad \text{or} \quad 5x + 4 = 0 \quad \text{or} \quad 5x - 4 = 0$$
$$5x = -4 \qquad 5x = 4$$
$$x = -\frac{4}{5} \qquad x = \frac{4}{5}$$
The solutions are $0, -\frac{4}{5}, \frac{4}{5}$.

49.
$$(x + 4)^2 - 16 = 0$$
$$[(x + 4) + 4][(x + 4) - 4] = 0$$
$$(x + 8)(x) = 0$$
$$x + 8 = 0 \quad \text{or} \quad x = 0$$
$$x = -8$$
The solutions are –8, 0.

51. $(x-7)(x+5) = -20$
$$x^2 - 2x - 35 = -20$$
$$x^2 - 2x - 15 = 0$$
$$(x-5)(x+3) = 0$$
$$x - 5 = 0 \quad \text{or} \quad x + 3 = 0$$
$$x = 5 \qquad\qquad x = -3$$
The solutions are 5, –3.

53. $6a^2 - 12 - 4a = 19a - 32$
$$6a^2 - 23a + 20 = 0$$
$$(3a-4)(2a-5) = 0$$
$$3a - 4 = 0 \quad \text{or} \quad 2a - 5 = 0$$
$$3a = 4 \qquad\qquad 2a = 5$$
$$a = \frac{4}{3} \qquad\qquad a = \frac{5}{2}$$
The solutions are $\dfrac{4}{3}, \dfrac{5}{2}$.

55. $(b-1)(3b+2) = 4b$
$$3b^2 - b - 2 = 4b$$
$$3b^2 - 5b - 2 = 0$$
$$(3b+1)(b-2) = 0$$
$$3b + 1 = 0 \quad \text{or} \quad b - 2 = 0$$
$$3b = -1 \qquad\qquad b = 2$$
$$b = -\frac{1}{3}$$
The solutions are $-\dfrac{1}{3}, 2$.

57. $2x^3 + 16x^2 + 30x = 0$
$$2x\left(x^2 + 8x + 15\right) = 0$$
$$2x(x+5)(x+3) = 0$$
$$2x = 0 \quad \text{or} \quad x + 5 = 0 \quad \text{or} \quad x + 3 = 0$$
$$x = 0 \qquad\qquad x = -5 \qquad\qquad x = -3$$
The solutions are 0, –5, –3.

59. $f(x) = 3x^2 + 7x + 9$
$$7 = 3a^2 + 7a + 9$$
$$0 = 3a^2 + 7a + 2$$
$$0 = (3a+1)(a+2)$$
$$3a + 1 = 0 \quad \text{or} \quad a + 2 = 0$$
$$a = -\frac{1}{3} \qquad\qquad a = -2$$
The values of a are $-\dfrac{1}{3}, -2$.

61. $g(x) = 10x^2 - 31x + 19$
$$4 = 10a^2 - 31a + 19$$
$$0 = 10a^2 - 31a + 15$$
$$0 = (5a-3)(2a-5)$$
$$5a - 3 = 0 \quad \text{or} \quad 2a - 5 = 0$$
$$a = \frac{3}{5} \qquad\qquad a = \frac{5}{2}$$
The values of a are $\dfrac{3}{5}, \dfrac{5}{2}$.

63. $r(x) = 4x^2 - 19x$
$$30 = 4a^2 - 19a$$
$$0 = 4a^2 - 19a - 30$$
$$0 = (4a+5)(a-6)$$
$$4a + 5 = 0 \quad \text{or} \quad a - 6 = 0$$
$$a = -\frac{5}{4} \qquad\qquad a = 6$$
The values of a are $-\dfrac{5}{4}, 6$.

65. $y = x^2 + 2x - 24$
Set $y = 0$ and solve the resulting equation.
$$x^2 + 2x - 24 = 0$$
$$(x+6)(x-4) = 0$$
$$x + 6 = 0 \quad \text{or} \quad x - 4 = 0$$
$$x = -6 \qquad\qquad x = 4$$
The x-intercepts are –6 and 4.

67. $y = x^2 + 14x + 49$

Set $y = 0$ and solve the resulting equation.

$x^2 + 14x + 49 = 0$

$(x+7)^2 = 0$

$x + 7 = 0$

$x = -7$

The x-intercept is -7.

69. $y = 6x^3 - 23x^2 + 20x$

Set $y = 0$ and solve the resulting equation.

$6x^3 - 23x^2 + 20x = 0$

$x(6x^2 - 23x + 20) = 0$

$x(3x-4)(2x-5) = 0$

$x = 0$　or　$3x - 4 = 0$　or　$2x - 5 = 0$

$\qquad\qquad\qquad 3x = 4 \qquad\qquad 2x = 5$

$\qquad\qquad\qquad x = \dfrac{4}{3} \qquad\qquad x = \dfrac{5}{2}$

The x-intercepts are 0, $\dfrac{4}{3}$, and $\dfrac{5}{2}$.

71. $y = x^2 - 5x + 6$

$0 = (x-2)(x-3)$

$x - 2 = 0$　or　$x - 3 = 0$

$\quad x = 2 \qquad\qquad x = 3$

The x-intercepts are 2 and 3.
Graph (d) matches this equation.

73. $y = x^2 + 5x + 6$

$0 = (x+2)(x+3)$

$x + 2 = 0$　or　$x + 3 = 0$

$\quad x = -2 \qquad\qquad x = -3$

The x-intercepts are -3 and -2.
Graph (b) matches this equation.

75. $y = (x-4)(x+3)$

$y = x^2 - x - 12$

77. $y = (2x+5)(x-3)$

$y = 2x^2 - x - 15$

79. Let w = width,
then $2w + 1$ = length
surface area = width $\cdot$ length

$10 = w(2w+1)$

$0 = 2w^2 + w - 10$

$0 = (2w+5)(w-2)$

$2w + 5 = 0$　or　$w - 2 = 0$

$w = -\dfrac{5}{2} \qquad\qquad w = 2$

Since width cannot be negative, the width is
2 feet and the length is $2(2) + 1 = 5$ feet.

81. Let b = length of base,
then $b + 6$ = height

area $= \dfrac{1}{2} \cdot$ base $\cdot$ height

$80 = \dfrac{1}{2} b(b+6)$

$160 = b(b+6)$

$0 = b^2 + 6b - 160$

$0 = (b+16)(b-10)$

$b + 16 = 0$　or　$b - 10 = 0$

$b = -16 \qquad\qquad b = 10$

Since the base length cannot be negative, the
base is 10 feet and the height is
$10 + 6 = 16$ feet.

83. Let x = width of frame.
picture area = total area – frame area

$414 = 28(23) - [2x(28 - x) + 2x(23 - x)]$

$414 = 644 - (56x - 2x^2 + 46x - 2x^2)$

$414 = 644 - 102x + 4x^2$

$0 = 4x^2 - 102x + 230$

$0 = 2(2x^2 - 51x + 115)$

$0 = 2(2x - 5)(x - 23)$

$2x - 5 = 0$　or　$x - 23 = 0$

$x = \dfrac{5}{2} \qquad\qquad x = 23$

The width cannot be 23 cm, therefore the
width is $\dfrac{5}{2}$ cm.

85. Let w = width of mulch border.
area of extra mulch = total area – garden area

$$2w(30+w)+2w(20+w)=936-20(30)$$
$$60w+2w^2+40w+2w^2=936-600$$
$$4w^2+100w=336$$
$$4w^2+100w-336=0$$
$$4\left(w^2+25w-84\right)=0$$
$$4(w-3)(w+28)=0$$
$$w-3=0 \quad \text{or} \quad w+28=0$$
$$w=3 \qquad \qquad w=-28$$

The width cannot be negative. Therefore the width is 3 feet.

87. $h(t)=-16t^2+32t$

$$0=-16t^2+32t$$
$$0=-16t(t-2)$$
$$-16t=0 \quad \text{or} \quad t-2=0$$
$$t=0 \qquad \qquad t=2$$

The spurt of water will return to the jet's height in 2 seconds.

89. Let x = height from the ground where the wire is attached, then $x + 8$ = length of the wire

$$c^2=a^2+b^2$$
$$(x+8)^2=x^2+12^2$$
$$x^2+16x+64=x^2+144$$
$$16x+64=144$$
$$16x=80$$
$$x=5$$
$$x+8=13$$

The height is 5 feet, so the wire is 13 feet.

91. $R(x)=C(x)$

$$3x^2-200x+450=x^2-75x+150$$
$$2x^2-125x+300=0$$
$$(2x-5)(x-60)=0$$
$$2x-5=0 \quad \text{or} \quad x-60=0$$
$$x=\frac{5}{2} \quad \text{or} \qquad x=60$$

Reject $\frac{5}{2}$ because it yields a negative revenue and cost. The company must sell 60 bicycles to break even.

93. Let x = length of side of original cardboard.
volume = length · width · height

$$162=(x-4)(x-4)\cdot 2$$
$$162=2\left(x^2-8x+16\right)$$
$$0=x^2-8x-65$$
$$0=(x-13)(x+5)$$
$$x-13=0 \quad \text{or} \quad x+5=0$$
$$x=13 \qquad \qquad x=-5$$

Disregard a negative length. The original cardboard measures 13 in. by 13 in.

95. a. $V=a^3-ab^2$

b. $V=a\left(a^2-b^2\right)$
$V=a(a-b)(a+b)$

c. $1620=12(12-b)(12+b)$
$$135=(12-b)(12+b)$$
$$135=144-b^2$$
$$b^2=9$$
$$b=\pm 3$$
Then, $b = 3$ in.

97. a. The x-intercepts are $x = -5$ and $x = -2$. The factors are $x + 5$ and $x + 2$. One possible representation for the function is $f(x)=(x+5)(x+2)=x^2+7x+10$.

b. The quadratic equation can be $x^2+7x+10=0$.

c. There are an infinite number. For this, express $f(x)$ as $f(x)=a(x^2+7x+10)$ where a is any real number except 0.

d. There are an infinite number. For this, use $a(x^2+7x+10)=0$ where a is any real number except 0. The solution is $x = -2$ or $x = -5$.

99. a. Answers will vary. One example is: No
　　　 x-intercepts

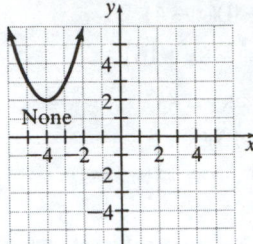

　　　 One x-intercept

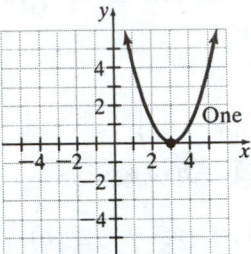

　　　 Two x-intercepts

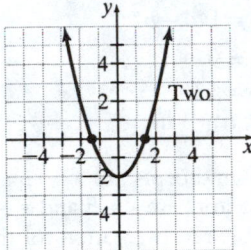

b. $ax^2 + bx + c$ could have no x-intercepts,
one x-intercept, or two x-intercepts. If
the graph does not cross the
x-axis, there are no x-intercepts (no real
solutions). If the vertex is located on the
x-axis, then there is one intercept (one
real solution). If the graph crosses the
x-axis at two different points, then there
are two x-intercepts (i.e., two real
solutions).

101. $d(s) = -0.31s^2 + 59.82s - 2180.22$

$545 = -0.31s^2 + 59.82s - 2180.22$

$y_1 = 545$

$y_2 = -0.31s^2 + 59.82s - 2180.22$

Intersection
X=73.721949　Y=545

60, 80, 2, 400, 600, 20

The intersection is approximately
$(73.721949, 545)$. The car was traveling
approximately 73.721949 mph.

103. $x^4 - 13x^2 = -36$

$\quad x^4 - 13x^2 + 36 = 0$

$\quad \left(x^2\right)^2 - 13x^2 + 36 = 0$

$\qquad y^2 - 13y + 36 = 0 \quad \leftarrow$ Replace x^2 with y

$\qquad (y - 9)(y - 4) = 0$

$\quad y - 9 = 0 \quad$ or $\quad y - 4 = 0$

$\quad x^2 - 9 = 0 \quad$ or $\quad x^2 - 4 = 0 \quad \leftarrow$ Replace y with x^2

$\quad (x - 3)(x + 3) = 0 \qquad (x - 2)(x + 2) = 0$

$\quad x - 3 = 0 \qquad x + 3 = 0 \qquad x - 2 = 0 \qquad x + 2 = 0$

$\qquad x = 3 \qquad\quad x = -3 \qquad\quad x = 2 \qquad\quad x = -2$

The solutions are $\pm 2, \pm 3$.

108. a. Let x be the rate for Carmen. Then $x + 1.2$ is the rate for Bob.

	Rate ×	**Time =**	**Distance**
Bob	$x + 1.2$	4	$4(x + 1.2)$
Carmen	x	5	$5x$

Since the distance is the same,
$$5x = 4(x + 1.2)$$
$$5x = 4x + 4.8$$
$$x = 4.8 \text{ mph}$$

Thus, the rate for Carmen is 4.8 mph and the rate for Bob is $4.8 + 1.2 = 6.0$ mph.

b. The distance is $5x = 5(4.8) = 24$ miles.

109.

x	$f(x) = x^3 + 3$
-2	$(-2)^3 + 3 = -5$
-1	$(-1)^3 + 3 = 2$
0	$0^3 + 3 = 3$
1	$1^3 + 3 = 4$
2	$2^3 + 3 = 11$

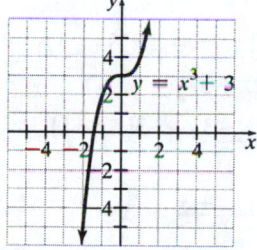

110. $3x + 5y = 9$
$2x - y = 6$

Multiply the second equation by 5 and then add the two equations.
$$3x + 5y = 9$$
$$5(2x - y = 6)$$
gives

$$\begin{array}{r} 3x + 5y = \ \ 9 \\ 10x - 5y = 30 \\ \hline \text{Add} \ \ 13x \ \ \ \ \ \ = 39 \end{array}$$

$$x = \frac{39}{13} = 3$$

Substitute 3 for x into the first equation and then solve for y.
$$3x + 5y = 9$$
$$3(3) + 5y = 9$$
$$9 + 5y = 9$$
$$5y = 0$$
$$y = \frac{0}{5} = 0$$

The solution is $(3, 0)$.

111. $2y > 6x + 12$
$$\frac{1}{2}y < \frac{3}{2}x + 2$$
$$y > 3x + 6$$
$$y < 3x + 4$$

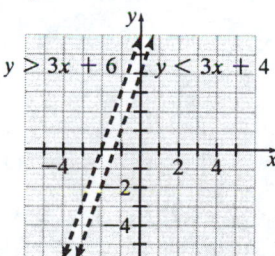

The two regions have no points in common. The solution is $\varnothing$. There is no solution.

112.

$$2x-3\overline{)6x^2-x-12}$$ quotient $3x+4$

$$\underline{6x^2-9x}\qquad \leftarrow 3x(2x-3)$$
$$8x-12$$
$$\underline{8x-12}\leftarrow 4(2x-3)$$
$$0\leftarrow \text{Remainder}$$

Review Exercises

1. $4x^2-3+5x$

   a. Trinomial (3 terms)

   b. $4x^2+5x-3$

   c. Second degree

2. x^2-y^2+xy

   a. Trinomial (3 terms)

   b. x^2+xy-y^2

   c. Second degree

3. $-3-9x^2y+6xy^3+2x^4$

   a. Polynomial (4 terms)

   b. $2x^4-9x^2y+6xy^3-3$

   c. Fourth degree

4. $3x^2+6x^{-1}+4$ is not a polynomial due to the negative exponent.

5. $\left(7x^2+3x-6\right)-\left(5x^2-7x-9\right)$
$$=7x^2-5x^2+3x+7x-6+9$$
$$=2x^2+10x+3$$

6. $4x\left(x^2+2x+3\right)$
$$=(4x)\left(x^2\right)+(4x)(2x)+(4x)(3)$$
$$=4x^3+8x^2+12x$$

7. $(x+5)^2=(x)^2+2(x)(5)+(5)^2$
$$=x^2+10x+25$$

8. $\dfrac{21y^3+6y}{3y}=\dfrac{21y^3}{3y}+\dfrac{6y}{3y}=7y^2+2$

9. $(5x+3)(5x-3)=(5x)^2-(3)^2$
$$=25x^2-9$$

10. $(3xy+1)(2x+3y)$
$$=(3xy)(2x)+(3xy)(3y)+(1)(2x)+(1)(3y)$$
$$=6x^2y+9xy^2+2x+3y$$

11.

$$3x-1\overline{)6x^2-11x+3}$$ quotient $2x-3$

$$\underline{6x^2-2x}\qquad \leftarrow 2x(3x-1)$$
$$-9x+3$$
$$\underline{-9x+3}\leftarrow -3(3x-1)$$
$$0$$

Thus, $\dfrac{6x^2-11x+3}{3x-1}=2x-3$

12. $\left(2x^3-4x^2-3x\right)-\left(4x^2-3x+9\right)$
$$=2x^3-4x^2-3x-4x^2+3x-9$$
$$=2x^3-4x^2-4x^2-3x+3x-9$$
$$=2x^3-8x^2-9$$

13. $-2xy^2\left(x^3+x^2y^5-6y\right)$
$$=\left(-2xy^2\right)\left(x^3\right)+\left(-2xy^2\right)\left(x^2y^5\right)+\left(-2xy^2\right)(-6y)$$
$$=-2x^4y^2-2x^3y^7+12xy^3$$

14. $(3x-2y)^2 = (3x)^2 - 2(3x)(2y) + (2y)^2$
$$= 9x^2 - 12xy + 4y^2$$

15. $(3x^2y + 6xy - 5y^2) - (4y^2 + 3xy)$
$$= 3x^2y + 6xy - 5y^2 - 4y^2 - 3xy$$
$$= 3x^2y + 6xy - 3xy - 5y^2 - 4y^2$$
$$= 3x^2y + 3xy - 9y^2$$

16. $(5xy - 6)(5xy + 6) = (5xy)^2 - (6)^2$
$$= 25x^2y^2 - 36$$

17. $\dfrac{9xy - 6y^2 + 3y}{3y} = \dfrac{9xy}{3y} - \dfrac{6y^2}{3y} + \dfrac{3y}{3y}$
$$= 3x - 2y + 1$$

18. $(2x - 5y^2)(2x + 5y^2) = (2x)^2 - (5y^2)^2$
$$= 4x^2 - 25y^4$$

19.
$$\begin{array}{r} x+4 \\ x-3 \overline{\smash{\big)}\, x^2 + x - 17} \\ \underline{x^2 - 3x} \\ 4x - 17 \\ \underline{4x - 12} \\ -5 \end{array}$$

Thus, $\dfrac{x^2 + x - 17}{x - 3} = x + 4 - \dfrac{5}{x - 3}$

20. $\dfrac{4x^3y^2 + 8x^2y^3 + 12xy^4}{8xy^3}$
$$= \dfrac{4x^3y^2}{8xy^3} + \dfrac{8x^2y^3}{8xy^3} + \dfrac{12xy^4}{8xy^3}$$
$$= \dfrac{x^2}{2y} + x + \dfrac{3y}{2}$$

21. $\big[(x + 3y) + 2\big]^2$
$$= (x + 3y)^2 + 2(x + 3y)(2) + (2)^2$$
$$= (x)^2 + 2(x)(3y) + (3y)^2 + 4(x + 3y) + 4$$
$$= x^2 + 6xy + 9y^2 + 4x + 12y + 4$$

22. $\big[(x + 3y) + 2\big]\big[(x + 3y) - 2\big]$
$$= (x + 3y)^2 - (2)^2$$
$$= (x)^2 + 2(x)(3y) + (3y)^2 - 4$$
$$= x^2 + 6xy + 9y^2 - 4$$

23. $(-6xy + 6y^2 - 3x) - (y^2 + 3xy + 6x)$
$$= -6xy + 6y^2 - 3x - y^2 - 3xy - 6x$$
$$= -6xy - 3xy + 6y^2 - y^2 - 3x - 6x$$
$$= -9xy + 5y^2 - 9x$$

24.
$$\begin{array}{r} 3x^2 + 4x - 6 \\ 2x - 3 \\ \hline -9x^2 - 12x + 18 \\ 6x^3 + 8x^2 - 12x \\ \hline 6x^3 - x^2 - 24x + 18 \end{array}$$

25.
$$\begin{array}{r} 2x^3 + x^2 - 3x - 4 \\ 2x - 1 \overline{\smash{\big)}\, 4x^4 + 0x^3 - 7x^2 - 5x + 4} \\ \underline{4x^4 - 2x^3} \\ 2x^3 - 7x^2 \\ \underline{2x^3 - x^2} \\ -6x^2 - 5x \\ \underline{-6x^2 + 3x} \\ -8x + 4 \\ \underline{-8x + 4} \\ 0 \end{array}$$

Thus,
$$\dfrac{4x^4 - 7x^2 - 5x + 4}{2x - 1} = 2x^3 + x^2 - 3x - 4$$

26.
$$\begin{array}{r} 4x^3 + 0x^2 + 6x - 5 \\ x + 3 \\ \hline 12x^3 + 0x^2 + 18x - 15 \\ 4x^4 + 0x^3 + 6x^2 - 5x \\ \hline 4x^4 + 12x^3 + 6x^2 + 13x - 15 \end{array}$$

27.

$$2x+3\overline{\smash{\big)}\,4x^3+12x^2+x-10}$$

$$2x^2+3x-4$$

$$\underline{4x^3+6x^2}$$

$$6x^2+x$$

$$\underline{6x^2+9x}$$

$$-8x-10$$

$$\underline{-8x-12}$$

$$2$$

Thus, $\dfrac{4x^3+12x^2+x-10}{2x+3}$

$$=2x^2+3x-4+\dfrac{2}{2x+3}$$

28.

$$x+y$$

$$x^2y+6xy+y^2$$

$$x^2y^2+6xy^2+y^3$$

$$x^3y+6x^2y \qquad +xy^2$$

$$x^3y+6x^2y+x^2y^2+7xy^2+y^3$$

29. $P(x)=3x^2-7x+9$

$$P(-3)=3(-3)^2-7(-3)+9$$

$$=3(9)-7(-3)+9$$

$$=27+21+9$$

$$=57$$

30. $P(x)=-x^3-7x^2+6x+3$

$$P(4)=-(4)^3-7(4)^2+6(4)+3$$

$$=-64-112+24+3$$

$$=-149$$

31. a. $t=2010-1997=13$

$$R(t)=0.78t^2+20.28t+385.0$$

$$R(13)=0.78(13)^2+20.28(13)+385.0$$

$$R(13)=780.46$$

The receipts in 2010 are estimated to be $780.46 billion.

b. Yes, the graph supports the answer.

32. a. $t=2010-1997=13$

$$G(t)=1.74t^2+7.32t+383.91$$

$$G(13)=1.74(13)^2+7.32(13)+383.91$$

$$G(13)=773.13$$

The outlays in 2010 are estimated to be $773.13 billion.

b. Yes, the graph supports the answer.

33. a. $(f \cdot g)(x)=f(x)\cdot g(x)$

$$=(x+2)(x-3)$$

$$=x^2+2x-3x-6$$

$$=x^2-x-6$$

b. $(f \cdot g)(3)=3^2-3-6$

$$=9-3-6$$

$$=0$$

34. a. $(f \cdot g)(x)=f(x)g(x)$

$$=(2x-4)(x^2-3)$$

$$=2x^3-6x-4x^2+12$$

$$=2x^3-4x^2-6x+12$$

b. $(f \cdot g)(3)=2(3)^3-4(3)^2-6(3)+12$

$$=54-36-18+12$$

$$=12$$

35. a. $(f \cdot g)(x)=f(x)\cdot g(x)$

$$=(x^2+x-3)(x-2)$$

$$=x^3-2x^2+x^2-2x-3x+6$$

$$=x^3-x^2-5x+6$$

b. $(f \cdot g)(3)=3^3-3^2-5(3)+6$

$$=27-9-15+6$$

$$=9$$

36. a. $(f \cdot g)(x)=f(x)\cdot g(x)$

$$=(x^2-2)(x^2+2)$$

$$=x^4-4$$

b. $(f \cdot g)(3)=3^4-4$

$$=77$$

37.

$$
\begin{array}{r|rrrr}
3 & 3 & -2 & 0 & 10 \\
  &   & 9 & 21 & 63 \\
\hline
  & 3 & 7 & 21 & 73
\end{array}
$$

Thus,

$$\frac{3x^3 - 2x^2 + 10}{x - 3} = 3x^2 + 7x + 21 + \frac{73}{x - 3}$$

38.

$$
\begin{array}{r|rrrrrr}
-1 & 2 & 0 & -10 & 0 & 1 & -1 \\
   &   & -2 & 2 & 8 & -8 & 7 \\
\hline
   & 2 & -2 & -8 & 8 & -7 & 6
\end{array}
$$

Thus, $\dfrac{2y^5 - 10y^3 + y - 1}{y + 1}$

$$= 2y^4 - 2y^3 - 8y^2 + 8y - 7 + \frac{6}{y + 1}$$

39.

$$
\begin{array}{r|rrrrrr}
2 & 1 & 0 & 0 & 0 & 0 & -20 \\
  &   & 2 & 4 & 8 & 16 & 32 \\
\hline
  & 1 & 2 & 4 & 8 & 16 & 12
\end{array}
$$

Thus,

$$\frac{x^5 - 20}{x - 2} = x^4 + 2x^3 + 4x^2 + 8x + 16 + \frac{12}{x - 2}$$

40.

$$
\begin{array}{r|rrrr}
\frac{1}{2} & 2 & 1 & 5 & -3 \\
  &   & 1 & 1 & 3 \\
\hline
  & 2 & 2 & 6 & 0
\end{array}
$$

Thus, $\dfrac{2x^3 + x^2 + 5x - 3}{x - \frac{1}{2}} = 2x^2 + 2x + 6$

41. To find the remainder, use synthetic division:

$$
\begin{array}{r|rrr}
3 & 1 & -4 & 6 \\
  &   & 3 & -3 \\
\hline
  & 1 & -1 & 3
\end{array}
$$

Thus, the remainder is 3.

42. To find the remainder, use synthetic division:

$$
\begin{array}{r|rrrr}
-4 & 2 & -6 & 3 & 0 \\
   &   & -8 & 56 & -236 \\
\hline
   & 2 & -14 & 59 & -236
\end{array}
$$

Thus, the remainder is –236.

43. To find the remainder, use synthetic division:

$$
\begin{array}{r|rrrr}
\frac{1}{3} & 3 & 0 & 0 & -6 \\
  &   & 1 & \frac{1}{3} & \frac{1}{9} \\
\hline
  & 3 & 1 & \frac{1}{3} & -\frac{53}{9}
\end{array}
\text{ or } -5.\overline{8}
$$

Thus, the remainder is $-\dfrac{53}{9}$ or $-5.\overline{8}$.

44. To find the remainder, use synthetic division:

$$
\begin{array}{r|rrrrr}
-2 & 2 & 0 & -6 & 0 & -8 \\
   &   & -4 & 8 & -4 & 8 \\
\hline
   & 2 & -4 & 2 & -4 & 0
\end{array}
$$

Since the remainder is 0, then $x + 2$ is a factor.

45. $12x^3 + 4x + 8 = 4\left(3x^2 + x + 2\right)$

46. $60x^4 + 6x^9 - 18x^5y^2 = 6x^4\left(10 + x^5 - 3xy^2\right)$

47. $4x(2x - 1) + 3(2x - 1)^2$
$= 4x(2x - 1) + 3(2x - 1) \cdot (2x - 1)$
$= (2x - 1)\left[4x + 3(2x - 1)\right]$
$= (2x - 1)\left[4x + 6x - 3\right]$
$= (2x - 1)(10x - 3)$

48. $12xy^4z^3 + 6x^2y^3z^2 - 15x^3y^2z^3$
$= 3xy^2z^2\left(4y^2z + 2xy - 5x^2z\right)$

49. $5x^2 - xy + 20xy - 4y^2$
$= x(5x - y) + 4y(5x - y)$
$= (x + 4y)(5x - y)$

50. $12x^2 - 8xy + 15xy - 10y^2$
$= 4x(3x - 2y) + 5y(3x - 2y)$
$= (4x + 5y)(3x - 2y)$

51. $(3x - y)(x + 2y) - (3x - y)(5x - 7y)$
$= (3x - y)\left[(x + 2y) - (5x - 7y)\right]$
$= (3x - y)(x + 2y - 5x + 7y)$
$= (3x - y)(-4x + 9y)$

52. $3a^4 - 12a^2b + 9a^2b - 36b^2$

$= 3\left(a^4 - 4a^2b + 3a^2b - 12b^2\right)$

$= 3\left[a^2\left(a^2 - 4b\right) + 3b\left(a^2 - 4b\right)\right]$

$= 3\left(a^2 + 3b\right)\left(a^2 - 4b\right)$

53. $x^2 + 8x + 15 = (x+5)(x+3)$ $\leftarrow$ Use 5 and 3: $(5)(3) = 15, 5 + 3 = 8$

54. $-x^2 + 12x + 45 = -\left(x^2 - 12x - 45\right) = -(x-15)(x+3)$ $\leftarrow$ Use -15 and 3: $(-15)(3) = -45, -15 + 3 = -12$

55. $x^2 - 15xy - 54y^2 = (x-18y)(x+3y)$ $\leftarrow$ Use -18 and 3: $(-18)(3) = -54, -18 + 3 = -15$

56. $8x^3 + 10x^2 - 25x = x\left(8x^2 + 10x - 25\right)$

$= x\left(8x^2 + 20x - 10x - 25\right)$ $\leftarrow$ Use $20x - 10x = 10x$

$= x\left[4x(2x+5) - 5(2x+5)\right]$

$= x(4x-5)(2x+5)$

57. $4x^3 - 9x^2 + 5x = x\left(4x^2 - 9x + 5\right)$

$= x\left(4x^2 - 4x - 5x + 5\right)$ $\leftarrow$ Use $-4x - 5x = -9x$

$= x\left[4x(x-1) - 5(x-1)\right]$

$= x(4x-5)(x-1)$

58. $12x^3 + 61x^2 + 5x = x\left(12x^2 + 61x + 5\right)$

$= x\left(12x^2 + 60x + x + 5\right)$ $\leftarrow$ Use $60x + x = 61x$

$= x\left[12x(x+5) + 1(x+5)\right]$

$= x(12x+1)(x+5)$

59. $x^4 - x^2 - 20 = y^2 - y - 20$ $\leftarrow$ Replace x^2 by y

$= (y-5)(y+4)$ $\leftarrow$ Use -5 and 4: $(-5)(4) = -20, -5 + 4 = -1$

$= \left(x^2 - 5\right)\left(x^2 + 4\right)$ $\leftarrow$ Replace y by x^2

60. $(x+5)^2 + 10(x+5) + 24 = w^2 + 10w + 24$ $\leftarrow$ Replace $x+5$ by w

$= (w+6)(w+4)$

$= \left[(x+5)+6\right]\left[(x+5)+4\right]$ $\leftarrow$ Replace w by $x+5$

$= (x+11)(x+9)$

61. $x^2 - 36 = x^2 - 6^2 = (x+6)(x-6)$

62. $x^4 - 81 = \left(x^2\right)^2 - 9^2$
$$= \left(x^2 + 9\right)\left(x^2 - 9\right)$$
$$= \left(x^2 + 9\right)\left(x^2 - 3^2\right)$$
$$= \left(x^2 + 9\right)(x+3)(x-3)$$

63. $(x-3)^2 - 4 = (x-3)^2 - 2^2$
$$= [(x-3)+2][(x-3)-2]$$
$$= (x-1)(x-5)$$

64. $4x^2 - 12x + 9 = (2x)^2 - 2(2x)(3) + (3)^2$
$$= (2x-3)^2$$

65. $9y^2 + 24y + 16 = (3y)^2 + 2(3y)(4) + (4)^2$
$$= (3y+4)^2$$

66. $w^4 - 16w^2 + 64 = \left(w^2\right)^2 - 2\left(w^2\right)(8) + (8)^2$
$$= \left(w^2 - 8\right)^2$$

67. $a^2 + 6ab + 9b^2 - 4c^2$
$$= (a+3b)^2 - (2c)^2$$
$$= [(a+3b)+2c][(a+3b)-2c]$$
$$= (a+3b+2c)(az+3b-2c)$$

68. $x^3 - 8 = x^3 - 2^3$
$$= (x-2)\left[x^2 + x(2) + 2^2\right]$$
$$= (x-2)\left(x^2 + 2x + 4\right)$$

69. $8x^3 + 27 = (2x)^3 + 3^3$
$$= (2x+3)\left[(2x)^2 - (2x)(3) + (3)^2\right]\text{‘}$$
$$= (2x+3)\left(4x^2 - 6x + 9\right)$$

70. $(x+1)^3 - 8$
$$= (x+1)^3 - 2^3$$
$$= [(x+1)-2]\left[(x+1)^2 + (x+1)(2) + (2)^2\right]$$
$$= (x-1)\left[x^2 + 2x + 1 + 2x + 2 + 4\right]$$
$$= (x-1)\left(x^2 + 4x + 7\right)$$

71. a. volume $= a^2 - 4b^2$

    b. $(a-2b)(a+2b)$

72. a. volume $= 4a^3 - 4ac^2$

    b. $4a\left(a^2 - c^2\right) = 4a(a+c)(a-c)$

73. $x^2y^2 - 2xy^2 - 15y^2 = y^2\left(x^2 - 2x - 15\right)$
$$= y^2(x+3)(x-5) \quad \leftarrow \text{ Use 3 and 5: } (3)(-5) = -15,\ 3+(-5) = -2$$

74. $3x^3 - 18x^2 + 24x = 3x\left(x^2 - 6x + 8\right)$
$$= 3x(x-4)(x-2) \quad \leftarrow \text{ Use } -4 \text{ and } -2: (-4)(-2) = 8,\ -4+(-2) = -6$$

75. $3x^3y^4 + 18x^2y^4 - 6x^2y^4 - 36xy^4 = 3xy^4\left(x^2 + 6x - 2x - 12\right)$
$$= 3xy^4\left[x(x+6) - 2(x+6)\right]$$
$$= 3xy^4(x-2)(x+6)$$

76. $3y^5 - 27y = 3y\left(y^4 - 9\right)$

$$= 3y\left[\left(y^2\right)^2 - 3^2\right]$$

$$= 3y\left(y^2 + 3\right)\left(y^2 - 3\right)$$

77. $2x^3y + 16y = 2y\left(x^3 + 8\right)$

$$= 2y\left(x^3 + 2^3\right)$$

$$= 2y(x+2)\left[x^2 - x(2) + 2^2\right]$$

$$= 2y(x+2)\left(x^2 - 2x + 4\right)$$

78. $5x^4y + 20x^3y + 20x^2y = 5x^2y\left(x^2 + 4x + 4\right)$

$$= 5x^2y\left[x^2 + 2(x)(2) + 2^2\right]$$

$$= 5x^2y(x+2)^2$$

79. $6x^3 - 21x^2 - 12x = 3x\left(2x^2 - 7x - 4\right)$

$$= 3x\left(2x^2 + x - 8x - 4\right) \qquad \leftarrow \text{Use } x - 8x \text{ for } -7x$$

$$= 3x\left[x(2x+1) - 4(2x+1)\right]$$

$$= 3x(x-4)(2x+1)$$

80. $x^2 + 10x + 25 - y^2 = (x+5)^2 - y^2$

$$= \left[(x+5) + y\right]\left[(x+5) - y\right]$$

$$= (x+5+y)(x+5-y)$$

81. $3x^3 + 24y^3 = 3\left(x^3 + 8y^3\right)$

$$= 3\left[x^3 + (2y)^3\right]$$

$$= 3(x+2y)\left[x^2 - x(2y) + (2y)^2\right]$$

$$= 3(x+2y)\left(x^2 - 2xy + 4y^2\right)$$

82. $x^2(x+4) - 3x(x+4) - 4(x+4) = (x+4)\left(x^2 + 3x - 4\right)$

$$= (x+4)(x+4)(x-1)$$

$$= (x+4)^2(x-1)$$

83.
$$
\begin{aligned}
4(2x+3)^2 - 12(2x+3) + 5 &= 4w^2 - 12w + 5 && \leftarrow \text{Replace } 2x+3 \text{ by } w \\
&= 4w^2 - 10w - 2w + 5 && \leftarrow \text{Use } -10w - 2w \text{ for } -12w \\
&= 2(2w-5) - 1(2w-5) \\
&= (2w-1)(2w-5) \\
&= [2(2x+3)-1][2(2x+3)-5] && \leftarrow \text{Replace } w \text{ by } 2x+3 \\
&= (4x+6-1)(4x+6-5) \\
&= (4x+5)(4x+1)
\end{aligned}
$$

84.
$$
\begin{aligned}
4x^4 + 4x^2 - 3 &= 4\left(x^2\right)^2 + 4x^2 - 3 \\
&= 4w^2 + 4w - 3 && \leftarrow \text{Replace } x^2 \text{ by } w \\
&= 4w^2 + 6w - 2w - 3 && \leftarrow \text{Use } 6w - 2w \text{ for } 4w \\
&= 2w(2w+3) - 1(2w+3) \\
&= (2w-1)(2w+3) \\
&= \left(2x^2-1\right)\left(2x^2+3\right) && \leftarrow \text{Replace } w \text{ by } x^2
\end{aligned}
$$

85.
$$
\begin{aligned}
(x-1)x^2 - (x-1)x - 2(x-1) &= (x-1)\left(x^2 - x - 2\right) \\
&= (x-1)(x-2)(x+1) && \leftarrow \text{Use } -2 \text{ and } 1: (-2)(1) = -2, \, -2 + (1) = -1
\end{aligned}
$$

86.
$$
\begin{aligned}
9ax - 3bx + 12ay - 4by &= 3x(3a-b) + 4y(3a-b) \\
&= (3x+4y)(3a-b)
\end{aligned}
$$

87.
$$
\begin{aligned}
6p^2q^2 - 5pq - 6 &= 6p^2q^2 - 9pq + 4pq - 6 && \leftarrow \text{Use } -9pq + 4pq \text{ for } -5pq \\
&= 3pq(2pq-3) + 2(2pq-3) \\
&= (3pq+2)(2pq-3)
\end{aligned}
$$

88.
$$
\begin{aligned}
9x^4 - 12x^2 + 4 &= \left(3x^2\right)^2 - 2\left(3x^2\right)(2) + (2)^2 \\
&= \left(3x^2 - 2\right)^2
\end{aligned}
$$

89.
$$
\begin{aligned}
4y^2 - \left(x^2 + 4x + 4\right) &= (2y)^2 - (x+2)^2 \\
&= [2y+(x+2)][2y-(x+2)] \\
&= (2y+x+2)(2y-x-2)
\end{aligned}
$$

90. $6(2a+3)^2 - 7(2a+3) - 3 = 6x^2 - 7x - 3$ ← Replace $2a+3$ by x

$\qquad\qquad = 6x^2 - 9x + 2x - 3$ ← Use $-9x + 2x$ for $-7x$

$\qquad\qquad = 3x(2x-3) + 1(2x-3)$

$\qquad\qquad = (3x+1)(2x-3)$

$\qquad\qquad = [3(2a+3)+1][2(2a+3)-3]$ ← Replace x by $2a+3$

$\qquad\qquad = (6a+9+1)(4a+6-3)$

$\qquad\qquad = (6a+10)(4a+3)$

$\qquad\qquad = 2(3a+5)(4a+3)$

91. $6x^4y^4 + 9x^3y^4 - 27x^2y^4$

$= 3x^2y^4\left(2x^2 + 3x - 9\right)$

$= 3x^2y^4\left(2x^2 + 6x - 3x - 9\right)$

$= 3x^2y^4[2x(x+3) - 3(x+3)]$

$= 3x^2y^4(2x-3)(x+3)$

92. $x^3 - \dfrac{8}{27}y^6$

$= x^3 - \left(\dfrac{2}{3}y^2\right)^3$

$= \left(x - \dfrac{2}{3}y^2\right)\left[x^2 + x\left(\dfrac{2}{3}y^2\right) + \left(\dfrac{2}{3}y^2\right)^2\right]$

$= \left(x - \dfrac{2}{3}y^2\right)\left(x^2 + \dfrac{2}{3}xy^2 + \dfrac{4}{9}y^4\right)$

93. a. The area of the large square is a^2. The sum of the areas of the four small squares is $4b^2$. The area of the shaded region is $a^2 - 4b^2$.

b. In factored form, the area is
$a^2 - 4b^2 = (a+2b)(a-2b)$.

94. a. The sum of the areas of the two large rectangles is $ab + ab = 2ab$. The sum of the areas of the two small squares is $b^2 + b^2 = 2b^2$. The area of the shaded region is $2ab + 2b^2$.

b. In factored form, the area is
$2ab + 2b^2 = 2b(a+b)$.

95. a. The sum of the areas of the three rectangles is
$a(a+3b) + a(a+3b) + b(a+3b)$
$= 2a(a+3b) + b(a+3b)$

b. In factored form, the area is
$2a(a+3b) + b(a+3b)$
$= (2a+b)(a+3b)$

96. a. The area of the large square is a^2. The area of the small square is b^2. The sum of the area of the two rectangles is $ab + ab = 2ab$. The area of the shaded region is $a^2 + 2ab + b^2$.

b. In factored form, the area is
$a^2 + 2ab + b^2 = (a+b)^2$.

97. $(x-5)(3x+2) = 0$

$x - 5 = 0$ or $3x + 2 = 0$

$\qquad x = 5$ $\qquad\qquad 3x = -2$

$\qquad\qquad\qquad\qquad\qquad x = -\dfrac{2}{3}$

The solutions are $-\dfrac{2}{3}, 5$.

98. $2x^2 = 3x$

$2x^2 - 3x = 0$

$x(2x - 3) = 0$

$x = 0$ or $2x - 3 = 0$

$2x = 3$

$x = \dfrac{3}{2}$

The solutions are $0, \dfrac{3}{2}$.

99. $15x^2 + 20x = 0$

$5x(3x + 4) = 0$

$5x = 0$ or $3x + 4 = 0$

$x = 0$ $3x = -4$

$x = -\dfrac{4}{3}$

The solutions are $-\dfrac{4}{3}, 0$.

100. $x^2 - 2x - 24 = 0$

$(x - 6)(x + 4) = 0$

$x - 6 = 0$ or $x + 4 = 0$

$x = 6$ $x = -4$

The solutions are $-4, 6$.

101. $x^2 + 8x + 15 = 0$

$(x + 5)(x + 3) = 0$

$x + 5 = 0$ or $x + 3 = 0$

$x = -5$ $x = -3$

The solutions are $-5, -3$.

102. $x^2 = -2x + 8$

$x^2 + 2x - 8 = 0$

$(x + 4)(x - 2) = 0$

$x + 4 = 0$ or $x - 2 = 0$

$x = -4$ $x = 2$

The solutions are $-4, 2$.

103. $3x^2 + 21x + 30 = 0$

$3\left(x^2 + 7x + 10\right) = 0$

$3(x + 5)(x + 2) = 0$

$x + 5 = 0$ or $x + 2 = 0$

$x = -5$ $x = -2$

The solutions are $-2, -5$.

104. $x^3 - 6x^2 + 8x = 0$

$x\left(x^2 - 6x + 8\right) = 0$

$x(x - 2)(x - 4) = 0$

$x = 0$ or $x - 2 = 0$ or $x - 4 = 0$

$x = 2$ $x = 4$

The solutions are $0, 2, 4$.

105. $12x^3 - 13x^2 - 4x = 0$

$x\left(12x^2 - 13x - 4\right) = 0$

$x(3x - 4)(4x + 1) = 0$

$x = 0$ or $3x - 4 = 0$ or $4x + 1 = 0$

$3x = 4$ $4x = -1$

$x = \dfrac{4}{3}$ $x = -\dfrac{1}{4}$

The solutions are $0, -\dfrac{1}{4}, \dfrac{4}{3}$.

106. $8x^2 - 3 = -10x$

$8x^2 + 10x - 3 = 0$

$(4x - 1)(2x + 3) = 0$

$4x - 1 = 0$ or $2x + 3 = 0$

$4x = 1$ $2x = -3$

$x = \dfrac{1}{4}$ $x = -\dfrac{3}{2}$

The solutions are $-\dfrac{3}{2}, \dfrac{1}{4}$.

107. $4x^2 = 16$

$$4x^2 - 16 = 0$$
$$4(x^2 - 4) = 0$$
$$4(x+2)(x-2) = 0$$
$$x + 2 = 0 \quad \text{or} \quad x - 2 = 0$$
$$x = -2 \qquad\qquad x = 2$$

The solutions are –2, 2.

108. $x(x+3) = 2(x+4) - 2$

$$x^2 + 3x = 2x + 8 - 2$$
$$x^2 + 3x = 2x + 6$$
$$x^2 + x - 6 = 0$$
$$(x+3)(x-2) = 0$$
$$x + 3 = 0 \quad \text{or} \quad x - 2 = 0$$
$$x = -3 \qquad\qquad x = 2$$

The solutions are –3, 2.

109. $y = 2x^2 - 2x - 60$

Set $y = 0$ and solve for x.

$$2x^2 - 2x - 60 = 0$$
$$x^2 - x - 30 = 0$$
$$(x-6)(x+5) = 0$$
$$x - 6 = 0 \quad \text{or} \quad x + 5 = 0$$
$$x = 6 \qquad\qquad x = -5$$

The x-intercepts are 6 and –5.

110. $y = 20x^2 - 49x + 30$

Set $y = 0$ and solve for x.

$$20x^2 - 49x + 30 = 0$$
$$(5x-6)(4x-5) = 0$$
$$5x - 6 = 0 \quad \text{or} \quad 4x - 5 = 0$$
$$5x = 6 \qquad\qquad 4x = 5$$
$$x = \frac{6}{5} \qquad\qquad x = \frac{5}{4}$$

The x-intercepts are $\dfrac{6}{5}$ and $\dfrac{5}{4}$.

111. $y = 14x^2 - 41x - 28$

Set $y = 0$ and solve for x.

$$14x^2 - 41x - 28 = 0$$
$$(2x-7)(7x+4) = 0$$

$$2x - 7 = 0 \quad \text{or} \quad 7x + 4 = 0$$
$$2x = 7 \qquad\qquad 7x = -4$$
$$x = \frac{7}{2} \qquad\qquad x = -\frac{4}{7}$$

The x-intercepts are $\dfrac{7}{2}$ and $-\dfrac{4}{7}$.

112. Let x be the width of the carpet. Then, the length is $x + 2$.

$$x(x+2) = 63$$
$$x^2 + 2x = 63$$
$$x^2 + 2x - 63 = 0$$
$$(x-7)(x+9) = 0$$
$$x - 7 = 0 \quad \text{or} \quad x + 9 = 0$$
$$x = 7 \qquad\qquad x = -9$$

Reject –9 for x since width cannot be negative. Thus, the width is 7 feet and the length is
$7 + 2 = 9$ feet.

113. Let x be the height. Then $2x + 3$ is the base

$$\frac{1}{2}bh = A$$
$$\frac{1}{2}(2x+3)x = 22$$
$$(2x+3)x = 44$$
$$2x^2 + 3x = 44$$
$$2x^2 + 3x - 44 = 0$$
$$(x-4)(2x+11) = 0$$
$$x - 4 = 0 \quad \text{or} \quad 2x + 11 = 0$$
$$x = 4 \qquad\qquad 2x = -11$$
$$x = -\frac{11}{2}$$

Reject $-\dfrac{11}{2}$ for x. Thus, the height is 4 feet and the base is $2(4) + 3 = 8 + 3 = 11$ feet.

114. Let x be the length of a side of the smaller square. Then $x + 4$ is the length of a side of the larger square.

$$(x+4)^2 = 81$$
$$(x+4)^2 - 81 = 0$$
$$(x+4)^2 - 9^2 = 0$$
$$(x+4-9)(x+4+9) = 0$$
$$(x-5)(x+13) = 0$$
$$x - 5 = 0 \quad \text{or} \quad x + 13 = 0$$
$$x = 5 \qquad\qquad x = -13$$

Reject -13 for x. Thus, $x = 5$ inches for the smaller square and $x + 4 = 5 + 4 = 9$ inches for the larger square.

115. $s = -16t^2 + 128t + 144$

Let $s = 0$

$$0 = -16t^2 + 128t + 144$$
$$0 = -16\left(t^2 - 8t - 9\right)$$
$$0 = -16(t-9)(t+1)$$
$$t - 9 = 0 \quad \text{or} \quad t + 1 = 0$$
$$t = 9 \qquad\qquad t = -1$$

Reject $t = -1$. Thus, $t = 9$ seconds.

116. $c^2 = a^2 + b^2$

$$(x+32)^2 = x^2 + (x+31)^2$$
$$x^2 + 64x + 1024 = x^2 + x^2 + 62x + 961$$
$$0 = x^2 - 2x - 63$$
$$0 = (x-9)(x+7)$$
$$x - 9 = 0 \quad \text{or} \quad x + 7 = 0$$
$$x = 9 \qquad\qquad x = -7$$

Disregard a negative length. x is 9.

Practice Test

1. a. Trinomial since it has three terms

b. $-6x^4 - 4x^2 + 2x$

c. Degree is 4

d. The leading coefficient of the polynomial is -6.

2.
$$\frac{12x^6 - 6xy^2 + 15}{3x} = \frac{12x^6}{3x} - \frac{6xy^2}{3x} + \frac{15}{3x}$$
$$= 4x^5 - 2y^2 + \frac{5}{x}$$

3. $(3x+y)(y-2x)$
$$= (3x)(y) + (3x)(-2x) + (y)(y) + (y)(-2x)$$
$$= 3xy - 6x^2 + y^2 - 2xy$$
$$= -6x^2 + xy + y^2$$

4.
$$\begin{array}{r} 2x^2 + 3xy - 6y^2 \\ 2x + y \\ \hline 2x^2 y + 3xy^2 - 6y^3 \\ 4x^3 + 6x^2 y - 12xy^2 \\ \hline 4x^3 + 8x^2 y - 9xy^2 - 6y^3 \end{array}$$

5.
$$\begin{array}{r} x - 5 \\ 2x+3\overline{\smash{\big)}\,2x^2 - 7x + 10} \\ \underline{2x^2 + 3x} \\ -10x + 10 \\ \underline{-10x - 15} \\ 25 \end{array}$$

Thus, $\dfrac{2x^2 - 7x + 10}{2x+3} = x - 5 + \dfrac{25}{2x+3}$

6. $\left(6x^2 y + 3y^2 + 5x\right) - \left(4x^2 y + 2x - 4y^2\right)$
$$= 6x^2 y + 3y^2 + 5x - 4x^2 y - 2x + 4y^2$$
$$= 2x^2 y + 3x + 7y^2$$

7.
$$\begin{array}{r|rrrr} 3 & 2 & -1 & 5 & -7 \\ & & 6 & 15 & 60 \\ \hline & 2 & 5 & 20 & 53 \end{array}$$

Thus,
$$\frac{2x^3 - x^2 + 5x - 7}{x - 3} = 2x^2 + 5x + 20 + \frac{53}{x-3}$$

8. $3x^2y^4\left(-2x^5y^2 + 6x^2y^3 - 3x\right) = 3x^2y^4\left(-2x^5y^2\right) + 3x^2y^4\left(6x^2y^3\right) + 3x^2y^4(-3x)$

$$= -6x^7y^6 + 18x^4y^7 - 9x^3y^4$$

9.

5	3	−12	0	−60	4
		15	15	75	75
	3	3	15	15	79

Thus, $\dfrac{3x^4 - 12x^3 - 60x + 4}{x - 5} = 3x^3 + 3x^2 + 15x + 15 + \dfrac{79}{x - 5}$

10.

−3	2	−6	−5	4
		−6	36	−93
	2	−12	31	−89

Thus, the remainder is −89.

11. $9x^3y^2 + 12x^2y^5 - 27xy^4 = 3xy^2\left(3x^2 + 4xy^3 - 9y^2\right)$ ← $3xy^2$ is a common factor

12. $3x^3 - 6x^2 - 9x = 3x\left(x^2 - 2x - 3\right)$ ← $3x$ is a common factor

$\qquad\qquad\qquad = 3x(x - 3)(x + 1)$ ← Use -3 and 1: $(-3)(1) = -3,\ -3 + 1 = -2$

13. $2x^2 + 4xy + 3xy + 6y^2 = 2x(x + 2y) + 3y(x + 2y)$ ← Factor by grouping

$\qquad\qquad\qquad\qquad = (2x + 3y)(x + 2y)$

14. $5(x - 2)^2 + 15(x - 2) = 5(x - 2) \cdot (x - 2) + 15(x - 2)$

$\qquad\qquad\qquad\qquad = 5(x - 2)[(x - 2) + 3]$ ← $5(x - 2)$ is a common factor

$\qquad\qquad\qquad\qquad = 5(x - 2)(x + 1)$ ← Simplify

15. $27x^3y^6 - 8y^6 = y^6\left(27x^3 - 8\right)$ ← y^6 is a common factor

$\qquad\qquad\quad = y^6\left[(3x)^3 - 2^3\right]$ ← Difference of two cubes

$\qquad\qquad\quad = y^6(3x - 2)\left[(3x)^2 + (3x)(2) + (2)^2\right]$

$\qquad\qquad\quad = y^6(3x - 2)\left(9x^2 + 6x + 4\right)$

16. $(x + 3)^2 + 2(x + 3) - 3$

$= w^2 + 2w - 3$

$= (w + 3)(w - 1)$

$= [(x + 3) + 3][(x + 3) - 1]$

$= (x + 6)(x + 2)$

17. $2x^4 + 5x^2 - 18 = 2w^2 + 5w - 18$

$\qquad\qquad\quad = 2w^2 - 4w + 9w - 18$

$\qquad\qquad\quad = 2w(w - 2) + 9(w - 2)$

$\qquad\qquad\quad = (2w + 9)(w - 2)$

$\qquad\qquad\quad = \left(2x^2 + 9\right)\left(x^2 - 2\right)$

18. $f(x) = 3x - 4$, $g(x) = x - 5$

    a. $(f \cdot g)(x) = f(x) \cdot g(x)$
$$= (3x - 4)(x - 5)$$
$$= 3x^2 - 19x + 20$$

    b. $(f \cdot g)(2) = 3(2)^2 - 19(2) + 20$
$$= 12 - 38 + 20$$
$$= -6$$

19. $4x^2 - 18 = 21x$
$$4x^2 - 21x - 18 = 0$$
$$(4x + 3)(x - 6) = 0$$
$$4x + 3 = 0 \quad \text{or} \quad x - 6 = 0$$
$$4x = -3 \qquad\qquad x = 6$$
$$x = -\frac{3}{4}$$

The solutions are $-\frac{3}{4}$, 6.

20. $x^3 + 4x^2 - 5x = 0$
$$x\left(x^2 + 4x - 5\right) = 0$$
$$x(x + 5)(x - 1) = 0$$
$$x = 0 \quad \text{or} \quad x + 5 = 0 \quad \text{or} \quad x - 1 = 0$$
$$x = -5 \qquad\qquad x = 1$$
The solutions are 0, –5, 1.

21. $b_n = b_1 + ac - c$
$$b_n - b_1 = ac - c$$
$$b_n - b_1 = c(a - 1)$$
$$\frac{b_n - b_1}{a - 1} = c$$
$$\text{or } c = \frac{b_n - b_1}{a - 1}$$

22. $y = 8x^2 + 10x - 3$
Set $y = 0$ and solve for x
$$8x^2 + 10x - 3 = 0$$
$$(4x - 1)(2x + 3) = 0$$

$$4x - 1 = 0 \quad \text{or} \quad 2x + 3 = 0$$
$$4x = 1 \qquad\qquad 2x = -3$$
$$x = \frac{1}{4} \qquad\qquad x = -\frac{3}{2}$$
The x-intercepts are $\frac{1}{4}$ and $-\frac{3}{2}$.

23. $y = (x - 2)(x - 6)$
$$y = x^2 - 6x - 2x + 12$$
$$y = x^2 - 8x + 12$$

24. Let x be the height of the triangle. Then, $3x + 2$ is the base.
$$\frac{1}{2}(\text{base})(\text{height}) = A$$
$$\frac{1}{2}(3x + 2)(x) = 28$$
$$(3x + 2)(x) = 56$$
$$3x^2 + 2x = 56$$
$$3x^2 + 2x - 56 = 0$$
$$(x - 4)(3x + 14) = 0$$
$$x - 4 = 0 \quad \text{or} \quad 3x + 14 = 0$$
$$x = 4 \qquad\qquad 3x = -14$$
$$x = -\frac{14}{3}$$

Reject $-\frac{14}{3}$. Thus, the height is 4 meters and the base is $3 \cdot 4 + 2 = 12 + 2 = 14$ meters.

25. $s = -16t^2 + 48t + 448$
Set $s = 0$ and solve for t
$$0 = -16t^2 + 48t + 448$$
$$0 = -16\left(t^2 - 3t - 28\right)$$
$$0 = -16(t - 7)(t + 4)$$
$$t - 7 = 0 \quad \text{or} \quad t + 4 = 0$$
$$t = 7 \qquad\qquad t = -4$$
Reject $t = -4$. The baseball strikes the ground in 7 seconds.

Cumulative Review Test

1. $\dfrac{\sqrt[3]{27}-\sqrt[3]{-8}+|-4|}{3^0-12\div3\div4-8}=\dfrac{3-(-2)+4}{1-12\div3\div4-8}$

$\qquad\qquad =\dfrac{3-(-2)+4}{1-4\div4-8}$

$\qquad\qquad =\dfrac{3-(-2)+4}{1-1-8}$

$\qquad\qquad =\dfrac{3+2+4}{1-1-8}$

$\qquad\qquad =\dfrac{9}{-8}\text{ or }-\dfrac{9}{8}$

2. $\left(\dfrac{8x^{-2}y^3}{4xy^{-1}}\right)^2=\left(\dfrac{8}{4}x^{-2-1}y^{3+1}\right)^2$

$\qquad\qquad =\left(2x^{-3}y^4\right)^2$

$\qquad\qquad =2^2x^{-3(2)}y^{4(2)}$

$\qquad\qquad =4x^{-6}y^8$

$\qquad\qquad =\dfrac{4y^8}{x^6}$

3. $\left(2p^4q^3\right)\left(3pq^4\right)^3=\left(2p^4q^3\right)\left[3^3p^3\left(q^4\right)^3\right]$

$\qquad\qquad =2p^4q^3\cdot27p^3q^{12}$

$\qquad\qquad =54p^{4+3}q^{3+12}$

$\qquad\qquad =54p^7q^{15}$

4. $\qquad\dfrac{1}{3}(x-6)=\dfrac{3}{4}(2x-1)$

$\qquad 12\left[\dfrac{1}{3}(x-6)\right]=12\left[\dfrac{3}{4}(2x-1)\right]$

$\qquad\qquad 4(x-6)=9(2x-1)$

$\qquad\qquad 4x-24=18x-9$

$\qquad 4x-4x-24=18x-4x-9$

$\qquad\qquad -24=14x-9$

$\qquad -24+9=14x-9+9$

$\qquad\qquad -15=14x$

$\qquad\qquad \dfrac{-15}{14}=\dfrac{14x}{14}$

$\qquad\qquad -\dfrac{15}{14}=x$

5. $\qquad 3P=\dfrac{2L-W}{4}$

$\qquad 4(3P)=4\left(\dfrac{2L-W}{4}\right)$

$\qquad 12P=2L-W$

$\qquad 12P+W=2L-W+W$

$\qquad 12P+W=2L$

$\qquad \dfrac{12P+W}{2}=\dfrac{2L}{2}$

$\qquad \dfrac{12P+W}{2}=L$

6. $f(x)=3x-4$

$\qquad y=3x-4$

x	y
0	-4
1	-1
2	2

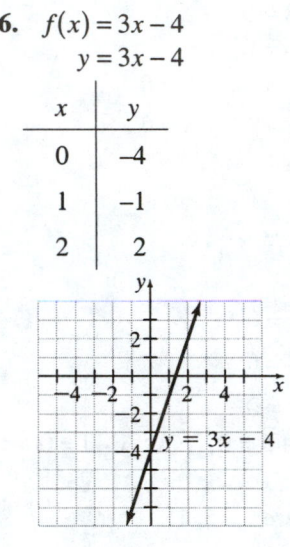

7. $2x-y\le6$

Graph the line $2x-y=6$ using a solid line.

$\qquad -y=-2x+6$

$\qquad y=2x-6$

For the test point, select $(0,0)$:

$\qquad 2x-y\le6$

$\qquad 2(0)-0\le6$

$\qquad 0-0\le6$

$\qquad 0\le6$

Since this is a true statement, shade the region which contains $(0,0)$.

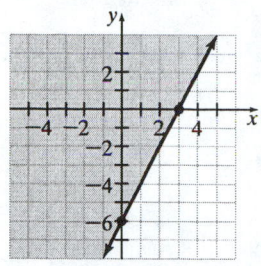

8. a. It is a function. For each value of x there is one value for y.

b. It is not a function since the ordered pairs
(1, 2) and (1, 0) have the same first coordinate but different second coordinates.

9. $3x - 2y = 8$
$2x - 5y = 10$
To eliminate x, multiply the first equation by 2 and the second equation by -3 and then add.
$2(3x - 2y = 8)$
$-3(2x - 5y = 10)$
gives
$$6x - 4y = 16$$
$$\underline{-6x + 15y = -30}$$
Add: $11y = -14$
$$y = -\frac{14}{11}$$

To find x, substitute $-\dfrac{14}{11}$ for y into the

second equation.

$$2x - 5y = 10$$
$$2x - 5\left(-\frac{14}{11}\right) = 10$$
$$2x + \frac{70}{11} = 10$$
$$2x = 10 - \frac{70}{11} = \frac{110}{11} - \frac{70}{11} = \frac{40}{11}$$
$$\frac{1}{2}(2x) = \frac{1}{2}\left(\frac{40}{11}\right)$$
$$x = \frac{20}{11}$$

The solution is $\left(\dfrac{20}{11}, -\dfrac{14}{11}\right)$.

10.

x	$y = \lvert x \rvert - 2$
-6	4
-4	2
-2	0
0	-2
1	-1

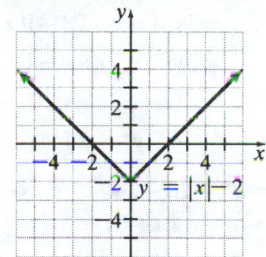

11. $3x^2 - 4x - 6 - \left(5x - 4x^2 - 6\right)$
$= 3x^2 - 4x - 6 - 5x + 4x^2 + 6$
$= 7x^2 - 9x$

12.
$$
\begin{array}{r}
x^2 - 3x - 6 \\
\underline{2x - 5} \\
-5x^2 + 15x + 30 \\
\underline{2x^3 - 6x^2 - 12x } \\
2x^3 - 11x^2 + 3x + 30
\end{array}
$$
$\leftarrow -5(x^2 - 3x - 6)$
$\leftarrow 2x(x^2 - 3x - 6)$
$\leftarrow$ Combine like terms

13. $\dfrac{9x^3y^5 - 8x^2y^4 - 12xy}{3x^2y}$

$= \dfrac{9x^3y^5}{3x^2y} - \dfrac{8x^2y^4}{3x^2y} - \dfrac{12xy}{3x^2y}$

$= 3xy^4 - \dfrac{8y^3}{3} - \dfrac{4}{x}$

14. $f(x) = 3x^3 - 6x^2 - 4x + 3$

$\quad f(2) = 3(2)^3 - 6(2)^2 - 4(2) + 3$

$\qquad\quad = 3(8) - 6(4) - 4(2) + 3$

$\qquad\quad = 24 - 24 - 8 + 3$

$\qquad\quad = -5$

15. $x^4 - 3x^3 + 2x^2 - 6x = x\left(x^3 - 3x^2 + 2x - 6\right)$

$\qquad\qquad\qquad\qquad\quad = x\left[x^2(x-3) + 2(x-3)\right]$

$\qquad\qquad\qquad\qquad\quad = x(x-3)\left(x^2 + 2\right)$

16. $12x^2y - 27xy + 6y = 3y\left(4x^2 - 9x + 2\right)$

$\qquad\qquad\qquad\qquad = 3y\left(4x^2 - 8x - x + 2\right) \quad \leftarrow \text{Use } -8x - x \text{ for } -9x$

$\qquad\qquad\qquad\qquad = 3y\left[4x(x-2) - 1(x-2)\right]$

$\qquad\qquad\qquad\qquad = 3y(x-2)(4x-1)$

17. $y^4 + 2y^2 - 24 = \left(y^2\right)^2 + 2y^2 - 24$

$\qquad\qquad\qquad = u^2 + 2u - 24 \qquad \leftarrow \text{Replace } y^2 \text{ with } u$

$\qquad\qquad\qquad = (u-4)(u+6)$

$\qquad\qquad\qquad = \left(y^2 - 4\right)\left(y^2 + 6\right) \qquad \leftarrow \text{Replace } u \text{ with } y^2$

$\qquad\qquad\qquad = (y+2)(y-2)\left(y^2 + 6\right)$

18. $8x^3 - 27y^6 = (2x)^3 - \left(3y^2\right)^3$

$\qquad\qquad\quad = \left(2x - 3y^2\right)\left[(2x)^2 + (2x)\left(3y^2\right) + \left(3y^2\right)^2\right]$

$\qquad\qquad\quad = \left(2x - 3y^2\right)\left(4x^2 + 6xy^2 + 9y^4\right)$

19. Let x be the number of pages.

$$0.15x + 0.05(6x) = 279$$
$$0.15x + 0.3x = 279$$
$$0.45x = 279$$
$$x = \frac{279}{0.45} = 620$$

The manuscript is 620 pages long.

20. $70 \leq \dfrac{68 + 72 + 90 + 86 + x}{5} < 80$

$350 \leq 316 + x < 400$

$34 \leq x < 84$

If Santo scores at least 34 points but less than 84 points, his average is in the 70's (and he will receive a grade of C).

Chapter 6

Exercise Set 6.1

1. a. A rational expression is an expression of the form $\dfrac{p}{q}$, p and q polynomials, $q \neq 0$.

b. Answers will vary.

3. a. a rational function is a function of the form $f(x) = \dfrac{p}{q}$, p and q polynomials, $q \neq 0$.

b. Answers will vary.

5. a. The domain of a rational function is the set of values that can replace the variable.

b. $\dfrac{3}{x^2 - 9} = \dfrac{3}{(x+3)(x-3)}$
The domain is
$\{x | x \text{ is a real number}, x \neq -3, x \neq 3\}$.

7. a. Answers will vary. One possible answer is:
Factor out (-1) from the numerator and then cancel the remaining factor with the denominator.

b. $\dfrac{3x^2 - 2x - 8}{-3x^2 + 2x + 8} = \dfrac{-(-3x^2 + 2x + 8)}{-3x^2 + 2x + 8}$
$= -1$

9. a. Answers will vary. One possible answer is:
Invert the second fraction and then multiply by factoring and cancelling common factors between the numerator and denominator.

b. $\dfrac{r+2}{r^2 + 7r + 12} \div \dfrac{(r+2)^2}{r^2 + 5r + 6}$
$= \dfrac{r+2}{r^2 + 7r + 12} \cdot \dfrac{r^2 + 5r + 6}{(r+2)^2}$
$= \dfrac{r+2}{(r+4)(r+3)} \cdot \dfrac{(r+3)(r+2)}{(r+2)(r+2)}$
$= \dfrac{1}{r+4}$

11. $\dfrac{3x}{2x-8} = \dfrac{3x}{2(x-4)}$
The excluded value is 4.

13. $\dfrac{4}{2x^2 - 15x + 25} = \dfrac{4}{(x-5)(2x-5)}$
The excluded values are 5 and $\dfrac{5}{2}$.

15. $\dfrac{x-3}{x^2 + 4}$
There are no values for which $x^2 + 4 = 0$.
Thus, there are no excluded values.

17. $f(p) = \dfrac{p+1}{p-4}$
The domain is
$\{p | p \text{ is a real number and } p \neq 4\}$

19. $y = \dfrac{5}{x^2 + x - 6}$
$y = \dfrac{5}{(x+3)(x-2)}$
The domain is
$\{x | x \text{ is a real number and } x \neq -3, x \neq 2\}$

21. $f(a) = \dfrac{3a^2 - 6a + 4}{2a^2 + 3a - 2}$
$2a^2 + 3a - 2 = 0$
$(2a - 1)(a + 2) = 0$
$2a - 1 = 0$ or $a + 2 = 0$
$2a = 1$ or $a = -2$
$a = \dfrac{1}{2}$
The domain is
$\left\{a \middle| a \text{ is a real number and } a \neq \dfrac{1}{2}, \ a \neq -2\right\}$

23. $\dfrac{x-xy}{x}=\dfrac{x(1-y)}{x}=1-y$

25. $\dfrac{5x^2-10xy}{25x}=\dfrac{5x(x-2y)}{5x\cdot5}=\dfrac{x-2y}{5}$

27. $\dfrac{5r-2}{2-5r}=\dfrac{-1(2-5r)}{2-5r}=-1$

29. $\dfrac{p^2-2p-24}{6-p}=\dfrac{(p-6)(p+4)}{-1(p-6)}$
$=\dfrac{p+4}{-1}$
$=-(p+4)$ or $-p-4$

31. $\dfrac{a^2-3a-10}{a^2+5a+6}=\dfrac{(a-5)(a+2)}{(a+3)(a+2)}=\dfrac{a-5}{a+3}$

33. $\dfrac{8x^3-125y^3}{2x-5y}$
$=\dfrac{(2x-5y)(4x^2+10xy+25y^2)}{2x-5y}$
$=4x^2+10xy+25y^2$

35. $\dfrac{(x+1)(x-3)+(x+1)(x-2)}{2(x+1)}$
$=\dfrac{(x+1)[(x-3)+(x-2)]}{2(x+1)}$
$=\dfrac{(x+1)(2x-5)}{2(x+1)}$
$=\dfrac{2x-5}{2}$

37. $\dfrac{xy-yw+xz-zw}{xy+yw+xz+zw}=\dfrac{y(x-w)+z(x-w)}{y(x+w)+z(x+w)}$
$=\dfrac{(y+z)(x-w)}{(y+z)(x+w)}$
$=\dfrac{x-w}{x+w}$

39. $\dfrac{a^3-b^3}{a^2-b^2}=\dfrac{(a-b)(a^2+ab+b^2)}{(a-b)(a+b)}$
$=\dfrac{a^2+ab+b^2}{a+b}$

41. $\dfrac{2x}{3y}\cdot\dfrac{y^3}{6}=\dfrac{x\cdot y^2}{3\cdot3}=\dfrac{xy^2}{9}$

43. $\dfrac{9x^3}{4}\div\dfrac{3}{16y^2}=\dfrac{9x^3}{4}\cdot\dfrac{16y^2}{3}$
$=\dfrac{3x^3\cdot4y^2}{1}$
$=\dfrac{12x^3y^2}{1}$
$=12x^3y^2$

45. $\dfrac{3-r}{r-3}\cdot\dfrac{r-5}{5-r}=\dfrac{-1(r-3)}{r-3}\cdot\dfrac{-1(5-r)}{5-r}$
$=(-1)(-1)$
$=1$

47. $\dfrac{p^2+3p-10}{p+5}\cdot\dfrac{1}{p-2}=\dfrac{(p+5)(p-2)}{p+5}\cdot\dfrac{1}{p-2}$
$=1$

49. $\dfrac{x^2+10x+21}{x+7}\div(x^2-5x-24)$
$=\dfrac{x^2+10x+21}{x+7}\cdot\dfrac{1}{x^2-5x-24}$
$=\dfrac{(x+3)(x+7)}{x+7}\cdot\dfrac{1}{(x+3)(x-8)}$
$=\dfrac{1}{x-8}$

51. $\dfrac{x^2-9x+14}{x^2-5x+6}\div\dfrac{x^2-5x-14}{x+2}$
$=\dfrac{x^2-9x+14}{x^2-5x+6}\cdot\dfrac{x+2}{x^2-5x-14}$
$=\dfrac{(x-7)(x-2)}{(x-2)(x-3)}\cdot\dfrac{x+2}{(x-7)(x+2)}$
$=\dfrac{1}{x-3}$

53. $\dfrac{a-b}{9a+9b} \div \dfrac{a^2-b^2}{a^2+2a+1} = \dfrac{a-b}{9a+9b} \cdot \dfrac{a^2+2a+1}{a^2-b^2}$

$\qquad\qquad\qquad\qquad = \dfrac{a-b}{9(a+b)} \cdot \dfrac{(a+1)(a+1)}{(a+b)(a-b)}$

$\qquad\qquad\qquad\qquad = \dfrac{(a+1)^2}{9(a+b)^2}$

55. $\dfrac{3x^2-x-4}{4x^2+5x+1} \cdot \dfrac{2x^2-5x-12}{6x^2+x-12} = \dfrac{(3x-4)(x+1)}{(4x+1)(x+1)} \cdot \dfrac{(2x+3)(x-4)}{(2x+3)(3x-4)}$

$\qquad\qquad\qquad\qquad\qquad = \dfrac{x-4}{4x+1}$

57. $\dfrac{x+2}{x^3-8} \cdot \dfrac{(x-2)^2}{x^2+4} = \dfrac{x+2}{(x-2)(x^2+2x+4)} \cdot \dfrac{(x-2)(x-2)}{x^2+4}$

$\qquad\qquad\qquad\qquad = \dfrac{(x+2)(x-2)}{(x^2+2x+4)(x^2+4)}$

59. $\dfrac{x^4-y^8}{x^2+y^4} \div \dfrac{x^2-y^4}{3x^2} = \dfrac{x^4-y^8}{x^2+y^4} \cdot \dfrac{3x^2}{x^2-y^4}$

$\qquad\qquad\qquad\qquad = \dfrac{(x^2+y^4)(x+y^2)(x-y^2)}{x^2+y^4} \cdot \dfrac{3x^2}{(x+y^2)(x-y^2)}$

$\qquad\qquad\qquad\qquad = 3x^2$

61. $\dfrac{2x^4+4x^2}{6x^2+14x+4} \div \dfrac{x^2+2}{3x^2+x}$

$\quad = \dfrac{2x^4+4x^2}{6x^2+14x+4} \cdot \dfrac{3x^2+x}{x^2+2}$

$\quad = \dfrac{2x^2(x^2+2)}{2(3x+1)(x+2)} \cdot \dfrac{x(3x+1)}{x^2+2}$

$\quad = \dfrac{x^3}{x+2}$

63. $\dfrac{r^2-9}{r^3-27} \div \dfrac{r^2+6r+9}{r^2+3r+9}$

$\quad = \dfrac{r^2-9}{r^3-27} \cdot \dfrac{r^2+3r+9}{r^2+6r+9}$

$\quad = \dfrac{(r-3)(r+3)}{(r-3)(r^2+3r+9)} \cdot \dfrac{r^2+3r+9}{(r+3)^2}$

$\quad = \dfrac{1}{r+3}$

65. $\dfrac{2x^3-7x^2+3x}{x^2+2x-3} \cdot \dfrac{x^2+3x}{(x-3)^2}$

$\quad = \dfrac{x(2x-1)(x-3)}{(x+3)(x-1)} \cdot \dfrac{x(x+3)}{(x-3)(x-3)}$

$\quad = \dfrac{x^2(2x-1)}{(x-1)(x-3)}$

67. $\dfrac{3r^2+17rs+10s^2}{6r^2+13rs-5s^2} \div \dfrac{6r^2+rs-2s^2}{6r^2-5rs+s^2}$

$= \dfrac{3r^2+17rs+10s^2}{6r^2+13rs-5s^2} \cdot \dfrac{6r^2-5rs+s^2}{6r^2+rs-2s^2}$

$= \dfrac{(3r+2s)(r+5s)}{(2r+5s)(3r-s)} \cdot \dfrac{(3r-s)(2r-s)}{(2r-s)(3r+2s)}$

$= \dfrac{r+5s}{2r+5s}$

69. $\dfrac{2p^2+2pq-pq^2-q^3}{p^3+p^2+pq^2+q^2} \div \dfrac{p^3+p+p^2q+q}{p^3+p+p^2+1}$

$= \dfrac{2p^2+2pq-pq^2-q^3}{p^3+p^2+pq^2+q^2} \cdot \dfrac{p^3+p+p^2+1}{p^3+p+p^2q+q}$

$= \dfrac{2p(p+q)-q^2(p+q)}{p^2(p+1)+q^2(p+1)} \cdot \dfrac{p(p^2+1)+1(p^2+1)}{p(p^2+1)+q(p^2+1)}$

$= \dfrac{(2p-q^2)(p+q)}{(p^2+q^2)(p+1)} \cdot \dfrac{(p+1)(p^2+1)}{(p+q)(p^2+1)}$

$= \dfrac{2p-q^2}{p^2+q^2}$

71. One possible answer is
$$\dfrac{1}{(x-2)(x+3)} \text{ or } \dfrac{1}{x^2+x-6}.$$

73. The numerator is never 0.

75. a. It is zero when the numerator is zero. That is, when $x-4=0$, or $x=4$.

   b. It is undefined when the denominator is zero. That is, when
$$x^2-4=0$$
$$(x+2)(x-2)=0$$
$$x+2=0 \quad \text{or} \quad x-2=0$$
$$x=-2 \qquad\quad x=2$$

77. One possible answer is
$$f(x)=\dfrac{x-2}{(x-3)(x+1)}.$$

79. $x^2+2x-15=(x-3)(x+5)$
$x+5$ is the factor missing from the denominator of the fraction on the right side.
$$\dfrac{1}{x-3} \cdot \dfrac{x+5}{x+5} = \dfrac{x+5}{(x-3)(x+5)}$$
$$= \dfrac{x+5}{x^2+2x-15}$$
The desired numerator is $x+5$.

81. $y^2-y-20=(y-5)(y+4)$
$y-5$ is the factor missing from the numerator of the fraction on the right side.
$$\dfrac{y+4}{y+1} \cdot \dfrac{y-5}{y-5} = \dfrac{(y+4)(y-5)}{(y+1)(y-5)} = \dfrac{y^2-y-20}{y^2-4y-5}.$$
The desired denominator is y^2-4y-5.

83. $\dfrac{x^2-x-12}{x^2+2x-3} \cdot \dfrac{?}{x^2-2x-8} = 1$

$\dfrac{(x-4)(x+3)}{(x+3)(x-1)} \cdot \dfrac{?}{(x-4)(x+2)} = 1$

$\dfrac{?}{(x-1)(x+2)} = 1$

The only way the left side can simplify to 1 is if the numerator is $(x-1)(x+2)$. The corresponding polynomial is x^2+x-2.

85. $\dfrac{x^2-9}{2x^2+3x-2} \div \dfrac{2x^2-9x+9}{?} = \dfrac{x+3}{2x-1}$

$\dfrac{x^2-9}{2x^2+3x-2} \cdot \dfrac{?}{2x^2-9x+9} = \dfrac{x+3}{2x-1}$

$\dfrac{(x+3)(x-3)}{(2x-1)(x+2)} \cdot \dfrac{?}{(2x-3)(x-3)} = \dfrac{x+3}{2x-1}$

$\dfrac{x+3}{(2x-1)(x+2)} \cdot \dfrac{?}{(2x-3)} = \dfrac{x+3}{2x-1}$

The only way the left side can simplify to the right side is if the missing numerator is $(2x-3)(x+2)$.
The corresponding polynomial is $2x^2+x-6$.

87. Area $= \dfrac{1}{2}$(base)(height)

$$a^2 + 2ab - 3b^2 = \frac{1}{2}(a + 3b)h$$

$$2(a^2 + 2ab - 3b^2) = 2\left[\frac{1}{2}(a + 3b)h\right]$$

$$2(a^2 + 2ab - 3b^2) = (a + 3b)h$$

$$\frac{2(a^2 + 2ab - 3b^2)}{a + 3b} = h$$

$$\frac{2(a + 3b)(a - b)}{a + 3b} = h$$

$$h = 2(a - b)$$

89. $\left(\dfrac{2x^2 - 3x - 14}{2x^2 - 9x + 7} \div \dfrac{6x^2 + x - 15}{3x^2 + 2x - 5}\right) \cdot \dfrac{6x^2 - 7x - 3}{2x^2 - x - 3} = \left(\dfrac{2x^2 - 3x - 14}{2x^2 - 9x + 7} \cdot \dfrac{3x^2 + 2x - 5}{6x^2 + x - 15}\right) \cdot \dfrac{6x^2 - 7x - 3}{2x^2 - x - 3}$

$$= \frac{(x + 2)(2x - 7)}{(x - 1)(2x - 7)} \cdot \frac{(x - 1)(3x + 5)}{(3x + 5)(2x - 3)} \cdot \frac{(3x + 1)(2x - 3)}{(x + 1)(2x - 3)}$$

$$= \frac{(x + 2)(3x + 1)}{(2x - 3)(x + 1)}$$

91. $\dfrac{5x^2(x - 1) - 3x(x - 1) - 2(x - 1)}{10x^2(x - 1) + 9x(x - 1) + 2(x - 1)} \cdot \dfrac{2x + 1}{x + 3}$

$$= \frac{(x - 1)(5x^2 - 3x - 2)}{(x - 1)(10x^2 + 9x + 2)} \cdot \frac{2x + 1}{x + 3}$$

$$= \frac{(x - 1)(5x + 2)(x - 1)}{(x - 1)(5x + 2)(2x + 1)} \cdot \frac{2x + 1}{x + 3}$$

$$= \frac{x - 1}{x + 3}$$

93. $\dfrac{(x - p)^n}{x^{-2}} \div \dfrac{(x - p)^{2n}}{x^{-4}}$

$$= x^2(x - p)^n \div x^4(x - p)^{2n}$$

$$= \frac{x^2(x - p)^n}{1} \cdot \frac{1}{x^4(x - p)^{2n}}$$

$$= \frac{x^2(x - p)^n}{1} \cdot \frac{1}{x^4(x - p)^n(x - p)^n}$$

$$= \frac{1}{x^2(x - p)^n}$$

95. $\dfrac{x^{5y} + 3x^{4y}}{3x^{3y} + x^{4y}} = \dfrac{x^y(x^{4y} + 3x^{3y})}{3x^{3y} + x^{4y}} = x^y$

97. a. $f(x) = \dfrac{1}{x - 2}$. The denominator cannot equal zero. Therefore, $x \ne 2$. The domain is $\{x | x$ is a real number, $x \ne 2\}$.

b.

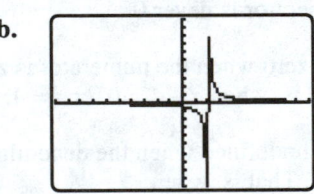

$-10, 10, 1, -10, 10, 1$

c. The function is decreasing as x gets closer to 2, approaching 2 from the left side.

d. The function is increasing as x gets closer to 2, approaching 2 from the right side.

99. a. $f(x) = \dfrac{x^2}{x-2}$. The denominator cannot

equal zero. Therefore, $x \neq 2$. The
domain is $\{x | x$ is a real number, $x \neq 2\}$.

b.

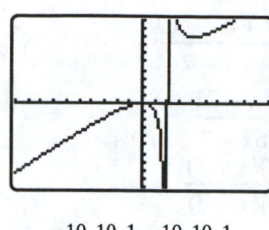

$-10, 10, 1, -10, 10, 1$

c. The function is decreasing as x gets
closer to 2, approaching 2 from the left
side.

d. The function is increasing as x gets
closer to 2, approaching 2 from the right
side.

103. $\dfrac{4x^{-3}y^4}{12x^{-2}y^3} = \dfrac{4}{12}x^{-3-(-2)}y^{4-3}$

$= \dfrac{1}{3}x^{-1}y^1$

$= \dfrac{y}{3x}$

104. $V = \dfrac{4}{3}\pi r^2 h$

$3(V) = 3\left(\dfrac{4}{3}\pi r^2 h\right)$

$3V = 4\pi r^2 h$

$\dfrac{3V}{4\pi r^2} = \dfrac{4\pi r^2 h}{4\pi r^2}$

$\dfrac{3V}{4\pi r^2} = h$

105. $-4 < 3x - 4 < 8$

$-4 + 4 < 3x - 4 + 4 < 8 + 4$

$0 < 3x < 12$

$\dfrac{0}{3} < \dfrac{3x}{3} < \dfrac{12}{3}$

$0 < x < 4$

The answer is the interval $(0, 4)$.

106. $3(y-4) = -(x-2)$

$3y - 12 = -x + 2$

$3y = -x + 14$

$y = -\dfrac{1}{3}x + \dfrac{14}{3}$

The slope is $-\dfrac{1}{3}$ and the y-intercept is $\dfrac{14}{3}$.

107. $x + 2y = 4$

$2y = 6x + 6$

Write the system in standard form.

$x + 2y = 4$

$-6x + 2y = 6$

To eliminate y, multiply the first equation by
-1 and then add.

$-1[x + 2y = 4]$

$-6x + 2y = 6$

gives

$-x - 2y = -4$

$-6x + 2y = 6$

Add: $-7x = 2$

$x = -\dfrac{2}{7}$

Substitute $-\dfrac{2}{7}$ for x in the first equation.

$x + 2y = 4$

$-\dfrac{2}{7} + 2y = 4$

$2y = 4 + \dfrac{2}{7}$

$2y = \dfrac{30}{7}$

$\dfrac{1}{2}(2y) = \dfrac{1}{2}\left(\dfrac{30}{7}\right)$

$y = \dfrac{15}{7}$

The solution is $\left(-\dfrac{2}{7}, \dfrac{15}{7}\right)$.

108. $3x^2y - 4xy + 2y^2 - (3xy + 6y^2 + 2x)$

$= 3x^2y - 4xy + 2y^2 - 3xy - 6y^2 - 2x$

$= 3x^2y - 4xy - 3xy + 2y^2 - 6y^2 - 2x$

$= 3x^2y - 7xy - 4y^2 - 2x$

Exercise Set 6.2

1. a. Answers will vary. One possible answer is: The LCD is the 'smallest' denominator into which the individual denominators divide.

b. Answers will vary. One possible answer is: Factor each denominator completely. Any factors that occur more than once should be expressed as powers. When the same factor appears in more than one denominator, write the factor with the highest power that appears. The LCD is the product of all of those factors found.

c. $\dfrac{5}{64x^2 - 121}, \dfrac{1}{8x^2 - 27x + 22}$

Factor denominators:

$64x^2 - 121 = (8x + 11)(8x - 11)$

$8x^2 - 27x + 22 = (8x - 11)(x - 2)$

LCD is $(8x + 11)(8x - 11)(x - 2)$.

3. a. The entire numerator was not subtracted.

b. $\dfrac{x^2 - 4x}{(x+3)(x-2)} - \dfrac{x^2 + x - 2}{(x+3)(x-2)}$

$= \dfrac{x^2 - 4x - (x^2 + x - 2)}{(x+3)(x-2)}$

$= \dfrac{x^2 - 4x - x^2 - x + 2}{(x+3)(x-2)}$

5. $\dfrac{5x-6}{x-2} + \dfrac{2x-5}{x-2} = \dfrac{5x-6+2x-5}{x-2} = \dfrac{7x-11}{x-2}$

7. $\dfrac{x^2 - 2}{x^2 + 6x - 7} - \dfrac{-4x + 19}{x^2 + 6x - 7}$

$= \dfrac{x^2 - 2 - (-4x + 19)}{x^2 + 6x - 7}$

$= \dfrac{x^2 - 2 + 4x - 19}{x^2 + 6x - 7}$

$= \dfrac{x^2 + 4x - 21}{x^2 + 6x - 7}$

$= \dfrac{(x - 3)(x + 7)}{(x - 1)(x + 7)}$

$= \dfrac{x - 3}{x - 1}$

9. $\dfrac{3r^2 + 15r}{r^3 + 2r^2 - 8r} + \dfrac{2r^2 + 5r}{r^3 + 2r^2 - 8r}$

$= \dfrac{3r^2 + 15r + 2r^2 + 5r}{r^3 + 2r^2 - 8r}$

$= \dfrac{5r^2 + 20r}{r^3 + 2r^2 - 8r}$

$= \dfrac{5r(r + 4)}{r(r + 4)(r - 2)}$

$= \dfrac{5}{r - 2}$

11. $\dfrac{3x^2 - x}{2x^2 - x - 21} + \dfrac{3x - 8}{2x^2 - x - 21} - \dfrac{x^2 - x + 27}{2x^2 - x - 21}$

$= \dfrac{3x^2 - x + (3x - 8) - (x^2 - x + 27)}{2x^2 - x - 21}$

$= \dfrac{3x^2 - x + 3x - 8 - x^2 + x - 27}{2x^2 - x - 21}$

$= \dfrac{2x^2 + 3x - 35}{2x^2 - x - 21}$

$= \dfrac{(2x - 7)(x + 5)}{(2x - 7)(x + 3)}$

$= \dfrac{x + 5}{x + 3}$

13. $\dfrac{4x}{x + 3} + \dfrac{6}{x + 2}$

The LCD is $(x + 3)(x + 2)$

15. $\dfrac{x+3}{16x^2y} - \dfrac{x^2}{3x^3}$

   Factor denominators: $16x^2y = 2^4 \cdot x^2 \cdot y$
                                            $3x^3 = 3 \cdot x^3$
   The LCD is $2^4 \cdot 3 \cdot x^3 \cdot y = 48x^3y$

17. $\dfrac{5z^2}{1} + \dfrac{9z}{z-4}$
   The LCD is $1(z-4) = z-4$

19. $\dfrac{a-2}{a^2-5a-24} + \dfrac{3}{a^2+11a+24}$

   Factor denominators: $a^2 - 5a - 24 = (a-8)(a+3)$
                                     $a^2 + 11a + 24 = (a+3)(a+8)$
   The LCD is $(a-8)(a+3)(a+8)$

21. $\dfrac{3}{x^2+3x-4} - \dfrac{4}{4x^2+5x-9} + \dfrac{x+2}{4x^2+25x+36}$
   Factor denominators:
$$x^2 + 3x - 4 = (x+4)(x-1)$$
$$4x^2 + 5x - 9 = (4x+9)(x-1)$$
$$4x^2 + 25x + 36 = (4x+9)(x+4)$$
   The LCD is $(x+4)(x-1)(4x+9)$

23. $\dfrac{2}{3x} + \dfrac{2}{x} = \dfrac{2}{3x} + \dfrac{2}{x} \cdot \dfrac{3}{3} \leftarrow$ The LCD is $3x$

      $= \dfrac{2}{3x} + \dfrac{6}{3x}$

      $= \dfrac{2+6}{3x}$

      $= \dfrac{8}{3x}$

25. $\dfrac{5}{6y} + \dfrac{3}{4y^2} = \dfrac{5}{6y} \cdot \dfrac{2y}{2y} + \dfrac{3}{4y^2} \cdot \dfrac{3}{3}$ $\leftarrow$ The LCD is $3 \cdot 2^2 \cdot y^2 = 12y^2$

        $= \dfrac{10y}{12y^2} + \dfrac{9}{12y^2}$

        $= \dfrac{10y+9}{12y^2}$

27. $\dfrac{5}{8x^4 y} - \dfrac{1}{5x^2 y^3} = \dfrac{5}{8x^4 y} \cdot \dfrac{5y^2}{5y^2} - \dfrac{1}{5x^2 y^3} \cdot \dfrac{8x^2}{8x^2}$ $\leftarrow$ The LCD is $40x^4 y^3$

$\qquad\qquad = \dfrac{25y^2}{40x^4 y^3} - \dfrac{8x^2}{40x^4 y^3}$

$\qquad\qquad = \dfrac{25y^2 - 8x^2}{40x^4 y^3}$

29. $\dfrac{4x}{3xy} + 2 = \dfrac{4}{3y} + \dfrac{2}{1}$ $\leftarrow$ Simplify first fraction

$\qquad\qquad = \dfrac{4}{3y} + \dfrac{2}{1} \cdot \dfrac{3y}{3y}$ $\leftarrow$ The LCD is $3y$

$\qquad\qquad = \dfrac{4}{3y} + \dfrac{6y}{3y} = \dfrac{4 + 6y}{3y}$

31. $\dfrac{2a}{a-b} - \dfrac{a}{b-a} = \dfrac{2a}{a-b} - \dfrac{a}{b-a} \cdot \dfrac{(-1)}{(-1)}$ $\leftarrow$ The LCD is $a - b$

$\qquad\qquad = \dfrac{2a}{a-b} - \dfrac{-a}{a-b}$

$\qquad\qquad = \dfrac{3a}{a-b}$

33. $\dfrac{x}{x^2 - 9} - \dfrac{4(x-3)}{x+3} = \dfrac{x}{(x+3)(x-3)} - \dfrac{4(x-3)}{x+3}$ $\leftarrow$ Factor denominators

$\qquad\qquad = \dfrac{x}{(x+3)(x-3)} - \dfrac{4(x-3)}{x+3} \cdot \dfrac{x-3}{x-3}$ $\leftarrow$ The LCD is $(x+3)(x-3)$

$\qquad\qquad = \dfrac{x}{(x+3)(x-3)} - \dfrac{4(x-3)^2}{(x+3)(x-3)}$

$\qquad\qquad = \dfrac{x}{(x+3)(x-3)} - \dfrac{4x^2 - 24x + 36}{(x+3)(x-3)}$

$\qquad\qquad = \dfrac{x - (4x^2 - 24x + 36)}{(x+3)(x-3)}$

$\qquad\qquad = \dfrac{x - 4x^2 + 24x - 36}{(x+3)(x-3)}$

$\qquad\qquad = \dfrac{-4x^2 + 25x - 36}{(x+3)(x-3)}$

35.

$$\frac{2m+1}{m-5} - \frac{4}{m^2 - 3m - 10} = \frac{2m+1}{m-5} - \frac{4}{(m-5)(m+2)}$$

$\leftarrow$ Factor denominators

$$= \frac{2m+1}{m-5} \cdot \frac{m+2}{m+2} - \frac{4}{(m-5)(m+2)}$$

$\leftarrow$ The LCD is $(m-5)(m+2)$

$$= \frac{(2m+1)(m+2)}{(m-5)(m+2)} - \frac{4}{(m-5)(m+2)}$$

$$= \frac{2m^2 + 5m + 2}{(m-5)(m+2)} - \frac{4}{(m-5)(m+2)}$$

$$= \frac{2m^2 + 5m + 2 - 4}{(m-5)(m+2)}$$

$$= \frac{2m^2 + 5m - 2}{(m-5)(m+2)}$$

37.

$$\frac{-x^2 + 5x}{(x-5)^2} + \frac{x+1}{x-5} = \frac{-x(x-5)}{(x-5)(x-5)} + \frac{x+1}{x-5}$$

$\leftarrow$ Factor first fraction and cancel common factor

$$= \frac{-x}{x-5} + \frac{x+1}{x-5}$$

$\leftarrow$ The LCD is now $x - 5$

$$= \frac{-x + x + 1}{x-5}$$

$$= \frac{1}{x-5}$$

39.

$$\frac{4}{(2p-3)(p+4)} - \frac{3}{(p+4)(p-4)}$$

$$= \frac{4}{(2p-3)(p+4)} \cdot \frac{p-4}{p-4} - \frac{3}{(p+4)(p-4)} \cdot \frac{2p-3}{2p-3}$$

$\leftarrow$ The LCD is $(2p-3)(p+4)(p-4)$

$$= \frac{4(p-4)}{(2p-3)(p+4)(p-4)} - \frac{3(2p-3)}{(2p-3)(p+4)(p-4)}$$

$$= \frac{4p-16}{(2p-3)(p+4)(p-4)} - \frac{6p-9}{(2p-3)(p+4)(p-4)}$$

$$= \frac{4p-16-(6p-9)}{(2p-3)(p+4)(p-4)}$$

$$= \frac{-2p-7}{(2p-3)(p+4)(p-4)}$$

41. $5 - \dfrac{x-1}{x^2+3x-10} = \dfrac{5}{1} \cdot \dfrac{x^2+3x-10}{x^2+3x-10} - \dfrac{x-1}{x^2+3x-10}$ $\leftarrow$ The LCD is $x^2+3x-10$ or $(x+5)(x-2)$

$\qquad = \dfrac{5x^2+15x-50}{x^2+3x-10} - \dfrac{x-1}{x^2+3x-10}$

$\qquad = \dfrac{5x^2+15x-50-(x-1)}{x^2+3x-10}$

$\qquad = \dfrac{5x^2+14x-49}{x^2+3x-10}$

$\qquad = \dfrac{5x^2+14x-49}{(x+5)(x-2)}$

43. $\dfrac{3a-4}{4a+1} + \dfrac{3a+6}{4a^2+9a+2} = \dfrac{3a-4}{4a+1} + \dfrac{3(a+2)}{(4a+1)(a+2)}$ $\leftarrow$ Factor the second fraction and simplify.

$\qquad = \dfrac{3a-4}{4a+1} + \dfrac{3}{4a+1}$

$\qquad = \dfrac{3a-4+3}{4a+1}$

$\qquad = \dfrac{3a-1}{4a+1}$

45. $\dfrac{x-y}{x^2-4xy+4y^2} + \dfrac{x-3y}{x^2-4y^2}$

$\qquad = \dfrac{x-y}{(x-2y)(x-2y)} + \dfrac{x-3y}{(x-2y)(x+2y)}$ $\leftarrow$ Factor denominators

$\qquad = \dfrac{x-y}{(x-2y)(x-2y)} \cdot \dfrac{x+2y}{x+2y} + \dfrac{x-3y}{(x-2y)(x+2y)} \cdot \dfrac{x-2y}{x-2y}$ $\leftarrow$ The LCD is $(x-2y)(x-2y)(x+2y)$

$\qquad = \dfrac{(x-y)(x+2y)}{(x-2y)^2(x+2y)} + \dfrac{(x-3y)(x-2y)}{(x-2y)^2(x+2y)}$

$\qquad = \dfrac{x^2+xy-2y^2}{(x-2y)^2(x+2y)} + \dfrac{x^2-5xy+6y^2}{(x-2y)^2(x+2y)}$

$\qquad = \dfrac{x^2+xy-2y^2+x^2-5xy+6y^2}{(x-2y)^2(x+2y)}$

$\qquad = \dfrac{2x^2-4xy+4y^2}{(x-2y)^2(x+2y)}$

47. $\dfrac{2r}{r-4} - \dfrac{2r}{r+4} + \dfrac{64}{r^2-16} = \dfrac{2r}{r-4} - \dfrac{2r}{r+4} + \dfrac{64}{(r+4)(r-4)}$ ← Factor third denominator

$$= \frac{2r}{r-4} \cdot \frac{r+4}{r+4} - \frac{2r}{r+4} \cdot \frac{r-4}{r-4} + \frac{64}{(r+4)(r-4)} \quad \leftarrow \text{The LCD is } (r+4)(r-4)$$

$$= \frac{2r(r+4)}{(r+4)(r-4)} - \frac{2r(r-4)}{(r+4)(r-4)} + \frac{64}{(r+4)(r-4)}$$

$$= \frac{2r^2+8r}{(r+4)(r-4)} - \frac{2r^2-8r}{(r+4)(r-4)} + \frac{64}{(r+4)(r-4)}$$

$$= \frac{2r^2+8r-(2r^2-8r)+64}{(r+4)(r-4)}$$

$$= \frac{16r+64}{(r+4)(r-4)}$$

$$= \frac{16(r+4)}{(r+4)(r-4)}$$

$$= \frac{16}{r-4}$$

49. $\dfrac{x^2+2}{x^2-x-2} + \dfrac{1}{x+1} - \dfrac{x}{x-2} = \dfrac{x^2+2}{(x+1)(x-2)} + \dfrac{1}{x+1} - \dfrac{x}{x-2}$ ← Factor first denominator

$$= \frac{x^2+2}{(x+1)(x-2)} + \frac{1}{x+1} \cdot \frac{x-2}{x-2} - \frac{x}{x-2} \cdot \frac{x+1}{x+1} \quad \leftarrow \text{The LCD is } (x+1)(x-2)$$

$$= \frac{x^2+2}{(x+1)(x-2)} + \frac{x-2}{(x+1)(x-2)} - \frac{x^2+x}{(x+1)(x-2)}$$

$$= \frac{x^2+2+x-2-(x^2+x)}{(x+1)(x-2)} = \frac{x^2+2+x-2-x^2-x}{(x+1)(x-2)}$$

$$= \frac{0}{(x+1)(x-2)}$$

$$= 0$$

51. $\dfrac{x}{3x+4} + \dfrac{3x+2}{x-5} - \dfrac{7x^2+24x+28}{3x^2-11x-20}$

$= \dfrac{x}{3x+4} + \dfrac{3x+2}{x-5} - \dfrac{7x^2+24x+28}{(x-5)(3x+4)}$ ← Factor third denominator

$= \dfrac{x}{3x+4} \cdot \dfrac{x-5}{x-5} + \dfrac{3x+2}{x-5} \cdot \dfrac{3x+4}{3x+4} - \dfrac{7x^2+24x+28}{(x-5)(3x+4)}$ ← The LCD is $(x-5)(3x+4)$

$= \dfrac{x(x-5)}{(x-5)(3x+4)} + \dfrac{(3x+2)(3x+4)}{(x-5)(3x+4)} - \dfrac{7x^2+24x+28}{(x-5)(3x+4)}$

$= \dfrac{x^2-5x}{(x-5)(3x+4)} + \dfrac{9x^2+18x+8}{(x-5)(3x+4)} - \dfrac{7x^2+24x+28}{(x-5)(3x+4)}$

$= \dfrac{x^2-5x+9x^2+18x+8-(7x^2+24x+28)}{(x-5)(3x+4)}$

$= \dfrac{3x^2-11x-20}{(x-5)(3x+4)}$

$= \dfrac{(x-5)(3x+4)}{(x-5)(3x+4)}$

$= 1$

53. $\dfrac{x}{x^2-10x+24} - \dfrac{3}{x-6} + 1$

$= \dfrac{x}{(x-6)(x-4)} - \dfrac{3}{x-6} + \dfrac{1}{1}$ ← Factor denominators

$= \dfrac{x}{(x-6)(x-4)} - \dfrac{3}{x-6} \cdot \dfrac{x-4}{x-4} + \dfrac{1}{1} \cdot \dfrac{(x-6)(x-4)}{(x-6)(x-4)}$ ← The LCD is $(x-6)(x-4)$

$= \dfrac{x}{(x-6)(x-4)} - \dfrac{3(x-4)}{(x-6)(x-4)} + \dfrac{1(x-6)(x-4)}{(x-6)(x-4)}$

$= \dfrac{x}{(x-6)(x-4)} - \dfrac{3x-12}{(x-6)(x-4)} + \dfrac{x^2-10x+24}{(x-6)(x-4)}$

$= \dfrac{x-(3x-12)+x^2-10x+24}{(x-6)(x-4)}$

$= \dfrac{x-3x+12+x^2-10x+24}{(x-6)(x-4)}$

$= \dfrac{x^2-12x+36}{(x-6)(x-4)}$

$= \dfrac{(x-6)(x-6)}{(x-6)(x-4)}$

$= \dfrac{x-6}{x-4}$

55.
$$\frac{3}{5x+6}+\frac{x^2-x}{5x^2-4x-12}-\frac{4}{x-2}$$

$$=\frac{3}{5x+6}+\frac{x^2-x}{(5x+6)(x-2)}-\frac{4}{x-2} \qquad \leftarrow \text{Factor second denominator}$$

$$=\frac{3}{5x+6}\cdot\frac{x-2}{x-2}+\frac{x^2-x}{(5x+6)(x-2)}-\frac{4}{x-2}\cdot\frac{5x+6}{5x+6} \qquad \leftarrow \text{The LCD is } (5x+6)(x-2)$$

$$=\frac{3(x-2)}{(5x+6)(x-2)}+\frac{x^2-x}{(5x+6)(x-2)}-\frac{4(5x+6)}{(5x+6)(x-2)}$$

$$=\frac{3x-6}{(5x+6)(x-2)}+\frac{x^2-x}{(5x+6)(x-2)}-\frac{20x+24}{(5x+6)(x-2)}$$

$$=\frac{3x-6+x^2-x-(20x+24)}{(5x+6)(x-2)}$$

$$=\frac{3x-6+x^2-x-20x-24}{(5x+6)(x-2)}$$

$$=\frac{x^2-18x-30}{(5x+6)(x-2)}$$

57.
$$\frac{m}{6m^2+13mn+6n^2}+\frac{2m}{4m^2+8mn+3n^2}$$

$$=\frac{m}{(2m+3n)(3m+2n)}+\frac{2m}{(2m+3n)(2m+n)} \qquad \leftarrow \text{Factor denominators}$$

$$=\frac{m}{(2m+3n)(3m+2n)}\cdot\frac{2m+n}{2m+n}+\frac{2m}{(2m+3n)(2m+n)}\cdot\frac{3m+2n}{3m+2n} \qquad \leftarrow \text{The LCD is } (2m+3n)(3m+2n)(2m+n)$$

$$=\frac{m(2m+n)}{(2m+3n)(3m+2n)(2m+n)}+\frac{2m(3m+2n)}{(2m+3n)(3m+2n)(2m+n)}$$

$$=\frac{2m^2+mn}{(2m+3n)(3m+2n)(2m+n)}+\frac{6m^2+4mn}{(2m+3n)(3m+2n)(2m+n)}$$

$$=\frac{2m^2+mn+6m^2+4mn}{(2m+3n)(3m+2n)(2m+n)}$$

$$=\frac{8m^2+5mn}{(2m+3n)(3m+2n)(2m+n)}$$

59. $\dfrac{5r-2s}{25r^2-4s^2} - \dfrac{2r-s}{10r^2-rs-2s^2} = \dfrac{5r-2s}{(5r+2s)(5r-2s)} - \dfrac{2r-s}{(5r+2s)(2r-s)}$ ← Factor denominators and simplify

$$= \frac{1}{5r+2s} - \frac{1}{5r+2s}$$

$$= \frac{1-1}{5r+2s}$$

$$= \frac{0}{5r+2s}$$

$$= 0$$

61. $\dfrac{2}{2x+3y} - \dfrac{4x^2-6xy+9y^2}{8x^3+27y^3}$ ← Factor second denominator and simplify

$$= \frac{2}{2x+3y} - \frac{4x^2-6xy+9y^2}{(2x+3y)(4x^2-6xy+9y^2)}$$

$$= \frac{2}{2x+3y} - \frac{1}{2x+3y}$$

$$= \frac{2-1}{2x+3y}$$

$$= \frac{1}{2x+3y}$$

63. No; they should be added first, then factored.

65. Yes, factor −1 from the numerator and the denominator.

67. a. $f(x) = \dfrac{x+2}{x-3}$

The denominator cannot equal zero. Therefore, $x \neq 3$. The domain is $\{x | x$ is a real number, $x \neq 3\}$

b. $g(x) = \dfrac{x}{x+4}$

The denominator cannot equal zero. Therefore, $x \neq -4$. The domain is $\{x | x$ is a real number, $x \neq -4\}$

c. $(f+g)(x) = f(x) + g(x)$

$$= \frac{x+2}{x-3} + \frac{x}{x+4}$$

$$= \frac{x+2}{x-3} \cdot \frac{x+4}{x+4} + \frac{x}{x+4} \cdot \frac{x-3}{x-3}$$

$$= \frac{(x+2)(x+4) + x(x-3)}{(x-3)(x+4)}$$

$$= \frac{x^2+6x+8+x^2-3x}{(x-3)(x+4)}$$

$$= \frac{2x^2+3x+8}{(x-3)(x+4)}$$

d. The denominator cannot equal zero. Therefore, $x \neq 3$ and $x \neq -4$. The domain is $\{x | x$ is a real number, $x \neq 3, x \neq -4\}$

69. The domain is $\{x | x \text{ is a real number}, x \neq 2\}$
The range is $\{y | y \text{ is a real number}, y \neq 1\}$

71. $(f + g)(x) = f(x) + g(x)$

$$= \frac{x}{x^2 - 4} + \frac{3}{x^2 + x - 6}$$

$$= \frac{x}{(x-2)(x+2)} + \frac{3}{(x-2)(x+3)}$$

$$= \frac{x}{(x-2)(x+2)} \cdot \frac{(x+3)}{(x+3)} + \frac{3}{(x-2)(x+3)} \cdot \frac{(x+2)}{(x+2)}$$

$$= \frac{x^2 + 3x + 3x + 6}{(x-2)(x+2)(x+3)}$$

$$= \frac{x^2 + 6x + 6}{(x-2)(x+2)(x+3)}$$

73. $(f \cdot g)(x) = f(x) \cdot g(x)$

$$= \frac{x}{x^2 - 4} \cdot \frac{3}{x^2 + x - 6}$$

$$= \frac{3x}{x^4 + x^3 - 6x^2 - 4x^2 - 4x + 24}$$

$$= \frac{3x}{x^4 + x^3 - 10x^2 - 4x + 24}$$

75. $\dfrac{a}{b} + \dfrac{c}{d} = \dfrac{a}{b} \cdot \dfrac{d}{d} + \dfrac{c}{d} \cdot \dfrac{b}{b}$

$$= \frac{ad}{bd} + \frac{cb}{db}$$

$$= \frac{ad + cb}{bd} \text{ or } \frac{ad + bc}{bd}$$

77. a.   Perimeter is

$$\frac{a+b}{a} + \frac{a+b}{a} + \frac{a-b}{a} + \frac{a-b}{a}$$

$$= \frac{a+b+a+b+a-b+a-b}{a}$$

$$= \frac{4a}{a}$$

$$= 4$$

b.   Area is $\left(\dfrac{a+b}{a}\right)\left(\dfrac{a-b}{a}\right) = \dfrac{(a+b)(a-b)}{a \cdot a} = \dfrac{a^2 - b^2}{a^2}$

79. Let $ax^2 + bx + c$ denote the missing numerator.

$$\frac{5x^2 - 6}{x^2 - x - 1} - \frac{ax^2 + bx + c}{x^2 - x - 1} = \frac{-2x^2 + 6x - 12}{x^2 - x - 1}$$

Since the denominators are the same, the fractions on the left side can be subtracted.

$$\frac{5x^2 - 6 - (ax^2 + bx + c)}{x^2 - x - 1} = \frac{-2x^2 + 6x - 12}{x^2 - x - 1}$$

Since the denominators are equal, the numerators must be equal for the fractions to be the same.

$$5x^2 - 6 - ax^2 - bx - c = -2x^2 + 6x - 12$$
$$(5 - a)x^2 + (-b)x + (-6 - c) = -2x^2 + 6x - 12$$

Thus,

$$\begin{array}{lll} 5 - a = -2 & -b = 6 & -6 - c = -12 \\ -a = -7 & b = -6 & -c = -6 \\ a = 7 & & c = 6 \end{array}$$

The missing numerator is $7x^2 - 6x + 6$.

81. $\left(3 + \dfrac{1}{x+3}\right)\left(\dfrac{x+3}{x-2}\right)$

$$= \left(\frac{3}{1} \cdot \frac{x+3}{x+3} + \frac{1}{x+3}\right)\left(\frac{x+3}{x-2}\right)$$

$$= \left(\frac{3x + 9 + 1}{x+3}\right)\left(\frac{x+3}{x-2}\right)$$

$$= \left(\frac{3x + 10}{x+3}\right)\left(\frac{x+3}{x-2}\right)$$

$$= \frac{3x + 10}{x-2}$$

83. $\left(\dfrac{5}{a-5} - \dfrac{2}{a+3}\right) \div (3a + 25)$

$$= \left(\frac{5}{a-5} - \frac{2}{a+3}\right) \cdot \frac{1}{3a+25}$$

$$= \left(\frac{5}{a-5} \cdot \frac{a+3}{a+3} - \frac{2}{a+3} \cdot \frac{a-5}{a-5}\right) \cdot \frac{1}{3a+25}$$

$$= \left(\frac{5a + 15}{(a-5)(a+3)} - \frac{2a - 10}{(a-5)(a+3)}\right) \cdot \frac{1}{3a+25}$$

$$= \left(\frac{5a + 15 - 2a + 10}{(a-5)(a+3)}\right) \cdot \frac{1}{3a+25}$$

$$= \frac{3a + 25}{(a-5)(a+3)} \cdot \frac{1}{3a+25}$$

$$= \frac{1}{(a-5)(a+3)}$$

85. $\left(\dfrac{x+5}{x-3}-x\right)\div\dfrac{1}{x-3}$

$=\left(\dfrac{x+5}{x-3}-x\right)(x-3)$

$=\dfrac{x+5}{x-3}\cdot(x-3)-x(x-3)$

$=x+5-x^2+3x$

$=-x^2+4x+5$

87. a. $a\left(\dfrac{x}{n}\right)+b\left(\dfrac{n-x}{n}\right)=\dfrac{ax}{n}+\dfrac{bn-bx}{n}$

$\qquad\qquad\qquad\qquad =\dfrac{ax+bn-bx}{n}$

b. $60\left(\dfrac{2}{5}\right)+92\left(\dfrac{3}{5}\right)=\dfrac{120}{5}+\dfrac{276}{5}$

$\qquad\qquad\qquad\qquad =\dfrac{396}{5}$

$\qquad\qquad\qquad\qquad =79.2$

89. $(a-b)^{-1}+(a-b)^{-2}$

$=\dfrac{1}{a-b}+\dfrac{1}{(a-b)^2}$

$=\dfrac{1}{a-b}\cdot\dfrac{a-b}{a-b}+\dfrac{1}{(a-b)^2}$

$=\dfrac{a-b}{(a-b)^2}+\dfrac{1}{(a-b)^2}$

$=\dfrac{a-b+1}{(a-b)^2}$

91. $y_1=\dfrac{x-3}{x+4}+\dfrac{x}{x^2-2x-24}$

$y_2=\dfrac{x^2-10x+18}{(x+4)(x-6)}$

$-10,10,1,-10,10,1$

No, the addition is not correct because the graphs are not the same.

93. a. $1+\dfrac{1}{x}=\dfrac{x}{x}+\dfrac{1}{x}=\dfrac{x+1}{x}$

b. $1+\dfrac{1}{x}+\dfrac{1}{x^2}=1\cdot\dfrac{x^2}{x^2}+\dfrac{1}{x}\cdot\dfrac{x}{x}+\dfrac{1}{x^2}$

$\qquad\qquad\qquad\quad =\dfrac{x^2}{x^2}+\dfrac{x}{x^2}+\dfrac{1}{x^2}$

$\qquad\qquad\qquad\quad =\dfrac{x^2+x+1}{x^2}$

c. $1+\dfrac{1}{x}+\dfrac{1}{x^2}+\dfrac{1}{x^3}+\dfrac{1}{x^4}+\dfrac{1}{x^5}$

$\qquad =\dfrac{x^5+x^4+x^3+x^2+x+1}{x^5}$

d. $1+\dfrac{1}{x}+\dfrac{1}{x^2}+\cdots+\dfrac{1}{x^n}$

$\qquad =\dfrac{x^n+x^{n-1}+x^{n-2}+\cdots+1}{x^n}$

94. Let x be the original price of the suit.

$\qquad x+0.20x-20=196$

$\qquad\quad 1.20x-20=196$

$\qquad 1.20x-20+20=196+20$

$\qquad\qquad\quad 1.20x=216$

$\qquad\qquad\dfrac{1.20x}{1.20}=\dfrac{216}{1.20}$

$\qquad\qquad\qquad x=180$

The original price was $180.

95. Let t be the time at the faster speed. Then 14 $-\,t$ is the time for the slower speed.

	rate	time	amount
fast	80	t	$80t$
slow	60	$14-t$	$60(14-t)$

The two amounts are the same.

$\qquad 80t=60(14-t)$

$\qquad 80t=840-60t$

$\quad 140t=840$

$\qquad\quad t=\dfrac{840}{140}=6$

a. The machine was used for 6 minutes at the faster setting.

b. While the machine was on the faster setting, $80(t) = 80(6) = 480$ bottles were filled. Since equal amounts were filled at each speed, the total number filled was $2(480) = 960$ bottles.

96. $\dfrac{9x^4y^6 - 3x^3y^2 + 5xy^5}{3xy^4}$

$= \dfrac{9x^4y^6}{3xy^4} - \dfrac{3x^3y^2}{3xy^4} + \dfrac{5xy^5}{3xy^4}$

$= 3x^3y^2 - \dfrac{x^2}{y^2} + \dfrac{5y}{3}$

97.

$$\begin{array}{r} 2x-1 \\ 3x+4\overline{\smash{\big)}\,6x^2+5x-4} \end{array}$$

$\quad\quad\quad \underline{6x^2 + 8x} \quad \leftarrow 2x(3x+4)$

$\quad\quad\quad\quad -3x - 4$

$\quad\quad\quad\quad \underline{-3x - 4} \leftarrow -1(3x+4)$

$\quad\quad\quad\quad\quad\quad 0 \leftarrow$ Remainder

$\dfrac{6x^2 + 5x - 4}{3x+4} = 2x - 1$

Exercise Set 6.3

1. A complex fraction is one that has a fractional expression in the numerator or the denominator or both the numerator and the denominator.

3. $\dfrac{1 - \frac{x}{y}}{x} = \dfrac{y\left(1 - \frac{x}{y}\right)}{y(x)} = \dfrac{y(1) - y\left(\frac{x}{y}\right)}{xy} = \dfrac{y - x}{xy}$

5. $\dfrac{\frac{15a}{b^2}}{\frac{b^3}{5}} = \dfrac{5b^2\left(\frac{15a}{b^2}\right)}{5b^2\left(\frac{b^3}{5}\right)} = \dfrac{75a}{b^5}$

or $\dfrac{\frac{15a}{b^2}}{\frac{b^3}{5}} = \dfrac{15a}{b^2} \div \dfrac{b^3}{5} = \dfrac{15a}{b^2} \cdot \dfrac{5}{b^3} = \dfrac{75a}{b^5}$

7. $\dfrac{\frac{36x^4}{5y^4z^5}}{\frac{9xy^2}{15z^5}} = \dfrac{15y^4z^5\left(\frac{36x^4}{5y^4z^5}\right)}{15y^4z^5\left(\frac{9xy^2}{15z^5}\right)} = \dfrac{108x^4}{9xy^6} = \dfrac{12x^3}{y^6}$

or

$\dfrac{\frac{36x^4}{5y^4z^5}}{\frac{9xy^2}{15z^5}} = \dfrac{36x^4}{5y^4z^5} \div \dfrac{9xy^2}{15z^5}$

$\quad\quad = \dfrac{36x^4}{5y^4z^5} \cdot \dfrac{15z^5}{9xy^2}$

$\quad\quad = \dfrac{12x^3}{y^6}$

9. $\dfrac{x - \frac{x}{y}}{\frac{1+x}{y}} = \dfrac{y\left(x - \frac{x}{y}\right)}{y\left(\frac{1+x}{y}\right)}$

$\quad = \dfrac{y(x) - y\left(\frac{x}{y}\right)}{y\left(\frac{1+x}{y}\right)}$

$\quad = \dfrac{xy - x}{1 + x}$ or $\dfrac{x(y-1)}{1+x}$

11. $\dfrac{\frac{2}{a} + \frac{1}{2a}}{a + \frac{a}{2}} = \dfrac{2a\left(\frac{2}{a} + \frac{1}{2a}\right)}{2a\left(a + \frac{a}{2}\right)}$

$\quad = \dfrac{2a\left(\frac{2}{a}\right) + 2a\left(\frac{1}{2a}\right)}{2a(a) + 2a\left(\frac{a}{2}\right)}$

$\quad = \dfrac{4 + 1}{2a^2 + a^2}$

$\quad = \dfrac{5}{3a^2}$

13. $\dfrac{\frac{x}{y}-\frac{y}{x}}{\frac{x+y}{x}}=\dfrac{xy\left(\frac{x}{y}-\frac{y}{x}\right)}{xy\left(\frac{x+y}{x}\right)}$

$=\dfrac{xy\left(\frac{x}{y}\right)-xy\left(\frac{y}{x}\right)}{xy\left(\frac{x+y}{x}\right)}$

$=\dfrac{x^2-y^2}{y(x+y)}$

$=\dfrac{(x+y)(x-y)}{y(x+y)}$

$=\dfrac{x-y}{y}$

15. $\dfrac{\frac{1}{m}+\frac{2}{m^2}}{2+\frac{1}{m^2}}=\dfrac{m^2\left(\frac{1}{m}+\frac{2}{m^2}\right)}{m^2\left(2+\frac{1}{m^2}\right)}$

$=\dfrac{m^2\left(\frac{1}{m}\right)+m^2\left(\frac{2}{m^2}\right)}{m^2(2)+m^2\left(\frac{1}{m^2}\right)}$

$=\dfrac{m+2}{2m^2+1}$

17. $\dfrac{\frac{x^2-y^2}{x}}{\frac{x+y}{x^3}}=\dfrac{x^3\left(\frac{x^2-y^2}{x}\right)}{x^3\left(\frac{x+y}{x^3}\right)}$

$=\dfrac{x^2(x^2-y^2)}{x+y}$

$=\dfrac{x^2(x+y)(x-y)}{x+y}$

$=x^2(x-y)$

19. $\dfrac{\frac{a}{a+1}-1}{\frac{2a+1}{a-1}}$

$=\dfrac{(a-1)(a+1)\left(\frac{a}{a+1}-1\right)}{(a-1)(a+1)\left(\frac{2a+1}{a-1}\right)}$

$=\dfrac{(a-1)(a+1)\left(\frac{a}{a+1}\right)-(a-1)(a+1)(1)}{(a-1)(a+1)\left(\frac{2a+1}{a-1}\right)}$

$=\dfrac{a(a-1)-(a-1)(a+1)}{(a+1)(2a+1)}$

$=\dfrac{(a-1)[a-(a+1)]}{(a+1)(2a+1)}$

$=\dfrac{(a-1)(-1)}{(a+1)(2a+1)}$

$=\dfrac{-a+1}{(a+1)(2a+1)}$

21. $\dfrac{1+\frac{x}{x+1}}{\frac{2x+1}{x-1}}$

$=\dfrac{(x-1)(x+1)\left(1+\frac{x}{x+1}\right)}{(x-1)(x+1)\left(\frac{2x+1}{x-1}\right)}$

$=\dfrac{(x-1)(x+1)(1)+(x-1)(x+1)\left(\frac{x}{x+1}\right)}{(x-1)(x+1)\left(\frac{2x+1}{x-1}\right)}$

$=\dfrac{(x-1)(x+1)+(x-1)(x)}{(x+1)(2x+1)}$

$=\dfrac{(x-1)(x+1+x)}{(x+1)(2x+1)}$

$=\dfrac{(x-1)(2x+1)}{(x+1)(2x+1)}$

$=\dfrac{x-1}{x+1}$

23. $\dfrac{\frac{a+1}{a-1}+\frac{a-1}{a+1}}{\frac{a+1}{a-1}-\frac{a-1}{a+1}} = \dfrac{(a-1)(a+1)\left(\frac{a+1}{a-1}+\frac{a-1}{a+1}\right)}{(a-1)(a+1)\left(\frac{a+1}{a-1}-\frac{a-1}{a+1}\right)}$

$\qquad = \dfrac{(a-1)(a+1)\left(\frac{a+1}{a-1}\right)+(a-1)(a+1)\left(\frac{a-1}{a+1}\right)}{(a-1)(a+1)\left(\frac{a+1}{a-1}\right)-(a-1)(a+1)\left(\frac{a-1}{a+1}\right)}$

$\qquad = \dfrac{(a+1)^2+(a-1)^2}{(a+1)^2-(a-1)^2}$

$\qquad = \dfrac{a^2+2a+1+a^2-2a+1}{a^2+2a+1-(a^2-2a+1)}$

$\qquad = \dfrac{2a^2+2}{4a}$

$\qquad = \dfrac{a^2+1}{2a}$

25. $\dfrac{\frac{5}{5-x}+\frac{6}{x-5}}{\frac{3}{x}+\frac{2}{x-5}} = \dfrac{x(x-5)\left(\frac{5}{5-x}+\frac{6}{x-5}\right)}{x(x-5)\left(\frac{3}{x}+\frac{2}{x-5}\right)}$

$\qquad = \dfrac{x(x-5)\left(\frac{5}{5-x}\right)+x(x-5)\left(\frac{6}{x-5}\right)}{x(x-5)\left(\frac{3}{x}\right)+x(x-5)\left(\frac{2}{x-5}\right)}$

$\qquad = \dfrac{-5x+6x}{3(x-5)+2x}$

$\qquad = \dfrac{x}{3x-15+2x}$

$\qquad = \dfrac{x}{5x-15} \text{ or } \dfrac{x}{5(x-3)}$

27. $\dfrac{\frac{3}{x^2}-\frac{1}{x}+\frac{2}{x-2}}{\frac{1}{x}} = \dfrac{x^2(x-2)\left(\frac{3}{x^2}-\frac{1}{x}+\frac{2}{x-2}\right)}{x^2(x-2)\left(\frac{1}{x}\right)}$

$\qquad = \dfrac{x^2(x-2)\left(\frac{3}{x^2}\right)-x^2(x-2)\left(\frac{1}{x}\right)+x^2(x-2)\left(\frac{2}{x-2}\right)}{x^2(x-2)\left(\frac{1}{x}\right)}$

$\qquad = \dfrac{3(x-2)-x(x-2)+2x^2}{x(x-2)}$

$\qquad = \dfrac{3x-6-x^2+2x+2x^2}{x(x-2)}$

$\qquad = \dfrac{x^2+5x-6}{x(x-2)}$

29.

$$\frac{\frac{1}{x^2+5x+4}+\frac{2}{x^2+2x-8}}{\frac{2}{x^2-x-2}+\frac{1}{x^2-5x+6}}$$

$$=\frac{\frac{1}{(x+4)(x+1)}+\frac{2}{(x+4)(x-2)}}{\frac{2}{(x-2)(x+1)}+\frac{1}{(x-3)(x-2)}}$$

$$=\frac{\frac{1(x-2)+2(x+1)}{(x+4)(x+1)(x-2)}}{\frac{2(x-3)+1(x+1)}{(x-2)(x+1)(x-3)}}$$

$$=\frac{x-2+2x+2}{(x+4)(x+1)(x-2)}\cdot\frac{(x-2)(x+1)(x-3)}{2x-6+x+1}$$

$$=\frac{3x(x-3)}{(x+4)(3x-5)}$$

31. $3a^{-2}+b=\dfrac{3}{a^2}+\dfrac{b}{1}=\dfrac{3}{a^2}+\dfrac{b}{1}\cdot\dfrac{a^2}{a^2}=\dfrac{3+a^2b}{a^2}$

33. $(a^{-1}+b^{-1})^{-1}=\left(\dfrac{1}{a}+\dfrac{1}{b}\right)^{-1}$

$$=\left(\dfrac{1}{a}\cdot\dfrac{b}{b}+\dfrac{1}{b}\cdot\dfrac{a}{a}\right)^{-1}$$

$$=\left(\dfrac{b}{ab}+\dfrac{a}{ab}\right)^{-1}$$

$$=\left(\dfrac{b+a}{ab}\right)^{-1}$$

$$=\dfrac{ab}{b+a}$$

35. $\dfrac{a^{-1}+1}{b^{-1}-1}=\dfrac{\frac{1}{a}+1}{\frac{1}{b}-1}$

$$=\dfrac{ab\left(\frac{1}{a}+1\right)}{ab\left(\frac{1}{b}-1\right)}$$

$$=\dfrac{ab\left(\frac{1}{a}\right)+ab(1)}{ab\left(\frac{1}{b}\right)-ab(1)}$$

$$=\dfrac{b+ab}{a-ab}\ \text{ or }\ \dfrac{b(1+a)}{a(1-b)}$$

37. $\dfrac{x^{-1}-y^{-1}}{x^{-1}+y^{-1}}=\dfrac{\frac{1}{x}-\frac{1}{y}}{\frac{1}{x}+\frac{1}{y}}$

$$=\dfrac{xy\left(\frac{1}{x}-\frac{1}{y}\right)}{xy\left(\frac{1}{x}+\frac{1}{y}\right)}$$

$$=\dfrac{xy\left(\frac{1}{x}\right)-xy\left(\frac{1}{y}\right)}{xy\left(\frac{1}{x}\right)+xy\left(\frac{1}{y}\right)}$$

$$=\dfrac{y-x}{y+x}$$

39. $\dfrac{a^{-1}+b^{-1}}{(a+b)^{-1}}=\dfrac{\frac{1}{a}+\frac{1}{b}}{\frac{1}{a+b}}=\dfrac{ab(a+b)\left(\frac{1}{a}+\frac{1}{b}\right)}{ab(a+b)\left(\frac{1}{a+b}\right)}$

$$=\dfrac{ab(a+b)\left(\frac{1}{a}\right)+ab(a+b)\left(\frac{1}{b}\right)}{ab(a+b)\left(\frac{1}{a+b}\right)}$$

$$=\dfrac{b(a+b)+a(a+b)}{ab}$$

$$=\dfrac{(a+b)(a+b)}{ab}$$

$$=\dfrac{(a+b)^2}{ab}$$

41. $2x^{-1}-(3y)^{-1}=\dfrac{2}{x}-\dfrac{1}{3y}$

$$=\dfrac{2}{x}\cdot\dfrac{3y}{3y}-\dfrac{1}{3y}\cdot\dfrac{x}{x}$$

$$=\dfrac{6y}{3xy}-\dfrac{x}{3xy}$$

$$=\dfrac{6y-x}{3xy}$$

43. $\dfrac{\frac{2}{xy}-\frac{3}{y}+\frac{5}{x}}{3x^{-1}-2y^{-2}}=\dfrac{\frac{2}{xy}-\frac{3}{y}+\frac{5}{x}}{\frac{3}{x}-\frac{2}{y^2}}$

$$=\dfrac{xy^2\left(\frac{2}{xy}-\frac{3}{y}+\frac{5}{x}\right)}{xy^2\left(\frac{3}{x}-\frac{2}{y^2}\right)}$$

$$=\dfrac{xy^2\left(\frac{2}{xy}\right)-xy^2\left(\frac{3}{y}\right)+xy^2\left(\frac{5}{x}\right)}{xy^2\left(\frac{3}{x}\right)-xy^2\left(\frac{2}{y^2}\right)}$$

$$=\dfrac{2y-3xy+5y^2}{3y^2-2x}$$

45. a.   Substitute $\frac{2}{3}$ for h.

$$E=\dfrac{\frac{1}{2}h}{h+\frac{1}{2}}$$

$$=\dfrac{\frac{1}{2}\left(\frac{2}{3}\right)}{\frac{2}{3}+\frac{1}{2}}$$

$$=\dfrac{\frac{1}{3}}{\frac{2}{3}+\frac{1}{2}}$$

$$=\dfrac{6\left(\frac{1}{3}\right)}{6\left(\frac{2}{3}+\frac{1}{2}\right)}\leftarrow\text{Multiply by LCD of 6}$$

$$=\dfrac{6\left(\frac{1}{3}\right)}{6\left(\frac{2}{3}\right)+6\left(\frac{1}{2}\right)}$$

$$=\dfrac{2}{4+3}$$

$$=\dfrac{2}{7}$$

b. Substitute $\frac{4}{5}$ for h.

$$E = \frac{\frac{1}{2}h}{h + \frac{1}{2}}$$

$$= \frac{\frac{1}{2}\left(\frac{4}{5}\right)}{\frac{4}{5} + \frac{1}{2}}$$

$$= \frac{\frac{2}{5}}{\frac{4}{5} + \frac{1}{2}}$$

$$= \frac{10\left(\frac{2}{5}\right)}{10\left(\frac{4}{5} + \frac{1}{2}\right)} \leftarrow \text{Multiply by LCD of 10}$$

$$= \frac{10\left(\frac{2}{5}\right)}{10\left(\frac{4}{5}\right) + 10\left(\frac{1}{2}\right)}$$

$$= \frac{4}{8 + 5}$$

$$= \frac{4}{13}$$

47. $R_T = \dfrac{R_1 R_2 R_3 (1)}{R_1 R_2 R_3 \left(\frac{1}{R_1} + \frac{1}{R_2} + \frac{1}{R_3}\right)}$

$$= \frac{R_1 R_2 R_3}{R_1 R_2 R_3 \left(\frac{1}{R_1}\right) + R_1 R_2 R_3 \left(\frac{1}{R_2}\right) + R_1 R_2 R_3 \left(\frac{1}{R_3}\right)}$$

$$= \frac{R_1 R_2 R_3}{R_2 R_3 + R_1 R_3 + R_1 R_2}$$

49. $f(x) = \dfrac{1}{x}$

$$f(a) = \frac{1}{a}$$

$$f(f(a)) = \frac{1}{\left(\frac{1}{a}\right)} = a$$

51. a. $f(x) = \dfrac{1}{x}$

$$f(x + h) = \frac{1}{x + h}$$

b. $f(x + h) - f(x) = \dfrac{1}{x + h} - \dfrac{1}{x}$

$$= \frac{x - (x + h)}{x(x + h)}$$

$$= \frac{-h}{x(x + h)}$$

c. $\dfrac{f(-x + h) - f(x)}{h} = \dfrac{-h}{hx(x + h)}$

$$= -\frac{1}{x(x + h)}$$

53. a. $f(x) = \dfrac{2}{x^2}$

$f(x+h) = \dfrac{2}{(x+h)^2}$

b. $f(x+h) - f(x)$

$= \dfrac{2}{(x+h)^2} - \dfrac{2}{x^2}$

$= \dfrac{2x^2 - 2(x+h)^2}{x^2(x+h)^2}$

$= \dfrac{2x^2 - 2x^2 - 4xh - 2h^2}{x^2(x+h)^2}$

$= \dfrac{-4xh - 2h^2}{x^2(x+h)^2}$

c. $\dfrac{f(x+h) - f(x)}{h} = \dfrac{-4xh - 2h^2}{h(x^2)(x+h)^2}$

$= \dfrac{h(-4x - 2h)}{h(x^2)(x+h)^2}$

$= \dfrac{-4x - 2h}{x^2(x+h)^2}$

55. $\dfrac{1}{2a + \dfrac{1}{2a + \frac{1}{2a}}} = \dfrac{1}{2a + \dfrac{1}{\frac{(2a)(2a)+1}{2a}}}$

$= \dfrac{1}{2a + \dfrac{1}{\frac{4a^2+1}{2a}}}$

$= \dfrac{1}{2a + \dfrac{2a}{4a^2+1}}$

$= \dfrac{1}{\dfrac{2a(4a^2+1)+2a}{4a^2+1}}$

$= \dfrac{1}{\dfrac{8a^3+2a+2a}{4a^2+1}}$

$= \dfrac{1}{\dfrac{8a^3+4a}{4a^2+1}}$

$= \dfrac{4a^2+1}{8a^3+4a}$

$= \dfrac{4a^2+1}{4a(2a^2+1)}$

57. $\dfrac{1}{2+\dfrac{1}{1+\frac{1}{2}}} = \dfrac{1}{2+\dfrac{1}{\frac{3}{2}}} = \dfrac{1}{2+\frac{2}{5}} = \dfrac{1}{\frac{12}{5}} = \dfrac{5}{12}$

58. $\dfrac{\left|-\frac{3}{9}\right| - \left(\frac{5}{-9}\right)\left|-\frac{3}{8}\right|}{|-5-(-3)|} = \dfrac{\left|-\frac{3}{9}\right| - \left(-\frac{5}{9}\right)\left|-\frac{3}{8}\right|}{|-2|}$

$= \dfrac{\frac{3}{9} + \frac{5}{9}\left(\frac{3}{8}\right)}{2}$

$= \dfrac{\frac{1}{3} + \frac{5}{24}}{2}$

$= \dfrac{24\left(\frac{1}{3} + \frac{5}{24}\right)}{24(2)}$

$= \dfrac{24\left(\frac{1}{3}\right) + 24\left(\frac{5}{24}\right)}{24(2)}$

$= \dfrac{8+5}{48}$

$= \dfrac{13}{48}$

59. $\left|\dfrac{4-2x}{3}\right| \geq 3$

$\dfrac{4-2x}{3} \leq -3$ or $\dfrac{4-2x}{3} \geq 3$

$4-2x \leq -9$ $\qquad$ $4-2x \geq 9$

$-2x \leq -13$ $\qquad$ $-2x \geq 5$

$x \geq \dfrac{13}{2}$ $\qquad$ $x \leq -\dfrac{5}{2}$

The solution is $\left\{ x \middle| x \leq -\dfrac{5}{2} \text{ or } x \geq \dfrac{13}{2} \right\}$

60. $6y - 3x < 12$

Graph the line $6y - 3x = 12$ using a dashed line. For the check point, select $(0,0)$:

$6y - 3x < 12$

$6(0) - 3(0) < 12$

$0 - 0 < 12$

$0 < 12$ True

Since this is a true statement, shade the region which contains the point $(0, 0)$.

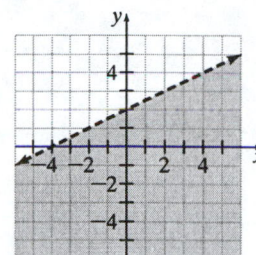

61.

$$
\begin{array}{r|rrrr}
2 & 1 & -7 & -13 & 9 \\
  &   & 2 & -10 & -46 \\
\hline
  & 1 & -5 & -23 & -37 \\
\end{array}
$$

$$\frac{x^3 - 7x^2 - 13x + 9}{x - 2} = x^2 - 5x - 23 - \frac{37}{x - 2}$$

Exercise Set 6.4

1. An extraneous root is a number obtained when solving an equation that is not a true solution to the original equation.

3. **a.** Multiply both sides of the equation by the LCD of 12. This removes fractions.

    b.
$$12\left(\frac{x}{4}\right) - 12\left(\frac{x}{3}\right) = 12(2)$$
$$3x - 4x = 24$$
$$-x = 24$$
$$x = \frac{24}{-1} = -24$$

    c. Write each term with the common denominator of 12. This allows the fractions to be added or subtracted.

13.
$$\frac{x}{3} - \frac{3x}{4} = -\frac{5x}{12}$$
$$12\left(\frac{x}{3}\right) - 12\left(\frac{3x}{4}\right) = 12\left(-\frac{5x}{12}\right) \quad \leftarrow \text{Multiply each term by the LCD of 12}$$
$$4x - 9x = -5x$$
$$-5x = -5x$$
Since this statement is true for all values of x, the solution is all real numbers.

    d.
$$\frac{x}{4} - \frac{x}{3} + 2 = \frac{x}{4} \cdot \frac{3}{3} - \frac{x}{3} \cdot \frac{4}{4} + \frac{2}{1} \cdot \frac{12}{12}$$
$$= \frac{3x}{12} - \frac{4x}{12} + \frac{24}{12}$$
$$= \frac{3x - 4x + 24}{12} = \frac{-x + 24}{12}$$

5. Similar figures are figures whose corresponding angles are the same and whose corresponding sides are in proportion.

7.
$$\frac{18}{3b} = \frac{-6}{2}$$
$$(3b)(-6) = (18)(2) \qquad \leftarrow \text{Cross multiply}$$
$$-18b = 36$$
$$b = \frac{36}{-18} = -2$$
This solution checks. The solution is –2.

9.
$$\frac{1}{4} = \frac{z + 1}{8}$$
$$4(z + 1) = 8 \qquad \leftarrow \text{Cross multiply}$$
$$4z + 4 = 8$$
$$4z = 4$$
$$z = \frac{4}{4} = 1$$
This solution checks. The solution is 1.

11.
$$\frac{6x + 7}{10} = \frac{2x + 9}{6}$$
$$6(6x + 7) = 10(2x + 9) \quad \leftarrow \text{Cross multiply}$$
$$36x + 42 = 20x + 90$$
$$16x + 42 = 90$$
$$16x = 48$$
$$x = \frac{48}{16} = 3$$

This solution checks. The solution is 3.

15. $\dfrac{3}{4} - x = 2x$

$$\dfrac{3}{4} = 3x$$

$$\dfrac{3}{4} = \dfrac{3x}{1}$$

$3(1) = 4(3x)$ ← Cross multiply

$$3 = 12x$$

$$\dfrac{3}{12} = x$$

$$\dfrac{1}{4} = x$$

This solution checks. The solution is $\dfrac{1}{4}$.

17. $\dfrac{3}{r} + \dfrac{5}{3r} = 1$

$3r\left(\dfrac{3}{r}\right) + 3r\left(\dfrac{5}{3r}\right) = 3r(1)$ ← Multiply each term by the LCD of $3x$

$$9 + 5 = 3r$$

$$14 = 3r$$

$$\dfrac{14}{3} = r$$

This solution checks. The solution is $\dfrac{14}{3}$.

19. $\dfrac{x-1}{x-5} = \dfrac{4}{x-5}$
Since the denominators are the same, the numerators must be equal.
$x - 1 = 4$
   $x = 5$
This does not check, since both denominators are 0 when $x = 5$. There is no solution.

21. $\dfrac{5y-2}{7} = \dfrac{15y-2}{28}$
$28(5y-2) = 7(15y-2)$ ← Cross multiply
$140y - 56 = 105y - 14$
   $35y = 42$
   $y = \dfrac{42}{35} = \dfrac{6}{5}$

This solution checks. The solution is $\dfrac{6}{5}$.

23. $\dfrac{5.6}{-p-6.2} = \dfrac{2}{p}$
$5.6(p) = 2(-p-6.2)$ ← Cross multiply
$5.6p = -2p - 12.4$
$7.6p = -12.4$
   $p = \dfrac{-12.4}{7.6} \approx -1.63$

This solution checks. The solution is ≈ -1.63.

25.

$$\frac{m+1}{m+10} = \frac{m-2}{m+4}$$

$(m+4)(m+1) = (m-2)(m+10) \quad \leftarrow \text{Cross multiply}$

$$m^2 + 5m + 4 = m^2 + 8m - 20$$

$$5m + 4 = 8m - 20$$

$$4 = 3m - 20$$

$$24 = 3m$$

$$\frac{24}{3} = m$$

$$8 = m$$

This solution checks. The solution is 8.

27.

$$x - \frac{4}{3x} = -\frac{1}{3}$$

$$3x(x) - 3x\left(\frac{4}{3x}\right) = 3x\left(-\frac{1}{3}\right)$$

$$3x^2 - 4 = -x$$

$$3x^2 + x - 4 = 0$$

$$(3x + 4)(x - 1) = 0$$

$$3x + 4 = 0 \quad \text{or} \quad x - 1 = 0$$

$$3x = -4 \qquad\qquad x = 1$$

$$x = -\frac{4}{3}$$

These solutions check. The solutions are $-\dfrac{4}{3}$ and 1.

29.

$$\frac{2x-1}{3} - \frac{x}{4} = \frac{7.4}{6}$$

$$12\left(\frac{2x-1}{3}\right) - 12\left(\frac{x}{4}\right) = 12\left(\frac{7.4}{6}\right) \quad \leftarrow \text{Multiply each term by the LCD of 12.}$$

$$4(2x-1) - 3x = 2(7.4)$$

$$8x - 4 - 3x = 14.8$$

$$5x - 4 = 14.8$$

$$5x = 18.8$$

$$x = \frac{18.8}{5} = 3.76$$

This solution checks. The solution is 3.76.

31.
$$x + \frac{6}{x} = -5$$

$$x(x) + x\left(\frac{6}{x}\right) = x(-5) \quad \leftarrow \text{ Multiply each term by the LCD of } x$$

$$x^2 + 6 = -5x$$

$$x^2 + 5x + 6 = 0$$

$$(x+3)(x+2) = 0$$

$$x + 3 = 0 \quad \text{or} \quad x + 2 = 0$$

$$x = -3 \qquad x = -2$$

These solutions check. The solutions are –3 and –2.

33.
$$2 - \frac{5}{2b} = \frac{2b}{b+1}$$

$$2[2b(b+1)] - 2b(b+1)\left(\frac{5}{2b}\right) = 2b(b+1)\left(\frac{2b}{b+1}\right) \quad \leftarrow \text{ Multiply each term by the LCD of } 2b(b+1)$$

$$4b(b+1) - 5(b+1) = 2b(2b)$$

$$4b^2 + 4b - 5b - 5 = 4b^2$$

$$4b^2 - b - 5 = 4b^2$$

$$-b - 5 = 0$$

$$b = -5$$

This solution checks. The solution is –5.

35.
$$\frac{1}{w-3} + \frac{1}{w+3} = \frac{-5}{w^2-9}$$

$$\frac{1}{w-3} + \frac{1}{w+3} = \frac{-5}{(w+3)(w-3)}$$

$$(w+3)(w-3)\left(\frac{1}{w-3}\right) + (w+3)(w-3)\left(\frac{1}{w+3}\right) = (w+3)(w-3)\left(\frac{-5}{(w+3)(w-3)}\right)$$

$$w + 3 + w - 3 = -5$$

$$2w = -5$$

$$w = \frac{-5}{2} = -\frac{5}{2}$$

This solution checks. The solution is $-\frac{5}{2}$.

37.

$$\frac{8}{x^2-9} = \frac{2}{x-3} - \frac{4}{x+3}$$

$$\frac{8}{(x-3)(x+3)} = \frac{2}{x-3} - \frac{4}{x+3}$$

$$(x-3)(x+3)\left(\frac{8}{(x-3)(x+3)}\right) = (x-3)(x+3)\left(\frac{2}{x-3}\right) - (x-3)(x+3)\left(\frac{4}{x+3}\right)$$

$$8 = 2(x+3) - 4(x-3)$$
$$8 = 2x+6-4x+12$$
$$8 = -2x+18$$
$$-10 = -2x$$
$$x = \frac{-10}{-2} = 5$$

This solution checks. The solution is 5.

39.

$$\frac{y}{2y+2} + \frac{2y-16}{4y+4} = \frac{2y-3}{y+1}$$

$$\frac{y}{2y+2} + \frac{2(y-8)}{4(y+1)} = \frac{2y-3}{y+1}$$

$$\frac{y}{2(y+1)} + \frac{y-8}{2(y+1)} = \frac{2y-3}{y+1}$$

$$\frac{y+y-8}{2(y+1)} = \frac{2y-3}{y+1}$$

$$\frac{2y-8}{2(y+1)} = \frac{2y-3}{y+1}$$

$$\frac{2(y-4)}{2(y+1)} = \frac{2y-3}{y+1}$$

$$\frac{y-4}{y+1} = \frac{2y-3}{y+1}$$

Since the denominators are the same, the numerators must be equal.
$$y-4 = 2y-3$$
$$-4 = y-3$$
$$-1 = y$$

This does not check since all the original denominators are 0 when $y = -1$. There is no solution.

41.

$$\frac{1}{x+2} + \frac{1}{x-2} = \frac{4}{x^2-4}$$

$$\frac{1}{x+2} + \frac{1}{x-2} = \frac{4}{(x+2)(x-2)}$$

$$(x+2)(x-2)\left(\frac{1}{x+2}\right) + (x+2)(x-2)\left(\frac{1}{x-2}\right) = (x+2)(x-2)\left(\frac{4}{(x+2)(x-2)}\right)$$

$$(x-2) + (x+2) = 4$$
$$2x = 4$$
$$x = \frac{4}{2} = 2$$

This does not check since the denominators $x-2$ and x^2-4 are both 0 when $x = 2$. There is no solution.

43.

$$\frac{5}{x^2+4x+3}+\frac{2}{x^2+x-6}=\frac{3}{x^2-x-2}$$

$$\frac{5}{(x+3)(x+1)}+\frac{2}{(x+3)(x-2)}=\frac{3}{(x-2)(x+1)}$$

$$(x+3)(x+1)(x-2)\left(\frac{5}{(x+3)(x+1)}\right)+(x+3)(x+1)(x-2)\left(\frac{2}{(x+3)(x-2)}\right)=(x+3)(x+1)(x-2)\left(\frac{3}{(x-2)(x+1)}\right)$$

$$5(x-2)+2(x+1)=3(x+3)$$
$$5x-10+2x+2=3x+9$$
$$7x-8=3x+9$$
$$7x=3x+17$$
$$4x=17$$
$$x=\frac{17}{4}$$

This solution checks. The solution is $\frac{17}{4}$.

45.

$$\frac{6x}{4}=\frac{6}{x}$$
$$x(6x)=4(6)$$
$$6x^2=24$$
$$6x^2-24=0$$
$$6(x^2-4)=0$$
$$6(x+2)(x-2)=0$$
$$x+2=0 \text{ or } x-2=0$$
$$x=-2 \qquad x=2$$

Reject $x=-2$ since x cannot be a negative number. Thus, $x=2$ and $6x$ is $6(2)=12$. The unknown lengths are 2 and 12.

47.

$$\frac{8}{2x+10}=\frac{x+3}{6}$$
$$(2x+10)(x+3)=8(6)$$
$$2x^2+16x+30=48$$
$$2x^2+16x-18=0$$
$$2(x^2+8x-9)=0$$
$$2(x+9)(x-1)=0$$
$$x+9=0 \quad \text{or} \quad x-1=0$$
$$x=-9 \qquad x=1$$

Reject $x=-9$ since x cannot be a negative number. Thus, $x=1$ so that $x+3$ is $1+3=4$ and $2x+10$ is $2\cdot1+10=2+10=12$. The unknown lengths are 4 and 12.

49.

$$f(x)=2x-\frac{4}{x}$$
$$f(a)=2a-\frac{4}{a}$$
$$-2=2a-\frac{4}{a}$$
$$a(-2)=a(2a)-a\left(\frac{4}{a}\right)$$
$$-2a=2a^2-4$$
$$0=2a^2+2a-4$$
$$0=2(a+2)(a-1)$$
$$a+2=0 \quad \text{or} \quad a-1=0$$
$$a=-2 \qquad a=1$$

$f(a)=-2$ when $a=-2$ and $a=1$.

51. $f(x) = 3x - \dfrac{5}{x}$

$f(a) = 3a - \dfrac{5}{a}$

$-14 = 3a - \dfrac{5}{a}$

$a(-14) = a(3a) - a\left(\dfrac{5}{a}\right)$

$-14a = 3a^2 - 5$

$0 = 3a^2 + 14a - 5$

$0 = (3a - 1)(a + 5)$

$3a - 1 = 0$ or $a + 5 = 0$

$a = \dfrac{1}{3}$ $a = -5$

$f(a) = -14$ when $a = -5$ and $a = \dfrac{1}{3}$.

53. $f(x) = \dfrac{x + 1}{x + 3}$

$f(a) = \dfrac{a + 1}{a + 3}$

$\dfrac{2}{3} = \dfrac{a + 1}{a + 3}$

$2(a + 3) = 3(a + 1)$

$2a + 6 = 3a + 3$

$3 = a$

$f(a) = \dfrac{2}{3}$ when $a = 3$.

55. $\dfrac{V_1}{V_2} = \dfrac{P_2}{P_1}$

$P_2 V_2 = V_1 P_1$

$P_2 = \dfrac{V_1 P_1}{V_2}$

57. $\dfrac{V_1}{V_2} = \dfrac{P_2}{P_1}$

$P_2 V_2 = V_1 P_1$

$V_2 = \dfrac{V_1 P_1}{P_2}$

59. $S = \dfrac{a}{1 - r}$

$S(1 - r) = a$

$S - Sr = a$

$Sr = S - a$

$r = \dfrac{S - a}{S}$ or $r = 1 - \dfrac{a}{S}$

61. $z = \dfrac{x - \bar{x}}{s}$

$zs = x - \bar{x}$

$s = \dfrac{x - \bar{x}}{z}$

63. $d = \dfrac{fl}{f + w}$

$d(f + w) = fl$

$df + dw = fl$

$dw = fl - df$

$w = \dfrac{fl - df}{d}$

65. $\dfrac{1}{p} + \dfrac{1}{q} = \dfrac{1}{f}$

$pqf\left(\dfrac{1}{p}\right) + pqf\left(\dfrac{1}{q}\right) = pqf\left(\dfrac{1}{f}\right)$

$qf + pf = pq$

$pf = pq - qf$

$pf = q(p - f)$

$\dfrac{pf}{p - f} = q$ or $q = \dfrac{pf}{p - f}$

67. $at_2 - at_1 + v_1 = v_2$

$at_2 - at_1 = v_2 - v_1$

$a(t_2 - t_1) = v_2 - v_1$

$a = \dfrac{v_2 - v_1}{t_2 - t_1}$

69. $a_n = a_1 + nd - d$

$a_n - a_1 = nd - d$

$a_n - a_1 = d(n - 1)$

$\dfrac{a_n - a_1}{n - 1} = d$ or $d = \dfrac{a_n - a_1}{n - 1}$

71. a.
$$\frac{2}{x-2} + \frac{3}{x^2-4} = \frac{2}{x-2} + \frac{3}{(x-2)(x+2)}$$

$$= \frac{2}{x-2} \cdot \frac{x+2}{x+2} + \frac{3}{(x-2)(x+2)}$$

$$= \frac{2(x+2)}{(x-2)(x+2)} + \frac{3}{(x-2)(x+2)}$$

$$= \frac{2x+4}{(x-2)(x+2)} + \frac{3}{(x-2)(x+2)}$$

$$= \frac{2x+7}{(x-2)(x+2)}$$

b.
$$\frac{2}{x-2} + \frac{3}{x^2-4} = 0$$

$$\frac{2x+7}{(x-2)(x+2)} = 0$$

$$2x+7 = 0$$

$$2x = -7$$

$$x = -\frac{7}{2}$$

73. a.
$$\frac{b+3}{b} - \frac{b+4}{b+5} - \frac{15}{b^2+5b} = \frac{b+3}{b} - \frac{b+4}{b+5} - \frac{15}{b(b+5)}$$

$$= \frac{b+3}{b} \cdot \frac{b+5}{b+5} - \frac{b+4}{b+5} \cdot \frac{b}{b} - \frac{15}{b(b+5)}$$

$$= \frac{(b+3)(b+5)}{b(b+5)} - \frac{b(b+4)}{b(b+5)} - \frac{15}{b(b+5)}$$

$$= \frac{b^2+8b+15}{b(b+5)} - \frac{b^2+4b}{b(b+5)} - \frac{15}{b(b+5)}$$

$$= \frac{b^2+8b+15-(b^2+4b)-15}{b(b+5)}$$

$$= \frac{4b}{b(b+5)}$$

$$= \frac{4}{b+5}$$

b.
$$\frac{b+3}{b} - \frac{b+4}{b+5} = \frac{15}{b(b+5)}$$

$$b(b+5)\left(\frac{b+3}{b}\right) - b(b+5)\left(\frac{b+4}{b+5}\right) = b(b+5)\left(\frac{15}{b(b+5)}\right)$$
$$(b+5)(b+3) - b(b+4) = 15$$
$$b^2 + 8b + 15 - b^2 - 4b = 15$$
$$4b + 15 = 15$$
$$4b = 0$$
$$b = 0$$

This does not check since the denominators b and $b(b+5)$ are 0 when $b = 0$.
There is no solution.

75. $c \neq 0$, since division by 0 is not defined.

77. $f(x)$ is graph **b)** and $g(x)$ is graph **a)**;
$f(x)$ is not defined for $x = 3$.

79. a.
$$I = \frac{AC}{0.80R}$$
$$I = \frac{50,000(10,000)}{0.80(100,000)}$$
$$I = 6250$$
The insurance company will pay $6250.

b.
$$I = \frac{AC}{0.80R}$$
$$I(0.80R) = AC$$
$$R = \frac{AC}{0.80I}$$

81. a.
$$a = \frac{v_2 - v_1}{t_2 - t_1}$$
$$a = \frac{60 - 20}{22 - 20}$$
$$a = 20$$

The average acceleration is 20 ft $/$ min^2.

b.
$$a = \frac{v_2 - v_1}{t_2 - t_1}$$
$$a(t_2 - t_1) = v_2 - v_1$$
$$at_2 - at_1 = v_2 - v_1$$
$$-at_1 = -at_2 + v_2 - v_1$$
$$at_1 = at_2 + v_1 - v_2$$
$$t_1 = \frac{at_2 + v_1 - v_2}{a}$$
$$t_1 = t_2 + \frac{v_1 - v_2}{a}$$

83. $Q = \dfrac{F + D}{R - V}$

$Q = \dfrac{2500 + 8000}{500 - 200}$

$Q = 35$

35 units must be rented.

85.

$\dfrac{1}{R_T} = \dfrac{1}{R_1} + \dfrac{1}{R_2} + \dfrac{1}{R_3}$

$\dfrac{1}{R_T} = \dfrac{1}{300} + \dfrac{1}{500} + \dfrac{1}{3000}$

$3000 R_T \left(\dfrac{1}{R_T} \right) = 3000 R_T \left(\dfrac{1}{300} \right) + 3000 R_T \left(\dfrac{1}{500} \right) + 3000 R_T \left(\dfrac{1}{3000} \right)$

$3000 = 10 R_T + 6 R_T + R_T$

$3000 = 17 R_T$

$\dfrac{3000}{17} = R_T$ or $R_T \approx 176.47$

The total resistance is about 176.47 ohms.

87. $\dfrac{1}{p} + \dfrac{1}{q} = \dfrac{1}{f}$

Solve for *p*.

$pqf \left(\dfrac{1}{p} + \dfrac{1}{q} \right) = pqf \left(\dfrac{1}{f} \right)$

$qf + pf = pq$

$pq - pf = qf$

$p(q - f) = qf$

$p = \dfrac{qf}{q - f}$

$p = \dfrac{7.5(0.1)}{7.5 - 0.1}$

$p \approx 0.101$

The lens should be about 0.101 m from the film.

89. a. T_a

$= \dfrac{T_f}{1 - [f + (s + c)(1 - f)]}$

$= \dfrac{0.0601}{1 - [0.33 + (0.046 + 0.03)(1 - 0.33)]}$

$= \dfrac{0.0601}{1 - [0.33 + (0.076)(0.67)]}$

$= \dfrac{0.0601}{1 - (0.33 + 0.05092)}$

$= \dfrac{0.0601}{1 - 0.38092}$

$= \dfrac{0.0601}{0.61908}$

≈ 0.0970795

Thus, the taxable equivalent is
$T_a \approx 9.71\%$.

b. Howard Levy should choose the Tax Free Money Market since $9.71\% > 7.68\%$.

91. Several answers are possible. One such equation is $\dfrac{1}{x - 4} + \dfrac{1}{x + 2} = 0$. Another one might be $\dfrac{1}{(x - 4)(x + 2)} = 0$.

93. Several answers are possible. One possible answer is $\frac{1}{x} + \frac{1}{x} = \frac{2}{x}$.

95. $-2 \le 4 - 2x < 6$

$-2 - 4 \le 4 - 2x - 4 < 6 - 4$

$-6 \le -2x < 2$

$\dfrac{-6}{-2} \ge \dfrac{-2x}{-2} > \dfrac{2}{-2}$

$3 \ge x > -1$

$-1 < x \le 3$

96. $f(x) = \frac{1}{2}x^2 - 3x + 4$

$f(5) = \dfrac{1}{2}(5)^2 - 3(5) + 4$

$\quad = \dfrac{25}{2} - 15 + 4$

$\quad = \dfrac{25}{2} - 11$

$\quad = \dfrac{25}{2} - \dfrac{22}{2}$

$\quad = \dfrac{25 - 22}{2}$

$\quad = \dfrac{3}{2}$

97.
$$\begin{array}{r} 3x^2 - 6x - 4 \\ -3x + 2 \\ \hline 6x^2 - 12x - 8 \\ -9x^3 + 18x^2 + 12x \\ \hline -9x^3 + 24x^2 + 0x - 8 \end{array}$$
or $-9x^3 + 24x^2 - 8$

98. $8x^3 - 64y^6$

$= 8(x^3 - 8y^6)$

$= 8[x^3 - (2y^2)^3]$

$= 8(x - 2y^2)[x^2 + x(2y^2) + (2y^2)^2]$

$= 8(x - 2y^2)(x^2 + 2xy^2 + 4y^4)$

Exercise Set 6.5

1. The total time needed will be equal to $\frac{1}{2}$ the time of each painting separately. In $\frac{1}{2}$ the time, each will complete $\frac{1}{2}$ the job.

3. Let x be the time to clean the carpet together.

	Rate	Time Worked	Part of carpet cleaned
Jason La Rue	$\dfrac{1}{3}$	x	$\dfrac{x}{3}$
Tom Lockhart	$\dfrac{1}{6}$	x	$\dfrac{x}{6}$

$$\frac{x}{3} + \frac{x}{6} = 1$$

$$6\left(\frac{x}{3}\right) + 6\left(\frac{x}{6}\right) = 6(1)$$

$$2x + x = 6$$

$$3x = 6$$

$$x = 2$$

Working together, they can shampoo the carpet in 2 hours.

5. Let x be the time needed for both of them to plow the section together.

	Rate	Time worked	Part of section plowed
Paula	$\dfrac{1}{4}$	x	$\dfrac{x}{4}$
Mike	$\dfrac{1}{6}$	x	$\dfrac{x}{6}$

$$\frac{x}{4} + \frac{x}{6} = 1$$

$$12\left(\frac{x}{4}\right) + 12\left(\frac{x}{6}\right) = 12(1)$$

$$3x + 2x = 12$$

$$5x = 12$$

$$x = \frac{12}{5} = 2.4$$

Working together, they can plow the section in 2.4 hours.

7. Let x be the time needed to fill the tank.

	Rate	Time	Part of tank filled or emptied
Belt	$\frac{1}{3}$	x	$\frac{x}{3}$
Valve	$\frac{1}{4}$	x	$\frac{x}{4}$

$$\frac{x}{3} - \frac{x}{4} = 1$$

$$12\left(\frac{x}{3}\right) - 12\left(\frac{x}{4}\right) = 12(1)$$

$$4x - 3x = 12$$

$$x = 12$$

The tank will be filled in 12 hours.

9. Let x be the time needed for O'Shea to tile the kitchen.

	Rate	Time	Part of floor tiled
Waunetta	$\frac{1}{12}$	8	$\frac{8}{12}$
O'Shea	$\frac{1}{x}$	8	$\frac{8}{x}$

$$\frac{8}{12} + \frac{8}{x} = 1$$

$$12x\left(\frac{8}{12}\right) + 12x\left(\frac{8}{x}\right) = 12x(1)$$

$$8x + 96 = 12x$$

$$96 = 4x$$

$$24 = x$$

It will take O'Shea 24 hours to tile the floor by himself.

11. Let x be the time needed for Mr. Guesner to shear the herd by himself.

	Rate	Time	Part of herd sheared
Mr. MacDonald	$\frac{1}{60}$	50	$\frac{50}{60}$
Mr. Guesner	$\frac{1}{x}$	50	$\frac{50}{x}$

$$\frac{50}{60} + \frac{50}{x} = 1$$

$$60x\left(\frac{50}{60}\right) + 60x\left(\frac{50}{x}\right) = 60x(1)$$

$$50x + 3000 = 60x$$

$$3000 = 10x$$

$$300 = x$$

It will take Mr. Guesner 300 minutes to sheer the herd by himself.

13. Let x be the time needed to fill the tub.

	Rate	Time worked	Part of tub filled or emptied
Cold Water valve	$\dfrac{1}{8}$	x	$\dfrac{x}{8}$
Hot water valve	$\dfrac{1}{12}$	x	$\dfrac{x}{12}$
Drain	$\dfrac{1}{7}$	x	$\dfrac{x}{7}$

$$\frac{x}{8} + \frac{x}{12} - \frac{x}{7} = 1$$
$$168\left(\frac{x}{8}\right) + 168\left(\frac{x}{12}\right) - 168\left(\frac{x}{7}\right) = 168(1)$$
$$21x + 14x - 24x = 168$$
$$11x = 168$$
$$x = \frac{168}{11} \approx 15.27$$

It takes about 15.27 minutes to fill the tub.

15. Let x be the time needed to empty the basement.

	Rate	Time worked	Part of job completed
First pump	$\dfrac{1}{6}$	x	$\dfrac{x}{6}$
Second pump	$\dfrac{1}{5}$	x	$\dfrac{x}{5}$
Third pump	$\dfrac{1}{4}$	x	$\dfrac{x}{4}$

The equation is $\frac{x}{6} + \frac{x}{5} + \frac{x}{4} = 1$.

$$60\left(\frac{x}{6}\right) + 60\left(\frac{x}{5}\right) + 60\left(\frac{x}{4}\right) = 60(1)$$
$$10x + 12x + 15x = 60$$
$$37x = 60$$
$$x = \frac{60}{37} \approx 1.62$$

It takes about 1.62 hours to empty the flooded basement.

17. Let x be the time needed for Anna to complete the job.

	Rate	Time worked	Part of job completed
Vic	$\dfrac{1}{15}$	6	$\dfrac{6}{15}$
Anna	$\dfrac{1}{20}$	x	$\dfrac{x}{20}$

$$\frac{6}{15} + \frac{x}{20} = 1$$
$$\frac{2}{5} + \frac{x}{20} = 1$$
$$20\left(\frac{2}{5}\right) + 20\left(\frac{x}{20}\right) = 20(1)$$
$$8 + x = 20$$
$$x = 12$$

It will take Anna 12 hours to complete the roofing job.

19. Let x be the unknown number.
$$\frac{4x}{3+x} = \frac{5}{2}$$
$$5(3 + x) = 4x(2)$$
$$15 + 5x = 8x$$
$$15 = 3x$$
$$\frac{15}{3} = x$$
$$5 = x$$

21. Let x and $2x$ be the two numbers. Their reciprocals are $\dfrac{1}{x}$ and $\dfrac{1}{2x}$.
$$\frac{1}{x} + \frac{1}{2x} = \frac{3}{4}$$
$$4x\left(\frac{1}{x}\right) + 4x\left(\frac{1}{2x}\right) = 4x\left(\frac{3}{4}\right)$$
$$4 + 2 = 3x$$
$$6 = 3x$$
$$\frac{6}{3} = x$$
$$2 = x$$

Thus, $x = 2$ and $2x = 2 \cdot 2 = 4$. The two numbers are 2 and 4.

23. Let x and $x + 2$ be the two consecutive even integers. Their reciprocals are $\dfrac{1}{x}$ and $\dfrac{1}{x+2}$.

$$\frac{1}{x} + \frac{1}{x+2} = \frac{5}{12}$$
$$12x(x+2)\left(\frac{1}{x}\right) + 12x(x+2)\left(\frac{1}{x+2}\right) = 12x(x+2)\left(\frac{5}{12}\right)$$
$$12(x+2) + 12x(1) = 5x(x+2)$$
$$12x + 24 + 12x = 5x^2 + 10x$$
$$24x + 24 = 5x^2 + 10x$$
$$0 = 5x^2 - 14x - 24$$
$$0 = (5x + 6)(x - 4)$$
$$5x + 6 = 0 \quad \text{or} \quad x - 4 = 0$$
$$5x = -6 \quad \text{or} \quad x = 4$$
$$x = -\frac{6}{5}$$

Reject $x = -\dfrac{6}{5}$ since x must be an integer. Thus, $x = 4$ and $x + 2 = 4 + 2 = 6$. The two integers are 4 and 6.

25. Let x be the unknown number.
$$\frac{2}{x} + 3 = \frac{31}{10}$$
$$10x\left(\frac{2}{x}\right) + 10x(3) = 10x\left(\frac{31}{10}\right)$$
$$20 + 30x = 31x$$
$$20 = x$$

27. Let x be the unknown number.
$$3x + \frac{2}{x} = 5$$
$$x(3x) + x\left(\frac{2}{x}\right) = 5(x)$$
$$3x^2 + 2 = 5x$$
$$3x^2 - 5x + 2 = 0$$
$$(3x - 2)(x - 1) = 0$$
$$3x - 2 = 0 \quad \text{or} \quad x - 1 = 0$$
$$3x = 2 \qquad\qquad x = 1$$
$$x = \frac{2}{3}$$

The two numbers that work are $\dfrac{2}{3}$ and 1.

29. Let x be the speed of the car. Then $40 + x$ is the speed of the train.

	d	r	$t = \dfrac{d}{r}$
Car	400	x	$\dfrac{400}{x}$
Train	600	$x + 40$	$\dfrac{600}{40 + x}$

The time is the same for both.
$$\frac{400}{x} = \frac{600}{x + 40}$$
$$400(x + 40) = 600(x)$$
$$400x + 16{,}000 = 600x$$
$$16{,}000 = 200x$$
$$\frac{16{,}000}{200} = x$$
$$80 = x$$

The speed of the car is 80 km/hr and the speed of the train is $80 + 40 = 120$ km/hr.

31. Let r be the rate of Nancy walking. Then $r + 2$ is the rate of Nancy walking on the moving sidewalk.

	d	r	$t = \dfrac{d}{r}$
Nancy on sidewalk	120	$r + 2$	$\dfrac{120}{r+2}$
Nancy	52	r	$\dfrac{52}{r}$

The time is the same for both.

$$\frac{120}{r+2} = \frac{52}{r}$$
$$120(r) = 52(r+2)$$
$$120r = 52r + 104$$
$$68r = 104$$
$$r \approx 1.53$$

Nancy walks at a rate of about 1.53 ft / sec.

33. Let x be Ruth's jogging rate.

	d	r	$t = \dfrac{d}{r}$
Jerry	5.7	$x + 2.9$	$\dfrac{5.7}{x+2.9}$
Ruth	3.4	x	$\dfrac{3.4}{x}$

$$\frac{5.7}{x+2.9} = \frac{3.4}{x}$$
$$5.7x = 3.4(x+2.9)$$
$$5.7x = 3.4x + 9.86$$
$$2.3x = 9.86$$
$$x \approx 4.29$$

Ruth's jogging rate is about 4.29 miles/hour.

35. Let x be the average speed of the boat.

	d	r	$t = \dfrac{d}{r}$
Boat	40	x	$\dfrac{40}{x}$

$$\frac{40}{x} + 2\left(\frac{40}{x}\right) = 4$$
$$x\left(\frac{40}{x}\right) + 2x\left(\frac{40}{x}\right) = 4$$
$$40 + 80 = 4x$$
$$120 = 4x$$
$$30 = x$$

The average speed of the boat is 30 kph.

37. Let x be the distance the ball traveled at 14.7 yards per second. Then $80 - x$ is the distance Tyrone traveled at 5.8 yards per second.

	d	r	$t = \dfrac{d}{r}$
Ball	x	14.7	$\dfrac{x}{14.7}$
Tyrone	$80 - x$	5.8	$\dfrac{80-x}{5.8}$

The total time of the play is 10.6 seconds.

$$\frac{x}{14.7} + \frac{80-x}{5.8} = 10.6$$
$$85.26\left(\frac{x}{14.7}\right) + 85.26\left(\frac{80-x}{5.8}\right) = 85.26(10.6)$$
$$5.8x + 14.7(80-x) = 903.756$$
$$5.8x + 1176 - 14.7x = 903.756$$
$$-8.9x = -272.244$$
$$x \approx 30.6$$

The ball traveled about 30.6 yards before Tyrone caught it.

39. Let x be the speed of the local train.

	d	r	$t = \dfrac{d}{r}$
local	$24.2 - 7.8 = 16.4$	x	$\dfrac{16.4}{x}$
express	24.2	$x + 5.2$	$\dfrac{24.2}{x+5.2}$

The time is the same.
$$\frac{16.4}{x} = \frac{24.2}{x+5.2}$$
$$16.4(x+5.2) = 24.2x$$
$$16.4x + 85.25 = 24.2x$$
$$85.28 = 7.8x$$
$$10.9 \approx x$$
$$16.1 \approx x + 5.2$$
The local train's speed is about 10.9 mph and the express train's speed is about 16.1 mph.

41. Let x be the distance to the State Fair.

	d	r	$t = \frac{d}{r}$
Train	x	70	$\frac{x}{70}$
Car	x	50	$\frac{x}{50}$

The train arrives 2 hours ahead of the car.
$$\frac{x}{70} + 2 = \frac{x}{50}$$
$$350\left(\frac{x}{70}\right) + 350(2) = 350\left(\frac{x}{50}\right)$$
$$5x + 700 = 7x$$
$$700 = 2x$$
$$\frac{700}{2} = x$$
$$350 = x$$
The distance is 350 miles.

43. Let x be the distance they swam.

	d	r	$t = \frac{d}{r}$
Jim	x	3.6	$\frac{x}{3.6}$
Pete	x	2.4	$\frac{x}{2.4}$

Jim arrives 0.2 hours ahead of Pete.
$$\frac{x}{3.6} + 0.2 = \frac{x}{2.4}$$
$$7.2\left(\frac{x}{3.6}\right) + 7.2(0.2) = 7.2\left(\frac{x}{2.4}\right)$$
$$2x + 1.44 = 3x$$
$$1.44 = x$$
The distance they swam was 1.44 miles.

45. Let r be the speed of Mary Ann's car.

	d	r	$t = \frac{d}{r}$
Mary Ann	600	r	$\frac{600}{r}$
Carla	600	$r-10$	$\frac{600}{r-10}$

$$\frac{600}{r} + 2 = \frac{600}{r-10}$$

$$r(r-10)\left(\frac{600}{r}\right) + 2r(r-10) = r(r-10)\left(\frac{600}{r-10}\right)$$

$$600(r-10) + 2r(r-10) = 600r$$

$$600r - 6000 + 2r^2 - 20r = 600r$$

$$2r^2 - 20r - 6000 = 0$$

$$2(r-60)(r+50) = 0$$

$$r - 60 = 0 \quad \text{or} \quad r + 50 = 0$$

$$r = 60 \qquad\qquad r = -50$$

Since a speed cannot be negative, $r = 60$.
Mary Ann's car travels at 60 mph.

47. Let x be the distance between the space station and NASA headquarters.

	d	r	$t = \dfrac{d}{r}$
First rocket	x	20,000	$\dfrac{x}{20,000}$
Second rocket	x	18,000	$\dfrac{x}{18,000}$

$$\frac{x}{20,000} + 0.6 = \frac{x}{18,000}$$

$$180,000\left(\frac{x}{20,000}\right) + 180,000(0.6) = 180,000\left(\frac{x}{18,000}\right)$$

$$9x + 108,000 = 10x$$

$$108,000 = x$$

The space station is 108,000 miles from NASA headquarters.

49. Answers will vary.

51. Answers will vary.

53. $\dfrac{1}{2}x^2 - 3x^2 + 2xy - \left(\dfrac{3}{5}xy + 6y^2\right) = \dfrac{1}{2}x^2 - 3x^2 + 2xy - \dfrac{3}{5}xy - 6y^2$

$$= \left(\frac{1}{2} - 3\right)x^2 + \left(2 - \frac{3}{5}\right)xy - 6y^2$$

$$= \left(\frac{1}{2} - \frac{6}{2}\right)x^2 + \left(\frac{10}{5} - \frac{3}{5}\right)xy - 6y^2$$

$$= -\frac{5}{2}x^2 + \frac{7}{5}xy - 6y^2$$

54.
$$4x^2 - 6x - 1$$
$$\underline{3x - 4}$$
$$-16x^2 + 24x + 4 \quad \leftarrow \; -4(4x^2 - 6x - 1)$$
$$\underline{12x^3 - 18x^2 - 3x} \quad \leftarrow \; 3x(4x^2 - 6x - 1)$$
$$12x^3 - 34x^2 + 21x + 4 \quad \leftarrow \; \text{Combine like terms}$$

55.
$$\begin{array}{r} 4x + 5 \\ 3x - 2 \overline{)\,12x^2 + 7x + 12} \\ \underline{12x^2 - 8x} \\ 15x + 12 \\ \underline{15x - 10} \\ 22 \end{array}$$

$$\frac{12x^2 + 7x + 12}{3x - 2} = 4x + 5 + \frac{22}{3x - 2}$$

56. Factor $8x^2 + 26x + 15$
Observe that $(8)(15) = 120$. Now, the two numbers whose product is 120 and whose sum is 26 are 20 and 6 since
$(20)(6) = 120$
$20 + 6 = 26$
Thus, express $26x$ as $20x + 6x$ and then complete the factorization
$$8x^2 + 26x + 15 = 8x^2 + 20x + 6x + 15$$
$$= 4x(2x + 5) + 3(2x + 5)$$
$$= (4x + 3)(2x + 5)$$

Exercise Set 6.6

1. a. As one increases, the other increases.

b. Answers will vary.

c. Answers will vary.

3. One quantity varies as a product of two or more quantities.

5. Direct

7. Inverse

9. Inverse

11. Direct

13. Inverse

15. Direct

17. Inverse

19. a. $x = ky$

b. Substitute 12 for y and 6 for k.
$x = 6(12)$
$x = 72$

21. a. $y = kR$

b. Substitute 180 for R and 1.7 for k.
$y = 1.7(180)$
$y = 306$

23. a. $R = \dfrac{k}{W}$

b. Substitute 160 for W and 8 for k.
$R = \dfrac{8}{160}$
$R = \dfrac{1}{20} = 0.05$

25. a. $A = \dfrac{kB}{C}$

b. Substitute 12 for B, 4 for C, and 3 for k.
$A = \dfrac{3(12)}{4} = \dfrac{36}{4} = 9$

27. a. $x = ky$

b. To find k, substitute 9 for x and 18 for y.

$$9 = k(18)$$
$$\frac{9}{18} = k$$
$$\frac{1}{2} = k$$

Thus, $x = \frac{1}{2}y$.

Now substitute 36 for y.

$$x = \frac{1}{2}(36) = 18$$

29. a. $y = kR^2$

b. To find k, substitute 5 for y and 5 for R.

$$5 = k(5)^2$$
$$5 = k(25)$$
$$\frac{5}{25} = k$$
$$\frac{1}{5} = k$$

Thus $y = \frac{1}{5}R^2$.

Now substitute 10 for R.

$$y = \frac{1}{5}(10)^2 = \frac{1}{5}(100) = 20$$

31. a. $C = \frac{k}{J}$

b. To find k, substitute 7 for C and 0.7 for J.

$$7 = \frac{k}{0.7}$$
$$7(0.7) = k$$
$$4.9 = k$$

Thus $C = \frac{4.9}{J}$.

Now substitute 12 for J.

$$C = \frac{4.9}{12} \approx 0.41$$

33. a. $F = \frac{kM_1M_2}{d}$

b. To find k, substitute 5 for M_1, 10 for M_2, 0.2 for d, and 20 for F.

$$20 = \frac{k(5)(10)}{0.2}$$
$$20 = k(250)$$
$$\frac{20}{250} = k$$
$$0.08 = k$$

Thus $F = \frac{0.08M_1M_2}{d}$.

Now substitute 10 for M_1, 20 for M_2, and 0.4 for d.

$$F = \frac{0.08(10)(20)}{0.4} = \frac{16}{0.4} = 40$$

35. $a = kb$
$k(2b) = 2(kb) = 2a$
If b is doubled, a is doubled.

37. $y = \frac{k}{x}$

$$\frac{k}{2x} = \frac{1}{2}\left(\frac{k}{x}\right) = \frac{1}{2}y$$

If x is doubled, y is halved.

39. The equation is $d = kw$. To find k, substitute 2376 for d and 132 for w.

$$2376 = k132$$
$$k = \frac{2376}{132}$$
$$k = 18$$

Thus $d = 18w$. Now substitute 172 for w.
$d = 18(172)$
$d = 3096$
The recommended dosage for Bill is 3096 mg.

41. The equation is $t = kl$. To find k substitute 62 for l and 96 for t.
$$96 = k(62)$$
$$k = \frac{96}{62}$$
Thus $t = \frac{96}{62}l$.
Now substitute 90 for l.
$$t = \frac{96}{62}(90)$$
$$t \approx 139.35$$
It will take him about 139.35 minutes.

43. The equation is $d = kt$. To find k, substitute 150 for d and 2.5 for t.
$$150 = k(2.5)$$
$$k = \frac{150}{2.5}$$
$$k = 60$$
Thus $d = 60t$.
Now substitute 4 for t.
$$d = 60(4)$$
$$d = 240$$
The car will travel 240 miles in 4 hours.

45. The equation is $V = \frac{k}{P}$. To find k, substitute 800 for V and 200 for P.
$$800 = \frac{k}{200}$$
$$800(200) = k$$
$$160,000 = k$$
Thus $V = \frac{160,000}{P}$.
Now substitute 25 for P.
$$V = \frac{160,000}{25} = 6400$$
The volume is 6400 cc.

47. The equation is $t = \frac{k}{s}$. To find k, substitute 122 for s and 0.21 for t.
$$0.21 = \frac{k}{122}$$
$$k = 0.21(122)$$
$$k = 25.62$$
Thus $t = \frac{25.62}{s}$.
Now substitute 80 for s.
$$t = \frac{25.62}{80} \approx 0.32$$
It takes about 0.32 seconds for the ball to hit the ground in the service box.

49. The equation is $w = km$. To find k, substitute 7040 for w and 3200 for m.
$$7040 = k(3200)$$
$$k = 2.2$$
Thus $w = 2.2m$.
Now substitute 18 for w.
$$18 = 2.2m$$
$$m = \frac{18}{2.2} \approx 8.18$$
The mass is about 8.18 kilograms.

51. The equation is $P = krm$. To find k, substitute 50,000 for m and 0.07 for r and 332.5 for P.
$$332.5 = k(0.07)(50,000)$$
$$332.5 = 3500k$$
$$k = \frac{332.5}{3500} = 0.095$$
Thus $P = 0.095rm$.
Now substitute 66,000 for m and 0.07 for r.
$$P = 0.095(0.07)(66,000)$$
$$P = 438.9$$
The payment is $438.90.

53. The equation is $R = \frac{kA}{P}$. To find k, substitute 400 for A, 2 for P, and 4600 for R.
$$4600 = \frac{k(400)}{2}$$
$$4600 = 200k$$
$$k = \frac{4600}{200} = 23$$

Thus $R = \dfrac{23A}{P}$. Now substitute 500 for A and 2.50 for P.

$R = \dfrac{23(500)}{2.50}$

$R = \dfrac{11,500}{2.50}$

$R = 4600$ tapes

They would rent 4600 tapes per week.

55. The equation is $R = \dfrac{kL}{A}$. To find k, substitute 0.2 for R, 200 for L, and 0.05 for A.

$0.02 = \dfrac{k(200)}{0.05}$

$0.2 = k(4000)$

$\dfrac{0.2}{4000} = k$

$0.00005 = k$

Thus $R = \dfrac{0.00005L}{A}$. Now substitute 5000 for L and 0.01 for A.

$R = \dfrac{0.00005(5000)}{0.01}$

$R = \dfrac{0.25}{0.01}$

$R = 25$

The resistance is 25 ohms.

57. The equation is $W = \dfrac{kTA\sqrt{F}}{R}$

To find k, substitute 78 for T, 5.6 for R, 4 for F, 1000 for A, and 68 for W.

$68 = \dfrac{k(78)(1000)\sqrt{4}}{5.6}$

$68 = \dfrac{156,000k}{5.6}$

$\dfrac{(68)(5.6)}{156,000} = k$

$0.002441 \approx k$

Thus $W = \dfrac{0.002441TA\sqrt{E}}{R}$

Now, substitute 78 for T, 5.6 for R, 6 for E, and 1500 for A.

$W = \dfrac{0.002441(78)(1500)\sqrt{6}}{5.6} \approx 124.92$

The water bill is about \$124.92.

59. a. The equation is $F = \dfrac{km_1m_2}{d^2}$.

b. If m_1 becomes $2m_1$, m_2 becomes $3m_2$ and

d becomes $\dfrac{d}{2}$, then

$F = \dfrac{k(2m_1)(3m_2)}{\left(\dfrac{d}{2}\right)^2}$

$= \dfrac{\dfrac{6km_1m_2}{d^2}}{4}$

$= 24\dfrac{km_1m_2}{d^2}$

The new force is 24 times the original force.

61. The slope of the line is $m = \dfrac{d-1}{-4-5} = \dfrac{d-1}{-9}$.

Set this equal to $\dfrac{2}{3}$ and solve.

$\dfrac{d-1}{-9} = \dfrac{2}{3}$

$-9\left(\dfrac{d-1}{-9}\right) = -9\left(\dfrac{2}{3}\right)$

$d-1 = -6$

$d = -5$

62. $2y < 3x - 6$

Graph the line $2y = 3x - 6$ with a dashed line. For the check point, select $(0, 0)$.

$2(0) < 3(0) - 6$

$2 < -6$

Since this is a false statement, shade the region which does not contain $(0, 0)$.

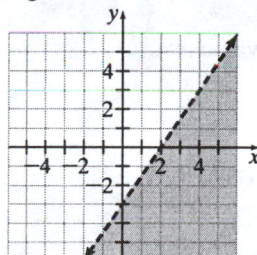

63. Let x be the base weekly salary and y be the commission rate. The system of equations is

$x + 5000y = 550$

$x + 8000y = 640$

To eliminate x, multiply the first equation by -1 and then add.

$-x - 5000y = -550$

$\underline{x + 8000y = 640}$

Add: $3000y = 90$

$y = \dfrac{90}{3000} = 0.03$

Now, substitute 0.03 for y in the first equation.

$x + 5000(0.03) = 550$

$x + 150 = 550$

$x = 400$

His base weekly salary is $400 and the commission rate is 3%.

64. $|x - 2| < 4$

$-4 < x - 2 < 4$

$-2 < x < 6$

For $|x - 2| < 4$, the graph is the region between the dashed lines $x = -2$ and $x = 6$.

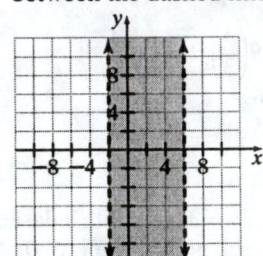

Review Exercises

1. $\dfrac{3}{x - 4}$

The excluded value is 4.

2. $\dfrac{x}{x + 1}$

The excluded value is -1.

3. $\dfrac{-2x}{x^2 + 5}$

There are no values for which $x^2 + 5 = 0$.

Thus, there are no excluded values.

4. $y = \dfrac{0}{(x + 3)^2}$

$x + 3 = 0$

$x = -3$

The domain is

$\{x | x$ is a real number and $x \neq -3\}$.

5. $f(x) = \dfrac{x + 6}{x^2}$

$x^2 = 0$

$x = 0$

The domain is

$\{x | x$ is a real number and $x \neq 0\}$.

6. $f(x) = \dfrac{x^2 - 2}{x^2 - 3x - 10}$

$x^2 - 3x - 10 = 0$

$(x - 5)(x + 2) = 0$

$x - 5 = 0$ or $x + 2 = 0$

$x = 5 x = -2$

The domain is

$\{x | x$ is a real number and $x \neq 5$ and $x \neq -2\}$.

7. $\dfrac{x^2 + xy}{x + y} = \dfrac{x(x + y)}{x + y} = \dfrac{x}{1} = x$

8. $\dfrac{x^2 - 9}{x + 3} = \dfrac{(x + 3)(x - 3)}{x + 3} = x - 3$

9. $\dfrac{4 - 5x}{5x - 4} = \dfrac{-(5x - 4)}{5x - 4} = -1$

10. $\dfrac{x^2 + 2x - 3}{x^2 + x - 6} = \dfrac{(x + 3)(x - 1)}{(x + 3)(x - 2)} = \dfrac{x - 1}{x - 2}$

11. $\dfrac{2x^2 - 6x + 5x - 15}{2x^2 + 7x + 5} = \dfrac{2x(x - 3) + 5(x - 3)}{(2x + 5)(x + 1)}$

$= \dfrac{(2x + 5)(x - 3)}{(2x + 5)(x + 1)}$

$= \dfrac{x - 3}{x + 1}$

12. $\dfrac{a^3 - 8b^3}{a^2 - 4b^2}$

$= \dfrac{(a - 2b)(a^2 + 2ab + 4b^2)}{(a - 2b)(a + 2b)}$

$= \dfrac{a^2 + 2ab + 4b^2}{a + 2b}$

13. Factor denominators: $x + 1$ is $x + 1$
 x is x.
The LCD is $x(x + 1)$.

14. Factor denominators: $x + y$ is $x + y$
$x^2 - y^2 = (x + y)(x - y)$
The LCD is $(x + y)(x - y)$.

15. Factor denominators:
$x^2 + 2x - 35 = (x + 7)(x - 5)$
$x^2 + 9x + 14 = (x + 7)(x + 2)$
The LCD is $(x + 7)(x - 5)(x + 2)$.

16. Factor denominators:
$(x + 2)^2 = (x + 2)(x + 2)$
$x^2 - 4 = (x + 2)(x - 2)$
$x + 1 = x + 1$
The LCD is $(x + 2)^2(x - 2)(x + 1)$.

17. $\dfrac{15x^2y^3}{3z} \cdot \dfrac{6z^3}{5xy^3} = \dfrac{6xz^2}{1} = 6xz^2$

18. $\dfrac{1}{x - 2} \cdot \dfrac{2 - x}{2} = \dfrac{1}{x - 2} \cdot \dfrac{-(x - 2)}{2} = \dfrac{-1}{2}$ or $-\dfrac{1}{2}$

19. $\dfrac{16x^2y^4}{xz^5} \div \dfrac{2x^2y^4}{x^4z^{10}} = \dfrac{16x^2y^4}{xz^5} \cdot \dfrac{x^4z^{10}}{2x^2y^4}$

$\phantom{\dfrac{16x^2y^4}{xz^5} \div \dfrac{2x^2y^4}{x^4z^{10}}} = 8x^3z^5$

20. $\dfrac{4}{2x} + \dfrac{x}{x^2} = \dfrac{2}{x} + \dfrac{1}{x} = \dfrac{3}{x}$

21. $\dfrac{4x + 4y}{x^2y} \cdot \dfrac{y^3}{8x} = \dfrac{4(x + y)}{x^2y} \cdot \dfrac{y^3}{8x} = \dfrac{(x + y)y^2}{2x^3}$

22. $\dfrac{4x^2 - 11x + 4}{x - 3} - \dfrac{x^2 - 4x + 10}{x - 3}$

$= \dfrac{4x^2 - 11x + 4 - (x^2 - 4x + 10)}{x - 3}$

$= \dfrac{4x^2 - 11x + 4 - x^2 + 4x - 10}{x - 3}$

$= \dfrac{3x^2 - 7x - 6}{x - 3}$

$= \dfrac{(3x + 2)(x - 3)}{x - 3}$

$= 3x + 2$

23. $\dfrac{a - 2}{a + 3} \cdot \dfrac{a^2 + 4a + 3}{a^2 - a - 2} = \dfrac{a - 2}{a + 3} \cdot \dfrac{(a + 3)(a + 1)}{(a - 2)(a + 1)}$

$\phantom{\dfrac{a - 2}{a + 3} \cdot \dfrac{a^2 + 4a + 3}{a^2 - a - 2}} = 1$

24. $\dfrac{x^2 - 5x - 2}{6x^2 - 11x - 35} - \dfrac{x^2 - 7x + 5}{6x^2 - 11x - 35}$

$= \dfrac{x^2 - 5x - 2 - (x^2 - 7x + 5)}{6x^2 - 11x - 35}$

$= \dfrac{x^2 - 5x - 2 - x^2 + 7x - 5}{6x^2 - 11x - 35}$

$= \dfrac{2x - 7}{6x^2 - 11x - 35}$

$= \dfrac{2x - 7}{(2x - 7)(3x + 5)}$

$= \dfrac{1}{3x + 5}$

25. $\dfrac{5x}{3xy} - \dfrac{4}{x^2} = \dfrac{5x}{3xy} \cdot \dfrac{x}{x} - \dfrac{4}{x^2} \cdot \dfrac{3y}{3y}$

$= \dfrac{5x^2}{3x^2y} - \dfrac{12y}{3x^2y}$

$= \dfrac{5x^2 - 12y}{3x^2y}$

26.
$$6 + \frac{x}{x+2} = \frac{6}{1} \cdot \frac{x+2}{x+2} + \frac{x}{x+2}$$
$$= \frac{6(x+2)}{x+2} + \frac{x}{x+2}$$
$$= \frac{6x+12}{x+2} + \frac{x}{x+2}$$
$$= \frac{6x+12+x}{x+2}$$
$$= \frac{7x+12}{x+2}$$

27.
$$5 - \frac{3}{x+3} = \frac{5}{1} \cdot \frac{x+3}{x+3} - \frac{3}{x+3}$$
$$= \frac{5(x+3)}{x+3} - \frac{3}{x-3}$$
$$= \frac{5x+15}{x+3} - \frac{3}{x+3}$$
$$= \frac{5x+15-3}{x+3}$$
$$= \frac{5x+12}{x+3}$$

28.
$$\frac{x^2-y^2}{x-y} \cdot \frac{x+y}{xy+x^2} = \frac{(x+y)(x-y)}{x-y} \cdot \frac{x+y}{x(x+y)}$$
$$= \frac{x+y}{x}$$

29.
$$\frac{1}{a^2+8a+15} \div \frac{3}{a+5} = \frac{1}{a^2+8a+15} \cdot \frac{a+5}{3}$$
$$= \frac{1}{(a+5)(a+3)} \cdot \frac{a+5}{3}$$
$$= \frac{1}{3(a+3)}$$

30.
$$\frac{a+c}{c} - \frac{a-c}{a} = \frac{a+c}{c} \cdot \frac{a}{a} - \frac{a-c}{a} \cdot \frac{c}{c}$$
$$= \frac{a(a+c)}{ac} - \frac{c(a-c)}{ac}$$
$$= \frac{a^2+ac}{ac} - \frac{ac-c^2}{ac}$$
$$= \frac{a^2+ac-(ac-c^2)}{ac}$$
$$= \frac{a^2+ac-ac+c^2}{ac}$$
$$= \frac{a^2+c^2}{ac}$$

31.
$$\frac{4x^2+8x-5}{2x+5} \cdot \frac{x+1}{4x^2-4x+1}$$
$$= \frac{(2x-1)(2x+5)}{2x+5} \cdot \frac{x+1}{(2x-1)(2x-1)}$$
$$= \frac{x+1}{2x-1}$$

32.
$$(x+3) \div \frac{x^2-4x-21}{x-7}$$
$$= \frac{x+3}{1} \cdot \frac{x-7}{x^2-4x-21}$$
$$= \frac{x+3}{1} \cdot \frac{x-7}{(x+3)(x-7)}$$
$$= 1$$

33.
$$\frac{x^2-3xy-10y^2}{6x} \div \frac{x+2y}{12x^2}$$
$$= \frac{x^2-3xy-10y^2}{6x} \cdot \frac{12x^2}{x+2y}$$
$$= \frac{(x+2y)(x-5y)}{6x} \cdot \frac{12x^2}{x+2y}$$
$$= 2x(x-5y)$$

34. $\dfrac{2}{3x} - \dfrac{3x}{3x-6}$

$= \dfrac{2}{3x} - \dfrac{3x}{3(x-2)}$

$= \dfrac{2}{3x} \cdot \dfrac{x-2}{x-2} - \dfrac{3x}{3(x-2)} \cdot \dfrac{x}{x}$

$= \dfrac{2(x-2)}{3x(x-2)} - \dfrac{3x(x)}{3x(x-2)}$

$= \dfrac{2x-4}{3x(x-2)} - \dfrac{3x^2}{3x(x-2)}$

$= \dfrac{2x-4-3x^2}{3x(x-2)}$ or $\dfrac{-3x^2+2x-4}{3x(x-2)}$

35. $\dfrac{x-4}{x-5} - \dfrac{3}{x+5}$

$= \dfrac{x-4}{x-5} \cdot \dfrac{x+5}{x+5} - \dfrac{3}{x+5} \cdot \dfrac{x-5}{x-5}$

$= \dfrac{(x-4)(x+5)}{(x-5)(x+5)} - \dfrac{3(x-5)}{(x-5)(x+5)}$

$= \dfrac{x^2+x-20}{(x-5)(x+5)} - \dfrac{3x-15}{(x-5)(x+5)}$

$= \dfrac{x^2+x-20-(3x-15)}{(x-5)(x+5)}$

$= \dfrac{x^2-2x-5}{(x-5)(x+5)}$

36. $\dfrac{x+3}{x^2-9} + \dfrac{2}{x+3}$

$= \dfrac{x+3}{(x+3)(x-3)} + \dfrac{2}{x+3}$

$= \dfrac{x+3}{(x+3)(x-3)} + \dfrac{2}{x+3} \cdot \dfrac{x-3}{x-3}$

$= \dfrac{x+3}{(x+3)(x-3)} + \dfrac{2(x-3)}{(x+3)(x-3)}$

$= \dfrac{x+3}{(x+3)(x-3)} + \dfrac{2x-6}{(x+3)(x-3)}$

$= \dfrac{x+3+2x-6}{(x+3)(x-3)}$

$= \dfrac{3x-3}{(x+3)(x-3)}$ or $\dfrac{3(x-1)}{(x+3)(x-3)}$

37. $\dfrac{1}{a-3} \cdot \dfrac{a^2-2a-3}{a^2+3a+2} = \dfrac{1}{a-3} \cdot \dfrac{(a-3)(a+1)}{(a+2)(a+1)}$

$\qquad\qquad\qquad = \dfrac{1}{a+2}$

38. $\dfrac{4x^2-16y^2}{9} \div \dfrac{(x+2y)^2}{12} = \dfrac{4x^2-16y^2}{9} \cdot \dfrac{12}{(x+2y)^2}$

$\qquad\qquad\qquad = \dfrac{4(x+2y)(x-2y)}{9} \cdot \dfrac{12}{(x+2y)(x+2y)}$

$\qquad\qquad\qquad = \dfrac{16(x-2y)}{3(x+2y)}$

39.

$$\frac{4}{(x+2)(x-3)} - \frac{4}{(x-2)(x+2)} = \frac{4}{(x+2)(x-3)} \cdot \frac{x-2}{x-2} - \frac{4}{(x+2)(x-2)} \cdot \frac{x-3}{x-3}$$

$$= \frac{4(x-2)}{(x+2)(x-3)(x-2)} - \frac{4(x-3)}{(x+2)(x-3)(x-2)}$$

$$= \frac{4x-8}{(x+2)(x-3)(x-2)} - \frac{4x-12}{(x+2)(x-3)(x-2)}$$

$$= \frac{4x-8-(4x-12)}{(x+2)(x-3)(x-2)}$$

$$= \frac{4}{(x+2)(x-3)(x-2)}$$

40.

$$\frac{2x^2+10x+12}{(x+2)^2} \cdot \frac{x+2}{x^3+5x^2+6x} = \frac{2(x+2)(x+3)}{(x+2)(x+2)} \cdot \frac{x+2}{x(x+2)(x+3)}$$

$$= \frac{2}{x(x+2)}$$

41.

$$\frac{x+2}{x^2-x-6} + \frac{x-3}{x^2-8x+15} = \frac{x+2}{(x+2)(x-3)} + \frac{x-3}{(x-3)(x-5)}$$

$$= \frac{1}{x-3} + \frac{1}{x-5}$$

$$= \frac{1}{x-3} \cdot \frac{x-5}{x-5} + \frac{1}{x-5} \cdot \frac{x-3}{x-3}$$

$$= \frac{x-5}{(x-3)(x-5)} + \frac{x-3}{(x-3)(x-5)}$$

$$= \frac{x-5+x-3}{(x-3)(x-5)}$$

$$= \frac{2x-8}{(x-3)(x-5)} \text{ or } \frac{2(x-4)}{(x-3)(x-5)}$$

42.
$$\frac{x+5}{x^2-15x+50} - \frac{x-2}{x^2-25} = \frac{x+5}{(x-5)(x-10)} - \frac{x-2}{(x-5)(x+5)}$$

$$= \frac{x+5}{(x-5)(x-10)} \cdot \frac{x+5}{x+5} - \frac{x-2}{(x-5)(x+5)} \cdot \frac{x-10}{x-10}$$

$$= \frac{(x+5)(x+5)}{(x-5)(x-10)(x+5)} - \frac{(x-2)(x-10)}{(x-5)(x-10)(x+5)}$$

$$= \frac{x^2+10x+25}{(x-5)(x-10)(x+5)} - \frac{x^2-12x+20}{(x-5)(x-10)(x+5)}$$

$$= \frac{x^2+10x+25-(x^2-12x+20)}{(x-5)(x-10)(x+5)}$$

$$= \frac{x^2+10x+25-x^2+12x-20}{(x-5)(x-10)(x+5)}$$

$$= \frac{22x+5}{(x-5)(x-10)(x+5)}$$

43.
$$\frac{y^4-x^6}{x^3-y^2} \div (y^2-x^3) = \frac{y^4-x^6}{x^3-y^2} \cdot \frac{1}{y^2-x^3}$$

$$= \frac{(y^2+x^3)(y^2-x^3)}{x^3-y^2} \cdot \frac{1}{y^2-x^3}$$

$$= \frac{x^3+y^2}{x^3-y^2}$$

44.
$$\frac{1}{x+3} - \frac{2}{x-3} + \frac{6}{x^2-9} = \frac{1}{x+3} - \frac{2}{x-3} + \frac{6}{(x+3)(x-3)}$$

$$= \frac{1}{x+3} \cdot \frac{x-3}{x-3} - \frac{2}{x-3} \cdot \frac{x+3}{x+3} + \frac{6}{(x+3)(x-3)}$$

$$= \frac{x-3}{(x+3)(x-3)} - \frac{2(x+3)}{(x+3)(x-3)} + \frac{6}{(x+3)(x-3)}$$

$$= \frac{x-3}{(x+3)(x-3)} - \frac{2x+6}{(x+3)(x-3)} + \frac{6}{(x+3)(x-3)}$$

$$= \frac{x-3-(2x+6)+6}{(x+3)(x-3)}$$

$$= \frac{x-3-2x-6+6}{(x+3)(x-3)}$$

$$= \frac{-x-3}{(x+3)(x-3)}$$

$$= \frac{-(x+3)}{(x+3)(x-3)}$$

$$= \frac{-1}{x-3} \text{ or } -\frac{1}{x-3}$$

45. $\dfrac{x^3+27}{4x^2-4} \div \dfrac{x^2-3x+9}{(x-1)^2} = \dfrac{x^3+27}{4x^2-4} \cdot \dfrac{(x-1)^2}{x^2-3x+9}$

$$= \frac{(x+3)(x^2-3x+9)}{4(x-1)(x+1)} \cdot \frac{(x-1)(x-1)}{x^2-3x+9}$$

$$= \frac{(x+3)(x-1)}{4(x+1)}$$

46. $\dfrac{x-4}{x-5} - \dfrac{3}{x+5} - \dfrac{10}{x^2-25} = \dfrac{x-4}{x-5} - \dfrac{3}{x+5} - \dfrac{10}{(x+5)(x-5)}$

$$= \frac{x-4}{x-5} \cdot \frac{x+5}{x+5} - \frac{3}{x+5} \cdot \frac{x-5}{x-5} - \frac{10}{(x+5)(x-5)}$$

$$= \frac{(x-4)(x+5)}{(x+5)(x-5)} - \frac{3(x-5)}{(x+5)(x-5)} - \frac{10}{(x+5)(x-5)}$$

$$= \frac{x^2+x-20}{(x+5)(x-5)} - \frac{3x-15}{(x+5)(x-5)} - \frac{10}{(x+5)(x-5)}$$

$$= \frac{x^2+x-20-(3x-15)-10}{(x+5)(x-5)}$$

$$= \frac{x^2+x-20-3x+15-10}{(x+5)(x-5)}$$

$$= \frac{x^2-2x-15}{(x+5)(x-5)}$$

$$= \frac{(x+3)(x-5)}{(x+5)(x-5)}$$

$$= \frac{x+3}{x+5}$$

47. $\left(\dfrac{x^2-8x+16}{2x^2-x-6} \cdot \dfrac{2x^2-7x-15}{x^2-2x-24} \right) \div \dfrac{x^2-9x+20}{x^2+2x-8} = \left(\dfrac{x^2-8x+16}{2x^2-x-6} \cdot \dfrac{2x^2-7x-15}{x^2-2x-24} \right) \cdot \dfrac{x^2+2x-8}{x^2-9x+20}$

$$= \frac{(x-4)(x-4)}{(2x+3)(x-2)} \cdot \frac{(2x+3)(x-5)}{(x-6)(x+4)} \cdot \frac{(x+4)(x-2)}{(x-5)(x-4)}$$

$$= \frac{x-4}{x-6}$$

48. $\left(\dfrac{x^2-x-56}{x^2+14x+49}\cdot\dfrac{x^2+4x-21}{x^2-9x+8}\right)+\dfrac{3}{x^2+8x-9}=\left(\dfrac{(x-8)(x+7)}{(x+7)(x+7)}\cdot\dfrac{(x+7)(x-3)}{(x-8)(x-1)}\right)+\dfrac{3}{(x-1)(x+9)}$

$$=\dfrac{x-3}{x-1}+\dfrac{3}{(x-1)(x+9)}$$

$$=\dfrac{x-3}{x-1}\cdot\dfrac{x+9}{x+9}+\dfrac{3}{(x-1)(x+9)}$$

$$=\dfrac{(x-3)(x+9)}{(x-1)(x+9)}+\dfrac{3}{(x-1)(x+9)}$$

$$=\dfrac{x^2+6x-27}{(x-1)(x+9)}+\dfrac{3}{(x-1)(x+9)}$$

$$=\dfrac{x^2+6x-27+3}{(x-1)(x+9)}$$

$$=\dfrac{x^2+6x-24}{(x-1)(x+9)}$$

49. a. $f(x)=\dfrac{x-4}{x+3}$

The denominator cannot equal zero.
Therefore $x\neq-3$.
The domain is
$\{x|x$ is a real number, $x\neq-3\}$

b. $g(x)=\dfrac{x}{x+5}$

The denominator cannot equal zero.
Therefore, $x\neq-5$.
The domain is
$\{x|x$ is a real number, $x\neq-5\}$

c. $(f+g)(x)=f(x)+g(x)$

$$=\dfrac{x-4}{x+3}+\dfrac{x}{x+5}$$

$$=\dfrac{x-4}{x+3}\cdot\dfrac{x+5}{x+5}+\dfrac{x}{x+5}\cdot\dfrac{x+3}{x+3}$$

$$=\dfrac{(x-4)(x+5)+x(x+3)}{(x+3)(x+5)}$$

$$=\dfrac{x^2+x-20+x^2+3x}{(x+3)(x+5)}$$

$$=\dfrac{2x^2+4x-20}{(x+3)(x+5)}$$

d. The denominator cannot equal zero.
Therefore, $x\neq-3$, $x\neq-5$.
The domain is
$\{x|x$ is a real number, $x\neq-3,x\neq-5\}$

50. a. $f(x)=\dfrac{x}{x^2-9}$ or $f(x)=\dfrac{x}{(x-3)(x+3)}$

The denominator cannot equal zero.
Therefore, $x\neq-3$, $x\neq3$.
The domain is
$\{x|x$ is a real number, $x\neq-3,x\neq3\}$

b. $g(x)=\dfrac{x+2}{x-3}$

The denominator cannot equal zero.
Therefore, $x\neq3$.
The domain is
$\{x|x$ is a real number, $x\neq3\}$

c. $(f+g)(x)$

$= f(x) + g(x)$

$= \dfrac{x}{x^2-9} + \dfrac{x+2}{x-3}$

$= \dfrac{x}{(x-3)(x+3)} + \dfrac{(x+2)}{(x-3)} \cdot \dfrac{(x+3)}{(x+3)}$

$= \dfrac{x + (x+2)(x+3)}{(x-3)(x+3)}$

$= \dfrac{x + x^2 + 5x + 6}{(x-3)(x+3)}$

$= \dfrac{x^2 + 6x + 6}{(x-3)(x+3)}$

d. The denominator cannot equal zero.
Therefore, $x \neq -3$ and $x \neq 3$.
The domain is
$\{x | x \text{ is a real number}, \ x \neq -3, x \neq 3\}$

51. $\dfrac{\frac{15xy}{6z}}{\frac{3x}{z^2}} = \dfrac{15xy}{6z} \cdot \dfrac{z^2}{3x} = \dfrac{5yz}{6}$

52. $\dfrac{x+\frac{1}{y}}{y^2} = \dfrac{y\left(x+\frac{1}{y}\right)}{y(y^2)} = \dfrac{y(x)+y\left(\frac{1}{y}\right)}{y(y^2)} = \dfrac{xy+1}{y^3}$

53. $\dfrac{\frac{4}{x}+\frac{2}{x^2}}{6-\frac{1}{x}} = \dfrac{x^2\left(\frac{4}{x}+\frac{2}{x^2}\right)}{x^2\left(6-\frac{1}{x}\right)}$

$= \dfrac{x^2\left(\frac{4}{x}\right)+x^2\left(\frac{2}{x^2}\right)}{x^2(6)-x^2\left(\frac{1}{x}\right)}$

$= \dfrac{4x+2}{6x^2-x} \text{ or } \dfrac{2(2x+1)}{x(6x-1)}$

54. $\dfrac{a^{-1}+2}{a^{-1}+\frac{1}{a}} = \dfrac{\frac{1}{a}+2}{\frac{1}{a}+\frac{1}{a}}$

$= \dfrac{\frac{1}{a}+2}{\frac{2}{a}}$

$= \left(\frac{1}{a}+2\right)\left(\frac{a}{2}\right)$

$= \dfrac{\left(\frac{1}{a}+2\right)a}{2}$

$= \dfrac{1+2a}{2}$

55. $\dfrac{x^{-2}+\frac{1}{x}}{\frac{1}{x^2}-\frac{1}{x}} = \dfrac{\frac{1}{x^2}+\frac{1}{x}}{\frac{1}{x^2}-\frac{1}{x}}$

$= \dfrac{x^2\left(\frac{1}{x^2}+\frac{1}{x}\right)}{x^2\left(\frac{1}{x^2}-\frac{1}{x}\right)}$

$= \dfrac{x^2\left(\frac{1}{x^2}\right)+x^2\left(\frac{1}{x}\right)}{x^2\left(\frac{1}{x^2}\right)-x^2\left(\frac{1}{x}\right)}$

$= \dfrac{1+x}{1-x} \text{ or } \dfrac{x+1}{-x+1}$

56. $\dfrac{\frac{1}{x^2-3x-18}+\frac{2}{x^2-2x-15}}{\frac{3}{x^2-11x+30}+\frac{1}{x^2-9x+20}}$

$= \dfrac{\frac{1}{(x-6)(x+3)}+\frac{2}{(x+3)(x-5)}}{\frac{3}{(x-5)(x-6)}+\frac{1}{(x-5)(x-4)}}$

$= \dfrac{\frac{(x-5)+2(x-6)}{(x-6)(x+3)(x-5)}}{\frac{3(x-4)+1(x-6)}{(x-5)(x-6)(x-4)}}$

$= \dfrac{x-5+2x-12}{(x-6)(x+3)(x-5)} \cdot \dfrac{(x-5)(x-6)(x-4)}{3x-12+x-6}$

$= \dfrac{(3x-17)(x-4)}{(4x-18)(x+3)}$

$= \dfrac{3x^2-29x+68}{4x^2-6x-54}$

57. $\dfrac{3}{x} = \dfrac{8}{24}$

$\dfrac{3}{x} = \dfrac{1}{3}$

$x = 3 \cdot 3$

$x = 9$

This solution checks. The solution is 9.

58. $\dfrac{5.6}{a} = \dfrac{14.6}{7.3}$

$14.6a = (5.6)(7.3)$

$14.6a = 40.88$

$a = \dfrac{40.88}{14.6}$

$a = 2.8$

This solution checks. The solution is 2.8.

59. $\dfrac{x+3}{5} = \dfrac{9}{5}$

$5 \cdot \dfrac{x+3}{5} = 5 \cdot \dfrac{9}{5}$

$x + 3 = 9$

$x = 6$

This solution checks. The solution is 6.

60. $\dfrac{x}{3.4} = \dfrac{x-4}{5.2}$

$5.2x = 3.4(x-4)$

$5.2x = 3.4x - 13.6$

$1.8x = -13.6$

$x = \dfrac{-13.6}{1.8}$

$x = -7.\overline{5}$

This solution checks. The solution is $-7.\overline{5}$.

61. $\dfrac{3x+4}{5} = \dfrac{2x-8}{3}$

$3(3x+4) = 5(2x-8)$

$9x + 12 = 10x - 40$

$9x + 52 = 10x$

$52 = x$

This solution checks. The solution is 52.

62. $\dfrac{x}{4.8} + \dfrac{x}{2} = 1.7$

$9.6\left(\dfrac{x}{4.8}\right) + 9.6\left(\dfrac{x}{2}\right) = 9.6(1.7)$

$2x + 4.8x = 16.32$

$6.8x = 16.32$

$x = \dfrac{16.32}{6.8} = 2.4$

This solution checks. The solution is 2.4.

63. $\dfrac{4}{x} - \dfrac{1}{6} = \dfrac{1}{x}$

$6x\left(\dfrac{4}{x}\right) - 6x\left(\dfrac{1}{6}\right) = 6x\left(\dfrac{1}{x}\right)$

$24 - x = 6$

$-x = -18$

$x = 18$

This solution checks. The solution is 18.

64.

$$\frac{1}{x-2}+\frac{1}{x+2}=\frac{1}{x^2-4}$$

$$\frac{1}{x-2}+\frac{1}{x+2}=\frac{1}{(x+2)(x-2)}$$

$$(x+2)(x-2)\left(\frac{1}{x-2}\right)+(x+2)(x-2)\left(\frac{1}{x+2}\right)=(x+2)(x-2)\left(\frac{1}{(x+2)(x-2)}\right)$$

$$x+2+x-2=1$$

$$2x=1$$

$$x=\frac{1}{2}$$

This solution checks. The solution is $\frac{1}{2}$.

65.

$$\frac{x-3}{x-2}+\frac{x+1}{x+3}=\frac{2x^2+x+1}{x^2+x-6}$$

$$\frac{x-3}{x-2}+\frac{x+1}{x+3}=\frac{2x^2+x+1}{(x+3)(x-2)}$$

$$(x+3)(x-2)\left(\frac{x-3}{x-2}\right)+(x+3)(x-2)\left(\frac{x+1}{x+3}\right)=(x+3)(x-2)\left(\frac{2x^2+x+1}{(x+3)(x-2)}\right)$$

$$(x+3)(x-3)+(x-2)(x+1)=2x^2+x+1$$

$$x^2-9+x^2-x-2=2x^2+x+1$$

$$2x^2-x-11=2x^2+x+1$$

$$-x-11=x+1$$

$$-11=2x+1$$

$$-12=2x$$

$$-6=x$$

This solution checks. The solution is -6.

66.

$$\frac{x}{x^2-9}+\frac{2}{x+3}=\frac{4}{x-3}$$

$$\frac{x}{(x+3)(x-3)}+\frac{2}{x+3}=\frac{4}{x-3}$$

$$(x+3)(x-3)\left(\frac{x}{(x+3)(x-3)}\right)+(x+3)(x-3)\left(\frac{2}{x+3}\right)=(x+3)(x-3)\left(\frac{4}{x-3}\right)$$

$$x+2(x-3)=4(x+3)$$

$$x+2x-6=4x+12$$

$$3x-6=4x+12$$

$$-6=x+12$$

$$-18=x$$

This solution checks. The solution is -18.

67.
$$I = \frac{nE}{R+nr}$$
$$I(R+nr) = nE$$
$$\frac{I(R+nr)}{n} = E \text{ or}$$
$$E = \frac{I(R+nr)}{n}$$

68.
$$S = \frac{a-ar^n}{1-r}$$
$$S(1-r) = a-ar^n$$
$$S(1-r) = a(1-r^n)$$
$$\frac{S(1-r)}{1-r^n} = a \text{ or}$$
$$a = \frac{S(1-r)}{1-r^n}$$

69. $\dfrac{1}{R_T} = \dfrac{1}{R_1} + \dfrac{1}{R_2} + \dfrac{1}{R_3}$

Substitute 200 for R_1, 400 for R_2, and 1200 for R_3.

$$\frac{1}{R_T} = \frac{1}{200} + \frac{1}{400} + \frac{1}{1200}$$
$$1200R_T\left(\frac{1}{R_T}\right) = 1200R_T\left(\frac{1}{200}\right) + 1200R_T\left(\frac{1}{400}\right) + 1200R_T\left(\frac{1}{1200}\right)$$
$$1200 = 6R_T + 3R_T + R_T$$
$$1200 = 10R_T$$
$$\frac{1200}{10} = R_T$$
$$R_T = 120$$

The total resistance is 120 ohms.

70. $\dfrac{1}{R_T} = \dfrac{1}{R_1} + \dfrac{1}{R_2}$

Let $x = R_1$ and $2x = R_2$. Also, substitute 600 for R_T.

$$\frac{1}{600} = \frac{1}{x} + \frac{1}{2x}$$
$$\frac{1}{600} = \frac{2}{2x} + \frac{1}{2x}$$
$$\frac{1}{600} = \frac{3}{2x}$$
$$2x = 3 \cdot 600$$
$$2x = 1800$$
$$x = 900$$

The two resistances should be 900 ohms and $2 \cdot 900 = 1800$ ohms.

71.
$$\frac{1}{p} + \frac{1}{q} = \frac{1}{f}$$
$$\frac{1}{12} + \frac{1}{4} = \frac{1}{f}$$
$$\frac{1}{12} + \frac{3}{12} = \frac{1}{f}$$
$$\frac{4}{12} = \frac{1}{f}$$
$$\frac{1}{3} = \frac{1}{f}$$
$$f = 3$$
The focal length is 3 centimeters.

72. Let $p = x$, then $q = 2x$.
$$\frac{1}{p} + \frac{1}{q} = \frac{1}{f}$$
$$\frac{1}{x} + \frac{1}{2x} = \frac{1}{10}$$
$$10x\left(\frac{1}{x}\right) + 10x\left(\frac{1}{2x}\right) = 10x\left(\frac{1}{10}\right)$$
$$10 + 5 = x$$
The object's distance is 15 cm.

73. Let x be time needed for both of them to mow the lawn.

	Rate	Time Worked	Part of lawn mowed
Dan	$\frac{1}{3}$	x	$\frac{x}{3}$
Kim	$\frac{1}{4}$	x	$\frac{x}{4}$

$$\frac{x}{3} + \frac{x}{4} = 1$$
$$12\left(\frac{x}{3}\right) + 12\left(\frac{x}{4}\right) = 12(1)$$
$$4x + 3x = 12$$
$$7x = 12$$
$$x = \frac{12}{7} \approx 1.71$$
Working together, they can mow the lawn in about 1.71 hours.

74. Let x be the time needed for Pete to edit the manuscript.

	Rate	Time Worked	Part of job completed
Pete	$\frac{1}{x}$	40	$\frac{40}{x}$
Annette	$\frac{1}{75}$	40	$\frac{40}{75}$

$$\frac{40}{75} + \frac{40}{x} = 1$$
$$75x\left(\frac{40}{75}\right) + 75x\left(\frac{40}{x}\right) = 75x(1)$$
$$40x + 3000 = 75x$$
$$3000 = 35x$$
$$\frac{3000}{35} = x$$
$$85.71 \approx x$$
Working alone, it takes Pete about 85.71 hours to edit the manuscript.

75. Let x be the unknown number.
$$\frac{5x}{x+8} = 1$$
$$5x = x + 8$$
$$4x = 8$$
$$x = 2$$
The desired number is 2.

76. Let x be the number.
$$1 - \frac{1}{2x} = \frac{1}{3x}$$
$$6x(1) - 6x\left(\frac{1}{2x}\right) = 6x\left(\frac{1}{3x}\right)$$
$$6x - 3 = 2$$
$$6x = 5$$
$$x = \frac{5}{6}$$
The desired number is $\frac{5}{6}$.

77. Let x be the speed of the current.

	d	r	$t = \dfrac{d}{r}$
With current	20	$15 + x$	$\dfrac{20}{15 + x}$
Against current	10	$15 - x$	$\dfrac{10}{15 - x}$

The times are the same.
$$\frac{20}{15+x} = \frac{10}{15-x}$$
$$20(15-x) = 10(15+x)$$
$$300 - 20x = 150 + 10x$$
$$300 = 150 + 30x$$
$$150 = 30x$$
$$5 = x$$
The speed of the current is 5 mph.

78. Let x be the speed of the car.

	d	r	$t = \dfrac{d}{r}$
Car	450	x	$\dfrac{450}{x}$
Plane	450	$3x$	$\dfrac{450}{3x}$

$$\frac{450}{3x} + 6 = \frac{450}{x}$$
$$\frac{150}{x} + 6 = \frac{450}{x}$$
$$x\left(\frac{150}{x}\right) + x(6) = x\left(\frac{450}{x}\right)$$
$$150 + 6x = 450$$
$$6x = 300$$
$$x = \frac{300}{6} = 50$$
The speed of the car is 50 mph and the speed of the plane is 3(50) = 150 mph.

79. The equation is $A = kC^2$. To find k, substitute 5 for A and 5 for C.
$$5 = k(5)^2$$
$$5 = 25k$$
$$\frac{5}{25} = k$$
$$\frac{1}{5} = k$$
Thus $A = \frac{1}{5}C^2$. Now substitute 10 for C.
$$A = \frac{1}{5}(10)^2 = \frac{1}{5}(100) = 20$$

80. The equation is $W = \dfrac{kL}{A}$. To find k, substitute 80 for W, 100 for L, and 20 for A.
$$80 = \frac{k(100)}{20}$$
$$80 = 5k$$
$$\frac{80}{5} = k$$
$$16 = k$$
Thus $W = \dfrac{16L}{A}$. Now substitute 50 for L and 40 for A.
$$W = \frac{16 \cdot 50}{40} = \frac{800}{40} = 20$$

81. The equation is $z = \dfrac{kxy}{r^2}$. To find k, substitute 12 for z, 20 for x, 8 for y, and 8 for r.
$$12 = \frac{k(20)(8)}{(8)^2}$$
$$12 = \frac{160k}{64}$$
$$12(64) = 160k$$
$$768 = 160k$$
$$\frac{768}{160} = k$$
$$k = 4.8$$
Thus $z = \dfrac{4.8xy}{r^2}$. Now substitute 10 for x, 80 for y, and 3 for r.
$$z = \frac{4.8(10)(80)}{3^2} \approx 426.7$$

82. Let x represent the distance on the map and h represent the actual distance. The equation is
$h = kx$.
To find k, substitute 1 for x and 60 for h.
$60 = k(1)$
$60 = k$
Thus $h = 60x$. Now substitute 300 for h.
$300 = 60x$
$\dfrac{300}{60} = x$
$5 = x$
The distance on the map is 5 inches.

83. Let B be the amount of the bill and H the number of kilowatt-hours used. The equation is
$B = 0.162H$.
Now substitute 740 for H.
$B = 0.162(740)$
$B = 119.88$
The bill is $119.88.

84. The equation is $d = kt^2$. To find k, substitute 16 for d and 1 for t.
$16 = k(1)^2$
$16 = k(1)$
$16 = k$
Thus $d = 16t^2$. Now substitute 5 for t.
$d = 16(5)^2 = 16(25) = 400$
An object will fall 400 feet.

85. The equation is $A = kr^2$. To find k, substitute 78.5 for A and 5 for r.
$78.5 = k(5)^2$
$78.5 = k(25)$
$\dfrac{78.5}{25} = k$
$3.14 = k$
Thus $A = 3.14r^2$. Now substitute 8 for r.
$A = 3.14(8)^2$
$A = 3.14(64)$
$A = 200.96$
The area is 200.96 square units.

86. The equation is $t = \dfrac{k}{w}$ where t is time and w is water temperature. To find k, substitute 1.7 for t and 70 for w.
$1.7 = \dfrac{k}{70}$
$(1.7)(70) = k$
$119 = k$
Thus $t = \dfrac{119}{w}$. Now substitute 50 for w.
$t = \dfrac{119}{50} = 2.38$
It takes the ice cube 2.38 minutes to melt.

Practice Test

1. $\dfrac{x+4}{x^2 - 3x - 28} = \dfrac{x+4}{(x+4)(x-7)}$
The denominator cannot equal zero.
Therefore,
the excluded values are –4 and 7.

2. $f(x) = \dfrac{x^2 - 4x}{8x^2 + 10x - 25} = \dfrac{x(x-4)}{(4x-5)(2x+5)}$
The denominator cannot equal zero.
Therefore,
$4x - 5 \neq 0$ or $2x + 5 \neq 0$
$x \neq \dfrac{5}{4}$ $x \neq -\dfrac{5}{2}$
The domain is
$\left\{ x \middle| x \text{ is a real number and } x \neq \dfrac{5}{4}, x \neq -\dfrac{5}{2} \right\}$

3. $\dfrac{8x^7 y^2 + 16x^2 y + 18x^3 y^3}{2x^2 y}$
$= \dfrac{8x^7 y^2}{2x^2 y} + \dfrac{16x^2 y}{2x^2 y} + \dfrac{18x^3 y^3}{2x^2 y}$
$= 4x^5 y + 8 + 9xy^2$

4. $\dfrac{4x^2 + 4xy - 15y^2}{4x^2 - 4xy - 3y^2} = \dfrac{(2x+5y)(2x-3y)}{(2x+y)(2x-3y)}$
$= \dfrac{2x+5y}{2x+y}$

5. $\dfrac{3xy^4}{6x^2y^3} \cdot \dfrac{2x^2y^4}{x^5y^7} = \dfrac{3 \cdot 2x^{1+2}y^{4+4}}{6 \cdot 1x^{2+5}y^{7+3}}$

$\phantom{\dfrac{3xy^4}{6x^2y^3} \cdot \dfrac{2x^2y^4}{x^5y^7}} = \dfrac{6x^3y^8}{6x^7y^{10}}$

$\phantom{\dfrac{3xy^4}{6x^2y^3} \cdot \dfrac{2x^2y^4}{x^5y^7}} = \dfrac{1}{x^4y^2}$

6. $\dfrac{a^2 - 9a + 14}{a - 2} \cdot \dfrac{a^2 - 4a - 21}{(a-7)^2}$

$= \dfrac{(a-7)(a-2)}{a-2} \cdot \dfrac{(a-7)(a+3)}{(a-7)(a-7)}$

$= a + 3$

7. $\dfrac{x^2 - 9y^2}{3x + 6y} \div \dfrac{x + 3y}{x + 2y}$

$= \dfrac{x^2 - 9y^2}{3x + 6y} \cdot \dfrac{x + 2y}{x + 3y}$

$= \dfrac{(x+3y)(x-3y)}{3(x+2y)} \cdot \dfrac{x + 2y}{x + 3y}$

$= \dfrac{x - 3y}{3}$

8. $\dfrac{x^3 + y^3}{x + y} \div \dfrac{x^2 - xy + y^2}{x^2 + y^2}$

$= \dfrac{x^3 + y^3}{x + y} \cdot \dfrac{x^2 + y^2}{x^2 - xy + y^2}$

$= \dfrac{(x+y)(x^2 - xy + y^2)}{x + y} \cdot \dfrac{x^2 + y^2}{x^2 - xy + y^2}$

$= x^2 + y^2$

9. $\dfrac{5}{x} + \dfrac{3}{2x^2} = \dfrac{5}{x} \cdot \dfrac{2x}{2x} + \dfrac{3}{2x^2}$

$\phantom{\dfrac{5}{x} + \dfrac{3}{2x^2}} = \dfrac{10x}{2x^2} + \dfrac{3}{2x^2}$

$\phantom{\dfrac{5}{x} + \dfrac{3}{2x^2}} = \dfrac{10x + 3}{2x^2}$

10. $\dfrac{x - 5}{x^2 - 16} - \dfrac{x - 2}{x^2 + 2x - 8}$

$= \dfrac{x - 5}{(x+4)(x-4)} - \dfrac{x - 2}{(x+4)(x-2)}$

$= \dfrac{x - 5}{(x+4)(x-4)} - \dfrac{1}{x + 4}$

$= \dfrac{x - 5}{(x+4)(x-4)} - \dfrac{1}{x + 4} \cdot \dfrac{x - 4}{x - 4}$

$= \dfrac{x - 5}{(x+4)(x-4)} - \dfrac{x - 4}{(x+4)(x-4)}$

$= \dfrac{(x-5) - (x-4)}{(x+4)(x-4)}$

$= \dfrac{x - 5 - x + 4}{(x+4)(x-4)}$

$= -\dfrac{1}{(x+4)(x-4)}$

11. $\dfrac{x+1}{4x^2-4x+1}+\dfrac{3}{2x^2+5x-3}=\dfrac{x+1}{(2x-1)(2x-1)}+\dfrac{3}{(2x-1)(x+3)}$

$$=\dfrac{x+1}{(2x-1)(2x-1)}\cdot\dfrac{x+3}{x+3}+\dfrac{3}{(2x-1)(x+3)}\cdot\dfrac{2x-1}{2x-1}$$

$$=\dfrac{(x+1)(x+3)}{(2x-1)(2x-1)(x+3)}+\dfrac{3(2x-1)}{(2x-1)(2x-1)(x+3)}$$

$$=\dfrac{x^2+4x+3}{(2x-1)(2x-1)(x+3)}+\dfrac{6x-3}{(2x-1)(2x-1)(x+3)}$$

$$=\dfrac{x^2+4x+3+6x-3}{(2x-1)(2x-1)(x+3)}$$

$$=\dfrac{x^2+10x}{(2x-1)(2x-1)(x+3)} \text{ or } \dfrac{x(x+10)}{(2x-1)^2(x+3)}$$

12. $\dfrac{m}{12m^2+4mn-5n^2}+\dfrac{2m}{12m^2+28mn+15n^2}$

$$=\dfrac{m}{(6m+5n)(2m-n)}+\dfrac{2m}{(6m+5n)(2m+3n)}$$

$$=\dfrac{m}{(6m+5n)(2m-n)}\cdot\dfrac{2m+3n}{2m+3n}+\dfrac{2m}{(6m+5n)(2m+3n)}\cdot\dfrac{2m-n}{2m-n}$$

$$=\dfrac{m(2m+3n)+2m(2m-n)}{(6m+5n)(2m-n)(2m+3n)}$$

$$=\dfrac{2m^2+3mn+4m^2-2mn}{(6m+5n)(2m-n)(2m+3n)}$$

$$=\dfrac{6m^2+mn}{(6m+5n)(2m-n)(2m+3n)} \text{ or } \dfrac{m(6m+n)}{(6m+5n)(2m-n)(2m+3n)}$$

13. $\dfrac{r^2-16}{r^3-64}\div\dfrac{r^2+2r-8}{r^2+4r+16}$

$$=\dfrac{r^2-16}{r^3-64}\cdot\dfrac{r^2+4r+16}{r^2+2r-8}$$

$$=\dfrac{(r-4)(r+4)}{(r-4)(r^2+4r+16)}\cdot\dfrac{r^2+4r+16}{(r+4)(r-2)}$$

$$=\dfrac{1}{r-2}$$

14. $(f+g)(x)=f(x)+g(x)$

$$=\dfrac{x-3}{x+5}+\dfrac{x}{2x+3}$$

$$=\dfrac{x-3}{x+5}\cdot\dfrac{2x+3}{2x+3}+\dfrac{x}{2x+3}\cdot\dfrac{x+5}{x+5}$$

$$=\dfrac{(x-3)(2x+3)+x(x+5)}{(x+5)(2x+3)}$$

$$=\dfrac{2x^2-3x-9+x^2+5x}{(x+5)(2x+3)}$$

$$=\dfrac{3x^2+2x-9}{(x+5)(2x+3)}$$

15.
$$(f-g)(x)$$
$$= f(x) - g(x)$$
$$= \frac{x-3}{x+5} - \frac{x}{2x+3}$$
$$= \frac{x-3}{x+5} \cdot \frac{2x+3}{2x+3} - \frac{x}{2x+3} \cdot \frac{x+5}{x+5}$$
$$= \frac{(x-3)(2x+3) - x(x+5)}{(x+5)(2x+3)}$$
$$= \frac{2x^2 - 3x - 9 - x^2 - 5x}{(x+5)(2x+3)}$$
$$= \frac{x^2 - 8x - 9}{(x+5)(2x+3)} \text{ or } \frac{(x-9)(x+1)}{(x+5)(2x+3)}$$

16. The denominator of $(f+g)(x)$ cannot be zero.

Therefore $x + 5 \neq 0$ or $2x + 3 \neq 0$

$$x \neq -5 \qquad\qquad x \neq -\frac{3}{2}$$

The domain is

$$\left\{ x \middle| x \text{ is a real number, } x \neq -5, x \neq -\frac{3}{2} \right\}$$

17.
$$\frac{\dfrac{1}{x} + \dfrac{1}{y}}{\dfrac{1}{x} - \dfrac{1}{y}} = \frac{xy\left(\dfrac{1}{x} + \dfrac{1}{y}\right)}{xy\left(\dfrac{1}{x} - \dfrac{1}{y}\right)} = \frac{xy\left(\dfrac{1}{x}\right) + xy\left(\dfrac{1}{y}\right)}{xy\left(\dfrac{1}{x}\right) - xy\left(\dfrac{1}{y}\right)} = \frac{y+x}{y-x}$$

18.
$$\frac{x + \dfrac{x}{y}}{x^{-1} + y^{-1}} = \frac{x + \dfrac{x}{y}}{\dfrac{1}{x} + \dfrac{1}{y}}$$

$$= \frac{xy\left(x + \dfrac{x}{y}\right)}{xy\left(\dfrac{1}{x} + \dfrac{1}{y}\right)}$$

$$= \frac{xy(x) + xy\left(\dfrac{x}{y}\right)}{xy\left(\dfrac{1}{x}\right) + xy\left(\dfrac{1}{y}\right)}$$

$$= \frac{x^2 y + x^2}{y + x} \text{ or } \frac{x^2(y+1)}{y+x}$$

19.
$$\frac{x}{3} - \frac{x}{4} = 5$$
$$12\left(\frac{x}{3}\right) - 12\left(\frac{x}{4}\right) = 12(5)$$
$$4x - 3x = 60$$
$$x = 60$$

20.

$$\frac{x}{x-8} + \frac{6}{x-2} = \frac{x^2}{x^2-10x+16}$$

$$\frac{x}{x-8} + \frac{6}{x-2} = \frac{x^2}{(x-8)(x-2)}$$

$$(x-8)(x-2)\left(\frac{x}{x-8}\right) + (x-8)(x-2)\left(\frac{6}{x-2}\right) = (x-8)(x-2)\left(\frac{x^2}{(x-8)(x-2)}\right)$$

$$x(x-2) + 6(x-8) = x^2$$

$$x^2 - 2x + 6x - 48 = x^2$$

$$x^2 + 4x - 48 = x^2$$

$$4x - 48 = 0$$

$$4x = 48$$

$$x = 12$$

This solution checks. The solution is 12.

21.
$$I = \frac{2R}{w+2s}$$
$$I(w+2s) = 2R$$
$$Iw + 2Is = 2R$$
$$2Is = 2R - Iw$$
$$s = \frac{2R - Iw}{2I}$$
$$s = \frac{2R}{2I} - \frac{Iw}{2I}$$
$$s = \frac{R}{I} - \frac{w}{2}$$

22. The equation is $W = kI^2R$.
To find k, substitute 10 for W, 1 for I, and 1000 for R.
$$10 = k \cdot 1^2 \cdot 1000$$
$$10 = 1000k$$
$$0.01 = k$$
Thus $W = 0.01I^2R$. Now substitute 0.5 for I and 300 for R.
$$W = 0.01(0.5)^2(300)$$
$$W = 0.75$$
The wattage is 0.75 watt.

23. The equation is $W = \frac{kPQ}{T^2}$.

To find k, substitute 6 for W, 20 for P, 8 for Q, and 4 for T.
$$6 = \frac{k(20)(8)}{4^2}$$
$$6 = \frac{160k}{16}$$
$$6 = 10k$$
$$\frac{6}{10} = k$$
$$0.6 = k$$

Thus $W = \frac{0.6PQ}{T^2}$. Now substitute 30 for P, 4 for Q, and 8 for T.
$$W = \frac{0.6(30)(4)}{8^2} = \frac{0.6(120)}{64} = \frac{72}{64} = 1.125$$

24. Let x be the amount of time needed for both of them to level the field.

	Rate	Time worked	Part of job
Kris	$\frac{1}{8}$	x	$\frac{x}{8}$
Heather	$\frac{1}{5}$	x	$\frac{x}{5}$

$$\frac{x}{8} + \frac{x}{5} = 1$$

$$40\left(\frac{x}{8}\right) + 40\left(\frac{x}{5}\right) = 40(1)$$

$$5x + 8x = 40$$

$$13x = 40$$

$$x = \frac{40}{13} \approx 3.08$$

Working together, they can level the field in about 3.08 hrs.

25. Let x be the length of the trail.

	d	r	$t = \dfrac{d}{r}$
Cameron	x	8	$\dfrac{x}{8}$
Ashley	x	5	$\dfrac{x}{5}$

$$\frac{x}{8} = \frac{x}{5} - \frac{1}{2}$$

$$40\left(\frac{x}{8}\right) = 40\left(\frac{x}{5}\right) - 40\left(\frac{1}{2}\right)$$

$$5x = 8x - 20$$

$$20 = 3x$$

$$\frac{20}{3} = x$$

$$x = 6\frac{2}{3}$$

The trail is $6\frac{2}{3}$ miles long.

Cumulative Review Test

1. $\left\{ x \,\middle|\, -\frac{5}{3} < x \le \frac{19}{4} \right\}$

2. $-3x^3 - 2x^2 y + \frac{1}{2} xy^2$

$$= -3(2)^3 - 2(2)^2\left(\frac{1}{2}\right) + \frac{1}{2}(2)\left(\frac{1}{2}\right)^2$$

$$= -3(8) - 2(4)\left(\frac{1}{2}\right) + \frac{1}{2}(2)\left(\frac{1}{4}\right)$$

$$= -24 - 4 + \frac{1}{4}$$

$$= -28 + \frac{1}{4}$$

$$= -27\frac{3}{4}$$

3. $\left(\dfrac{4a^3 b^{-2}}{2a^{-2} b^{-3}}\right)^{-2} = \left(\dfrac{4}{2} \cdot a^{3-(-2)} b^{-2-(-3)}\right)^{-2}$

$$= (2a^5 b)^{-2}$$

$$= \left(\frac{1}{2a^5 b}\right)^2$$

$$= \frac{1}{2^2 a^{5\cdot 2} b^2}$$

$$= \frac{1}{4a^{10} b^2}$$

4. a. $(1.631 \times 10^{12})(0.459)$

$$= (1.631 \times 10^{12})(4.59 \times 10^{-1})$$

$$= (1.631)(4.59) \times 10^{12-1}$$

$$= 7.48629 \times 10^{11}$$

About 7.486×10^{11} was collected from individuals.

b. $(1.631 \times 10^{12})(0.118)$

$$= (1.631 \times 10^{12})(1.18 \times 10^{-1})$$

$$= (1.631)(1.18) \times 10^{12-1}$$

$$= 1.92458 \times 10^{11}$$

About 1.925×10^{11} was collected from corporations.

c. $7.486 \times 10^{11} - 1.95 \times 10^{11}$

$$= 5.561 \times 10^{11}$$

The difference is about 5.561×10^{11}.

5. $4[3x - 2(2x - 4)] = -[6x - 4 - (3x + 8)]$
$4(3x - 4x + 8) = -(6x - 4 - 3x - 8)$
$4(-x + 8) = -(3x - 12)$
$-4x + 32 = -3x + 12$
$20 = x$

6. $\dfrac{-b + \sqrt{b^2 - 4ac}}{2a}$

$= \dfrac{-(-14) + \sqrt{(-14)^2 - 4(1)(48)}}{2(1)}$

$= \dfrac{14 + \sqrt{196 - 192}}{2}$

$= \dfrac{14 + \sqrt{4}}{2}$

$= \dfrac{14 + 2}{2}$

$= \dfrac{16}{2}$

$= 8$

7. a. Let x = client's assets
   Plan 1: $0.06x$
   Plan 2: $500 + 0.04x$
   $0.06x = 500 + 0.04x$
   $0.02x = 500$
   $x = 25,000$
   They are equal at $25,000

b. Plan 2 would give the lower fee because $60,000 is greater than $25,000.

8. Let x be the number of bottles filled per hour at the faster speed.
   $3.6x + 4.4(x - 400) = 25,440$
   $3.6x + 4.4x - 1760 = 25,440$
   $8x = 27,200$
   $x = 3400$
   The machine fills 3400 bottles per hour at the faster speed.

9. $y = x^2 - 2$

10. $y = |x| + 2$

11. a) and **c)** are functions.

12. $3x - 5y = 7$
   $-5y = -3x + 7$
   $y = \dfrac{3}{5}x - \dfrac{7}{5}$

   $m = \dfrac{3}{5}$, y-intercept = -2

   $y = \dfrac{3}{5}x - 2$

13. $4x - 3y = 12$
   $-3y = -4x + 12$
   $y = \dfrac{4}{3}x - 4$

   $m = \dfrac{4}{3}$

   The slope of a perpendicular line will be
   $m = -\dfrac{3}{4}$.

$$y - y_1 = m(x - x_1)$$
$$y - (-2) = -\frac{3}{4}(x - 3)$$
$$y + 2 = -\frac{3}{4}x + \frac{9}{4}$$
$$y = -\frac{3}{4}x + \frac{1}{4}$$

14. a. $(G + B)(96) = G(96) + B(96)$
      $= 60,703 + 97,312$
      $= 158,015$

 b. $(B - G)(96) = B(96) - G(96)$
      $= 97,312 - 60,703$
      $= 36,609$

 c. Answers will vary.

15. $2a - b - 2c = -1$ (1)
   $a - 2b - c = 1$ (2)
   $a + b + c = 4$ (3)
To eliminate a and c between equations (1)
and (2), multiply equation (2) by –2 and add.
 $2a - b - 2c = -1$
$-2[a - 2b - c = 1]$
gives
    $2a - b - 2c = -1$
    $\underline{-2a + 4b + 2c = -2}$
Add: $3b = -3$
     $b = -1$
To eliminate c between equations (2) and
(3), simply add.
   $a - 2b - c = 1$
   $\underline{a + b + c = 4}$
Add: $2a - b\quad = 5$ (4)
Substitute –1 for b in equation (4).
  $2a - b = 5$
 $2a - (-1) = 5$
  $2a + 1 = 5$
   $2a = 4$
   $a = 2$
Substitute 2 for a and –1 for b in equation
(3).
  $a + b + c = 4$
 $2 + (-1) + c = 4$
   $1 + c = 4$
    $c = 3$
The solution is (2, –1, 3).

16. Let x be the amount of 10% sulfuric acid,
then $4 - x$ is the amount of 20% sulfuric
acid.
 $0.10x + 0.20(4 - x) = 4(0.12)$
  $0.10x + 0.8 - 0.2x = 0.48$
      $-0.1x = -0.32$
       $x = 3.2$
     $4 - x = 0.8$
John should mix 3.2 liters of 10% solution
with 0.8 liter of 20% solution.

17. a. $(f \cdot g)(x)$
  $= f(x) \cdot g(x)$
  $= (x^2 - 11x + 30)(x - 5)$
  $= x^3 - 5x^2 - 11x^2 + 55x + 30x - 150$
  $= x^3 - 16x^2 + 85x - 150$

 b. $\left(\dfrac{f}{g}\right)(x) = \dfrac{f(x)}{g(x)}$

   $= \dfrac{x^2 - 11x + 30}{x - 5}$

   $= \dfrac{(x - 5)(x - 6)}{x - 5}$

   $= x - 6$

 c. $\left(\dfrac{f}{g}\right)(x) = \dfrac{x^2 - 11x + 30}{x - 5}$

  The denominator cannot equal zero.
  Therefore $x \neq 5$.
  The domain is
  $\{x | x$ is a real number, $x \neq 5\}$

18. $24p^3q + 16p^2q - 30pq$
  $= 2pq(12p^2 + 8p - 15)$
  $= 2pq(2p + 3)(6p - 5)$

19.
$$\frac{5}{c} = 2 - \frac{2c}{c+1}$$

$$c(c+1)\left(\frac{5}{c}\right) = c(c+1)(2) - c(c+1)\left(\frac{2c}{c+1}\right)$$

$$5(c+1) = 2c(c+1) - 2c^2$$

$$5c + 5 = 2c^2 + 2c - 2c^2$$

$$5c + 5 = 2c$$

$$3c = -5$$

$$c = -\frac{5}{3}$$

20.
$$\frac{4}{r+5} + \frac{1}{r+3} = \frac{2}{r^2 + 8r + 15}$$

$$\frac{4}{r+5} + \frac{1}{r+3} = \frac{2}{(r+5)(r+3)}$$

$$(r+5)(r+3)\left(\frac{4}{r+5} + \frac{1}{r+3}\right) = (r+5)(r+3)\left(\frac{2}{(r+5)(r+3)}\right)$$

$$4(r+3) + (r+5) = 2$$

$$4r + 12 + r + 5 = 2$$

$$5r = -15$$

$$r = -3$$

When $r = -3$, $\frac{1}{r+3} = \frac{1}{0}$, which is undefined.

Therefore, there is no solution.

Chapter 7

Exercise Set 7.1

1. **a.** Every real number has two square roots; a positive or principal square root and a negative square root.

   b. The square roots of 36 are 6 and –6.

   c. When we say square root, we are referring to the principal square root.

   d. $\sqrt{36} = 6$

3. There is no real number which, when squared, results in –49.

5. No. If the number under the radical is negative, the answer is not a real number.

7. $\sqrt{49} = 7$ since $7^2 = 49$

9. $\sqrt[3]{-64} = -4$ since $(-4)^3 = -64$

11. $\sqrt[3]{125} = 5$ since $5^3 = 125$

13. $\sqrt{-9}$ is not a real number.

15. $\sqrt[3]{216} = 6$ since $6^3 = 216$

17. $\sqrt[3]{-343} = -7$ since $(-7)^3 = -343$

19. $\sqrt[4]{16} = 2$ since $2^4 = 16$

21. $\sqrt{-36}$ is not a real number.

23. $-\sqrt[3]{102.4} \approx -4.68$

25. $\sqrt{\dfrac{25}{9}} = \dfrac{5}{3}$ since $\left(\dfrac{5}{3}\right)^2 = \dfrac{25}{9}$

27. $\sqrt[4]{-81}$ is not a real number.

29. $\sqrt[5]{16.2} \approx 1.75$

31. $\sqrt{4^2} = |4| = 4$

33. $\sqrt{(-1)^2} = |-1| = 1$

35. $\sqrt{(43)^2} = |43| = 43$

37. $\sqrt{(235.23)^2} = |235.23| = 235.23$

39. $\sqrt{(-0.03)^2} = |-0.03| = 0.03$

41. $\sqrt{\left(-\dfrac{162}{5}\right)^2} = \left|-\dfrac{162}{5}\right| = \dfrac{162}{5}$ or 32.4

43. $\sqrt{(a-9)^2} = |a-9|$

45. $\sqrt{(x-3)^2} = |x-3|$

47. $\sqrt{(3x+5)^2} = |3x+5|$

49. $\sqrt{(6-3x)^2} = |6-3x|$

51. $\sqrt{(y^2-4y+3)^2} = |y^2-4y+3|$

53. $\sqrt{(8a-b)^2} = |8a-b|$

55. $\sqrt{a^8} = \sqrt{(a^4)^2} = |a^4|$

57. $\sqrt{a^{24}} = \sqrt{(a^{12})^2} = |a^{12}|$

59. $\sqrt{a^2-8a+16} = \sqrt{(a-4)^2} = |a-4|$

61. $\sqrt{x^2-8x+16} = \sqrt{(x-4)^2} = |x-4|$

345

63. $f(x) = \sqrt{5x - 6}$
$f(2) = \sqrt{5 \cdot 2 - 6}$
$\quad\quad = \sqrt{10 - 6}$
$\quad\quad = \sqrt{4}$
$\quad\quad = 2$

65. $f(x) = \sqrt{64 - 8x}$
$f(-3) = \sqrt{64 - 8(-3)}$
$\quad\quad\quad = \sqrt{64 + 24}$
$\quad\quad\quad = \sqrt{88}$
$\quad\quad\quad \approx 9.381$

67. $f(c) = \sqrt[3]{9c^2 - 4}$
$f(4) = \sqrt[3]{9(4)^2 - 4}$
$\quad\quad = \sqrt[3]{144 - 4}$
$\quad\quad = \sqrt[3]{140}$
$\quad\quad \approx 5.192$

69. $g(x) = \sqrt[3]{-3x^2 + 6x - 1}$
$g(-3) = \sqrt[3]{-3(-3)^2 + 6(-3) - 1}$
$\quad\quad\quad = \sqrt[3]{-27 - 18 - 1}$
$\quad\quad\quad = \sqrt[3]{-46}$
$\quad\quad\quad \approx -3.583$

71. Select $x = 0$.
$\sqrt{(2(0) - 1)^2} \neq 2(0) - 1$
$\sqrt{(-1)^2} \neq -1$
$\sqrt{1} \neq -1$
$1 \neq -1$
This is true for all $x < \dfrac{1}{2}$.

73. $\sqrt{(x - 1)^2} = x - 1$ for all $x \geq 1$. The expression $\sqrt{(x - 1)^2} = x - 1$, when $(x - 1)$ is equal to zero or positive. Therefore, $x - 1 \geq 0$ and $x \geq 1$.

75. $\sqrt{(2x - 6)^2} = 2x - 6$ for all $x \geq 3$. The expression $\sqrt{(2x - 6)^2} = 2x - 6$, when $(2x - 6)$ is positive or equal to 0. Therefore,
$2x - 6 \geq 0$
$\quad\quad x \geq 3$

77. a. $\sqrt{a^2} = |a|$ for all real values

b. $\sqrt{a^2} = a$ when $a \geq 0$

79. If n is even, we are finding the even root of a positive number. If n is odd, the expression is also real.

81. $\dfrac{\sqrt{x + 3}}{\sqrt[3]{x + 3}}$ The denominator cannot equal zero.
$\sqrt[3]{x + 3} \neq 0$
$\quad\quad x \neq -3$
The numerator must be greater than or equal to zero.
$\sqrt{x + 3} \geq 0$
$\quad\quad x \geq -3$
Therefore the domain is
$\{x | x \text{ is a real number } x > -3\}$

83. $f(x) = \sqrt{x}$ matches graph d).
The x-intercept is 0 and the domain is $x \geq 0$.

85. $f(x) = \sqrt{x - 5}$ matches graph a). The x-intercept is 5.

87. One answer is $f(x) = \sqrt{x - 4}$

89. $f(x) = -\sqrt{x}$

a. No

b. Yes

c. Yes

Answers will vary.

91. $V = \sqrt{64.4h}$

a. $V = \sqrt{64.4(20)}$
$\quad\quad = \sqrt{1288}$
$\quad\quad \approx 35.89$
The velocity will be about 35.89 ft/sec.

b. $V = \sqrt{64.4(40)}$
      $= \sqrt{2576}$
      ≈ 50.75

The velocity will be about 50.75 ft/sec.

93. $f(x) = \sqrt{x+1}$

x	$f(x)$
-1	0
0	1
3	2
8	3

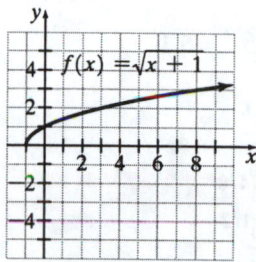

95. $g(x) = \sqrt{x} + 1$

x	$g(x)$
0	1
4	3
9	4

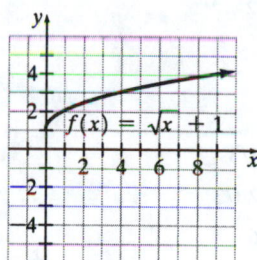

97. $y_1 = \sqrt{x+1}$

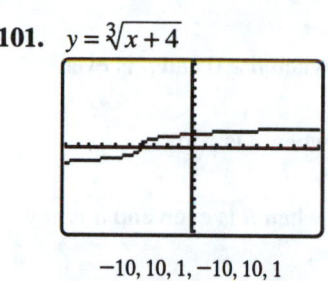

$-10, 10, 1, -10, 10, 1$

99. $y_1 = \dfrac{\sqrt{x+3}}{\sqrt[3]{x+3}}$

$-10, 10, 1, -10, 10, 1$

101. $y = \sqrt[3]{x+4}$

$-10, 10, 1, -10, 10, 1$

106. $3y^2 - 18y + 27 - 3z^2$
      $= 3(y^2 - 6y + 9 - z^2)$
      $= 3[(y^2 - 6y + 9) - z^2]$
      $= 3[(y-3)^2 - z^2]$
      $= 3[(y-3) + z][(y-3) - z]$
      $= 3(y - 3 + z)(y - 3 - z)$

107. $x^3 + \dfrac{1}{27} = x^3 + \left(\dfrac{1}{3}\right)^3$
      $= \left(x + \dfrac{1}{3}\right)\left(x^2 - \dfrac{1}{3}x + \dfrac{1}{9}\right)$

108. $(x+2)^2 - (x+2) - 12$
      $= y^2 - y - 12$ ← Use y for $x + 2$
      $= (y - 4)(y + 3)$ ← Factor
      $= [(x+2) - 4][(x+2) + 3]$ ← Use $x + 2$ for y
      $= (x - 2)(x + 5)$ ← Simplify

109. $2x^4 - 3x^3 - 6x^2 + 9x$
      $= x[2x^3 - 3x^2 - 6x + 9]$
      $= x[x^2(2x - 3) - 3(2x - 3)]$
      $= x(2x - 3)(x^2 - 3)$

Exercise Set 7.2

1. a. $\sqrt[n]{a}$ is a real number when n is even and $a \geq 0$, or n is odd.

b. $\sqrt[n]{a}$ can be expressed with rational exponents as $a^{1/n}$.

3. a. $\sqrt[n]{a^n}$ is always real

b. $\sqrt[n]{a^n} = a$ when $a \geq 0$ and n is even

c. $\sqrt[n]{a^n} = a$ when n is odd

d. $\sqrt[n]{a^n} = |a|$ when n is even and a is any real number

5. $\sqrt{x^5} = x^{5/2}$

7. $\sqrt{8^5} = 8^{5/2}$

9. $\sqrt[5]{x^4} = x^{4/5}$

11. $\left(\sqrt{x}\right)^3 = x^{3/2}$

13. $\sqrt[4]{a^3 b^5} = (a^3 b^5)^{1/4}$

15. $\sqrt[8]{5r^4 s^9 t^{12}} = (5r^4 s^9 t^{12})^{1/8}$

17. $x^{1/2} = \sqrt{x}$

19. $z^{3/2} = \sqrt{z^3}$

21. $(24y^2)^{1/2} = \sqrt{24y^2}$

23. $(19x^2 y^4)^{-1/2} = \dfrac{1}{\sqrt{19x^2 y^4}}$

25. $(2m^2 n^3)^{2/5} = \left(\sqrt[5]{2m^2 n^3}\right)^2$

27. $(a^2 - 4b^2)^{-2/3} = \dfrac{1}{\left(\sqrt[3]{a^2 - 4b^2}\right)^2}$

29. $\sqrt{y^6} = y^{6/2} = y^3$

31. $\sqrt{z^8} = z^{8/2} = z^4$

33. $\sqrt[3]{x^9} = x^{9/3} = x^3$

35. $\sqrt[10]{z^5} = z^{5/10} = z^{1/2} = \sqrt{z}$

37. $(\sqrt{5.1})^2 = (5.1)^{2/2} = 5.1$

39. $\sqrt[6]{y^6} = y^{6/6} = y^1 = y$

41. $(\sqrt[8]{xyz})^4 = (xyz)^{4/8}$
$= (xyz)^{1/2}$
$= \sqrt{xyz}$

43. $(\sqrt[3]{xy^2})^9 = (xy^2)^{9/3}$
$= (xy^2)^3$
$= x^3 y^6$

45. $\sqrt{\sqrt{x}} = (\sqrt{x})^{1/2}$
$= (x^{1/2})^{1/2}$
$= x^{1/4}$
$= \sqrt[4]{x}$

47. $\sqrt[3]{\sqrt{x^5}} = (\sqrt{x^5})^{1/3}$
$= (x^5)^{1/2 \cdot 1/3}$
$= (x^5)^{1/6}$
$= \sqrt[6]{x^5}$

49. $4^{1/2} = \sqrt{4} = 2$

51. $(-4)^{1/2} = \sqrt{-4}$ Not a real number

53. $\left(\dfrac{9}{25}\right)^{1/2} = \sqrt{\dfrac{9}{25}} = \dfrac{3}{5}$

55. $-16^{1/2} = -\sqrt{16} = -4$

57. $-64^{1/3} = -\sqrt[3]{64} = -4$

59. $64^{-1/3} = \dfrac{1}{64^{1/3}} = \dfrac{1}{\sqrt[3]{64}} = \dfrac{1}{4}$

61. $4^{-3/2} = \dfrac{1}{4^{3/2}} = \dfrac{1}{\left(\sqrt{4}\right)^3} = \dfrac{1}{2^3} = \dfrac{1}{8}$

63. $-\left(\dfrac{4}{49}\right)^{-1/2} = -\dfrac{1}{\sqrt{4/49}} = -\dfrac{1}{\frac{2}{7}} = -\dfrac{7}{2}$ or

     $-\left(\dfrac{4}{49}\right)^{-1/2} = -\left(\dfrac{49}{4}\right)^{1/2} = -\sqrt{\dfrac{49}{4}} = -\dfrac{7}{2}$

65. $25^{1/2} + 169^{1/2} = \sqrt{25} + \sqrt{169} = 5 + 13 = 18$

67. $343^{-1/3} + 9^{-1/2} = \dfrac{1}{343^{1/3}} + \dfrac{1}{9^{1/2}}$

     $= \dfrac{1}{\sqrt[3]{343}} + \dfrac{1}{\sqrt{9}} = \dfrac{1}{7} + \dfrac{1}{3} = \dfrac{3}{21} + \dfrac{7}{21} = \dfrac{10}{21}$

69. $x^5 \cdot x^{1/2} = x^{5+1/2} = x^{11/2}$

71. $\dfrac{x^{1/2}}{x^{1/3}} = x^{1/2-1/3} = x^{3/6-2/6} = x^{1/6}$

73. $(x^{1/2})^{-2} = x^{1/2(-2)} = x^{-1} = \dfrac{1}{x}$

75. $(6^{-1/3})^0 = 6^{-1/3(0)} = 6^0 = 1$

77. $\dfrac{5y^{-1/3}}{60y^{-2}} = \dfrac{1}{12}y^{-1/3-(-2)} = \dfrac{1}{12}y^{5/3} = \dfrac{y^{5/3}}{12}$

79. $4x^{5/3} \cdot 2x^{-7/2} = 4 \cdot 2 \cdot x^{5/3} \cdot x^{-7/2}$

     $= 8x^{5/3-7/2}$

     $= 8x^{10/6-21/6}$

     $= 8x^{-11/6}$

     $= \dfrac{8}{x^{11/6}}$

81. $\left(\dfrac{8}{64x}\right)^{1/3} = \dfrac{8^{1/3}}{64^{1/3}x^{1/3}} = \dfrac{2}{4x^{1/3}} = \dfrac{1}{2x^{1/3}}$

     or $\left(\dfrac{8}{64x}\right)^{1/3} = \left(\dfrac{1}{8x}\right)^{1/3} = \dfrac{1}{8^{1/3}x^{1/3}} = \dfrac{1}{2x^{1/3}}$

83. $\left(\dfrac{22x^{3/7}}{2x^{1/2}}\right)^2 = (11x^{3/7-1/2})^2$

     $= (11x^{6/14-7/14})^2$

     $= (11x^{-1/14})^2$

     $= (11)^2(x^{-1/14})^2$

     $= 121x^{-1/7}$

     $= \dfrac{121}{x^{1/7}}$

85. $\left(\dfrac{y^4}{4y^{-2/5}}\right)^{-3} = \dfrac{y^{-12}}{4^{-3}y^{6/5}}$

     $= 4^3 y^{-12-6/5}$

     $= 64y^{-66/5}$

     $= \dfrac{64}{y^{66/5}}$

87. $\left(\dfrac{x^{3/4}y^{-2}}{x^{1/2}y^2}\right)^4 = \left(x^{3/4-1/2}y^{-2-2}\right)^4$

     $= (x^{1/4}y^{-4})^4$

     $= (x^{1/4})^4(y^{-4})^4$

     $= xy^{-16}$

     $= \dfrac{x}{y^{16}}$

89. $3z^{-1/2}(2z^4 - z^{1/2})$

     $= 3z^{-1/2} \cdot 2z^4 - 3z^{-1/2}z^{1/2}$

     $= 6z^{-1/2+4} - 3z^{-1/2+1/2}$

     $= 6z^{7/2} - 3z^0$

     $= 6z^{7/2} - 3$

91. $5x^{-1}(x^{-4} + 2x^{-1/2})$
$= 5x^{-1} \cdot x^{-4} + 5x^{-1} \cdot 2x^{-1/2}$
$= 5x^{-1-4} + 10x^{-1-1/2}$
$= 5x^{-5} + 10x^{-3/2}$
$= \dfrac{5}{x^5} + \dfrac{10}{x^{3/2}}$

93. $-4x^{5/3}(-2x^{1/2} + x^{1/3})$
$= (-4x^{5/3})(-2x^{1/2}) + (-4x^{5/3})(x^{1/3})$
$= 8x^{5/3+1/2} - 4x^{5/3+1/3}$
$= 8x^{13/6} - 4x^{6/3}$
$= 8x^{13/6} - 4x^2$

95. $\sqrt{120} \approx 10.95$

97. $\sqrt[5]{402.83} \approx 3.32$

99. $45^{2/3} \approx 12.65$

101. $1000^{-1/2} \approx 0.03$

103. $\sqrt[n]{a^n} = \left(\sqrt[n]{a}\right)^n = a$ when n is odd, or n is even with $a \ge 0$.

105. To show $(a^{1/2} + b^{1/2})^2 \ne a + b$, use $a = 9$ and $b = 16$. Then $(a^{1/2} + b^{1/2})^2$ becomes $(9^{1/2} + 16^{1/2})^2 = (3+4)^2 = 7^2 = 49$ whereas $a + b$ becomes $9 + 16 = 25$. Since $49 \ne 25$, then $(a^{1/2} + b^{1/2})^2 \ne a + b$. Answers will vary.

107. $x^{3/2} + x^{1/2} = x^{1/2} \cdot x^1 + x^{1/2}$
$= x^{1/2}(x+1)$

109. $y^{1/3} - y^{4/3} = y^{1/3} - y^{1/3}y^1$
$= y^{1/3}(1-y)$

111. $y^{-3/5} + y^{2/5}$
$= y^{-3/5} + y^{-3/5}y^1$
$= y^{-3/5}(1+y)$
$= \dfrac{1+y}{y^{3/5}}$

113. a. $E(t) = 2^{10} \cdot 2^t$
$E(0) = 2^{10} \cdot 2^0$
$= 2^{10} \cdot 1$
$= 2^{10}$
Initially, there are 2^{10} bacteria.

b. $E\left(\dfrac{1}{2}\right) = 2^{10} \cdot 2^{1/2}$
$= 2^{10}\sqrt{2}$
≈ 1448.15
After $\dfrac{1}{2}$ hour there are about 1448 bacteria.

115. $A(t) = 2.69t^{3/2}$

a. $t = 1990 - 1983 = 7$
$A(7) = 2.69(7)^{3/2}$
≈ 49.82
In 1990, there was about \$49.82 billion in total assets in the U.S. in 401(k) plans.

b. $t = 1997 - 1983 = 14$
$A(14) = 2.69(14)^{3/2}$
≈ 140.91
In 1997, there was about \$140.91 billion in total assets in the U.S. in 401(k) plans.

117. $(3^{\sqrt{2}})^{\sqrt{2}} = 3^{\sqrt{2} \cdot \sqrt{2}} = 3^2 = 9$

119. $f(x) = (x-2)^{1/2}(x+3)^{-1/2}$
$= \dfrac{(x-2)^{1/2}}{(x+3)^{1/2}}$
$= \dfrac{\sqrt{x-2}}{\sqrt{x+3}}$
The denominator must be greater than zero.

$\sqrt{x+3} > 0$

$x > -3$

The numerator must be greater than or equal to zero.

$\sqrt{x-2} \geq 0$

$x \geq 2$

Therefore, the domain is $\{x | x$ is a real number $x \geq 2\}$

121. Let a be the unknown index in the shaded area.

$\sqrt[4]{\sqrt[a]{\sqrt{x}}} = x^{1/24}$

$\left(\left(x^{1/2} \right)^{1/a} \right)^{1/4} = x^{1/24}$

$x^{1/2 \cdot 1/a \cdot 1/4} = x^{1/24}$

$x^{1/8a} = x^{1/24}$ ← Equate exponents

$\dfrac{1}{8a} = \dfrac{1}{24}$

$8a = 24$

$a = 3$

123. a. $f(x) = \sqrt{2x+3} = (2x+3)^{1/2}$

  b. $y_1 = \sqrt{2x+3}, \ y_2 = (2x+3)^{1/2}$

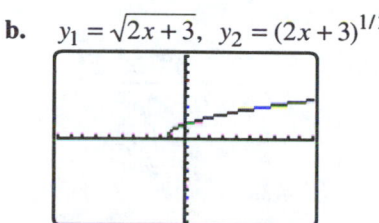

$-10, 10, 1, -10, 10, 1$

  a) is correct because the graphs are the same.

124. a. The graph is a relation but not a function.

  b. The graph is a relation but not a function.

  c. The graph is both a relation and a function.

125. $\dfrac{a^{-2} + ab^{-1}}{ab^{-2} - a^{-2}b^{-1}} = \dfrac{\dfrac{1}{a^2} + \dfrac{a}{b}}{\dfrac{a}{b^2} - \dfrac{1}{a^2 b}}$

$= \dfrac{a^2 b^2 \left(\dfrac{1}{a^2} \right) + a^2 b^2 \left(\dfrac{a}{b} \right)}{a^2 b^2 \left(\dfrac{a}{b^2} \right) - a^2 b^2 \left(\dfrac{1}{a^2 b} \right)}$

$= \dfrac{b^2 + a^3 b}{a^3 - b}$

126. $\dfrac{3x-2}{x+4} = \dfrac{2x+1}{3x-2}$

$(3x-2)(3x-2) = (2x+1)(x+4)$

$9x^2 - 12x + 4 = 2x^2 + 9x + 4$

$7x^2 - 21x = 0$

$7x(x-3) = 0$

$7x = 0 \quad \text{or} \quad x-3 = 0$

$x = 0 \qquad\qquad x = 3$

The solution is 0 or 3.

127. Let y be the speed of the plane in still air. The table is

	d	r	$t = \dfrac{d}{r}$
With wind	560	$y + 25$	$\dfrac{560}{y + 25}$
Against wind	500	$y - 25$	$\dfrac{500}{y - 25}$

Since the time is the same for both parts of the
trip. The equation is

$$\frac{560}{y+25} = \frac{500}{y-25}$$
$$560(y - 25) = 500(y + 25)$$
$$560y - 14,000 = 500y + 12,500$$
$$560y = 500y + 26,500$$
$$60y = 26,500$$
$$y = \frac{26,500}{60}$$
$$\approx 441.67 \text{ mph}$$

The speed of the plane in still air is about 441.67 mph.

Exercise Set 7.3

1. a. Square the natural numbers.

b. $1^2 = 1, \ 2^2 = 4, \ 3^2 = 9,$
$4^2 = 16, \ 5^2 = 25, \ 6^2 = 36$

3. a. Raise natural numbers to the fifth power.

b. $1^5 = 1, \ 2^5 = 32, \ 3^5 = 243,$
$4^5 = 1024, \ 5^5 = 3125$

5. If n is even and a or b is negative, the numbers are not real numbers and the rule does not apply.

7. $\sqrt{75} = \sqrt{25 \cdot 3} = \sqrt{25}\sqrt{3} = 5\sqrt{3}$

9. $\sqrt{32} = \sqrt{16 \cdot 2} = \sqrt{16}\sqrt{2} = 4\sqrt{2}$

11. $\sqrt[3]{16} = \sqrt[3]{8 \cdot 2} = \sqrt[3]{8}\sqrt[3]{2} = 2\sqrt[3]{2}$

13. $\sqrt[3]{54} = \sqrt[3]{27 \cdot 2} = \sqrt[3]{27}\sqrt[3]{2} = 3\sqrt[3]{2}$

15. $-\sqrt{x^3} = -\sqrt{x^2 \cdot x} = -\sqrt{x^2}\sqrt{x} = -x\sqrt{x}$

17. $7\sqrt{x^{11}} = 7\sqrt{x^{10} \cdot x} = 7\sqrt{x^{10}}\sqrt{x} = 7x^5\sqrt{x}$

19. $\sqrt{b^{27}} = \sqrt{b^{26} \cdot b} = \sqrt{b^{26}}\sqrt{b} = b^{13}\sqrt{b}$

21. $\sqrt[4]{b^{23}} = \sqrt[4]{b^{20} \cdot b^3} = \sqrt[4]{b^{20}}\sqrt[4]{b^3} = b^5\sqrt[4]{b^3}$

23. $3\sqrt[3]{24c^{11}} = 3\sqrt[3]{8 \cdot 3 \cdot c^9 \cdot c^2}$
$= 3\sqrt[3]{8c^9 \cdot 3c^2}$
$= 3\sqrt[3]{8c^9}\sqrt[3]{3c^2}$
$= 6c^3\sqrt[3]{3c^2}$

25. $\sqrt{x^3y^7} = \sqrt{x^2 \cdot x \cdot y^6 \cdot y}$
$= \sqrt{x^2y^6 \cdot xy}$
$= \sqrt{x^2y^6}\sqrt{xy}$
$= xy^3\sqrt{xy}$

27. $\sqrt[3]{81a^6b^8} = \sqrt[3]{27 \cdot 3 \cdot a^6 \cdot b^6 \cdot b^2}$
$= \sqrt[3]{27a^6b^6 \cdot 3b^2}$
$= \sqrt[3]{27a^6b^6}\sqrt[3]{3b^2}$
$= 3a^2b^2\sqrt[3]{3b^2}$

29. $4\sqrt[3]{54x^{12}y^{13}} = 4\sqrt[3]{27 \cdot 2 \cdot x^{12} \cdot y^{12} \cdot y}$
$= 4\sqrt[3]{27x^{12}y^{12} \cdot 2y}$
$= 4\sqrt[3]{27x^{12}y^{12}}\sqrt[3]{2y}$
$= 12x^4y^4\sqrt[3]{2y}$

31. $-\sqrt[5]{64x^{12}y^7} = -\sqrt[5]{32 \cdot 2 \cdot x^{10} \cdot x^2 \cdot y^5 \cdot y^2}$
$= -\sqrt[5]{32x^{10}y^5 \cdot 2x^2y^2}$
$= -\sqrt[5]{32x^{10}y^5}\sqrt[5]{2x^2y^2}$
$= -2x^2y\sqrt[5]{2x^2y^2}$

33. $\sqrt[4]{32x^8y^9z^{19}} = \sqrt[4]{16 \cdot 2 \cdot x^8 \cdot y^8 \cdot y \cdot z^{16} \cdot z^3}$
$$= \sqrt[4]{16x^8y^8z^{16} \cdot 2yz^3}$$
$$= \sqrt[4]{16x^8y^8z^{16}} \sqrt[4]{2yz^3}$$
$$= 2x^2y^2z^4\sqrt[4]{2yz^3}$$

35. $\sqrt{50}\sqrt{2} = \sqrt{50 \cdot 2} = \sqrt{100} = 10$

37. $\sqrt[3]{2}\sqrt[3]{28} = \sqrt[3]{56} = \sqrt[3]{8 \cdot 7} = 2\sqrt[3]{7}$

39. $\sqrt{15xy^4}\sqrt{6xy^3} = \sqrt{15xy^4 \cdot 6xy^3}$
$$= \sqrt{90x^2y^7}$$
$$= \sqrt{9 \cdot 10 \cdot x^2 \cdot y^6 \cdot y}$$
$$= \sqrt{9x^2y^6 \cdot 10y}$$
$$= \sqrt{9x^2y^6}\sqrt{10y}$$
$$= 3xy^3\sqrt{10y}$$

41. $\sqrt{9m^3n^7}\sqrt{3mn^4} = \sqrt{9m^3n^7 \cdot 3mn^4}$
$$= \sqrt{27m^4n^{11}}$$
$$= \sqrt{9 \cdot 3 \cdot m^4 \cdot n^{10} \cdot n}$$
$$= \sqrt{9m^4n^{10} \cdot 3n}$$
$$= \sqrt{9m^4n^{10}}\sqrt{3n}$$
$$= 3m^2n^5\sqrt{3n}$$

43. $\sqrt[3]{9x^7y^{12}}\sqrt[3]{6x^4y} = \sqrt[3]{9x^7y^{12} \cdot 6x^4y}$
$$= \sqrt[3]{54x^{11}y^{13}}$$
$$= \sqrt[3]{27 \cdot 2 \cdot x^9 \cdot x^2 \cdot y^{12} \cdot y}$$
$$= \sqrt[3]{27x^9y^{12} \cdot 2x^2y}$$
$$= \sqrt[3]{27x^9y^{12}}\sqrt[3]{2x^2y}$$
$$= 3x^3y^4\sqrt[3]{2x^2y}$$

45. $\left(\sqrt[3]{5x^2y^6}\right)^2 = \sqrt[3]{(5x^2y^6)^2}$
$$= \sqrt[3]{25x^4y^{12}}$$
$$= \sqrt[3]{25 \cdot x^3 \cdot x \cdot y^{12}}$$
$$= \sqrt[3]{x^3y^{12} \cdot 25x}$$
$$= \sqrt[3]{x^3y^{12}}\sqrt[3]{25x}$$
$$= xy^4\sqrt[3]{25x}$$

47. $\sqrt[5]{x^{24}y^{30}z^9}\sqrt[5]{x^{13}y^8z^7}$
$$= \sqrt[5]{x^{24}y^{30}z^9 \cdot x^{13}y^8z^7}$$
$$= \sqrt[5]{x^{37}y^{38}z^{16}}$$
$$= \sqrt[5]{x^{35} \cdot x^2 \cdot y^{35} \cdot y^3 \cdot z^{15} \cdot z}$$
$$= \sqrt[5]{x^{35}y^{35}z^{15} \cdot x^2y^3z}$$
$$= \sqrt[5]{x^{35}y^{35}z^{15}}\sqrt[5]{x^2y^3z}$$
$$= x^7y^7z^3\sqrt[5]{x^2y^3z}$$

49. $\sqrt{2}\left(\sqrt{6} + \sqrt{2}\right) = \left(\sqrt{2}\right)\left(\sqrt{6}\right) + \left(\sqrt{2}\right)\left(\sqrt{2}\right)$
$$= \sqrt{12} + \sqrt{4}$$
$$= \sqrt{4}\sqrt{3} + 2$$
$$= 2\sqrt{3} + 2$$

51. $\sqrt{3}\left(\sqrt{12} - \sqrt{6}\right) = \left(\sqrt{3}\right)\left(\sqrt{12}\right) - \left(\sqrt{3}\right)\left(\sqrt{6}\right)$
$$= \sqrt{36} - \sqrt{18}$$
$$= 6 - \sqrt{9}\sqrt{2}$$
$$= 6 - 3\sqrt{2}$$

53. $\sqrt{2}\left(\sqrt{18} + \sqrt{8}\right) = \left(\sqrt{2}\right)\left(\sqrt{18}\right) + \left(\sqrt{2}\right)\left(\sqrt{8}\right)$
$$= \sqrt{36} + \sqrt{16}$$
$$= 6 + 4$$
$$= 10$$

55. $\sqrt{3y}\left(\sqrt{27y^2} - \sqrt{y}\right)$

$= \left(\sqrt{3y}\right)\left(\sqrt{27y^2}\right) - \left(\sqrt{3y}\right)\left(\sqrt{y}\right)$

$= \sqrt{81y^3} - \sqrt{3y^2}$

$= \sqrt{81y^2}\sqrt{y} - \sqrt{y^2}\sqrt{3}$

$= 9y\sqrt{y} - y\sqrt{3}$

57. $\sqrt[4]{8x^3y^5}\left(\sqrt[4]{4x^5y^7} - \sqrt[4]{3x^7y^6}\right)$

$= \left(\sqrt[4]{8x^3y^5}\right)\left(\sqrt[4]{4x^5y^7}\right) - \left(\sqrt[4]{8x^3y^5}\right)\left(\sqrt[4]{3x^7y^6}\right)$

$= \sqrt[4]{32x^8y^{12}} - \sqrt[4]{24x^{10}y^{11}}$

$= \sqrt[4]{16x^8y^{12}}\sqrt[4]{2} - \sqrt[4]{x^8y^8}\sqrt[4]{24x^2y^3}$

$= 2x^2y^3\sqrt[4]{2} - x^2y^2\sqrt[4]{24x^2y^3}$

59. $(f \cdot g)(x) = f(x) \cdot g(x)$

$= \sqrt{2x}(\sqrt{8x} - \sqrt{32})$

$= \sqrt{2x} \cdot \sqrt{8x} - \sqrt{2x} \cdot \sqrt{32}$

$= \sqrt{16x^2} - \sqrt{64x}$

$= 4x - 8\sqrt{x}$

61. $(f \cdot g)(x) = f(x) \cdot g(x)$

$= \sqrt[3]{xy^2}\left(\sqrt[3]{x^5y^2} + \sqrt[3]{x^2y^2}\right)$

$= \sqrt[3]{xy^2}\sqrt[3]{x^5y^2} + \sqrt[3]{xy^2}\sqrt[3]{x^2y^2}$

$= \sqrt[3]{x^6y^4} + \sqrt[3]{x^3y^4}$

$= \sqrt[3]{x^6y^3}\sqrt[3]{y} + \sqrt[3]{x^3y^3}\sqrt[3]{y}$

$= x^2y\sqrt[3]{y} + xy\sqrt[3]{y}$

63. $(f \cdot g)(x) = f(x) \cdot g(x)$

$= \sqrt[4]{3x^2y}\left(\sqrt[4]{9x^4y} - \sqrt[4]{x^7}\right)$

$= \sqrt[4]{3x^2y}\sqrt[4]{9x^4y} - \sqrt[4]{3x^2y}\sqrt[4]{x^7}$

$= \sqrt[4]{27x^6y^2} - \sqrt[4]{3x^9y}$

$= \sqrt[4]{x^4}\sqrt[4]{27x^2y^2} - \sqrt[4]{x^8}\sqrt[4]{3xy}$

$= x\sqrt[4]{27x^2y^2} - x^2\sqrt[4]{3xy}$

65. $(f \cdot g)(x)$

$= f(x) \cdot g(x)$

$= \sqrt[5]{8x^4y^6}\left(\sqrt[5]{4x^6y^9} - \sqrt[5]{10xy^7}\right)$

$= \sqrt[5]{8x^4y^6}\sqrt[5]{4x^6y^9} - \sqrt[5]{8x^4y^6}\sqrt[5]{10xy^7}$

$= \sqrt[5]{32x^{10}y^{15}} - \sqrt[5]{80x^5y^{13}}$

$= 2x^2y^3 - \sqrt[5]{x^5y^{10}}\sqrt[5]{80y^3}$

$= 2x^2y^3 - xy^2\sqrt[5]{80y^3}$

67. $\sqrt{24} = \sqrt{4 \cdot 6} = \sqrt{4}\sqrt{6} = 2\sqrt{6}$

69. $\sqrt[3]{32} = \sqrt[3]{8 \cdot 4} = \sqrt[3]{8}\sqrt[3]{4} = 2\sqrt[3]{4}$

71. $\sqrt[3]{y^{13}} = \sqrt[3]{y^{12} \cdot y} = y^4\sqrt[3]{y}$

73. $\sqrt[3]{80x^{11}} = \sqrt[3]{8x^9 \cdot 10x^2} = 2x^3\sqrt[3]{10x^2}$

75. $\sqrt[6]{128ab^{17}c^9} = \sqrt[6]{64b^{12}c^6 \cdot 2ab^5c^3}$

$= \sqrt[6]{64b^{12}c^6} \cdot \sqrt[6]{2ab^5c^3}$

$= 2b^2c\sqrt[6]{2ab^5c^3}$

77. $\sqrt[4]{8}\sqrt[4]{10} = \sqrt[4]{8 \cdot 10} = \sqrt[4]{80} = \sqrt[4]{16 \cdot 5} = 2\sqrt[4]{5}$

79. $\sqrt{20xy^4}\sqrt{6x^5y^7} = \sqrt{20xy^4 \cdot 6x^5y^7}$

$= \sqrt{120x^6y^{11}}$

$= \sqrt{4x^6y^{10} \cdot 30y}$

$= \sqrt{4x^6y^{10}}\sqrt{30y}$

$= 2x^3y^5\sqrt{30y}$

81. $\sqrt{x}\left(\sqrt{x} + 3\right) = \left(\sqrt{x}\right)\left(\sqrt{x}\right) + \left(\sqrt{x}\right)(3)$

$= \sqrt{x^2} + 3\sqrt{x}$

$= x + 3\sqrt{x}$

83. $\sqrt[3]{y}\left(2\sqrt[3]{y} - \sqrt[3]{y^8}\right) = \left(\sqrt[3]{y}\right)\left(2\sqrt[3]{y}\right) - \left(\sqrt[3]{y}\right)\left(\sqrt[3]{y^8}\right)$

$= 2\sqrt[3]{y^2} - \sqrt[3]{y^9}$

$= 2\sqrt[3]{y^2} - y^3$

85. $\sqrt[3]{3ab^2}\left(\sqrt[3]{4a^4b^3} - \sqrt[3]{8a^5b^4}\right) = \left(\sqrt[3]{3ab^2}\right)\left(\sqrt[3]{4a^4b^3}\right) - \left(\sqrt[3]{3ab^2}\right)\left(\sqrt[3]{8a^5b^4}\right)$

$$= \sqrt[3]{12a^5b^5} - \sqrt[3]{24a^6b^6}$$
$$= \sqrt[3]{a^3b^3}\sqrt[3]{12a^2b^2} - \sqrt[3]{8a^6b^6}\sqrt[3]{3}$$
$$= ab\sqrt[3]{12a^2b^2} - 2a^2b^2\sqrt[3]{3}$$

87. $f(x) = \sqrt{2x+5}\sqrt{2x+5}$
$$= \sqrt{(2x+5)(2x+5)}$$
$$= \sqrt{(2x+5)^2}$$
$$= |2x+5|$$

89. $h(r) = \sqrt{4r^2 - 32r + 64}$
$$= \sqrt{4(r^2 - 8r + 16)}$$
$$= \sqrt{4(r-4)^2}$$
$$= 2|r-4|$$

91. a. $\sqrt[n]{(x-4)^{2n}} = (x-4)^{2n/n}$
$$= (x-4)^2$$

   b. $\sqrt[n]{(x-4)^{2n}} = (x-4)^{2n/n}$
$$= (x-4)^2$$

93. $\left(\sqrt{3} + \sqrt{2}\right)\left(\sqrt{3} - \sqrt{2}\right) = \left(\sqrt{3}\right)\left(\sqrt{3}\right) + \left(\sqrt{3}\right)\left(-\sqrt{2}\right) + \left(\sqrt{2}\right)\left(\sqrt{3}\right) + \left(\sqrt{2}\right)\left(-\sqrt{2}\right)$

$$= 3 - \sqrt{6} + \sqrt{6} - 2$$
$$= 1$$

95. $\left(5\sqrt{a} - 4\sqrt{b}\right)\left(2\sqrt{a} + 3\sqrt{b}\right) = \left(5\sqrt{a}\right)\left(2\sqrt{a}\right) + \left(5\sqrt{a}\right)\left(3\sqrt{b}\right) + \left(-4\sqrt{b}\right)\left(2\sqrt{a}\right) + \left(-4\sqrt{b}\right)\left(3\sqrt{b}\right)$

$$= 10a + 15\sqrt{ab} - 8\sqrt{ab} - 12b$$
$$= 10a + 7\sqrt{ab} - 12b$$

97. a. $s = \sqrt{30FB}$
$s = \sqrt{30(0.85)(80)} \approx 45.17$
The car's speed was about 45.17 mph.

   b. $s = \sqrt{30FB}$
$s = \sqrt{30(0.52)(80)} \approx 35.33$
The car's speed was about 35.33 mph.

99. $R = 28d^2\sqrt{P}$
$R = 28(2.5)^2\sqrt{80}$
$R \approx 1565.25$
The flow rate is about 1565.25 gallons per minute.

101. $\sigma = \sqrt{npq}$

$\sigma = \sqrt{600(0.93)(0.07)}$

$\sigma = 6.25$

In this case the standard deviation is about 6.25.

102. A rational number is a number that can be expressed as a quotient of two integers in which the denominator is not equal to zero.

103. A real number is a number that can be represented on the real number line.

104. An irrational number is a real number that cannot be expressed as the quotient of two integers.

105. $|a| = \begin{cases} a & a \ge 0 \\ -a & a < 0 \end{cases}$

106. $E = \dfrac{1}{2}mv^2$ Solve for m.

$2E = mv^2$

$\dfrac{2E}{v^2} = m$ or $m = \dfrac{2E}{v^2}$

107. $-4 < 2x - 3 \le 5$

$-1 < 2x \le 8$

$-\dfrac{1}{2} < x \le 4$

a.

b. $\left(-\dfrac{1}{2},\ 4\right]$

c. $\left\{ x \middle| -\dfrac{1}{2} < x \le 4 \right\}$

Exercise Set 7.4

1. Answers will vary. Possible answer:

$\dfrac{\sqrt[n]{a}}{\sqrt[n]{b}} = \sqrt[n]{\dfrac{a}{b}},\ \ a \ge 0,\ \ b > 0$

3. To remove radicals from a denominator

5. a. Answers will vary. Possible answer: Multiply the numerator and denominator by the conjugate of the denominator.

b. $\dfrac{\sqrt{2}+\sqrt{5}}{\sqrt{2}-\sqrt{5}} = \dfrac{\sqrt{2}+\sqrt{5}}{\sqrt{2}-\sqrt{5}} \cdot \dfrac{\left(\sqrt{2}+\sqrt{5}\right)}{\left(\sqrt{2}+\sqrt{5}\right)}$

$= \dfrac{\sqrt{4} + 2\left(\sqrt{2}\right)\left(\sqrt{5}\right) + \sqrt{25}}{\left(\sqrt{2}\right)^2 - \left(\sqrt{5}\right)^2}$

$= \dfrac{2 + 2\sqrt{10} + 5}{2 - 5}$

$= \dfrac{7 + 2\sqrt{10}}{-3} = -\dfrac{7 + 2\sqrt{10}}{3}$

7. $\sqrt{\dfrac{36}{4}} = \sqrt{9} = 3$

9. $\dfrac{\sqrt{3}}{\sqrt{27}} = \sqrt{\dfrac{3}{27}} = \sqrt{\dfrac{1}{9}} = \dfrac{\sqrt{1}}{\sqrt{9}} = \dfrac{1}{3}$

11. $\sqrt[3]{\dfrac{2}{16}} = \sqrt[3]{\dfrac{1}{8}} = \dfrac{\sqrt[3]{1}}{\sqrt[3]{8}} = \dfrac{1}{2}$

13. $\dfrac{-\sqrt{24}}{\sqrt{3}} = -\sqrt{\dfrac{24}{3}} = -\sqrt{8} = -\sqrt{4}\sqrt{2} = -2\sqrt{2}$

15. $\sqrt{\dfrac{r^4}{25}} = \dfrac{\sqrt{r^4}}{\sqrt{25}} = \dfrac{r^2}{5}$

17. $\sqrt{\dfrac{27x^6}{3x^2}} = \sqrt{9x^4} = 3x^2$

19. $\sqrt[3]{\dfrac{7xy}{8x^{13}}} = \sqrt[3]{\dfrac{7y}{8x^{12}}} = \dfrac{\sqrt[3]{7y}}{\sqrt[3]{8x^{12}}} = \dfrac{\sqrt[3]{7y}}{2x^4}$

21. $\sqrt[4]{\dfrac{20x^4}{81x^{-8}}} = \sqrt[4]{\dfrac{20x^{12}}{81}}$

$= \dfrac{\sqrt[4]{20x^{12}}}{\sqrt[4]{81}}$

$= \dfrac{\sqrt[4]{x^{12}}\,\sqrt[4]{20}}{\sqrt[4]{81}}$

$= \dfrac{x^3\sqrt[4]{20}}{3}$

23. $\dfrac{1}{\sqrt{5}} = \dfrac{1}{\sqrt{5}} \cdot \dfrac{\sqrt{5}}{\sqrt{5}} = \dfrac{\sqrt{5}}{5}$

25. $\dfrac{\sqrt{m}}{\sqrt{2}} = \dfrac{\sqrt{m}}{\sqrt{2}} \cdot \dfrac{\sqrt{2}}{\sqrt{2}} = \dfrac{\sqrt{2m}}{2}$

27. $\dfrac{\sqrt{x}}{\sqrt{y}} = \dfrac{\sqrt{x}}{\sqrt{y}} \cdot \dfrac{\sqrt{y}}{\sqrt{y}} = \dfrac{\sqrt{xy}}{y}$

29. $\sqrt{\dfrac{5m}{8}} = \dfrac{\sqrt{5m}}{\sqrt{8}}$

$= \dfrac{\sqrt{5m}}{2\sqrt{2}}$

$= \dfrac{\sqrt{5m}}{2\sqrt{2}} \cdot \dfrac{\sqrt{2}}{\sqrt{2}}$

$= \dfrac{\sqrt{10m}}{2\cdot 2} = \dfrac{\sqrt{10m}}{4}$

31. $\dfrac{2n}{\sqrt{18n}} = \dfrac{2n}{\sqrt{9}\sqrt{2n}}$

$= \dfrac{2n}{3\sqrt{2n}}$

$= \dfrac{2n}{3\sqrt{2n}} \cdot \dfrac{\sqrt{2n}}{\sqrt{2n}}$

$= \dfrac{2n\sqrt{2n}}{3\cdot 2n} = \dfrac{\sqrt{2n}}{3}$

33. $\sqrt{\dfrac{8x^6y}{2xz}} = \sqrt{\dfrac{4x^5y}{z}}$

$= \dfrac{\sqrt{4x^4}\,\sqrt{xy}}{\sqrt{z}}$

$= \dfrac{2x^2\sqrt{xy}}{\sqrt{z}} \cdot \dfrac{\sqrt{z}}{\sqrt{z}}$

$= \dfrac{2x^2\sqrt{xyz}}{z}$

35. $\sqrt{\dfrac{20y^4z^3}{3xy^{-2}}} = \sqrt{\dfrac{20y^6z^3}{3x}}$

$= \dfrac{\sqrt{20y^6z^3}}{\sqrt{3x}}$

$= \dfrac{\sqrt{4y^6z^2}\,\sqrt{5z}}{\sqrt{3x}}$

$= \dfrac{2y^3z\sqrt{5z}}{\sqrt{3x}}$

$= \dfrac{2y^3z\sqrt{5z}}{\sqrt{3x}} \cdot \dfrac{\sqrt{3x}}{\sqrt{3x}}$

$= \dfrac{2y^3z\sqrt{15xz}}{3x}$

37. $\sqrt{\dfrac{18x^4y^3}{2z^3}} = \sqrt{\dfrac{9x^4y^3}{z^3}} = \dfrac{\sqrt{9x^4y^3}}{\sqrt{z^2\cdot z}}$

$= \dfrac{\sqrt{9x^4y^2}\,\sqrt{y}}{\sqrt{z^2}\,\sqrt{z}}$

$= \dfrac{3x^2y\sqrt{y}}{z\sqrt{z}}$

$= \dfrac{3x^2y\sqrt{y}}{z\sqrt{z}} \cdot \dfrac{\sqrt{z}}{\sqrt{z}}$

$= \dfrac{3x^2y\sqrt{yz}}{z^2}$

39. $\dfrac{1}{\sqrt[3]{2}} = \dfrac{1}{\sqrt[3]{2}} \cdot \dfrac{\sqrt[3]{4}}{\sqrt[3]{4}} = \dfrac{\sqrt[3]{4}}{\sqrt[3]{8}} = \dfrac{\sqrt[3]{4}}{2}$

41. $\dfrac{1}{\sqrt[3]{5}} = \dfrac{1}{\sqrt[3]{5}} \cdot \dfrac{\sqrt[3]{25}}{\sqrt[3]{25}} = \dfrac{\sqrt[3]{25}}{5}$

43. $\sqrt[3]{\dfrac{5x}{y}} = \dfrac{\sqrt[3]{5x}}{\sqrt[3]{y}} = \dfrac{\sqrt[3]{5x}}{\sqrt[3]{y}} \cdot \dfrac{\sqrt[3]{y^2}}{\sqrt[3]{y^2}} = \dfrac{\sqrt[3]{5xy^2}}{y}$

45. $-\sqrt[3]{\dfrac{5c}{9y^2}} = -\dfrac{\sqrt[3]{5c}}{\sqrt[3]{3^2 y^2}} \cdot \dfrac{\sqrt[3]{3y}}{\sqrt[3]{3y}} = -\dfrac{\sqrt[3]{15cy}}{3y}$

47. $\dfrac{5m}{\sqrt[4]{2}} = \dfrac{5m}{\sqrt[4]{2}} \cdot \dfrac{\sqrt[4]{2^3}}{\sqrt[4]{2^3}} = \dfrac{5m\sqrt[4]{8}}{2}$

49. $\sqrt[4]{\dfrac{2x^3}{4y^2}} = \sqrt[4]{\dfrac{x^3}{2y^2}}$

$= \dfrac{\sqrt[4]{x^3}}{\sqrt[4]{2y^2}}$

$= \dfrac{\sqrt[4]{x^3}}{\sqrt[4]{2y^2}} \cdot \dfrac{\sqrt[4]{8y^2}}{\sqrt[4]{8y^2}} = \dfrac{\sqrt[4]{8x^3y^2}}{2y}$

51. $\sqrt[3]{\dfrac{15x^6 y^7}{2z^2}} = \dfrac{\sqrt[3]{15x^6 y^7}}{\sqrt[3]{2z^2}}$

$= \dfrac{\sqrt[3]{x^6 y^6}\sqrt[3]{15y}}{\sqrt[3]{2z^2}}$

$= \dfrac{x^2 y^2 \sqrt[3]{15y}}{\sqrt[3]{2z^2}} \cdot \dfrac{\sqrt[3]{2^2 z}}{\sqrt[3]{2^2 z}}$

$= \dfrac{x^2 y^2 \sqrt[3]{60yz}}{2z}$

53. $\sqrt[6]{\dfrac{r^4 s^9}{2r^5}} = \sqrt[6]{\dfrac{s^9}{2r}}$

$= \dfrac{\sqrt[6]{s^9}}{\sqrt[6]{2r}}$

$= \dfrac{s\sqrt[6]{s^3}}{\sqrt[6]{2r}} \cdot \dfrac{\sqrt[6]{32r^5}}{32r^5}$

$= \dfrac{s\sqrt[6]{32r^5 s^3}}{2r}$

55. $(5 + \sqrt{5})(5 - \sqrt{5}) = 5^2 - (\sqrt{5})^2$
$= 25 - 5$
$= 20$

57. $(7 - \sqrt{5})(7 + \sqrt{5}) = 7^2 - (\sqrt{5})^2$
$= 49 - 5$
$= 44$

59. $(\sqrt{6} + x)(\sqrt{6} - x) = (\sqrt{6})^2 - x^2$
$= 6 - x^2$

61. $(\sqrt{x} + \sqrt{y})(\sqrt{x} - \sqrt{y}) = (\sqrt{x})^2 - (\sqrt{y})^2$
$= x - y$

63. $(2\sqrt{3} - \sqrt{2})(2\sqrt{3} + \sqrt{2}) = (2\sqrt{3})^2 - (\sqrt{2})^2$
$= 4 \cdot 3 - 2$
$= 12 - 2$
$= 10$

65. $\dfrac{5}{\sqrt{2} + 1} = \dfrac{5}{\sqrt{2} + 1} \cdot \dfrac{(\sqrt{2} - 1)}{(\sqrt{2} - 1)}$

$= \dfrac{5\sqrt{2} - 5}{(\sqrt{2})^2 - 1^2}$

$= \dfrac{5\sqrt{2} - 5}{2 - 1}$

$= \dfrac{5\sqrt{2} - 5}{1}$

$= 5\sqrt{2} - 5$

67. $\dfrac{2}{5 - \sqrt{6}} = \dfrac{2}{5 - \sqrt{6}} \cdot \dfrac{(5 - \sqrt{6})}{(5 + \sqrt{6})}$

$= \dfrac{10 + 2\sqrt{6}}{25 - (\sqrt{6})^2}$

$= \dfrac{10 + 2\sqrt{6}}{25 - 6}$

$= \dfrac{10 + 2\sqrt{6}}{19}$

69. $\dfrac{2}{\sqrt{2}+\sqrt{3}} = \dfrac{2}{\sqrt{2}+\sqrt{3}} \cdot \dfrac{\left(\sqrt{2}-\sqrt{3}\right)}{\left(\sqrt{2}-\sqrt{3}\right)}$

$= \dfrac{2\left(\sqrt{2}-\sqrt{3}\right)}{2-3}$

$= \dfrac{2\left(\sqrt{2}-\sqrt{3}\right)}{-1}$

$= -2\left(\sqrt{2}-\sqrt{3}\right)$

$= -2\sqrt{2}+2\sqrt{3} = 2\sqrt{3}-2\sqrt{2}$

71. $\dfrac{1}{\sqrt{17}-\sqrt{8}} = \dfrac{1}{\sqrt{17}-\sqrt{8}} \cdot \dfrac{\left(\sqrt{17}+\sqrt{8}\right)}{\left(\sqrt{17}+\sqrt{8}\right)}$

$= \dfrac{\sqrt{17}+\sqrt{8}}{17-8}$

$= \dfrac{\sqrt{17}+\sqrt{4}\sqrt{2}}{9}$

$= \dfrac{\sqrt{17}+2\sqrt{2}}{9}$

73. $\dfrac{3\sqrt{5}}{\sqrt{a}-3} = \dfrac{3\sqrt{5}}{\sqrt{a}-3} \cdot \dfrac{\left(\sqrt{a}+3\right)}{\left(\sqrt{a}+3\right)}$

$= \dfrac{3\sqrt{5a}+9\sqrt{5}}{a-9}$

75. $\dfrac{\sqrt{8x}}{x+\sqrt{y}} = \dfrac{\sqrt{8x}}{x+\sqrt{y}} \cdot \dfrac{\left(x-\sqrt{y}\right)}{\left(x-\sqrt{y}\right)}$

$= \dfrac{2\sqrt{2x}\left(x-\sqrt{y}\right)}{x^2-y}$

$= \dfrac{2x\sqrt{2x}-2\sqrt{2xy}}{x^2-y}$

77. $\dfrac{\sqrt{c}-\sqrt{2d}}{\sqrt{c}-\sqrt{d}}$

$= \dfrac{\sqrt{c}-\sqrt{2d}}{\sqrt{c}-\sqrt{d}} \cdot \dfrac{\left(\sqrt{c}+\sqrt{d}\right)}{\left(\sqrt{c}+\sqrt{d}\right)}$

$= \dfrac{\left(\sqrt{c}\right)^2+\sqrt{c}\sqrt{d}-\sqrt{2d}\sqrt{c}-\sqrt{2d}\sqrt{d}}{\left(\sqrt{c}\right)^2-\left(\sqrt{d}\right)^2}$

$= \dfrac{c+\sqrt{cd}-\sqrt{2cd}-d\sqrt{2}}{c-d}$

79. $\dfrac{2\sqrt{xy}-\sqrt{xy}}{\sqrt{x}+\sqrt{y}} = \dfrac{\sqrt{xy}}{\sqrt{x}+\sqrt{y}}$

$= \dfrac{\sqrt{xy}}{\sqrt{x}+\sqrt{y}} \cdot \dfrac{\sqrt{x}-\sqrt{y}}{\sqrt{x}-\sqrt{y}}$

$= \dfrac{\sqrt{xy}\left(\sqrt{x}-\sqrt{y}\right)}{\left(\sqrt{x}\right)^2-\left(\sqrt{y}\right)^2}$

$= \dfrac{\sqrt{x^2y}-\sqrt{xy^2}}{x-y}$

$= \dfrac{x\sqrt{y}-y\sqrt{x}}{x-y}$

81. $\sqrt{\dfrac{x}{9}} = \dfrac{\sqrt{x}}{\sqrt{9}} = \dfrac{\sqrt{x}}{3}$

83. $\sqrt{\dfrac{2}{5}} = \dfrac{\sqrt{2}}{\sqrt{5}} = \dfrac{\sqrt{2}}{\sqrt{5}} \cdot \dfrac{\sqrt{5}}{\sqrt{5}} = \dfrac{\sqrt{10}}{5}$

85. $\left(\sqrt{5}+\sqrt{6}\right)\left(\sqrt{5}-\sqrt{6}\right) = \left(\sqrt{5}\right)^2-\left(\sqrt{6}\right)^2$

$= 5-6 = -1$

87. $\sqrt{\dfrac{24x^3y^6}{5z}} = \dfrac{\sqrt{24x^3y^6}}{\sqrt{5z}}$

$= \dfrac{\sqrt{4x^2y^6}\sqrt{6x}}{\sqrt{5z}}$

$= \dfrac{2xy^3\sqrt{6x}}{\sqrt{5z}} \cdot \dfrac{\sqrt{5z}}{\sqrt{5z}} = \dfrac{2xy^3\sqrt{30xz}}{5z}$

89. $\sqrt{\dfrac{12xy^4}{2x^3y^4}} = \sqrt{\dfrac{6}{x^2}} = \dfrac{\sqrt{6}}{\sqrt{x^2}} = \dfrac{\sqrt{6}}{x}$

91. $\left(\sqrt{x}+3\right)\left(\sqrt{x}-3\right) = \left(\sqrt{x}\right)^2 - 3^2 = x - 9$

93. $-\dfrac{7\sqrt{x}}{\sqrt{98}} = -\dfrac{7\sqrt{x}}{7\sqrt{2}}$

$\qquad = -\dfrac{\sqrt{x}}{\sqrt{2}} \cdot \dfrac{\sqrt{2}}{\sqrt{2}}$

$\qquad = -\dfrac{\sqrt{2x}}{\sqrt{4}}$

$\qquad = -\dfrac{\sqrt{2x}}{2}$

95. $\sqrt[4]{\dfrac{3y^2}{2x}} = \dfrac{\sqrt[4]{3y^2}}{\sqrt[4]{2x}} \cdot \dfrac{\sqrt[4]{8x^3}}{\sqrt[4]{8x^3}}$

$\qquad = \dfrac{\sqrt[4]{24x^3y^2}}{\sqrt[4]{16x^4}} = \dfrac{\sqrt[4]{24x^3y^2}}{2x}$

97. $\sqrt[3]{\dfrac{32y^{12}z^{10}}{2x}} = \sqrt[3]{\dfrac{16y^{12}z^{10}}{x}}$

$\qquad = \dfrac{\sqrt[3]{16y^{12}z^{10}}}{\sqrt[3]{x}}$

$\qquad = \dfrac{\sqrt[3]{8y^{12}z^9}\sqrt[3]{2z}}{\sqrt[3]{x}}$

$\qquad = \dfrac{2y^4z^3\sqrt[3]{2z}}{\sqrt[3]{x}} \cdot \dfrac{\sqrt[3]{x^2}}{\sqrt[3]{x^2}}$

$\qquad = \dfrac{2y^4z^3\sqrt[3]{2x^2z}}{\sqrt[3]{x^3}}$

$\qquad = \dfrac{2y^4z^3\sqrt[3]{2x^2z}}{x}$

99. $\dfrac{\sqrt{ar}}{\sqrt{a}-2\sqrt{r}} \cdot \dfrac{\left(\sqrt{a}+2\sqrt{r}\right)}{\left(\sqrt{a}+2\sqrt{r}\right)} = \dfrac{\sqrt{ar}\left(\sqrt{a}+2\sqrt{r}\right)}{\left(\sqrt{a}\right)^2 - \left(2\sqrt{r}\right)^2}$

$\qquad\qquad = \dfrac{a\sqrt{r}+2r\sqrt{a}}{a-4r}$

101. $\dfrac{\sqrt[3]{6x}}{\sqrt[3]{5xy}} = \sqrt[3]{\dfrac{6x}{5xy}}$

$\qquad = \sqrt[3]{\dfrac{6}{5y}}$

$\qquad = \dfrac{\sqrt[3]{6}}{\sqrt[3]{5y}} \cdot \dfrac{\sqrt[3]{25y^2}}{\sqrt[3]{25y^2}}$

$\qquad = \dfrac{\sqrt[3]{150y^2}}{5y}$

103. $\sqrt[4]{\dfrac{2x^7y^{12}z^4}{3x^9}} = \sqrt[4]{\dfrac{2y^{12}z^4}{3x^2}}$

$\qquad = \dfrac{\sqrt[4]{2y^{12}z^4}}{\sqrt[4]{3x^2}}$

$\qquad = \dfrac{\sqrt[4]{y^{12}z^4}\sqrt[4]{2}}{\sqrt[4]{3x^2}}$

$\qquad = \dfrac{y^3z\sqrt[4]{2}}{\sqrt[4]{3x^2}} \cdot \dfrac{\sqrt[4]{27x^2}}{\sqrt[4]{27x^2}}$

$\qquad = \dfrac{y^3z\sqrt[4]{54x^2}}{\sqrt[4]{81x^4}}$

$\qquad = \dfrac{y^3z\sqrt[4]{54x^2}}{3x}$

105. $\dfrac{\sqrt{(a+b)^4}}{\sqrt[3]{a+b}} = \dfrac{(a+b)^{4/2}}{(a+b)^{1/3}}$

$\qquad = (a+b)^{6/3-1/3}$

$\qquad = (a+b)^{5/3}$

$\qquad = \sqrt[3]{(a+b)^5}$

107. $\dfrac{\sqrt[5]{(a+2b)^4}}{\sqrt[3]{(a+2b)^2}} = \dfrac{(a+2b)^{4/5}}{(a+2b)^{2/3}}$

$\qquad = (a+2b)^{4/5-2/3}$

$\qquad = (a+2b)^{2/15}$

$\qquad = \sqrt[15]{(a+2b)^2}$

109. $\dfrac{\sqrt[3]{r^2 s^4}}{\sqrt{rs}} = \dfrac{(r^2 s^4)^{1/3}}{(rs)^{1/2}}$

$\phantom{\dfrac{\sqrt[3]{r^2 s^4}}{\sqrt{rs}}} = \dfrac{r^{2/3} s^{4/3}}{r^{1/2} s^{1/2}}$

$\phantom{\dfrac{\sqrt[3]{r^2 s^4}}{\sqrt{rs}}} = r^{2/3 - 1/2} s^{4/3 - 1/2}$

$\phantom{\dfrac{\sqrt[3]{r^2 s^4}}{\sqrt{rs}}} = r^{1/6} s^{5/6}$

$\phantom{\dfrac{\sqrt[3]{r^2 s^4}}{\sqrt{rs}}} = (rs^5)^{1/6}$

$\phantom{\dfrac{\sqrt[3]{r^2 s^4}}{\sqrt{rs}}} = \sqrt[6]{rs^5}$

111. $\dfrac{\sqrt[5]{x^4 y^6}}{\sqrt[3]{(xy)^2}} = \dfrac{(x^4 y^6)^{1/5}}{(xy)^{2/3}}$

$\phantom{\dfrac{\sqrt[5]{x^4 y^6}}{\sqrt[3]{(xy)^2}}} = \dfrac{x^{4/5} y^{6/5}}{x^{2/3} y^{2/3}}$

$\phantom{\dfrac{\sqrt[5]{x^4 y^6}}{\sqrt[3]{(xy)^2}}} = x^{4/5 - 2/3} y^{6/5 - 2/3}$

$\phantom{\dfrac{\sqrt[5]{x^4 y^6}}{\sqrt[3]{(xy)^2}}} = x^{2/15} y^{8/15}$

$\phantom{\dfrac{\sqrt[5]{x^4 y^6}}{\sqrt[3]{(xy)^2}}} = (x^2 y^8)^{1/15}$

$\phantom{\dfrac{\sqrt[5]{x^4 y^6}}{\sqrt[3]{(xy)^2}}} = \sqrt[15]{x^2 y^8}$

113. $d = \sqrt{\dfrac{72}{I}}$

$d = \sqrt{\dfrac{72}{5.3}} \approx 3.69$

The person is about 3.69 m from the light source.

115. $r = \sqrt[3]{\dfrac{3V}{4\pi}}$

$r = \sqrt[3]{\dfrac{3(7238.23)}{4\pi}} = 12$

The radius of the tank is 12 inches.

117. $N(t) = \dfrac{6.21}{\sqrt[4]{t}}$

    a. $t = 1950 - 1949 = 1$

$N(1) = \dfrac{6.21}{\sqrt[4]{1}} = 6.21$

The number of farms in 1950 was 6.21 million.

    b. $t = 1997 - 1949 = 48$

$N(48) = \dfrac{6.21}{\sqrt[4]{48}} \approx 2.36$

The number of farms in 1997 was about 2.36 million.

119. If n is even and a or b is negative, the root of the numbers is not real and this rule does not apply.

121. $\dfrac{2}{\sqrt{2}} = \dfrac{2}{\sqrt{2}} \cdot \dfrac{\sqrt{2}}{\sqrt{2}} = \dfrac{2\sqrt{2}}{\sqrt{4}} = \dfrac{2\sqrt{2}}{2} = \sqrt{2}$

$\dfrac{3}{\sqrt{3}} = \dfrac{3}{\sqrt{3}} \cdot \dfrac{\sqrt{3}}{\sqrt{3}} = \dfrac{3\sqrt{3}}{\sqrt{9}} = \dfrac{3\sqrt{3}}{3} = \sqrt{3}$

Since $3 > 2$, then $\sqrt{3} > \sqrt{2}$ and we conclude that $\dfrac{3}{\sqrt{3}} > \dfrac{2}{\sqrt{2}}$.

123. $f(x) = x^{a/2}$, $g(x) = x^{b/3}$

    a. $x^{4/2} = x^2$

$x^{12/2} = x^6$

$x^{8/2} = x^4$

Therefore $x^{a/2}$ is a perfect square when $a = 4, 8, 12$.

    b. $x^{9/3} = x^3$

$x^{18/3} = x^6$

$x^{27/3} = x^9$

Therefore, $x^{b/3}$ is a perfect cube when $b = 9, 18, 27$.

    c. $(f \cdot g)(x) = f(x) \cdot g(x)$

$= x^{a/2} \cdot x^{b/3}$

$= x^{a/2 + b/3}$

$= x^{3a/6 + 2b/6}$

$= x^{(3a + 2b)/6}$

d. $\left(\dfrac{f}{g}\right)(x) = \dfrac{f(x)}{g(x)}$

$= \dfrac{x^{a/2}}{x^{b/3}}$

$= x^{a/2 - b/3}$

$= x^{3a/6 - 2b/6}$

$= x^{(3a-2b)/6}$

125. $\dfrac{3}{\sqrt{2a-3b}} = \dfrac{3}{\sqrt{2a-3b}} \cdot \dfrac{\sqrt{2a-3b}}{\sqrt{2a-3b}}$

$= \dfrac{3\sqrt{2a-3b}}{2a-3b}$

127. $\dfrac{5-\sqrt{5}}{6} = \dfrac{5-\sqrt{5}}{6} \cdot \dfrac{5+\sqrt{5}}{5+\sqrt{5}}$

$= \dfrac{25-5}{30+6\sqrt{5}}$

$= \dfrac{20}{2(15+3\sqrt{5})}$

$= \dfrac{10}{15+3\sqrt{5}}$

129. $\dfrac{\sqrt{x+h}-\sqrt{x}}{h} = \dfrac{\sqrt{x+h}-\sqrt{x}}{h} \cdot \dfrac{\sqrt{x+h}+\sqrt{x}}{\sqrt{x+h}+\sqrt{x}}$

$= \dfrac{x+h-x}{h\left(\sqrt{x+h}+\sqrt{x}\right)}$

$= \dfrac{h}{h\left(\sqrt{x+h}+\sqrt{x}\right)}$

$= \dfrac{1}{\sqrt{x+h}+\sqrt{x}}$

131. $A = \dfrac{1}{2}h(b_1 + b_2)$

$2A = h(b_1 + b_2)$

$\dfrac{2A}{h} = b_1 + b_2$

$\dfrac{2A}{h} - b_1 = b_2$

$b_2 = \dfrac{2A}{h} - b_1$

132. Let r be the rate of the slower car and $r + 10$ be the rate of the faster.
Distance the first traveled plus distance the second traveled is 270 miles.

$3r + 3(r + 10) = 270$
$3r + 3r + 30 = 270$
$6r + 30 = 270$
$6r = 240$
$r = 40$

The rate of the slower car is 40 mph and the rate of the faster car is $r + 10 = 50$ mph.

133.

$$\begin{array}{r} 4x^2 - 3x - 2 \\ 2x - 3 \\ \hline -12x^2 + 9x + 6 \\ 8x^3 - 6x^2 - 4x \\ \hline 8x^3 - 18x^2 + 5x + 6 \end{array}$$

134.
$$(2x-3)(x-2)=4x-6$$
$$2x^2-4x-3x+6=4x-6$$
$$2x^2-11x+12=0$$
$$2x^2-8x-3x+12=0 \quad \leftarrow \text{ Use } -8x-3x \text{ for } -11x$$
$$2x(x-4)-3(x-4)=0$$
$$(2x-3)(x-4)=0$$
$$2x=3 \quad \text{or} \quad x-4=0$$
$$x=\frac{3}{2} \qquad\qquad x=4$$

The solution is $\dfrac{3}{2}$ or 4.

Exercise Set 7.5

1. Like radicals are radicals with the same radicands and index.

3. $\sqrt{3}+3\sqrt{2} \approx 1.732+3(1.414)$
$$\approx 1.732+4.242$$
$$\approx 5.974 \text{ or } 5.97$$

5. No. To see this, let $a=16$ and $b=9$. Then, the left side is
$$\sqrt{a}+\sqrt{b}=\sqrt{16}+\sqrt{9}=4+3=7$$
whereas the right side is
$$\sqrt{a+b}=\sqrt{16+9}=\sqrt{25}=5.$$

7. $6\sqrt{3}-2\sqrt{3}=4\sqrt{3}$

9. $2\sqrt{3}-2\sqrt{3}-4\sqrt{3}+5=5-4\sqrt{3}$

11. $3\sqrt[4]{y}-6\sqrt[4]{y}=-3\sqrt[4]{y}$

13. $3\sqrt{5}-\sqrt[3]{x}+4\sqrt{5}+3\sqrt[3]{x}=7\sqrt{5}+2\sqrt[3]{x}$

15. $6+4\sqrt[3]{a}-7\sqrt[3]{a}=6-3\sqrt[3]{a}$

17. $\sqrt{8}-\sqrt{12}=\sqrt{4}\cdot\sqrt{2}-\sqrt{4}\cdot\sqrt{3}$
$$=2\sqrt{2}-2\sqrt{3}$$
$$=2\left(\sqrt{2}-\sqrt{3}\right)$$

19. $-6\sqrt{75}+4\sqrt{125}=-6\sqrt{25}\cdot\sqrt{3}+4\sqrt{25}\cdot\sqrt{5}$
$$=-6\left(5\sqrt{3}\right)+4\left(5\sqrt{5}\right)$$
$$=-30\sqrt{3}+20\sqrt{5}$$

21. $-4\sqrt{90}+3\sqrt{40}+2\sqrt{10}$
$$=-4\sqrt{9}\cdot\sqrt{10}+3\sqrt{4}\cdot\sqrt{10}+2\sqrt{10}$$
$$=-4\left(3\sqrt{10}\right)+3\left(2\sqrt{10}\right)+2\left(\sqrt{10}\right)$$
$$=-12\sqrt{10}+6\sqrt{10}+2\sqrt{10}$$
$$=-4\sqrt{10}$$

23. $\sqrt{500xy^2}+y\sqrt{320x}$
$$=\sqrt{100y^2}\cdot\sqrt{5x}+y\sqrt{64}\sqrt{5x}$$
$$=10y\sqrt{5x}+8y\sqrt{5x}$$
$$=18y\sqrt{5x}$$

25. $5\sqrt{8}+2\sqrt{50}-3\sqrt{72}$
$$=5\sqrt{4}\cdot\sqrt{2}+2\sqrt{25}\cdot\sqrt{2}-3\sqrt{36}\cdot\sqrt{2}$$
$$=5(2\sqrt{2})+2(5\sqrt{2})-3(6\sqrt{2})$$
$$=10\sqrt{2}+10\sqrt{2}-18\sqrt{2}$$
$$=2\sqrt{2}$$

27. $3\sqrt{27c^2}-2\sqrt{108c^2}-\sqrt{48c^2}$
$$=3\sqrt{9c^2}\cdot\sqrt{3}-2\sqrt{36c^2}\cdot\sqrt{3}-\sqrt{16c^2}\cdot\sqrt{3}$$
$$=3(3c\sqrt{3})-2(6c\sqrt{3})-4c\sqrt{3}$$
$$=9c\sqrt{3}-12c\sqrt{3}-4c\sqrt{3}$$
$$=-7c\sqrt{3}$$

29. $\sqrt[3]{54} - \sqrt[3]{16} = \sqrt[3]{27} \cdot \sqrt[3]{2} - \sqrt[3]{8} \cdot \sqrt[3]{2}$
$$= 3\sqrt[3]{2} - 2\sqrt[3]{2}$$
$$= \sqrt[3]{2}$$

31. $\sqrt[3]{108} + 2\sqrt[3]{32} = \sqrt[3]{27} \cdot \sqrt[3]{4} + 2\sqrt[3]{8} \cdot \sqrt[3]{4}$
$$= 3\sqrt[3]{4} + 2(2\sqrt[3]{4})$$
$$= 3\sqrt[3]{4} + 4\sqrt[3]{4}$$
$$= 7\sqrt[3]{4}$$

33. $\sqrt[3]{27} - 5\sqrt[3]{8} = 3 - 5 \cdot 2 = 3 - 10 = -7$

35. $2\sqrt[3]{a^4 b^2} + 3a\sqrt[3]{ab^2} = 2\sqrt[3]{a^3} \cdot \sqrt[3]{ab^2} + 3a\sqrt[3]{ab^2}$
$$= 2a\sqrt[3]{ab^2} + 3a\sqrt[3]{ab^2}$$
$$= 5a\sqrt[3]{ab^2}$$

37. $\sqrt{4r^7 s^5} + 3r^2\sqrt{r^3 s^5} - 2rs\sqrt{r^5 s^3} = \sqrt{4r^6 s^4} \cdot \sqrt{rs} + 3r^2\sqrt{r^2 s^4} \cdot \sqrt{rs} - 2rs\sqrt{r^4 s^2} \cdot \sqrt{rs}$
$$= 2r^3 s^2 \sqrt{rs} + 3r^2(rs^2\sqrt{rs}) - 2rs(r^2 s\sqrt{rs})$$
$$= 2r^3 s^2\sqrt{rs} + 3r^3 s^2\sqrt{rs} - 2r^3 s^2\sqrt{rs}$$
$$= 3r^3 s^2\sqrt{rs}$$

39. $\sqrt[3]{128x^9 y^{10}} - 2x^2 y^3\sqrt[3]{16x^3 y^7}$
$$= \sqrt[3]{64x^9 y^9}\sqrt[3]{2y} - 2x^2 y^3\sqrt[3]{8x^3 y^6}\sqrt[3]{2y}$$
$$= 4x^3 y^3\sqrt[3]{2y} - 2x^2 y\left(2xy^2\sqrt[3]{2y}\right)$$
$$= 4x^3 y^3\sqrt[3]{2y} - 4x^3 y^3\sqrt[3]{2y}$$
$$= 0$$

41. $\dfrac{1}{\sqrt{2}} + \dfrac{\sqrt{2}}{2} = \dfrac{1}{\sqrt{2}} \cdot \dfrac{\sqrt{2}}{\sqrt{2}} + \dfrac{\sqrt{2}}{2}$
$$= \dfrac{\sqrt{2}}{\sqrt{4}} + \dfrac{\sqrt{2}}{2}$$
$$= \dfrac{\sqrt{2}}{2} + \dfrac{\sqrt{2}}{2}$$
$$= \dfrac{2\sqrt{2}}{2}$$
$$= \sqrt{2}$$

43. $\sqrt{3} - \dfrac{1}{\sqrt{3}} = \sqrt{3} - \dfrac{1}{\sqrt{3}} \cdot \dfrac{\sqrt{3}}{\sqrt{3}}$
$$= \sqrt{3} - \dfrac{\sqrt{3}}{3}$$
$$= \dfrac{3\sqrt{3}}{3} - \dfrac{\sqrt{3}}{3}$$
$$= \dfrac{2\sqrt{3}}{3}$$

45. $3\sqrt{2} - \dfrac{2}{\sqrt{8}} + \sqrt{50}$

$= 3\sqrt{2} - \dfrac{2}{\sqrt{8}} \cdot \dfrac{\sqrt{8}}{\sqrt{8}} + \sqrt{25}\sqrt{2}$

$= 3\sqrt{2} - \dfrac{2\sqrt{4}\sqrt{2}}{\sqrt{64}} + 5\sqrt{2}$

$= \dfrac{3\sqrt{2}}{1} - \dfrac{4\sqrt{2}}{8} + \dfrac{5\sqrt{2}}{1}$

$= \dfrac{6\sqrt{2}}{2} - \dfrac{\sqrt{2}}{2} + \dfrac{10\sqrt{2}}{2}$

$= \dfrac{(6 - 1 + 10)\sqrt{2}}{2}$

$= \dfrac{15\sqrt{2}}{2}$

47. $4\sqrt{x} + \dfrac{1}{\sqrt{x}} + \sqrt{\dfrac{1}{x}}$

$= 4\sqrt{x} + \dfrac{1}{\sqrt{x}} \cdot \dfrac{\sqrt{x}}{\sqrt{x}} + \dfrac{\sqrt{1}}{\sqrt{x}} \cdot \dfrac{\sqrt{x}}{\sqrt{x}}$

$= 4\sqrt{x} + \dfrac{\sqrt{x}}{\sqrt{x^2}} + \dfrac{\sqrt{x}}{\sqrt{x^2}}$

$= 4\sqrt{x} + \dfrac{\sqrt{x}}{x} + \dfrac{\sqrt{x}}{x}$

$= 4\sqrt{x} + \dfrac{2\sqrt{x}}{x}$

$= 2\sqrt{x}\left(2 + \dfrac{1}{x}\right)$

49. $\dfrac{1}{2}\sqrt{18} - \dfrac{3}{\sqrt{2}} - 3\sqrt{50}$

$= \dfrac{1}{2}\left(3\sqrt{2}\right) - \dfrac{3}{\sqrt{2}} - 3\left(5\sqrt{2}\right)$

$= \dfrac{3\sqrt{2}}{2} - \dfrac{3}{\sqrt{2}} \cdot \dfrac{\sqrt{2}}{\sqrt{2}} - 15\sqrt{2}$

$= \dfrac{3\sqrt{2}}{2} - \dfrac{3\sqrt{2}}{2} - 15\sqrt{2}$

$= 0 - 15\sqrt{2}$

$= -15\sqrt{2}$

51. $-2\sqrt{\dfrac{x}{y}} + 3\sqrt{\dfrac{y}{x}} = -2\dfrac{\sqrt{x}}{\sqrt{y}} + 3\dfrac{\sqrt{y}}{\sqrt{x}}$

$= -2\dfrac{\sqrt{x}}{\sqrt{y}} \cdot \dfrac{\sqrt{y}}{\sqrt{y}} + 3\dfrac{\sqrt{y}}{\sqrt{x}} \cdot \dfrac{\sqrt{x}}{\sqrt{x}}$

$= -2\dfrac{\sqrt{xy}}{y} + 3\dfrac{\sqrt{xy}}{x}$

$= \left(-\dfrac{2}{y} + \dfrac{3}{x}\right)\sqrt{xy}$

53. $\left(\sqrt{3} + 4\right)\left(\sqrt{3} + 5\right) = \sqrt{9} + 5\sqrt{3} + 4\sqrt{3} + 20$

$= 3 + 9\sqrt{3} + 20$

$= 23 + 9\sqrt{3}$

55. $\left(1 + \sqrt{5}\right)\left(6 + \sqrt{5}\right) = 6 + \sqrt{5} + 6\sqrt{5} + \left(\sqrt{5}\right)^2$

$= 6 + 7\sqrt{5} + 5$

$= 11 + 7\sqrt{5}$

57. $\left(4 - \sqrt{2}\right)\left(5 + \sqrt{2}\right) = 20 + 4\sqrt{2} - 5\sqrt{2} - \sqrt{4}$

$= 20 - \sqrt{2} - 2$

$= 18 - \sqrt{2}$

59. $\left(4\sqrt{3} + \sqrt{2}\right)\left(\sqrt{3} - \sqrt{2}\right)$

$= 4\sqrt{9} - 4\sqrt{6} + \sqrt{6} - \sqrt{4}$

$= 4 \cdot 3 - 3\sqrt{6} - 2$

$= 12 - 3\sqrt{6} - 2$

$= 10 - 3\sqrt{6}$

61. $\left(2\sqrt{5} - 3\right)^2 = \left(2\sqrt{5} - 3\right)\left(2\sqrt{5} - 3\right)$

$= 4\sqrt{25} - 6\sqrt{5} - 6\sqrt{5} + 9$

$= 4 \cdot 5 - 12\sqrt{5} + 9$

$= 20 + 9 - 12\sqrt{5}$

$= 29 - 12\sqrt{5}$

63. $\left(2\sqrt{3x} - \sqrt{y}\right)\left(3\sqrt{3x} + \sqrt{y}\right)$

$= 6\left(\sqrt{3x}\right)^2 + 2\sqrt{3x}\sqrt{y} - 3\sqrt{3x}\sqrt{y} - \left(\sqrt{y}\right)^2$

$= 6(3x) + 2\sqrt{3xy} - 3\sqrt{3xy} - y$

$= 18x - \sqrt{3xy} - y$

65. $\left(\sqrt[3]{4}-\sqrt[3]{6}\right)\left(\sqrt[3]{2}-\sqrt[3]{36}\right)$

$=\sqrt[3]{4}\sqrt[3]{2}-\sqrt[3]{4}\sqrt[3]{36}-\sqrt[3]{6}\sqrt[3]{2}+\sqrt[3]{6}\sqrt[3]{36}$

$=\sqrt[3]{8}-\sqrt[3]{144}-\sqrt[3]{12}+\sqrt[3]{216}$

$=2-2\sqrt[3]{18}-\sqrt[3]{12}+6$

$=8-2\sqrt[3]{18}-\sqrt[3]{12}$

67. $\sqrt{5}+2\sqrt{5}=3\sqrt{5}$

69. $\sqrt{125}+\sqrt{20}=5\sqrt{5}+2\sqrt{5}=7\sqrt{5}$

71. $\dfrac{\sqrt{6}}{2}+\dfrac{1}{\sqrt{6}}=\dfrac{\sqrt{6}}{2}+\dfrac{1}{\sqrt{6}}\cdot\dfrac{\sqrt{6}}{\sqrt{6}}$

$=\dfrac{\sqrt{6}}{2}+\dfrac{\sqrt{6}}{6}$

$=\dfrac{3\sqrt{6}}{6}+\dfrac{\sqrt{6}}{6}$

$=\dfrac{4\sqrt{6}}{6}$

$=\dfrac{2\sqrt{6}}{3}$

73. $2\sqrt[3]{81}+4\sqrt[3]{24}=2\sqrt[3]{27}\cdot\sqrt[3]{3}+4\sqrt[3]{8}\cdot\sqrt[3]{3}$

$=2\left(3\sqrt[3]{3}\right)+4\left(2\sqrt[3]{3}\right)$

$=6\sqrt[3]{3}+8\sqrt[3]{3}$

$=14\sqrt[3]{3}$

75. $\left(3\sqrt{2}-4\right)\left(\sqrt{2}+5\right)$

$=3\left(\sqrt{2}\right)^2+15\sqrt{2}-4\sqrt{2}-20$

$=6+11\sqrt{2}-20$

$=-14+11\sqrt{2}$

77. $4\sqrt{3x^3}-\sqrt{12x}=4x\sqrt{3x}-2\sqrt{3x}$

$=(4x-2)\sqrt{3x}$

79. $\dfrac{3}{\sqrt{y}}-\sqrt{\dfrac{9}{y}}+\sqrt{y}=\dfrac{3}{\sqrt{y}}-\dfrac{3}{\sqrt{y}}+\sqrt{y}$

$=0+\sqrt{y}$

$=\sqrt{y}$

81. $\left(\sqrt[3]{x^2}-\sqrt[3]{y}\right)\left(\sqrt[3]{x}-2\sqrt[3]{y^2}\right)$

$=\sqrt[3]{x^2}\sqrt[3]{x}-2\sqrt[3]{x^2}\sqrt[3]{y^2}-\sqrt[3]{y}\sqrt[3]{x}+2\sqrt[3]{y}\sqrt[3]{y^2}$

$=\sqrt[3]{x^3}-2\sqrt[3]{x^2y^2}-\sqrt[3]{xy}+2\sqrt[3]{y^3}$

$=x-2\sqrt[3]{x^2y^2}-\sqrt[3]{xy}+2y$

83. $\left(\sqrt[3]{a}+5\right)\left(\sqrt[3]{a^2}-3\right)$

$=\sqrt[3]{a}\sqrt[3]{a^2}-3\sqrt[3]{a}+5\sqrt[3]{a^2}-15$

$=\sqrt[3]{a^3}-3\sqrt[3]{a}+5\sqrt[3]{a^2}-15$

$=a-3\sqrt[3]{a}+5\sqrt[3]{a^2}-15$

85. $2\sqrt{\dfrac{8}{3}}-4\sqrt{\dfrac{100}{6}}=2\sqrt{\dfrac{8}{3}}-4\sqrt{\dfrac{50}{3}}$

$=2\dfrac{\sqrt{8}}{\sqrt{3}}-4\dfrac{\sqrt{50}}{\sqrt{3}}$

$=\dfrac{2\sqrt{4}\sqrt{2}}{\sqrt{3}}-\dfrac{4\sqrt{25}\sqrt{2}}{\sqrt{3}}$

$=\dfrac{2\left(2\sqrt{2}\right)}{\sqrt{3}}-\dfrac{4\left(5\sqrt{2}\right)}{\sqrt{3}}$

$=\dfrac{4\sqrt{2}}{\sqrt{3}}-\dfrac{20\sqrt{2}}{\sqrt{3}}$

$=\dfrac{-16\sqrt{2}}{\sqrt{3}}\cdot\dfrac{\sqrt{3}}{\sqrt{3}}$

$=-\dfrac{16\sqrt{6}}{3}$

87. Perimeter $= \sqrt{45} + \sqrt{45} + \sqrt{80} + \sqrt{80}$
$$= 2\sqrt{45} + 2\sqrt{80}$$
$$= 2\sqrt{9}\sqrt{5} + 2\sqrt{16}\sqrt{5}$$
$$= 2(3\sqrt{5}) + 2(4\sqrt{5})$$
$$= 6\sqrt{5} + 8\sqrt{5}$$
$$= 14\sqrt{5}$$

Area $= \sqrt{45}\sqrt{80}$
$$= 3\sqrt{5} \cdot 4\sqrt{5}$$
$$= 12(\sqrt{5})^2$$
$$= 12 \cdot 5$$
$$= 60$$

89. Perimeter $= \sqrt{245} + \sqrt{180} + \sqrt{80}$
$$= \sqrt{49}\sqrt{5} + \sqrt{36}\sqrt{5} + \sqrt{16}\sqrt{5}$$
$$= 7\sqrt{5} + 6\sqrt{5} + 4\sqrt{5}$$
$$= 17\sqrt{5}$$

Area $= \dfrac{1}{2}\sqrt{245}\sqrt{45}$
$$= \dfrac{1}{2}\sqrt{49}\sqrt{5}\sqrt{9}\sqrt{5}$$
$$= \dfrac{1}{2} \cdot 7 \cdot 3(\sqrt{5})^2$$
$$= \dfrac{21}{2} \cdot 5$$
$$= 52.5$$

91. $f(t) = 3\sqrt{t} + 19$

   a. $t = 36$
$$f(36) = 3\sqrt{36} + 19$$
$$= 3(6) + 19$$
$$= 18 + 19$$
$$= 37$$
The length at 36 months is 37 inches.

   b. $t = 40$
$$f(40) = 3\sqrt{40} + 19$$
$$= 3\sqrt{4}\sqrt{10} + 19$$
$$= 3 \cdot 2\sqrt{10} + 19$$
$$= 6\sqrt{10} + 19$$
$$\approx 37.97$$
The length at 40 months is about 37.97 inches.

93. $\dfrac{1}{\sqrt{3}+2} = \dfrac{1}{\sqrt{3}+2} \cdot \dfrac{\sqrt{3}-2}{\sqrt{3}-2}$
$$= \dfrac{\sqrt{3}-2}{(\sqrt{3})^2 - 2^2}$$
$$= \dfrac{\sqrt{3}-2}{3-4}$$
$$= \dfrac{\sqrt{3}-2}{-1}$$
$$= -\sqrt{3}+2$$
$$= 2-\sqrt{3}$$
$$2+\sqrt{3} > 2-\sqrt{3}$$
Therefore, $2+\sqrt{3} > \dfrac{1}{\sqrt{3}+2}$.

95. Yes. Answers will vary.

97. No; $-\sqrt{2} + \sqrt{2} = 0$

99. $\sqrt{10} + \sqrt{5} = \sqrt{5 \cdot 2} + \sqrt{5}$
$$= \sqrt{5}\sqrt{2} + \sqrt{5}$$
$$= \sqrt{5}(\sqrt{2} + 1)$$

101. **a.** $x^2 - 3 = x^2 - (\sqrt{3})^2$

   b. $x^2 - (\sqrt{3})^2 = (x+\sqrt{3})(x-\sqrt{3})$

103. **a.**
$$f(x) = \sqrt{x}$$
$$g(x) = 2$$
$$(f+g)(x) = f(x) + g(x)$$
$$= \sqrt{x} + 2$$

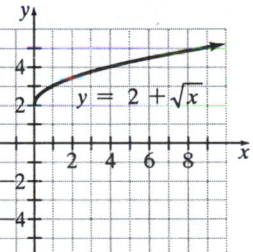

   b.   It raises the graph 2 units.

105. a. $(f - g)(x) = f(x) - g(x)$
$$= \sqrt{x} - \left(\sqrt{x} - 2\right)$$
$$= \sqrt{x} - \sqrt{x} + 2$$
$$= 2$$

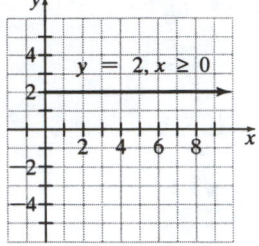

$y = 2, x \geq 0$

b. $\sqrt{x} \geq 0$, so $x \geq 0$
The domain is
$\{x | x$ is a real number, $x \geq 0\}$.

107. $f(x) = \sqrt{x^2}$

x	$f(x)$
-3	3
-2	2
0	0
1	1
2	2
3	3

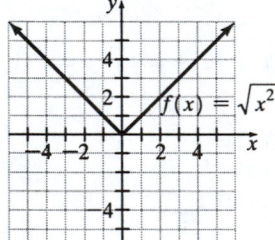

$f(x) = \sqrt{x^2}$

111. $\dfrac{(2x^{-2}y^3)^2}{(xy^2)^{-2}} = \dfrac{4x^{-4}y^6}{x^{-2}y^{-4}}$
$$= 4x^{-4-(-2)}y^{6-(-4)}$$
$$= 4x^{-2}y^{10}$$
$$= \dfrac{4y^{10}}{x^2}$$

112.
$$20x^2 + 3x - 9 = 0$$
$$20x^2 + 15x - 12x - 9 = 0$$
$$5x(4x + 3) - 3(4x + 3) = 0$$
$$(5x - 3)(4x + 3) = 0$$
$$5x - 3 = 0 \quad \text{or} \quad 4x + 3 = 0$$
$$5x = 3 \qquad\qquad 4x = -3$$
$$x = \dfrac{3}{5} \qquad\qquad x = -\dfrac{3}{4}$$

113. $\left(\dfrac{x^{3/4}y^{2/3}}{x^{1/2}y}\right)^2 = (x^{3/4 - 1/2}y^{2/3 - 1})^2$
$$= (x^{1/4})^2(y^{-1/3})^2$$
$$= x^{1/2}y^{-2/3}$$
$$= \dfrac{x^{1/2}}{y^{2/3}}$$

114. $\sqrt[3]{3x^2y}\left(\sqrt[3]{9x^4y^3} - \sqrt[3]{x^{10}y^7}\right)$
$$= \sqrt[3]{27x^6y^4} - \sqrt[3]{3x^{12}y^8}$$
$$= \sqrt[3]{27x^6y^3}\sqrt[3]{y} - \sqrt[3]{x^{12}y^6}\sqrt[3]{3y^2}$$
$$= 3x^2y\sqrt[3]{y} - x^4y^2\sqrt[3]{3y^2}$$

Exercise Set 7.6

1. a. Answers will vary.

b.
$$\sqrt{2x + 26} - 2 = 4$$
$$\sqrt{2x + 26} - 2 + 2 = 4 + 2$$
$$\sqrt{2x + 26} = 6$$
$$\left(\sqrt{2x + 26}\right)^2 = 6^2$$
$$2x + 26 = 36$$
$$2x = 10$$
$$x = \dfrac{10}{2}$$
$$x = 5$$

3. 0 is the only solution to the equation. For all other values, the left side of the equation is negative and the right side is positive.

5. Answers will vary. Possible answer:
The equation has no solution since the left side is a positive number whereas the right side is 0. A positive number is never equal to 0.
Also, the equation can be written as
$\sqrt{x-3} = -3$ for which the left side is positive and the right side is negative. It is impossible for $\sqrt{x-3}$ to equal a negative number.

7. $\sqrt{x} = 4$
$$\left(\sqrt{x}\right)^2 = 4^2$$
$$x = 16$$

9. $\sqrt[3]{x} = 2$
$$\left(\sqrt[3]{x}\right)^3 = 2^3$$
$$x = 8$$

11. $\sqrt[4]{x} = 3$
$$\left(\sqrt[4]{x}\right)^4 = 3^4$$
$$x = 81$$

13. $-\sqrt{2x+4} = -6$
$$\left(-\sqrt{2x+4}\right)^2 = (-6)^2$$
$$2x+4 = 36$$
$$2x = 32$$
$$x = 16$$
Check: $\quad -\sqrt{2x+4} = -6$
$$-\sqrt{2(16)+4} \stackrel{?}{=} -6$$
$$-\sqrt{32+4} \stackrel{?}{=} -6$$
$$-\sqrt{36} \stackrel{?}{=} -6$$
$$-6 = -6 \quad \text{True}$$

15. $\sqrt[3]{2x+11} = 3$
$$\left(\sqrt[3]{2x+11}\right)^3 = 3^3$$
$$2x+11 = 27$$
$$2x = 16$$
$$x = 8$$

17. $\sqrt[3]{3x} + 4 = 7$
$$\sqrt[3]{3x} = 3$$
$$\left(\sqrt[3]{3x}\right)^3 = 3^3$$
$$3x = 27$$
$$x = 9$$

19. $\sqrt{a-8} = 2\sqrt{3a-2}$
$$\left(\sqrt{a-8}\right)^2 = \left(2\sqrt{3a-2}\right)^2$$
$$a-8 = 4(3a-2)$$
$$a-8 = 12a-8$$
$$a = 12a$$
$$0 = 11a$$
$$0 = a$$
Check: $\sqrt{a-8} = 2\sqrt{3a-2}$
$$\sqrt{0-8} \stackrel{?}{=} 2\sqrt{3(0)-2}$$
$$\sqrt{-8} = 2 \quad \text{False}$$
Thus, 0 is not a solution to this equation and we conclude that there is no solution.

21. $\sqrt{5b+10} = -\sqrt{3b+8}$
$$\left(\sqrt{5b+10}\right)^2 = \left(-\sqrt{3b+8}\right)^2$$
$$5b+10 = 3b+8$$
$$2b = -2$$
$$b = -1$$
Check: $\quad \sqrt{5b+10} = -\sqrt{3b+8}$
$$\sqrt{5(-1)+10} \stackrel{?}{=} -\sqrt{3(-1)+8}$$
$$\sqrt{-5+10} \stackrel{?}{=} -\sqrt{-3+8}$$
$$\sqrt{5} = -\sqrt{5} \quad \text{False}$$
Thus, -1 is not a solution to this equation and we conclude that there is no solution.

23. $\sqrt[3]{6x+1} = \sqrt[3]{2x+5}$
$$\left(\sqrt[3]{6x+1}\right)^3 = \left(\sqrt[3]{2x+5}\right)^3$$
$$6x+1 = 2x+5$$
$$4x = 4$$
$$x = 1$$

25.
$$\sqrt{x^2 + 9x + 3} = -x$$
$$\left(\sqrt{x^2 + 9x + 3}\right)^2 = (-x)^2$$
$$x^2 + 9x + 3 = x^2$$
$$9x + 3 = 0$$
$$9x = -3$$
$$x = \frac{-3}{9} = \frac{-1}{3} = -\frac{1}{3}$$

Check:
$$\sqrt{x^2 + 9x + 3} = -x$$
$$\sqrt{\left(-\frac{1}{3}\right)^2 + 9\left(-\frac{1}{3}\right) + 3} \stackrel{?}{=} -\left(-\frac{1}{3}\right)$$
$$\sqrt{\frac{1}{9} - 3 + 3} \stackrel{?}{=} \frac{1}{3}$$
$$\sqrt{\frac{1}{9}} \stackrel{?}{=} \frac{1}{3}$$
$$\frac{1}{3} = \frac{1}{3} \quad \text{True}$$

27.
$$\sqrt{m^2 + 6m - 4} = m$$
$$\left(\sqrt{m^2 + 6m - 4}\right)^2 = (m)^2$$
$$m^2 + 6m - 4 = m^2$$
$$6m - 4 = 0$$
$$6m = 4$$
$$m = \frac{2}{3}$$

29.
$$\sqrt{x^2 - 2} = x + 4$$
$$\left(\sqrt{x^2 - 2}\right)^2 = (x + 4)^2$$
$$x^2 - 2 = x^2 + 8x + 16$$
$$-2 = 8x + 16$$
$$-18 = 8x$$
$$-\frac{18}{8} = x$$
$$-\frac{9}{4} = x$$

Check:
$$\sqrt{x^2 - 2} = x + 4$$
$$\sqrt{\left(-\frac{9}{4}\right)^2 - 2} \stackrel{?}{=} -\frac{9}{4} + 4$$
$$\sqrt{\frac{81}{16} - 2} \stackrel{?}{=} \frac{7}{4}$$
$$\sqrt{\frac{49}{16}} \stackrel{?}{=} \frac{7}{4}$$
$$\frac{7}{4} = \frac{7}{4} \quad \text{True}$$

31.
$$-\sqrt{x} = 2x - 1$$
$$\left(-\sqrt{x}\right)^2 = (2x - 1)^2$$
$$x = 4x^2 - 4x + 1$$
$$0 = 4x^2 - 5x + 1$$
$$0 = (x - 1)(4x - 1)$$
$$x - 1 = 0 \quad \text{or} \quad 4x - 1 = 0$$
$$x = 1 \qquad\qquad 4x = 1$$
$$x = \frac{1}{4}$$

Check: Check:
$$-\sqrt{x} = 2x - 1 \qquad -\sqrt{x} = 2x - 1$$
$$-\sqrt{1} \stackrel{?}{=} 2(1) - 1 \qquad -\sqrt{\frac{1}{4}} \stackrel{?}{=} 2\left(\frac{1}{4}\right) - 1$$
$$-1 = 1 \qquad\qquad -\frac{1}{2} \stackrel{?}{=} \frac{1}{2} - 1$$
$$\text{False} \qquad\qquad -\frac{1}{2} \stackrel{?}{=} \frac{1}{2} - \frac{2}{2}$$
$$-\frac{1}{2} = -\frac{1}{2} \quad \text{True}$$

Only $\frac{1}{4}$ is a solution. 1 is an extraneous root.

33.
$$\sqrt{5x+6} = 2x-6$$
$$\left(\sqrt{5x+6}\right)^2 = (2x-6)^2$$
$$5x+6 = 4x^2 - 24x + 36$$
$$0 = 4x^2 - 29x + 30$$
$$0 = (4x-5)(x-6)$$
$$4x-5=0 \quad \text{or} \quad x-6=0$$
$$x = \frac{5}{4} \qquad\qquad x = 6$$

Check: $\sqrt{5x+6} = 2x-6$
$$\sqrt{5\left(\frac{5}{4}\right)+6} \stackrel{?}{=} 2\left(\frac{5}{4}\right)-6$$
$$\sqrt{\frac{49}{4}} \stackrel{?}{=} -\frac{14}{4}$$
$$\frac{7}{2} = -\frac{7}{2} \quad \text{False}$$

Check: $\sqrt{5x+6} = 2x-6$
$$\sqrt{5(6)+6} \stackrel{?}{=} 2(6)-6$$
$$6 = 6 \qquad \text{True}$$

Only 6 is a solution. $\dfrac{5}{4}$ is an extraneous solution.

35.
$$\sqrt[3]{x-12} = \sqrt[3]{5x+16}$$
$$\left(\sqrt[3]{x-12}\right)^3 = \left(\sqrt[3]{5x+16}\right)^3$$
$$x-12 = 5x+16$$
$$-28 = 4x$$
$$-7 = x$$
Check: $\sqrt[3]{x-12} = \sqrt[3]{5x+16}$
$$\sqrt[3]{-7-12} \stackrel{?}{=} \sqrt[3]{5x+16}$$
$$\sqrt[3]{-19} = \sqrt[3]{-19} \qquad \text{True}$$

37.
$$\sqrt{8b-15} + b = 10$$
$$\sqrt{8b-15} = 10 - b$$
$$\left(\sqrt{8b-15}\right)^2 = (10-b)^2$$
$$8b-15 = 100 - 20b + b^2$$
$$b^2 - 28b + 115 = 0$$
$$(b-5)(b-23) = 0$$
$$b-5=0 \quad \text{or} \quad b-23=0$$
$$b = 5 \qquad\qquad b = 23$$

Check: $\sqrt{8b-15} + b = 10$
$$\sqrt{8\cdot5-15} + 5 \stackrel{?}{=} 10$$
$$\sqrt{40-15} + 5 \stackrel{?}{=} 10$$
$$\sqrt{25} + 5 \stackrel{?}{=} 10$$
$$5+5 \stackrel{?}{=} 10$$
$$10 = 10 \quad \text{True}$$

Check: $\sqrt{8b-15} + b = 10$
$$\sqrt{8\cdot23-15} + 23 \stackrel{?}{=} 10$$
$$\sqrt{184-15} + 23 \stackrel{?}{=} 10$$
$$\sqrt{169} + 23 \stackrel{?}{=} 10$$
$$13+23 \stackrel{?}{=} 10$$
$$36 = 10 \quad \text{False}$$

Only 5 is a solution. 23 is an extraneous solution.

39.
$$(2a+9)^{1/2} - a + 3 = 0$$
$$(2a+9)^{1/2} = a - 3$$
$$[(2a+9)^{1/2}]^2 = (a-3)^2$$
$$2a+9 = a^2 - 6a + 9$$
$$0 = a^2 - 8a$$
$$0 = a(a-8)$$
$$a = 0 \quad \text{or} \quad a = 8$$

Check: $(2a+9)^{1/2} - a + 3 = 0$
$$(2\cdot0+9)^{1/2} - 0 + 3 \stackrel{?}{=} 0$$
$$3+3 \stackrel{?}{=} 0$$
$$6 = 0 \quad \text{False}$$

Check: $(2a+9)^{1/2} - a + 3 = 0$
$$(2\cdot8+9)^{1/2} - 8 + 3 \stackrel{?}{=} 0$$
$$5-8+3 \stackrel{?}{=} 0$$
$$0 = 0 \quad \text{True}$$

Only 8 is a solution. 0 is an extraneous solution.

41. $(r+2)^{1/3} = (3r+8)^{1/3}$

$[(r+2)^{1/3}]^3 = [(3r+8)^{1/3}]^3$

$r+2 = 3r+8$

$2 = 2r+8$

$-6 = 2r$

$-3 = r$

Check: $(r+2)^{1/3} = (3r+8)^{1/3}$

$(-3+2)^{1/3} \overset{?}{=} [3(-3)+8]^{1/3}$

$(-1)^{1/3} \overset{?}{=} (-1)^{1/3}$

$-1 = -1$ True

43. $(5x+18)^{1/4} = (9x+2)^{1/4}$

$[(5x+18)^{1/4}]^4 = [(9x+2)^{1/4}]^4$

$5x+18 = 9x+2$

$18 = 4x+2$

$16 = 4x$

$4 = x$

Check:

$(5x+18)^{1/4} = (9x+2)^{1/4}$

$(5 \cdot 4+18)^{1/4} \overset{?}{=} (9 \cdot 4+2)^{1/4}$

$(20+18)^{1/4} \overset{?}{=} (36+2)^{1/4}$

$(38)^{1/4} = (38)^{1/4}$ True

45. $(x^2+4x+4)^{1/2} - x - 3 = 0$

$(x^2+4x+4)^{1/2} = x+3$

$[(x^2+4x+4)^{1/2}]^2 = (x+3)^2$

$x^2+4x+4 = x^2+6x+9$

$4x+4 = 6x+9$

$4 = 2x+9$

$-5 = 2x$

$-\dfrac{5}{2} = x$

47. $\sqrt{2a-3} = \sqrt{2a}-1$

$\left(\sqrt{2a-3}\right)^2 = \left(\sqrt{2a}-1\right)^2$

$2a-3 = 2a-2\sqrt{2a}+1$

$-4 = -2\sqrt{2a}$

$-2 = -\sqrt{2a}$

$(-2)^2 = \left(-\sqrt{2a}\right)^2$

$4 = 2a$

$2 = a$

49. $\sqrt{x+1} = 2 - \sqrt{x}$

$\left(\sqrt{x+1}\right)^2 = \left(2-\sqrt{x}\right)^2$

$x+1 = 4-4\sqrt{x}+x$

$-3 = -4\sqrt{x}$

$(-3)^2 = \left(-4\sqrt{x}\right)^2$

$9 = 16x$

$\dfrac{9}{16} = x$

Check: $\sqrt{x+1} = 2 - \sqrt{x}$

$\sqrt{\dfrac{9}{16}+1} \overset{?}{=} 2 - \sqrt{\dfrac{9}{16}}$

$\sqrt{\dfrac{9}{16}+\dfrac{16}{16}} \overset{?}{=} 2 - \dfrac{3}{4}$

$\sqrt{\dfrac{25}{16}} \overset{?}{=} \dfrac{8}{4} - \dfrac{3}{4}$

$\dfrac{5}{4} = \dfrac{5}{4}$ True

51. $\sqrt{x+7} = 5 - \sqrt{x-8}$

$\left(\sqrt{x+7}\right)^2 = \left(5-\sqrt{x-8}\right)^2$

$x+7 = 25-10\sqrt{x-8}+x-8$

$7 = 17-10\sqrt{x-8}$

$-10 = -10\sqrt{x-8}$

$1 = \sqrt{x-8}$

$(1)^2 = \left(\sqrt{x-8}\right)^2$

$1 = x-8$

$9 = x$

Check:

$\sqrt{x+7} = 5 - \sqrt{x-8}$

$\sqrt{9+7} \overset{?}{=} 5 - \sqrt{9-8}$

$\sqrt{16} \overset{?}{=} 5 - \sqrt{1}$

$4 \overset{?}{=} 5 - 1$

$4 = 4$ True

53.
$$\sqrt{b-3} = 4 - \sqrt{b+5}$$
$$\left(\sqrt{b-3}\right)^2 = \left(4 - \sqrt{b+5}\right)^2$$
$$b - 3 = 16 - 8\sqrt{b+5} + b + 5$$
$$-3 = 21 - 8\sqrt{b+5}$$
$$-24 = -8\sqrt{b+5}$$
$$3 = \sqrt{b+5}$$
$$(3)^2 = \left(\sqrt{b+5}\right)^2$$
$$9 = b + 5$$
$$4 = b$$
Check: $\sqrt{b-3} = 4 - \sqrt{b+5}$
$$\sqrt{4-3} \overset{?}{=} 4 - \sqrt{4+5}$$
$$\sqrt{1} \overset{?}{=} 4 - \sqrt{9}$$
$$1 \overset{?}{=} 4 - 3$$
$$1 = 1 \quad \text{True}$$

55. $\sqrt{r+10} + 3 + \sqrt{r-5} = 0$

Without going any further, it is clear that there is no solution to this problem since all the quantitites on the left side are positive and the right side is 0. A positive quantity is never 0.

57. $\sqrt{2x+4} - \sqrt{x+3} - 1 = 0$
$$\sqrt{2x+4} = \sqrt{x+3} + 1$$
$$\left(\sqrt{2x+4}\right)^2 = \left(\sqrt{x+3} + 1\right)^2$$
$$2x + 4 = (x+3) + 2\sqrt{x+3} + 1$$
$$2x + 4 = x + 4 + 2\sqrt{x+3}$$
$$x = 2\sqrt{x+3}$$
$$x^2 = \left(2\sqrt{x+3}\right)^2$$
$$x^2 = 4(x+3)$$
$$x^2 = 4x + 12$$
$$x^2 - 4x - 12 = 0$$
$$(x-6)(x+2) = 0$$
$$x - 6 = 0 \quad \text{or} \quad x + 2 = 0$$
$$x = 6 \qquad\qquad x = -2$$

A check shows that –2 is not a solution and that 6 is a solution.

59.
$$f(x) = g(x)$$
$$\sqrt{x+4} = \sqrt{2x-2}$$
$$\left(\sqrt{x+4}\right)^2 = \left(\sqrt{2x-2}\right)^2$$
$$x + 4 = 2x - 2$$
$$6 = x$$

61.
$$f(x) = g(x)$$
$$\sqrt[3]{5x-12} = \sqrt[3]{6x-16}$$
$$\left(\sqrt[3]{5x-12}\right)^3 = \left(\sqrt[3]{6x-16}\right)^3$$
$$5x - 12 = 6x - 16$$
$$x = 4$$

63.
$$f(x) = g(x)$$
$$2(8x+24)^{1/3} = 4(2x-2)^{1/3}$$
$$[2(8x+24)^{1/3}]^3 = [4(2x-2)^{1/3}]^3$$
$$8(8x+24) = 64(2x-2)$$
$$64x + 192 = 128x - 128$$
$$64x = 320$$
$$x = 5$$

65.
$$p = \sqrt{2v}$$
$$p^2 = \left(\sqrt{2v}\right)^2$$
$$p^2 = 2v$$
$$\frac{p^2}{2} = v$$

67.
$$v = \sqrt{2gh}$$
$$v^2 = \left(\sqrt{2gh}\right)^2$$
$$v^2 = 2gh$$
$$g = \frac{v^2}{2h}$$

69.
$$v = \sqrt{\frac{FR}{m}}$$
$$v^2 = \left(\sqrt{\frac{FR}{m}}\right)^2$$
$$v^2 = \frac{FR}{m}$$
$$mv^2 = FR$$
$$F = \frac{mv^2}{R}$$

71. $x = \sqrt{\dfrac{m}{k}} V_0$

$$x^2 = \left(\sqrt{\dfrac{m}{k}} V_0 \right)^2$$

$$x^2 = \dfrac{m V_0^2}{k}$$

$$x^2 k = m V_0^2$$

$$m = \dfrac{x^2 k}{V_0^2}$$

73. $l = \sqrt{20^2 + 40^2}$

$l = \sqrt{400 + 1600}$

$= \sqrt{2000}$

≈ 44.7 ft

75. $s = \sqrt{A}$

$s = \sqrt{144}$

$= 12$ feet

77. $T = 2\pi \sqrt{\dfrac{l}{32}}$

   a.   Let $l = 6$

$$T = 2\pi \sqrt{\dfrac{l}{32}}$$

$$= 2\pi \sqrt{\dfrac{6}{32}}$$

$$= 2\pi \sqrt{0.1875}$$

$$\approx 2\pi(0.433012702)$$

$$\approx 2.72 \text{ seconds}$$

   b.   Replace l with $2l$:

$$T_D = 2\pi \sqrt{\dfrac{2l}{32}}$$

$$= 2\pi \sqrt{2} \sqrt{\dfrac{l}{32}}$$

$$= \sqrt{2} \left(2\pi \sqrt{\dfrac{l}{32}} \right)$$

$$= \sqrt{2} T$$

   c.   This part must be solved in two phases. First, we need to find the length of the pendulum:

$$T = 2\pi \sqrt{\dfrac{l}{g}}$$

$$2 = 2\pi \sqrt{\dfrac{l}{32}}$$

$$\dfrac{1}{\pi} = \sqrt{\dfrac{l}{32}}$$

$$\left(\dfrac{1}{\pi} \right)^2 = \dfrac{l}{32}$$

$$l = \dfrac{32}{\pi^2}$$

Now, find T using $g = \dfrac{32}{6}$ and $l = \dfrac{32}{\pi^2}$

$$T = 2\pi \sqrt{\dfrac{l}{g}}$$

$$= 2\pi \sqrt{\dfrac{\frac{32}{\pi^2}}{\frac{32}{6}}}$$

$$= 2\pi \sqrt{\dfrac{6}{\pi^2}}$$

$$= 2\pi \dfrac{\sqrt{6}}{\sqrt{\pi^2}}$$

$$= 2\pi \dfrac{\sqrt{6}}{\pi}$$

$$= 2\sqrt{6}$$

$$\approx 4.90 \text{ seconds}$$

79. $r = \sqrt[4]{\dfrac{8\mu l}{\pi R}}$

$$r^4 = \left(\sqrt[4]{\dfrac{8\mu l}{\pi R}} \right)^4$$

$$r^4 = \dfrac{8\mu l}{\pi R^4}$$

$$\pi R r^4 = 8\mu l$$

$$R = \dfrac{8\mu l}{\pi r^4}$$

81. $N = 0.2\left(\sqrt{R}\right)^3$

$\quad N = 0.2\left(\sqrt{149.4}\right)^3$

$\quad\quad = 0.2(12.223)^3$

$\quad\quad = 0.2(1826.106)$

$\quad\quad \approx 365.2$ days

83. $R = \sqrt{F_1^2 + F_2^2}$

$\quad R = \sqrt{600^2 + 800^2}$

$\quad\quad = \sqrt{1,000,000}$

$\quad\quad = 1000$ lb

85. $c = \sqrt{gH} = \sqrt{32 \cdot 10} = \sqrt{320} \approx 17.89$ ft / sec

87. $x = \dfrac{-b \pm \sqrt{b^2 - 4ac}}{2a}$

$\quad x = \dfrac{-0 \pm \sqrt{0^2 - 4(1)(-4)}}{2(1)} = \dfrac{\pm\sqrt{16}}{2}$

$\quad$ Now, $x = \dfrac{\sqrt{16}}{2} = \dfrac{4}{2} = 2$ or

$\quad x = -\dfrac{\sqrt{16}}{2} = -\dfrac{4}{2} = -2$

89. $x = \dfrac{-b \pm \sqrt{b^2 - 4ac}}{2a}$

$\quad x = \dfrac{-5 \pm \sqrt{5^2 - 4(2)(-12)}}{2(2)}$

$\quad\quad = \dfrac{-5 \pm \sqrt{121}}{4}$

$\quad\quad = \dfrac{-5 \pm 11}{4}$

$\quad$ Now, $x = \dfrac{-5 + 11}{4} = \dfrac{6}{4} = \dfrac{3}{2}$ or

$\quad x = \dfrac{-5 - 11}{4} = \dfrac{-16}{4} = -4$

91. $f(x) = \sqrt{x - 5}$

$\quad 4 = \sqrt{x - 5}$

$\quad 4^2 = \left(\sqrt{x - 5}\right)^2$

$\quad 16 = x - 5$

$\quad 21 = x$

93. $f(x) = \sqrt{3x^2 - 11} + 4$

$\quad 12 = \sqrt{3x^2 - 11} + 4$

$\quad 8 = \sqrt{3x^2 - 11}$

$\quad 8^2 = \left(\sqrt{3x^2 - 11}\right)^2$

$\quad 64 = 3x^2 - 11$

$\quad 75 = 3x^2$

$\quad 25 = x^2$

$\quad \pm 5 = x$

95. a. $y = \sqrt{4x - 12}, y = x - 3$

The points of intersection are $(3, 0)$ and $(7, 4)$. The x-values are 3 and 7.

b. $\sqrt{4x - 12} = x - 3$

For $x = 3$:	For $x = 7$:
$\sqrt{4 \cdot 3 - 12} = 3 - 3$	$\sqrt{4x - 12} = x - 3$
$\sqrt{12 - 12} = 0$	$\sqrt{4 \cdot 7 - 12} = 7 - 3$
$\sqrt{0} = 0$	$\sqrt{16} = 4$
$0 = 0$ True	$4 = 4$ True

c. $\sqrt{4x - 12} = x - 3$

$\quad \left(\sqrt{4x - 12}\right)^2 = (x - 3)^2$

$\quad 4x - 12 = x^2 - 6x + 9$

$\quad 0 = x^2 - 10x + 21$

$\quad 0 = (x - 3)(x - 7)$

$\quad x - 3 = 0 \quad$ or $\quad x - 7 = 0$

$\quad\quad x = 3 \quad\quad\quad\quad x = 7$

97. At $x = 4$, $g(x) = 0$ or $y = 0$. Therefore, the graph must have an x-intercept at 4.

99. $L_1 = p - 1.96\sqrt{\dfrac{p(1-p)}{n}}$

$L_1 = 0.60 - 1.96\sqrt{\dfrac{0.60(1-0.60)}{36}}$

$\approx 0.60 - 0.16$

≈ 0.44

$L_2 = p + 1.96\sqrt{\dfrac{p(1-p)}{n}}$

$L_2 = 0.60 + 1.96\sqrt{\dfrac{0.60(1-0.60)}{36}}$

$= 0.60 + 0.16$

≈ 0.76

101. $\sqrt{x^2+9} = (x^2+9)^{1/2}$

$\left(\sqrt{x^2+9}\right)^2 = [(x^2+9)^{1/2}]^2$

$x^2+9 = x^2+9$

$9 = 9$

All real numbers, x, satisfy this equation.

103. Graph:

$y_1 = \sqrt{x+8}$

$y_2 = \sqrt{3x+5}$

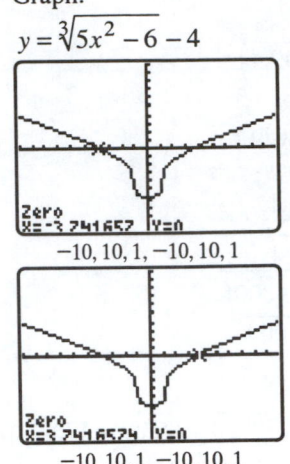

Intersection
X=1.5   Y=3.082207

$-10, 10, 1, -10, 10, 1$

The graphs of the equations intersect when $x = 1.5$.

105. Graph:

$y = \sqrt[3]{5x^2 - 6} - 4$

Zero
X=-3.741657   Y=0

$-10, 10, 1, -10, 10, 1$

Zero
X=3.7416574   Y=0

$-10, 10, 1, -10, 10, 1$

The graph of the equation crosses the x-axis at $x \approx -3.74$ and $x \approx 3.74$.

107. $\sqrt{\sqrt{x+25} - \sqrt{x}} = 5$

$\left(\sqrt{\sqrt{x+25} - \sqrt{x}}\right)^2 = 5^2$

$\sqrt{x+25} - \sqrt{x} = 25$

$\sqrt{x+25} = 25 + \sqrt{x}$

$\left(\sqrt{x+25}\right)^2 = \left(25 + \sqrt{x}\right)^2$

$x + 25 = 625 + 50\sqrt{x} + x$

$-600 = 50\sqrt{x}$

$-12 = \sqrt{x}$

$(-12)^2 = \left(\sqrt{x}\right)^2$

$144 = x$

Check: $\quad \sqrt{\sqrt{x+25} - \sqrt{x}} = 5$

$\sqrt{\sqrt{144+25} - \sqrt{144}} \overset{?}{=} 5$

$\sqrt{\sqrt{169} - \sqrt{144}} \overset{?}{=} 5$

$\sqrt{13 - 12} \overset{?}{=} 5$

$\sqrt{1} \overset{?}{=} 5$

$1 = 5$ False

Thus, 144 is not a solution and we conclude that there is no solution.

109.
$$(3p-1)^{2/3} = (5p^2-p)^{1/3}$$
$$[(3p-1)^{2/3}]^3 = [(5p^2-p)^{1/3}]^3$$
$$(3p-1)^2 = (5p^2-p)^1$$
$$9p^2-6p+1 = 5p^2-p$$
$$4p^2-5p+1 = 0$$
$$(4p-1)(p-1) = 0$$
$$4p-1=0 \quad \text{or} \quad p-1=0$$
$$4p=1 \qquad\qquad p=1$$
$$p = \frac{1}{4}$$

Check: $p = \dfrac{1}{4}$
$$(3p-1)^{2/3} = (5p^2-p)^{1/3}$$
$$\left(3\cdot\frac{1}{4}-1\right)^{2/3} \overset{?}{=} \left[5\left(\frac{1}{4}\right)^2-\frac{1}{4}\right]^{1/3}$$
$$\left(\frac{3}{4}-1\right)^{2/3} \overset{?}{=} \left(\frac{5}{16}-\frac{4}{16}\right)^{1/3}$$
$$\left(-\frac{1}{4}\right)^{2/3} \overset{?}{=} \left(\frac{1}{16}\right)^{1/3}$$
$$\left(\frac{1}{16}\right)^{1/3} = \left(\frac{1}{16}\right)^{1/3} \quad \text{True}$$

$p = 1$
$$(3p-1)^{2/3} = (5p^2-p)^{1/3}$$
$$(3\cdot1-1)^{2/3} \overset{?}{=} [5(1)^2-1]^{1/3}$$
$$(2)^{2/3} \overset{?}{=} (4)^{1/3}$$
$$4^{1/3} = 4^{1/3} \quad \text{True}$$

111.
$$z = \frac{p'-p}{\sqrt{\dfrac{pq}{n}}}$$
$$z\sqrt{\frac{pq}{n}} = p'-p$$
$$\left(z\sqrt{\frac{pq}{n}}\right)^2 = (p'-p)^2$$
$$\frac{z^2 pq}{n} = (p'-p)^2$$
$$z^2 pq = n(p'-p)^2$$
$$\frac{z^2 pq}{(p'-p)^2} = n \text{ or } n = \frac{z^2 pq}{(p'-p)^2}$$

113.
$$P_1P_2 - P_1P_3 = P_2P_3 \quad \text{Solve for } P_2.$$
$$P_1P_2 - P_1P_2 - P_1P_3 = P_2P_3 - P_1P_2$$
$$-P_1P_3 = P_2P_3 - P_1P_2$$
$$-P_1P_3 = P_2(P_3 - P_1)$$
$$\frac{-P_1P_3}{P_3-P_1} = \frac{P_2(P_3-P_1)}{P_3-P_1}$$
$$\text{or } \frac{P_1P_3}{P_1-P_3} = P_2 \text{ or } P_2 = \frac{P_1P_3}{P_1-P_3}$$

114.
$$\frac{(2x+3)(3x-4)-(2x+3)(5x-1)}{2x+3}$$
$$= \frac{(2x+3)[(3x-4)-(5x-1)]}{2x+3}$$
$$= \frac{(2x+3)[3x-4-5x+1]}{2x+3}$$
$$= \frac{(2x+3)(-2x-3)}{2x+3}$$
$$= -2x-3$$

115.
$$\frac{4a^2-9b^2}{4a^2+12ab+9b^2} \cdot \frac{6a^2b}{8a^2b^2-12ab^3}$$
$$= \frac{(2a-3b)(2a+3b)}{(2a+3b)(2a+3b)} \cdot \frac{6a^2b}{4ab^2(2a-3b)}$$
$$= \frac{3a}{2b(2a+3b)}$$

116.
$$(t^2-t-12) \div \frac{t^2-9}{t^2-3t}$$
$$= (t^2-t-12) \cdot \frac{t^2-3t}{t^2-9}$$
$$= (t-4)(t+3) \cdot \frac{t(t-3)}{(t+3)(t-3)}$$
$$= t(t-4)$$

117. $\dfrac{2}{x+3} - \dfrac{1}{x-3} + \dfrac{2x}{x^2-9} = \dfrac{2}{x+3} - \dfrac{1}{x-3} + \dfrac{2x}{(x+3)(x-3)}$

$\qquad = \dfrac{2}{x+3} \cdot \dfrac{x-3}{x-3} - \dfrac{1}{x-3} \cdot \dfrac{x+3}{x+3} + \dfrac{2x}{(x+3)(x-3)}$

$\qquad = \dfrac{2(x-3)}{(x+3)(x-3)} - \dfrac{x+3}{(x+3)(x-3)} + \dfrac{2x}{(x+3)(x-3)}$

$\qquad = \dfrac{2x-6}{(x+3)(x-3)} - \dfrac{x+3}{(x+3)(x-3)} + \dfrac{2x}{(x+3)(x-3)}$

$\qquad = \dfrac{2x-6-(x+3)+2x}{(x+3)(x-3)}$

$\qquad = \dfrac{2x-6-x-3+2x}{(x+3)(x-3)}$

$\qquad = \dfrac{3x-9}{(x+3)(x-3)}$

$\qquad = \dfrac{3(x-3)}{(x+3)(x-3)}$

$\qquad = \dfrac{3}{x+3}$

118.

$$2 + \frac{3x}{x-1} = \frac{8}{x-1}$$

$$(x-1)(2) + (x-1)\left(\frac{3x}{x-1}\right) = (x-1)\left(\frac{8}{x-1}\right)$$

$$2(x-1) + 3x = 8$$

$$2x - 2 + 3x = 8$$

$$5x - 2 = 8$$

$$5x = 10$$

$$x = 2$$

Exercise Set 7.7

1. $i = \sqrt{-1}$

3. Yes

5. Yes

7. The conjugate of $a + bi$ is $a - bi$.

9. Answers will vary. Possible answers:

    a. $\sqrt{2}$

    b. 2

    c. $\sqrt{-3}$

    d. 6

    e. Every number we have studied is a complex number.

11. $3 = 3 + 0i$

13. $3 + \sqrt{-4} = 3 + \sqrt{4}\sqrt{-1}$
$\qquad\qquad\quad = 3 + 2i$

15. $6 + \sqrt{3} = \left(6 + \sqrt{3}\right) + 0i$

17. $\sqrt{-25} = 0 + \sqrt{25}\sqrt{-1}$
$\qquad\quad = 0 + 5i$

19. $4 + \sqrt{-12} = 4 + \sqrt{12}\sqrt{-1}$
$\qquad\qquad\quad = 4 + \sqrt{4}\sqrt{3}\sqrt{-1}$
$\qquad\qquad\quad = 4 + 2i\sqrt{3}$

21. $\sqrt{-25} - 2i = 0 + \sqrt{25}\sqrt{-1} - 2i$
$\qquad\qquad\quad = 0 + 5i - 2i$
$\qquad\qquad\quad = 0 + 3i$

23. $9 - \sqrt{-9} = 9 - \sqrt{9}\sqrt{-1}$
$\qquad\qquad\quad = 9 - 3i$

25. $2i - \sqrt{-80} = 0 + 2i - \sqrt{16}\sqrt{5}\sqrt{-1}$
$$= 0 + 2i - 4i\sqrt{5}$$
$$= 0 + \left(2 - 4\sqrt{5}\right)i$$

27. $(12 - 6i) + (3 + 2i) = 12 - 6i + 3 + 2i$
$$= 15 - 4i$$

29. $\left(12 + \dfrac{5}{9}i\right) - \left(4 - \dfrac{3}{4}i\right) = 12 + \dfrac{5}{9}i - 4 + \dfrac{3}{4}i$
$$= 8 + \dfrac{5}{9}i + \dfrac{3}{4}i$$
$$= 8 + \dfrac{20}{36}i + \dfrac{27}{36}i$$
$$= 8 + \dfrac{47}{36}i$$

31. $\left(13 - \sqrt{-4}\right) - \left(-5 + \sqrt{-9}\right)$
$$= 13 - \sqrt{-4} + 5 - \sqrt{-9}$$
$$= 13 - \sqrt{4}\sqrt{-1} + 5 - \sqrt{9}\sqrt{-1}$$
$$= 13 - 2i + 5 - 3i$$
$$= 18 - 5i$$

33. $\left(\sqrt{3} + \sqrt{2}\right) + \left(3\sqrt{2} - \sqrt{-8}\right)$
$$= \sqrt{3} + \sqrt{2} + 3\sqrt{2} - \sqrt{8}\sqrt{-1}$$
$$= \sqrt{3} + \sqrt{2} + 3\sqrt{2} - \sqrt{4}\sqrt{2}\sqrt{-1}$$
$$= \sqrt{3} + \sqrt{2} + 3\sqrt{2} - 2i\sqrt{2}$$
$$= \left(4\sqrt{2} + \sqrt{3}\right) - 2i\sqrt{2}$$

35. $\left(19 + \sqrt{-147}\right) + \left(\sqrt{-75}\right)$
$$= 19 + \sqrt{147}\sqrt{-1} + \sqrt{75}\sqrt{-1}$$
$$= 19 + \sqrt{49}\sqrt{3}\sqrt{-1} + \sqrt{25}\sqrt{3}\sqrt{-1}$$
$$= 19 + 7i\sqrt{3} + 5i\sqrt{3}$$
$$= 19 + 12i\sqrt{3}$$

37. $\left(\sqrt{12} + \sqrt{-49}\right) - \left(\sqrt{49} - \sqrt{-12}\right)$
$$= \sqrt{12} + \sqrt{49}\sqrt{-1} - \sqrt{49} + \sqrt{12}\sqrt{-1}$$
$$= \sqrt{4}\sqrt{3} + \sqrt{49}\sqrt{-1} - \sqrt{49} + \sqrt{4}\sqrt{3}\sqrt{-1}$$
$$= 2\sqrt{3} + 7i - 7 + 2i\sqrt{3}$$
$$= (2\sqrt{3} - 7) + \left(7 + 2\sqrt{3}\right)i$$

39. $2(-3 - 2i) = -6 - 4i$

41. $i(6 + i) = i(6) + i(i)$
$$= 6i + i^2$$
$$= 6i - 1 \text{ or } -1 + 6i$$

43. $-3.5i(6.4 - 1.8i) = -3.5i(6.4) - 3.5i(-1.8i)$
$$= -22.4i + 6.3i^2$$
$$= -22.4i + 6.3(-1)$$
$$= -6.3 - 22.4i$$

45. $\sqrt{-4}\left(\sqrt{3} + 2i\right) = \sqrt{4}\sqrt{-1}\left(\sqrt{3} + 2i\right)$
$$= 2i\left(\sqrt{3} + 2i\right)$$
$$= 2i\sqrt{3} + 4i^2$$
$$= 2i\sqrt{3} + 4(-1)$$
$$= 2i\sqrt{3} - 4 \text{ or } -4 + 2i\sqrt{3}$$

47. $\sqrt{-6}\left(\sqrt{3} + \sqrt{-6}\right) = \sqrt{6}\sqrt{-1}\left(\sqrt{3} + \sqrt{6}\sqrt{-1}\right)$
$$= i\sqrt{6}\left(\sqrt{3} + i\sqrt{6}\right)$$
$$= i\sqrt{18} + i^2\sqrt{36}$$
$$= i\sqrt{9}\cdot\sqrt{2} - 1\sqrt{36}$$
$$= 3i\sqrt{2} - 6 \text{ or } -6 + 3i\sqrt{2}$$

49. $(3 + 2i)(1 + i) = 3(1) + 3(i) + 2i(1) + 2i(i)$
$$= 3 + 3i + 2i + 2i^2$$
$$= 3 + 3i + 2i + 2(-1)$$
$$= 3 + 3i + 2i - 2$$
$$= 1 + 5i$$

51. $(4 - 6i)(3 - i) = 4(3) - 4(i) - 6i(3) - 6i(-i)$
$$= 12 - 4i - 18i + 6i^2$$
$$= 12 - 4i - 18i + 6(-1)$$
$$= 12 - 4i - 18i - 6$$
$$= 6 - 22i$$

53. $\left(\frac{1}{4}+\sqrt{-3}\right)\left(2+\sqrt{3}\right)$

$=\frac{1}{4}(2)+\frac{1}{4}\sqrt{3}+2\sqrt{-3}+\sqrt{-3}\sqrt{3}$

$=\frac{1}{2}+\frac{\sqrt{3}}{4}+2i\sqrt{3}+i\sqrt{3}\sqrt{3}$

$=\frac{1}{2}+\frac{\sqrt{3}}{4}+2i\sqrt{3}+3i$

$=\left(\frac{1}{2}+\frac{\sqrt{3}}{4}\right)+\left(2\sqrt{3}+3\right)i$

55. $\left(5-\sqrt{-8}\right)\left(\frac{1}{4}+\sqrt{-2}\right)$

$=(5-2i\sqrt{2})\left(\frac{1}{4}+i\sqrt{2}\right)$

$=\frac{5}{4}+5i\sqrt{2}-\frac{2i\sqrt{2}}{4}-2i^2\sqrt{4}$

$=\frac{5}{4}+5i\sqrt{2}-\frac{i\sqrt{2}}{2}-4i^2$

$=\frac{5}{4}+5i\sqrt{2}-\frac{i\sqrt{2}}{2}-4(-1)$

$=\frac{5}{4}+5i\sqrt{2}-\frac{i\sqrt{2}}{2}+4$

$=\left(\frac{5}{4}+\frac{16}{4}\right)+\left(5\sqrt{2}-\frac{\sqrt{2}}{2}\right)i$

$=\frac{21}{4}+\frac{9\sqrt{2}}{2}i$ or $\frac{21}{4}+\frac{9}{2}i\sqrt{2}$

57. $\frac{-5}{-3i}=\frac{-5}{-3i}\cdot\frac{i}{i}=\frac{-5i}{-3i^2}=\frac{-5i}{-3(-1)}=-\frac{5i}{3}$

59. $\frac{2+3i}{2i}=\frac{2+3i}{2i}\cdot\frac{-i}{-i}$

$=\frac{(2+3i)(-i)}{-2i^2}$

$=\frac{-2i-3i^2}{-2i^2}$

$=\frac{-2i-3(-1)}{-2(-1)}$

$=\frac{-2i+3}{2}$ or $\frac{3-2i}{2}$

61. $\frac{2+5i}{5i}=\frac{2+5i}{5i}\cdot\frac{-i}{-i}$

$=\frac{(2+5i)(-i)}{-5i^2}$

$=\frac{-2i-5i^2}{-5i^2}$

$=\frac{-2i-5(-1)}{-5(-1)}$

$=\frac{-2i+5}{5}$ or $\frac{5-2i}{5}$

63. $\frac{7}{7-2i}=\frac{7}{7-2i}\cdot\frac{7+2i}{7+2i}$

$=\frac{7(7+2i)}{49-4i^2}$

$=\frac{49+14i}{49-4(-1)}$

$=\frac{49+14i}{49+4}$

$=\frac{49+14i}{53}$

65. $\frac{6-3i}{4+2i}=\frac{6-3i}{4+2i}\cdot\frac{4-2i}{4-2i}$

$=\frac{(6-3i)(4-2i)}{16-4i^2}$

$=\frac{24-12i-12i+6i^2}{16-4i^2}$

$=\frac{24-12i-12i-6}{16+4}$

$=\frac{18-24i}{20}$

$=\frac{2(9-12i)}{20}$

$=\frac{9-12i}{10}$

67. $\dfrac{4}{6-\sqrt{-4}} = \dfrac{4}{6-\sqrt{4}\sqrt{-1}}$

$= \dfrac{4}{6-2i} \cdot \dfrac{6+2i}{6+2i}$

$= \dfrac{4(6+2i)}{36-4i^2}$

$= \dfrac{24+8i}{36-4(-1)}$

$= \dfrac{24+8i}{36+4}$

$= \dfrac{8(3+i)}{40}$

$= \dfrac{3+i}{5}$

69. $\dfrac{\sqrt{6}}{\sqrt{3}-\sqrt{-9}} = \dfrac{\sqrt{6}}{\sqrt{3}-\sqrt{9}\sqrt{-1}}$

$= \dfrac{\sqrt{6}}{\sqrt{3}-3i} \cdot \dfrac{\sqrt{3}+3i}{\sqrt{3}+3i}$

$= \dfrac{\sqrt{6}\left(\sqrt{3}+3i\right)}{\sqrt{9}-9i^2}$

$= \dfrac{\sqrt{18}+3i\sqrt{6}}{3-9(-1)}$

$= \dfrac{\sqrt{9}\sqrt{2}+3i\sqrt{6}}{3+9}$

$= \dfrac{3\sqrt{2}+3i\sqrt{6}}{12}$

$= \dfrac{3\left(\sqrt{2}+i\sqrt{6}\right)}{12}$

$= \dfrac{\sqrt{2}+i\sqrt{6}}{4}$

71. $\dfrac{\sqrt{10}+\sqrt{-3}}{5-\sqrt{-20}}$

$= \dfrac{\sqrt{10}+\sqrt{3}\sqrt{-1}}{5-\sqrt{4}\sqrt{5}\sqrt{-1}}$

$= \dfrac{\sqrt{10}+i\sqrt{3}}{5-2i\sqrt{5}} \cdot \dfrac{5+2i\sqrt{5}}{5+2i\sqrt{5}}$

$= \dfrac{(\sqrt{10}+i\sqrt{3})(5+2i\sqrt{5})}{5^2-4i^2\sqrt{5}^2}$

$= \dfrac{5\sqrt{10}+2i\sqrt{50}+5i\sqrt{3}+2i^2\sqrt{15}}{5^2-4(-1)(5)}$

$= \dfrac{5\sqrt{10}+2i\sqrt{25}\sqrt{2}+5i\sqrt{3}+2(-1)\sqrt{15}}{25+20}$

$= \dfrac{(5\sqrt{10}-2\sqrt{15})+(10\sqrt{2}+5\sqrt{3})i}{45}$

73. $\dfrac{\sqrt{-60}}{\sqrt{-2}} = \dfrac{\sqrt{60}\sqrt{-1}}{\sqrt{2}\sqrt{-1}} = \dfrac{i\sqrt{60}}{i\sqrt{2}} = \sqrt{\dfrac{60}{2}} = \sqrt{30}$

75. $\dfrac{\sqrt{-80}\sqrt{-5}}{\sqrt{-2}} = \dfrac{\sqrt{80}\sqrt{-1}\cdot\sqrt{5}\sqrt{-1}}{\sqrt{2}\sqrt{-1}}$

$= \dfrac{(i\sqrt{80})(i\sqrt{5})}{i\sqrt{2}}$

$= \dfrac{i\sqrt{80}\sqrt{5}}{\sqrt{2}}$

$= i\sqrt{\dfrac{80\cdot 5}{2}}$

$= i\sqrt{200}$

$= i\sqrt{100}\sqrt{2}$

$= 10i\sqrt{2}$

77. $(4-2i)+(3-5i) = 4+3-2i-5i = 7-7i$

79. $(8-\sqrt{-6})-(2-\sqrt{-24})$

$= (8-i\sqrt{6})-(2-2i\sqrt{6})$

$= 8-i\sqrt{6}-2+2i\sqrt{6}$

$= 6+i\sqrt{6}$

81. $5.2(4-3.2i) = 5.2(4)-5.2(3.2i)$

$\qquad = 20.8-16.64\,i$

83. $(5+2i)(3-5i) = 15 - 25i + 6i - 10i^2$
$$= 15 - 19i - 10(-1)$$
$$= 15 + 10 - 19i$$
$$= 25 - 19i$$

85. $\dfrac{5-4i}{2i} \cdot \dfrac{-2i}{-2i} = \dfrac{(5-4i)(-2i)}{-4i^2}$
$$= \dfrac{-10i + 8i^2}{-4i^2}$$
$$= \dfrac{-10i + 8(-1)}{-4(-1)}$$
$$= \dfrac{-8 - 10i}{4} = \dfrac{-4 - 5i}{2}$$

87. $\dfrac{4}{\sqrt{3} - \sqrt{-4}} = \dfrac{4}{\sqrt{3} - 2i}$
$$= \dfrac{4}{\sqrt{3} - 2i} \cdot \dfrac{\sqrt{3} + 2i}{\sqrt{3} + 2i}$$
$$= \dfrac{4(\sqrt{3} + 2i)}{(\sqrt{3})^2 - (2i)^2}$$
$$= \dfrac{4\sqrt{3} + 8i}{3 - 4i^2}$$
$$= \dfrac{4\sqrt{3} + 8i}{3 - 4(-1)}$$
$$= \dfrac{4\sqrt{3} + 8i}{7}$$

89. $\left(5 - \dfrac{5}{9}i\right) - \left(2 - \dfrac{3}{5}i\right) = 5 - \dfrac{5}{9}i - 2 + \dfrac{3}{5}i$
$$= 3 - \dfrac{5}{9}i + \dfrac{3}{5}i$$
$$= 3 - \dfrac{25}{45}i + \dfrac{27}{45}i$$
$$= 3 + \dfrac{2}{45}i$$

91. $\left(\dfrac{2}{3} - \dfrac{1}{5}i\right)\left(\dfrac{3}{5} - \dfrac{3}{4}i\right)$
$$= \left(\dfrac{2}{3}\right)\left(\dfrac{3}{5}\right) - \dfrac{2}{3}\left(\dfrac{3}{4}i\right) - \left(\dfrac{1}{5}i\right)\left(\dfrac{3}{5}\right) + \left(\dfrac{1}{5}i\right)\left(\dfrac{3}{4}i\right)$$
$$= \dfrac{2}{5} - \dfrac{1}{2}i - \dfrac{3}{25}i + \dfrac{3}{20}i^2$$
$$= \dfrac{2}{5} - \dfrac{1}{2}i - \dfrac{3}{25}i + \dfrac{3}{20}(-1)$$
$$= \left(\dfrac{2}{5} - \dfrac{3}{20}\right) + \left(-\dfrac{1}{2} - \dfrac{3}{25}\right)i$$
$$= \left(\dfrac{8}{20} - \dfrac{3}{20}\right) + \left(-\dfrac{25}{50} - \dfrac{6}{50}\right)i$$
$$= \dfrac{5}{20} - \dfrac{31}{50}i$$
$$= \dfrac{1}{4} - \dfrac{31}{50}i$$

93. $\dfrac{\sqrt{-96}}{\sqrt{-24}} = \dfrac{\sqrt{96}\sqrt{-1}}{\sqrt{24}\sqrt{-1}}$
$$= \dfrac{i\sqrt{96}}{i\sqrt{24}}$$
$$= \dfrac{\sqrt{96}}{\sqrt{24}}$$
$$= \sqrt{\dfrac{96}{24}}$$
$$= \sqrt{4}$$
$$= 2$$

95. $(5.23 - 6.41i) - (8.56 - 4.5i) - 7.1i$
$$= 5.23 - 6.41i - 8.56 + 4.5i - 7.1i$$
$$= -3.33 - 9.01i$$

97. $i^{10} = i^8 i^2$
$$= (i^4)^2 i^2$$
$$= 1^2 \cdot i^2$$
$$= 1(i^2)$$
$$= i^2$$
$$= -1$$

99. $i^{200} = (i^4)^{50} = 1^{50} = 1$

101. $i^{93} = i^{92} \cdot i^1 = (i^4)^{23}i = 1^{23} \cdot i = 1(i) = i$

103.
$$i^{907} = i^{904}i^3$$
$$= (i^4)^{226}i^3$$
$$= 1^{236} \cdot i^3$$
$$= 1(i^3)$$
$$= i^3$$
$$= -i$$

105. True. The product of two pure imaginary numbers is always a real number. Consider two pure imaginary numbers bi and di where b, d are non-zero real numbers whose product

$$(bi)(di) = bdi^2$$
$$= bd(-1)$$
$$= -bd$$

which is a real number.

107. False. The product of two complex numbers is not always a real number. For example,

$$(1+i)(1+i) = 1 + i + i + i^2$$
$$= 1 + 2i + (-1)$$
$$= 0 + 2i$$

which is not a real number.

109. Even values of n will result in i^n being a real number. $i^2 = -1$, $i^{2n} = (i^2)^n = (-1)^n$

111.
$$f(x) = x^2$$
$$f(i) = i^2 = -1$$

113.
$$f(x) = x^4 - 2x$$
$$f(2i) = (2i)^4 - 2(2i)$$
$$= 2^4 i^4 - 4i$$
$$= 16(1) - 4i$$
$$= 16 - 4i$$

115.
$$f(x) = x^2 - x$$
$$f(3+i) = (3+i)^2 - (3+i)$$
$$= 9 + 6i + i^2 - 3 - i$$
$$= 9 + 6i - 1 - 3 - i$$
$$= 5 + 5i$$

117.
$$x^2 - 2x + 5$$
$$= (1-2i)^2 - 2(1-2i) + 5$$
$$= 1^2 - 2(1)(2i) + (2i)^2 - 2 + 4i + 5$$
$$= 1 - 4i - 4 - 2 + 4i + 5$$
$$= 0 + 0i$$
$$= 0$$

119.
$$x^2 + 2x + 8$$
$$= (-1 + i\sqrt{5})^2 + 2(-1 + i\sqrt{5}) + 8$$
$$= (-1)^2 - 2(1)(i\sqrt{5}) + (i\sqrt{5})^2 - 2 + 2i\sqrt{5} + 8$$
$$= 1 - 2i\sqrt{5} - 5 - 2 + 2i\sqrt{5} + 8$$
$$= 2 + 0i$$
$$= 2$$

121.
$$x^2 - 4x + 5 = 0$$
$$(2-i)^2 - 4(2-i) + 5 \stackrel{?}{=} 0$$
$$2^2 - 2(2)(i) + (i)^2 - 8 + 4i + 5 \stackrel{?}{=} 0$$
$$4 - 4i - 1 - 8 + 4i + 5 \stackrel{?}{=} 0$$
$$0 + 0i \stackrel{?}{=} 0$$
$$0 = 0 \quad \text{True}$$

$2 - i$ is a solution.

123.
$$x^2 - 6x + 12 = 0$$
$$(-3 + i\sqrt{3})^2 - 6(-3 + i\sqrt{3}) + 12 \stackrel{?}{=} 0$$
$$(-3)^2 - 2(3)(i\sqrt{3}) + (i\sqrt{3})^2 + 18 - 6i\sqrt{3} + 12 \stackrel{?}{=} 0$$
$$9 - 6i\sqrt{3} - 3 + 18 - 6i\sqrt{3} + 12 \stackrel{?}{=} 0$$
$$36 - 12i\sqrt{3} = 0 \quad \text{False}$$
$$-3 + i\sqrt{3} \text{ is not a solution.}$$

125. $Z = \dfrac{V}{I}$

$Z = \dfrac{1.8 + 0.5i}{0.6i}$

$= \dfrac{1.8 + 0.5i}{0.6i} \cdot \dfrac{-0.6i}{-0.6i}$

$= \dfrac{-1.08i - 0.3i^2}{-0.36i^2}$

$= \dfrac{-1.08i - 0.3(-1)}{-0.36(-1)}$

$= \dfrac{0.3 - 1.08i}{0.36}$

$\approx 0.83 - 3i$

127. $Z_T = \dfrac{Z_1 Z_2}{Z_1 + Z_2}$

$= \dfrac{(2-i)(4+i)}{(2-i)+(4+i)}$

$= \dfrac{8 + 2i - 4i - i^2}{6}$

$= \dfrac{8 - 2i - (-1)}{6}$

$= \dfrac{9 - 2i}{6}$

$\approx 1.5 - 0.33i$

129. $i^{-1} = \dfrac{1}{i} = \dfrac{1}{i} \cdot \dfrac{i}{i} = \dfrac{i}{i^2} = \dfrac{i}{-1} = -i$

131. $x^2 - 4x + 6 = 0$

$a = 1, b = -4, c = 6$

$x = \dfrac{-(-4) \pm \sqrt{(-4)^2 - 4(1)(6)}}{2(1)}$

$= \dfrac{4 \pm \sqrt{16 - 24}}{2}$

$= \dfrac{4 \pm \sqrt{-8}}{2}$

$= \dfrac{4 \pm 2i\sqrt{2}}{2}$

$= \dfrac{2\left(2 \pm i\sqrt{2}\right)}{2}$

$= 2 \pm i\sqrt{2}$

133. $a + b = 3 + 2i\sqrt{3} + 1 + i\sqrt{3}$

$= 3 + 1 + 2i\sqrt{3} + i\sqrt{3}$

$= 4 + 3i\sqrt{3}$

135. ab

$= (3 + 2i\sqrt{3})(1 + i\sqrt{3})$

$= 3(1) + (3)(i\sqrt{3}) + (2i\sqrt{3})(1) + (2i\sqrt{3})(i\sqrt{3})$

$= 3 + 3i\sqrt{3} + 2i\sqrt{3} + 2i^2(\sqrt{3})^2$

$= 3 + 3i\sqrt{3} + 2i\sqrt{3} - 6$

$= -3 + 5i\sqrt{3}$

137. This problem can be solved using a single variable. To do this, let x be the amount that is \$5.50 per pound. Then $40 - x$ is the amount that is \$6.30 per pound and the equation is

$5.50(x) + 6.30(40 - x) = 6(40)$

$5.5x + 252 - 6.3x = 240$

$252 - 0.8x = 240$

$-0.8x = -12$

$x = 15$

Thus, combine 15 lb of the \$5.50 per pound coffee with $40 - 15 = 25$ lb of the \$6.30 per pound coffee to obtain 40 lb of \$6.00 per pound coffee.

138. $4x^4 + 12x^2 + 9$

$= (2x^2)^2 + 2(2x^2)(3) + (3)^2$

$= (2x^2 + 3)^2$

139. $15r^2 s^2 + rs - 6$

Observe that $(15)(-6) = -90$. The two numbers whose product is -90 and whose sum is 1 are -9 and 10 since $(-9)(10) = -90$ and $-9 + 10 = 1$.

Write the middle term, rs, as $-9rs + 10rs$ and the factorization is

$15r^2 s^2 + rs - 6 = 15r^2 s^2 - 9rs + 10rs - 6$

$= 3rs(5rs - 3) + 2(5rs - 3)$

$= (3rs + 2)(5rs - 3)$

Review Exercises

1. $\sqrt{9} = 3$

2. $\sqrt[3]{-27} = -3$

3. $\sqrt[4]{256} = 4$

4. $\sqrt[3]{-64} = -4$

5. $\sqrt{(-7)^2} = |-7| = 7$

6. $\sqrt{(-93.4)^2} = |-93.4| = 93.4$

7. $\sqrt{x^2} = |x|$

8. $\sqrt{(x-2)^2} = |x-2|$

9. $\sqrt{(x-y)^2} = |x-y|$

10. $\sqrt{(x^2-4x+12)^2} = |x^2-4x+12|$

11. $\sqrt{x^5} = x^{5/2}$

12. $\sqrt[3]{x^5} = x^{5/3}$

13. $\left(\sqrt[4]{y}\right)^{15} = y^{15/4}$

14. $\sqrt[7]{5^2} = 5^{2/7}$

15. $a^{1/2} = \sqrt{a}$

16. $y^{3/5} = \sqrt[5]{y^3}$

17. $(2m^2n)^{9/5} = \left(\sqrt[5]{2m^2n}\right)^9$

18. $(2a+3b)^{-3/4} = \dfrac{1}{\left(\sqrt[4]{2a+3b}\right)^3}$

19. $\sqrt[3]{3^6} = 3^{6/3} = 3^2 = 9$

20. $\sqrt{x^{10}} = x^{10/2} = x^5$

21. $\left(\sqrt[3]{4}\right)^6 = 4^{6/3} = 4^2 = 16$

22. $\sqrt[20]{x^4} = x^{4/20} = x^{1/5} = \sqrt[5]{x}$

23. $-25^{1/2} = -\sqrt{25} = -5$

24. $(-25)^{1/2}$ is not a real number.

25. $\left(\dfrac{8}{27}\right)^{-1/3} = \left(\dfrac{27}{8}\right)^{1/3} = \sqrt[3]{\dfrac{27}{8}} = \dfrac{\sqrt[3]{27}}{\sqrt[3]{8}} = \dfrac{3}{2}$

26. $36^{-1/2} - 8^{-2/3} = \dfrac{1}{36^{1/2}} - \dfrac{1}{8^{2/3}}$

$= \dfrac{1}{\sqrt{36}} - \dfrac{1}{\left(\sqrt[3]{8}\right)^2}$

$= \dfrac{1}{6} - \dfrac{1}{4}$

$= \dfrac{2}{12} - \dfrac{3}{12}$

$= -\dfrac{1}{12}$

27. $x^{3/5}x^{-1/3} = x^{3/5-1/3} = x^{9/15-5/15} = x^{4/15}$

28. $\left(\dfrac{64}{y^6}\right)^{1/3} = \sqrt[3]{\dfrac{64}{y^6}} = \dfrac{\sqrt[3]{64}}{\sqrt[3]{y^6}} = \dfrac{4}{y^2}$

29. $\left(\dfrac{y^{-3/5}}{y^{1/5}}\right)^{2/3} = \dfrac{y^{-3/5\cdot2/3}}{y^{1/5\cdot2/3}}$

$= \dfrac{y^{-2/5}}{y^{2/15}}$

$= y^{-2/5-2/15}$

$= y^{-6/15-2/15}$

$= y^{-8/15}$

$= \dfrac{1}{y^{8/15}}$

30. $\left(\dfrac{30x^4y^{-2}}{5y^{1/2}}\right)^2 = \left(\dfrac{6x^4}{y^{5/2}}\right)^2$

$= \dfrac{6^2x^{4\cdot2}}{y^{5/2\cdot2}}$

$= \dfrac{36x^8}{y^5}$

31. $z^{1/3}(2z^{5/3} - 4z) = z^{1/3}(2z^{5/3}) - z^{1/3}(4z)$
$$= 2z^{6/3} - 4z^{4/3}$$
$$= 2z^2 - 4z^{4/3}$$

32. $\frac{3}{4}r^{-2/3}\left(4r^{-3/2} + \frac{4}{3}r^{2/3}\right)$
$$= \frac{3}{4}r^{-2/3}(4r^{-3/2}) + \frac{3}{4}r^{-2/3}\left(\frac{4}{3}r^{2/3}\right)$$
$$= 3r^{-2/3-3/2} + r^{-2/3+2/3}$$
$$= 3r^{-13/6} + r^0$$
$$= 3r^{-13/6} + 1$$
$$= \frac{3}{r^{13/6}} + 1$$

33. $x^{1/5} + x^{6/5} = x^{1/5} + x^{1/5}x^1$
$$= x^{1/5}(1 + x)$$

34. $x^{-1/2} + x^{2/3} = x^{-3/6} + x^{4/6}$
$$= x^{-3/6}(1 + x^{7/6})$$
$$= \frac{1 + x^{7/6}}{x^{1/2}}$$

35. $f(x) = \sqrt{6x + 13}$
$$f(6) = \sqrt{6(6) + 13}$$
$$= \sqrt{36 + 13}$$
$$= \sqrt{49}$$
$$= 7$$

36. $g(x) = \sqrt[3]{9r - 17}$
$$g(4) = \sqrt[3]{9(4) - 17}$$
$$= \sqrt[3]{36 - 17}$$
$$= \sqrt[3]{19}$$
$$\approx 2.668$$

37. $f(x) = \sqrt{x}$

x	$f(x)$
0	0
1	1
4	2
9	3

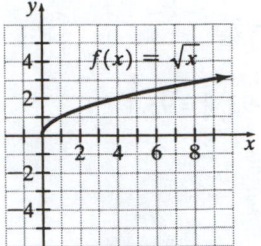

38. $f(x) = \sqrt{x} - 4$

x	$f(x)$
0	-4
1	-3
4	-2
9	-1
16	0

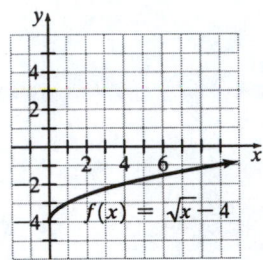

39. $\left(\dfrac{3r^2p^{1/3}}{r^{1/2}p^{4/3}}\right)^3 = (3r^{2-1/2}p^{1/3-4/3})^3$
$$= (3r^{3/2}p^{-1})^3$$
$$= 3^3(r^{3/2})^3(p^{-1})^3$$
$$= 27r^{9/2}p^{-3}$$
$$= \frac{27r^{9/2}}{p^3}$$

40. $\left(\dfrac{4y^{2/5}z^{1/3}}{x^{-1}y^{3/5}}\right)^{-1} = \left(\dfrac{4y^{2/5-3/5}z^{1/3}}{x^{-1}}\right)^{-1}$

$\qquad = \left(\dfrac{4y^{-1/5}z^{1/3}}{x^{-1}}\right)^{-1}$

$\qquad = \left(\dfrac{4xz^{1/3}}{y^{1/5}}\right)^{-1}$

$\qquad = \dfrac{y^{1/5}}{4xz^{1/3}}$

41. $\sqrt{80} = \sqrt{16}\sqrt{5} = 4\sqrt{5}$

42. $\sqrt[3]{54} = \sqrt[3]{27}\sqrt[3]{2} = 3\sqrt[3]{2}$

43. $\sqrt{50x^3y^7} = \sqrt{25x^2y^6}\sqrt{2xy} = 5xy^3\sqrt{2xy}$

44. $\sqrt[3]{125x^7y^{10}} = \sqrt[3]{125x^6y^9}\sqrt[3]{xy} = 5x^2y^3\sqrt[3]{xy}$

45. $\left(\sqrt[6]{x^{12}y^7z^{17}}\right)^{42} = (x^{12}y^7z^{17})^{42/6}$

$\qquad = (x^{12}y^7z^{17})^7$

$\qquad = x^{84}y^{49}z^{119}$

46. $\sqrt{20}\sqrt{5} = \sqrt{100} = 10$

47. $\sqrt{5x}\sqrt{8x^5} = \sqrt{40x^6}$

$\qquad = \sqrt{4x^6}\sqrt{10}$

$\qquad = 2x^3\sqrt{10}$

48. $\sqrt[3]{2x^4y^5}\sqrt[3]{16x^4y^4} = \sqrt[3]{32x^8y^9}$

$\qquad = \sqrt[3]{8x^6y^9}\sqrt[3]{4x^2}$

$\qquad = 2x^2y^3\sqrt[3]{4x^2}$

49. $\left(\sqrt[3]{5x^2y^3}\right)^2 = \sqrt[3]{(5x^2y^3)^2}$

$\qquad = \sqrt[3]{25x^4y^6}$

$\qquad = \sqrt[3]{x^3y^6}\sqrt[3]{25x}$

$\qquad = xy^2\sqrt[3]{25x}$

50. $\sqrt[4]{8x^4y^7}\sqrt[4]{2x^5y^9} = \sqrt[4]{16x^9y^{16}}$

$\qquad = \sqrt[4]{16x^8y^{16}}\sqrt[4]{x}$

$\qquad = 2x^2y^4\sqrt[4]{x}$

51. $\sqrt{3x}\left(\sqrt{12x} - \sqrt{20}\right) = \sqrt{36x^2} - \sqrt{60x}$

$\qquad = \sqrt{36x^2} - \sqrt{4}\sqrt{15x}$

$\qquad = 6x - 2\sqrt{15x}$

52. $\left(\sqrt[5]{a^7b^{12}c^9}\right)^{35} = (a^7b^{12}c^9)^{35/5}$

$\qquad = (a^7b^{12}c^9)^7$

$\qquad = a^{49}b^{84}c^{63}$

53. $\sqrt[3]{2x^2y}\left(\sqrt[3]{4x^4y^7} + \sqrt[3]{9x}\right)$

$\qquad = \sqrt[3]{8x^6y^8} + \sqrt[3]{18x^3y}$

$\qquad = \sqrt[3]{8x^6y^6}\sqrt[3]{y^2} + \sqrt[3]{x^3}\sqrt[3]{18y}$

$\qquad = 2x^2y^2\sqrt[3]{y^2} + x\sqrt[3]{18y}$

54. $\sqrt{\sqrt{x^2y}} = \left(\sqrt{x^2y}\right)^{1/2}$

$\qquad = [(x^2y)^{1/2}]^{1/2}$

$\qquad = (x^2y)^{1/4}$

$\qquad = \sqrt[4]{x^2y}$

55. $\sqrt{\sqrt[3]{x^4y}} = \left(\sqrt[3]{x^4y}\right)^{1/2}$

$\qquad = [(x^4y)^{1/2}]^{1/3}$

$\qquad = (x^4y)^{1/6}$

$\qquad = \sqrt[6]{x^4y}$

56. $\sqrt{\dfrac{36}{25}} = \dfrac{\sqrt{36}}{\sqrt{25}} = \dfrac{6}{5}$

57. $\sqrt[3]{\dfrac{x^3}{8}} = \dfrac{\sqrt[3]{x^3}}{\sqrt[3]{8}} = \dfrac{x}{2}$

58. $\dfrac{\sqrt[3]{2x^9}}{\sqrt[3]{16x^6}} = \sqrt[3]{\dfrac{2x^9}{16x^6}} = \sqrt[3]{\dfrac{x^3}{8}} = \dfrac{\sqrt[3]{x^3}}{\sqrt[3]{8}} = \dfrac{x}{2}$

59. $\sqrt{\dfrac{32x^2y^5}{2x^4y}} = \sqrt{\dfrac{16y^4}{x^2}} = \dfrac{\sqrt{16y^4}}{\sqrt{x^2}} = \dfrac{4y^2}{x}$

60. $\sqrt[3]{\dfrac{108x^3y^6}{2y^3}} = \sqrt[3]{54x^3y^3} = 3xy\sqrt[3]{2}$

61. $\dfrac{x}{\sqrt{7}} = \dfrac{x}{\sqrt{7}} \cdot \dfrac{\sqrt{7}}{\sqrt{7}} = \dfrac{x\sqrt{7}}{7}$

62. $\sqrt{\dfrac{2}{5}} = \dfrac{\sqrt{2}}{\sqrt{5}} \cdot \dfrac{\sqrt{5}}{\sqrt{5}} = \dfrac{\sqrt{10}}{5}$

63. $\sqrt{\dfrac{12x}{5y}} = \dfrac{\sqrt{12x}}{\sqrt{5y}}$

$\phantom{\sqrt{\dfrac{12x}{5y}}} = \dfrac{2\sqrt{3x}}{\sqrt{5y}}$

$\phantom{\sqrt{\dfrac{12x}{5y}}} = \dfrac{2\sqrt{3x}}{\sqrt{5y}} \cdot \dfrac{\sqrt{5y}}{\sqrt{5y}}$

$\phantom{\sqrt{\dfrac{12x}{5y}}} = \dfrac{2\sqrt{15xy}}{5y}$

64. $\dfrac{2}{\sqrt[3]{x}} = \dfrac{2}{\sqrt[3]{x}} \cdot \dfrac{\sqrt[3]{x^2}}{\sqrt[3]{x^2}} = \dfrac{2\sqrt[3]{x^2}}{\sqrt[3]{x^3}} = \dfrac{2\sqrt[3]{x^2}}{x}$

65. $\sqrt[3]{\dfrac{3x}{5y}} = \dfrac{\sqrt[3]{3x}}{\sqrt[3]{5y}} \cdot \dfrac{\sqrt[3]{25y^2}}{\sqrt[3]{25y^2}} = \dfrac{\sqrt[3]{75xy^2}}{5y}$

66. $\sqrt{\dfrac{3x^2}{y}} = \dfrac{\sqrt{3x^2}}{\sqrt{y}}$

$\phantom{\sqrt{\dfrac{3x^2}{y}}} = \dfrac{\sqrt{x^2}\sqrt{3}}{\sqrt{y}}$

$\phantom{\sqrt{\dfrac{3x^2}{y}}} = \dfrac{x\sqrt{3}}{\sqrt{y}} \cdot \dfrac{\sqrt{y}}{\sqrt{y}}$

$\phantom{\sqrt{\dfrac{3x^2}{y}}} = \dfrac{x\sqrt{3y}}{y}$

67. $\sqrt{\dfrac{18x^4y^5}{3z}} = \dfrac{\sqrt{18x^4y^5}}{\sqrt{3z}}$

$\phantom{\sqrt{\dfrac{18x^4y^5}{3z}}} = \dfrac{\sqrt{9x^4y^4}\sqrt{2y}}{\sqrt{3z}}$

$\phantom{\sqrt{\dfrac{18x^4y^5}{3z}}} = \dfrac{3x^2y^2\sqrt{2y}}{\sqrt{3z}}$

$\phantom{\sqrt{\dfrac{18x^4y^5}{3z}}} = \dfrac{3x^2y^2\sqrt{2y}}{\sqrt{3z}} \cdot \dfrac{\sqrt{3z}}{\sqrt{3z}}$

$\phantom{\sqrt{\dfrac{18x^4y^5}{3z}}} = \dfrac{3x^2y^2\sqrt{6yz}}{3z}$

$\phantom{\sqrt{\dfrac{18x^4y^5}{3z}}} = \dfrac{x^2y^2\sqrt{6yz}}{z}$

68. $\sqrt{\dfrac{125x^2y^5}{3z}} = \dfrac{\sqrt{125x^2y^5}}{\sqrt{3z}}$

$\phantom{\sqrt{\dfrac{125x^2y^5}{3z}}} = \dfrac{\sqrt{25x^2y^4}\sqrt{5y}}{\sqrt{3z}}$

$\phantom{\sqrt{\dfrac{125x^2y^5}{3z}}} = \dfrac{5xy^2\sqrt{5y}}{\sqrt{3z}} \cdot \dfrac{\sqrt{3z}}{\sqrt{3z}}$

$\phantom{\sqrt{\dfrac{125x^2y^5}{3z}}} = \dfrac{5xy^2\sqrt{15yz}}{3z}$

69. $\sqrt[4]{\dfrac{2x^2y^6}{8x^3}} = \sqrt[4]{\dfrac{y^6}{4x}}$

$\phantom{\sqrt[4]{\dfrac{2x^2y^6}{8x^3}}} = \dfrac{\sqrt[4]{y^6}}{\sqrt[4]{4x}}$

$\phantom{\sqrt[4]{\dfrac{2x^2y^6}{8x^3}}} = \dfrac{y\sqrt[4]{y^2}}{\sqrt[4]{4x}}$

$\phantom{\sqrt[4]{\dfrac{2x^2y^6}{8x^3}}} = \dfrac{y\sqrt[4]{y^2}}{\sqrt[4]{4x}} \cdot \dfrac{\sqrt[4]{4x^3}}{\sqrt[4]{4x^3}}$

$\phantom{\sqrt[4]{\dfrac{2x^2y^6}{8x^3}}} = \dfrac{y\sqrt[4]{4x^3y^2}}{2x}$

70. $\sqrt[3]{\dfrac{4x^5y^3}{x^6}} = \sqrt[3]{\dfrac{4y^3}{x}}$

$$= \dfrac{\sqrt[3]{4y^3}}{\sqrt[3]{x}}$$

$$= \dfrac{\sqrt[3]{y^3}\,\sqrt[3]{4}}{\sqrt[3]{x}}$$

$$= \dfrac{y\sqrt[3]{4}}{\sqrt[3]{x}} \cdot \dfrac{\sqrt[3]{x^2}}{\sqrt[3]{x^2}}$$

$$= \dfrac{y\sqrt[3]{4x^2}}{x}$$

71. $\sqrt[3]{\dfrac{y^6}{2x^2}} = \dfrac{\sqrt[3]{y^6}}{\sqrt[3]{2x^2}}$

$$= \dfrac{y^2}{\sqrt[3]{2x^2}} \cdot \dfrac{\sqrt[3]{4x}}{\sqrt[3]{4x}}$$

$$= \dfrac{y^2\sqrt[3]{4x}}{2x}$$

72. $\left(3-\sqrt{2}\right)\left(3+\sqrt{2}\right) = 3^2 - \left(\sqrt{2}\right)^2 = 9 - 2 = 7$

73. $\left(\sqrt{x}+y\right)\left(\sqrt{x}-y\right) = \left(\sqrt{x}\right)^2 - y^2 = x - y^2$

74. $\left(x-\sqrt{y}\right)\left(x+\sqrt{y}\right) = x^2 - \left(\sqrt{y}\right)^2 = x^2 - y$

75. $\left(\sqrt{3}+5\right)^2 = \left(\sqrt{3}\right)^2 + 2(5)\left(\sqrt{3}\right) + 5^2$

$$= 3 + 10\sqrt{3} + 25$$

$$= 28 + 10\sqrt{3}$$

76. $\left(\sqrt{x}-\sqrt{3y}\right)\left(\sqrt{x}+\sqrt{5y}\right)$

$$= \left(\sqrt{x}\right)^2 + \sqrt{x}\sqrt{5y} - \sqrt{x}\sqrt{3y} - \sqrt{3y}\sqrt{5y}$$

$$= x + \sqrt{5xy} - \sqrt{3xy} - \sqrt{15y^2}$$

$$= x + \sqrt{5xy} - \sqrt{3xy} - y\sqrt{15}$$

77. $\left(\sqrt[3]{2x}-\sqrt[3]{3y}\right)\left(\sqrt[3]{3x}-\sqrt[3]{2y}\right) = \sqrt[3]{2x}\left(\sqrt[3]{3x}\right) - \left(\sqrt[3]{2x}\right)\left(\sqrt[3]{2y}\right) - \sqrt[3]{3y}\left(\sqrt[3]{3x}\right) + \sqrt[3]{3y}\sqrt[3]{2y}$

$$= \sqrt[3]{6x^2} - \sqrt[3]{4xy} - \sqrt[3]{9xy} + \sqrt[3]{6y^2}$$

78. $\dfrac{5}{2+\sqrt{5}} = \dfrac{5}{2+\sqrt{5}} \cdot \dfrac{2-\sqrt{5}}{2-\sqrt{5}}$

$$= \dfrac{5\left(2-\sqrt{5}\right)}{2^2 - \left(\sqrt{5}\right)^2}$$

$$= \dfrac{10 - 5\sqrt{5}}{4 - 5}$$

$$= \dfrac{10 - 5\sqrt{5}}{-1}$$

$$= -10 + 5\sqrt{5}$$

79. $\dfrac{x}{3+\sqrt{x}} = \dfrac{x}{3+\sqrt{x}} \cdot \dfrac{3-\sqrt{x}}{3-\sqrt{x}}$

$\qquad = \dfrac{x\left(3-\sqrt{x}\right)}{3^2 - \left(\sqrt{x}\right)^2}$

$\qquad = \dfrac{3x - x\sqrt{x}}{9-x}$

80. $\dfrac{\sqrt{x}}{\sqrt{x}+\sqrt{y}} = \dfrac{\sqrt{x}}{\sqrt{x}+\sqrt{y}} \cdot \dfrac{\sqrt{x}-\sqrt{y}}{\sqrt{x}-\sqrt{y}}$

$\qquad = \dfrac{\sqrt{x}\left(\sqrt{x}-\sqrt{y}\right)}{\left(\sqrt{x}\right)^2 - \left(\sqrt{y}\right)^2}$

$\qquad = \dfrac{\sqrt{x^2} - \sqrt{xy}}{x-y}$

$\qquad = \dfrac{x - \sqrt{xy}}{x-y}$

81. $\dfrac{\sqrt{x}-2\sqrt{y}}{\sqrt{x}-\sqrt{y}} = \dfrac{\sqrt{x}-2\sqrt{y}}{\sqrt{x}-\sqrt{y}} \cdot \dfrac{\sqrt{x}+\sqrt{y}}{\sqrt{x}+\sqrt{y}}$

$\qquad = \dfrac{x + \sqrt{xy} - 2\sqrt{xy} - 2y}{\left(\sqrt{x}\right)^2 - \left(\sqrt{y}\right)^2}$

$\qquad = \dfrac{x - \sqrt{xy} - 2y}{x-y}$

82. $\dfrac{4}{\sqrt{y+2}-3} = \dfrac{4}{\sqrt{y+2}-3} \cdot \dfrac{\sqrt{y+2}+3}{\sqrt{y+2}+3}$

$\qquad = \dfrac{4\sqrt{y+2}+12}{\left(\sqrt{y+2}\right)^2 - 3^2}$

$\qquad = \dfrac{4\sqrt{y+2}+12}{y+2-9}$

$\qquad = \dfrac{4\sqrt{y+2}+12}{y-7}$

83. $\sqrt[3]{x} + 3\sqrt[3]{x} - 2\sqrt[3]{x} = 2\sqrt[3]{x}$

84. $\sqrt{3} + \sqrt{27} - \sqrt{192} = \sqrt{3} + 3\sqrt{3} - \sqrt{64}\sqrt{3}$

$\qquad = \sqrt{3} + 3\sqrt{3} - 8\sqrt{3}$

$\qquad = -4\sqrt{3}$

85. $\sqrt[3]{16} - 5\sqrt[3]{54} + 2\sqrt[3]{64}$

$\qquad = \sqrt[3]{8}\sqrt[3]{2} - 5\sqrt[3]{27}\sqrt[3]{2} + 2\sqrt[3]{64}$

$\qquad = 2\sqrt[3]{2} - 5\left(3\sqrt[3]{2}\right) + 2(4)$

$\qquad = 2\sqrt[3]{2} - 15\sqrt[3]{2} + 8$

$\qquad = 8 - 13\sqrt[3]{2}$

86. $4\sqrt{2} - \dfrac{3}{\sqrt{32}} + \sqrt{50}$

$\qquad = 4\sqrt{2} - \dfrac{3}{4\sqrt{2}} + 5\sqrt{2}$

$\qquad = 4\sqrt{2} - \dfrac{3}{4\sqrt{2}} \cdot \dfrac{\sqrt{2}}{\sqrt{2}} + 5\sqrt{2}$

$\qquad = 4\sqrt{2} - \dfrac{3\sqrt{2}}{8} + 5\sqrt{2}$

$\qquad = \dfrac{8}{8}\left(4\sqrt{2}\right) - \dfrac{3\sqrt{2}}{8} + \left(\dfrac{8}{8}\right)5\sqrt{2}$

$\qquad = \dfrac{32\sqrt{2}}{8} - \dfrac{3\sqrt{2}}{8} + \dfrac{40\sqrt{2}}{8}$

$\qquad = \dfrac{69\sqrt{2}}{8}$

87. $3\sqrt{x^5 y^6} - \sqrt{16x^7 y^8}$

$\qquad = 3\sqrt{x^4 y^6}\sqrt{x} - \sqrt{16x^6 y^8}\sqrt{x}$

$\qquad = 3x^2 y^3 \sqrt{x} - 4x^3 y^4 \sqrt{x}$

$\qquad = \left(3x^2 y^3 - 4x^3 y^4\right)\sqrt{x}$

88. $2\sqrt[3]{x^7 y^8} - \sqrt[3]{x^4 y^2} + 3\sqrt[3]{x^{10} y^2}$

$\qquad = 2\sqrt[3]{x^6 y^6}\sqrt[3]{xy^2} - \sqrt[3]{x^3}\sqrt[3]{xy^2} + 3\sqrt[3]{x^9}\sqrt[3]{xy^2}$

$\qquad = 2x^2 y^2 \sqrt[3]{xy^2} - x\sqrt[3]{xy^2} + 3x^3 \sqrt[3]{xy^2}$

$\qquad = (2x^2 y^2 - x + 3x^3)\sqrt[3]{xy^2}$

89. $(f \cdot g)(x) = f(x) \cdot g(x)$

$\qquad = \sqrt{3x} \cdot \left(\sqrt{6x} - \sqrt{10}\right)$

$\qquad = \sqrt{3x}\sqrt{6x} - \sqrt{3x}\sqrt{10}$

$\qquad = \sqrt{18x^2} - \sqrt{30x}$

$\qquad = \sqrt{9x^2}\sqrt{2} - \sqrt{30x}$

$\qquad = 3x\sqrt{2} - \sqrt{30x}$

90. $(f \cdot g)(x) = f(x) \cdot g(x)$
$$= \sqrt[3]{2x^2}\left(\sqrt[3]{4x^4} + \sqrt[3]{8x^5}\right)$$
$$= \sqrt[3]{2x^2}\sqrt[3]{4x^4} + \sqrt[3]{2x^2}\sqrt[3]{8x^5}$$
$$= \sqrt[3]{8x^6} + \sqrt[3]{16x^7}$$
$$= \sqrt[3]{8x^6} + \sqrt[3]{8x^6}\sqrt[3]{2x}$$
$$= 2x^2 + 2x^2\sqrt[3]{2x}$$

91. $f(x) = \sqrt{2x+4}\sqrt{2x+4}$
$$= \sqrt{(2x+4)^2}$$
$$= |2x+4|$$

92. $g(a) = \sqrt{20a^2 + 50a + 125}$
$$= \sqrt{5(4a^2 + 10a + 25)}$$
$$= \sqrt{5(2a+5)^2}$$
$$= \sqrt{5}|2a+5|$$

93. $\dfrac{\sqrt[3]{(x+5)^5}}{\sqrt{(x+5)^3}} = \dfrac{(x+5)^{5/3}}{(x+5)^{3/2}}$
$$= (x+5)^{5/3 - 3/2}$$
$$= (x+5)^{1/6}$$
$$= \sqrt[6]{x+5}$$

94. $\dfrac{\sqrt[3]{a^3b^2}}{\sqrt[4]{a^4b}} = \dfrac{a\sqrt[3]{b^2}}{a\sqrt[4]{b}}$
$$= \dfrac{\sqrt[3]{b^2}}{\sqrt[4]{b}}$$
$$= \dfrac{b^{2/3}}{b^{1/4}}$$
$$= b^{2/3 - 1/4}$$
$$= b^{5/12}$$
$$= \sqrt[12]{b^5}$$

95. a. $f(x) = \sqrt{x} + 2$
$g(x) = -3$
$(f+g)(x) = f(x) + g(x)$
$$= \sqrt{x} + 2 - 3$$
$$= \sqrt{x} - 1$$

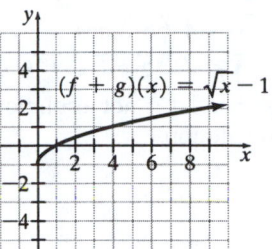

b. $\sqrt{x} \geq 0, \;\; x \geq 0$
The domain is $\{x | x$ is a real number, $x \geq 0\}$

96. a. $f(x) = -\sqrt{x}$
$g(x) = \sqrt{x} + 2$
$(f+g)(x) = f(x) + g(x)$
$$= -\sqrt{x} + \sqrt{x} + 2$$
$$= 2$$

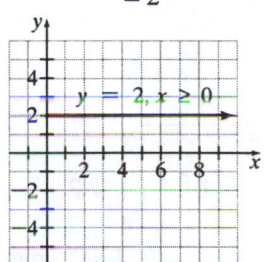

b. $\sqrt{x} \geq 0, \;\; x \geq 0$
The domain is $\{x | x$ is a real number, $x \geq 0\}$

97. $\sqrt[3]{x} = 4$
$$\left(\sqrt[3]{x}\right)^3 = 4^3$$
$$x = 64$$
Check: $\sqrt[3]{x} = 4$
$$\sqrt[3]{64} \overset{?}{=} 4$$
$$4 = 4 \text{ True}$$
64 is a solution.

98. $\sqrt{3x+4} = \sqrt{5x+12}$

$\left(\sqrt{3x+4}\right)^2 = \left(\sqrt{5x+12}\right)^2$

$3x+4 = 5x+12$

$-8 = 2x$

$-4 = x$

Check: $\sqrt{3x+4}$ becomes

$\sqrt{3(-4)+4} = \sqrt{-12+4} = \sqrt{-8}$ which is not a real number. -4 is not a solution and there is no real solution.

99. $2+\sqrt[3]{x} = 4$

$\sqrt[3]{x} = 2$

$\left(\sqrt[3]{x}\right)^3 = 2^3$

$x = 8$

Check: $2+\sqrt[3]{8} \overset{?}{=} 4$

$2+2 \overset{?}{=} 4$

$4 = 4$ True

8 is a solution.

100. $\sqrt{x^2+2x-4} = x$

$\left(\sqrt{x^2+2x-4}\right)^2 = (x)^2$

$x^2+2x-4 = x^2$

$2x-4 = 0$

$x = 2$

Check: $\sqrt{2^2+2\cdot2-4} \overset{?}{=} 2$

$\sqrt{4+4-4} \overset{?}{=} 2$

$\sqrt{4} \overset{?}{=} 2$

$2 = 2$ True

2 is a solution.

101. $\sqrt[3]{x-9} = \sqrt[3]{5x+3}$

$\left(\sqrt[3]{x-9}\right)^3 = \left(\sqrt[3]{5x+3}\right)^3$

$x-9 = 5x+3$

$-4x = 12$

$x = -3$

Check: $\sqrt[3]{-3-9} \overset{?}{=} \sqrt[3]{5(-3)+3}$

$\sqrt[3]{-12} \overset{?}{=} \sqrt[3]{-15+3}$

$\sqrt[3]{-12} = \sqrt[3]{-12}$ True

-3 is a solution.

102. $(x^2+5)^{1/2} = x+1$

$\left[(x^2+5)^{1/2}\right]^2 = (x+1)^2$

$x^2+5 = x^2+2x+1$

$5 = 2x+1$

$4 = 2x$

$x = 2$

Check: $(2^2+5)^{1/2} \overset{?}{=} 2+1$

$(4+5)^{1/2} \overset{?}{=} 3$

$9^{1/2} \overset{?}{=} 3$

$3 = 3$ True

2 is a solution.

103. $\sqrt{x}+3 = \sqrt{3x+9}$

$\left(\sqrt{x}+3\right)^2 = \left(\sqrt{3x+9}\right)^2$

$\left(\sqrt{x}+3\right)\left(\sqrt{x}+3\right) = 3x+9$

$x+6\sqrt{x}+9 = 3x+9$

$6\sqrt{x} = 2x$

$\left(6\sqrt{x}\right)^2 = (2x)^2$

$36x = 4x^2$

$4x^2-36x = 0$

$4x(x-9) = 0$

$4x = 0$ or $x-9 = 0$

$x = 0$ $\qquad$ $x = 9$

A check shows that 0 and 9 are solutions.

104. $\sqrt{6x-5} - \sqrt{2x+6} - 1 = 0$

$$\sqrt{6x-5} = \sqrt{2x+6} + 1$$

$$\left(\sqrt{6x-5}\right)^2 = \left(\sqrt{2x+6} + 1\right)^2$$

$$6x - 5 = 2x + 6 + 2\sqrt{2x+6} + 1$$

$$6x - 5 = 2x + 7 + 2\sqrt{2x+6}$$

$$4x - 12 = 2\sqrt{2x+6}$$

$$\frac{4x}{2} - \frac{12}{2} = \frac{2}{2}\sqrt{2x+6}$$

$$2x - 6 = \sqrt{2x+6}$$

$$(2x-6)^2 = \left(\sqrt{2x+6}\right)^2$$

$$4x^2 - 24x + 36 = 2x + 6$$

$$4x^2 - 26x + 30 = 0$$

$$2x^2 - 13x + 15 = 0$$

Express the middle term, $-13x$, as $-10x - 3x$.

$$2x^2 - 10x - 3x + 15 = 0$$
$$2x(x-5) - 3(x-5) = 0$$
$$(2x-3)(x-5) = 0$$
$$2x - 3 = 0 \quad \text{or} \quad x - 5 = 0$$
$$x = \frac{3}{2} \qquad\qquad x = 5$$

The solution $x = 5$ checks in the original equation but $x = \frac{3}{2}$ does not check. Therefore the only solution is $x = 5$.

105. $f(x) = g(x)$

$$\sqrt{3x+4} = 2\sqrt{2x-4}$$

$$\left(\sqrt{3x+4}\right)^2 = \left(2\sqrt{2x-4}\right)^2$$

$$3x + 4 = 4(2x-4)$$
$$3x + 4 = 8x - 16$$
$$20 = 5x$$
$$4 = x$$

106. $f(x) = g(x)$

$$(4x+3)^{1/3} = (6x-9)^{1/3}$$
$$[(4x+3)^{1/3}]^3 = [(6x-9)^{1/3}]^3$$
$$4x + 3 = 6x - 9$$
$$12 = 2x$$
$$6 = x$$

107. $V = \sqrt{\dfrac{2L}{w}}$ Solve for L.

$$V^2 = \frac{2L}{w}$$

$$V^2 w = 2L$$

$$\frac{V^2 w}{2} = L \text{ or } L = \frac{V^2 w}{2}$$

108. $r = \sqrt{\dfrac{A}{\pi}}$ Solve for A.

$$r^2 = \left(\sqrt{\frac{A}{\pi}}\right)^2$$

$$r^2 = \frac{A}{\pi}$$

$$\pi r^2 = A \text{ or } A = \pi r^2$$

109.
$$l = \sqrt{a^2 + b^2}$$
$$= \sqrt{5^2 + 2^2}$$
$$= \sqrt{29}$$
$$\approx 5.39 \text{ m}$$

110.
$$v = \sqrt{2gh}$$
$$= \sqrt{2(32)(20)}$$
$$= \sqrt{1280} \approx 35.78 \text{ ft/sec}$$

111.
$$T = 2\pi\sqrt{\frac{L}{32}}$$
$$= 2\pi\sqrt{\frac{64}{32}}$$
$$= 2\pi\sqrt{2}$$
$$\approx 8.89 \text{ sec}$$

112.
$$V = \sqrt{\frac{2K}{m}}$$
$$= \sqrt{\frac{2(45)}{0.145}}$$
$$\approx 25 \text{ meters per second}$$

113.
$$m = \frac{m_0}{\sqrt{1 - \frac{v^2}{c^2}}}$$
$$= \frac{m_0}{\sqrt{1 - \frac{(0.98c)^2}{c^2}}}$$
$$= \frac{m_0}{\sqrt{1 - \frac{0.9604c^2}{c^2}}}$$
$$= \frac{m_0}{\sqrt{1 - 0.9604}}$$
$$= \frac{m_0}{\sqrt{0.0396}}$$
$$\approx 5m_0$$

It is about 5 times its original mass.

114. $5 = 5 + 0i$

115. $-6 = -6 + 0i$

116.
$$2 - \sqrt{-256} = 2 - \sqrt{-1}\sqrt{256}$$
$$= 2 - 16i$$

117.
$$3 + \sqrt{-16} = 3 + \sqrt{16}\sqrt{-1}$$
$$= 3 + 4i$$

118. $(3 + 2i) + (4 - i) = 7 + i$

119.
$$(4 - 6i) - (3 - 4i) = 4 - 6i - 3 + 4i$$
$$= 1 - 2i$$

120.
$$\left(\sqrt{3} + \sqrt{-5}\right) + \left(2\sqrt{3} - \sqrt{-7}\right)$$
$$= \sqrt{3} + \sqrt{5}\sqrt{-1} + 2\sqrt{3} - \sqrt{7}\sqrt{-1}$$
$$= \sqrt{3} + i\sqrt{5} + 2\sqrt{3} - i\sqrt{7}$$
$$= 3\sqrt{3} + \left(\sqrt{5} - \sqrt{7}\right)i$$

121.
$$\sqrt{-6}\left(\sqrt{6} + \sqrt{-6}\right) = \sqrt{6}\sqrt{-1}\left(\sqrt{6} + \sqrt{6}\sqrt{-1}\right)$$
$$= i\sqrt{6}\left(\sqrt{6} + i\sqrt{6}\right)$$
$$= i\sqrt{36} + i^2\sqrt{36}$$
$$= 6i + 6(-1) = -6 + 6i$$

122.
$$(4 + 3i)(2 - 3i) = 8 - 12i + 6i - 9i^2$$
$$= 8 - 6i - 9(-1)$$
$$= 8 - 6i + 9$$
$$= 17 - 6i$$

123.
$$\left(6 + \sqrt{-3}\right)\left(4 - \sqrt{-15}\right)$$
$$= \left(6 + \sqrt{3}\sqrt{-1}\right)\left(4 - \sqrt{15}\sqrt{-1}\right)$$
$$= \left(6 + i\sqrt{3}\right)\left(4 - i\sqrt{15}\right)$$
$$= 24 - 6i\sqrt{15} + 4i\sqrt{3} - i^2\sqrt{45}$$
$$= 24 - 6i\sqrt{15} + 4i\sqrt{3} - (-1)\sqrt{9}\sqrt{5}$$
$$= \left(24 + 3\sqrt{5}\right) + \left(4\sqrt{3} - 6\sqrt{15}\right)i$$

124.
$$\frac{2}{3i} = \frac{2}{3i} \cdot \frac{-3i}{-3i}$$
$$= \frac{-6i}{-9i^2}$$
$$= \frac{-6i}{-9(-1)}$$
$$= \frac{-6i}{9}$$
$$= \frac{-2i}{3}$$

125. $\dfrac{2+\sqrt{3}}{2i} = \dfrac{2+\sqrt{3}}{2i} \cdot \dfrac{-2i}{-2i}$

$\qquad = \dfrac{-4i - 2i\sqrt{3}}{-4i^2}$

$\qquad = \dfrac{-4i - 2i\sqrt{3}}{-4(-1)}$

$\qquad = \dfrac{2\left(-2i - i\sqrt{3}\right)}{4}$

$\qquad = \dfrac{-2i - i\sqrt{3}}{2}$

$\qquad = \dfrac{\left(-2 - \sqrt{3}\right)i}{2}$

126. $\dfrac{5}{3+2i} = \dfrac{5}{3+2i} \cdot \dfrac{3-2i}{3-2i}$

$\qquad = \dfrac{5(3-2i)}{9 - 4i^2}$

$\qquad = \dfrac{15 - 10i}{9 - 4(-1)}$

$\qquad = \dfrac{15 - 10i}{9 + 4}$

$\qquad = \dfrac{15 - 10i}{13}$

127. $\dfrac{\sqrt{3}}{5 - \sqrt{-6}} = \dfrac{\sqrt{3}}{5 - i\sqrt{6}}$

$\qquad = \dfrac{\sqrt{3}}{\left(5 - i\sqrt{6}\right)} \cdot \dfrac{5 + i\sqrt{6}}{5 + i\sqrt{6}}$

$\qquad = \dfrac{5\sqrt{3} + i\sqrt{18}}{(5)^2 - \left(i\sqrt{6}\right)^2}$

$\qquad = \dfrac{5\sqrt{3} + 3i\sqrt{2}}{(5)^2 + \left(\sqrt{6}\right)^2}$

$\qquad = \dfrac{5\sqrt{3} + 3i\sqrt{2}}{25 + 6}$

$\qquad = \dfrac{5\sqrt{3} + 3i\sqrt{2}}{31}$

128. $x^2 - 2x + 9$

$\qquad = \left(1 + 2i\sqrt{2}\right)^2 - 2\left(1 + 2i\sqrt{2}\right) + 9$

$\qquad = 1^2 + 2(1)\left(2i\sqrt{2}\right) + \left(2i\sqrt{2}\right)^2 - 2 - 4i\sqrt{2} + 9$

$\qquad = 1 + 4i\sqrt{2} - 8 - 2 - 4i\sqrt{2} + 9$

$\qquad = 0 + 0i$

$\qquad = 0$

129. $x^2 - 2x + 12$

$\qquad = (1 - 2i)^2 - 2(1 - 2i) + 12$

$\qquad = 1^2 - 2(1)(2i) + (2i)^2 - 2 + 4i + 12$

$\qquad = 1 - 4i - 4 - 2 + 4i + 12$

$\qquad = 7 + 0i$

$\qquad = 7$

130. $i^{53} = i^{52} i = (i^4)^{13} = 1^{13} \cdot i = i$

131. $i^{19} = i^{16} i^3$

$\qquad = (i^4)^4 i^3$

$\qquad = 1^4 \cdot i^3$

$\qquad = 1(i^3)$

$\qquad = i^3$

$\qquad = -i$

132. $i^{404} = (i^4)^{101} = 1^{101} = 1$

133. $i^{5326} = i^{5324} i^2$

$\qquad = (i^4)^{1331} i^2$

$\qquad = 1^{1331} \cdot i^2$

$\qquad = 1(i^2)$

$\qquad = i^2$

$\qquad = -1$

Practice Test

1. $\sqrt{(-26)^2} = |-26| = 26$

2. $\sqrt{(3x-4)^2} = |3x - 4|$

3. $\left(\dfrac{y^{2/3}\cdot y^{-1}}{y^{1/4}}\right)^2 = \dfrac{y^{4/3}y^{-2}}{y^{2/4}}$

$= y^{4/3-2-1/2}$

$= y^{(8/6-12/6-3/6)}$

$= y^{-7/6}$

$= \dfrac{1}{y^{7/6}}$

4. $x^{3/5} + x^{-2/5} = x^{-2/5}(x^{5/5}) + x^{-2/5}(1)$

$= x^{-2/5}(x+1)$

5. $f(x) = \sqrt{x}$

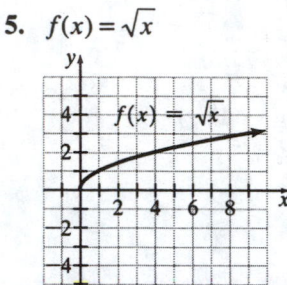

6. $\sqrt{50x^5y^8} = \sqrt{25x^4y^8}\sqrt{2x}$

$= 5x^2y^4\sqrt{2x}$

7. $\sqrt[3]{4x^5y^2}\sqrt[3]{10x^6y^8} = \sqrt[3]{40x^{11}y^{10}}$

$= \sqrt[3]{8x^9y^9}\cdot\sqrt[3]{5x^2y}$

$= 2x^3y^3\sqrt[3]{5x^2y}$

8. $\sqrt{\dfrac{2x^4y^5}{8z}} = \dfrac{\sqrt{2x^4y^5}}{\sqrt{8z}}$

$= \dfrac{\sqrt{x^4y^4}\sqrt{2y}}{\sqrt{4}\sqrt{2z}}$

$= \dfrac{x^2y^2\sqrt{2y}}{2\sqrt{2z}}\cdot\dfrac{\sqrt{2z}}{\sqrt{2z}}$

$= \dfrac{x^2y^2\sqrt{4}\sqrt{yz}}{2(2z)}$

$= \dfrac{2x^2y^2\sqrt{yz}}{4z}$

$= \dfrac{x^2y^2\sqrt{yz}}{2z}$

9. $\sqrt[3]{\dfrac{1}{x}} = \dfrac{\sqrt[3]{1}}{\sqrt[3]{x}}\cdot\dfrac{\sqrt[3]{x^2}}{\sqrt[3]{x^2}} = \dfrac{\sqrt[3]{x^2}}{x}$

10. $\dfrac{\sqrt{2}}{2+\sqrt{8}}\cdot\dfrac{2-\sqrt{8}}{2-\sqrt{8}} = \dfrac{\sqrt{2}\left(2-\sqrt{8}\right)}{4-8}$

$= \dfrac{2\sqrt{2}-\sqrt{16}}{-4}$

$= \dfrac{2\sqrt{2}-4}{-4}$

$= \dfrac{2\left(\sqrt{2}-2\right)}{-4}$

$= \dfrac{\sqrt{2}-2}{-2}$

$= \dfrac{-\sqrt{2}+2}{2}$ or $\dfrac{2-\sqrt{2}}{2}$

11. $\sqrt{27} + 2\sqrt{3} - 5\sqrt{75}$

$= \sqrt{9}\sqrt{3} + 2\sqrt{3} - 5\sqrt{25}\sqrt{3}$

$= 3\sqrt{3} + 2\sqrt{3} - 25\sqrt{3}$

$= -20\sqrt{3}$

12. $\sqrt[3]{8x^3y^5} + 2\sqrt[3]{x^6y^8}$

$= \sqrt[3]{8x^3y^3}\sqrt[3]{y^2} + 2\sqrt[3]{x^6y^6}\sqrt[3]{y^2}$

$= 2xy\sqrt[3]{y^2} + 2x^2y^2\sqrt[3]{y^2}$

$= (2xy + 2x^2y^2)\sqrt[3]{y^2}$

13. $\left(\sqrt{5}-3\right)\left(2-\sqrt{8}\right)$

$= \sqrt{5}(2) - \sqrt{5}\sqrt{8} - 3(2) + 3\sqrt{8}$

$= 2\sqrt{5} - \sqrt{40} - 6 + 3\sqrt{8}$

$= 2\sqrt{5} - \sqrt{4}\sqrt{10} - 6 + 3\sqrt{4}\sqrt{2}$

$= 2\sqrt{5} - 2\sqrt{10} - 6 + 6\sqrt{2}$

14. $\sqrt[3]{\sqrt{x^4 y^2}} = \sqrt[3]{(x^4 y^2)^{1/2}}$

$= [(x^4 y^2)^{1/2}]^{1/3}$

$= (x^4 y^2)^{1/6}$

$= \sqrt[6]{x^4 y^2}$

15. $\dfrac{\sqrt[4]{(7x+2)^5}}{\sqrt[3]{(7x+2)^2}} = \dfrac{(7x+2)^{5/4}}{(7x+2)^{2/3}}$

$= (7x+2)^{5/4 - 2/3}$

$= (7x+2)^{7/12}$

$= \sqrt[12]{(7x+2)^7}$

16. $\sqrt{4x-3} = 7$

$\left(\sqrt{4x-3}\right)^2 = 7^2$

$4x - 3 = 49$

$4x = 52$

$x = 13$

Check: $\sqrt{4 \cdot 13 - 3} \overset{?}{=} 7$

$\sqrt{52 - 3} \overset{?}{=} 7$

$\sqrt{49} \overset{?}{=} 7$

$7 = 7$ True

13 is a solution.

17. $\sqrt{x^2 - x - 12} = x + 3$

$\left(\sqrt{x^2 - x - 12}\right)^2 = (x+3)^2$

$x^2 - x - 12 = x^2 + 6x + 9$

$-x - 12 = 6x + 9$

$-12 = 7x + 9$

$-21 = 7x$

$x = -3$

Check: $\sqrt{(-3)^2 - (-3) - 12} \overset{?}{=} -3 + 3$

$\sqrt{9 + 3 - 12} \overset{?}{=} -3 + 3$

$\sqrt{0} \overset{?}{=} 0$

$0 = 0$

-3 is a solution.

18. $\sqrt{x-15} = \sqrt{x} - 3$

$\left(\sqrt{x-15}\right)^2 = \left(\sqrt{x} - 3\right)^2$

$x - 15 = x - 6\sqrt{x} + 9$

$-15 = -6\sqrt{x} + 9$

$-24 = -6\sqrt{x}$

$4 = \sqrt{x}$

$(4)^2 = \left(\sqrt{x}\right)^2$

$16 = x$

Check: $\sqrt{16 - 15} \overset{?}{=} \sqrt{16} - 3$

$\sqrt{1} \overset{?}{=} 4 - 3$

$1 = 1$ True

16 is a solution.

19. $f(x) = g(x)$

$(9x + 37)^{1/3} = 2(2x + 2)^{1/3}$

$[(9x + 37)^{1/3}]^3 = [2(2x + 2)^{1/3}]^3$

$9x + 37 = 8(2x + 2)$

$9x + 37 = 16x + 16$

$21 = 7x$

$3 = x$

20. $w = \dfrac{\sqrt{2gh}}{4}$ Solve for h.

$$4w = \sqrt{2gh}$$
$$(4w)^2 = \left(\sqrt{2gh}\right)^2$$
$$16w^2 = 2gh$$
$$\frac{16w^2}{2g} = \frac{2gh}{2g}$$
$$\frac{8w^2}{g} = h$$

21. Let x be the length of the ladder.
$$x = \sqrt{12^2 + 5^2}$$
$$= 169$$
$$= 13 \text{ feet}$$

22. $T = 2\pi\sqrt{\dfrac{m}{k}}$

$$T = 2\pi\sqrt{\frac{1400}{65,000}}$$
$$\approx 0.92 \text{ sec}$$

23. $\left(6 - \sqrt{-4}\right)\left(3 + \sqrt{-2}\right)$
$$= (6 - 2i)\left(3 + i\sqrt{2}\right)$$
$$= 18 + 6i\sqrt{2} - 6i - 2i^2\sqrt{2}$$
$$= 18 + 6i\sqrt{2} - 6i - 2(-1)\sqrt{2}$$
$$= 18 + 6i\sqrt{2} - 6i + 2\sqrt{2}$$
$$= \left(18 + 2\sqrt{2}\right) + \left(6\sqrt{2} - 6\right)i$$

24. $\dfrac{\sqrt{5}}{2 - \sqrt{-8}} = \dfrac{\sqrt{5}}{2 - 2i\sqrt{2}}$

$$= \frac{\sqrt{5}}{2 - 2i\sqrt{2}} \cdot \frac{2 + 2i\sqrt{2}}{2 + 2i\sqrt{2}}$$
$$= \frac{2\sqrt{5} + 2i\sqrt{10}}{2^2 + \left(2\sqrt{2}\right)^2}$$
$$= \frac{2\sqrt{5} + 2i\sqrt{10}}{4 + 4 \cdot 2}$$
$$= \frac{2\sqrt{5} + 2i\sqrt{10}}{4 + 8}$$
$$= \frac{2\sqrt{5} + 2i\sqrt{10}}{12}$$
$$= \frac{2\left(\sqrt{5} + i\sqrt{10}\right)}{12}$$
$$= \frac{\sqrt{5} + i\sqrt{10}}{6}$$

25. $x^2 + 6x + 12$
$$= (-3 + i)^2 + 6(-3 + i) + 12$$
$$= (-3)^2 - 2(3)(i) + (i)^2 - 18 + 6i + 12$$
$$= 9 - 6i - 1 - 18 + 6i + 12$$
$$= 2 + 0i$$
$$= 2$$

Cumulative Review Test

1. $\dfrac{1}{5}(x - 3) = \dfrac{3}{4}(x + 3) - x$

$$20\left[\frac{1}{5}(x - 3)\right] = 20\left[\frac{3}{4}(x + 3)\right] - 20x$$
$$4(x - 3) = 5(3)(x + 3) - 20x$$
$$4x - 12 = 15x + 45 - 20x$$
$$4x - 12 = -5x + 45$$
$$9x - 12 = 45$$
$$9x = 57$$
$$x = \frac{57}{9}$$

2. a.   Relation: any set of ordered pairs

b.   Function: a correspondence such that each element of the domain corresponds with exactly one member of the range.

3. **a.**   Domain: $\{x | x \geq 0\}$ because $\sqrt{x} \geq 0$

     b. $y = f(x) = \sqrt{x} + 2$

x	y
0	2
1	3
4	4
9	5

The graph is

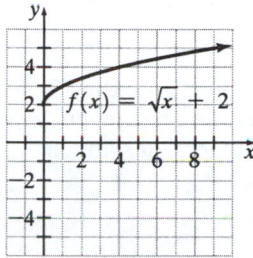

4. $x - 4y = 6$
    $3x - y = 2$
    To solve, use the substitution method by solving the first equation for x:
    $x - 4y = 6$
        $x = 4y + 6$

Now, substitute $4y + 6$ for x into the second equation and then solve for y:
$$3x - y = 2$$
$$3(4y + 6) - y = 2$$
$$12y + 18 - y = 2$$
$$11y + 18 = 2$$
$$11y = -16$$
$$y = -\frac{16}{11}$$

Finally, substitute $-\dfrac{16}{11}$ for y into $x = 4y + 6$.

$$x = 4\left(-\frac{16}{11}\right) + 6$$
$$= -\frac{64}{11} + 6$$
$$= -\frac{64}{11} + \frac{66}{11}$$
$$= \frac{2}{11}$$

The solution is $\left(\dfrac{2}{11}, \ -\dfrac{16}{11}\right)$.

5. $\left(\dfrac{3x^2 y^{-2}}{x^4 y^{-5}}\right)^{-1}$

$= \left(\dfrac{3y^{-2-(-5)}}{x^{4-2}}\right)^{-1}$ ← Simplify within parentheses

$= \left(\dfrac{3y^3}{x^2}\right)^{-1}$ ← Simplify

$= \left(\dfrac{x^2}{3y^3}\right)^{1}$ ← Negative exponent rule

$= \dfrac{x^2}{3y^3}$

6. $3x^2 - 4x - 6$
             $2x - 5$
   _____
   $-15x^2 + 20x + 30$ ← $-5(3x^2 - 4x - 6)$
   $6x^3 - 8x^2 - 12x$ ← $2x(3x^2 - 4x - 6)$
   _____
   $6x^3 - 23x^2 + 8x + 30$ ← Combine like terms

7. Using synthetic division:

$$
\begin{array}{r|rrr}
-2 & 3 & 10 & 10 \\
   &   & -6 & -8 \\
\hline
   & 3 & 4 & 2
\end{array}
$$

Thus, $\dfrac{3x^2 + 10x + 10}{x+2} = 3x + 4 + \dfrac{2}{x+2}$.

8. $y = f(x) = x^3 - 2$

x	y
-2	-10
-1	-3
0	-2
1	-1
2	6

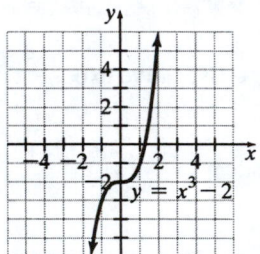

9. $2x^2 - 12x + 18 - 2y^2$
$= 2(x^2 - 6x + 9 - y^2)$
$= 2[(x^2 - 6x + 9) - y^2]$
$= 2[(x-3)^2 - y^2]$
$= 2[(x-3) - y][(x-3) + y]$
$= 2(x - 3 - y)(x - 3 + y)$

10. $\dfrac{x+2}{3x-5}$

The domain is found by setting $3x - 5$ equal to zero and solving. The domain will be all real numbers except that solution.
$$3x - 5 = 0$$
$$x = \frac{5}{3}$$

Domain: $\left\{ x \middle| x \ne \dfrac{5}{3} \right\}$

11. $\dfrac{(x+2)(x-4) + (x-1)(x-4)}{3(x-4)}$

$= \dfrac{(x-4)[(x+2) + (x-1)]}{3(x-4)}$

$= \dfrac{(x-4)(2x+1)}{3(x-4)}$

$= \dfrac{2x+1}{3}$

12. $\dfrac{4x^2 + 8x + 3}{2x^2 - x - 1} \cdot \dfrac{x^2 - 1}{4x^2 + 12x + 9}$

$= \dfrac{(2x+1)(2x+3)}{(2x+1)(x-1)} \cdot \dfrac{(x-1)(x+1)}{(2x+3)(2x+3)}$

$= \dfrac{x+1}{2x+3}$

13. $\dfrac{x+1}{x^2+2x-3}-\dfrac{x}{2x^2+11x+15}$

$=\dfrac{x+1}{(x+3)(x-1)}-\dfrac{x}{(2x+5)(x+3)}$

$=\dfrac{x+1}{(x+3)(x-1)}\cdot\dfrac{2x+5}{2x+5}-\dfrac{x}{(2x+5)(x+3)}\cdot\dfrac{x-1}{x-1}$

$=\dfrac{(x+1)(2x+5)}{(2x+5)(x+3)(x-1)}-\dfrac{x(x-1)}{(2x+5)(x+3)(x-1)}$

$=\dfrac{2x^2+7x+5}{(2x+5)(x+3)(x-1)}-\dfrac{x^2-x}{(2x+5)(x+3)(x-1)}$

$=\dfrac{2x^2+7x+5-(x^2-x)}{(2x+5)(x+3)(x-1)}$

$=\dfrac{x^2+8x+5}{(2x+5)(x+3)(x-1)}$

14. $\qquad 4-\dfrac{5}{y}=\dfrac{4y}{y+1}$

$y(y+1)4-y(y+1)\left(\dfrac{5}{y}\right)=y(y+1)\left(\dfrac{4y}{y+1}\right)$

$\qquad 4y(y+1)-5(y+1)=y(4y)$

$\qquad 4y^2+4y-5y-5=4y^2$

$\qquad\qquad 4y-5y-5=0$

$\qquad\qquad\qquad -y-5=0$

$\qquad\qquad\qquad -5=y$

15. $\left(\dfrac{x^2 y^{1/2}}{x^{1/4}}\right)^2=(x^{2-1/4}y^{1/2})^2$

$\qquad\qquad\quad =(x^{7/4}y^{1/2})^2$

$\qquad\qquad\quad =(x^{7/4})^2(y^{1/2})^2$

$\qquad\qquad\quad =x^{7/4\cdot 2}y^{1/2\cdot 2}$

$\qquad\qquad\quad =x^{7/2}y$

16. $\sqrt[3]{4x^{10}y^{20}}\sqrt[3]{4x^3y^9}=\sqrt[3]{16x^{13}y^{29}}$

$\qquad\qquad\qquad\qquad =\sqrt[3]{8x^{12}y^{27}}\sqrt[3]{2xy^2}$

$\qquad\qquad\qquad\qquad =2x^4y^9\sqrt[3]{2xy^2}$

17. $\qquad \sqrt{2x^2+7}+3=8$

$\qquad\qquad \sqrt{2x^2+7}=5$

$\qquad\qquad \left(\sqrt{2x^2+7}\right)^2=5^2$

$\qquad\qquad\qquad 2x^2+7=25$

$\qquad\qquad\qquad 2x^2-18=0$

$\qquad\qquad\qquad 2(x^2-9)=0$

$\qquad\qquad 2(x+3)(x-3)=0$

$\qquad x+3=0\quad\text{or}\quad x-3=0$

$\qquad\quad x=-3\qquad\qquad x=3$

Check for $x=-3$ Check for $x=3$

$\sqrt{2(-3)^2+7}+3\overset{?}{=}8$ $\sqrt{2(3)^2+7}+3\overset{?}{=}8$

$\sqrt{18+7}+3\overset{?}{=}8$ $\sqrt{18+7}+3\overset{?}{=}8$

$\sqrt{25}+3\overset{?}{=}8$ $\sqrt{25}+3\overset{?}{=}8$

$5+3\overset{?}{=}8$ $5+3\overset{?}{=}8$

$8=8$ $8=8$

      True   True

Solutions are -3 and 3.

18.
$$\frac{2}{3+\sqrt{-6}} = \frac{2}{3+\sqrt{-1}\sqrt{6}}$$
$$= \frac{2}{3+i\sqrt{6}}$$
$$= \frac{2}{3+i\sqrt{6}} \cdot \frac{3-i\sqrt{6}}{3-i\sqrt{6}}$$
$$= \frac{2\left(3-i\sqrt{6}\right)}{3^2 + \left(\sqrt{6}\right)^2}$$
$$= \frac{6-2i\sqrt{6}}{9+6}$$
$$= \frac{6-2i\sqrt{6}}{15}$$

19. Let x be the amount of time for them to paint the room together.

Person	Rate	Time	Part completed
Jim	$\dfrac{1}{2}$	x	$\dfrac{x}{2}$
Mike	$\dfrac{1}{3}$	x	$\dfrac{x}{3}$

$$\frac{x}{2} + \frac{x}{3} = 1$$
$$6\left(\frac{x}{2}\right) + 6\left(\frac{x}{3}\right) = 6(1)$$
$$3x + 2x = 6$$
$$5x = 6$$
$$x = \frac{6}{5} \text{ or } 1\frac{1}{5} \text{ hr}$$

20. Let x be the length of the wire.
$$x = \sqrt{30^2 + 20^2}$$
$$= \sqrt{900 + 400}$$
$$= \sqrt{1300} \approx 36.1 \text{ feet}$$

Chapter 8

Exercise Set 8.1

1. The two square roots of 36 are $\pm\sqrt{36} = \pm 6$.

3. The square root property is: If $x^2 = a$, where a is a real number, then $x = \pm\sqrt{a}$.

5. A trinomial, $x^2 + bx + c$, is a perfect square trinomial if $\left(\dfrac{b}{2}\right)^2 = c$.

7. $x^2 = 49$
   $x = \pm\sqrt{49} = \pm 7$

9. $y^2 = 48$
   $y = \pm\sqrt{48} = \pm 4\sqrt{3}$

11. $z^2 + 12 = 40$
    $z^2 = 28$
    $z = \pm\sqrt{28} = \pm 2\sqrt{7}$

13. $(p-4)^2 = 16$
    $p - 4 = \pm\sqrt{16}$
    $p - 4 = \pm 4$
    $p = 4 \pm 4$
    $p = 4 + 4 \quad$ or $\quad P = 4 - 4$
    $P = 8 \qquad\qquad P = 0$

15. $\left(z + \dfrac{1}{3}\right)^2 = \dfrac{4}{9}$
    $z + \dfrac{1}{3} = \pm\sqrt{\dfrac{4}{9}}$
    $z + \dfrac{1}{3} = \pm\dfrac{2}{3}$
    $z = -\dfrac{1}{3} \pm \dfrac{2}{3}$
    $z = -\dfrac{1}{3} + \dfrac{2}{3} \quad$ or $\quad z = -\dfrac{1}{3} - \dfrac{2}{3}$
    $z = \dfrac{1}{3} \qquad\qquad\quad z = -\dfrac{3}{3}$
    $\qquad\qquad\qquad\qquad\quad z = -1$

17. $(x + 1.8)^2 = 0.81$
    $x + 1.8 = \pm\sqrt{0.81}$
    $x + 1.8 = \pm 0.9$
    $x = -1.8 \pm 0.9$
    $x = -1.8 + 0.9 \quad$ or $\quad x = -1.8 - 0.9$
    $x = -0.9 \qquad\qquad\quad x = -2.7$

19. $(2a - 5)^2 = 12$
    $2a - 5 = \pm\sqrt{12}$
    $2a - 5 = \pm 2\sqrt{3}$
    $2a = 5 \pm 2\sqrt{3}$
    $a = \dfrac{5 \pm 2\sqrt{3}}{2}$

21. $\left(2y+\dfrac{1}{2}\right)^2 = \dfrac{4}{25}$

$\quad 2y+\dfrac{1}{2} = \pm\sqrt{\dfrac{4}{25}}$

$\quad 2y+\dfrac{1}{2} = \pm\dfrac{2}{5}$

$2y+\dfrac{1}{2} = \dfrac{2}{5} \quad$ or $\quad 2y+\dfrac{1}{2} = -\dfrac{2}{5}$

$\quad 2y = -\dfrac{1}{2}+\dfrac{2}{5} \qquad\quad 2y = -\dfrac{1}{2}-\dfrac{2}{5}$

$\quad 2y = -\dfrac{1}{10} \qquad\qquad\quad 2y = -\dfrac{9}{10}$

$\quad\quad y = -\dfrac{1}{20} \qquad\qquad\quad y = -\dfrac{9}{20}$

23. $x^2 + 2x - 15 = 0$

$\quad x^2 + 2x = 15$

$\quad x^2 + 2x + 1 = 15 + 1$

$\quad (x+1)^2 = 16$

$\quad x+1 = \pm 4$

$\quad x = \pm 4 - 1$

$x = 4 - 1 \quad$ or $\quad x = -4 - 1$

$x = 3 \qquad\qquad x = -5$

25. $x^2 - 4x - 5 = 0$

$\quad x^2 - 4x = 5$

$\quad x^2 - 4x + 4 = 5 + 4$

$\quad (x-2)^2 = 9$

$\quad x - 2 = \pm 3$

$\quad x = \pm 3 + 2$

$x = 3 + 2 \quad$ or $\quad x = -3 + 2$

$x = 5 \qquad\qquad x = -1$

27. $a^2 + 3a + 2 = 0$

$\quad a^2 + 3a = -2$

$\quad a^2 + 3a + \dfrac{9}{4} = -2 + \dfrac{9}{4}$

$\quad \left(a+\dfrac{3}{2}\right)^2 = \dfrac{1}{4}$

$\quad a + \dfrac{3}{2} = \pm\dfrac{1}{2}$

$\quad a = \pm\dfrac{1}{2} - \dfrac{3}{2}$

$a = -\dfrac{1}{2} - \dfrac{3}{2} \quad$ or $\quad a = \dfrac{1}{2} - \dfrac{3}{2}$

$a = -\dfrac{4}{2} \qquad\qquad\quad a = -\dfrac{2}{2}$

$a = -2 \qquad\qquad\quad\quad a = -1$

29. $r^2 - 8r + 15 = 0$

$\quad r^2 - 8r = -15$

$\quad r^2 - 8r + 16 = -15 + 16$

$\quad (r-4)^2 = 1$

$\quad r - 4 = \pm 1$

$\quad r = \pm 1 + 4$

$r = 1 + 4 \quad$ or $\quad r = -1 + 4$

$r = 5 \qquad\qquad r = 3$

31. $x^2 + 2x + 12 = 0$

$\quad x^2 + 2x = -12$

$\quad x^2 + 2x + 1 = -12 + 1$

$\quad (x+1)^2 = -11$

$\quad x + 1 = \pm\sqrt{-11}$

$\quad x + 1 = \pm i\sqrt{11}$

$\quad x = -1 \pm i\sqrt{11}$

33.
$$-z^2 + 9z - 20 = 0$$
$$z^2 - 9z + 20 = 0$$
$$z^2 - 9z = -20$$
$$z^2 - 9z + \frac{81}{4} = -20 + \frac{81}{4}$$
$$\left(z - \frac{9}{2}\right)^2 = \frac{1}{4}$$
$$z - \frac{9}{2} = \pm\frac{1}{2}$$
$$z = \frac{9}{2} \pm \frac{1}{2}$$
$$z = \frac{9}{2} + \frac{1}{2} \quad \text{or} \quad z = \frac{9}{2} - \frac{1}{2}$$
$$z = \frac{10}{2} \qquad\qquad z = \frac{8}{2}$$
$$z = 5 \qquad\qquad\quad z = 4$$

35.
$$b^2 = 3b + 28$$
$$b^2 - 3b = 28$$
$$b^2 - 3b + \frac{9}{4} = \frac{112}{4} + \frac{9}{4}$$
$$\left(b - \frac{3}{2}\right)^2 = \frac{121}{4}$$
$$b - \frac{3}{2} = \pm\frac{11}{2}$$
$$b = \pm\frac{11}{2} + \frac{3}{2}$$
$$b = -\frac{11}{2} + \frac{3}{2} \quad \text{or} \quad b = \frac{11}{2} + \frac{3}{2}$$
$$b = -\frac{8}{2} \qquad\qquad b = \frac{14}{2}$$
$$b = -4 \qquad\qquad b = 7$$

37.
$$x^2 + 3x + 6 = 0$$
$$x^2 + 3x = -6$$
$$x^2 + 3x + \frac{9}{4} = -\frac{24}{4} + \frac{9}{4}$$
$$\left(x + \frac{3}{2}\right)^2 = \frac{-15}{4}$$
$$x + \frac{3}{2} = \pm\sqrt{\frac{-15}{4}}$$
$$x + \frac{3}{2} = \pm\frac{i\sqrt{15}}{2}$$
$$x = -\frac{3}{2} \pm \frac{i\sqrt{15}}{2}$$
$$x = \frac{-3 \pm i\sqrt{15}}{2}$$

39.
$$-s^2 + 5s = -8$$
$$s^2 - 5s = 8$$
$$s^2 - 5s + \frac{25}{4} = 8 + \frac{25}{4}$$
$$\left(s - \frac{5}{2}\right)^2 = \frac{32}{4} + \frac{25}{4}$$
$$\left(s - \frac{5}{2}\right)^2 = \frac{57}{4}$$
$$s - \frac{5}{2} = \pm\frac{\sqrt{57}}{2}$$
$$s = \frac{5}{2} \pm \frac{\sqrt{57}}{2}$$
$$s = \frac{5 \pm \sqrt{57}}{2}$$

41.
$$-\frac{1}{4}a^2 - \frac{1}{2}a = 0$$
$$-4\left(-\frac{1}{4}a^2 - \frac{1}{2}a = 0\right)$$
$$a^2 + 2a = 0$$
$$a^2 + 2a + 1 = 0 + 1$$
$$(a + 1)^2 = 1$$
$$a + 1 = \pm 1$$
$$a + 1 = 1 \quad \text{or} \quad a + 1 = -1$$
$$a = 0 \qquad\qquad a = -2$$

43.
$$12a^2 - 4a = 0$$
$$a^2 - \frac{1}{3}a = 0$$
$$a^2 - \frac{1}{3}a + \frac{1}{36} = 0 + \frac{1}{36}$$
$$\left(a - \frac{1}{6}\right)^2 = \frac{1}{36}$$
$$a - \frac{1}{6} = \pm\frac{1}{6}$$
$$a = \pm\frac{1}{6} + \frac{1}{6}$$
$$x = -\frac{1}{6} + \frac{1}{6} \quad \text{or} \quad x = \frac{1}{6} + \frac{1}{6}$$
$$x = 0 \qquad\qquad x = \frac{2}{6}$$
$$x = \frac{1}{3}$$

45.
$$-\frac{1}{2}p^2 - p + \frac{3}{2} = 0$$
$$-2\left(-\frac{1}{2}p^2 - p + \frac{3}{2} = 0\right)$$
$$p^2 + 2p - 3 = 0$$
$$p^2 + 2p = 3$$
$$p^2 + 2p + 1 = 3 + 1$$
$$(p+1)^2 = 4$$
$$p + 1 = \pm 2$$
$$p = -1 \pm 2$$
$$p = -1 + 2 \quad \text{or} \quad p = -1 - 2$$
$$p = 1 \qquad\qquad p = -3$$

47.
$$2x^2 + 18x + 4 = 0$$
$$x^2 + 9x + 2 = 0$$
$$x^2 + 9x = -2$$
$$x^2 + 9x + \frac{81}{4} = -2 + \frac{81}{4}$$
$$\left(x + \frac{9}{2}\right)^2 = -\frac{8}{4} + \frac{81}{4}$$
$$\left(x + \frac{9}{2}\right)^2 = \frac{73}{4}$$
$$x + \frac{9}{2} = \pm\frac{\sqrt{73}}{2}$$
$$x = -\frac{9}{2} \pm \frac{\sqrt{73}}{2}$$
$$x = \frac{-9 \pm \sqrt{73}}{2}$$

49.
$$3x^2 + 33x + 72 = 0$$
$$x^2 + 11x + 24 = 0$$
$$x^2 + 11x = -24$$
$$x^2 + 11x + \frac{121}{4} = -24 + \frac{121}{4}$$
$$\left(x + \frac{11}{2}\right)^2 = -\frac{96}{4} + \frac{121}{4}$$
$$\left(x - \frac{11}{2}\right)^2 = \frac{25}{4}$$
$$x + \frac{11}{2} = \pm\frac{5}{2}$$
$$x = \pm\frac{5}{2} - \frac{11}{2}$$
$$x = -\frac{5}{2} - \frac{11}{2} \quad \text{or} \quad x = \frac{5}{2} - \frac{11}{2}$$
$$x = -\frac{16}{2} \qquad \text{or} \quad x = -\frac{6}{2}$$
$$x = -8 \qquad\quad \text{or} \quad x = -3$$

51.

$$\frac{2}{3}x^2 + \frac{4}{3}x + 1 = 0$$

$$\frac{3}{2}\left(\frac{2}{3}x^2 + \frac{4}{3}x + 1\right) = 0$$

$$x^2 + 2x + \frac{3}{2} = 0$$

$$x^2 + 2x = -\frac{3}{2}$$

$$x^2 + 2x + 1 = -\frac{3}{2} + 1$$

$$(x+1)^2 = -\frac{1}{2}$$

$$x + 1 = \pm\sqrt{-\frac{1}{2}}$$

$$x + 1 = \pm\frac{i\sqrt{2}}{2}$$

$$x = -1 \pm \frac{i\sqrt{2}}{2}$$

$$x = \frac{-2 \pm i\sqrt{2}}{2}$$

53.

$$-3x^2 + 6x = 6$$

$$x^2 - 2x = -2$$

$$x^2 - 2x + 1 = -2 + 1$$

$$(x-1)^2 = -1$$

$$x - 1 = \pm\sqrt{-1}$$

$$x - 1 = \pm i$$

$$x = 1 \pm i$$

55.

$$\frac{5}{2}x^2 + \frac{3}{2}x - \frac{5}{4} = 0$$

$$\frac{2}{5}\left[\frac{5}{2}x^2 + \frac{3}{2}x - \frac{5}{4} = 0\right]$$

$$x^2 + \frac{3}{5}x - \frac{1}{2} = 0$$

$$x^2 + \frac{3}{5}x = \frac{1}{2}$$

$$x^2 + \frac{3}{5}x + \frac{9}{100} = \frac{1}{2} + \frac{9}{100}$$

$$\left(x + \frac{3}{10}\right)^2 = \frac{59}{100}$$

$$x + \frac{3}{10} = \pm\frac{\sqrt{59}}{10}$$

$$x = -\frac{3}{10} \pm \frac{\sqrt{59}}{10}$$

$$x = \frac{-3 \pm \sqrt{59}}{10}$$

57.

$$x^2 + 2ax + a^2 = k$$

$$x^2 + 2ax = k - a^2$$

$$x^2 + 2ax + a^2 = k - a^2 + a^2$$

$$(x+a)^2 = k$$

$$x + a = \pm\sqrt{k}$$

$$x = -a \pm \sqrt{k}$$

59. Let x be the first integer. Then $x + 2$ is the next consecutive odd integer.

$$x(x+2) = 63$$

$$x^2 + 2x = 63$$

$$x^2 + 2x + 1 = 63 + 1$$

$$(x+1)^2 = 64$$

$$x + 1 = \pm 8$$

$$x = -1 \pm 8$$

$$x = -1 + 8 \quad \text{or} \quad x = -1 - 8$$

$$x = 7 \qquad\qquad x = -9$$

Since it was given that the integers are positive, one integer is 7 and the other is $7 + 2 = 9$.

61. Let x be the width of the rectangle. Then $2x + 2$ is the length.
Use length · width = area.

$$(2x+2)x = 60$$
$$2x^2 + 2x = 60$$
$$x^2 + x = 30$$
$$x^2 + x + \frac{1}{4} = 30 + \frac{1}{4}$$
$$\left(x+\frac{1}{2}\right)^2 = \frac{120}{4} + \frac{1}{4}$$
$$\left(x+\frac{1}{2}\right)^2 = \frac{121}{4}$$
$$x+\frac{1}{2} = \pm\frac{11}{2}$$
$$x = -\frac{1}{2} \pm \frac{11}{2}$$
$$x = -\frac{1}{2}+\frac{11}{2} \quad \text{or} \quad x = -\frac{1}{2}-\frac{11}{2}$$
$$x = \frac{10}{2} = 5 \qquad x = -\frac{12}{2} = 6$$

Since the width cannot be negative, the width is 5 ft.
Length $= 2(5)+2 = 10+2 = 12$ ft.
The rectangle is 5 ft by 12 ft.

63. Let s be the length of the side. Then $s + 6$ is the length of the diagonal (d). Use $s^2+s^2 = d^2$.
$$2s^2 = (s+6)^2$$
$$2s^2 = s^2 + 12s + 36$$
$$s^2 = 12s + 36$$
$$s^2 - 12s = 36$$
$$s^2 - 12s + 36 = 36 + 36$$
$$(s-6)^2 = 72$$
$$s-6 = \pm 6\sqrt{2}$$
$$s = 6 \pm 6\sqrt{2}$$
Length is never negative. Thus,
$s = 6 + 6\sqrt{2} \approx 14.49$.
The room is about 14.49 ft by 14.49 ft.

65. Since the radius is 10 inches, the diameter (d) is 20 inches. Use the formula $s^2+s^2=d^2$ to find the length (s) of the other two sides.

$$s^2 + s^2 = d^2$$
$$2s^2 = 20^2$$
$$2s^2 = 400$$
$$s^2 = 200$$
$$s = \pm\sqrt{200} = \pm 10\sqrt{2}$$
Length is never negative.
Thus $s = 10\sqrt{2} \approx 14.14$ inches.

67.
$$A = \pi r^2$$
$$24\pi = \pi r^2$$
$$24 = r^2$$
$$\pm\sqrt{24} = r$$
$$\pm 2\sqrt{6} = r$$
Length is never negative.
Thus $r = 2\sqrt{6} \approx 4.90$ feet.

69. $A = P\left(1+\frac{r}{n}\right)^{nt}$
$$551.25 = 500\left(1+\frac{r}{1}\right)^{1(2)}$$
$$551.25 = 500(1+r)^2$$
$$1.1025 = (1+r)^2$$
$$\pm 1.05 = 1+r$$
$$-1 \pm 1.05 = r$$
An interest rate is never negative. Thus $r = -1 + 1.05 = 0.05 = 5\%$.

71. $A = P\left(1+\frac{r}{n}\right)^{nt}$
$$1432.86 = 1200\left(1+\frac{r}{2}\right)^{2(3)}$$
$$1432.86 = 1200\left(1+\frac{r}{2}\right)^6$$
$$1.19405 = \left(1+\frac{r}{2}\right)^6$$
$$\pm 1.03 \approx 1+\frac{r}{2}$$
$$-1 \pm 1.03 \approx \frac{r}{2}$$
$$-2 \pm 2.06 \approx r$$
An interest rate is never negative.
Thus the annual interest rate is
$-2 + 2.06 = 0.06 = 6\%$.

73.
$$x^2 + 6x + y^2 - 10y = -18$$
$$x^2 + 6x + 9 + y^2 - 10y = -18 + 9$$
$$(x+3)^2 + y^2 - 10y = -9$$
$$(x+3)^2 + y^2 - 10y + 25 = -9 + 25$$
$$(x+3)^2 + (y-5)^2 = 16$$

75. a. To find the surface area, we must first determine the radius. Use $V = \pi r^2 h$ with $V = 160$ and $h = 10$ to get
$$160 = \pi r^2 (10)$$
$$16 = \pi r^2$$
$$\frac{16}{\pi} = r^2$$
$$\frac{4}{\sqrt{\pi}} = r$$

Since the radius equals $\dfrac{4}{\sqrt{\pi}}$, use the formula $S = 2\pi r^2 + 2\pi rh$ to calculate the surface area.
$$S = 2\pi \left(\frac{4}{\sqrt{\pi}}\right)^2 + 2\pi \left(\frac{4}{\sqrt{\pi}}\right)(10)$$
$$= 2\pi \left(\frac{16}{\pi}\right) + \frac{80\pi}{\sqrt{\pi}}$$
$$= 32 + 80\sqrt{\pi}$$
$$\approx 173.80$$
The surface area is about 173.80 square inches.

b. Use $V = \pi r^2 h$ with $V = 160$ and $h = 10$ to obtain $160 = \pi r^2 (10)$. In part (a) this was solved for r to get
$$r = \frac{4}{\sqrt{\pi}} = \frac{4}{\sqrt{\pi}} \cdot \frac{\sqrt{\pi}}{\sqrt{\pi}} = \frac{4\sqrt{\pi}}{\pi} \approx 2.26$$
The radius is about 2.26 inches.

c. Use $S = 2\pi r^2 + 2\pi rh$ with $S = 160$ and $h = 10$.

$$160 = 2\pi r^2 + 2\pi r(10)$$
$$160 = 2\pi r^2 + 20\pi r$$
$$\frac{160}{2\pi} = \frac{2\pi r^2}{2\pi} + \frac{20\pi r}{2\pi}$$
$$\frac{80}{\pi} = r^2 + 10r$$
$$\frac{80}{\pi} + 25 = r^2 + 10r + 25$$
$$\frac{80 + 25\pi}{\pi} = (r+5)^2$$
$$\pm\sqrt{\frac{80 + 25\pi}{\pi}} = r + 5$$
$$\pm\sqrt{\frac{80 + 25\pi}{\pi}} - 5 = r$$
The radius is never negative.
Thus $r \approx 2.1$ inches.

77.
$$2xy - 3yz = -xy + z$$
$$2xy + xy = 3yz + z$$
$$3xy = z(3y+1)$$
$$\frac{3xy}{3y+1} = z$$

78. $\sqrt{\left(x^2 - 4x\right)^2} = \left|x^2 - 4x\right|$

79. $25^{-1/2} = \dfrac{1}{\sqrt{25}} = \dfrac{1}{5}$

80.
$$\frac{x^{3/4}y^{1/2}}{x^{1/4}y^2} = x^{(3/4 - 1/4)}y^{(1/2 - 2)}$$
$$= x^{2/4}y^{-3/2}$$
$$= \frac{x^{1/2}}{y^{3/2}}$$

Exercise Set 8.2

1. The quadratic formula is
$$x = \frac{-b \pm \sqrt{b^2 - 4ac}}{2a}$$ which gives the values of x where $ax^2 + bx + c = 0$.

3. a. Yes, the equations will have the same solutions since
$$12x^2 - 15x - 6 = 3(4x^2 - 5x - 2).$$

b. $12x^2 - 15x - 6 = 0$. Use the quadratic formula, $x = \dfrac{-b \pm \sqrt{b^2 - 4ac}}{2a}$, with $a = 12$, $b = -15$, and $c = -6$.

$$x = \frac{15 \pm \sqrt{(-15)^2 - 4(12)(-6)}}{2(12)}$$

$$= \frac{15 \pm \sqrt{225 + 288}}{24}$$

$$= \frac{15 \pm \sqrt{513}}{24}$$

$$= \frac{15 \pm 3\sqrt{57}}{24}$$

$$= \frac{5 \pm \sqrt{57}}{8}$$

The solutions are $\dfrac{5 + \sqrt{57}}{8}$ and $\dfrac{5 - \sqrt{57}}{8}$.

c. $3(4x^2 - 5x - 2) = 0$. Use the quadratic formula, $x = \dfrac{-b \pm \sqrt{b^2 - 4ac}}{2a}$, with $a = 4$, $b = -5$ and $c = -2$.

$$x = \frac{5 \pm \sqrt{(-5)^2 - 4(4)(-2)}}{2(4)}$$

$$= \frac{5 \pm \sqrt{25 + 32}}{8}$$

$$= \frac{5 \pm \sqrt{57}}{8}$$

The solutions are $\dfrac{5 + \sqrt{57}}{8}$ and $\dfrac{5 - \sqrt{57}}{8}$.

5. If $b^2 - 4ac > 0$, then the quadratic equation will have two distinct real solutions. Since there is a positive number under the radical sign in the quadratic formula, the value of the radical will be real and there will be two real solutions. If $b^2 - 4ac = 0$, then the equation has the single real solution $\dfrac{-b}{2a}$. If $b^2 - 4ac < 0$, the expression under the radical sign in the quadratic formula is negative. Thus the equation has no real solution.

7. $x^2 - 5x + 6 = 0$
$$b^2 - 4ac = (-5)^2 - 4(1)(6)$$
$$= 25 - 24$$
$$= 1$$
Since $1 > 0$, there is two real solution.

9. $2a^2 - 4a + 7 = 0$
$$b^2 - 4ac = (-4)^2 - 4(2)(7)$$
$$= 16 - 56$$
$$= -40$$
Since $-40 < 0$ there is no real solution.

11. $5p^2 + 3p - 7 = 0$
$$b^2 - 4ac = 3^2 - 4(5)(-7)$$
$$= 9 + 140$$
$$= 149$$
Since $149 > 0$ there are two real solutions.

13. $-5x^2 + 5x - 6 = 0$
$$b^2 - 4ac = 5^2 - 4(-5)(-6)$$
$$= 25 - 120$$
$$= -95$$
Since $-95 < 0$, there is no real solution.

15. $x^2 + 10.2x + 26.01 = 0$
$$b^2 - 4ac = (10.2)^2 - 4(1)(26.01)$$
$$= 104.04 - 104.04$$
$$= 0$$
There is one real solution.

17. $b^2 = -3b - \dfrac{9}{4}$

$\quad b^2 + 3b + \dfrac{9}{4} = 0$

$\quad\quad b^2 - 4ac = 3^2 - 4(1)\left(\dfrac{9}{4}\right)$

$\quad\quad\quad\quad\quad = 9 - 9$

$\quad\quad\quad\quad\quad = 0$

There is one real solution.

19. $x^2 + 9x + 20 = 0$

$\quad x = \dfrac{-9 \pm \sqrt{9^2 - 4(1)(20)}}{2(1)}$

$\quad\quad = \dfrac{-9 \pm \sqrt{81 - 80}}{2}$

$\quad\quad = \dfrac{-9 \pm \sqrt{1}}{2}$

$\quad\quad = \dfrac{-9 \pm 1}{2}$

$x = \dfrac{-9 + 1}{2} \quad$ or $\quad x = \dfrac{-9 - 1}{2}$

$\quad = \dfrac{-8}{2} \quad\quad\quad\quad = \dfrac{-10}{2}$

$\quad = -4 \quad\quad\quad\quad\quad = -5$

The solutions are -4 and -5.

21. $x^2 + 2x - 3 = 0$

$\quad x = \dfrac{-2 \pm \sqrt{2^2 - 4(1)(-3)}}{2(1)}$

$\quad\quad = \dfrac{-2 \pm \sqrt{4 + 12}}{2}$

$\quad\quad = \dfrac{-2 \pm \sqrt{16}}{2}$

$\quad\quad = \dfrac{-2 \pm 4}{2}$

$x = \dfrac{-2 + 4}{2} \quad$ or $\quad x = \dfrac{-2 - 4}{2}$

$\quad = \dfrac{2}{2} \quad\quad\quad\quad = \dfrac{-6}{2}$

$\quad = 1 \quad\quad\quad\quad\quad = -3$

The solutions are 1 and -3.

23. $a^2 - 6a = -5$

$\quad a^2 - 6a + 5 = 0$

$\quad a = \dfrac{-(-6) \pm \sqrt{(-6)^2 - 4(1)(5)}}{2(1)}$

$\quad\quad = \dfrac{6 \pm \sqrt{36 - 20}}{2}$

$\quad\quad = \dfrac{6 \pm \sqrt{16}}{2}$

$\quad\quad = \dfrac{6 \pm 4}{2}$

$a = \dfrac{6 - 4}{2} \quad$ or $\quad a = \dfrac{6 + 4}{2}$

$\quad = \dfrac{2}{2} \quad\quad\quad\quad\quad = \dfrac{10}{2}$

$\quad = 1 \quad\quad\quad\quad\quad\quad = 5$

The solutions are 1 and 5.

25. $r^2 - 81 = 0$

$\quad r = \dfrac{0 \pm \sqrt{0^2 - 4(1)(-81)}}{2(1)}$

$\quad\quad = \dfrac{\pm \sqrt{324}}{2}$

$\quad\quad = \pm \dfrac{18}{2}$

$\quad\quad = \pm 9$

The solutions are 9 and -9.

27. $x^2 - 4x = 0$

$\quad x = \dfrac{4 \pm \sqrt{(-4)^2 - 4(1)(0)}}{2(1)}$

$\quad\quad = \dfrac{4 \pm \sqrt{16}}{2}$

$\quad\quad = \dfrac{4 \pm 4}{2}$

$x = \dfrac{4 + 4}{2} \quad$ or $\quad x = \dfrac{4 - 4}{2}$

$\quad = \dfrac{8}{2} \quad\quad\quad\quad\quad = \dfrac{0}{2}$

$\quad = 4 \quad\quad\quad\quad\quad\quad = 0$

The solutions are 4 and 0.

29. $3w^2 - 4w + 5 = 0$

$$w = \frac{-(-4) \pm \sqrt{(-4)^2 - 4(3)(5)}}{2(3)}$$

$$= \frac{4 \pm \sqrt{16 - 60}}{6}$$

$$= \frac{4 \pm \sqrt{-44}}{6}$$

$$= \frac{4 \pm 2i\sqrt{11}}{6}$$

$$= \frac{2(2 \pm i\sqrt{11})}{6}$$

$$= \frac{2 \pm i\sqrt{11}}{3}$$

The solutions are $\dfrac{2 - i\sqrt{11}}{3}$ and $\dfrac{2 + i\sqrt{11}}{3}$.

31. $x^2 + 6x = -2$

$x^2 + 6x + 2 = 0$

$$x = \frac{-6 \pm \sqrt{6^2 - 4(1)(2)}}{2(1)}$$

$$= \frac{-6 \pm \sqrt{36 - 8}}{2}$$

$$= \frac{-6 \pm \sqrt{28}}{2}$$

$$= \frac{-6 \pm 2\sqrt{7}}{2}$$

$$= -3 \pm \sqrt{7}$$

The solutions are $-3 + \sqrt{7}$ and $-3 - \sqrt{7}$.

33. $-6x^2 + 21x = -27$

$$\frac{-6x^2}{-3} + \frac{21}{-3}x = \frac{-27}{-3}$$

$$2x^2 - 7x = 9$$

$$2x^2 - 7x - 9 = 0$$

$$x = \frac{-(-7) \pm \sqrt{(-7^2) - 4(2)(-9)}}{2(2)}$$

$$= \frac{7 \pm \sqrt{49 + 72}}{4}$$

$$= \frac{7 \pm \sqrt{121}}{4}$$

$$= \frac{7 \pm 11}{4}$$

$$x = \frac{7 + 11}{4} \quad \text{or} \quad x = \frac{7 - 11}{4}$$

$$= \frac{18}{4} \qquad\qquad = \frac{-4}{4}$$

$$= \frac{9}{2} \qquad\qquad = -1$$

The solutions are $\dfrac{9}{2}$ or -1.

35. $(2a + 3)(3a - 1) = 2$

$6a^2 + 7a - 3 = 2$

$6a^2 + 7a - 5 = 0$

$$a = \frac{-(7) \pm \sqrt{(7)^2 - 4(6)(-5)}}{2(6)}$$

$$= \frac{-7 \pm \sqrt{49 + 120}}{12}$$

$$= \frac{-7 \pm \sqrt{169}}{12}$$

$$= \frac{-7 \pm 13}{12}$$

$$a = \frac{-7 - 13}{12} \quad \text{or} \quad a = \frac{-7 + 13}{12}$$

$$= \frac{-20}{12} \qquad\qquad = \frac{6}{12}$$

$$= -\frac{5}{3} \qquad\qquad = \frac{1}{2}$$

The solutions are $\dfrac{1}{2}$ and $-\dfrac{5}{3}$.

37.
$$-2a^2 = a + 3$$
$$-2a^2 - a - 3 = 0$$
$$\frac{-2a^2}{-1} - \frac{a}{-1} - \frac{3}{-1} = \frac{0}{-1}$$
$$2a^2 + a + 3 = 0$$
$$a = \frac{-1 \pm \sqrt{(1)^2 - 4(2)(3)}}{2(2)}$$
$$= \frac{-1 \pm \sqrt{1 - 24}}{4}$$
$$= \frac{-1 \pm \sqrt{-23}}{4}$$
$$= \frac{-1 \pm i\sqrt{23}}{4}$$

The solutions are $\dfrac{-1 + i\sqrt{23}}{4}$ and

$\dfrac{-1 - i\sqrt{23}}{4}$.

39.
$$2x^2 + 6x = 0$$
$$x = \frac{-6 \pm \sqrt{6^2 - 4(2)(0)}}{2(2)}$$
$$= \frac{-6 \pm \sqrt{36 - 0}}{4}$$
$$= \frac{-6 \pm 6}{4}$$
$$x = \frac{-6 + 6}{4} \quad \text{or} \quad x = \frac{-6 - 6}{4}$$
$$= \frac{0}{4} \qquad\qquad = \frac{-12}{4}$$
$$= 0 \qquad\qquad\quad = -3$$

The solutions are 0 and –3.

41.
$$m = \frac{-m + 6}{m - 4}$$
$$m(m - 4) = -m + 6$$
$$m^2 - 4m = -m + 6$$
$$m^2 - 3m - 6 = 0$$

$$m = \frac{-(-3) \pm \sqrt{(-3)^2 - 4(1)(-6)}}{2(1)}$$
$$= \frac{3 \pm \sqrt{9 + 24}}{2}$$
$$= \frac{3 \pm \sqrt{33}}{2}$$

The solutions are $\dfrac{3 + \sqrt{33}}{2}$ and $\dfrac{3 - \sqrt{33}}{2}$.

43. $\dfrac{1}{2}x^2 + 2x + \dfrac{2}{3} = 0$

$$6\left(\frac{1}{2}x^2 + 2x + \frac{2}{3} = 0\right)$$
$$3x^2 + 12x + 4 = 0$$
$$x = \frac{-12 \pm \sqrt{(12)^2 - 4(3)(4)}}{2(3)}$$
$$= \frac{-12 \pm \sqrt{144 - 48}}{6}$$
$$= \frac{-12 \pm \sqrt{96}}{6}$$
$$= \frac{-12 \pm 4\sqrt{6}}{6}$$
$$= \frac{2(-6 \pm 2\sqrt{6})}{2(3)}$$
$$= \frac{-6 \pm 2\sqrt{6}}{3}$$

The solutions are
$\dfrac{-6 + 2\sqrt{6}}{3}$ and $\dfrac{-6 - 2\sqrt{6}}{3}$.

45. $-x^2 + \dfrac{11}{3}x + \dfrac{10}{3} = 0$

$$-3\left(-x^2 + \frac{11}{3}x + \frac{10}{3} = 0\right)$$
$$3x^2 - 11x - 10 = 0$$
$$x = \frac{-(-11) \pm \sqrt{(-11)^2 - 4(3)(-10)}}{2(3)}$$
$$= \frac{11 \pm \sqrt{121 + 120}}{6}$$
$$= \frac{11 \pm \sqrt{241}}{6}$$

The solutions are

$$\frac{11+\sqrt{241}}{6} \text{ and } \frac{11-\sqrt{241}}{6}.$$

47. $0.1x^2 + 0.6x - 1.2 = 0$

$10(0.1x^2 + 0.6x - 1.2 = 0)$

$x^2 + 6x - 12 = 0$

$$x = \frac{-6 \pm \sqrt{6^2 - 4(1)(-12)}}{2(1)}$$

$$= \frac{-6 \pm \sqrt{36 + 48}}{2}$$

$$= \frac{-6 \pm \sqrt{84}}{2}$$

$$= \frac{-6 \pm 2\sqrt{21}}{2}$$

$$= -3 \pm \sqrt{21}$$

The solutions are $-3 + \sqrt{21}$ or $-3 - \sqrt{21}$.

49. $-1.62x^2 - 0.94x + 4.85 = 0$

$-100(-1.62x^2 - 0.94x + 4.85 = 0)$

$162x^2 + 94x - 485 = 0$

$$x = \frac{-94 \pm \sqrt{(94)^2 - 4(162)(-485)}}{2(162)}$$

$$= \frac{-94 \pm \sqrt{8836 + 314,280}}{324}$$

$$= \frac{-94 \pm \sqrt{323,116}}{324}$$

$$= \frac{-94 \pm 2\sqrt{80,779}}{324}$$

$$= \frac{-47 \pm \sqrt{80,779}}{162}$$

The solutions are $\dfrac{-47 - \sqrt{80,779}}{162}$ and

$\dfrac{-47 + \sqrt{80,779}}{162}.$

51. $f(x) = x^2 - 3x + 3, \ f(x) = 3$

$x^2 - 3x + 3 = 3$

$x^2 - 3x = 0$

$$x = \frac{3 \pm \sqrt{(-3)^2 - 4(1)(0)}}{2(1)}$$

$$= \frac{3 \pm \sqrt{9}}{2}$$

$$= \frac{3 \pm 3}{2}$$

$x = \dfrac{3+3}{2}$ or $x = \dfrac{3-3}{2}$

$\quad = \dfrac{6}{2} \qquad\qquad = \dfrac{0}{2}$

$\quad = 3 \qquad\qquad\ = 0$

The values of x are 3 and 0.

53. $h(t) = 2t^2 - 7t + 1, \ h(t) = -3$

$2t^2 - 7t + 1 = -3$

$2t^2 - 7t + 4 = 0$

$$t = \frac{7 \pm \sqrt{(-7)^2 - 4(2)(4)}}{2(2)}$$

$$= \frac{7 \pm \sqrt{49 - 32}}{4}$$

$$= \frac{7 \pm \sqrt{17}}{4}$$

The values of t are $\dfrac{7+\sqrt{17}}{4}$ and $\dfrac{7-\sqrt{17}}{4}.$

55. $g(a) = 2a^2 - 3a + 16, \ g(a) = 14$

$2a^2 - 3a + 16 = 14$

$2a^2 - 3a + 2 = 0$

$$a = \frac{3 \pm \sqrt{(-3)^2 - 4(2)(2)}}{2(2)}$$

$$= \frac{3 \pm \sqrt{9 - 16}}{4}$$

$$= \frac{3 \pm \sqrt{-7}}{4}$$

There are no real values of a for which $g(a) = 14.$

57. If 4 and 6 are solutions, the factors must be $(x-4)$ and $(x-6)$.
$$f(x) = (x-4)(x-6)$$
$$f(x) = x^2 - 6x - 4x + 24$$
$$f(x) = x^2 - 10x + 24$$

59. If 3 and –4 are solutions, the factors must be $(x-3)$ and $(x+4)$.
$$f(x) = (x-3)(x+4)$$
$$f(x) = x^2 + 4x - 3x - 12$$
$$f(x) = x^2 + x - 12$$

61. If $\dfrac{1}{2}$ and 3 are solutions, the factors must be $(2x-1)$ and $(x-3)$.
$$f(x) = (2x-1)(x-3)$$
$$f(x) = 2x^2 - 6x - x + 3$$
$$f(x) = 2x^2 - 7x + 3$$

63. If $-\dfrac{3}{5}$ and $\dfrac{2}{3}$ are solutions, the factors must be $(5x+3)$ and $(3x-2)$.
$$f(x) = (5x+3)(3x-2)$$
$$f(x) = 15x^2 - 10x + 9x - 6$$
$$f(x) = 15x^2 - x - 6$$

65. If $\sqrt{3}$ and $-\sqrt{3}$ are solutions, the factors must be $\left(x - \sqrt{3}\right)$ and $\left(x + \sqrt{3}\right)$.
$$f(x) = \left(x - \sqrt{3}\right)\left(x + \sqrt{3}\right)$$
$$f(x) = x^2 - 3$$

67. $2i$ and $-2i$ are solutions, the factors must be $(x-2i)$ and $(x+2i)$.
$$f(x) = (x-2i)(x+2i)$$
$$f(x) = x^2 - 4i^2$$
$$f(x) = x^2 + 4$$

69. If $3 + \sqrt{2}$ and $3 - \sqrt{2}$ are solutions, the factors must be $\left(x - \left(3 + \sqrt{2}\right)\right)$ and $\left(x - \left(3 - \sqrt{2}\right)\right)$.
$$f(x) = \left(x - \left(3 + \sqrt{2}\right)\right)\left(x - \left(3 - \sqrt{2}\right)\right)$$
$$f(x) = \left(x - 3 - \sqrt{2}\right)\left(x - 3 + \sqrt{2}\right)$$
$$f(x) = (x-3)^2 - \left(\sqrt{2}\right)^2$$
$$f(x) = x^2 - 6x + 9 - 2$$
$$f(x) = x^2 - 6x + 7$$

71. If $2 + 3i$ and $2 - 3i$ are solutions, the factors must be $(x - (2+3i))$ and $(x - (2-3i))$.
$$f(x) = (x-(2+3i))(x-(2-3i))$$
$$f(x) = (x - 2 - 3i)(x - 2 + 3i)$$
$$f(x) = (x-2)^2 - 9i^2$$
$$f(x) = x^2 - 4x + 4 + 9$$
$$f(x) = x^2 - 4x + 13$$

73. Any quadratic equation for which the discriminant is a non-negative perfect square can be solved by factoring. Any quadratic equation for which the discriminant is a positive number but not a perfect square can be solved by the quadratic formula but not by factoring over the set of integers.

75. Yes. If the discriminant is a perfect square, the simplified expression will not contain a radical and the quadratic equation can be solved by factoring.

77. Let x be the number.
$$2x^2 + 3x = 14$$
$$2x^2 + 3x - 14 = 0$$

$$x = \frac{-3 \pm \sqrt{3^2 - 4(2)(-14)}}{2(2)}$$

$$= \frac{-3 \pm \sqrt{9 + 112}}{4}$$

$$= \frac{-3 \pm \sqrt{121}}{4}$$

$$= \frac{-3 \pm 11}{4}$$

$$x = \frac{-3 + 11}{4} \text{ since } x \text{ is positive}$$

$$x = \frac{8}{4}$$

$$x = 2$$

79. Let x be the width. Then $3x - 2$ is the length. Use $A = (\text{length})(\text{width})$.

$$21 = (3x - 2)(x)$$

$$21 = 3x^2 - 2x$$

$$3x^2 - 2x - 21 = 0$$

$$x = \frac{-(-2) \pm \sqrt{(-2)^2 - 4(3)(-21)}}{2(3)}$$

$$= \frac{2 \pm \sqrt{4 + 252}}{6}$$

$$= \frac{2 \pm \sqrt{256}}{6}$$

$$= \frac{2 \pm 16}{6}$$

Since width is positive, use

$$x = \frac{2 + 16}{6} = \frac{18}{6} = 3$$

$$3x - 2 = 3(3) - 2 = 9 - 2 = 7$$

The width is 3 feet and the length is 7 feet.

81. Let x be the amount by which each side is to be reduced.
Then $6 - x$ is the new width
and $8 - x$ is the new length

$$\text{new area} = \frac{1}{2} \text{ (old area)}$$

$$= \frac{1}{2}(6 \cdot 8)$$

$$= \frac{1}{2}(48)$$

$$= 24$$

$$\text{new area} = (\text{new width})(\text{new length})$$

$$24 = (6 - x)(8 - x)$$

$$0 = 48 - 14x + x^2 - 24$$

$$0 = x^2 - 14x + 24$$

$$x = \frac{-(-14) \pm \sqrt{(-14)^2 - 4(1)(24)}}{2(1)}$$

$$= \frac{14 \pm \sqrt{196 - 96}}{2}$$

$$= \frac{14 \pm \sqrt{100}}{2}$$

$$= \frac{14 \pm 10}{2}$$

$$x = \frac{14 + 10}{2} \quad \text{or} \quad x = \frac{14 - 10}{2}$$

$$= \frac{24}{2} \qquad\qquad = \frac{4}{2}$$

$$= 12 \qquad\qquad\quad = 2$$

We reject $x = 12$, since this would give negative values for width and length. The only meaningful value is $x = 2$ inches since this gives positive values for the new width and length.

83. a. $P = 0.2(1)^2 + 1.5(1) - 1.2 = 0.5$
The profit is 0.5 thousand dollars or $500.

b. $P = 0.2(5)^2 + 1.5(5) - 1.2 = 11.3$
The profit is 11.3 thousand dollars or $11,300.

c. $1.6 = 0.2t^2 + 1.5t - 1.2$

$$16 = 2t^2 + 15t - 12$$

$$0 = 2t^2 + 15t - 28$$

$$t = \frac{-15 \pm \sqrt{15^2 - 4(2)(-28)}}{2(2)}$$

$$t = \frac{-15 \pm \sqrt{449}}{4}$$

$$t = \frac{-15 + \sqrt{449}}{4} \quad \text{or} \quad t = \frac{-15 - \sqrt{449}}{4}$$

$$t \approx 1.55 \qquad\qquad\qquad t \approx -9.05$$

The company should charge $1.55 per tape.

85. a. $A(t) = 0.03t^2 + 0.02t + 2.69$

In 1990, $t = 3$.

$A(3) = 0.03(3^2) + 0.02(3) + 2.69$

$\qquad = 0.27 + 0.06 + 2.69$

$\qquad = 3.02$

The NASCAR attendance in 1990 was 3.02 million.

b. $A(t) = 0.03t^2 + 0.02t + 2.69$, $A(t) = 5.3$

$5.3 = 0.03t^2 + 0.02t + 2.69$

$0 = 0.03t^2 + 0.02t - 2.61$

$t = \dfrac{-0.02 \pm \sqrt{(0.02)^2 - 4(0.03)(-2.61)}}{2(0.03)}$

$\quad = \dfrac{-0.02 \pm \sqrt{0.0004 + 0.3132}}{0.06}$

$\quad = \dfrac{-0.2 \pm \sqrt{0.3136}}{0.06}$

$\quad = \dfrac{-0.02 \pm 0.56}{0.06}$

$t = \dfrac{-0.02 + 0.56}{0.06} \quad$ or $\quad t = \dfrac{-0.02 - 0.56}{0.06}$

$\quad = \dfrac{0.54}{0.06} \qquad\qquad\quad = -\dfrac{0.58}{0.06}$

$\quad = 9 \qquad\qquad\qquad\quad \approx -9.67$

Since $1 \le t \le 20$, the only solution is $t = 9$. Thus NASCAR attendance was 5.3 million 9 years after 1987, or in the year 1996.

87. a. $m(t) = 0.05t^2 - 0.32t + 3.15$

In 1985, $t = 3$.

$m(3) = 0.05(3^2) - 0.32(3) + 3.15$

$\qquad = 0.45 - 0.96 + 3.15$

$\qquad = 2.64$

Veterinary bills for dogs in 1985 amounted to about $2.64 billion.

b. $m(t) = 0.05t^2 - 0.32t + 3.15$, $m(t) = 12$

$12 = 0.05t^2 - 0.32t + 3.15$

$0 = 0.05t^2 - 0.32t - 8.85$

$t = \dfrac{0.32 \pm \sqrt{(0.32)^2 - 4(0.05)(-8.85)}}{2(0.05)}$

$\quad = \dfrac{0.32 \pm \sqrt{0.1024 + 1.77}}{0.1}$

$\quad = \dfrac{0.32 \pm \sqrt{1.8724}}{0.1}$

$\quad \approx \dfrac{0.32 \pm 1.3683567}{0.1}$

$t \approx \dfrac{0.32 + 1.3683567}{0.1}$

$\quad = \dfrac{1.6883567}{0.1} \qquad$ or

$\quad \approx 16.9$

$t \approx \dfrac{0.32 - 1.3683567}{0.1}$

$\quad = \dfrac{-1.0483567}{0.1}$

$\quad \approx -10.5$

Since $1 \le t \le 18$, the only solution is $t \approx 16.9$. Thus $12 billion was spent on veterinary bills for dogs approximately 16.9 years after 1982 or in 1998.

89. a. $s(t) = 0.02t^2 - 0.02t + 1.20$

In 1995, $t = 3$.

$s(3) = 0.02(3^2) - 0.02(3) + 1.20$

$\qquad = 0.18 - 0.06 + 1.20$

$\qquad = 1.32$

The amount spent on indigestion remedies in 1995 was $1.32 billion.

b. $s(t) = 0.02t^2 - 0.02t + 1.20$, $s(t) = 2.0$

$2.0 = 0.02t^2 - 10.02t + 1.20$

$0 = 0.02t^2 - 0.02t - 0.80$

$t = \dfrac{0.02 \pm \sqrt{(-0.02)^2 - 4(0.02)(-0.80)}}{2(0.02)}$

$\quad = \dfrac{0.02 \pm \sqrt{0.0004 + 0.064}}{0.04}$

$\quad = \dfrac{0.02 \pm \sqrt{0.0644}}{0.04}$

$\quad \approx \dfrac{0.02 \pm 0.2537716}{0.04}$

$$t \approx \frac{0.02 + 0.2537716}{0.04} \quad \text{or} \quad t \approx \frac{0.02 - 0.2537716}{0.04}$$

$$\approx \frac{0.2737716}{0.04} \qquad\qquad \approx \frac{-0.2337716}{0.04}$$

$$\approx 6.84 \qquad\qquad\qquad \approx -5.84$$

Since $1 \le t \le 9$, the only solution is $t \approx 6.84$. U.S. residents spent \$20 billion on indigestion remedies approximately 6.84 years after 1992 or in 1998.

91. a. $0 = -3.3t^2 - 2.3t + 62$

$$t = \frac{-(-2.3) \pm \sqrt{(-2.3)^2 - 4(-3.3)(62)}}{2(-3.3)}$$

$$t = \frac{2.3 \pm 28.7}{-6.6}$$

$t \approx -4.7$ or $t = 4$

Since time must be positive, the car takes 4 seconds for the drop.

b. $s = 6.74(4) + 2.3 = 29.26$
The speed is 29.26 feet per second.

93. Since gravity is one-sixth that on Earth a in this context will equal $\frac{1}{6}(-32) = \frac{-16}{3}$.

$$0 = \frac{1}{2}\left(-\frac{16}{3}\right)t^2 + 40t + 60$$

$$0 = -\frac{8}{3}t^2 + 40t + 60$$

$$\frac{3}{4}\left(0 = -\frac{8}{3}t^2 + 40t + 60\right)$$

$$0 = -2t^2 + 30t + 45$$

$$t = \frac{-30 \pm \sqrt{30^2 - 4(-2)(45)}}{2(-2)}$$

$$t = \frac{-30 \pm \sqrt{1260}}{-4}$$

$$t = \frac{-30 + \sqrt{1260}}{-4} \quad \text{or} \quad t = \frac{-30 - \sqrt{1260}}{-4}$$

$$t \approx -1.37 \qquad\qquad\qquad t \approx 16.37$$

Neil Armstrong will land on the ground after about 16.37 sec.

95. $x^2 + 5\sqrt{6}x + 36 = 0$, $a = 1$,
$b = 5\sqrt{6}$, $c = 36$

$$x = \frac{-5\sqrt{6} \pm \sqrt{\left(5\sqrt{6}\right)^2 - 4(1)(36)}}{2(1)}$$

$$= \frac{-5\sqrt{6} \pm \sqrt{150 - 144}}{2}$$

$$= \frac{-5\sqrt{6} \pm \sqrt{6}}{2}$$

$$x = \frac{-5\sqrt{6} - \sqrt{6}}{2} \quad \text{or} \quad x = \frac{-5\sqrt{6} + \sqrt{6}}{2}$$

$$= \frac{-6\sqrt{6}}{2} \qquad\qquad = \frac{-4\sqrt{6}}{2}$$

$$= -3\sqrt{6} \qquad\qquad\quad = -2\sqrt{6}$$

The solutions are $-3\sqrt{6}$ and $-2\sqrt{6}$.

97.
$$x^6 = 1$$
$$x^6 - 1 = 0$$
$$(x^3)^2 - 1^2 = 0$$
$$(x^3 - 1)(x^3 + 1) = 0$$
$$(x - 1)(x^2 + x + 1)(x + 1)(x^2 - x + 1) = 0$$

$$x - 1 = 0$$
$$x = 1$$

$$x^2 + x + 1 = 0$$

$$x = \frac{-1 \pm \sqrt{1^2 - 4(1)(1)}}{2(1)}$$

$$= \frac{-1 \pm \sqrt{-3}}{2}$$

$$= \frac{-1 \pm i\sqrt{3}}{2}$$

$$x + 1 = 0$$
$$x = -1$$

$$x^2 - x + 1 = 0$$

$$x = \frac{-(-1) \pm \sqrt{(-1)^2 - 4(1)(1)}}{2(1)}$$

$$= \frac{1 \pm \sqrt{-3}}{2}$$

$$= \frac{1 \pm i\sqrt{3}}{2}$$

The 6 solutions to $x^6 = 1$ are $x = \pm 1$,
$$x = \frac{-1 \pm i\sqrt{3}}{2}, \ x = \frac{1 \pm i\sqrt{3}}{2}.$$

100. $\sqrt[5]{64x^9y^{12}z^{20}} = \sqrt[5]{32x^5y^{10}z^{20} \cdot 2x^4y^2}$

$$= \sqrt[5]{32x^5y^{10}z^{20}} \ \sqrt[5]{2x^4y^2}$$

$$= 2xy^2z^4 \ \sqrt[5]{2x^4y^2}$$

101. $\sqrt[3]{4x^2y^8}\left(\sqrt[3]{2x^5y^4} + \sqrt[3]{6xy}\right)$

$$= \sqrt[3]{8x^7y^{12}} + \sqrt[3]{24x^3y^9}$$

$$= \sqrt[3]{8x^6y^{12} \cdot x} + \sqrt[3]{8x^3y^9 \cdot 3}$$

$$= \sqrt[3]{8x^6y^{12}} \ \sqrt[3]{x} + \sqrt[3]{8x^3y^9} \ \sqrt[3]{3}$$

$$= 2x^2y^4\sqrt[3]{x} + 2xy^3\sqrt[3]{3}$$

102. $\dfrac{x + \sqrt{y}}{x - \sqrt{y}} = \dfrac{x + \sqrt{y}}{x - \sqrt{y}} \cdot \dfrac{x + \sqrt{y}}{x + \sqrt{y}}$

$$= \frac{x^2 + x\sqrt{y} + x\sqrt{y} + \left(\sqrt{y}\right)^2}{x^2 - \left(\sqrt{y}\right)^2}$$

$$= \frac{x^2 + 2x\sqrt{y} + y}{x^2 - y}$$

103.
$$\sqrt{2x + 4} - 1 = \sqrt{x + 3}$$
$$\left(\sqrt{2x + 4} - 1\right)^2 = \left(\sqrt{x + 3}\right)^2$$
$$(2x + 4) - 2\sqrt{2x + 4} + 1 = x + 3$$
$$-2\sqrt{2x + 4} = -x - 2$$
$$\left(2\sqrt{2x + 4}\right)^2 = (x + 2)^2$$
$$4(2x + 4) = x^2 + 4x + 4$$
$$8x + 16 = x^2 + 4x + 4$$
$$0 = x^2 - 4x - 12$$
$$0 = (x - 6)(x + 2)$$

$$x - 6 = 0 \quad \text{or} \quad x + 2 = 0$$
$$x = 6 \qquad\qquad x = -2$$
$$\text{True} \qquad\qquad \text{False}$$

The only solution is $x = 6$; $x = -2$ is an extraneous solution.

Exercise Set 8.3

1. Answers will vary.

3. $A = s^2$, for s.
$$\sqrt{A} = s$$

5. $E = i^2 r$, for i
$$\frac{E}{r} = i^2$$
$$\sqrt{\frac{E}{r}} = i$$

7. $V = \pi r^2 h$, for r
$$\frac{V}{\pi h} = r^2$$
$$\sqrt{\frac{V}{\pi h}} = r$$

9. $a^2 + b^2 = c^2$, for b
$$b^2 = c^2 - a^2$$
$$b = \sqrt{c^2 - a^2}$$

11. $E = \dfrac{1}{2}mv^2$, for v

$$2E = mv^2$$

$$\frac{2E}{m} = v^2$$

$$\sqrt{\frac{2E}{m}} = v$$

13. $a = \dfrac{v_2^2 - v_1^2}{2d}$, for v_1

$$2ad = v_2^2 - v_1^2$$

$$2ad + v_1^2 = v_2^2$$

$$v_1^2 = v_2^2 - 2ad$$

$$v_1 = \sqrt{v_2^2 - 2ad}$$

15. $v' = \sqrt{c^2 - v^2}$, for c

$$(v')^2 = c^2 - v^2$$

$$(v')^2 + v^2 = c^2$$

$$c = \sqrt{(v')^2 + v^2}$$

17. Let x be Julie's rate going uphill so $x + 2$ is her rating going downhill. Using $\dfrac{d}{r} = t$ gives

$$t_{\text{uphill}} + t_{\text{downhill}} = 1.75$$

$$\frac{6}{x} + \frac{6}{x+2} = 1.75$$

$$x(x+2)\left(\frac{6}{x}\right) + x(x+2)\left(\frac{6}{x+2}\right) = x(x+2)(1.75)$$

$$6(x+2) + 6x = 1.75x(x+2)$$

$$6x + 12 + 6x = 1.75x^2 + 3.5x$$

$$0 = 1.75x^2 - 8.5x - 12$$

$$x = \frac{-(-8.5) \pm \sqrt{(-8.5)^2 - 4(1.75)(-12)}}{2(1.75)}$$

$$= \frac{8 \pm \sqrt{156.25}}{3.5}$$

$$= \frac{8.5 \pm 12.5}{3.5}$$

$$x = 6 \quad \text{or} \quad x \approx -1.14$$

Since the time must be positive, Julie's uphill rate is 6 mph and her downhill rate is
$x + 2 = 8$ mph.

19. Let r be the rate at which the present equipment drills.

	d	r	$t = \dfrac{d}{r}$
present equipment	64	r	$\dfrac{64}{r}$
new equipment	64	$r + 1$	$\dfrac{64}{r+1}$

They would have hit water in 3.2 hours less time with the new equipment.

$$\frac{64}{r+1} = \frac{64}{r} - 3.2$$

$$r(r+1)\left(\frac{64}{r+1}\right) = r(r+1)\left(\frac{64}{r}\right) - r(r+1)(3.2)$$

$$64r = 64(r+1) - 3.2r(r+1)$$

$$64r = 64r + 64 - 3.2r^2 - 3.2r$$

$$0 = 64 - 3.2r^2 - 3.2r$$

$$3.2r^2 + 3.2r - 64 = 0$$

$$r^2 + r - 20 = 0$$

$$(r+5)(r-4) = 0$$

$$r + 5 = 0 \quad \text{or} \quad r - 4 = 0$$

$$r = -5 \qquad\qquad r = 4$$

Use the positive value. The present equipment drills at a rate of 4 ft/hr.

21. Let r be the average speed of the car.

	d	r	$t = \dfrac{d}{r}$
car	500	r	$\dfrac{500}{r}$
train	800	$r + 20$	$\dfrac{800}{r+20}$

The train traveled $1\dfrac{2}{3}$ or $\dfrac{5}{3}$ hours longer.

$$\frac{800}{r+20} = \frac{500}{r} + \frac{5}{3}$$

$$3r(r+20)\left(\frac{800}{r+20}\right) = 3r(r+20)\left(\frac{500}{r}\right) + 3r(r+20)\left(\frac{5}{3}\right)$$

$$3r(800) = 3(r+20)(500) + 5r(r+20)$$

$$2400r = 1500r + 30,000 + 5r^2 + 100r$$

$$2400r = 5r^2 + 1600r + 30,000$$

$$0 = 5r^2 - 800r + 30,000$$

$$0 = r^2 - 160r + 6000$$

$$0 = (r-60)(r-100)$$

$$r - 60 = 0 \quad \text{or} \quad r - 100 = 0$$

$$r = 60 \qquad\qquad r = 100$$

A car traveling 100 mph is not reasonable. Thus the car's average rate was 60 mph and the train's average rate was 80 mph.

23. Let x be the time of the experienced mechanic then $x + 1$ is the time of the inexperienced mechanic.

$$\frac{6}{x} + \frac{6}{x+1} = 1$$

$$x(x+1)\left(\frac{6}{x}\right) + x(x+1)\left(\frac{6}{x+1}\right) = x(x+1)(1)$$

$$6(x+1) + 6x = x^2 + x$$

$$6x + 6 + 6x = x^2 + x$$

$$0 = x^2 - 11x - 6$$

$$x = \frac{-(-11) \pm \sqrt{(-11)^2 - 4(1)(-6)}}{2(1)}$$

$$= \frac{11 \pm \sqrt{121 + 24}}{2}$$

$$= \frac{11 \pm \sqrt{145}}{2}$$

$$x = \frac{11 + \sqrt{145}}{2} \quad \text{or} \quad x = \frac{11 - \sqrt{145}}{2}$$

$$\approx 11.52 \qquad\qquad \approx -0.52$$

Since the time must be positive, it takes Bonita about 11.52 hours and Pamela about 12.52 hours to rebuild the engine.

25. Let t be the number of hours for Chris to clean alone. Then $t + 0.5$ is the number of hours for John to clean alone.

	Rate of work	Time worked	Part of Task completed
Chris	$\dfrac{1}{t}$	6	$\dfrac{6}{t}$
John	$\dfrac{1}{t+0.5}$	6	$\dfrac{6}{t+0.5}$

$$\frac{6}{t} + \frac{6}{t+0.5} = 1$$

$$t(t+0.5)\left(\frac{6}{t}\right) + t(t+0.5)\left(\frac{6}{t+0.5}\right) = t(t+0.5)(1)$$

$$6(t+0.5) + 6t = t(t+0.5)$$

$$6t + 3 + 6t = t^2 + 0.5t$$

$$12t + 3 = t^2 + 0.5t$$

$$0 = t^2 - 11.5t - 3$$

$$t = \frac{11.5 \pm \sqrt{(-11.5)^2 - 4(1)(-3)}}{2(1)}$$

$$= \frac{11.5 \pm \sqrt{132.25 + 12}}{2}$$

$$= \frac{11.5 \pm \sqrt{144.25}}{2}$$

$$t = \frac{11.5 + \sqrt{144.25}}{2} \quad \text{or} \quad t = \frac{11.5 - \sqrt{144.25}}{2}$$

$$\approx 11.76 \qquad\qquad\qquad \approx -0.26$$

Since the time must be positive, it takes Chris about 11.76 hours and John about $11.76 + 0.5 = 12.26$ hours to clean alone.

27. Let x be the speed of the trip from Lubbock to Plainview. Then $x - 10$ is the speed from Plainview to Amarillo.

	d	r	t
first part	60	x	$\dfrac{60}{x}$
second part	100	$x - 10$	$\dfrac{100}{x-10}$

Including the 2.5 hours she spent in Plainview, the entire trip took Lisa 5.5 hours.

$$\frac{60}{x} + 2.5 + \frac{100}{x-10} = 5.5$$

$$\frac{60}{x} + \frac{100}{x-10} = 3$$

$$x(x-10)\left(\frac{60}{x}\right) + x(x-10)\left(\frac{100}{x-10}\right) = x(x-10)(3)$$

$$60(x-10) + 100x = 3(x^2 - 10x)$$

$$60x - 600 + 100x = 3x^2 - 30x$$

$$-600 + 160x = 3x^2 - 30x$$

$$0 = 3x^2 - 190x + 600$$

$$0 = (x-60)(3x-10)$$

$$x - 60 = 0 \quad \text{or} \quad 3x - 10 = 0$$
$$x = 60 \qquad\qquad x = \frac{10}{3}$$

$\frac{10}{3}$ miles per hour is too slow for a car, so the speed of the trip from Lubbock to Plainview was 60 mph.

29. Answers will vary.

31.
$$S = 2\pi rh + 2\pi r^2$$
$$S - 2\pi r^2 = 2\pi rh$$
$$\frac{S - 2\pi r^2}{2\pi r} = \frac{2\pi rh}{2\pi r}$$
$$\frac{S - 2\pi r^2}{2\pi r} = h$$

32. No. In a function each x-value must have a unique y-value. This is not the case with the points $(2, 3)$ and $(2, 1)$.

33. If a vertical line can be drawn to intersect the graph at more than one point, the graph is not a function.

34. $f(x) = \frac{3}{2}x^3 - \frac{1}{2}x^2 + 3$

$$f\left(\frac{1}{3}\right) = \frac{3}{2}\left(\frac{1}{3}\right)^3 - \frac{1}{2}\left(\frac{1}{3}\right)^2 + 3$$
$$= \frac{3}{2}\left(\frac{1}{27}\right) - \frac{1}{2}\left(\frac{1}{9}\right) + 3$$
$$= \frac{1}{18} - \frac{1}{18} + 3$$
$$= 3$$

35. The domain is the set of values that can be used for the independent variable.

36. The range is the set of values that are obtained for the dependent variable.

37. No. $f(x) = \frac{1}{x}$ is not a polynomial since a polynomial function must have whole number exponents on the variable and, in this case, $f(x) = \frac{1}{x} = x^{-1}$ and -1 is not a whole number.

Exercise Set 8.4

1. A given equation can be expressed as an equation in quadratic form if the equation can be written in the form $au^2 + bu + c = 0$.

3. $x^4 - 10x^2 + 9 = 0$

$\left(x^2\right)^2 - 10x^2 + 9 = 0$

$u^2 - 10u + 9 = 0 \quad \leftarrow \text{Replace } x^2 \text{ with } u$

$(u - 9)(u - 1) = 0$

$u - 9 = 0 \qquad\qquad \text{or} \quad u - 1 = 0$

$\quad u = 9 \qquad\qquad\qquad u = 1$

$\quad x^2 = 9 \qquad\qquad\qquad x^2 = 1 \quad \leftarrow \text{Replace } u \text{ with } x^2$

$\quad x = \pm\sqrt{9} = \pm 3 \qquad x = \pm\sqrt{1} = \pm 1$

The solutions are 3, –3, 1, and –1.

5. $x^4 - 7x^2 + 12 = 0$

$\left(x^2\right)^2 - 7x^2 + 12 = 0$

$u^2 - 7u + 12 = 0 \quad \leftarrow \text{Replace } x^2 \text{ with } u$

$(u - 4)(u - 3) = 0$

$u - 4 = 0 \qquad\qquad \text{or} \quad u - 3 = 0$

$\quad u = 4 \qquad\qquad\qquad u = 3$

$\quad x^2 = 4 \qquad\qquad\qquad x^2 = 3 \quad \leftarrow \text{Replace } u \text{ with } u^2$

$\quad x = \pm\sqrt{4} = \pm 2 \qquad x = \pm\sqrt{3}$

The solutions are 2, –2, $\sqrt{3}$, and $-\sqrt{3}$.

7. $r^4 - 8r^2 = -15$

$\quad r^4 - 8r^2 + 15 = 0$

$\left(r^2\right)^2 - 8r^2 + 15 = 0$

$u^2 - 8u + 15 = 0 \quad \leftarrow \text{Replace } r^2 \text{ with } u$

$(u - 3)(u - 5) = 0$

$u - 3 = 0 \qquad\qquad u - 5 = 0$

$\quad u = 3 \qquad\qquad\qquad u = 5$

$\quad r^2 = 3 \qquad\qquad\qquad r^2 = 5 \quad \leftarrow \text{Replace } u \text{ with } r^2$

$\quad r = \pm\sqrt{3} \qquad\qquad r = \pm\sqrt{5}$

The solutions are $\sqrt{3},\ -\sqrt{3},\ \sqrt{5},$ and $-\sqrt{5}$.

9. $x^4 + 2x^2 = 8$

$\quad x^4 + 2x^2 - 8 = 0$

$\left(x^2\right)^2 + 2x^2 - 8 = 0$

$u^2 + 2u - 8 = 0 \quad \leftarrow \text{Replace } x^2 \text{ with } u$

$(u - 2)(u + 4) = 0$

$$u - 2 = 0 \quad \text{or} \quad u + 4 = 0$$
$$u = 2 \qquad\qquad u = -4$$
$$x^2 = 2 \qquad\qquad x^2 = -4 \leftarrow \text{Replace } u \text{ with } x^2$$
$$x = \pm\sqrt{2} \qquad\qquad x = \pm\sqrt{-4} = \pm 2i$$

The solutions are $\sqrt{2}, -\sqrt{2}, 2i,$ and $-2i$.

11. $x + 4\sqrt{x} - 12 = 0$

$$\left(x^{1/2}\right)^2 + 4x^{1/2} - 12 = 0$$
$$u^2 + 4u - 12 = 0 \leftarrow \text{Replace } x^{1/2} \text{ with } u$$
$$(u - 2)(u + 6) = 0$$
$$u - 2 = 0 \quad \text{or} \quad u + 6 = 0$$
$$u = 2 \quad \text{or} \qquad u = -6$$
$$x^{1/2} = 2 \quad \text{or} \quad x^{1/2} = -6 \leftarrow \text{Replace } u \text{ with } x^{1/2}$$
$$x = 2^2 = 4$$

$x^{1/2} = -6$ has no solution since there is no value of x for which $x^{1/2} = -6$.
The solution is 4.

13. $x + \sqrt{x} = 6$

$$x + \sqrt{x} - 6 = 0$$
$$\left(x^{1/2}\right)^2 + x^{1/2} - 6 = 0$$
$$u^2 + u - 6 = 0 \leftarrow \text{Replace } x^{1/2} \text{ with } u$$
$$(u - 2)(u + 3) = 0$$
$$u - 2 = 0 \quad \text{or} \quad u + 3 = 0$$
$$u = 2 \qquad\qquad u = -3$$
$$x^{1/2} = 2 \qquad x^{1/2} = -3 \leftarrow \text{Replace } u \text{ with } x^{1/2}$$
$$x = 4$$

$x^{1/2} = -3$ has no solution since there is no value of x for which $x^{1/2} = -3$. The solution is 4.

15. $\qquad\qquad 9x + 3\sqrt{x} = 2$

$$9x + 3\sqrt{x} - 2 = 0$$
$$9\left(x^{1/2}\right)^2 + 3x^{1/2} - 2 = 0$$
$$9u^2 + 3u - 2 = 0 \leftarrow \text{Replace } x^{1/2} \text{ with } u$$
$$(3u - 1)(3u + 2) = 0$$

$$3u - 1 = 0 \quad \text{or} \quad 3u + 2 = 0$$

$$u = \frac{1}{3} \qquad\qquad u = -\frac{2}{3}$$

$$x^{1/2} = \frac{1}{3} \qquad\qquad x^{1/2} = -\frac{2}{3} \quad \leftarrow \text{Replace } u \text{ with } x^{1/2}$$

$$x = \frac{1}{9}$$

$x^{1/2} = -\frac{2}{3}$ has no solution since there is no value of x for which $x^{1/2} = -\frac{2}{3}$.

The solution is $\frac{1}{9}$.

17.
$$(x+3)^2 + 2(x+3) = 24$$

$$(x+3)^2 + 2(x+3) - 24 = 0$$

$$u^2 + 2u - 24 = 0 \quad \leftarrow \text{Replace } x+3 \text{ with } u$$

$$(u-4)(u+6) = 0$$

$$u - 4 = 0 \quad \text{or} \quad u + 6 = 0$$

$$u = 4 \qquad\qquad u = -6$$

$$x + 3 = 4 \qquad x + 3 = -6 \quad \leftarrow \text{Replace } u \text{ with } x+3$$

$$x = 1 \qquad\qquad x = -9$$

The solutions are 1 and –9.

19.
$$6(x-2)^2 = -19(x-2) - 10$$

$$6(x-2)^2 + 19(x-2) + 10 = 0$$

$$6u^2 + 19u + 10 = 0 \quad \leftarrow \text{Replace } x-2 \text{ with } u$$

$$(3u + 2)(2u + 5) = 0$$

$$3u + 2 = 0 \quad \text{or} \quad 2u + 5 = 0$$

$$u = -\frac{2}{3} \qquad\qquad u = -\frac{5}{2}$$

$$x - 2 = -\frac{2}{3} \qquad x - 2 = -\frac{5}{2} \quad \leftarrow \text{Replace } u \text{ with } x-2$$

$$x = \frac{4}{3} \qquad\qquad x = -\frac{1}{2}$$

The solutions are $\frac{4}{3}$ and $-\frac{1}{2}$.

21. $(x^2-1)^2-(x^2-1)-6=0$

$\qquad u^2-u-6=0 \quad \leftarrow$ Replace x^2-1 with u

$\qquad (u+2)(u-3)=0$

$u+2=0 \qquad$ or $\quad u-3=0$

$\quad u=-2 \qquad\qquad\quad u=3$

$x^2-1=-2 \qquad\qquad x^2-1=3 \quad \leftarrow$ Replace u with x^2-1

$\quad x^2=-1 \qquad\qquad\quad x^2=4$

$\quad x=\sqrt{-1}=\pm i \qquad\quad x=\pm\sqrt{4}=\pm 2$

The solutions are i, $-i$, 2, and -2.

23. $18(x^2-5)^2+27(x^2-5)+10=0$

$\qquad 18u^2+27u+10=0 \quad \leftarrow$ Replace x^2-5 with u

$\qquad (3u+2)(6u+5)=0$

$3u+2=0 \qquad\qquad$ or $\quad 6u+5=0$

$\quad u=-\dfrac{2}{3} \qquad\qquad\qquad u=-\dfrac{5}{6}$

$x^2-5=-\dfrac{2}{3} \qquad\qquad x^2-5=-\dfrac{5}{6} \leftarrow$ Replace u with x^2-5

$\quad x^2=\dfrac{13}{3} \qquad\qquad\quad x^2=\dfrac{25}{6}$

$\quad x=\pm\sqrt{\dfrac{13}{3}} \qquad\qquad x=\pm\sqrt{\dfrac{25}{6}}$

$\qquad =\pm\dfrac{\sqrt{13}}{\sqrt{3}}\cdot\dfrac{\sqrt{3}}{\sqrt{3}} \qquad\qquad =\pm\dfrac{5}{\sqrt{6}}\cdot\dfrac{\sqrt{6}}{\sqrt{6}}$

$\qquad =\pm\dfrac{\sqrt{39}}{3} \qquad\qquad\qquad =\pm\dfrac{5\sqrt{6}}{6}$

The solutions are $\dfrac{\sqrt{39}}{3}$, $-\dfrac{\sqrt{39}}{3}$, $\dfrac{5\sqrt{6}}{6}$, and $-\dfrac{5\sqrt{6}}{6}$.

25. $r^{-2}+6r^{-1}+9=0$

$\quad \dfrac{1}{r^2}+\dfrac{6}{r}+9=0$

$\quad r^2\left(\dfrac{1}{r^2}+\dfrac{6}{r}+9\right)=r^2(0)$

$\quad 1+6r+9r^2=0$

$\quad (3r+1)^2=0$

$\quad 3r+1=0$

$\quad 3r=-1$

$\quad r=-\dfrac{1}{3}$

The solution is $-\dfrac{1}{3}$.

27. $\qquad 2b^{-2}=7b^{-1}-3$

$\quad 2b^{-2}-7b^{-1}+3=0$

$\qquad \dfrac{2}{b^2}-\dfrac{7}{b}+3=0$

$\quad b^2\left(\dfrac{2}{b^2}-\dfrac{7}{b}+3\right)=b^2(0)$

$\qquad 2-7b+3b^2=0$

$\quad (3b-1)(b-2)=0$

$3b - 1 = 0$ or $b - 2 = 0$

$\qquad 3b = 1 \qquad\qquad b = 2$

$\qquad b = \dfrac{1}{3}$

The solutions are $\dfrac{1}{3}$ and 2.

29. $x^{-2} + 6x^{-1} - 16 = 0$

$\qquad \dfrac{1}{x^2} + \dfrac{6}{x} - 16 = 0$

$\qquad x^2 \left(\dfrac{1}{x^2} + \dfrac{6}{x} - 16 \right) = x^2(0)$

$\qquad 1 + 6x - 16x^2 = 0$

$\qquad -16x^2 + 6x + 1 = 0$

$\qquad 16x^2 - 6x - 1 = 0$

$\qquad (8x + 1)(2x - 1) = 0$

$\quad 8x + 1 = 0 \qquad$ or $\quad 2x - 1 = 0$

$\qquad 8x = -1 \qquad\qquad 2x = 1$

$\qquad x = -\dfrac{1}{8} \qquad\qquad x = \dfrac{1}{2}$

The solutions are $-\dfrac{1}{8}$ and $\dfrac{1}{2}$.

31. $x^{2/3} - 5x^{1/3} + 6 = 0$

$\quad \left(x^{1/3}\right)^2 - 5x^{1/3} + 6 = 0$

$\qquad u^2 - 5u + 6 = 0 \;\leftarrow$ Replace $x^{1/3}$ with u

$\qquad (u - 3)(u - 2) = 0$

$\quad u - 3 = 0 \quad$ or $\quad u - 2 = 0$

$\qquad u = 3 \qquad\qquad u = 2$

$\qquad x^{1/3} = 3 \qquad\quad x^{1/3} = 2 \;\leftarrow$ Replace u with $x^{1/3}$

$\qquad x = 3^3 \qquad\qquad x = 2^3$

$\qquad = 27 \qquad\qquad = 8$

The solutions are 27 and 8.

33. $c^{2/5} - 9c^{1/5} = -18$

$\quad c^{2/5} - 9c^{1/5} + 18 = 0$

$\quad \left(c^{1/5}\right)^2 - 9c^{1/5} + 18 = 0$

$\qquad u^2 - 9u + 18 = 0 \;\leftarrow$ Replace $c^{1/5}$ with u

$\qquad (u - 3)(u - 6) = 0$

$\quad u - 3 = 0 \qquad\quad u - 6 = 0$

$\qquad u = 3 \qquad\qquad u = 6$

$\qquad c^{1/5} = 3 \qquad\quad c^{1/5} = 6 \;\leftarrow$ Replace u with $c^{1/5}$

$\qquad c = 3^5 \qquad\qquad c = 6^5$

$\qquad = 243 \qquad\qquad = 7776$

The solutions are 243 and 7776.

35. $-2a - 5a^{1/2} + 3 = 0$

$-2\left(a^{1/2}\right)^2 - 5a^{1/2} + 3 = 0$

$-2u^2 - 5u + 3 = 0 \quad \leftarrow$ Replace $a^{1/2}$ with u

$2u^2 + 5u - 3 = 0$

$(2u - 1)(u + 3) = 0$

$2u - 1 = 0 \qquad$ or $\quad u + 3 = 0$

$u = \dfrac{1}{2} \qquad\qquad\quad u = -3$

$a^{1/2} = 2 \qquad\qquad a^{1/2} = -3 \quad \leftarrow$ Replace u with $a^{1/2}$

$a = \left(\dfrac{1}{2}\right)^2$

$ = \dfrac{1}{4}$

$a^{1/2} = -3$ has no solution since there is no value of a for which $a^{\frac{1}{2}} = -3$.

The solution is $\dfrac{1}{4}$.

37. $f(x) = x - 5\sqrt{x} + 4$, $f(x) = 0$

$0 = \left(x^{1/2}\right)^2 - 5x^{1/2} + 4$

$0 = u^2 - 5u + 4 \quad \leftarrow$ Replace $x^{1/2}$ with u

$0 = (u - 1)(u - 4)$

$u - 1 = 0 \quad$ or $\quad u - 4 = 0$

$u = 1 \qquad\qquad u = 4$

$x^{1/2} = 1 \qquad\quad x^{1/2} = 4 \quad \leftarrow$ Replace u with $x^{1/2}$

$x = 1 \qquad\qquad x = 16$

The x-intercepts are $(1, 0)$ and $(16, 0)$.

39. $g(x) = 4x^{-2} - 19x^{-1} - 5$, $g(x) = 0$

$0 = 4\left(x^{-1}\right)^2 - 19x^{-1} - 5$

$0 = 4u^2 - 19u - 5 \quad \leftarrow$ Replace x^{-1} with u

$0 = (4u + 1)(u - 5)$

$4u + 1 = 0 \qquad$ or $\quad u - 5 = 0$

$u = -\dfrac{1}{4} \qquad\qquad u = 5$

$x^{-1} = -\dfrac{1}{4} \qquad\quad x^{-1} = 5 \quad \leftarrow$ Replace u with x^{-1}

$x = -4 \qquad\qquad x = \dfrac{1}{5}$

The x-intercepts are $(-4, 0)$ and $\left(\dfrac{1}{5}, 0\right)$.

41. $f(x) = x^{2/3} + x^{1/3} - 6$, $f(x) = 0$

$0 = \left(x^{1/3}\right)^2 + x^{1/3} - 6$

$0 = u^2 + u - 6 \;\leftarrow$ Replace $x^{1/3}$ with u

$0 = (u+3)(u-2)$

$u + 3 = 0 \quad$ or $\quad u - 2 = 0$

$\quad u = -3 \qquad\qquad u = 2$

$x^{1/3} = -3 \qquad\quad x^{1/3} = 2 \;\leftarrow$ Replace u with $x^{1/3}$

$\quad x = -27 \qquad\qquad x = 8$

The x-intercepts are $(-27, 0)$ and $(8, 0)$.

43. $g(x) = \left(x^2 - 3x\right)^2 + 2\left(x^2 - 3x\right) - 24$, $g(x) = 0$

$0 = u^2 + 2u - 24 \;\leftarrow$ Replace $x^2 - 3x$ with u

$0 = (u+6)(u-4)$

$\quad\quad u + 6 = 0 \quad$ or $\qquad\quad u - 4 = 0$

$x^2 - 3x + 6 = 0 \qquad\quad x^2 - 3x - 4 = 0 \;\leftarrow$ Replace u with $x^2 - 3x$

$\qquad\qquad\qquad\qquad (x-4)(x+1) = 0$

$\qquad\qquad\qquad\qquad\quad x - 4 = 0 \text{ or } x + 1 = 0$

$\qquad\qquad\qquad\qquad\qquad x = 4 \qquad\quad x = -1$

There are no x-intercepts for $x^2 - 3x + 6 = 0$ since $b^2 - 4ac = (-3)^2 - 4(1)(6) = 9 - 24 = -15$. The x-intercepts are $(4, 0)$ or $(-1, 0)$.

45. When solving an equation of the form $ax^4 + bx^2 + c = 0$, let $u = x^2$.

47. If the solutions are ± 2 and ± 4, the factors must be $(x-2)$, $(x+2)$, $(x-4)$ and $(x+4)$.

$0 = (x-2)(x+2)(x-4)(x+4)$

$0 = \left(x^2 - 4\right)\left(x^2 - 16\right)$

$0 = x^4 - 20x^2 + 64$

49. No. An equation of the form $ax^4 + bx^2 + c = 0$ can have no real solutions, two real solutions, or four real solutions.

51. a. $\quad \dfrac{3}{x^2} - \dfrac{3}{x} = 60$ The LCD is x^2

$x^2\left(\dfrac{3}{x^2}\right) - x^2\left(\dfrac{3}{x}\right) = x^2(60)$

$3 - 3x = 60x^2$

$0 = 60x^2 + 3x - 3$

$0 = 3\left(20x^2 + x - 1\right)$

$0 = 3(5x - 1)(4x + 1)$

$5x - 1 = 0 \quad$ or $\quad 4x + 1 = 0$

$\quad x = \dfrac{1}{5} \qquad\qquad\quad x = -\dfrac{1}{4}$

The solutions are $\dfrac{1}{5}$ and $-\dfrac{1}{4}$.

b.
$$\frac{3}{x^2} - \frac{3}{x} = 60$$
$$3x^{-2} - 3x^{-1} = 60$$
$$3\left(x^{-1}\right)^2 - 3x^{-1} - 60 = 0$$
$$3u^2 - 3u - 60 = 0 \quad \leftarrow \text{ Replace } x^{-1} \text{ with } u$$
$$3\left(u^2 - u - 20\right) = 0$$
$$3(u - 5)(u + 4) = 0$$

$$u - 5 = 0 \quad \text{or} \quad u + 4 = 0$$
$$u = 5 \qquad\qquad u = -4$$
$$x^{-1} = 5 \qquad\quad x^{-1} = -4 \quad \leftarrow \text{ Replace } u \text{ with } x^{-1}$$
$$x = \frac{1}{5} \qquad\qquad x = -\frac{1}{4}$$

The solutions are $\dfrac{1}{5}$ and $-\dfrac{1}{4}$.

53.
$$15(r + 2) + 22 = -\frac{8}{r + 2}$$
$$15(r + 2)(r + 2) + 22(r + 2) = -\frac{8}{r + 2}(r + 2)$$
$$15(r + 2)^2 + 22(r + 2) = -8$$
$$15(r + 2)^2 + 22(r + 2) + 8 = 0$$
$$15u^2 + 22u + 8 = 0 \quad \leftarrow \text{ Replace } r + 2 \text{ with } u$$
$$(5u + 4)(3u + 2) = 0$$

$$5u + 4 = 0 \qquad \text{or} \quad 3u + 2 = 0$$
$$u = -\frac{4}{5} \qquad\qquad u = -\frac{2}{3}$$
$$r + 2 = -\frac{4}{5} \qquad\quad r + 2 = -\frac{2}{3} \quad \leftarrow \text{ Replace } u \text{ with } r + 2$$
$$r = -\frac{14}{5} \qquad\qquad r = -\frac{8}{3}$$

The solutions are $-\dfrac{14}{5}$ and $-\dfrac{8}{3}$.

55. $4 - (x-2)^{-1} = 3(x-2)^{-2}$

$\qquad 4 - u^{-1} = 3u^{-2} \quad \leftarrow$ Replace $x-2$ by u

$\qquad 4 - \dfrac{1}{u} = \dfrac{3}{u^2}$

$\qquad u^2\left(4 - \dfrac{1}{u}\right) = u^2\left(\dfrac{3}{u^2}\right)$

$\qquad 4u^2 - u = 3$

$\quad 4u^2 - u - 3 = 0$

$\quad (4u+3)(u-1) = 0$

$\quad 4u+3 = 0 \qquad \text{or} \quad u-1 = 0$

$\qquad u = -\dfrac{3}{4} \qquad\qquad u = 1$

$\quad x-2 = -\dfrac{3}{4} \qquad x-2 = 1 \quad \leftarrow$ Replace u by $x-2$

$\qquad x = \dfrac{5}{4} \qquad\qquad x = 3$

The solutions are $\dfrac{5}{4}$ and 3.

57. $\qquad x^6 - 9x^3 + 8 = 0$

$\qquad \left(x^3\right)^2 - 9x^3 + 8 = 0$

$\qquad\quad u^2 - 9u + 8 = 0 \quad \leftarrow$ Replace x^3 with u

$\qquad\quad (u-8)(u-1) = 0$

$\quad u-8 = 0 \quad \text{or} \quad u-1 = 0$

$\qquad u = 8 \qquad\qquad u = 1$

$\qquad x^3 = 8 \qquad\qquad x^3 = 1 \quad \leftarrow$ Replace u with x^3

$\qquad x = 2 \qquad\qquad x = 1$

The solutions are 2 and 1.

59. $\left(x^2 + 2x - 2\right)^2 - 7\left(x^2 + 2x - 2\right) + 6 = 0$

$\qquad\qquad\qquad u^2 - 7u + 6 = 0 \quad \leftarrow$ Replace $x^2 + 2x - 2$ with u

$\qquad\qquad\qquad (u-6)(u-1) = 0$

$\qquad\qquad\qquad u-6 = 0 \qquad \text{or} \qquad\qquad u-1 = 0$

$\qquad\qquad\qquad\quad u = 6 \qquad\qquad\qquad\qquad u = 1$

$\qquad\qquad x^2 + 2x - 2 = 6 \qquad\qquad x^2 + 2x - 2 = 1 \quad \leftarrow$ Replace u with $x^2 + 2x - 2$

$\qquad\qquad x^2 + 2x - 8 = 0 \qquad\qquad x^2 + 2x - 3 = 0$

$\qquad\qquad (x+4)(x-2) = 0 \qquad\qquad (x+3)(x-1) = 0$

$x+4 = 0 \quad \text{or} \quad x-2 = 0 \qquad x+3 = 0 \quad \text{or} \quad x-1 = 0$

$\quad x = -4 \qquad\quad x = 2 \qquad\qquad x = -3 \qquad\quad x = 1$

The solutions are –4, 2, –3, and 1.

61. $2n^4 - 6n^2 - 3 = 0$

$\quad 2\left(n^2\right)^2 - 6n^2 - 3 = 0$

$\qquad 2u^2 - 6u - 3 = 0 \leftarrow$ Replace n^2 with u

$\quad u = \dfrac{6 \pm \sqrt{(-6)^2 - 4(2)(-3)}}{2(2)}$

$\qquad = \dfrac{6 \pm \sqrt{60}}{4}$

$\qquad = \dfrac{6 \pm 2\sqrt{15}}{4}$

$\qquad = \dfrac{3 \pm \sqrt{15}}{2}$

$\quad n^2 = \dfrac{3 \pm \sqrt{15}}{2} \leftarrow$ Replace u with n^2

$\quad n = \pm\sqrt{\dfrac{3 \pm \sqrt{15}}{2}}$

63. $P_1 = \dfrac{T_1 P_2}{T_2}$ for T_2

$\quad P_1 T_2 = T_1 P_2$

$\quad T_2 = \dfrac{T_1 P_2}{P_1}$

64. $a + 2b + 2c = 1$ (1)

$\quad 2a - b + c = 3$ (2)

$\quad 4a + b + 2c = 0$ (3)

To eliminate b between equations (1) and (2), multiply equation (2) by 2 and add.

$\quad a + 2b + 2c = 1$

$\quad 2[2a - b + c = 3]$

gives

$\qquad a + 2b + 2c = 1$

$\quad \underline{4a - 2b + 2c = 6}$

Add: $5a \qquad + 4c = 7$ (4)

To eliminate b between equations (2) and (3), simply add.

$\quad 2a - b + \; c = 3$

$\quad \underline{4a + b + 2c = 0}$

Add: $\quad 6a + 3c = 3$ (5)

Equations (4) and (5) are a system of two equations in two unknowns.

$5a + 4c = 7$

$6a + 3c = 3$

To eliminate c between equations (4) and (5), multiply equation (4) by -3 and equation (5) by 4 and add.

$\quad -3[5a + 4c = 7]$

$\quad \; 4[6a + 3c = 3]$

gives

$\qquad -15a - 12c = -21$

$\qquad \underline{24a + 12c = \;\; 12}$

Add: $\;\; 9a \qquad = -9$

$\qquad a \qquad = -1$

Substitute -1 for a in equation (4).

$\qquad 5a + 4c = 7$

$\quad 5(-1) + 4c = 7$

$\qquad \quad 4c = 12$

$\qquad \quad \; c = 3$

Substitute -1 for a and 3 for c in equation (1).

$\qquad a + 2b + 2c = 1$

$\quad -1 + 2b + 2(3) = 1$

$\qquad \quad 2b + 5 = 1$

$\qquad \qquad 2b = -4$

$\qquad \qquad \; b = -2$

The solution is $(-1, -2, 3)$.

65. a.   1: $a \cdot a = a^2$

$\qquad$ 2: $a \cdot b = ab$

$\qquad$ 3: $b \cdot a = ab$

$\qquad$ 4: $b \cdot b = b^2$

b. $a^2 + ab + ab + b^2 = a^2 + 2ab + b^2$

$\qquad\qquad\qquad\qquad\; = (a + b)^2$

66. $\dfrac{6x^2 + 7x - 20}{4x^2 - 25}$

$\quad = \dfrac{(3x - 4)(2x + 5)}{(2x - 5)(2x + 5)}$

$\quad = \dfrac{3x - 4}{2x - 5}$

Exercise Set 8.5

1. The graph of a quadratic equation is called a parabola.

3. The axis of symmetry of a parabola is the line where, if the graph is folded, the two sides overlap.

5. For $f(x) = ax^2 + bx + c$, the vertex of the graph is $\left(-\dfrac{b}{2a}, \dfrac{4ac - b^2}{4} \right)$.

7. a. For $f(x) = ax^2 + bx + c$, $f(x)$ will have a minimum if $a > 0$ since the graph opens upward.

   b. For $f(x) = ax^2 + bx + c$, $f(x)$ will have a maximum if $a < 0$ since the graph opens downward.

9. To find the *y*-intercepts of the graph of a quadratic function, set $x = 0$ and solve for *y*.

11. a. For $f(x) = ax^2$, the general shape of $f(x)$ if $a > 0$ is

   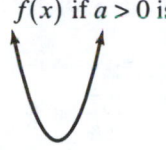

   b. For $f(x) = ax^2$, the general shape of $f(x)$ if $a < 0$ is

13. a. Since $a = 1$, the parabola opens upward.

   b. $y = 0^2 + 8(0) + 15 = 15$
   The *y*-intercept is (0, 15).

c. $x = -\dfrac{b}{2a} = -\dfrac{8}{2(1)} = -\dfrac{8}{2} = -4$

   $y = \dfrac{4ac - b^2}{4a}$

   $= \dfrac{4(1)(15) - 8^2}{4(1)}$

   $= \dfrac{60 - 64}{4}$

   $= \dfrac{-4}{4}$

   $= -1$

   The vertex is (–4, –1).

d. $0 = x^2 + 8x + 15 = (x + 5)(x + 3)$
   $x + 5 = 0 \quad$ or $\quad x + 3 = 0$
   $\quad x = -5 \qquad\qquad x = -3$
   The *x*-intercepts are (–5, 0) and (–3, 0).

e. $f(x) = x^2 + 8x + 15$

   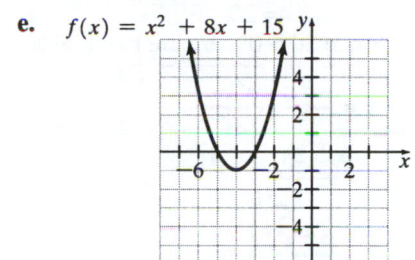

15. $f(x) = x^2 - 6x + 4$

   a. Since $a = 1$ the parabola opens upward.

   b. $y = 0^2 - 6(0) + 4 = 4$
   The *y*-intercept is (0, 4).

c. $x = -\dfrac{b}{2a} = -\dfrac{-6}{2(1)} = \dfrac{6}{2} = 3$

$$y = \dfrac{4ac - b^2}{4a}$$

$$= \dfrac{4(1)(4) - (-6)^2}{4(1)}$$

$$= \dfrac{16 - 36}{4}$$

$$= \dfrac{-20}{4}$$

$$= -5$$

The vertex is (3, –5).

d. $0 = x^2 - 6x + 4$

Since this is not factorable check the discriminant.

$$b^2 - 4ac = (-6)^2 - 4(1)(4)$$

$$= 36 - 16$$

$$= 20$$

Since 20 > 0 there are two real roots.

$$x = \dfrac{-b \pm \sqrt{b^2 - 4ac}}{2a}$$

$$= \dfrac{-(-6) \pm \sqrt{20}}{2(1)}$$

$$= \dfrac{6 \pm 2\sqrt{5}}{2}$$

$$= 3 \pm \sqrt{5}$$

The *x*-intercepts are

$(3 + \sqrt{5},\ 0)$ and $(3 - \sqrt{5},\ 0)$.

e.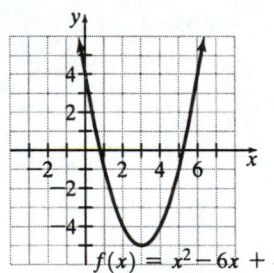

17. $f(x) = x^2 + 6x + 9$

a. Since *a* = 1 the parabola opens upward.

b. $y = 0^2 + 6(0) + 9 = 9$

The *y*-intercept is (0, 9).

c. $x = -\dfrac{b}{2a} = -\dfrac{6}{2(1)} = -\dfrac{6}{2} = -3$

$$y = \dfrac{4ac - b^2}{4a}$$

$$= \dfrac{4(1)(9) - 6^2}{4(1)}$$

$$= \dfrac{36 - 36}{0}$$

$$= 0$$

The vertex is (–3, 0).

d. $0 = x^2 + 6x + 9$

$$= (x + 3)(x + 3)$$

$$= (x + 3)^2$$

$$x + 3 = 0$$

$$x = -3$$

The *x*-intercept is (–3, 0).

e. $f(x) = x^2 + 6x + 9$

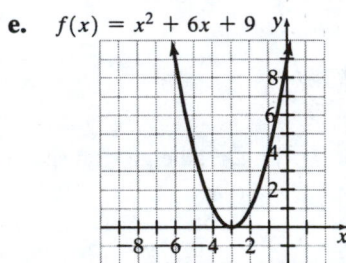

19. $y = 2x^2 - x - 6$

a. Since *a* = 2 the parabola opens upward.

b. $y = 2(0)^2 - 0 - 6 = -6$

The *y*-intercept is (0, –6).

c. $x = -\dfrac{b}{2a} = -\dfrac{-1}{2(2)} = \dfrac{1}{4}$

$y = \dfrac{4ac - b^2}{4a}$

$= \dfrac{4(2)(-6) - (-1)^2}{4(2)}$

$= \dfrac{-48 - 1}{8}$

$= -\dfrac{49}{8}$

The vertex is $\left(\dfrac{1}{4}, -\dfrac{49}{8}\right)$.

d. $0 = 2x^2 - x - 6 = (2x + 3)(x - 2)$

$2x + 3 = 0 \quad \text{or} \quad x - 2 = 0$

$x = -\dfrac{3}{2} \qquad\qquad x = 2$

The *x*-intercepts are $\left(-\dfrac{3}{2}, 0\right)$ and $(2, 0)$.

e.

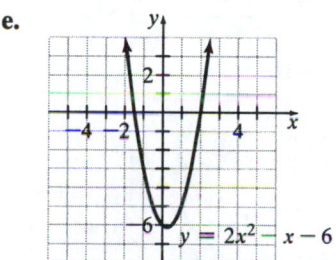

21. $y = 3x^2 + 4x + 3$

a. Since $a = 3$ the parabola opens upward.

b. $y = 3(0) + 4(0) + 3 = 3$
The *y*-intercept is $(0, 3)$.

c. $x = -\dfrac{b}{2a} = -\dfrac{4}{2(3)} = -\dfrac{4}{6} = -\dfrac{2}{3}$

$y = \dfrac{4ac - b^2}{4a}$

$= \dfrac{4(3)(3) - 4^2}{4(3)}$

$= \dfrac{36 - 16}{12}$

$= \dfrac{20}{12}$

$= \dfrac{5}{3}$

The vertex is $\left(-\dfrac{2}{3}, \dfrac{5}{3}\right)$.

d. $0 = 3x^2 + 4x + 3$
Since this is not factorable, check the discriminant.

$b^2 - 4ac = 4^2 - 4(3)(3)$

$= 16 - 36$

$= -20$

Since $-20 < 0$ there are no real roots.
Thus, there are no *x*-intercepts.

e.

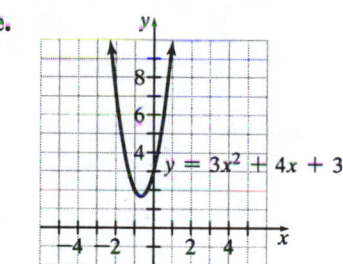

23. $f(x) = -2x^2 - 6x + 4$

a. Since $a = -2$ the parabola opens downward.

b. $y = -2(0)^2 - 6(0) + 4 = 4$
The *y*-intercept is $(0, 4)$.

c. $x = -\dfrac{b}{2a} = -\dfrac{(-6)}{2(-2)} = \dfrac{6}{-4} = -\dfrac{3}{2}$

$y = \dfrac{4ac - b^2}{4a}$

$= \dfrac{4(-2)(4) - (-6)^2}{4(-2)}$

$= \dfrac{-32 - 36}{-8}$

$= \dfrac{-68}{-8}$

$= \dfrac{17}{2}$

The vertex is $\left(-\dfrac{3}{2}, \dfrac{17}{2}\right)$.

d. $0 = -2x^2 - 6x + 4 = -2(x^2 + 3x - 2)$
Since this is not factorable, check the discriminant.

$b^2 - 4ac = (-6)^2 - 4(-2)(4)$

$\qquad\qquad = 36 + 32$

$\qquad\qquad = 68$

There are two real roots.

$x = \dfrac{-b \pm \sqrt{b^2 - 4ac}}{2a}$

$= \dfrac{-(-6) \pm \sqrt{68}}{2(-2)}$

$= \dfrac{6 \pm 2\sqrt{17}}{-4}$

$= \dfrac{3 \pm \sqrt{17}}{-2}$

$= \dfrac{-3 \pm \sqrt{17}}{2}$

The x-intercepts are

$\left(\dfrac{-3 + \sqrt{17}}{2},\ 0\right)$ and $\left(\dfrac{-3 - \sqrt{17}}{2}, 0\right)$.

e.

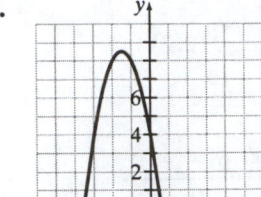

$f(x) = -2x^2 - 6x + 4$

25. $y = x^2 + 4x$

a. Since $a = 1$ the parabola opens upward.

b. $y = 0^2 + 4(0) = 0$
The y-intercept is $(0, 0)$.

c. $x = -\dfrac{b}{2a} = -\dfrac{4}{2(1)} = -\dfrac{4}{2} = -2$

$y = \dfrac{4ac - b^2}{4a}$

$= \dfrac{4(1)(0) - 4^2}{4(1)}$

$= \dfrac{-16}{4}$

$= -4$

The vertex is $(-2, -4)$.

d. $0 = x^2 + 4x = x(x + 4)$

$x = 0$ or $x + 4 = 0$

$x = 0$ $\qquad\qquad x = -4$

The x-intercepts are $(0, 0)$ and $(-4, 0)$.

e.

$y = x^2 + 4x$

27. $f(x) = 3x^2 + 10x$

a. Since $a = 3$ the parabola opens upward.

b. $y = 3(0)^2 + 10(0) = 0$
The y-intercept is $(0, 0)$.

c. $x = -\dfrac{b}{2a} = -\dfrac{10}{2(3)} = -\dfrac{10}{6} = -\dfrac{5}{3}$

$y = \dfrac{4ac - b^2}{4a}$

$= \dfrac{4(3)(0) - 10^2}{4(3)}$

$= \dfrac{-100}{12}$

$= -\dfrac{25}{3}$

The vertex is $\left(-\dfrac{5}{3}, -\dfrac{25}{3}\right)$.

d. $0 = 3x^2 + 10x = x(3x + 10)$
$x = 0$ or $3x + 10 = 0$
$x = 0$ $\qquad$ $x = -\dfrac{10}{3}$
The x-intercepts are $(0, 0)$ and
$\left(-\dfrac{10}{3}, 0\right)$.

e.
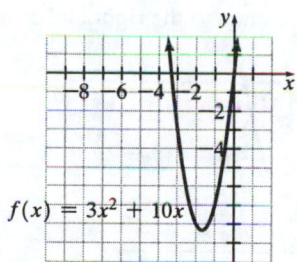
$f(x) = 3x^2 + 10x$

29. $f(x) = -x^2 + 3x - 5$

a. Since $a = -1$ the parabola opens downward.

b. $y = -0^2 + 3(0) - 5 = -5$
The y-intercept is $(0, -5)$.

c. $x = -\dfrac{b}{2a} = -\dfrac{3}{2(-1)} = -\dfrac{3}{-2} = \dfrac{3}{2}$

$y = \dfrac{4ac - b^2}{4a} = \dfrac{4(-1)(-5) - 3^2}{4(-1)} = -\dfrac{11}{4}$

The vertex is $\left(\dfrac{3}{2}, -\dfrac{11}{4}\right)$.

d. $0 = -x^2 + 3x - 5$
Since this is not factorable, check the discriminant.
$b^2 - 4ac = 3^2 - 4(-1)(-5)$
$\qquad\qquad = 9 - 20$
$\qquad\qquad = -11$
Since $-11 < 0$ there are no real roots. Thus, there are no x-intercepts.

e.
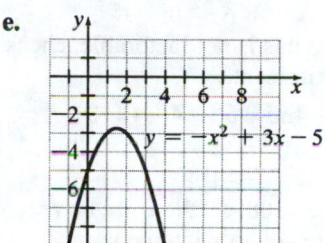
$y = -x^2 + 3x - 5$

31. $f(x) = -4x^2 + 6x - 9$

a. Since $a = -4$ the parabola opens downward.

b. $y = -4(0)^2 + 6(0) - 9 = -9$
The y-intercept is $(0, -9)$.

c. $x = -\dfrac{b}{2a} = -\dfrac{6}{2(-4)} = -\dfrac{6}{-8} = \dfrac{3}{4}$

$y = \dfrac{4ac - b^2}{4a}$

$= \dfrac{4(-4)(-9) - 6^2}{4(-4)}$

$= \dfrac{144 - 36}{-16}$

$= \dfrac{108}{-16}$

$= -\dfrac{27}{4}$

The vertex is $\left(\dfrac{3}{4}, -\dfrac{27}{4}\right)$.

d. $0 = -4x^2 + 6x - 9$
Since this is not factorable, check the discriminant.
$b^2 - 4ac = 6^2 - 4(-4)(-9)$
$= 36 - 144$
$= -108$
Since $-108 < 0$ there are no real roots.
There are no x-intercepts.

e.

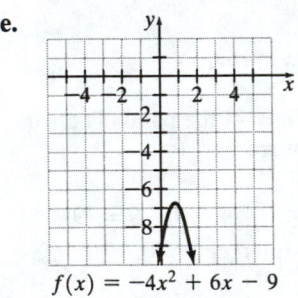

$f(x) = -4x^2 + 6x - 9$

33. In the function $f(x) = (x - 3)^2$, h has a value of 3. The graph of $f(x)$ is the graph of $g(x) = x^2$ shifted 3 units to the right.

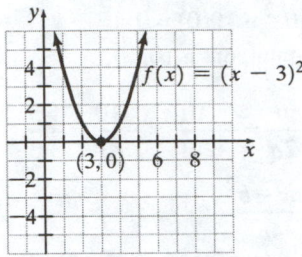

35. In the function $f(x) = x^2 + 3$, k has a value of 3. The graph $f(x)$ will be the graph of $g(x) = x^2$ shifted 3 units up.

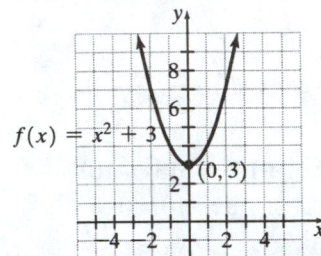

37. In the function $f(x) = (x - 2)^2 + 3$, h has a value of 2 and k has a value of 3. The graph of $f(x)$ will be the graph of $g(x) = x^2$ shifted 2 units to the right and 3 units up.

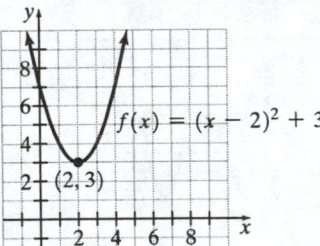

39. In the function $g(x) = -(x + 3)^2 - 2$, a has the value -1, h has the value -3, and k has the value -2. Since $a < 0$, the parabola opens downward. The graph of $g(x)$ will be the graph of $f(x) = -x^2$ shifted 3 units to the left and 2 units down.

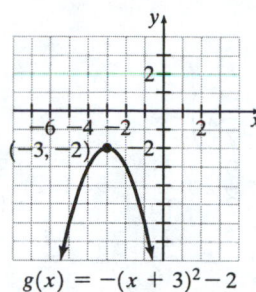

$$g(x) = -(x + 3)^2 - 2$$

41. In the function $f(x) = (x+4)^2 + 4$, h has a value of -4 and k has a value of 4. The graph of $f(x)$ will be the graph of $g(x) = x^2$ shifted 4 units to the left and 4 units up.

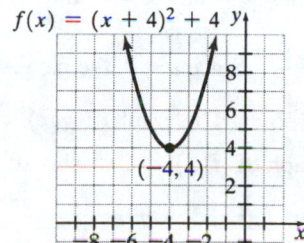

$$f(x) = (x + 4)^2 + 4$$

43. In the function $h(x) = -2(x+1)^2 - 3$, a has a value of -2, h has a value of -1, and k has a value of -3. Since $a < 0$, the parabola opens downward. The graph of $h(x)$ will be the graph of $f(x) = -2x^2$ shifted 1 unit left and 3 units down.

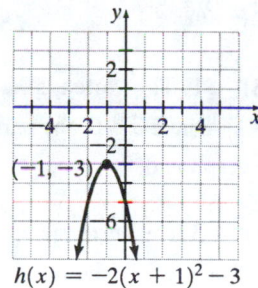

$$h(x) = -2(x + 1)^2 - 3$$

45. In the function $y = -2(x-2)^2 + 2$, a has a value of -2, h has a value of 2, and k has a value of 2. The graph of y will be the graph of $g(x) = -2x^2$ shifted 2 units to the right and 2 units up.

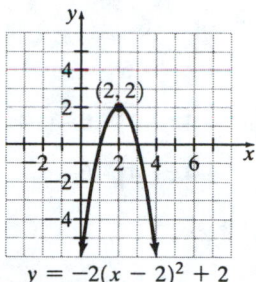

$$y = -2(x - 2)^2 + 2$$

47. a.
$$f(x) = x^2 - 6x + 8$$
$$= \left(x^2 - 6x + 9\right) + 8 - 9$$
$$f(x) = (x - 3)^2 - 1$$

b. Since $h = 3$ and $k = -1$, the vertex is $(3, -1)$

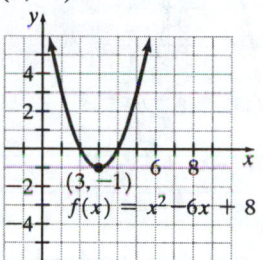

$$f(x) = x^2 - 6x + 8$$

49. a.
$$g(x) = x^2 - x - 12$$
$$= \left(x^2 - x + \frac{1}{4}\right) - 12 - \frac{1}{4}$$
$$g(x) = \left(x - \frac{1}{2}\right)^2 - \frac{49}{4}$$

b. Since $h = \frac{1}{2}$ and $k = -\frac{49}{4}$, the vertex is $\left(\frac{1}{2}, -\frac{49}{4}\right)$.

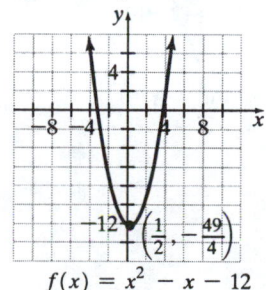

$$f(x) = x^2 - x - 12$$

51. a. $f(x) = -x^2 - 4x - 6$

$= -\left(x^2 + 4x\right) - 6$

$= -\left(x^2 + 4x + 4\right) - 6 - (-4)$

$= -\left(x^2 + 4x + 4\right) - 6 + 4$

$f(x) = -(x+2)^2 - 2$

b. Since $h = -2$ and $k = -2$, the vertex is $(-2, -2)$. Since $a < 0$, the parabola opens downward.

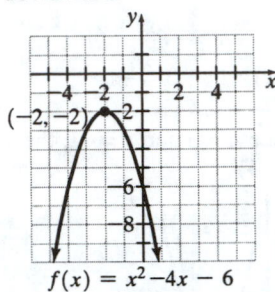

$f(x) = x^2 - 4x - 6$

53. a. $h(x) = -x^2 + 4x - 12$

$= -\left(x^2 - 4x\right) - 12$

$= -\left(x^2 - 4x + 4\right) - 12 - (-4)$

$= -\left(x^2 - 4x + 4\right) - 12 + 4$

$h(x) = -(x-2)^2 - 8$

b. Since $h = 2$ and $k = -8$, the vertex is $(2, -8)$. Since $a < 0$, the parabola opens downward.

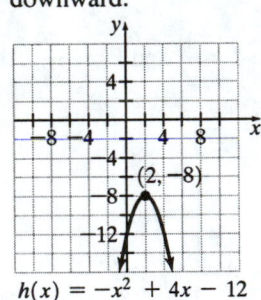

$h(x) = -x^2 + 4x - 12$

55. a. $g(x)$

$= -2x^2 - 9x - 9$

$= -2\left(x^2 + \frac{9}{2}x\right) - 9$

$= -2\left(x^2 + \frac{9}{2}x + \frac{81}{16}\right) - 9 - (-2)\left(\frac{81}{16}\right)$

$= -2\left(x^2 + \frac{9}{2}x + \frac{81}{16}\right) - 9 + \frac{81}{8}$

$g(x)$

$= -2\left(x + \frac{9}{4}\right)^2 + \frac{9}{8}$

b. Since $h = -\frac{9}{4}$ and $k = \frac{9}{8}$, the vertex is $\left(-\frac{9}{4}, \frac{9}{8}\right)$. Since $a < 0$, the parabola opens down. The graph of $g(x)$ will be the graph of $f(x) = -2x^2$ shifted $\frac{9}{4}$ units left and $\frac{9}{8}$ units up.

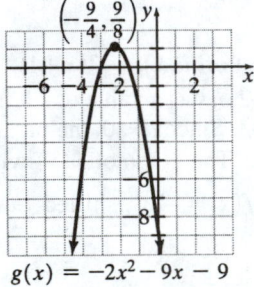

$g(x) = -2x^2 - 9x - 9$

57. $f(x) = 2(x+3)^2 - 1$. The vertex is $(h, k) = (-3, -1)$. Since $a > 0$, the parabola opens up. The graph is d).

59. $f(x) = 2(x-1)^2 + 3$ The vertex is $(h, k) = (1, 3)$. Since $a > 0$, the parabola opens up. The graph is b).

61. For $f(x) = (x-2)^2 + \dfrac{5}{2}$, the vertex is

$\left(2, \dfrac{5}{2}\right)$. For $g(x) = (x-2)^2 - \dfrac{3}{2}$, the vertex

is $\left(2, -\dfrac{3}{2}\right)$. These points are on the vertical

line $x = 2$. The distance between the two

points is $\dfrac{5}{2} - \left(-\dfrac{3}{2}\right) = \dfrac{5}{2} + \dfrac{3}{2} = \dfrac{8}{2} = 4$ units.

63. For $f(x) = 2(x+4)^2 - 3$, the vertex is

$(-4, -3)$. For $g(x) = -(x+1)^2 - 3$, the vertex
is $(-1, -3)$. These points are on the
horizontal line $y = -3$. The distance between
the two points is
$-1 - (-4) = -1 + 4 = 3$ units.

65. A function that has the shape of $f(x) = 2x^2$

will have the form $f(x) = 2(x-h)^2 + k$. If

$(h, k) = (3, -2)$, the function is

$f(x) = 2(x-3)^2 - 2$.

67. A function that has the shape of

$f(x) = -4x^2$ will have the form

$f(x) = -4(x-h)^2 + k$. If

$(h, k) = \left(-\dfrac{3}{5}, -\sqrt{2}\right)$, the function is

$f(x) = -4\left(x + \dfrac{3}{5}\right)^2 - \sqrt{2}$.

69. a. The graphs will have the same
   x-intercepts but $f(x) = x^2 - 8x + 12$
   will open up and $g(x) = -x^2 + 8x - 12$
   will open down.

   b. Yes, because the x-intercepts are located
   by setting $x^2 - 8x + 12$ and
   $-x^2 + 8x - 12$ equal to zero. They have
   the same solution set, therefore the
   x-intercepts are equal.

   c. No. The vertex for $y = x^2 - 8x + 12$ has
   the same x-coordinate as the vertex for
   $y = -x^2 + 8x - 12$ but the y-coordinates
   will differ by sign.

   d.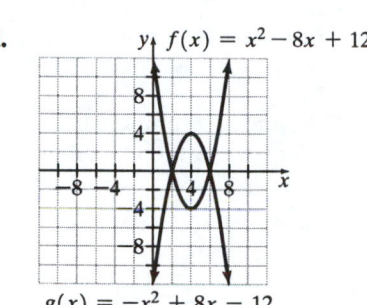

71. a. The vertex $x = -\dfrac{b}{2a} = -\dfrac{24}{2(-1)} = 12$

$I = -(12)^2 + 24(12) - 44$
$ = -144 + 288 - 44$
$ = 100$
The vertex is at $(12, 100)$. To find the
roots set $I = 0$.
$0 = -x^2 + 24x - 44$
$ = -1(x^2 - 24x + 44)$
$ = -1(x-2)(x-22)$
$x - 2 = 0 \quad$ or $\quad x - 22 = 0$
$ x = 2 \qquad\qquad x = 22$
The roots are 2 and 22.

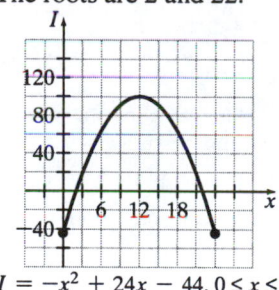

$I = -x^2 + 24x - 44,\ 0 \le x \le 24$

   b. The minimum cost will be $2 since the
   smaller root is 2.

   c. The maximum cost is $22 since the
   larger root is 22.

d. The maximum value will occur at the vertex of the parabola, (12, 100). Therefore, they should charge $12.

e. The maximum value will occur at the vertex of the parabola, (12, 100). Since I is in hundreds of dollars the maximum income is 100($100) = $10,000.

73. a. The number of bird feeders sold for the maximum profit will be the x-coordinate of the vertex.
$$f(x) = -0.4x^2 + 80x - 200$$
$$x = -\frac{b}{2a} = -\frac{80}{2(-0.4)} = -\frac{80}{-0.8} = 100$$
The company must sell 100 bird feeders for maximum profit.

b. The maximum profit will be the y-coordinate of the vertex, $y = f(100)$.
$$f(100) = -0.4\left(100^2\right) + 80(100) - 200$$
$$= -0.4(10,000) + 8000 - 200$$
$$= -4000 + 8000 - 200$$
$$= 3800$$
The maximum profit will be $3800.

75. The time the ball reaches the maximum height is t, the x-coordinate of the vertex, and the maximum height is h, the y-coordinate.
$$h = 32t - 16t^2 = -16t^2 + 32t$$
$$t = -\frac{b}{2a} = -\frac{32}{2(-16)} = -\frac{32}{-32} = 1$$
The ball will reach the maximum height after 1 second.
$$h(1) = 32(1) - 16\left(1^2\right) = 32 - 16 = 16$$
The maximum height is 16 feet.

77. If the perimeter of the room is 60 ft., then $60 = 2l + 2w$, where l is the length and w is the width. Then $60 = 2(l + w)$ and $30 = l + w$. Therefore $l = 30 - w$. The area of the room is

$$A = lw$$
$$= (30 - w)w$$
$$= 30w - w^2$$
$$= -w^2 + 30w$$
The maximum area is the y-coordinate of the vertex.
$$A = \frac{4ac - b^2}{4a}$$
$$= \frac{4(-1)(0) - 30^2}{4(-1)}$$
$$= \frac{0 - 900}{-4}$$
$$= 225$$
The maximum area is 225 ft^2.

79. If two numbers differ by 8 and x is one of the numbers, then $x + 8$ is the other number. The product is $f(x) = x(x + 8) = x^2 + 8x$. The maximum product is the y-coordinate of the vertex.
$$x = -\frac{b}{2a} = -\frac{8}{2(1)} = -4$$
$$y = f(-4)$$
$$= (-4)^2 + 8(-4)$$
$$= 16 - 32$$
$$= -16$$
The maximum product is -16. The numbers are -4 and $-4 + 8 = 4$.

81. If two numbers add to 40 and x is one of the numbers, then $40 - x$ is the other number. The product is
$$f(x) = x(40 - x) = 40x - x^2 = -x^2 + 40x.$$
The maximum product is the y-coordinate of the vertex.
$$x = -\frac{b}{2a} = -\frac{40}{2(-1)} = -\frac{40}{-2} = 20$$
$$y = f(20)$$
$$= -20^2 + 40(20)$$
$$= -400 + 800$$
$$= 400$$
The maximum product is 400. The numbers are 20 and $40 - 20 = 20$.

83. a. The number of years after 1989 in which the maximum percent of workers felt insecure is the x-coordinate of the vertex of

$$p(x) = -0.007x^2 + 0.099x + 0.090$$

$$x = -\frac{b}{2a}$$

$$= -\frac{0.099}{2(-0.007)}$$

$$= -\frac{0.099}{-0.014}$$

$$\approx 7.07$$

The year in which the maximum percent of workers felt insecure was about 7.07 years after 1989 or in 1996.

b. The maximum percent is the y-coordinate of the vertex.

$$y = p(7.07)$$

$$= -0.007(7.07)^2 + 0.099(7.07) + 0.090$$

$$= -0.007(49.9849) + 0.6993 + 0.090$$

$$= 0.4400357$$

$$\approx 44\%$$

The maximum percent of workers that felt insecure on the job was approximately 44%.

85. a. The number of years after 1986, of greatest emigration is the x-coordinate of the vertex of

$$N(x) = -1.82x^2 + 20.08x + 11.86.$$

$$x = -\frac{b}{2a}$$

$$= -\frac{20.08}{2(-1.82)}$$

$$= -\frac{20.08}{-3.64}$$

$$\approx 5.52$$

The greatest emigration was 5.52 years after 1986 or in 1991.

b. The number of people, in thousands, who emigrated is $N(x)$, the y-coordinate of the vertex.

$$N(5.52)$$

$$= -1.82(5.52)^2 + 20.08(5.52) + 11.86$$

$$= -55.456128 + 110.8416 + 11.86$$

$$= 67.245472$$

$$\approx 67.245$$

The number of people who emigrated was about 67.245472 thousand or 67,245.

87. $C(x) = 2000 + 40x,\ R(x) = 800x - x^2$

$$P(x) = R(x) - C(x)$$

$$P(x) = \left(800x - x^2\right) - (2000 + 40x)$$

$$= 800x - x^2 - 2000 - 40x$$

$$= -x^2 + 760x - 2000$$

The maximum profit is $P(x)$, the y-coordinate of the vertex. The number of items that must be produced and sold to obtain maximum profit is the x coordinate of the vertex.

$$x = -\frac{b}{2a} = -\frac{760}{2(-1)} = 380$$

a. $P(380) = -380^2 + 760(380) - 2000$

$$= -144,400 + 288800 - 2000$$

$$= 142,400$$

The maximum profit is $142,400.

b. The number of items that must be produced and sold to obtain maximum profit is 380.

89. a.
$$h(t) = -4.9t^2 + 24.5t + 9.8$$
$$= -4.9\left(t^2 - 5t\right) + 9.8$$
$$= -4.9\left(t^2 - 5t + 6.25\right) + 9.8 - (-4.9)(6.25)$$
$$= -4.9(t - 2.5)^2 + 9.8 + 30.625$$
$$h(t) = -4.9(t - 2.5)^2 + 40.425$$

b. The maximum height is 40.425 m, the time of maximum height is 2.5 seconds.

c. Yes, it is the same.

91. The radius of the outer circle is $r = 15$ ft. The area $A = \pi r^2 = \pi\left(15^2\right) = 225\pi$ ft^2. The radius of the inner circle is $r = 5$ ft. The area is $A = \pi r^2 = \pi\left(5^2\right) = 25\pi$ ft^2. The area shaded blue is
225π ft^2 $- 25\pi$ ft^2 $= 200\pi$ ft^2.

92. Mr. Duncan's salary plus commission is his take home pay. Let s represent his salary and r represent his commission rate.

	salary	commission	take home
1st week	s	$6000r$	1300
2nd week	s	$4000r$	1000

$s + 6000r = 1300$
$s + 4000r = 1000$
Multiply the second equation by -1 and add the two equations.
$$\begin{aligned} s + 6000r &= 1300 \\ -s - 4000r &= -1000 \\ \hline 2000r &= 300 \\ r &= 0.15 \end{aligned}$$
Substitute 0.15 for r in the first equation.
$$s + 6000(0.15) = 1300$$
$$s + 900 = 1300$$
$$s = 400$$
Mr. Duncan's salary is \$400 and his commission rate is $0.15 = 15\%$.

93.
$$x^3 + 2x - 5x^2 - 10 = x^3 - 5x^2 + 2x - 10$$
$$= x^2(x - 5) + 2(x - 5)$$
$$= (x - 5)\left(x^2 + 2\right)$$

94.
$$\frac{3a^{-1} - b^{-1}}{(a - b)^{-1}} = \frac{\frac{3}{a} - \frac{1}{b}}{\frac{1}{a-b}}$$
$$= \frac{\frac{3b - a}{ab}}{\frac{1}{a-b}}$$
$$= \frac{3b - a}{ab} \cdot a - b$$
$$= \frac{(3b - a)(a - b)}{ab}$$

Exercise Set 8.6

1. a. For $f(x) = x^2 - 7x + 10$, $f(x) > 0$ when the graph is above the x-axis. The solution is $x < 2$ or $x > 5$.

b. For $f(x) = x^2 - 7x + 10$ $f(x) < 0$ when the graph is below the x-axis. The solution is $2 < x < 5$.

3. $x^2 - 3x - 10 \geq 0$
$(x + 2)(x - 5) \geq 0$

5. $x^2 + 4x > 0$
$x(x+4) > 0$

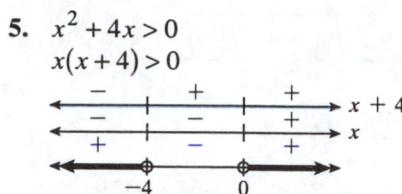

7. $x^2 - 16 < 0$
$(x+4)(x-4) < 0$

9. $2x^2 + 5x - 3 \geq 0$
$(2x-1)(x+3) \geq 0$

11. $5x^2 + 19x \leq 4$
$5x^2 + 19x - 4 \leq 0$
$(x+4)(5x-1) \leq 0$

13. $2x^2 - 12x + 9 \leq 0$
$2x^2 - 12x + 9 = 0$

$x = \dfrac{-(-12) \pm \sqrt{(-12)^2 - 4(2)(9)}}{2(2)}$

$= \dfrac{12 \pm \sqrt{144 - 72}}{4}$

$= \dfrac{12 \pm \sqrt{72}}{4}$

$= \dfrac{12 \pm 6\sqrt{2}}{4}$

$= \dfrac{6 \pm 3\sqrt{2}}{2}$

$\dfrac{6 - 3\sqrt{2}}{2}$ $\dfrac{6 + 3\sqrt{2}}{2}$

15. $(x-1)(x+1)(x+4) > 0$
$x-1=0$ $x+1=0$ $x+4=0$
$x=1$ $x=-1$ $x=-4$

$(-4, -1) \cup (1, \infty)$

17. $x(x-3)(2x+6) \geq 0$
$x=0$ $x-3=0$ $2x+6=0$
$x=0$ $x=3$ $2x=-6$
$x=-3$

$[-3, 0] \cup [3, \infty)$

19. $(2c+5)(3c-6)(c+6) > 0$
$2c+5=0$ $3c-6=0$ $c+6=0$
$2c=-5$ $3c=6$ $c=-6$
$c=-\dfrac{5}{2}$ $c=2$

$\left(-6, -\dfrac{5}{2}\right) \cup (2, \infty)$

21. $(x+2)(x+2)(3x-8) \geq 0$
$x+2=0$ $3x-8=0$
$x=-2$ $x=\dfrac{8}{3}$

$$\begin{array}{ccc}+&+&+\\-&-&+\\-&-&+\end{array}$$
$(x+2)^2$
$3x-8$

$$\xrightarrow[\hspace{2cm}\bullet\hspace{1.2cm}]{}$$
$-2 \qquad \dfrac{8}{3}$

$\left[\dfrac{8}{3},\ \infty\right)$

23. $x^3 - 4x^2 + 4x < 0$

$x\left(x^2 - 4x - 4\right) < 0$

$x(x-2)^2 < 0$

$x = 0 \quad x - 2 = 0$

$\qquad x = 2$

$$\begin{array}{ccc}-&+&+\\+&+&+\\-&+&+\end{array}$$
x
$(x-2)^2$

$$\xleftarrow{\hspace{2.5cm}}\underset{0}{\circ}\hspace{0.8cm}\underset{2}{\circ}$$

$(-\infty, 0)$

25. $f(x) = x^2 + 4x,\ f(x) \geq 0$

$x^2 + 4x \geq 0$

$x(x+4) \geq 0$

$x = 0 \quad x + 4 = 0$

$\qquad x = -4$

$$\begin{array}{ccc}-&-&+\\-&+&+\\+&-&+\end{array}$$
x
$x+4$

$$\xleftarrow{\hspace{1cm}}\underset{-4}{\bullet}\hspace{1.2cm}\underset{0}{\bullet}\xrightarrow{\hspace{1cm}}$$

27. $f(x) = x^2 - 14x + 48,\ f(x) < 0$

$x^2 - 14x + 48 < 0$

$(x-6)(x-8) < 0$

$x - 6 = 0 \quad x - 8 = 0$

$x = 6 \qquad x = 8$

$$\begin{array}{ccc}-&+&+\\-&-&+\\+&-&+\end{array}$$
$x-6$
$x-8$

$$\xleftarrow{\hspace{2cm}}\underset{6}{\circ}\hspace{1cm}\underset{8}{\circ}\xrightarrow{\hspace{1cm}}$$

29. $f(x) = 2x^2 + 9x - 4,\ f(x) \leq 2$

$2x^2 + 9x - 4 \leq 2$

$2x^2 + 9x - 6 \leq 0$

$x = \dfrac{-9 \pm \sqrt{9^2 - 4(2)(-6)}}{2(2)}$

$= \dfrac{-9 \pm \sqrt{129}}{4}$

$x = \dfrac{-9 - \sqrt{129}}{4} \qquad x = \dfrac{-98 + \sqrt{129}}{4}$

$$\begin{array}{ccc}-&+&+\\-&-&+\\+&-&+\end{array}$$
$x - \left(\dfrac{-9 - \sqrt{129}}{4}\right)$
$x - \left(\dfrac{-9 + \sqrt{129}}{4}\right)$

$$\xleftarrow{\hspace{1.5cm}}\underset{}{\bullet}\hspace{1cm}\underset{}{\bullet}\xrightarrow{\hspace{1.5cm}}$$
$\dfrac{-9 - \sqrt{129}}{4} \quad \dfrac{-9 + \sqrt{129}}{4}$

31. $f(x) = 2x^3 + 9x^2 - 35x,\ f(x) \geq 0$

$2x^3 + 9x^2 - 35x \geq 0$

$x\left(2x^2 + 9x - 35\right) \geq 0$

$x(2x-5)(x+7) \geq 0$

$x = 0 \quad 2x - 5 = 0 \quad x + 7 = 0$

$\qquad x = \dfrac{5}{2} \qquad x = -7$

$$\begin{array}{cccc}-&+&+&+\\-&-&+&+\\-&-&-&+\\-&+&-&+\end{array}$$
$x + 7$
x
$2x - 5$

$$\xleftarrow{}\underset{-7}{\bullet}\hspace{0.8cm}\underset{0}{\bullet}\hspace{0.8cm}\underset{\frac{5}{2}}{\bullet}\xrightarrow{}$$

33. $\dfrac{x+3}{x-1} > 0$

$x \neq 1$

$$\begin{array}{ccc}-&+&+\\-&-&+\\+&-&+\end{array}$$
$x + 3$
$x - 1$

$$\xleftarrow{\hspace{1.5cm}}\underset{-3}{\circ}\hspace{1cm}\underset{1}{\circ}\xrightarrow{\hspace{1.5cm}}$$

$\{x \mid x < -3 \text{ or } x > 1\}$

35. $\dfrac{y-4}{y-1} \leq 0$

$y \neq 1$

$$\begin{array}{ccc}-&+&+\\-&-&+\\+&-&+\end{array}$$
$y - 1$
$y - 4$

$$\xleftarrow{\hspace{2cm}}\underset{1}{\circ}\hspace{1cm}\underset{4}{\bullet}\xrightarrow{\hspace{1cm}}$$

$\{y \mid 1 < y \leq 4\}$

37. $\dfrac{2x-4}{x-1} < 0$

$x \neq 1$

$\{x \mid 1 < x < 2\}$

39. $\dfrac{3a+6}{2a-1} \geq 0$

$a \neq \dfrac{1}{2}$

$\left\{ a \mid a \leq -2 \text{ or } a > \dfrac{1}{2} \right\}$

41. $\dfrac{x+4}{x-4} \leq 0$

$x \neq 4$

$\{x \mid -4 \leq x < 4\}$

43. $\dfrac{4x-2}{2x-4} > 0$

$x \neq 2$

$\left\{ x \mid x < \dfrac{1}{2} \text{ or } x > 2 \right\}$

45. $\dfrac{(x+2)(x-4)}{x+6} < 0$

$x \neq -6$

$(-\infty, -6) \cup (-2, 4)$

47. $\dfrac{(w-6)(w-1)}{w-3} \geq 0$

$w \neq 3$

$[1, 3) \cup [6, \infty)$

49. $\dfrac{x-6}{(x+4)(x-1)} \leq 0$

$x \neq -4,\ x \neq 1$

$(-\infty, -4) \cup (1, 6]$

51. $\dfrac{(x-3)(2x+5)}{x-6} > 0$

$x \neq 6$

$\left(-\dfrac{5}{2}, 3\right) \cup (6, \infty)$

53. $\dfrac{2}{x-3} \geq -1$

$$\dfrac{2}{x-3} + 1 \geq 0$$

$$\dfrac{2}{x-3} + \dfrac{1(x-3)}{x-3} \geq 0$$

$$\dfrac{2+x-3}{x-3} \geq 0$$

$$\dfrac{x-1}{x-3} \geq 0$$

$$x \neq 3$$

55. $\dfrac{4}{x-2} \geq 2$

$$\dfrac{4}{x-2} - 2 \geq 0$$

$$\dfrac{4}{x-2} - \dfrac{2(x-2)}{x-2} \geq 0$$

$$\dfrac{4-2x+4}{x-2} \geq 0$$

$$\dfrac{8-2x}{x-2} \geq 0$$

$$x \neq 2$$

57. $\dfrac{2p-5}{p-4} \leq 1$

$$\dfrac{2p-5}{p-4} - 1 \leq 0$$

$$\dfrac{2p-5}{p-4} - \dfrac{1(p-4)}{p-4} \leq 0$$

$$\dfrac{2p-5-p+4}{p-4} \leq 0$$

$$\dfrac{p-1}{p-4} \leq 0$$

$$p \neq 4$$

59. $\dfrac{w}{3w-2} > -2$

$$\dfrac{w}{3w-2} + 2 > 0$$

$$\dfrac{w}{3w-2} + \dfrac{2(3w-2)}{3w-2} > 0$$

$$\dfrac{w+6w-4}{3w-2} > 0$$

$$\dfrac{7w-4}{3w-2} > 0$$

$$w \neq \dfrac{2}{3}$$

61. $\dfrac{x+8}{x+2} > 1$

$$\dfrac{x+8}{x+2} - 1 > 0$$

$$\dfrac{x+8}{x+2} - \dfrac{1(x+2)}{x+2} > 0$$

$$\dfrac{x+8-x-2}{x+2} > 0$$

$$\dfrac{6}{x+2} > 0$$

$$x \neq -2$$

63. a. $y = \dfrac{x^2-4x+4}{x-4} > 0$ where the graph of y is above the x-axis, on the interval $(4, \infty)$.

b. $y = \dfrac{x^2-4x+4}{x-4} < 0$ where the graph of y is below the x-axis, on the interval $(-\infty, 2) \cup (2, 4)$.

65. A quadratic inequality with the union of the two outer regions of the number line as its solution, not including the boundary values, will be of the form $ax^2 + bx + c > 0$ with $a > 0$. Since the boundary values are $x = -4$ and $x = 2$, the factors are $x + 4$ and $x - 2$. Therefore one quadratic inequality is

$(x+4)(x-2) > 0$ or $x^2 + 2x - 8 > 0$.

67. Since the solution set is $x \le -3$ and $x > 4$, the factors are $x + 3$ and $x - 4$. Because -3 is included in the solution set, $x + 3$ is the numerator. Since 4 is not included in the solution set, $x - 4$ is the denominator. The inequality symbol will be $\ge$ because the union of the outer regions of number line is the solution set. Therefore, the rational inequality is $\dfrac{x+3}{x-4} \ge 0$.

69. $(x+3)^2(x-1)^2 \ge 0$

The solution is all real numbers since any nonzero number squared is positive and zero squared is zero.

71. $\dfrac{x^2}{(x+1)^2} \ge 0$

This statement is true for all real numbers, except -1, since any nonzero number squared is negative. It is undefined when $x = -1$. Therefore, -1 is not a solution.

73. If $f(x) = ax^2 + bx + c$ and $a > 0$, the graph of $f(x)$ opens upward. If the discriminant is negative, the graph of $f(x)$ has no x-intercepts. Therefore, the graph lies above the x-axis and $f(x) < 0$ has no solution.

75. $(x+1)(x-3)(x+5)(x+9) \ge 0$

77. One possible answer is: Use a parabola that opens upward and has x-intercepts of $(0, 0)$ and $(3, 0)$. The x-values for which the parabola lies above the x-axis are $(-\infty, 0) \cup (3, \infty)$.

$x^2 - 3x > 0$

79. One possible answer is: Use a parabola that opens upward and has its vertex on or above the x-axis. Then there are no x-values for which the parabola lies below the x-axis.

$x^2 < 0$

83. $f(x) = \dfrac{3}{x^2 - 4}$

$x^2 - 4 = 0$

$(x+2)(x-2) = 0$

$x + 2 = 0 \quad x - 2 = 0$

$x = -2 \quad\quad x = 2$

Domain:

$\{x | x \text{ is a real number and } x \ne -2,\ x \ne 2\}$

84. $f(x) = \sqrt{x-4}$

$x - 4 \ge 0$

$x \ge 4$

The domain is

$\{x | x \text{ is a real number and } x \ge 4\}$.

85. $\dfrac{ab^{-2} - a^{-1}b}{a^{-2} + ab^{-1}} = \dfrac{\frac{a}{b^2} - \frac{b}{a}}{\frac{1}{a^2} + \frac{a}{b}}$

$= \dfrac{b^2 a^2 \left(\frac{a}{b^2}\right) - b^2 a^2 \left(\frac{b}{a}\right)}{b^2 a^2 \left(\frac{1}{a^2}\right) + b^2 a^2 \left(\frac{a}{b}\right)}$

$= \dfrac{a^3 - ab^3}{b^2 + a^3 b}$

86. $\left(\sqrt{-8} + \sqrt{2}\right)\left(\sqrt{-2} - \sqrt{8}\right)$

$= \left(2i\sqrt{2} + \sqrt{2}\right)\left(i\sqrt{2} - 2\sqrt{2}\right)$

$= 2\left(\sqrt{2}\right)^2 i^2 - 4\left(\sqrt{2}\right)^2 i + \left(\sqrt{2}\right)^2 i - 2\left(\sqrt{2}\right)^2$

$= 2(2)(-1) - 4(2)i + 2i - 2(2)$

$= -4 - 8i + 2i - 4$

$= -8 - 6i$

Review Exercises

1. $(x-4)^2 = 20$

   $x - 4 = \pm\sqrt{20}$

   $x - 4 = \pm 2\sqrt{5}$

   $x = 4 \pm 2\sqrt{5}$

   $x = 4 + 2\sqrt{5}$ or $x = 4 - 2\sqrt{5}$

2. $(3x-4)^2 = 60$

   $3x - 4 = \pm\sqrt{60}$

   $3x - 4 = \pm 2\sqrt{15}$

   $3x = 4 \pm 2\sqrt{15}$

   $x = \dfrac{4 \pm 2\sqrt{15}}{3}$

   $x = \dfrac{4 + 2\sqrt{15}}{3}$ or $x = \dfrac{4 - 2\sqrt{15}}{3}$

3. $\left(x - \dfrac{2}{3}\right)^2 = \dfrac{1}{9}$

   $x - \dfrac{2}{3} = \pm\sqrt{\dfrac{1}{9}}$

   $x - \dfrac{2}{3} = \pm\dfrac{1}{3}$

   $x = \dfrac{2}{3} \pm \dfrac{1}{3}$

   $x = \dfrac{2}{3} + \dfrac{1}{3}$ or $x = \dfrac{2}{3} - \dfrac{1}{3}$

   $\quad = \dfrac{3}{3} \qquad\qquad = \dfrac{1}{3}$

   $\quad = 1$

   $x = 1$ or $x = \dfrac{1}{3}$

4. $\left(2x - \dfrac{1}{2}\right)^2 = 4$

   $2x - \dfrac{1}{2} = \pm\sqrt{4}$

   $2x - \dfrac{1}{2} = \pm 2$

   $2x = \dfrac{1}{2} \pm 2$

   $2x = \dfrac{1 \pm 4}{2}$

   $x = \dfrac{1 \pm 4}{4}$

   $x = \dfrac{1+4}{4}$ or $x = \dfrac{1-4}{4}$

   $\quad = \dfrac{5}{4} \qquad\qquad = -\dfrac{3}{4}$

   $x = \dfrac{5}{4}$ or $x = -\dfrac{3}{4}$

5. $x^2 - 8x + 15 = 0$

   $x^2 - 8x = -15$

   $x^2 - 8x + 16 = -15 + 16$

   $x^2 - 8x + 16 = 1$

   $(x-4)^2 = 1$

   $x - 4 = \pm 1$

   $x = 4 \pm 1$

   $x = 4 + 1$ or $x = 4 - 1$

   $\quad = 5 \qquad\qquad = 3$

   $x = 5$ or $x = 3$

6. $x^2 - 3x - 54 = 0$

   $x^2 - 3x = 54$

   $x^2 - 3x + \dfrac{9}{4} = \dfrac{216}{4} + \dfrac{9}{4}$

   $\left(x - \dfrac{3}{2}\right)^2 = \dfrac{225}{4}$

   $x - \dfrac{3}{2} = \pm\dfrac{15}{2}$

   $x = \dfrac{3}{2} \pm \dfrac{15}{2}$

$$x = \frac{3}{2} + \frac{15}{2} \quad \text{or} \quad x = \frac{3}{2} - \frac{15}{2}$$
$$= \frac{18}{2} \qquad\qquad = \frac{-12}{2}$$
$$= 9 \qquad\qquad\quad = -6$$
$$x = 9 \text{ or } x = -6$$

7. $x^2 = -5x + 6$
$$x^2 + 5x = 6$$
$$x^2 + 5x + \frac{25}{4} = 6 + \frac{25}{4}$$
$$\left(x + \frac{5}{2}\right)^2 = \frac{24}{4} + \frac{25}{4}$$
$$\left(x + \frac{5}{2}\right)^2 = \frac{49}{4}$$
$$x + \frac{5}{2} = \pm\frac{7}{2}$$
$$x = -\frac{5}{2} \pm \frac{7}{2}$$
$$x = -\frac{5}{2} + \frac{7}{2} \quad \text{or} \quad x = -\frac{5}{2} - \frac{7}{2}$$
$$= \frac{2}{2} \qquad\qquad = -\frac{12}{2}$$
$$= 1 \qquad\qquad\quad = -6$$
$$x = 1 \text{ or } x = -6$$

8. $x^2 + 2x - 5 = 0$
$$x^2 + 2x = 5$$
$$x^2 + 2x + 1 = 5 + 1$$
$$(x + 1)^2 = 6$$
$$x + 1 = \pm\sqrt{6}$$
$$x = -1 \pm \sqrt{6}$$
$$x = -1 + \sqrt{6} \quad \text{or} \quad x = -1 - \sqrt{6}$$

9. $-a^2 - 6a + 10 = 0$
$$a^2 + 6a = 10$$
$$a^2 + 6a + 9 = 10 + 9$$
$$(a + 3)^2 = 19$$
$$a + 3 = \pm\sqrt{19}$$
$$a = -3 \pm \sqrt{19}$$
$$a = -3 + \sqrt{19} \quad \text{or} \quad a = -3 - \sqrt{19}$$

10. $2r^2 - 8r = -64$
$$r^2 - 4r = -32$$
$$r^2 - 4r + 4 = -32 + 4$$
$$(r - 2)^2 = -28$$
$$r - 2 = \pm\sqrt{-28}$$
$$r = 2 \pm \sqrt{4}\sqrt{7}\sqrt{-1}$$
$$r = 2 \pm 2i\sqrt{7}$$
$$r = 2 + 2i\sqrt{7} \quad \text{or} \quad r = 2 - 2i\sqrt{7}$$

11. $3x^2 - 4x - 20 = 0 \qquad a = 3, b = -4, c = -20$
$$b^2 - 4ac = (-4)^2 - 4(3)(-20)$$
$$= 16 + 240$$
$$= 256$$
Since the discriminant is positive, this equation has two distinct real solutions.

12. $-3x^2 + 4x = 9$
$$-3x^2 + 4x - 9 = 0 \qquad a = -3, b = 4, c = -9$$
$$b^2 - 4ac = (4)^2 - 4(-3)(-9)$$
$$= 16 - 108$$
$$= -92$$
Since the discriminant is negative, this equation has no real solutions.

13. $2x^2 + 6x + 7 = 0 \qquad a = 2, b = 6, c = 7$
$$b^2 - 4ac = 6^2 - 4(2)(7)$$
$$= 36 - 56$$
$$= -20$$
Since the discriminant is negative, this equation has no real solutions.

14. $n^2 - 12n = -36$
$$n^2 - 12n + 36 = 0 \qquad a = 1, b = -12, c = 36$$
$$b^2 - 4ac = (-12)^2 - 4(1)(36)$$
$$= 144 - 144$$
$$= 0$$
Since the discriminant is 0, the equation has one real solution.

15. $-3x^2 - 4x + 8 = 0 \qquad a = -3, b = -4, c = 8$
$$b^2 - 4ac = (-4)^2 - 4(-3)(8)$$
$$= 16 + 96$$
$$= 112$$
Since the discriminant is positive, this equation has two real solutions.

16. $x^2 - 9x + 6 = 0$ $a = 1, b = -9, c = 6$

$b^2 - 4ac = (-9)^2 - 4(1)(6)$

$\qquad = 81 - 24$

$\qquad = 57$

Since the discriminant is positive, this equation has two distinct real solutions.

17. $5x^2 - 7x = 6$

$5x^2 - 7x - 6 = 0$ $a = 5, b = -7, c = -6$

$x = \dfrac{-b \pm \sqrt{b^2 - 4ac}}{2a}$

$x = \dfrac{-(-7) \pm \sqrt{(-7)^2 - 4(5)(-6)}}{2(5)}$

$= \dfrac{7 \pm \sqrt{49 + 120}}{10}$

$= \dfrac{7 \pm \sqrt{169}}{10}$

$= \dfrac{7 \pm 13}{10}$

$x = \dfrac{7 + 13}{10}$ or $x = \dfrac{7 - 13}{10}$

$= \dfrac{20}{10}$ $= \dfrac{-6}{10}$

$= 2$ $= -\dfrac{3}{5}$

18. $x^2 - 18 = 7x$

$x^2 - 7x - 18 = 0$ $a = 1, b = -7, c = -18$

$x = \dfrac{-b \pm \sqrt{b^2 - 4ac}}{2a}$

$x = \dfrac{-(-7) \pm \sqrt{(-7)^2 - 4(1)(-18)}}{2(1)}$

$= \dfrac{7 \pm \sqrt{49 + 72}}{2}$

$= \dfrac{7 \pm \sqrt{121}}{2}$

$= \dfrac{7 \pm 11}{2}$

$x = \dfrac{7 + 11}{2}$ or $x = \dfrac{7 - 11}{2}$

$= \dfrac{18}{2}$ $= \dfrac{-4}{2}$

$= 9$ $= -2$

19. $x^2 - x + 30 = 0$ $a = 1, b = -1, c = 30$

$x = \dfrac{-b \pm \sqrt{b^2 - 4ac}}{2a}$

$x = \dfrac{-(-1) \pm \sqrt{(-1)^2 - 4(1)(30)}}{2(1)}$

$= \dfrac{1 \pm \sqrt{1 - 120}}{2}$

$= \dfrac{1 \pm \sqrt{-119}}{2}$

$= \dfrac{1 \pm \sqrt{119}\sqrt{-1}}{2}$

$= \dfrac{1 \pm i\sqrt{119}}{2}$

$x = \dfrac{1 + i\sqrt{119}}{2}$ or $x = \dfrac{1 - i\sqrt{119}}{2}$

20. $6d^2 + d - 15 = 0$ $a = 6, b = 1, c = -15$

$d = \dfrac{-b \pm \sqrt{b^2 - 4ac}}{2a}$

$d = \dfrac{-1 \pm \sqrt{1^2 - 4(6)(-15)}}{2(6)}$

$= \dfrac{-1 \pm \sqrt{1 + 360}}{12}$

$= \dfrac{-1 \pm \sqrt{361}}{12}$

$= \dfrac{-1 \pm 19}{12}$

$d = \dfrac{-1 + 19}{12}$ or $d = \dfrac{-1 - 19}{12}$

$= \dfrac{18}{12}$ $= \dfrac{-20}{12}$

$= \dfrac{3}{2}$ $= -\dfrac{5}{3}$

21. $2x^2 + 4x - 3 = 0$ $a = 2, b = 4, c = -3$

$$x = \frac{-b \pm \sqrt{b^2 - 4ac}}{2a}$$

$$x = \frac{-4 \pm \sqrt{4^2 - 4(2)(-3)}}{2(2)}$$

$$= \frac{-4 \pm \sqrt{16 + 24}}{4}$$

$$= \frac{-4 \pm \sqrt{40}}{4}$$

$$= \frac{-4 \pm \sqrt{4}\sqrt{10}}{4}$$

$$= \frac{-4 \pm 2\sqrt{10}}{4}$$

$$= \frac{2(-2 \pm \sqrt{10})}{2(2)}$$

$$x = \frac{-2 + \sqrt{10}}{2} \quad \text{or} \quad x = \frac{-2 - \sqrt{10}}{2}$$

22. $-2x^2 + 3x + 6 = 0$ $a = -2, b = 3, c = 6$

$$x = \frac{-b \pm \sqrt{b^2 - 4ac}}{2a}$$

$$x = \frac{-3 \pm \sqrt{3^2 - 4(-2)(6)}}{2(-2)}$$

$$= \frac{-3 \pm \sqrt{9 + 48}}{-4}$$

$$= \frac{3 \pm \sqrt{57}}{4}$$

$$x = \frac{3 + \sqrt{57}}{4} \quad \text{or} \quad x = \frac{3 - \sqrt{57}}{4}$$

23. $x^2 - 6x + 7 = 0$ $a = 1, b = -6, c = 7$

$$x = \frac{-b \pm \sqrt{b^2 - 4ac}}{2a}$$

$$x = \frac{-(-6) \pm \sqrt{(-6)^2 - 4(1)(7)}}{2(1)}$$

$$= \frac{6 \pm \sqrt{36 - 28}}{2}$$

$$= \frac{6 \pm \sqrt{8}}{2}$$

$$= \frac{6 \pm \sqrt{4}\sqrt{2}}{2}$$

$$= \frac{6 \pm 2\sqrt{2}}{2}$$

$$= \frac{2(3 \pm \sqrt{2})}{2}$$

$$x = 3 + \sqrt{2} \quad \text{or} \quad x = 3 - \sqrt{2}$$

24. $3x^2 - 6x - 8 = 0$ $a = 3, b = -6, c = -8$

$$x = \frac{-b \pm \sqrt{b^2 - 4ac}}{2a}$$

$$x = \frac{-(-6) \pm \sqrt{(-6)^2 - 4(3)(-8)}}{2(3)}$$

$$= \frac{6 \pm \sqrt{36 + 96}}{6}$$

$$= \frac{6 \pm \sqrt{132}}{6}$$

$$= \frac{6 \pm \sqrt{4}\sqrt{33}}{6}$$

$$= \frac{6 \pm 2\sqrt{33}}{6}$$

$$= \frac{2(3 \pm \sqrt{33})}{6}$$

$$x = \frac{3 + \sqrt{33}}{3} \quad \text{or} \quad x = \frac{3 - \sqrt{33}}{3}$$

25. $2x^2 - 5x = 0$ $\qquad$ $a = 2, b = -5, c = 0$

$$x = \frac{-b \pm \sqrt{b^2 - 4ac}}{2a}$$

$$x = \frac{-(-5) \pm \sqrt{(-5)^2 - 4(2)(0)}}{2(2)}$$

$$= \frac{5 \pm \sqrt{25 - 0}}{4}$$

$$= \frac{5 \pm 5}{4}$$

$$x = \frac{5 + 5}{4} \quad \text{or} \quad x = \frac{5 - 5}{4}$$

$$= \frac{10}{4} \qquad\qquad = \frac{0}{4}$$

$$= \frac{5}{2} \qquad\qquad = 0$$

$$x = \frac{-(-5) \pm \sqrt{(-5)^2 - 4(6)(-25)}}{2(6)}$$

$$x = \frac{5 \pm \sqrt{25 + 600}}{12}$$

$$= \frac{5 \pm \sqrt{625}}{12}$$

$$= \frac{5 \pm 25}{12}$$

$$x = \frac{5 + 25}{12} \quad \text{or} \quad x = \frac{5 - 25}{12}$$

$$= \frac{30}{12} \qquad\qquad = \frac{-20}{12}$$

$$= \frac{5}{2} \qquad\qquad = -\frac{5}{3}$$

26. $1.2r^2 + 5.7r = 2.3$

$1.2r^2 + 5.7r - 2.3 = 0$

$a = 1.2, b = 5.7, c = -2.3$

$$r = \frac{-b \pm \sqrt{b^2 - 4ac}}{2a}$$

$$r = \frac{-5.7 \pm \sqrt{(5.7)^2 - 4(1.2)(-2.3)}}{2(1.2)}$$

$$= \frac{-5.7 \pm \sqrt{32.49 + 11.04}}{2.4}$$

$$= \frac{-5.7 \pm \sqrt{43.53}}{2.4}$$

$$x = \frac{-5.7 + \sqrt{43.53}}{2.4} \quad \text{or} \quad x = \frac{-5.7 - \sqrt{43.53}}{2.4}$$

28. $x^2 + \frac{5x}{4} = \frac{3}{8}$

$8x^2 + 10x = 3$

$8x^2 + 10 - 3 = 0$

$a = 8, b = 10, c = -3$

$$x = \frac{-b \pm \sqrt{b^2 - 4ac}}{2a}$$

$$= \frac{-10 \pm \sqrt{10^2 - 4(8)(-3)}}{2(8)}$$

$$= \frac{-10 \pm \sqrt{100 + 96}}{16}$$

$$= \frac{-10 \pm \sqrt{196}}{16}$$

$$= \frac{-10 \pm 14}{16}$$

$$x = \frac{-10 + 14}{16} \quad \text{or} \quad x = \frac{-10 - 14}{16}$$

$$= \frac{4}{16} \qquad\qquad = \frac{-24}{16}$$

$$= \frac{1}{4} \qquad\qquad = -\frac{3}{2}$$

27. $x^2 = \frac{5}{6}x + \frac{25}{6}$

$6x^2 = 5x + 25$

$6x^2 - 5x - 25 = 0$ $\qquad$ $a = 6, b = -5, c = -25$

$$x = \frac{-b \pm \sqrt{b^2 - 4ac}}{2a}$$

29. $f(x) = x^2 - 4x - 45$, $f(x) = 15$

$x^2 - 4x - 45 = 15$

$x^2 - 4x - 60 = 0$

$(x - 10)(x + 6) = 0$

$x - 10 = 0$ or $x + 6 = 0$

$x = 10$ $\quad x = -6$

The solutions are 10 and –6.

30. $g(x) = 6x^2 + 5x$, $g(x) = 6$

$6x^2 + 5x = 6$

$6x^2 + 5x - 6 = 0$

$(2x + 3)(3x - 2) = 0$

$2x + 3 = 0$ or $3x - 2 = 0$

$x = -\dfrac{3}{2}$ $\quad x = \dfrac{2}{3}$

The solutions are $-\dfrac{3}{2}$ and $\dfrac{2}{3}$.

31. $h(r) = 5r^2 - 7r - 6$, $h(r) = -4$

$5r^2 - 7r - 6 = -4$

$5r^2 - 7r - 2 = 0$

$r = \dfrac{7 \pm \sqrt{(-7)^2 - 4(5)(-2)}}{2(5)}$

$= \dfrac{7 \pm \sqrt{49 + 40}}{10}$

$= \dfrac{7 \pm \sqrt{89}}{10}$

The solutions are $\dfrac{7 + \sqrt{89}}{10}$ and $\dfrac{7 - \sqrt{89}}{10}$.

32. $f(x) = -2x^2 + 6x + 5$, $f(x) = -4$

$-2x^2 + 6x + 5 = -4$

$-2x^2 + 6x + 9 = 0$

$x = \dfrac{-6 \pm \sqrt{6^2 - 4(-2)(9)}}{2(-2)}$

$= \dfrac{-6 \pm \sqrt{36 + 72}}{-4}$

$= \dfrac{-6 \pm \sqrt{108}}{-4}$

$= \dfrac{-6 \pm 6\sqrt{3}}{-4}$

$= \dfrac{3 \pm 3\sqrt{3}}{2}$

The solutions are $\dfrac{3 + 3\sqrt{3}}{2}$ and $\dfrac{3 - 3\sqrt{3}}{2}$.

33. Solutions are 3 and 2.

Factors are $(x - 3)$ and $(x + 2)$.

$f(x) = (x - 3)(x + 2)$

$f(x) = x^2 - x - 6$

34. Solutions are $\dfrac{2}{3}$ and –3.

Factors are $(3x - 2)$ and $(x + 3)$.

$f(x) = (3x - 2)(x + 3)$

$f(x) = 3x^2 + 7x - 6$

35. Solutions are $x = -\sqrt{5}$ and $x = \sqrt{5}$. Factors are $\left(x + \sqrt{5}\right)$ and $\left(x - \sqrt{5}\right)$.

$f(x) = \left(x + \sqrt{5}\right)\left(x - \sqrt{5}\right)$

$f(x) = x^2 - 5$

36. Solutions are $x = 3 - 2i$ and $x = 3 + 2i$.

Factors are $x - (3 - 2i) = x - 3 + 2i$ and

$x - (3 + 2i) = x - 3 - 2i$.

$f(x) = (x - 3 + 2i)(x - 3 - 2i)$

$= (x - 3)^2 - (2i)^2$

$= x^2 - 6x + 9 - 4i^2$

$= x^2 - 6x + 9 + 4$

$f(x) = x^2 - 6x + 13$

37. Let w = the width. Then the length is $w + 2$.

$A = lw$ so

$63 = (w + 2)w$

$63 = w^2 + 2w$

$0 = w^2 + 2w - 63$

$0 = (w + 9)(w - 7)$

$w + 9 = 0 \qquad w - 7 = 0$

$\quad w = -9 \qquad\quad w = 7$

Since length is never negative, the width is 7 ft and length is 9 ft.

38. Using the Pythagorean Theorem,

$a^2 + b^2 = c^2$

$8^2 + 8^2 = x^2$

$64 + 64 = x^2$

$\qquad 128 = x^2$

$\quad \sqrt{128} = x$

$\qquad x = 8\sqrt{2} \approx 11.31$

39. $A = P\left(1 + \dfrac{r}{n}\right)^{nt}$

$882 = 800\left(1 + \dfrac{r}{1}\right)^{1(2)}$

$882 = 800(1 + r)^2$

$1.1025 = (1 + r)^2$

$\pm 1.05 = 1 + r$

$\quad r = -1 \pm 1.05$

Since the rate must be positive,

$r = -1 + 1.05 = 0.05$.

The annual interest is 5%.

40. Let x be the smaller positive number. Then $x + 4$ is the larger positive number.

$x(x + 4) = 45$

$x^2 + 4x - 45 = 0$

$(x + 9)(x - 5) = 0$

$x = -9 \quad$ or $\quad x = 5$

Since x must be positive, $x = 5$ and $x + 4 = 9$.

41. Let x be the width. Then $2x - 1$ is the length and the equation is $A = lw$.

$66 = (2x - 1)(x)$

$66 = 2x^2 - x$

$0 = 2x^2 - x - 66$

$0 = (2x + 11)(x - 6)$

$\quad 0 = 2x + 11 \quad$ or $\quad 0 = x - 6$

$-11 = 2x \qquad\qquad\quad 6 = x$

$-\dfrac{11}{2} = x$

Since the width must be positive, $x = 6$. The width is 6 inches and the length $2x - 1$ is

$2(6) - 1 = 12 - 1 = 11$ inches.

42. $V = 12d - 0.05d^2$, $d = 50$

$V = 12(50) - 0.05(50)^2$

$V = 12(50) - 0.05(2500)$

$V = 600 - 125$

$V = 475$

The value is $475.

43. $d = -16t^2 + 1800$

a. $d = -16(3)^2 + 1800$

$d = -16(9) + 1800$

$d = -144 + 1800$

$d = 1656$

The object is 1656 feet from the ground 3 seconds after being dropped.

b. $\quad 0 = -16t^2 + 1800$

$16t^2 = 1800$

$\quad t^2 = 112.5$

$\quad t = \pm\sqrt{112.5}$

$\quad t \approx \pm 10.61$

Since the time must be positive,

$t = 10.61$ seconds.

44. $h = -16t^2 + 16t + 100$

a. $h = -16(2)^2 + 16(2) + 100$

$= -16(4) + 32 + 100$

$= -64 + 32 + 100$

$= 68$

The height is 68 feet.

b. $0 = -16t^2 + 16t + 100$

$0 = 4t^2 - 4t - 25$

$t = \dfrac{-(-4) \pm \sqrt{(-4)^2 - 4(4)(-25)}}{2(4)}$

$= \dfrac{4 \pm \sqrt{16 + 400}}{8}$

$= \dfrac{4 \pm \sqrt{416}}{8}$

Since the time must be positive,

$t = \dfrac{4 + \sqrt{416}}{8}$

≈ 3.05 seconds

The object hits the ground in about 3.05 seconds.

45. $L = 0.0004t^2 + 0.16t + 20$

a. $L = 0.0004(100)^2 + 0.16(100) + 20$

$= 40$

40 milliliters will leak out at 100°C.

b. $53 = 0.0004t^2 + 0.16t + 20$

$0 = 0.0004t^2 + 0.16t - 33$

$t = \dfrac{-0.16 \pm \sqrt{(0.16)^2 - 4(0.0004)(-33)}}{2(0.0004)}$

$= \dfrac{-0.16 \pm \sqrt{0.0784}}{0.0008}$

$t = \dfrac{-0.16 + \sqrt{0.0784}}{0.0008} = 150$

or $t = \dfrac{-0.16 - \sqrt{0.0784}}{0.0008} = -550$

Since the temperature must be positive,

$t = 150$. The operating temperature is 150°C.

46. Let x be the time in which the smaller machine can do the job then $x - 1$ is the time for the larger machine.

$$\frac{12}{x} + \frac{12}{x-1} = 1$$

$$x(x-1)\left(\frac{12}{x}\right) + x(x-1)\left(\frac{12}{x-1}\right) = x(x-1)$$

$$12(x-1) + 12x = x^2 - x$$

$$12x - 12 + 12x = x^2 - x$$

$$-12 + 24x = x^2 - x$$

$$0 = x^2 - 25x + 12$$

$$x = \frac{-(-25) \pm \sqrt{(-25)^2 - 4(1)(12)}}{2(1)}$$

$$= \frac{25 \pm \sqrt{577}}{2}$$

$$x = \frac{25 + \sqrt{577}}{2} \approx 24.5 \text{ or}$$

$$x = \frac{25 - \sqrt{577}}{2} \approx 0.49$$

x cannot equal 0.49 since this would mean the smaller machine could do the work in 0.49 hours and the larger can do the work in $x - 1$ or –0.51 hours. Therefore the smaller machine does the work in 24.51 hours and the larger machine does the work in 23.51 hours.

47. Let x be the speed (in miles per hour) for the first 20 miles. Then, the speed for the next 30 miles is $x + 10$. The time for the first 20 miles is $\dfrac{d}{r} = \dfrac{20}{x}$ and the time for the next 30 miles is $\dfrac{d}{r} = \dfrac{30}{x+10}$. The total time is 0.9 hours.

$$\frac{20}{x} + \frac{30}{x+10} = 0.9$$

$$x(x+10)\left(\frac{20}{x}\right) + x(x+10)\left(\frac{30}{x+10}\right) = x(x+10)(0.9)$$

$$20(x+10) + x(30) = \left(x^2 + 10x\right)(0.9)$$

$$50x + 200 = 0.9x^2 + 9x$$

$$0 = 0.9x^2 - 41x - 200$$

$$0 = 9x^2 - 410x - 2000$$

$$0 = (x - 50)(9x + 40)$$

$$x - 50 = 0 \quad \text{or} \quad 9x + 40 = 0$$

$$x = 50 \quad \text{or} \quad x = -\frac{40}{9}$$

Since the speed must be positive, $x = 50$.
Thus, the speed was 50 mph.

48. Let r be the speed (in miles per hour) of the canoe in still water. For the trip downstream, the rate is $r + 0.4$ and the distance 3 miles so that the time is $t = \dfrac{3}{r + 0.4}$. For the trip upstream the rate is $r - 0.4$ and the distance is 3 miles so that the time is $t = \dfrac{3}{r - 0.4}$. The total time is 4 hours.

$$\frac{3}{r + 0.4} + \frac{3}{r - 0.4} = 4$$

$$(r + 0.4)(r - 0.4)\left[\frac{3}{r + 0.4} + \frac{3}{r - 0.4} = 4\right]$$

$$3(r - 0.4) + 3(r + 0.4) = 4(r + 0.4)(r - 0.4)$$

$$3r - 1.2 + 3r + 1.2 = 4\left(r^2 - 0.16\right)$$

$$6r = 4r^2 - 0.64$$

$$0 = 4r^2 - 6r - 0.64$$

$$0 = 2r^2 - 3r - 0.32$$

$$r = \frac{-(-3) \pm \sqrt{(-3)^2 - 4(2)(-0.32)}}{2(2)}$$

$$= \frac{3 \pm \sqrt{9 + 2.56}}{4}$$

$$= \frac{3 \pm \sqrt{11.56}}{4}$$

$$= \frac{3 \pm 3.4}{4}$$

$$r = \frac{3 + 3.4}{4} \quad \text{or} \quad r = \frac{3 - 3.4}{4}$$

$$= \frac{6.4}{4} \qquad\qquad\quad = \frac{-0.4}{4}$$

$$= 1.6 \qquad\qquad\qquad = -0.1$$

Since the rate must be positive, $r = 1.6$. Rachel canoes 1.6 miles per hour in still water.

49. $V_x^2 + V_y^2 = V^2$ for V_y

$$V_y^2 = V^2 - V_x^2$$

$$V_y = \sqrt{V^2 - V_x^2}$$

50. $a = \dfrac{v_2^2 - v_1^2}{2d}$ for v_2

$$2ad = v_2^2 - v_1^2$$

$$2ad + v_1^2 = v_2^2$$

$$v_2^2 = \sqrt{v_1^2 + 2ad}$$

51. $p^4 - 5p^2 = 24$

$$p^4 - 5p^2 - 24 = 0$$

$$(p^2)^2 - 5p^2 - 24 = 0$$

$$u^2 - 5u - 24 = 0$$

$$(u - 8)(u + 3) = 0$$

$$u - 8 = 0 \qquad \text{or} \quad u + 3 = 0$$

$$u = 8 \qquad\qquad\qquad u = -3$$

$$p^2 = 8 \qquad\qquad\qquad p^2 = -3$$

$$p = \pm\sqrt{8} \qquad\qquad y = \pm\sqrt{-3}$$

$$= \pm 2\sqrt{2} \qquad\qquad = \pm i\sqrt{3}$$

The solutions are $\pm 2\sqrt{2}$ and $\pm i\sqrt{3}$.

52. $6m^{-2} + 11m^{-1} - 10 = 0$

$$\frac{6}{m^2} + \frac{11}{m} - 10 = 0$$

$$m^2\left(\frac{6}{m^2} + \frac{11}{m} - 10\right) = m^2(0)$$

$$6 + 11m - 10m^2 = 0$$

$$10m^2 - 11m - 6 = 0$$

$$(2m - 3)(5m + 2) = 0$$

$$2m - 3 = 0 \quad \text{or} \quad 5m + 2 = 0$$

$$2m = 3 \qquad\qquad\quad 5m = -2$$

$$m = \frac{3}{2} \qquad\qquad\quad m = -\frac{2}{5}$$

The solutions are $\dfrac{3}{2}$ and $-\dfrac{2}{5}$.

53. $4x + 23\sqrt{x} - 6 = 0$

$$4\left(x^{1/2}\right)^2 + 23x^{1/2} - 6 = 0$$

$$4u^2 + 23u - 6 = 0$$

$$(4u - 1)(u + 6) = 0$$

$4u - 1 = 0$ or $u + 6 = 0$

$4u = 1$ $u = -6$

$u = \dfrac{1}{4}$

$x^{1/2} = \dfrac{1}{4}$ $x^{1/2} = -6$

$x = \left(\dfrac{1}{4}\right)^2$

$= \dfrac{1}{16}$

There are no solutions for $x^{1/2} = -6$ since there is no real number x for which $x^{1/2} = -6$.

The solution is $\dfrac{1}{16}$.

54. $2m^{2/3} - 7m^{1/3} = -6$

$2m^{2/3} - 7m^{1/3} + 6 = 0$

$2\left(m^{1/3}\right)^2 - 7m^{1/3} + 6 = 0$

$2u^2 - 7u + 6 = 0$

$(2u - 3)(u - 2) = 0$

$2u - 3 = 0$ or $u - 2 = 0$

$u = \dfrac{3}{2}$ $u = 2$

$m^{1/3} = \dfrac{3}{2}$ $m^{1/3} = 2$

$m = \left(\dfrac{3}{2}\right)^3$ $m = 2^3$

$= \dfrac{27}{8}$ $= 8$

The solutions are $\dfrac{27}{8}$ and 8.

55. $10(p + 2) + 7 = \dfrac{12}{p + 2}$

$10(p + 2)^2 + 7(p + 2) = 12$

$10(p + 2)^2 + 7(p + 2) - 12 = 0$

$10u^2 + 7u - 12 = 0$

$(5u - 4)(2u + 3) = 0$

$5u - 4 = 0$ or $2u + 3 = 0$

$u = \dfrac{4}{5}$ $u = -\dfrac{3}{2}$

$p + 2 = \dfrac{4}{5}$ $p + 2 = -\dfrac{3}{2}$

$p = -\dfrac{6}{5}$ $p = -\dfrac{7}{2}$

The solutions are $-\dfrac{6}{5}$ and $-\dfrac{7}{2}$.

56. $6(x - 2)^{-2} = -13(x - 2)^{-1} + 8$

$6\left[(x - 2)^{-1}\right]^2 = -13(x - 2)^{-1} + 8$

$6u^2 = -13u + 8$

$6u^2 + 13u - 8 = 0$

$(2u - 1)(3u + 8) = 0$

$2u - 1 = 0$ or $3u + 8 = 0$

$u = \dfrac{1}{2}$ $u = -\dfrac{8}{3}$

$(x - 2)^{-1} = \dfrac{1}{2}$ $(x - 2)^{-1} = -\dfrac{8}{3}$

$x - 2 = 2$ $x - 2 = -\dfrac{3}{8}$

$x = 4$ $x = \dfrac{13}{8}$

The solutions are 4 and $\dfrac{13}{8}$.

57. $f(x) = 30x + 13\sqrt{x} - 10$

To find the x-intercepts, set $f(x) = 0$.

$0 = 30x + 13\sqrt{x} - 10$

$0 = 30\left(\sqrt{x}\right)^2 + 13\sqrt{x} - 10$

$0 = 30u^2 + 13u - 10$

$0 = (6u + 5)(5u - 2)$

$6u + 5 = 0$ $5u - 2 = 0$

$u = -\dfrac{5}{6}$ $u = \dfrac{2}{5}$

$\sqrt{x} = -\dfrac{5}{6}$ $\sqrt{x} = \dfrac{2}{5}$

$x = \dfrac{4}{25}$

Since $\sqrt{x}$ cannot be negative, the solution is $\dfrac{4}{25}$. The only x-intercept is $\left(\dfrac{4}{25}, 0\right)$.

58. $g(x) = 12(x^2 - 4x)^2 + 16(x^2 - 4x) - 3$

To find the x-intercepts, set $g(x) = 0$.

$0 = 12(x^2 - 4x)^2 + 16(x^2 - 4x) - 3$

$0 = 12u^2 + 16u - 3$

$0 = (2u + 3)(6y - 1)$

$$
\begin{array}{lll}
2u + 3 = 0 & \text{or} & 6u - 1 = 0 \\
u = -\dfrac{3}{2} & & u = \dfrac{1}{6} \\
x^2 - 4x = -\dfrac{3}{2} & & x^2 - 4x = \dfrac{1}{6} \\
2x^2 - 8x = -3 & & 6x^2 - 24x = 1 \\
2x^2 - 8x + 3 = 0 & & 6x^2 - 24x - 1 = 0
\end{array}
$$

$$
\begin{array}{ll}
x = \dfrac{8 \pm \sqrt{(-8)^2 - 4(2)(3)}}{2(2)} & x = \dfrac{24 \pm \sqrt{(-24)^2 - 4(6)(-1)}}{2(6)} \\[3mm]
= \dfrac{8 \pm \sqrt{40}}{4} & = \dfrac{24 \pm \sqrt{600}}{12} \\[3mm]
= \dfrac{8 \pm 2\sqrt{10}}{4} & = \dfrac{24 \pm 10\sqrt{6}}{12} \\[3mm]
= \dfrac{4 \pm \sqrt{10}}{2} & = \dfrac{12 \pm 5\sqrt{6}}{6}
\end{array}
$$

The solutions are $\dfrac{4 + \sqrt{10}}{2}$, $\dfrac{4 - \sqrt{10}}{2}$, $\dfrac{12 + 5\sqrt{6}}{6}$, and $\dfrac{12 - 5\sqrt{6}}{6}$.

The x-intercepts are $\left(\dfrac{4 + \sqrt{10}}{2}, 0\right)$, $\left(\dfrac{4 - \sqrt{10}}{2}, 0\right)$, $\left(\dfrac{12 + 5\sqrt{6}}{6}, 0\right)$ and $\left(\dfrac{12 - 5\sqrt{6}}{6}, 0\right)$.

59. $y = x^2 + 6x$

a. Since $a = 1$ the parabola opens upward.

b. $y = 0^2 + 6(0) = 0$
The y-intercept is $(0, 0)$.

c. $x = -\dfrac{b}{2a} = -\dfrac{6}{2(1)} = -\dfrac{6}{2} = -3$

$y = \dfrac{4ac - b^2}{4a}$

$= \dfrac{4(1)(0) - 6^2}{4(1)}$

$= \dfrac{-36}{4} = -9$

The vertex is $(-3, -9)$.

d. $0 = x^2 + 6x$
$0 = x(x + 6)$
$0 = x$ or $0 = x + 6$
$x = 0$ $x = -6$
The x-intercepts are $(0, 0)$ and $(-6, 0)$.

e.

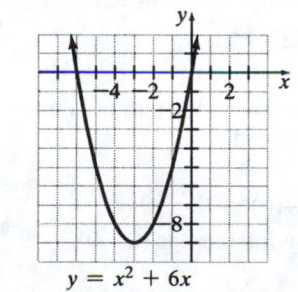

$y = x^2 + 6x$

60. $y = x^2 + 2x - 8$

a. Since $a = 1$ the parabola opens upward.

b. $y = (0)^2 + 2(0) - 8 = -8$
The y-intercept is $(0, -8)$.

c. $x = -\dfrac{b}{2a} = -\dfrac{2}{2(1)} = -\dfrac{2}{2} = -1$

$y = \dfrac{4ac - b^2}{4a}$

$= \dfrac{4(1)(-8) - (2)^2}{4(1)}$

$= \dfrac{-32 - 4}{4}$

$= \dfrac{-36}{4}$

$= -9$

The vertex is $(-1, -9)$.

d. $0 = x^2 + 2x - 8$
$0 = (x + 4)(x - 2)$
$0 = x + 4$ or $0 = x - 2$
$-4 = x$ or $2 = x$
The x-intercepts are $(-4, 0)$ and $(2, 0)$.

e.

61. $y = -x^2 - 9$

a. Since $a = -1$ the parabola opens downward.

b. $y = -(0)^2 - 9 = -9$
The y-intercept is $(0, -9)$.

c. $x = -\dfrac{b}{2a} = -\dfrac{0}{2(-1)} = -\dfrac{0}{-2} = 0$

$y = \dfrac{4ac - b^2}{4a}$

$= \dfrac{4(-1)(-9) - 0^2}{4(-1)}$

$= \dfrac{36}{-4}$

$= -9$

The vertex is $(0, -9)$.

d. $0 = -x^2 - 9$
$x^2 = -9$
$x = \pm\sqrt{-9}$ or $\pm 3i$
There are no real roots. Thus, there are no x-intercepts.

e.
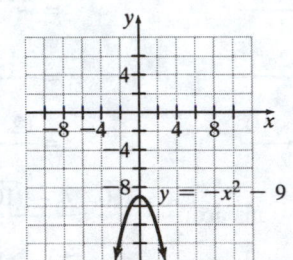

62. $y = -2x^2 - x + 15$

a. Since $a = -2$ the parabola opens downward.

b. $y = -2(0)^2 - 0 + 15 = 15$
The y-intercept is $(0, 15)$.

c. $x = -\dfrac{b}{2a} = -\dfrac{-1}{2(-2)} = \dfrac{1}{-4} = -\dfrac{1}{4}$

$y = \dfrac{4ac - b^2}{4a}$

$= \dfrac{4(-2)(15) - (-1)^2}{4(-2)}$

$= \dfrac{121}{8}$

The vertex is $\left(-\dfrac{1}{4}, \dfrac{121}{8}\right)$.

d.
$$0 = -1\left(2x^2 + x - 15\right)$$
$$= -1(2x - 5)(x + 3)$$
$$0 = (2x - 5) \quad \text{or} \quad 0 = x + 3$$
$$5 = 2x$$
$$\frac{5}{2} = x \qquad\qquad -3 = x$$

The x-intercepts are $(-3, 0)$ and $\left(\dfrac{5}{2}, 0\right)$.

e.

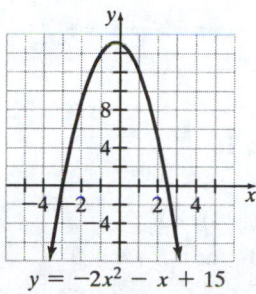

$$y = -2x^2 - x + 15$$

63. a. $I = -x^2 + 22x - 30,\ 2 \le x \le 20$

The x-coordinate of the vertex will be the cost per ticket to maximize profit.

$$x = -\frac{b}{2a} = -\frac{22}{2(-1)} = 11$$

They should charge $11 per ticket.

b. The maximum profit in hundreds is the y-coordinate of the vertex.

$$I(11) = -11^2 + 22(11) - 30$$
$$= -121 + 242 - 30$$
$$= 91$$

The maximum profit is $91 hundred or $9100.

64. a. $s(t) = -16t^2 + 80t + 60$

The ball will attain maximum height at the x-coordinate of the vertex.

$$t = -\frac{b}{2a} = -\frac{80}{2(-16)} = -\frac{80}{-32} = 2.5$$

The ball will attain maximum height 2.5 seconds after being thrown.

b. The maximum height is the y-coordinate of the vertex.

$$s(2.5) = -16(2.5)^2 + 80(2.5) + 60$$
$$= -100 + 200 + 60$$
$$= 160$$

The maximum height is 160 feet.

65. The graph of $f(x) = (x - 3)^2$ has vertex $(3, 0)$. The graph will be $g(x) = x^2$ shifted right 3 units.

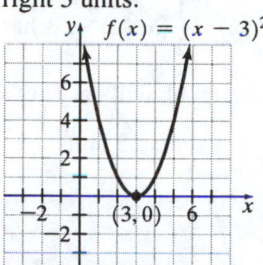

66. The graph of $f(x) = -(x + 2)^2 - 3$ has vertex $(-2, -3)$. Since $a < 0$, the parabola opens downward. The graph will be $g(x) = -x^2$ shifted left 2 units and down 3 units.

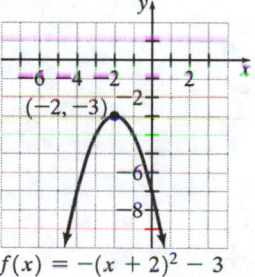

$$f(x) = -(x + 2)^2 - 3$$

67. The graph of $g(x) = -2(x + 4)^2 - 1$ has vertex $(-4, -1)$. Since $a < 0$, the parabola opens downward. The graph will be $f(x) = -2x^2$ shifted left 4 units and down 1 unit.

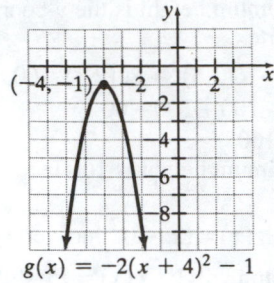

$g(x) = -2(x+4)^2 - 1$

68. The graph of $h(x) = \dfrac{1}{2}(x-1)^2 + 3$ has

vertex $(1, 3)$. The graph will be $f(x) = \dfrac{1}{2}x^2$

shifted right 1 unit and up 3 units.

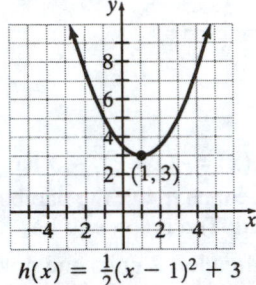

$h(x) = \frac{1}{2}(x-1)^2 + 3$

69. $x^2 + 6x + 5 \geq 0$

$(x+1)(x+5) \geq 0$

$$x = \frac{-(-11) \pm \sqrt{(-11)^2 - 4(1)(20)}}{2(1)}$$

$$= \frac{11 \pm \sqrt{121 - 80}}{2}$$

$$= \frac{11 \pm \sqrt{41}}{2}$$

$\underset{\text{False}}{+} \quad \underset{\text{True}}{-} \quad \underset{\text{False}}{+}$

$\dfrac{11 - \sqrt{41}}{2} \quad \dfrac{11 + \sqrt{41}}{2}$

72. $3x^2 + 8x > 16$

$3x^2 + 8x - 16 > 0$

$(3x-4)(x+4) > 0$

$3x - 4 = 0 \quad$ or $\quad x + 4 = 0$

$x = \dfrac{4}{3} \qquad\qquad x = -4$

$-4 \qquad \dfrac{4}{3}$

73. $4x^2 - 9 \leq 0$

$(2x-3)(2x+3) \leq 0$

$2x - 3 = 0 \quad$ or $\quad 2x + 3 = 0$

$x = \dfrac{3}{2} \qquad\qquad x = -\dfrac{3}{2}$

$-\dfrac{3}{2} \qquad \dfrac{3}{2}$

70. $x^2 + 2x - 15 \leq 0$

$(x+5)(x-3) \leq 0$

$-5 \qquad 3$

74. $5x^2 - 25 > 0$

$5(x^2 - 5) > 0$

$5(x+\sqrt{5})(x-\sqrt{5}) > 0$

$x + \sqrt{5} = 0 \qquad$ or $\quad x - \sqrt{5} = 0$

$x = -\sqrt{5} \qquad\qquad x = \sqrt{5}$

$-\sqrt{5} \qquad \sqrt{5}$

71. $x^2 \leq 11x - 20$

$x^2 - 11x + 20 \leq 0$

$x^2 - 11x + 20 = 0$

75. $\dfrac{x+2}{x-3} > 0$

$x \neq 3$

$$\begin{array}{ccccc} - & + & & + & x+2 \\ - & & - & + & x-3 \\ + & & - & + & \end{array}$$

$\xleftarrow{\hspace{2cm}} \overset{-2}{\circ} \quad \overset{3}{\circ} \xrightarrow{\hspace{2cm}}$

$\{x \mid x < -2 \text{ or } x > 3\}$

76. $\dfrac{x-5}{x+2} \leq 0$

$x = -2$

$$\begin{array}{ccccc} - & - & & + & x-5 \\ - & & + & + & x+2 \\ + & & - & + & \end{array}$$

$\{x \mid -2 < x \leq 5\}$

77. $\dfrac{2x-4}{x+1} \geq 0$

$\dfrac{2(x-2)}{x+1} \geq 0$

$x \neq -1$

$$\begin{array}{ccccc} - & - & & + & x-2 \\ - & & + & + & x+1 \\ + & & - & + & \end{array}$$

$\{x \mid x < -1 \text{ or } x \geq 2\}$

78. $\dfrac{3x+5}{x-6} < 0$

$x \neq 6$

$$\begin{array}{ccccc} - & + & & + & 3x+5 \\ - & & - & + & x-6 \\ + & & - & + & \end{array}$$

$\left\{x \mid -\dfrac{5}{3} < x < 6\right\}$

79. $(x+3)(x+1)(x-2) > 0$

$$\begin{array}{ccccccc} - & + & & + & & + & x+3 \\ - & & - & & + & + & x+1 \\ - & & - & & - & + & x-2 \\ - & + & & - & & + & \end{array}$$

$\{x \mid -3 < x < -1 \text{ or } x > 2\}$

80. $x(x-3)(x-5) \leq 0$

$$\begin{array}{ccccccc} - & + & & + & & + & x \\ - & & - & & + & + & x-3 \\ - & & - & & - & + & x-5 \\ - & + & & - & & + & \end{array}$$

$\{x \mid x \leq 0 \text{ or } 3 \leq x \leq 5\}$

81. $(3x+4)(x-1)(x-3) \geq 0$

$$\begin{array}{ccccccc} - & + & & + & & + & 3x+4 \\ - & & - & & + & + & x-1 \\ - & & - & & - & + & x-3 \\ - & + & & - & & + & \end{array}$$

$\left[-\dfrac{4}{3}, 1\right] \cup [3, \infty)$

82. $2x(x+2)(x+5) < 0$

$$\begin{array}{ccccccc} - & - & & - & & + & x \\ - & & - & & + & + & x+2 \\ - & & + & & + & + & x+5 \\ - & + & & - & & + & \end{array}$$

$(-\infty, -5) \cup (-2, 0)$

83. $\dfrac{x(x-4)}{x+2} > 0$

$x \neq -2$

$$\begin{array}{ccccccc} - & - & & + & & + & x \\ - & & - & & - & + & x-4 \\ - & & + & & + & + & x+2 \\ - & & + & - & & + & \end{array}$$

$(-2, 0) \cup (4, \infty)$

84. $\dfrac{(x-2)(x-5)}{x+3} < 0$

$x \neq -3$

$(-\infty, -3) \cup (2, 5)$

85. $\dfrac{x-3}{(x+2)(x-5)} \geq 0$

$x \neq -2, \, x \neq 5$

$(-2, 3] \cup (5, \infty)$

86. $\dfrac{x(x-5)}{x+3} \leq 0$

$x \neq -3$

$(-\infty, -3) \cup [0, 5]$

87. $\dfrac{3}{x+4} \geq -1$

$\dfrac{3}{x+4} + 1 \geq 0$

$\dfrac{3+1(x+4)}{x+4} \geq 0$

$\dfrac{3+x+4}{x+4} \geq 0$

$\dfrac{x+7}{x+4} \geq 0$

$x \neq -4$

88. $\dfrac{2x}{x-2} \leq 1$

$\dfrac{2x}{x-2} - 1 \leq 0$

$\dfrac{2x}{x-2} - \dfrac{1(x-2)}{x-2} \leq 0$

$\dfrac{2x-x+2}{x-2} \leq 0$

$\dfrac{x+2}{x-2} \leq 0$

$x \neq 2$

89. $\dfrac{2x+3}{3x-5} < 4$

$\dfrac{2x+3}{3x-5} - 4 < 0$

$\dfrac{2x+3-4(3x-5)}{3x-5} < 0$

$\dfrac{2x+3-12x+20}{3x-5} < 0$

$\dfrac{-10x+23}{3x-5} < 0$

Practice Test

1. $x^2 = -x + 12$

$x^2 + x = 12$

$x^2 + x + \dfrac{1}{4} = \dfrac{48}{4} + \dfrac{1}{4}$

$\left(x + \dfrac{1}{2}\right)^2 = \dfrac{49}{4}$

$x + \dfrac{1}{2} = \pm \dfrac{7}{2}$

$x = -\dfrac{1}{2} \pm \dfrac{7}{2}$

$$x = -\frac{1}{2} + \frac{7}{2} \quad \text{or} \quad x = -\frac{1}{2} - \frac{7}{2}$$
$$= \frac{6}{2} \qquad\qquad = -\frac{8}{2}$$
$$= 3 \qquad\qquad = -4$$

2. $4x^2 + 8x = -12$

$$\frac{4x^2}{4} + \frac{8x}{4} = \frac{-12}{4}$$
$$x^2 + 2x = -3$$
$$x^2 + 2x + 1 = -3 + 1$$
$$(x+1)^2 = -2$$
$$x + 1 = \pm\sqrt{-2}$$
$$x + 1 = \pm i\sqrt{2}$$
$$x = -1 \pm i\sqrt{2}$$
$$x = -1 + i\sqrt{2} \quad \text{or} \quad x = -1 - i\sqrt{2}$$

3. $x^2 - 5x - 6 = 0 \qquad a = 1, b = -5, c = -6$

$$x = \frac{-b \pm \sqrt{b^2 - 4ac}}{2a}$$
$$x = \frac{-(-5) \pm \sqrt{(-5)^2 - 4(1)(-6)}}{2(1)}$$
$$= \frac{5 \pm \sqrt{25 + 24}}{2}$$
$$= \frac{5 \pm \sqrt{49}}{2}$$
$$= \frac{5 \pm 7}{2}$$
$$x = \frac{5 + 7}{2} \quad \text{or} \quad x = \frac{5 - 7}{2}$$
$$= \frac{12}{2} \qquad\qquad = \frac{-2}{2}$$
$$= 6 \qquad\qquad = -1$$

4. $x^2 + 5 = -8x$

$$x^2 + 8x + 5 = 0$$
$$a = 1, b = 8, c = 5$$
$$x = \frac{-b \pm \sqrt{b^2 - 4ac}}{2a}$$

$$x = \frac{-8 \pm \sqrt{8^2 - 4(1)(5)}}{2(1)}$$
$$= \frac{-8 \pm \sqrt{64 - 20}}{2}$$
$$= \frac{-8 \pm \sqrt{44}}{2}$$
$$= \frac{-8 \pm 2\sqrt{11}}{2}$$
$$= \frac{2\left(-4 \pm \sqrt{11}\right)}{2}$$
$$= -4 \pm \sqrt{11}$$
$$x = -4 + \sqrt{11} \quad \text{or} \quad x = -4 - \sqrt{11}$$

5. $3x^2 - 5x = 0$

$$x(3x - 5) = 0$$
$$x = 0 \quad \text{or} \quad 3x - 5 = 0$$
$$3x = 5$$
$$x = \frac{5}{3}$$

6. $-2x^2 = 9x - 5$

$$0 = 2x^2 + 9x - 5$$
$$0 = (2x - 1)(x + 5)$$
$$2x - 1 = 0 \quad \text{or} \quad x + 5 = 0$$
$$x = \frac{1}{2} \qquad\qquad x = -5$$

7. x-intercepts are 3 and $-\dfrac{5}{2}$

Factors are $(x - 3)$ and $(2x + 5)$
$$f(x) = (x - 3)(2x + 5)$$
$$f(x) = 2x^2 - x - 15$$

8. $K = \dfrac{1}{2}mv^2$ for v

$$2K = mv^2$$
$$\frac{2K}{m} = v^2$$
$$v = \sqrt{\frac{2K}{m}}$$

9. a. $c = -0.01s^2 + 78s + 22,000$

$c(2000)$

$= -0.01(2000)^2 + 78(2000) + 22,000$

$= -40,000 + 156,000 + 22,000$

$= 138,000$

The cost is $138,000.

b. $160,000 = -0.01s^2 + 78s + 22,000$

$0 = -0.01s^2 + 78s - 138,000$

$s = \dfrac{-78 \pm \sqrt{78^2 - 4(-0.01)(-138,000)}}{2(-0.01)}$

$= \dfrac{-78 \pm \sqrt{564}}{-0.02}$

$s = \dfrac{-78 + \sqrt{564}}{-0.02} \approx 2712.57$

$s = \dfrac{-78 - \sqrt{564}}{-0.02} \approx 5087.43$

Since $1300 \le s \le 3900$, the house should have about 2712.57 square feet.

10. The formula $d = rt$ can be written $t = \dfrac{d}{r}$.

Let $r =$ his actual rate.

	distance	rate	time $= \dfrac{d}{r}$
actual trip	520	r	$\dfrac{520}{r}$
faster trip	520	$r + 15$	$\dfrac{520}{r+15}$

The faster trip would have taken 2.4 hours less time than the actual trip.

$\dfrac{520}{r+15} = \dfrac{520}{r} - 2.4$

$r(r+15)\left(\dfrac{520}{r+15}\right) = r(r+15)\left(\dfrac{520}{r} - 2.4\right)$

$520r = 520(r+15) - 2.4r(r+15)$

$520r = 520r + 7800 - 2.4r^2 - 36r$

$0 = -2.4r^2 - 36r + 7800$

$0 = r^2 + 15r - 3250$

$0 = (r+65)(r-50)$

$r + 65 = 0 \qquad r - 50 = 0$

$r = -65 \qquad r = 50$

Since speed is never negative, Jacob drove an average speed of 50 mph.

11. $10m^4 + 21m^2 = 10$

$10m^4 + 21m^2 - 10 = 0$

$10(m^2)^2 + 21m^2 - 10 = 0$

$10u^2 + 21u - 10 = 0$

$(5u - 2)(2u + 5) = 0$

$5u - 2 = 0 \quad$ or $\quad 2u = -5$

$u = \dfrac{2}{5} \qquad\qquad u = -\dfrac{5}{2}$

$m^2 = \dfrac{2}{5} \qquad\qquad m^2 = -\dfrac{5}{2}$

$m = \pm\sqrt{\dfrac{2}{5}} \qquad\qquad m = \pm\sqrt{-\dfrac{5}{2}}$

$= \pm\dfrac{\sqrt{2}}{\sqrt{5}} \cdot \dfrac{\sqrt{5}}{\sqrt{5}} \qquad = \pm\dfrac{i\sqrt{5}}{\sqrt{2}} \cdot \dfrac{\sqrt{2}}{\sqrt{2}}$

$= \pm\dfrac{\sqrt{10}}{5} \qquad\qquad = \pm\dfrac{i\sqrt{10}}{2}$

12. $3r^{2/3} + 11r^{1/3} - 42 = 0$

$3\left(r^{1/3}\right)^2 + 11r^{1/3} - 42 = 0$

$3u^2 + 11u - 42 = 0$

$(3u - 7)(u + 6) = 0$

$3u - 7 = 0$ or $u + 6 = 0$

$u = \dfrac{7}{3}$ $u = -6$

$r^{1/3} = \dfrac{7}{3}$ $r^{1/3} = -6$

$r = \left(\dfrac{7}{3}\right)^3$ $r = (-6)^3$

$= \dfrac{343}{27}$ $r = -216$

13. $f(x) = 16x - 40\sqrt{x} + 25$

$0 = 16\left(\sqrt{x}\right)^2 - 40\sqrt{x} + 25$

$0 = 16u^2 - 40u + 25$

$0 = (4u - 5)^2$

$4u - 5 = 0$

$u = \dfrac{5}{4}$

$\sqrt{x} = \dfrac{5}{4}$

$x = \dfrac{25}{16}$

The x-intercept is $\left(\dfrac{25}{16}, 0\right)$.

14. $f(x) = (x - 3)^2 + 2$

The vertex is $(3, -2)$. The graph will be the graph of $g(x) = x^2$ shifted 3 units right and 2 units down.

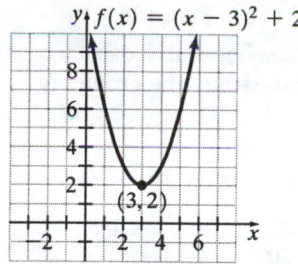

15. $h(x) = -\dfrac{1}{2}(x - 2)^2 - 2$

The vertex is $(2, -2)$. The graph will be the graph of $g(x) = -\dfrac{1}{2}x^2$ shifted 2 units right and 2 units down.

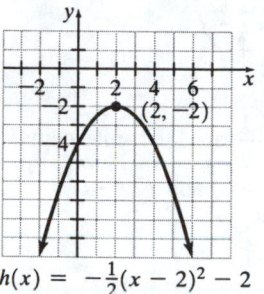

$h(x) = -\frac{1}{2}(x - 2)^2 - 2$

16. Begin by rewriting the equation in standard form.

$5x^2 - 4x - 2 = 0$ $a = 5, b = -4, c = -2$

The discriminant is

$b^2 - 4ac = (-4)^2 - 4(5)(-2)$

$= 56$

Since the discriminant is greater than 0, the quadratic equation has two distinct real solutions.

17. $y = x^2 - 2x - 8$

a. Since $a = 1$ the parabola opens upward.

b. $y = 0^2 - 2(0) - 8$

$y = -8$

The y-intercept is $(0, -8)$.

c. $x = -\dfrac{b}{2a} = -\dfrac{-2}{2(1)} = 1$

$y = 1^2 - 2(1) - 8 = -9$

The vertex is $(1, -9)$.

d. The x-intercepts occur when $y = 0$.

$0 = x^2 - 2x - 8$

$0 = (x - 4)(x + 2)$

$x - 4 = 0$ or $x + 2 = 0$

$x = 4$ $x = -2$

The x-intercepts are $(-2, 0)$ and $(4, 0)$.

e.

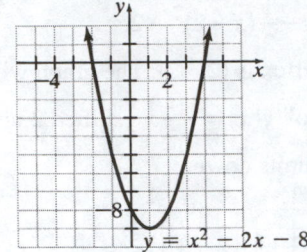

$y = x^2 - 2x - 8$

18. Since -6 and $\dfrac{1}{2}$ are the x-intercepts, the factors are $(x + 6)$ and $(2x - 1)$.
$$f(x) = (x + 6)(2x - 1)$$
$$f(x) = 2x^2 + 11x - 6$$

19.
$$x^2 - x \geq 42$$
$$x^2 - x - 42 \geq 0$$
$$(x - 7)(x + 6) \geq 0$$
$$x - 7 = 0 \quad \text{or} \quad x + 6 = 0$$
$$x = 7 \qquad\qquad x = -6$$

20. $\dfrac{(x + 3)(x - 4)}{x + 1} \geq 0$
$$x \neq -1$$
$$x + 3 = 0 \quad \text{or} \quad x - 4 = 0 \quad \text{or} \quad x + 1 = 0$$
$$x = -3 \qquad\qquad x = 4 \qquad\qquad x = -1$$

21.
$$\frac{x + 3}{x + 2} \leq -1$$
$$\frac{x + 3}{x + 2} + 1 \leq 0$$
$$\frac{x + 3}{x + 2} + \frac{x + 2}{x + 2} \leq 0$$
$$\frac{2x + 5}{x + 2} \leq 0$$
$$x \neq -2$$

a. $\left[-\dfrac{5}{2}, -2 \right)$

b. $\left\{ x \middle| -\dfrac{5}{2} \leq x < -2 \right\}$

22. Let x be the width of the carpet. Then $2x + 4$ is the length $A = lw$.
$$48 = x(2x + 4)$$
$$48 = 2x^2 + 4x$$
$$0 = 2x^2 + 4x - 48$$
$$0 = x^2 + 2x - 24$$
$$0 = (x - 4)(x + 6)$$
$$x - 4 = 0 \quad \text{or} \quad x + 6 = 0$$
$$x = 4 \qquad\qquad x = -6$$
The width is 4 feet and the length is $2 \cdot 4 + 4 = 8 + 4 = 12$ feet.

23. $d = -16t^2 + 64t + 80$
$d = 0$ when the ball strikes the ground.
$$0 = -16t^2 + 64t + 80$$
$$0 = -16\left(t^2 - 4t - 5 \right)$$
$$0 = -16(t - 5)(t + 1)$$
$$t - 5 = 0 \quad \text{or} \quad t + 1 = 0$$
$$t = 5 \qquad\qquad t = -1$$
The time must be positive, so $t = 5$. Thus, the ball strikes the ground in 5 seconds.

24. a. $f(x) = -1.4x^2 + 56x - 60$
$$x = -\frac{b}{2a} = -\frac{56}{2(-1.4)} = -\frac{56}{-2.8} = 20$$
The company must sell 20 carvings.

b. $f(20) = -1.4(20)^2 + 56(20) - 60$
           $= -560 + 1120 - 60$
           $= 500$
The maximum weekly profit is $500.

25. a.   Profit = Revenue − Cost
       $P(x) = \left(500x - x^2\right) - (8000 + 20x)$
           $= 500x - x^2 - 8000 - 20x$
           $= -x^2 + 480x - 8000$
The x coordinate of the vertex is
       $x = -\dfrac{b}{2a} = -\dfrac{480}{2(-1)} = 240$

To maximize profit, they must sell 240 items.

b. $P(240) = -240^2 + 480(240) - 8000$
           $= -57,600 + 115,200 - 8000$
           $= 49,600$
The maximum profit is $49,600.

Cumulative Review Test

1. $\left(|-3| - 5\right) - \left(4 \cdot |-6|\right) = (3 - 5) - (4 \cdot 6)$
                      $= -2 - 24$
                      $= -26$

2. a. $\left(9.96 \times 10^7\right)(0.56) = 5.5776 \times 10^7$

b. $\left(9.96 \times 10^7\right)(0.121) = 1.20516 \times 10^7$

c. $5.5776 \times 10^7 - 1.20516 \times 10^7$
      $= 4.37244 \times 10^7$

3. $-2(x + 5) = 3\left\{[5 - (x - 5)] - 6x\right\}$
    $-2x - 10 = 3\left\{[5 - x + 5] - 6x\right\}$
    $-2x - 10 = 3\left\{10 - x - 6x\right\}$
    $-2x - 10 = 3\left\{10 - 7x\right\}$
    $-2x - 10 = 30 - 21x$
     $19x - 10 = 30$
        $19x = 40$
          $x = \dfrac{40}{19}$

4. $\left|\dfrac{x - 5}{3}\right| < 8$
     $-8 < \dfrac{x - 5}{3} < 8$
     $-24 < x - 5 < 24$
     $-19 < x < 29$

5. $\left|\dfrac{1}{3}a + 3\right| = \left|\dfrac{2}{3}a - 1\right|$

   $\dfrac{1}{3}a + 3 = \dfrac{2}{3}a - 1$ or $\dfrac{1}{3}a + 3 = -\left(\dfrac{2}{3}a - 1\right)$

    $-\dfrac{1}{3}a + 3 = -1$ $\dfrac{1}{3}a + 3 = -\dfrac{2}{3}a + 1$

     $-\dfrac{1}{3}a = -4$ $a + 3 = 1$

        $a = 12$ $a = -2$

The solution set is $\{12, -2\}$.

6.   Yes, the set of ordered pairs is a function since no two ordered pairs have the same first coordinate.

7. a.   No, the graph is not a function since each
       x-value does not have a unique y-value.

b.   The domain is the set of x-values,
       Domain: $\{x | x \geq -2\}$. The range is the
       set of y-values, Range: $\mathbb{R}$

8. a. $x = -4$ is a vertical line.

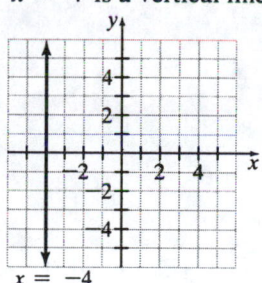

$x = -4$

b. $y = 2$ is a horizontal line.

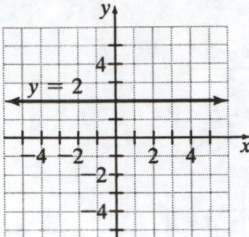

9. a. $\dfrac{1}{2}y = 2(x-3)+4$

$y = 4(x-3)+8$

$y = 4x - 12 + 8$

$y = 4x - 4$

$4x - y = 4$

b. The graph of $\dfrac{1}{2}y = 2(x-3)+4$ is the graph of $4x - y = 4$.

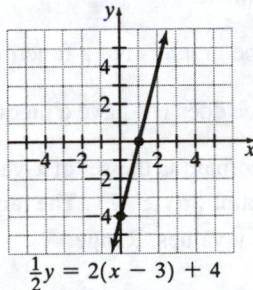

$\frac{1}{2}y = 2(x-3)+4$

10. $4y = -3x + 7$

$y = -\dfrac{3}{4}x + \dfrac{7}{4}$

The slope of the line is $-\dfrac{3}{4}$. A line that passes through $(2, -1)$ perpendicular to this line will have slope $\dfrac{4}{3}$.

$y - y_1 = m(x - x_1)$

$y - (-1) = \dfrac{4}{3}(x - 2)$

$y + 1 = \dfrac{4}{3}(x - 2)$

11. $x - 2y = 8$

$2x + y = 6$

$\begin{bmatrix} 1 & -2 & | & 8 \\ 2 & 1 & | & 6 \end{bmatrix}$

$\begin{bmatrix} 1 & -2 & | & 8 \\ 0 & 5 & | & -10 \end{bmatrix} 2R_1 + R_2$

$\begin{bmatrix} 1 & -2 & | & 8 \\ 0 & 1 & | & -2 \end{bmatrix} \frac{1}{5}R_2$

The system is

$x - 2y = 8$

$y = -2$

Substitute -2 for y in the first equation.

$x - 2y = 8$

$x - 2(-2) = 8$

$x + 4 = 8$

$x = 4$

The solution to the system is $(4, -2)$.

12. $\begin{vmatrix} 4 & 0 & -2 \\ 3 & 5 & 1 \\ 1 & -1 & 7 \end{vmatrix} = 4\begin{vmatrix} 5 & 1 \\ -1 & 7 \end{vmatrix} - 3\begin{vmatrix} 0 & -2 \\ -1 & 7 \end{vmatrix} + 1\begin{vmatrix} 0 & -2 \\ 5 & 1 \end{vmatrix}$

$= 4[5(7) - (-1)(1)] - 3[0(7) - (-1)(-2)] + 1[0(1) - 5(-2)]$

$= 4(35 + 1) - 3(0 - 2) + 1(0 + 10)$

$= 4(36) - 3(-2) + 1(10)$

$= 144 + 6 + 10$

$= 160$

13. $3p^4 - 12p^2q + 9p^2q - 36q^2 = 3\left(p^4 - 4p^2q + 3p^2q - 12q^2\right)$

$\qquad\qquad\qquad\qquad\qquad = 3\left[p^2\left(p^2 - 4q\right) + 3q\left(p^2 - 4q\right)\right]$

$\qquad\qquad\qquad\qquad\qquad = 3\left(p^2 - 4q\right)\left(p^2 + 3q\right)$

14. $f(x) = x^3 - 13x - 12$, $g(x) = x - 4$

    a. $\left(\dfrac{f}{g}\right)(x) = \left(x^3 - 13x - 12\right) \div (x - 4)$

$$
\begin{array}{r}
x^2 + 4x + 3 \\
x-4\overline{\smash{)}x^3 - 0x^2 - 13x - 12} \\
\underline{x^3 - 4x^2} \\
4x^2 - 13x \\
\underline{4x^2 - 16x} \\
3x - 12 \\
\underline{3x - 12} \\
0
\end{array}
$$

$\qquad\qquad \left(\dfrac{f}{g}\right)(x) = x^2 + 4x + 3,\ x \ne 4$

    b. $\left(\dfrac{f}{g}\right)(6) = 6^2 + 4(6) + 3$

$\qquad\qquad\qquad = 36 + 24 + 3$

$\qquad\qquad\qquad = 63$

15. $f(x) = 2x^2 + 4x - 6$, $g(x) = 3x - 4$

$(f \cdot g)(x) = \left(2x^2 + 4x - 6\right)(3x - 4)$

$\qquad\qquad = 2x^2(3x - 4) + 4x(3x - 4) - 6(3x - 4)$

$\qquad\qquad = 6x^3 - 8x^2 + 12x^2 - 16x - 18x + 24$

$\qquad\qquad = 6x^3 + 4x^2 - 34x + 24$

16. Let $x =$ the length of the side of the second square. Then $x + 4 =$ the length of the side of the first and larger square. Use $A = s^2$.

$121 = (x + 4)^2$

$121 = x^2 + 8x + 16$

$0 = x^2 + 8x - 105$

$0 = (x + 15)(x - 7)$

$x + 15 = 0 \qquad x - 7 = 0$

$\qquad x = -15 \qquad\quad x = 7$

Since length is never negative, $x = 7$ and $x + 4 = 11$. The squares have sides 11 in. and 7 in.

17.

$$\frac{1}{a-2} = \frac{4a-1}{a^2+5a-14} + \frac{2}{a+7}$$

$$\frac{1}{a-2} = \frac{4a-1}{(a+7)(a-2)} + \frac{2}{a+7}$$

$$(a+7)(a-2)\left(\frac{1}{a-2}\right) = (a+7)(a-2)\left[\frac{4a-1}{(a+7)(a-2)} + \frac{2}{a+7}\right]$$

$$a+7 = 4a-1+2(a-2)$$
$$a+7 = 4a-1+2a-4$$
$$a+7 = 6a-5$$
$$-5a+7 = -5$$
$$-5a = -12$$
$$a = \frac{12}{5}$$

18. $w = kI^2R,\ w = 12,\ I = 2,\ R = 100$

$$12 = k(2^2)(100)$$
$$12 = 400k$$
$$\frac{12}{400} = k$$
$$k = \frac{3}{100}$$
$$w = \frac{3}{100}I^2R,\ I = 0.8,\ R = 600$$
$$w = \frac{3}{100}(0.8)^2(600)$$
$$w = 11.52$$

The wattage is 11.52 watts.

19.

$$\frac{3-4i}{2+3i} = \left(\frac{3-4i}{2+3i}\right)\left(\frac{2-3i}{2-3i}\right)$$

$$= \frac{6-17i+12i^2}{4-9i^2}$$

$$= \frac{6-17i-12}{4+9}$$

$$= \frac{-6-17i}{13}$$

20.

$$4x^2 = -3x-12$$
$$4x^2+3x+12 = 0$$

$$x = \frac{-3 \pm \sqrt{3^2-4(4)(12)}}{2(4)}$$

$$= \frac{-3 \pm \sqrt{9-192}}{8}$$

$$= \frac{-3 \pm \sqrt{-183}}{8}$$

$$= \frac{-3 \pm i\sqrt{183}}{8}$$

The solutions are $\dfrac{-3+i\sqrt{183}}{8}$ and $\dfrac{-3-i\sqrt{183}}{8}$.

Chapter 9

Exercise Set 9.1

1. To find $(f \circ g)(x)$, substitute $g(x)$ for x in $f(x)$.

3. a. Each y has a unique x in a one-to-one function.

    b. Use the horizontal line test to determine whether a graph is one-to-one.

5. a. Yes; each first coordinate is paired with only one second coordinate.

    b. Yes; each second coordinate is paired with only one first coordinate.

    c. $\{(5, 3), (2, 4), (3, -1), (-2, 0)\}$
Reverse each ordered pair.

7. The domain of f is the range of f^{-1} and the range of f is the domain of f^{-1}.

9. $f(x) = x^2 + 5$, $g(x) = x - 4$

    a. $(f \circ g)(x) = (x - 4)^2 + 5$
$$= x^2 - 8x + 16 + 5$$
$$= x^2 - 8x + 21$$

    b. $(f \circ g)(4) = 4^2 - 8(4) + 21 = 5$

    c. $(g \circ f)(x) = (x^2 + 5) - 4 = x^2 + 1$

    d. $(g \circ f)(4) = 4^2 + 1 = 17$

11. $f(x) = x + 2$, $g(x) = x^2 + 4x - 2$

    a. $(f \circ g)(x) = (x^2 + 4x - 2) + 2 = x^2 + 4x$

    b. $(f \circ g)(4) = 4^2 + 4(4) = 32$

    c. $(g \circ f)(x) = (x + 2)^2 + 4(x + 2) - 2$
$$= x^2 + 4x + 4 + 4x + 8 - 2$$
$$= x^2 + 8x + 10$$

    d. $(g \circ f)(4) = 4^2 + 8(4) + 10 = 58$

13. $f(x) = \dfrac{3}{x}$, $g(x) = x^2 + 1$

    a. $(f \circ g)(x) = \dfrac{3}{x^2 + 1}$

    b. $(f \circ g)(4) = \dfrac{3}{4^2 + 1} = \dfrac{3}{17}$

    c. $(g \circ f)(x) = \left(\dfrac{3}{x}\right)^2 + 1$
$$= \dfrac{9}{x^2} + \dfrac{x^2}{x^2}$$
$$= \dfrac{9 + x^2}{x^2}$$

    d. $(g \circ f)(4) = \dfrac{9 + 4^2}{4^2} = \dfrac{25}{16}$

15. $f(x) = x - 4$, $g(x) = \sqrt{x + 5}$, $x \geq -5$

    a. $(f \circ g)(x) = \sqrt{x + 5} - 4$

    b. $(f \circ g)(4) = \sqrt{4 + 5} - 4$
$$= \sqrt{9} - 4$$
$$= 3 - 4$$
$$= -1$$

    c. $(g \circ f)(x) = \sqrt{(x - 4) + 5} = \sqrt{x + 1}$

    d. $(g \circ f)(4) = \sqrt{4 + 1} = \sqrt{5}$

17. This function is not a one-to-one function since it does not pass the horizontal line test.

19. This function is a one-to-one function since it passes both the vertical line test and the horizontal line test.

21. Yes, the ordered pairs represent a one-to-one function. For each value of x there is a unique value for y and each y-value has a unique x-value.

23. No, the ordered pairs do not represent a one-to-one function. For each value of x there is a unique y, but for each y-value there is not a unique x since $(-4, 2)$ and $(0, 2)$ are ordered pairs in the given set.

25. $y = 2x - 4$ is a line with a slope of 2 and having a y-intercept of -4. It is a one-to-one function since it passes both the vertical line test and the horizontal line test.

27. $y = x^2 - 4$ is a parabola with vertex at $(0, -4)$. It is not a one-to-one function since it does not pass the horizontal line test. Horizontal lines above $y = -4$ intersect the graph in 2 different points.

29. $y = x^2 - 4$, $x \geq 0$ is the right side of the parabola from Exercise 27. It is a one-to-one function since it passes both the vertical line test and the horizontal line test.

31. $y = |x|$. It is not a one-to-one function since it does not pass the vertical line test and the horizontal line test.

33. For $f(x)$: Domain: $\{-2, -1, 2, 4, 9\}$
Range: $\{0, 3, 4, 6, 7\}$
For $f^{-1}(x)$: Domain:$\{0, 3, 4, 6, 7\}$
Range: $\{-2, -1, 2, 4, 9\}$

35. For $f(x)$: Domain:$\{-1, 1, 2, 4\}$
Range: $\{-3, -1, 0, 2\}$
For $f^{-1}(x)$: Domain: $\{-3, -1, 0, 2\}$
Range: $\{-1, 1, 2, 4\}$

37. For $f(x)$: Domain: $\{x | x \geq 2\}$
Range: $\{y | y \geq 0\}$
For $f^{-1}(x)$: Domain: $\{x | x \geq 0\}$
Range: $\{y | y \geq 2\}$

39. a. Yes, $f(x) = x - 5$ is a one-to-one function.

b. $$y = x - 5$$
$$x = y - 5$$
$$x + 5 = y$$
$$y = x + 5$$
$$f^{-1}(x) = x + 5$$

41. a. Yes, $h(x) = 4x$ is a one-to-one function.

b. $$y = 4x$$
$$x = 4y$$
$$y = \frac{x}{4}$$
$$h^{-1}(x) = \frac{x}{4}$$

43. a. Yes; $g(x) = \frac{1}{x}$ is a one-to-one function.

b. $$y = \frac{1}{x}$$
$$x = \frac{1}{y}$$
$$y = \frac{1}{x}$$
$$g^{-1}(x) = \frac{1}{x}$$

45. a. No; $f(x) = x^2 + 4$ is not a one-to-one function.

b. Does not exist

47. a. Yes, $g(x) = x^3 - 5$ is a one-to-one function.

b.
$$y = x^3 - 5$$
$$x = y^3 - 5$$
$$x + 5 = y^3$$
$$\sqrt[3]{x+5} = y$$
$$g^{-1}(x) = \sqrt[3]{x+5}$$

49. a. Yes, $g(x) = \sqrt{x+2}$, $x \geq -2$ is a one-to-one function.

b.
$$y = \sqrt{x+2}$$
$$x = \sqrt{y+2}$$
$$x^2 = y + 2$$
$$x^2 - 2 = y$$
$$g^{-1}(x) = x^2 - 2, \ x \geq 0$$

51. a. Yes, $h(x) = x^2 - 4$, $x \geq 0$ is a one-to-one function.

b.
$$y = x^2 - 4$$
$$x = y^2 - 4$$
$$x + 4 = y^2$$
$$y = \sqrt{x+4}$$
$$h^{-1}(x) = \sqrt{x+4}, \ x \geq -4$$

53. $f(x) = 2x + 8$

a.
$$y = 2x + 8$$
$$x = 2y + 8$$
$$x - 8 = 2y$$
$$\frac{x-8}{2} = y$$
$$f^{-1}(x) = \frac{x-8}{2}$$

b.

x	$f(x)$
0	8
-4	0

x	$f^{-1}(x)$
0	-4
8	0

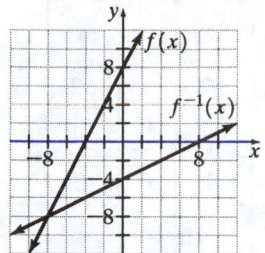

55. $f(x) = \sqrt{x}$, $x \geq 0$

a.
$$y = \sqrt{x}$$
$$x = \sqrt{y}$$
$$x^2 = \left(\sqrt{y}\right)^2$$
$$x^2 = y$$
$$f^{-1}(x) = x^2 \text{ for } x \geq 0$$

b.

x	$f(x)$
0	0
1	1
4	2

x	$f^{-1}(x)$
0	0
1	1
2	4

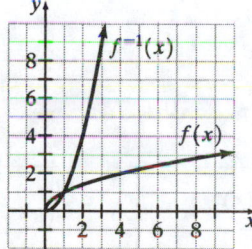

57. $f(x) = \sqrt[3]{x}$

a.
$$y = \sqrt[3]{x}$$
$$x = \sqrt[3]{y}$$
$$x^3 = \left(\sqrt[3]{y}\right)^3$$
$$x^3 = y$$
$$f^{-1}(x) = x^3$$

b.

x	$f(x)$
−8	−2
−1	−1
0	0
1	1
8	2

x	$f^{-1}(x)$
−2	−8
−1	−1
0	0
1	1
2	8

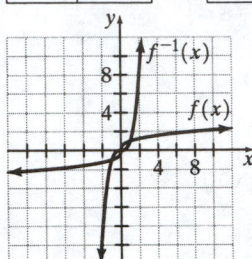

59. $f(x) = \dfrac{1}{x}, \; x > 0$

a. $y = \dfrac{1}{x}$

$x = \dfrac{1}{y}$

$xy = 1$

$y = \dfrac{1}{x}$

$f^{-1}(x) = \dfrac{1}{x}, \; x > 0$

b.

x	$f(x)$
$\dfrac{1}{2}$	2
1	1
3	$\dfrac{1}{3}$

x	$f^{-1}(x)$
2	$\dfrac{1}{2}$
1	1
$\dfrac{1}{3}$	3

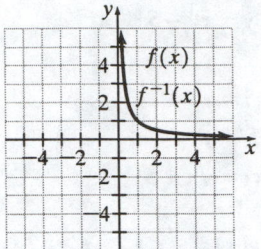

61. $(f \circ f^{-1})(x) = (x + 4) - 4 = x$

$(f^{-1} \circ f)(x) = (x - 4) + 4 = x$

63. $(f \circ f^{-1})(x) = \sqrt[3]{(x^3 + 2) - 2}$

$= \sqrt[3]{x^3}$

$= x$

$(f^{-1} \circ f)(x) = \left(\sqrt[3]{x - 2}\right)^3 + 2$

$= x - 2 + 2$

$= x$

65. $(f \circ f^{-1})(x) = \dfrac{2}{\frac{2}{x}} = 2 \cdot \dfrac{x}{2} = x$

$(f^{-1} \circ f)(x) = \dfrac{2}{\frac{2}{x}} = 2 \cdot \dfrac{x}{2} = x$

67. No, composition of functions is not commutative.

Let $f(x) = x^2$ and $g(x) = x + 1$.

Then $(f \circ g)(x) = (x + 1)^2 = x^2 + 2x + 1$

while $(g \circ f)(x) = x^2 + 1$.

69. **a.** $(f \circ g)(x) = f[g(x)]$

$= \left(\sqrt[3]{x - 2}\right)^3 + 2$

$= x - 2 + 2$

$= x$

$(g \circ f)(x) = g[f(x)]$

$= \sqrt[3]{(x^3 + 2) - 2}$

$= \sqrt[3]{x^3}$

$= x$

b. The domain of *f* is all real numbers and the domain of *g* is all real numbers. The domains of $(f \circ g)(x)$ and $(g \circ f)(x)$ are also all real numbers.

71. The range of $f^{-1}(x)$ is the domain of $f(x)$.

73. $f(x) = 3x$ converts yards, *x*, into feet, *y*.

$$y = 3x$$
$$x = 3y$$
$$\frac{x}{3} = y$$
$$f^{-1}(x) = \frac{x}{3}$$

Here, *x* is feet and $f^{-1}(x)$ is yards. The inverse function converts feet to yards.

75. $f(x) = \frac{5}{9}(x - 32)$

$$y = \frac{5}{9}(x - 32)$$
$$x = \frac{5}{9}(y - 32)$$
$$\frac{9}{5}x = \frac{9}{5}\left[\frac{5}{9}(y - 32)\right]$$
$$\frac{9}{5}x = y - 32$$
$$\frac{9}{5}x + 32 = y$$
$$f^{-1}(x) = \frac{9}{5}x + 32$$

Here, *x* is degrees Celsius and $f^{-1}(x)$ is degrees Fahrenheit. The inverse function converts Celsius to Fahrenheit.

77.

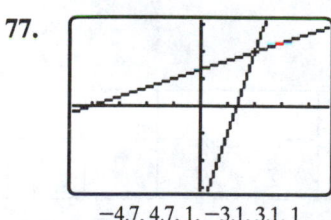

$$-4.7, 4.7, 1, -3.1, 3.1, 1$$

Yes, the functions are inverses.

79.

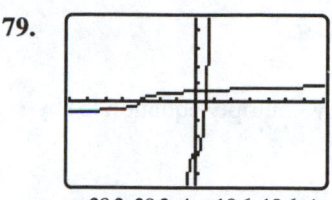

$$-28.2, 28.2, 4, -18.6, 18.6, 4$$

Yes, the functions are inverses.

81. a. $r(3) = 2(3) = 6$
The radius is 6 feet.

b. $A = \pi r^2$
$$A = \pi(6)^2$$
$$A = 36\pi \approx 113.10$$
The surface area is $36\pi \approx 113.10$ square feet.

c. $(A \circ r)(t) = \pi(2t)^2 = \pi(4t^2) = 4\pi t^2$

d. $4\pi(3)^2 = 4\pi(9) = 36\pi$

e. The answers agree.

84. $2x + 3y - 4z = 18$ (1)
$$x - y - z = 3 \quad (2)$$
$$x - 2y - 2z = 2 \quad (3)$$

To eliminate *y* and *z* between equations (2) and (3), multiply equation (2) by −2 and then add.
$$-2[x - y - z = 3]$$
$$x - 2y - 2z = 2$$
gives
$$-2x + 2y + 2z = -6$$
$$\underline{x - 2y - 2x = 2}$$
Add: $-x \quad\quad = -4$
$$x = 4$$

Now substitute 4 for *x* in equations (1) and (2).
Equation (1) becomes
$$2x + 3y - 4z = 18$$
$$2(4) + 3y - 4z = 18$$
$$3y - 4z = 10 \quad (4)$$
Equation (2) becomes
$$x - y - z = 3$$
$$4 - y - z = 3$$
$$-y - z = -1 \quad (5)$$
Equations (4) and (5) are a system in two

unknowns.
$3y - 4z = 10$
$\quad y - z = -1$
To eliminate y, multiply equation (5) by 3
and then add.
$\quad 3y - 4z = 10$
$3[-y - z = -1]$
gives
$$3y - 4z = 10$$
$$-3y - 3z = -3$$
Add: $\quad \overline{-7z = 7}$
$$z = \frac{7}{-7} = -1$$

Finally, substitute -1 for z in equation (5).
$$-y - z = -1$$
$$-y - (-1) = -1$$
$$-y + 1 = -1$$
$$-y = -2$$
$$y = 2$$
The solution is $(4, 2, -1)$.

85.

$$\begin{array}{r|rrrr} -2 & 1 & 6 & 6 & -8 \\ & & -2 & -8 & 4 \\ \hline & 1 & 4 & -2 & -4 \end{array}$$

$$\frac{x^3 + 6x^2 + 6x - 8}{x + 2} = x^2 + 4x - 2 - \frac{4}{x+2}$$

86. $\sqrt{\dfrac{24x^3 y^2}{3xy^3}} = \sqrt{\dfrac{8x^2}{y}}$

$$= \frac{\sqrt{8x^2}}{\sqrt{y}}$$

$$= \frac{\sqrt{4x^2 \cdot 2}}{\sqrt{y}}$$

$$= \frac{2x\sqrt{2}}{\sqrt{y}} \cdot \frac{\sqrt{y}}{\sqrt{y}}$$

$$= \frac{2x\sqrt{2y}}{\left(\sqrt{y}\right)^2}$$

$$= \frac{2x\sqrt{2y}}{y}$$

87. $x^2 + 2x - 6 = 0$
$$x^2 + 2x = 6$$
$$x^2 + 2x + 1 = 6 + 1$$
$$(x+1)^2 = 7$$
$$x + 1 = \pm\sqrt{7}$$
$$x = -1 \pm \sqrt{7}$$

Exercise Set 9.2

1. Exponential functions are functions of the form $f(x) = a^x, a > 0, \ a \neq 1$.

3. a. $y = \left(\dfrac{1}{2}\right)^x$; as x increases, y decreases.

b. No, y can never be zero because $\left(\dfrac{1}{2}\right)^x$ can never be 0.

c. No, y can never be negative because $\left(\dfrac{1}{2}\right)^x$ is never negative.

5. $y = 2^x$ and $y = 3^x$

a. Let $x = 0$
$$y = 2^0 \quad y = 3^0$$
$$y = 1 \quad \ y = 1$$
They have the same y-intercepts at $(0, 1)$.

b. $y = 3^x$ will be steeper than $y = 2^x$ for $x > 0$.

7. $y = 2^x$

x	-2	-1	0	1	2
y	$\frac{1}{4}$	$\frac{1}{2}$	1	2	4

Domain: $\mathbb{R}$

R: $\{y|y > 0\}$

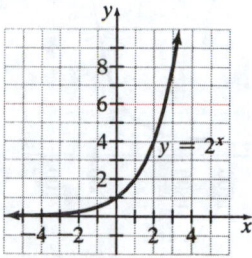

9. $y = \left(\frac{1}{2}\right)^x$

x	-2	-1	0	1	2
y	4	2	1	$\frac{1}{2}$	$\frac{1}{4}$

Domain: $\mathbb{R}$

R: $\{y|y > 0\}$

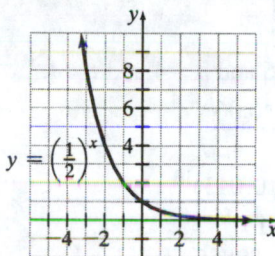

11. $y = 4^x$

x	-2	-1	0	1	2
y	$\frac{1}{16}$	$\frac{1}{4}$	1	4	16

Domain: $\mathbb{R}$

R: $\{y|y > 0\}$

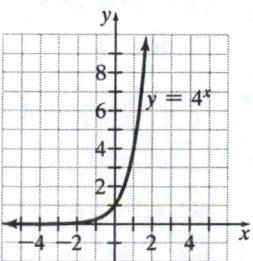

13. $y = 3^{-x} = \frac{1}{3^x} = \left(\frac{1}{3}\right)^x$

x	-2	-1	0	1	2
y	9	3	1	$\frac{1}{3}$	$\frac{1}{9}$

Domain: $\mathbb{R}$

R: $\{y|y > 0\}$

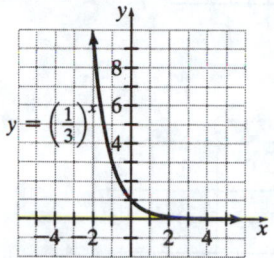

15. $y = 2^{x-1}$

x	-2	0	2	4	6
y	$\frac{1}{8}$	$\frac{1}{2}$	2	8	32

Domain: $\mathbb{R}$

Range: $\{y|y > 0\}$

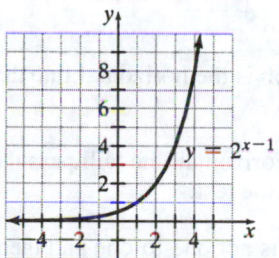

17. $y = 2^x - 1$

x	-2	-1	0	1	2	3
y	$-\frac{3}{4}$	$-\frac{1}{2}$	0	1	3	7

Domain: $\mathbb{R}$

Range: $\{y|y > -1\}$

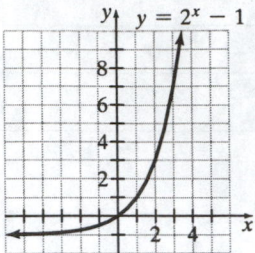

19. $y = 2^{2x} - 4$

x	-1	0	1	2
y	$-\frac{15}{4}$	-3	0	12

Domain: $\mathbb{R}$

R: $\{y|y > -4\}$

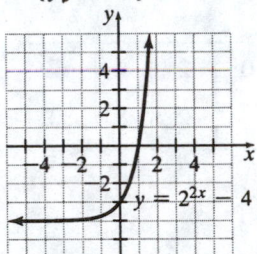

21.　a. The graph is the horizontal line through $y = 1$.

　　b. Yes. A horizontal line will pass the vertical line test.

　　c. No. $f(x)$ is not one-to-one and therefore does not have an inverse function.

23. $y = a^x - k$ will have the same basic shape as the graph $y = a^x$. However, $y = a^x - k$ will be k units lower than $y = a^x$.

25. The graph of $y = a^{x+2}$ is the graph of $y = a^x$ shifted 2 units to the left.

27.　a.

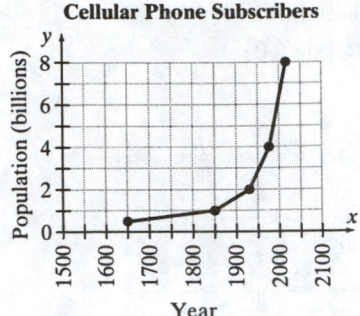

　　b. 200 years

　　c. 40 years

29. $g = 2^n$, $n = 8$

$g = 2^8 = 256$

The plant has 256 gametes.

31. $A = p\left(1 + \dfrac{r}{n}\right)^{nt}$.

Use $p = 5000$,

$r = 6\% = 0.06$ and $n = 4$ and $t = 4$.

$A = 5000\left(1 + \dfrac{0.06}{4}\right)^{4 \cdot 4}$

$A = 5000(1 + 0.015)^{16}$

$A = 5000(1.015)^{16}$

$A \approx 5000(1.2689855)$

$A \approx 6344.93$

He has \$6344.93 after 4 years.

33. $A = A_0 2^{-t/5600}$

Use $A_0 = 12$ and $t = 1000$.

$A = 12(2^{-1000/5600})$

$A \approx 12(2^{-0.18})$

$A \approx 12(0.88)$

$A \approx 10.6$ grams

There are about 10.6 grams left.

35. $y = 80(2)^{-0.4t}$

   a. $t = 10$

   $y = 80(2)^{-0.4(10)}$

   $y = 80(2)^{-4} = 80\left(\dfrac{1}{16}\right)$

   $y = 5$

   After 10 years, 5 grams remain.

   b. $t = 100$

   $y = 80(2)^{-0.4(100)}$

   $y = 80(2)^{-40}$

   $y \approx 80(9.094947 \times 10^{-13})$

   $y \approx 7.28 \times 10^{-11}$

   After 100 years, about 7.28×10^{-11} grams are left.

37. $y = 5000(3)^{x}$

   a. $x = 5$

   $y = 5000(3)^{5}$

   $y = 5000(243)$

   $y = 1,215,000$

   After 5 days, the number of bacteria is expected to be 1,215,000.

   b. $x = 7$

   $y = 5000(3)^{7}$

   $y = 5000(2187)$

   $y = 10,935,000$

   After 7 days, the number of bacteria is expected to be 10,935,000.

39. a. Answers will vary. One possible answer is: Since $\dfrac{2}{3}$ are being recycled, then the previous year's amount is multiplied by $\dfrac{2}{3}$.

   b. $n = 2001 - 1998 = 3$

   $A(3) = 190,000,000\left(\dfrac{2}{3}\right)^{3}$

   $= 190,000,000\left(\dfrac{8}{27}\right)$

   ≈ 56.3

   About 56.3 million cans are expected to be made from recycled 1998 cans.

41. $A = 41.97(0.996)^{x}$

   $A(389) = 41.97(0.996)^{389}$

   $A \approx 8.83$

   The altitude at the top of Mt. Everest is about 8.83 kilometers.

43. $x = 1998 - 1964 = 34$

   $f(x) = 3318.10(1.08)^{x}$

   $f(34) = 3318.10(1.08)^{34} = 45,425.23$

   The cost of a one-year stay was expected to be $45,425.23 in 1998.

45. $2^{3x+2} = 16$

   $2^{3x+2} = 2^{4}$

   $3x + 2 = 4$

   $3x = 2$

   $x = \dfrac{2}{3}$

47. a. $A = p\left(1 + \dfrac{r}{n}\right)^{nt}$

   $A = 100\left(1 + \dfrac{0.07}{365}\right)^{365 \cdot 10}$

   $A \approx 100(1.0001918)^{3650}$

   $A = 201.36$

   The amount is $201.36.

   b. For simple interest,

   $A = 100 + 100(0.07)t$

   $A = 100 + 100(0.07)(10)$

   $A = 100 + 70$

   $A = 170$

   $201.36 - $170 = $31.36

49. a. $y_1 = 3^{x-5}$

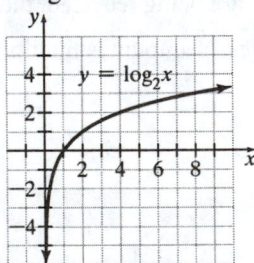

$-10, 10, 1, -10, 10, 1$

b. $4 = 3^{x-5}$ when $x \approx 6.26$.

53. a. $2.3x^4y - 6.2x^6y^2 + 9.2x^5y^2$
$= -6.2x^6y^2 + 9.2x^5y^2 + 2.3x^4y$

b. $-6.2x^6y^2$ is the leading term.
$6 + 2 = 8$ is the degree of the polynomial.

c. $-6.2x^6y^2$ is the leading term, so -6.2 is the leading coefficient.

54. $(f \cdot g)(x) = f(x) \cdot g(x)$
$= (x+3)(x^2 - 2x + 4)$
$= x^3 - 2x^2 + 4x + 3x^2 - 6x + 12$
$= x^3 + x^2 - 2x + 12$

55. $\sqrt{a^2 - 8a + 16} = \sqrt{(a-4)^2} = |a - 4|$

56.

$\sqrt[4]{\dfrac{32x^5y^9}{2y^3z}} = \sqrt[4]{\dfrac{16x^5y^6}{z}}$

$= \dfrac{\sqrt[4]{16x^5y^6}}{\sqrt[4]{z}}$

$= \dfrac{\sqrt[4]{16x^4y^4 \cdot xy^2}}{\sqrt[4]{z}}$

$= \dfrac{2xy\sqrt[4]{xy^2}}{\sqrt[4]{z}} \cdot \dfrac{\sqrt[4]{z^3}}{\sqrt[4]{z^3}}$

$= \dfrac{2xy\sqrt[4]{xy^2z^3}}{\sqrt[4]{z^4}}$

$= \dfrac{2xy\sqrt[4]{xy^2z^3}}{z}$

Exercise Set 9.3

1. $y = \log_a x$

a.   The base a must be positive and must not be equal to one.

b.   The argument x represents a number that is greater than 0. Thus, the domain is $\{x | x \text{ is a real number and } x > 0\}$.

c. $\mathbb{R}$

3.   The functions $f(x) = a^x$ and $g(x) = \log_a x$ are inverse functions. Therefore, some of the points on the function are $g(x) = \log_a x$ are $\left(\dfrac{1}{27}, -3\right)\left(\dfrac{1}{9}, -2\right)$, $\left(\dfrac{1}{3}, -1\right)(1, \ 0)$, $(3, 1)$, $(9, 2)$, and $(27, 3)$.

5.   The functions $y = a^x$ and $y = \log_a x$ for $a \neq 1$ are inverses of each other, thus the graphs are symmetric with respect to the line $y = x$. For each ordered pair (x, y) on the graph of $y = a^x$, the ordered pair (y, x) is on the graph of $y = \log_a x$.

7. $y = \log_2 x$
Convert to exponential form.
$2^y = x$

x	$\frac{1}{4}$	$\frac{1}{2}$	1	2	4
y	-2	-1	0	1	2

Domain: $\{x | x > 0\}$
Range: $\mathbb{R}$

9.　$y = \log_{1/2} x$
Convert to exponential form.

$$x = \left(\frac{1}{2}\right)^y$$

x	4	2	1	$\frac{1}{2}$	$\frac{1}{4}$
y	-2	-1	0	1	2

Domain: $\{x \mid x > 0\}$
Range: $\mathbb{R}$

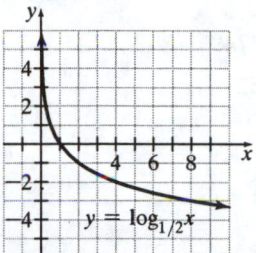

11.　$y = \log_5 x$
Convert to the exponential form.

$$x = 5^y$$

x	$\frac{1}{25}$	$\frac{1}{5}$	1	5	25
y	-2	-1	0	1	2

Domain: $\{x \mid x > 0\}$

Range: $\mathbb{R}$

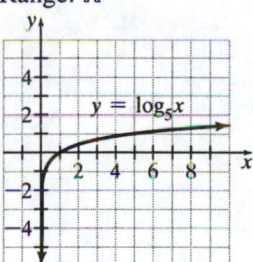

13.　$y = 2^x$

x	-2	-1	0	1	2
y	$\frac{1}{4}$	$\frac{1}{2}$	1	2	4

$y = \log_{1/2} x$
Convert to exponential form.

$$x = \left(\frac{1}{2}\right)^y$$

x	4	2	1	$\frac{1}{2}$	$\frac{1}{4}$
y	-2	-1	0	1	2

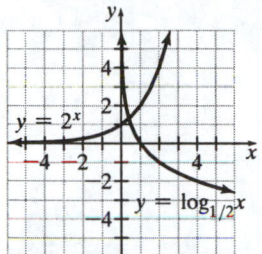

15.　$y = 2^x$

x	-2	-1	0	1	2
y	$\frac{1}{4}$	$\frac{1}{2}$	1	2	4

$y = \log_2 x$
Convert to exponential form.

$$x = 2^y$$

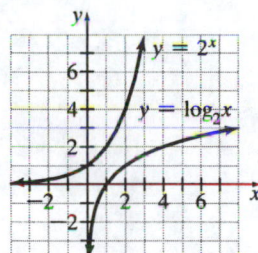

17.　　$2^3 = 8$
　　　$\log_2 8 = 3$

19.　　$9^{1/2} = 3$
　　　$\log_9 3 = \dfrac{1}{2}$

21.　　$\left(\dfrac{1}{2}\right)^5 = \dfrac{1}{32}$
　　　$\log_{1/2}\left(\dfrac{1}{32}\right) = 5$

23. $2^{-3} = \dfrac{1}{8}$

     $\log_2 \dfrac{1}{8} = -3$

25. $4^{-3} = \dfrac{1}{64}$

     $\log_4 \dfrac{1}{64} = -3$

27. $16^{-1/2} = \dfrac{1}{4}$

     $\log_{16} \dfrac{1}{4} = -\dfrac{1}{2}$

29. $8^{-1/3} = \dfrac{1}{2}$

     $\log_8 \dfrac{1}{2} = -\dfrac{1}{3}$

31. $10^{0.6990} = 5$

     $\log_{10} 5 = 0.6990$

33. $e^2 = 7.3891$

   $\log_e 7.3891 = 2$

35. $c^b = w$

   $\log_c w = b$

37. $\log_2 8 = 3$

     $2^3 = 8$

39. $\log_{1/3} \dfrac{1}{9} = 2$

     $\left(\dfrac{1}{3}\right)^2 = \dfrac{1}{9}$

41. $\log_5 \dfrac{1}{125} = -3$

     $5^{-3} = \dfrac{1}{125}$

43. $\log_{125} 5 = \dfrac{1}{3}$

     $125^{1/3} = 5$

45. $\log_{27} \dfrac{1}{3} = -\dfrac{1}{3}$

     $27^{-1/3} = \dfrac{1}{3}$

47. $\log_{10} 1000 = 3$

       $10^3 = 1000$

49. $\log_{10} 8 = 0.9031$

   $10^{0.9031} = 8$

51. $\log_e 6.52 = 1.8749$

     $e^{1.8749} = 6.52$

53. $\log_r c = -a$

     $r^{-a} = c$

55. $\log_4 16 = y$

     $4^y = 16$

     $4^y = 4^2$

     $y = 2$

57. $\log_2 x = 5$

     $2^5 = x$

     $32 = x$

59. $\log_a \dfrac{1}{27} = -3$

     $a^{-3} = \dfrac{1}{27}$

     $a^{-3} = 3^{-3}$

     $a = 3$

61. $\log_a \dfrac{1}{64} = 3$

     $a^3 = \dfrac{1}{64}$

     $a^3 = \left(\dfrac{1}{4}\right)^3$

     $a = \dfrac{1}{4}$

63. $\log_{10} 100 = 2$ because $10^2 = 100$

65. $\log_{10} 1 = 0$ because $10^0 = 1$

67. $\log_{10} 10,000 = 4$ because $10^4 = 10,000$

69. $\log_4 64 = 3$ because $4^3 = 64$

71. $\log_8 \dfrac{1}{64} = -2$ because $8^{-2} = \dfrac{1}{8^2} = \dfrac{1}{64}$

73. $\log_9 9 = 1$ because $9^1 = 9$

75. $\log_5 1 = 0$ because $5^0 = 1$

77. $\log_{10} 425$ lies between 2 and 3 since 425 lies between $10^2 = 100$ and $10^3 = 1000$.

79. $\log_3 62$ lies between 3 and 4 since 62 lies between $3^3 = 27$ and $3^4 = 81$.

81. For $x > 1$, 2^x will grow faster than $\log_{10} x$. Note that when $x = 10$, $2^x = 1024$ while $\log_{10} x = 1$.

83. $\quad x = \log_{10} 10^5$
$\quad 10^x = 10^5$
$\quad\quad x = 5$

85. $\quad x = \log_b b^3$
$\quad b^x = b^3$
$\quad\quad x = 3$

87. $\quad\quad x = 10^{\log_{10} 8}$
$\log_{10} x = \log_{10} 8$
$\quad\quad x = 8$

89. $\quad\quad x = b^{\log_b 9}$
$\log_b x = \log_b 9$
$\quad\quad x = 9$

91. $\quad R = \log_{10} I$
$\quad 7 = \log_{10} I$
$\quad 10^7 = I$
$\quad\quad I = 10,000,000$

The earthquake is 10,000,000 times more intense than the smallest measurable activity.

$\dfrac{1000}{10} = 100$

An earthquake that measures 3 is 100 times more intense than one that measures 1.

93. $y = \log_2(x-1)$ or $2^y = x - 1$

x	$1\frac{1}{4}$	$1\frac{1}{2}$	2	3	5
y	-2	-1	0	1	2

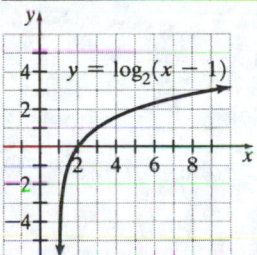

95. $24x^2 - 6xy + 16xy - 4y^2$
$= 2[12x^2 - 3xy + 8xy - 2y^2]$
$= 2[3x(4x - y) + 2y(4x - y)]$
$= 2(3x + 2y)(4x - y)$

96. $2(a-3)^2 + 7(a-3) - 15 = 2y^2 + 7y - 15$ $\leftarrow$ Replace $a - 3$ with y
$\quad\quad\quad\quad\quad\quad\quad\quad = 2y^2 + 10y - 3y - 15$
$\quad\quad\quad\quad\quad\quad\quad\quad = 2y(y + 5) - 3(y + 5)$
$\quad\quad\quad\quad\quad\quad\quad\quad = (2y - 3)(y + 5)$
$\quad\quad\quad\quad\quad\quad\quad\quad = [2(a-3) - 3][(a-3) + 5]$ $\leftarrow$ Replace y with $a - 3$
$\quad\quad\quad\quad\quad\quad\quad\quad = (2a - 9)(a + 2)$

97. $4x^4 - 36x^2 = 4x^2(x^2 - 9)$
$= 4x^2(x+3)(x-3)$

98. $8x^3 + \dfrac{1}{27}$

$= (2x)^3 + \left(\dfrac{1}{3}\right)^3$

$= \left(2x + \dfrac{1}{3}\right)\left[(2x)^2 - (2x)\left(\dfrac{1}{3}\right) + \left(\dfrac{1}{3}\right)^2\right]$

$= \left(2x + \dfrac{1}{3}\right)\left(4x^2 - \dfrac{2}{3}x + \dfrac{1}{9}\right)$

Exercise Set 9.4

5. $\log_4 6 \cdot 9 = \log_4 6 + \log_4 9$

7. $\log_8 7(x+3) = \log_8 7 + \log_8(x+3)$

9. $\log_6 \dfrac{27}{5} = \log_6 27 - \log_6 5$

11. $\log_{10} \dfrac{\sqrt{x}}{x-9} = \log_{10} \dfrac{x^{1/2}}{x-9}$

$= \log_{10} x^{1/2} - \log_{10}(x-9)$

$= \dfrac{1}{2}\log_{10} x - \log_{10}(x-9)$

13. $\log_8 x^4 = 4\log_8 x$

15. $\log_{10} 3(8^2) = \log_{10} 3 + \log_{10} 8^2$
$= \log_{10} 3 + 2\log_{10} 8$

17. $\log_4 \sqrt{\dfrac{a^5}{a+4}} = \log_4 \left(\dfrac{a^5}{a+4}\right)^{1/2}$

$= \dfrac{1}{2}\log_4 \dfrac{a^5}{a+4}$

$= \dfrac{1}{2}[\log_4 a^5 - \log_4(a+4)]$

$= \dfrac{1}{2}[5\log_4 a - \log_4(a+4)]$

$= \dfrac{5}{2}\log_4 a - \dfrac{1}{2}\log_4(a+4)$

19. $\log_{10} \dfrac{d^4}{(a+2)^3} = \log_{10} d^4 - \log_{10}(a+2)^3$

$= 4\log_{10} d - 3\log_{10}(a+2)$

21. $\log_8 \dfrac{y(y+2)}{y^3}$

$= \log_8 y + \log_8(y+2) - \log_8 y^3$
$= \log_8 y + \log_8(y+2) - 3\log_8 y$
$= -2\log_8 y + \log_8(y+2)$

23. $\log_{10} \dfrac{2m}{3n}$

$= \log_{10} 2m - \log_{10} 3n$
$= \log_{10} 2 + \log_{10} m - (\log_{10} 3 + \log_{10} n)$
$= \log_{10} 2 + \log_{10} m - \log_{10} 3 - \log_{10} n$

25. $2\log_{10} x - \log_{10}(x-5)$

$= \log_{10} x^2 - \log_{10}(x-5)$

$= \log_{10} \dfrac{x^2}{x-5}$

27. $2(\log_5 a - \log_5 3) = 2\log_5 \dfrac{a}{3}$

$= \log_5 \left(\dfrac{a}{3}\right)^2$

29. $\log_{10} n + \log_{10}(n-3) - \log_{10}(n+1)$
$= \log_{10} n(n-3) - \log_{10}(n+1)$

$= \log_{10} \dfrac{n(n-3)}{n+1}$

31. $\dfrac{1}{2}[\log_5(x-4) - \log_5 x]$

$= \dfrac{1}{2}\log_5 \dfrac{x-4}{x}$

$= \log_5 \left(\dfrac{x-4}{x}\right)^{1/2}$

$= \log_5 \sqrt{\dfrac{x-4}{x}}$

33. $2\log_9 5 + \dfrac{1}{3}\log_9 (r-6) - \dfrac{1}{2}\log_9 r$

$= \log_9 5^2 + \log_9 (r-6)^{1/3} - \log_9 r^{1/2}$

$= \log_9 25 + \log_9 \sqrt[3]{r-6} - \log_9 \sqrt{r}$

$= \log_9 25\sqrt[3]{r-6} - \log_9 \sqrt{r}$

$= \log_9 \dfrac{25\sqrt[3]{r-6}}{\sqrt{r}}$

35. $4\log_6 3 - [2\log_6 (x+3) + 4\log_6 x]$

$= \log_6 3^4 - [\log_6 (x+3)^2 + \log_6 x^4]$

$= \log_6 3^4 - \log_6 (x+3)^2 x^4$

$= \log_6 \dfrac{3^4}{(x+3)^2 x^4}$

37. $\log_{10} 10 = \log_{10}(2)(5)$

$= \log_{10} 2 + \log_{10}(5)$

$= 0.3010 + 0.6990$

$= 1$

39. $\log_{10} 2.5 = \log_{10} \dfrac{5}{2}$

$= \log_{10} 5 - \log_{10} 2$

$= 0.6990 - 0.3010$

$= 0.3980$

41. $\log_{10} 25 = \log_{10} 5^2$

$= 2(\log_{10} 5)$

$= 2(0.6990)$

$= 1.3980$

43. $5^{\log_5 10} = 10$

45. $(2^3)^{\log_8 5} = 8^{\log_8 5} = 5$

47. $\log_3 27 = \log_3 3^3 = 3$

49. $5\left(\sqrt[3]{27}\right)^{\log_3 5} = 5(3)^{\log_3 5}$

$= 5(5)$

$= 25$

51. Yes

53. $\log_a \dfrac{x}{y} = \log_a xy^{-1}$

$= \log_a x + \log_a y^{-1}$

$= \log_a x + \log_a \dfrac{1}{y}$

55. $\log_a (x^2 - 4) - \log_a (x+2)$

$= \log_a \dfrac{x^2 - 4}{x+2}$

$= \log_a \dfrac{(x+2)(x-2)}{x+2}$

$= \log_a (x-2)$

57. Yes; $\log_a (x^2 + 8x + 16) = \log_a (x+4)^2$

$= 2\log_a (x+4)$

59. $\log_{10} x^2 = 2\log_{10} x$

$= 2(0.4320)$

$= 0.8640$

61. $\log_{10} \sqrt[4]{x} = \log_{10} x^{1/4}$

$= \dfrac{1}{4}\log_{10} x$

$= \dfrac{1}{4}(0.4320) = 0.1080$

63. $\log_{10} xy = \log_{10} x + \log_{10} y$

$= 0.5000 + 0.2000$

$= 0.7000$

65. No; answers will vary.

67. $\log_2 \dfrac{\sqrt[4]{xy}\,\sqrt[3]{a}}{\sqrt[5]{a-b}} = \log_2 \sqrt[4]{xy}\,\sqrt[3]{a} - \log_2 \sqrt[5]{a-b}$

$\qquad\qquad = \log_2 (xy)^{1/4} + \log_2 a^{1/3} - \log_2 (a-b)^{1/5}$

$\qquad\qquad = \dfrac{1}{4}\log_2 xy + \dfrac{1}{3}\log_2 a - \dfrac{1}{5}\log_2 (a-b)$

$\qquad\qquad = \dfrac{1}{4}\log_2 x + \dfrac{1}{4}\log_2 y + \dfrac{1}{3}\log_2 a - \dfrac{1}{5}\log_2 (a-b)$

69. Let $\log_a x = m$ and $\log_a y = n$. Then $a^m = x$ and $a^n = y$, so $\dfrac{x}{y} = \dfrac{a^m}{a^n} = a^{m-n}$.

Thus, $\log_a \dfrac{x}{y} = m - n = \log_a x - \log_a y$.

72. $\dfrac{2x+5}{x^2-7x+12} \div \dfrac{x-4}{2x^2-x-15} = \dfrac{2x+5}{x^2-7x+12} \cdot \dfrac{2x^2-x-15}{x-4}$

$\qquad\qquad\qquad = \dfrac{2x+5}{(x-4)(x-3)} \cdot \dfrac{(2x+5)(x-3)}{x-4}$

$\qquad\qquad\qquad = \dfrac{(2x+5)^2}{(x-4)^2}$

73. $\dfrac{2x+5}{x^2-7x+12} - \dfrac{x-4}{2x^2-x-15} = \dfrac{2x+5}{(x-4)(x-3)} - \dfrac{(x-4)}{(2x+5)(x-3)}$

$\qquad\qquad\qquad = \dfrac{2x+5}{(x-4)(x-3)} \cdot \dfrac{2x+5}{2x+5} - \dfrac{x-4}{(2x+5)(x-3)} \cdot \dfrac{x-4}{x-4}$

$\qquad\qquad\qquad = \dfrac{(4x^2+20x+25)-(x^2-8x+16)}{(x-4)(x-3)(2x+5)}$

$\qquad\qquad\qquad = \dfrac{4x^2+20x+25-x^2+8x-16}{(x-4)(x-3)(2x+5)}$

$\qquad\qquad\qquad = \dfrac{3x^2+28x+9}{(x-4)(x-3)(2x+5)}$

$\qquad\qquad\qquad = \dfrac{(3x+1)(x+9)}{(x-4)(x-3)(2x+5)}$

74. Let x equal the time needed to paint the house if they work together.

$$\frac{x}{4}+\frac{x}{5}=1$$
$$20\left(\frac{x}{4}\right)+20\left(\frac{x}{5}\right)=20(1)$$
$$5x+4x=20$$
$$9x=20$$
$$x=\frac{20}{9}=2\frac{2}{9}$$

It would take them $2\frac{2}{9}$ days working together.

75. $\sqrt[3]{4x^4y^7}\cdot\sqrt[3]{12x^7y^{10}}=\sqrt[3]{48x^{11}y^{17}}$
$$=\sqrt[3]{8x^9y^{15}\cdot 6x^2y^2}$$
$$=\sqrt[3]{8x^9y^{15}}\cdot\sqrt[3]{6x^2y^2}$$
$$=2x^3y^5\sqrt[3]{6x^2y^2}$$

Exercise Set 9.5

1. Common logarithms are logarithms with base 10.

3. Antilogarithms are numbers obtained by taking 10 to the power of the logarithm.

5. $\log 45 = 1.6532$

7. $\log 19,200 = 4.2833$

9. $\log 0.0000857 = -4.0670$

11. $\log 100 = 2.0000$

13. $\log 3.75 = 0.5740$

15. $\log 0.000472 = -3.3261$

17. antilog $0.6325 = 4.29$

19. antilog $4.6283 = 42,500$

21. antilog$(-1.0585) = 0.0874$

23. antilog $0.0000 = 1.00$

25. antilog $2.5011 = 317$

27. antilog$(-0.1543) = 0.701$

29. $\log N = 2.0000$
$N = $ antilog 2.000
$N = 100$

31. $\log N = -2.103$
$N = $ antilog (-2.103)
$N = 0.00789$

33. $\log N = 4.5202$
$N = $ antilog 4.5202
$N = 33,100$

35. $\log N = -1.06$
$N = $ antilog (-1.06)
$N = 0.0871$

37. $\log N = -0.3686$
$N = $ antilog (-0.3686)
$N = 0.428$

39. $\log N = -0.3936$
$N = $ antilog (-0.3936)
$N = 0.404$

41. $\log 3560 = 3.5514$
Therefore, $10^{3.5514}\approx 3560$.

43. $\log 0.0727 = -1.1385$
Therefore, $10^{-1.1385}\approx 0.0727$

45. $\log 102 = 2.0086$
Therefore, $10^{2.0086}\approx 102$.

47. $\log 0.00128 = -2.8928$
Therefore, $10^{-2.8928}\approx 0.00128$.

49. $10^{2.4360} = 273$

51. $10^{-0.158} = 0.695$

53. $10^{-1.6091} = 0.0246$

55. $10^{1.3503} = 22.4$

57. $\log 1 = x$
$10^x = 1$
$10^x = 10^0$
$x = 0$
Therefore, $\log 1 = 0$.

59. $\log 0.1 = x$
$10^x = 0.1$
$10^x = \dfrac{1}{10}$
$10^x = 10^{-1}$
$x = -1$
Therefore, $\log 0.1 = -1$.

61. $\log 0.01 = x$
$10^x = 0.01$
$10^x = \dfrac{1}{100}$
$10^x = 10^{-2}$
$x = -2$
Therefore, $\log 0.01 = -2$.

63. $\log 0.001 = x$
$10^x = 0.001$
$10^x = \dfrac{1}{1000}$
$10^x = 10^{-3}$
$x = -3$
Therefore, $\log 0.001 = -3$.

65. $\log 10^5 = 5$

67. $10^{\log 7} = 7$

69. $6 \log 10^{5.2} = 6(5.2) = 31.2$

71. $5(10^{\log 9.4}) = 5(9.4) = 47$

73. No; $10^2 = 100$ and since $462 > 100$, $\log 462$ must be greater than 2.

75. No; $10^0 = 1$ and $10^{-1} = 0.1$ and, since $\log 0.163$ must be between 0 and -1.

77. No;
$$\log \dfrac{y}{3x} = \log y - \log 3x$$
$$= \log y - (\log 3 + \log x)$$
$$= \log y - \log 3 - \log x$$

79. $\log 125 = \log 5^3$
$$= 3 \log 5$$
$$= 3(0.6990)$$
$$= 2.0970$$

81. $\log 30$ is not possible given this information.

83. $\log \dfrac{1}{25} = \log 25^{-1}$
$$= -\log 25$$
$$= -1(1.3979)$$
$$= -1.3979$$

85. $R = \log I$

a. $R = \log 12,000 = 4.08$
An earthquake that is 12,000 times more intense than the minimum level has a Richter number of 4.08.

b. $4.29 = \log I$
$10^{4.29} = I$
$I = 19,500$
An earthquake with a Richter number of 4.29 is 19,500 times more intense than the minimum level.

87. $f(x) = 30 - 5 \log x$

a. $f(1) = 30 - 5 \log 1$
$$= 30 - 5(0)$$
$$= 30$$
The box office receipts during the first week were $30 million.

b. $f(8) = 30 - 5 \log 8$
$$= 30 - 5(0.90309)$$
$$\approx 25.48$$
The box office receipts during the eighth week were about $25.48 million.

89. $\log E = 11.8 + 1.5 m_s$

    a. $\log E = 11.8 + 1.5(6)$
              $\log E = 20.8$
              $10^{20.8} = E$
                  $E = 6.31 \times 10^{20}$
     The energy released is 6.31×10^{20}.

    b. $\log(1.2 \times 10^{15}) = 11.8 + 1.5 m_s$
         $15.07918125 = 11.8 + 1.5 m_s$
           $3.27918125 = 1.5 m_s$
                $m_s \approx 2.19$
     The surface wave has magnitude 2.19.

91. $M = \dfrac{\log E - 11.8}{1.5}$

     $M = \dfrac{\log(1.259 \times 10^{21}) - 11.8}{1.5}$

       $= \dfrac{\log 1.259 + \log 10^{21} - 11.8}{1.5}$

       $= \dfrac{\log 1.259 + 21 - 11.8}{1.5}$

       $= \dfrac{\log 1.259 + 9.2}{1.5}$

       $\approx \dfrac{0.1000 + 9.2}{1.5}$

       $\approx \dfrac{9.3}{1.5}$

       ≈ 6.2

    The magnitude is about 6.2.

94. $-3x^2 - 4x - 8 = 0, \quad a = -3, \ b = -4, \ c = -8$

$$x = \frac{-(-4) \pm \sqrt{(-4)^2 - 4(-3)(-8)}}{2(-3)}$$

$$= \frac{4 \pm \sqrt{16 - 96}}{-6}$$

$$= \frac{4 \pm \sqrt{-80}}{-6}$$

$$= \frac{4 \pm \sqrt{16}\sqrt{-1}\sqrt{5}}{-6}$$

$$= \frac{4 \pm 4i\sqrt{5}}{-6}$$

$$= \frac{-2\left(-2 \pm 2i\sqrt{5}\right)}{-6}$$

$$= \frac{-2 \pm 2i\sqrt{5}}{3}$$

95. Let r equal the rate of the boat in still water.

	d	r	t
Down river	15	$r + 5$	$\frac{15}{r+5}$
Up river	15	$r - 5$	$\frac{15}{r-5}$

The total time was 4 hours.

$$\frac{15}{r+5} + \frac{15}{r-5} = 4$$

$$(r+5)(r-5)\left(\frac{15}{r+5}\right) + (r+5)(r-5)\left(\frac{15}{r-5}\right) = 4(r+5)(r-5)$$

$$15(r-5) + 15(r+5) = 4(r^2 - 25)$$

$$15r - 75 + 15r + 75 = 4r^2 - 100$$

$$30r = 4r^2 - 100$$

$$0 = 4r^2 - 30r - 100$$

$$0 = 2(2r^2 - 15r - 50)$$

$$0 = 2(2r+5)(r-10)$$

$$2r + 5 = 0 \quad \text{or} \quad r - 10 = 0$$
$$r = -\frac{5}{2} \qquad\qquad r = 10$$

Since the speed must be positive, the speed of the boat in still water is 10 miles per hour.

96. $\frac{2x-3}{5x+10} < 0$

$$2x - 3 = 0 \qquad 5x + 10 = 0$$
$$x = \frac{3}{2} \qquad\qquad x = -2$$

$x \neq -2$

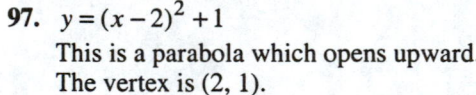

97. $y = (x-2)^2 + 1$

This is a parabola which opens upward.
The vertex is (2, 1).

x	y
–1	10
0	5
1	2
2	1
4	5

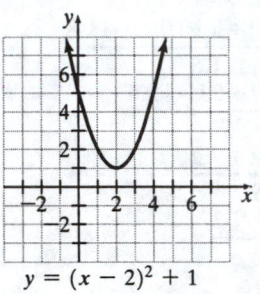

$y = (x - 2)^2 + 1$

Exercise Set 9.6

1. $a^m = a^n,\ \log m = \log n$

3. $r = s$

5. Check for extraneous roots.

7. $5^x = 125$
$$5^x = 5^3$$
$$x = 3$$

9. $16^x = \dfrac{1}{4}$

$(4^2)^x = 4^{-1}$

$4^{2x} = 4^{-1}$

$2x = -1$

$x = -\dfrac{1}{2}$

11. $2^{3x-2} = 16$

$2^{3x-2} = 2^4$

$3x - 2 = 4$

$3x = 6$

$x = 2$

13. $27^x = 3^{2x+3}$

$3^{3x} = 3^{2x+3}$

$3x = 2x + 3$

$x = 3$

15. $7^x = 50$

$\log 7^x = \log 50$

$x \log 7 = \log 50$

$x = \dfrac{\log 50}{\log 7}$

$x \approx 2.01$

17. $4^{x-1} = 20$

$\log 4^{x-1} = \log 20$

$(x-1) \log 4 = \log 20$

$x - 1 = \dfrac{\log 20}{\log 4}$

$x = \dfrac{\log 20}{\log 4} + 1$

$x \approx 3.16$

19. $1.63^{x+1} = 25$

$\log 1.63^{x+1} = \log 25$

$(x+1) \log 1.63 = \log 25$

$x + 1 = \dfrac{\log 25}{\log 1.63}$

$x + 1 \approx 6.59$

$x \approx 5.59$

21. $3^{x+4} = 6^x$

$\log 3^{x+4} = \log 6^x$

$(x + 4) \log 3 = x \log 6$

$x \log 3 + 4 \log 3 = x \log 6$

$4 \log 3 = x \log 6 - x \log 3$

$4 \log 3 = x(\log 6 - \log 3)$

$\dfrac{4 \log 3}{\log 6 - \log 3} = x$

$6.34 \approx x$

23. $\log_9 x = \dfrac{1}{2}$

$9^{1/2} = x$

$\sqrt{9} = x$

$3 = x$

25. $\log_5 x = -2$

$5^{-2} = x$

$\dfrac{1}{5^2} = x$

$\dfrac{1}{25} = x$

27. $\log_2(5 - 3x) = 3$

$2^3 = 5 - 3x$

$8 = 5 - 3x$

$3x = -3$

$x = -1$

29. $\log_5(x + 2)^3 = 3$

$(x + 2)^3 = 5^3$

$x + 2 = 5$

$x = 3$

31. $\log_2(x + 4)^2 = 4$

$(x + 4)^2 = 2^4$

$x^2 + 8x + 16 = 16$

$x^2 + 8x = 0$

$x(x + 8) = 0$

$x = 0 \quad \text{or} \quad x + 8 = 0$

$x = -8$

33. $\log(2x-3)^3 = 3$
$3\log(2x-3) = 3$
$\log(2x-3) = 1$
$2x-3 = 10^1$
$2x-3 = 10$
$2x = 13$
$x = \dfrac{13}{2}$

35. $\log(r+2) = \log(3r-1)$
$r+2 = 3r-1$
$3 = 2r$
$\dfrac{3}{2} = r$

37. $\log(2x+1) + \log 4 = \log(7x+8)$
$\log(8x+4) = \log(7x+8)$
$8x+4 = 7x+8$
$x = 4$

39. $\log n + \log(3n-5) = \log 2$
$\log(3n^2 - 5n) = \log 2$
$3n^2 - 5n = 2$
$3n^2 - 5n - 2 = 0$
$(3n+1)(n-2) = 0$

$3n+1 = 0$ or $n-2 = 0$
$3n = -1 \qquad\qquad n = 2$
$n = -\dfrac{1}{3}$

Check: $n = -\dfrac{1}{3}$
$\log n + \log(3n-5) = \log 2$
$\log\left(-\dfrac{1}{3}\right) + \log\left[3\left(\dfrac{-1}{3}\right)-5\right] = \log 2$
Logarithms of negative numbers are not real numbers.
Check: $n = 2$
$\log n + \log(3n-5) = \log 2$
$\log 2 + \log[3(2)-5] = \log 2$
$\log 2 + \log 1 = \log 2$
$\log(2 \cdot 1) = \log 2$
$\log 2 = \log 2$
2 is the only solution.
$-\dfrac{1}{3}$ is an extraneous solution.

41. $\log 5 + \log y = 0.72$
$\log 5y = 0.72$
$5y \approx 5.2481$
$y \approx 1.05$

43. $2\log x - \log 4 = 2$
$\log x^2 - \log 4 = 2$
$\log \dfrac{x^2}{4} = 2$
$\dfrac{x^2}{4} = \text{antilog } 2$
$\dfrac{x^2}{4} = 100$
$x^2 = 400$
$x^2 - 400 = 0$
$(x+20)(x-20) = 0$

$x+20 = 0$ or $x-20 = 0$
$x = -20 \qquad\qquad x = 20$
Check: $x = -20$
$2\log x - \log 4 = 2$
$2\log(-20) - \log 4 = 2$
Logarithms of negative numbers are not real numbers.
Check: $x = 20$
$2\log x - \log 4 = 2$
$2\log 20 - \log 4 = 2$
$\log \dfrac{400}{4} = 2$
$\log 100 = 2$
$100 = \text{antilog } 2$
$100 = 100$
Thus, 20 is the only solution.
-20 is an extraneous solution.

45. $\log x + \log(x-3) = 1$
$\log(x^2 - 3x) = 1$
$x^2 - 3x = \text{antilog } 1$
$x^2 - 3x = 10$
$x^2 - 3x - 10 = 0$
$(x-5)(x+2) = 0$
$x-5 = 0$ or $x+2 = 0$
$x = 5 \qquad\qquad x = -2$
A check shows that 5 is the only solution.
-2 is an extraneous solution.

47. $\log x = \dfrac{1}{3}\log 27$

$\log x = \log 27^{1/3}$
$\log x = \log 3$
$x = 3$

49. $\log_8 x = 3\log_8 2 - \log_8 4$

$\log_8 x = \log_8 2^3 - \log_8 4$
$\log_8 x = \log_8 \dfrac{8}{4}$
$\log_8 x = \log_8 2$
$x = 2$

51. $\log_5(x+3) + \log_5(x-2) = \log_5 6$

$\log_5(x+3)(x-2) = \log_5 6$
$\log_5(x^2 + x - 6) = \log_5 6$
$x^2 + x - 6 = 6$
$x^2 + x - 12 = 0$
$(x+4)(x-3) = 0$

$x = -4$ or $x = 3$
Disregard $x = -4$ since
$\log(-4 + 3) = \log(-1)$.
Therefore, $x = 3$ is the only solution.

53. $\log_2(x+3) - \log_2(x-6) = \log_2 4$

$\log_2 \dfrac{x+3}{x-6} = \log_2 4$
$\dfrac{x+3}{x-6} = 4$
$x + 3 = 4x - 24$
$27 = 3x$
$9 = x$

55. $A = P(1+r)^n$

$A = 1200(1 + 0.06)^5$
$= 1200(1.338226)$
$= \$1605.87$
A total of $\$1605.87$ accumulates in 5 years.

57. Let x be the number of bacteria present initially.

$2224 = x \cdot 2^4$
$2224 = 16x$
$139 = x$
There were 139 bacteria present initially.

59. $f(x) = 26 - 12\log x$

a. $x = 1960 - 1959 = 1$
$f(1) = 26 - 12\log 1 = 26$
In 1960, the rate was 26 deaths per 1000 live births.

b. $x = 1996 - 1959 = 37$
$f(37) = 26 - 12\log 37 \approx 7.18$
In 1996, the rate was 7.18 deaths per 1000 live births.

61. $c = 50{,}000$, $n = 12$, $r = 0.15$.

$S = c(1-r)^n$
$S = 50{,}000(1 - 0.15)^{12}$
$S = 50{,}000(0.85)^{12}$
$S \approx 7112.09$
The scrap value is about $\$7112.09$.

63. $P_{out} = 12.6$ and $P_{in} = 0.146$

$P = 10\log\left(\dfrac{12.6}{0.146}\right)$
$P \approx 10\log 86.30137$
$P \approx 10(1.936)$
$P \approx 19.36$
The power gain is about 19.36.

65. a. $d = 120$

$d = 10\log I$
$120 = 10\log I$
$12 = \log I$
$I = \text{antilog } 12$
$I = 10^{12}$
$I = 1{,}000{,}000{,}000{,}000$
The intensity is 1,000,000,000,000 times the minimum intensity of audible sound.

b. $d = 70$

$d = 10\log I$
$70 = 10\log I$
$7 = \log I$
$I = \text{antilog } 7$
$I = 10^7$
$I = 10{,}000{,}000$
$\dfrac{1{,}000{,}000{,}000{,}000}{10{,}000{,}000} = 100{,}000$

The sound of an airplane engine is 100,000 times more intense than the noise in a busy city street.

67. $8^x = 16^{x-2}$

$2^{3x} = 2^{4(x-2)}$

$3x = 4(x-2)$

$3x = 4x - 8$

$8 = x$

69. $2^{2x} - 6(2^x) + 8 = 0$

$(2^x)^2 - 6(2^x) + 8 = 0$

$y^2 - 6y + 8 = 0 \leftarrow$ Replace 2^x with y

$(y-4)(y-2) = 0$

$y - 4 = 0$ or $y - 2 = 0$

　　$y = 4$　　　　$y = 2$

$2^x = 4$　　　$2^x = 2 \leftarrow$ Replace y with 2^x

$2^x = 2^2$　　　$2^x = 2^1$

　$x = 2$　　　　$x = 1$

The solutions are $x = 2$ and $x = 1$.

71.　　$2^x = 8^y$

$x + y = 4$

The first equation simplifies to

$2^x = (2^3)^y$

$2^x = 2^{3y}$

　$x = 3y$

The system becomes

　$x = 3y$

$x + y = 4$

Substitute $3y$ for x in the second equation.

　$x + y = 4$

$3y + y = 4$

　　$4y = 4$

　　　$y = 1$

Now, substitute 1 for y in the first equation.

$x = 3y$

$x = 3(1) = 3$

The solution is (3, 1).

73. $\log(x + y) = 2$

　　　$x - y = 8$

The first equation can be written as

$x + y = 10^2$

$x + y = 100$

The system becomes

$x + y = 100$

　　　$x + y = 100$

　　　$\underline{x - y = 8}$

Add: $2x = 108$

　　$x = 54$

Substitute 54 for x in the first equation.

　$x + y = 100$

$54 + y = 100$

　　　$y = 46$

The solution is (54, 46).

75.

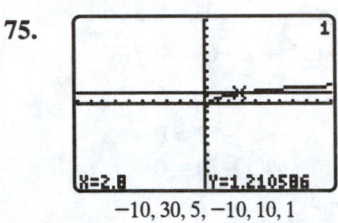

$-10, 30, 5, -10, 10, 1$

The solution is $x \approx 2.8$.

77.

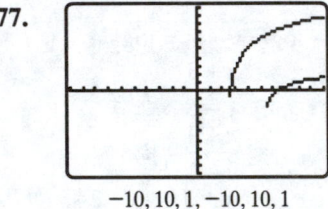

$-10, 10, 1, -10, 10, 1$

There is no real-number solution.

79.　　　$\dfrac{x-4}{2} - \dfrac{2x-5}{5} > 3$

$10\left(\dfrac{x-4}{2}\right) - 10\left(\dfrac{2x-5}{5}\right) > 10(3)$

　$5(x-4) - 2(2x-5) > 30$

　　$5x - 20 - 4x + 10 > 30$

　　　　　　　$x - 10 > 30$

　　　　　　　　　$x > 40$

a. The solution is $\{x | x > 40\}$.

b. The solution is $(40, \infty)$.

80. b) and c) are functions; only b) is one-to-one.

81.

$$x^2 - 4x + 3$$

$$\frac{2x - 3}{-3x^2 + 12x - 9}$$

$$\frac{2x^3 - 8x^2 + 6x}{2x^3 - 11x^2 + 18x - 9}$$

82. Using synthetic division:

$$\begin{array}{r|rrr} -4 & 2 & 11 & 15 \\ & & -8 & -12 \\ \hline & 2 & 3 & 3 \end{array}$$

$$\frac{2x^2 + 11x + 15}{x + 4} = 2x + 3 + \frac{3}{x + 4}.$$

Exercise Set 9.7

1. a. The base in the natural exponential function is e.

b. The approximate value of e is 2.7183.

3. The domain of $\ln x$ is
$\{x | x$ is a real number and $x > 0\}$.

5. $\log_a x = \dfrac{\log_b x}{\log_b a}$

7. $\ln e^x = x$

9. P increases when t increases for $k > 0$.

11. $\ln 35 = 3.5553$

13. $\ln 302 = 5.7104$

15. $\ln N = 16$
$e^{\ln N} = e^{1.6}$
$N = e^{1.6} \approx 4.95$

17. $\ln N = -2.72$
$e^{\ln N} = e^{-2.72}$
$N = e^{-2.72} \approx 0.0659$

19. $\ln 60 = \dfrac{\log 60}{\log e} \approx 4.0943$

21. $\ln 0.046 = \dfrac{\log 0.046}{\log e} \approx -3.0791$

23. $\log_3 25 = \dfrac{\log 25}{\log 3} \approx 2.9300$

25. $\log_2 32 = \dfrac{\log 32}{\log 2} \approx 5.000$

27. $\ln 2700 = \dfrac{\log 2700}{\log e} \approx 7.9010$

29. $\log_3 0.0049 = \dfrac{\log 0.0049}{\log 3} \approx -4.8411$

31. $\ln x + \ln(x - 1) = \ln 12$
$\ln x(x - 1) = \ln 12$
$e^{\ln[x(x-1)]} = e^{\ln 12}$
$x(x - 1) = 12$
$x^2 - x - 12 = 0$
$(x - 4)(x + 3) = 0$
$x - 4 = 0 \qquad x + 3 = 0$
$x = 4 \qquad\quad x = -3$

Only $x = 4$ checks. $x = -3$ is an extraneous solution since $\ln x$ becomes $\ln(-3)$ which is not a real number.

33. $\ln x = 5\ln 2 - \ln 8$
$\ln x = \ln 2^5 - \ln 8$
$\ln x = \ln \dfrac{32}{8}$
$\ln x = \ln 4$
$e^{\ln x} = e^{\ln 4}$
$x = 4$
$x = 4$ checks.

35. $\ln(x^2 - 4) - \ln(x + 2) = \ln 1$

$\ln(x^2 - 4) - \ln(x + 2) = 0$

$\ln(x^2 - 4) = \ln(x + 2)$

$e^{\ln(x^2 - 4)} = e^{\ln(x+2)}$

$x^2 - 4 = x + 2$

$x^2 - x - 6 = 0$

$(x - 3)(x + 2) = 0$

$x - 3 = 0$ or $x + 2 = 0$

$x = 3$ $x = -2$

Only $x = 3$ checks. $x = -2$ is an extraneous solution since $\ln(x + 2)$ becomes $\ln(-2 + 2) = \ln(0)$ which is not a real number.

37. $P = 700e^{(1.7)(1.2)}$

$P = 700e^{2.04}$

$P \approx 5383.43$

39. $50 = P_0 e^{-0.05(3)}$

$50 = P_0 e^{-0.15}$

$\dfrac{50}{e^{-0.15}} = P_0$

$P_0 \approx 58.09$

41. $90 = 30e^{1.4t}$

$3 = e^{1.4t}$

$\ln 3 = \ln e^{1.4t}$

$\ln 3 = 1.4t$

$t = \dfrac{\ln 3}{1.4}$

$t \approx 0.7847$

43. $80 = 40e^{k(3)}$

$2 = e^{3k}$

$\ln 2 = \ln e^{3k}$

$\ln 2 = 3k$

$k = \dfrac{\ln 2}{3}$

$k \approx 0.2310$

45. $20 = 40e^{k(2.4)}$

$0.5 = e^{2.4k}$

$\ln 0.5 = \ln e^{2.4k}$

$\ln 0.5 = 2.4k$

$k = \dfrac{\ln 0.5}{2.4}$

$k \approx -0.2888$

47. $A = 6000e^{-0.08(3)}$

$A = 6000e^{-0.24}$

$A \approx 4719.77$

49. $V = V_0 e^{kt}$

$\dfrac{V}{e^{kt}} = V_0$ or $V_0 = \dfrac{V}{e^{kt}}$

51. $P = 150e^{4t}$

$\dfrac{P}{150} = e^{4t}$

$\ln \dfrac{P}{150} = \ln e^{4t}$

$\ln \dfrac{P}{150} = 4t$

$\dfrac{\ln P - \ln 150}{4} = t$ or $t = \dfrac{\ln P - \ln 150}{4}$

53. $A = A_0 e^{kt}$

$\dfrac{A}{A_0} = e^{kt}$

$\ln \dfrac{A}{A_0} = \ln e^{kt}$

$\ln A - \ln A_0 = kt$

$\dfrac{\ln A - \ln A_0}{t} = k$ or $k = \dfrac{\ln A - \ln A_0}{t}$

55. $\ln y - \ln x = 2.3$

$\ln \dfrac{y}{x} = 2.3$

$e^{\ln(y/x)} = e^{2.3}$

$\dfrac{y}{x} = e^{2.3}$

$y = xe^{2.3}$

57. $\ln y - \ln(x+3) = 6$

$$\ln \frac{y}{x+3} = 6$$

$$e^{\ln \frac{y}{x+3}} = e^6$$

$$\frac{y}{x+3} = e^6$$

$$y = (x+3)e^6$$

59. $e^x = 12.183$

Take the natural logarithm of both sides of the equation.

$$\ln e^x = \ln 12.183$$
$$x = \ln 12.183 \approx 2.5000$$

61. $P = P_0 e^{kt}$

  a. $P = 5000 e^{0.08(2)}$

$$= 5000 e^{0.16}$$
$$\approx 5867.55$$

The amount will be \$5867.55.

  b. If the amount in the account is to double, then $P = 2(5000) = 10,000$.

$$10,000 = 5000 e^{0.08t}$$
$$2 = e^{0.08t}$$
$$\ln 2 = \ln e^{0.08t}$$
$$\ln 2 = 0.08t$$
$$\frac{\ln 2}{0.08} = t$$
$$8.66 \approx t$$

It would take about 8.66 years for the value to double.

63. $f(x) = 30 - 15 \ln 0.4x$

  a. $x = 1989 - 1984 = 5$
$f(5) = 30 - 15 \ln 0.4(5) \approx 19.60$
In 1989, about 19.60 thousand tons were available.

  b. $x = 2000 - 1984 = 16$
$f(16) = 30 - 15 \ln 0.4(16) \approx 2.16$
In 2000, about 2.16 tons are available.

65. $f(t) = 1 - e^{-0.04t}$

  a. $f(t) = 1 - e^{-0.04(50)} = 1 - e^{-2} \approx 0.8647$
About 86.47% of the target market buys the drink after 50 days of advertising.

  b.
$$0.75 = 1 - e^{-0.04t}$$
$$-0.25 = -e^{-0.04t}$$
$$0.25 = e^{-0.04t}$$
$$\ln 0.25 = \ln e^{-0.04t}$$
$$\ln 0.25 = -0.04t$$
$$t = \frac{\ln 0.25}{-0.04}$$
$$t \approx 34.66$$

About 34.66 days of advertising are needed if 75% of the target market is to buy the soft drink.

67. $f(P) = 0.37 \ln P + 0.05$

  a. $f(972,000) = 0.37 \ln(972,000) + 0.05$
$$\approx 5.1012311 + 0.05$$
$$\approx 5.15$$
The average walking speed in Nashville, Tennessee is 5.15 feet per second.

  b. $f(8,567,000)$
$$= 0.37 \ln(8,567,000) + 0.05$$
$$\approx -5.906 + 0.05$$
$$\approx 5.96$$
The average walking speed in New York City is 5.96 feet per second.

  c.
$$5 = 0.37 \ln P + 0.05$$
$$4.95 = 0.37 \ln P$$
$$13.378378 = \ln P$$
$$e^{13.378378} = e^{\ln P}$$
$$P = e^{13.378378}$$
$$P \approx 646,000$$
The population is about 646,000.

69. $P(t) = 5.98e^{0.0133t}$

   a. $t = 2006 - 1999 = 7$

$$P(7) = 5.98e^{0.0133(7)}$$
$$= 5.98e^{0.0931}$$
$$\approx 6.56$$

The world's population in 2006 is expected to be about 6.56 billion.

   b. $2(5.98) = 11.96$

$$11.96 = 5.98e^{0.0133t}$$
$$2 = e^{0.0133t}$$
$$\ln 2 = \ln e^{0.0133t}$$
$$\ln 2 = 0.0133t$$
$$\frac{\ln 2}{0.0133} = t$$
$$52.12 \approx t$$

The world's population will double in about 52.12 years.

71. **a.** $f(t) = v_0 e^{-0.0001205t}$

Use $f(t) = 9$ and $v_0 = 20$.

$$9 = 20e^{-0.0001205t}$$
$$0.45 = e^{-0.0001205t}$$
$$\ln 0.45 = -0.0001205t$$
$$\frac{\ln 0.45}{-0.0001205} = t$$
$$t \approx 6626.62$$

The bone is about 6626.62 years old.

   b.   Let x equal the original amount of carbon 14 then $0.5x$ equals the remaining amount.

$$0.5x = xe^{-0.0001205t}$$
$$\frac{0.5x}{x} = \frac{xe^{-0.0001205t}}{x}$$
$$0.5 = e^{-0.0001205t}$$
$$\ln 0.5 = \ln e^{-0.0001205t}$$
$$\ln 0.5 = -0.0001205t$$
$$\frac{\ln 0.5}{-0.0001205} = t$$
$$t \approx 5752.26$$

If 50% of the carbon 14 remains, the item is about 5752.26 years old.

73. The equation is $P = P_0^{kt}$. Let $P = 2P_0$ and $t = 6$ to obtain

$$P = P_0^{kt}$$
$$2P_0 = P_0 e^{k \cdot 6}$$
$$2P_0 = P_0 e^{6k}$$
$$2 = e^{6k}$$
$$\ln 2 = \ln e^{6k}$$
$$\ln 2 = 6k$$
$$\frac{\ln 2}{6} = k$$
$$0.1155 \approx k \text{ or}$$
$$k \approx 11.55\%$$

The rate needed is about 11.55%.

75. $P = 50e^{-0.002t}$

   a.   Use $t = 100$.

$$P = 50e^{-0.002(100)}$$
$$P = 50e^{-0.2}$$
$$P \approx 40.94$$

After 100 days, about 40.94 watts remain.

   b.   Use $P = 10$.

$$10 = 50e^{-0.002t}$$
$$0.2 = e^{-0.002t}$$
$$\ln 0.2 = \ln e^{-0.002t}$$
$$\ln 0.2 = -0.002t$$
$$\frac{\ln 0.2}{-0.002} = t$$
$$t \approx 804.72$$

There will be 10 watts of power remaining after about 804.72 days.

77. $t = \dfrac{t_h}{0.693} \ln\left(\dfrac{N_0}{N}\right)$

$$= \frac{4.5 \times 10^9}{0.693} \ln\left(\frac{5 \times 10^{12}}{4 \times 10^{12}}\right)$$
$$\approx 1.45 \times 10^9$$

The rock is about 1.45×10^9 years old.

79.

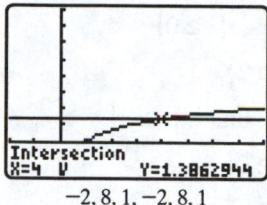

Intersection
X=4   Y=1.3862944

$$-2, 8, 1, -2, 8, 1$$

81. $\ln(4-x) = 2\ln x + \ln 2.4$
$$y_1 = \ln(4-x)$$
$$y_2 = 2\ln x + \ln 2.4$$

Intersection
X=1.0993629   Y=1.0649304

$$-10, 10, 1, -10, 10, 1$$

The intersection is approximately $(1.099, 1.065)$. Therefore, $x \approx 1.099$.

83. $x = k(\ln I_0 - \ln I)$
$$\frac{x}{k} = \ln\frac{I_0}{I}$$
$$e^{x/k} = e^{\ln(I_0/I)}$$
$$e^{x/k} = \frac{I_0}{I}$$
$$Ie^{x/k} = I_0 \text{ or } I_0 = Ie^{x/k}$$

85. $\ln M = \ln Q - \ln(1-Q)$
$$\ln M = \ln\frac{Q}{1-Q}$$
$$e^{\ln M} = e^{\ln[Q/(1-Q)]}$$
$$M = \frac{Q}{1-Q}$$
$$M(1-Q) = Q$$
$$M - MQ = Q$$
$$M = Q + MQ$$
$$M = Q(1+M)$$
$$\frac{M}{1+M} = Q \text{ or } Q = \frac{M}{1+M}$$

87. $(x^2 y^{-2})^{-1} (4xy^3)^2$
$$= x^{-2} y^2 \cdot 16 x^2 y^6$$
$$= 16 x^{-2+2} y^{2+6}$$
$$= 16 x^0 y^8$$
$$= 16 y^8$$

88. $\dfrac{\dfrac{3}{x^2} - \dfrac{2}{x}}{\dfrac{x}{4}} = \dfrac{4x^2\left(\dfrac{3}{x^2}\right) - 4x^2\left(\dfrac{2}{x}\right)}{4x^2\left(\dfrac{x}{4}\right)}$

$$= \frac{12 - 8x}{x^3}$$

89. $\dfrac{1}{f} = \dfrac{1}{p} + \dfrac{1}{q}$
$$fpq\left(\frac{1}{f}\right) = fpq\left(\frac{1}{p}\right) + fpq\left(\frac{1}{q}\right)$$
$$pq = fq + fp$$
$$pq - fq = fp$$
$$q(p-f) = fp$$
$$q = \frac{fp}{p-f}$$

90. $\sqrt[3]{128 x^7 y^9 z^{13}} = \sqrt[3]{64 x^6 y^9 z^{12} \cdot 2xz}$
$$= \sqrt[3]{64 x^6 y^9 z^{12}} \cdot \sqrt[3]{2xz}$$
$$= 4 x^2 y^3 z^4 \sqrt[3]{2xz}$$

Review Exercises

1. $(f \circ g)(x) = (2x-5)^2 - 3(2x-5) + 4$
$$= 4x^2 - 20x + 25 - 6x + 15 + 4$$
$$= 4x^2 - 26x + 44$$

2. $(f \circ g)(x) = 4x^2 - 26x + 44$
$$(f \circ g)(2) = 4(2)^2 - 26(2) + 44$$
$$= 16 - 52 + 44$$
$$= 8$$

3. $(g \circ f)(x) = 2(x^2 - 3x + 4) - 5$
$$= 2x^2 - 6x + 8 - 5$$
$$= 2x^2 - 6x + 3$$

4. $(g \circ f)(x) = 2x^2 - 6x + 3$
    $(g \circ f)(-3) = 2(-3)^2 - 6(-3) + 3$
    $\qquad\qquad = 18 + 18 + 3$
    $\qquad\qquad = 39$

5. $(f \circ g)(x) = 3\sqrt{x - 4} + 2$

6. $(g \circ f)(x) = \sqrt{(3x + 2) - 4}$
    $\qquad\qquad = \sqrt{3x - 2}$
    $\qquad x \geq \dfrac{2}{3}$

7. This function is one-to-one since it passes both the vertical line test and the horizontal line test.

8. The function is not one-to-one since the graph does not pass the horizontal line test.

9. Yes, the ordered pairs represent a one-to-one function. For each value of x, there is a unique value for y and each y-value has a unique x-value.

10. No, the ordered pairs do not represent a one-to-one function since the pairs $(0, -2)$ and $(3, -2)$ have different x-values but the same y-values.

11. Yes, $y = \sqrt{x + 1}$, $x \geq -1$, is a one-to-one function since it passes both the vertical line test and the horizontal line test.

12. No, $y = x^2 - 9$ is a parabola with vertex at $(0, -9)$. It is not a one-to-one function since it does not pass the horizontal line test. Horizontal lines above $y = -9$ intersect the graph in two points.

13. $f(x)$: Domain: $\{-4, 0, 5, 6\}$
    Range: $\{-3, 2, 3, 7\}$
    $f^{-1}(x)$: Domain: $\{-3, 2, 3, 7\}$
    Range: $\{-4, 0, 5, 6\}$

14. $f(x)$: Domain: $\{x \mid x \geq 0\}$
    Range: $\{y \mid y \geq 2\}$
    $f^{-1}(x)$: Domain: $\{x \mid x \geq 2\}$
    Range: $\{y \mid y \geq 0\}$

15. $y = f(x) = 4x - 2$
    $\qquad x = 4y - 2$
    $\qquad x + 2 = 4y$
    $\qquad \dfrac{x + 2}{4} = y$
    $\qquad f^{-1}(x) = \dfrac{x + 2}{4}$

x	$f(x)$
0	–2
$\frac{1}{2}$	0

x	$f^{-1}(x)$
0	$\frac{1}{2}$
–2	0

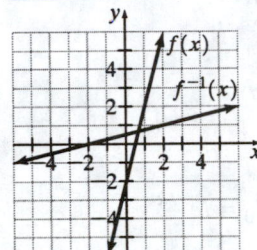

16. $y = f(x) = \sqrt[3]{x - 1} = (x - 1)^{1/3}$
    $\qquad x = (y - 1)^{1/3} \qquad$ a
    $\qquad x^3 = [(y - 1)^{1/3}]^3$
    $\qquad x^3 = y - 1$
    $\qquad x^3 + 1 = y$
    $\qquad f^{-1}(x) = x^3 + 1$

x	$f(x)$
–7	–2
0	–1
1	0
2	1
9	2

x	$f^{-1}(x)$
–2	–7
–1	0
0	1
1	2
2	9

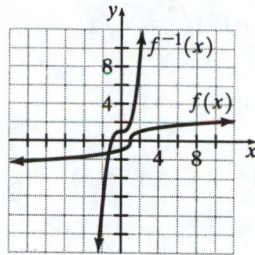

17. $y = 2^x$

x	-2	-1	0	1	2	3
y	$\frac{1}{4}$	$\frac{1}{2}$	1	2	4	8

Domain: $\mathbb{R}$

Range: $\{y \mid y > 0\}$

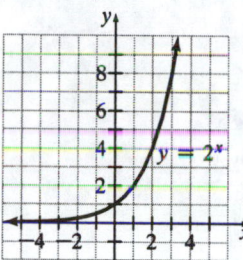

18. $y = \left(\frac{1}{2}\right)^x$

x	-2	-1	0	1	2
y	4	2	1	$\frac{1}{2}$	$\frac{1}{4}$

Domain: $\mathbb{R}$

Range: $\{y \mid y > 0\}$

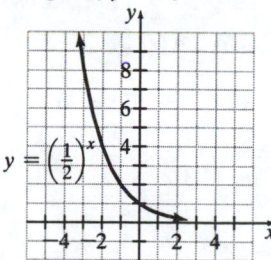

19. a. Appears to be linear since it resembles a straight line.

b. Appears to be exponential

c. 8 measurements were used

d. 1 year

e. Less than 1 year, possibly 9 months

f. 35 million drives were in use in mid-1994.

20. $\quad 4^2 = 16$
$$\log_4 16 = 2$$

21. $\quad 8^{1/3} = 2$
$$\log_8 2 = \frac{1}{3}$$

22. $\quad 6^{-2} = \frac{1}{36}$
$$\log_6 \frac{1}{36} = -2$$

23. $\log_5 25 = 2$
$$5^2 = 25$$

24. $\log_{1/3} \frac{1}{9} = 2$
$$\left(\frac{1}{3}\right)^2 = \frac{1}{9}$$

25. $\log_3 \frac{1}{9} = -2$
$$3^{-2} = \frac{1}{9}$$

26. $3 = \log_4 x$
$$x = 4^3$$
$$x = 64$$

27. $\quad 3 = \log_a 8$
$$a^3 = 8$$
$$a^3 = 2^3$$
$$a = 2$$

28. $-3 = \log_{1/4} x$

$x = \left(\dfrac{1}{4}\right)^{-3}$

$x = \dfrac{1}{\left(\frac{1}{4}\right)^3}$

$x = \dfrac{1}{\frac{1}{64}}$

$x = 64$

29. $y = \log_2 x$

$x = 2^y$

x	$\frac{1}{4}$	$\frac{1}{2}$	1	2	4	8
y	-2	-1	0	1	2	3

Domain: $\{x \mid x > 0\}$

Range: $\mathbb{R}$

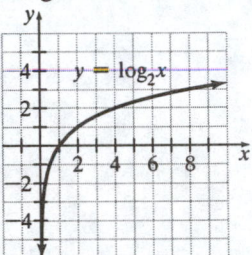

30. $y = \log_{1/2} x$

$x = \left(\dfrac{1}{2}\right)^y$

x	4	2	1	$\frac{1}{2}$	$\frac{1}{4}$
y	-2	-1	0	1	2

Domain: $\{x \mid x > 0\}$

Range: $\mathbb{R}$

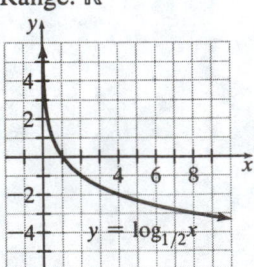

31. $\log_8 \sqrt{12} = \log_8 (12)^{1/2} = \dfrac{1}{2} \log_8 12$

32. $\log(x-8)^5 = 5\log(x-8)$

33. $\log \dfrac{2(x-3)}{x} = \log 2 + \log(x-3) - \log x$

34. $\log \dfrac{x^4}{39(2x+8)}$

$= \log x^4 - \log 39(2x+8)$

$= 4\log x - [\log 39 + \log(2x-8)]$

$= 4\log x - \log 39 - \log(2x+8)$

35. $2\log x - 3\log(x+1)$

$= \log x^2 - \log(x+1)^3$

$= \log \dfrac{x^2}{(x+1)^3}$

36. $3(\log 2 + \log x) - \log y$

$= 3(\log 2x) - \log y$

$= \log(2x)^3 - \log y$

$= \log \dfrac{(2x)^3}{y}$

37. $\dfrac{1}{2}[\ln x - \ln(x+2)] - \ln 2$

$= \dfrac{1}{2}\left(\ln \dfrac{x}{x+2}\right) - \ln 2$

$= \ln\left(\dfrac{x}{x+2}\right)^{1/2} - \ln 2$

$= \ln\left(\dfrac{\sqrt{\frac{x}{x+2}}}{2}\right)$

38. $3\ln x + \dfrac{1}{2}\ln(x+1) - 3\ln(x+4)$

$= \ln x^3 + \ln(x+1)^{1/2} - \ln(x+4)^3$

$= \ln \dfrac{x^3\sqrt{x+1}}{(x+4)^3}$

39. $8^{\log_8 9} = 9$

40. $\log_4 4^5 = 5$

41. $3\log_7 49 = 3\log_7 7^2 = 3(2) = 6$

42. $4^{\log_8 \sqrt{8}} = 4^{\log_8 8^{1/2}}$
$= 4^{1/2}$
$= \sqrt{4}$
$= 2$

43. $\log 8200 = 3.9138$

44. $\log 0.000716 = -3.1451$

45. antilog $2.9186 = 829$

46. antilog$(-1.3747) = 0.0422$

47. $\log N = 2.3304$
$N = $ antilog 2.3304
$N = 214$

48. $\log N = -1.2262$
$N = $ antilog (-1.2262)
$N = 0.0594$

49. $\log 10^4 = 4$

50. $10^{\log 3} = 3$

51. $7.5\log 10^{4.2} = 7.5(4.2) = 31.5$

52. $3(10^{\log 1.7}) = 3(1.7) = 5.1$

53. $9 = 3^x$
$3^2 = 3^x$
$2 = x$

54. $49^x = \dfrac{1}{7}$
$(7^2)^x = 7^{-1}$
$7^{2x} = 7^{-1}$
$2x = -1$
$x = -\dfrac{1}{2}$

55. $2^{2x+3} = 32$
$2^{2x+3} = 2^5$
$2x + 3 = 5$
$2x = 2$
$x = 1$

56. $27^x = 3^{2x+5}$
$(3^3)^x = 3^{2x+5}$
$3^{3x} = 3^{2x+5}$
$3x = 2x + 5$
$x = 5$

57. $4^x = 37$
$\log 4^x = \log 37$
$x \log 4 = \log 37$
$x = \dfrac{\log 37}{\log 4}$
$x \approx 2.605$

58. $3.2^x = 187$
$\log 3.2^x = \log 187$
$x \log 3.2 = \log 187$
$x = \dfrac{\log 187}{\log 3.2}$
$x \approx 4.497$

59. $10.9^{x+1} = 492$
$\log 10.9^{x+1} = \log 492$
$(x+1)\log 10.9 = \log 492$
$x + 1 = \dfrac{\log 492}{\log 10.9}$
$x = \dfrac{\log 492}{\log 10.9} - 1$
$x \approx 1.595$

60.

$$3^{x+2} = 8^x$$
$$\log 3^{x+2} = \log 8^x$$
$$(x+2)\log 3 = x \log 8$$
$$x \log 3 + 2 \log 3 = x \log 8$$
$$2 \log 3 = x \log 8 - x \log 3$$
$$2 \log 3 = x(\log 8 - \log 3)$$
$$\frac{2 \log 3}{\log 8 - \log 3} = x$$
$$2.240 \approx x$$

61. $\log_5(x+2) = 3$

$$5^3 = x + 2$$
$$125 = x + 2$$
$$123 = x$$

62. $\log x - \log(3x - 5) = \log 2$

$$\log \frac{x}{3x-5} = \log 2$$
$$\frac{x}{3x-5} = 2$$
$$x = 6x - 10$$
$$0 = 5x - 10$$
$$x = 2$$

63. $\log_3 x + \log_3(2x+1) = 1$

$$\log_3 x(2x+1) = 1$$
$$3^1 = x(2x+1)$$
$$0 = 2x^2 + x - 3$$
$$0 = (2x+3)(x-1)$$
$$2x + 3 = 0 \quad \text{or} \quad x - 1 = 0$$
$$x = -\frac{3}{2} \qquad\qquad x = 1$$

Only $x = 1$ checks. $x = -\frac{3}{2}$ is an extraneous solution since $\log_3 x$ becomes $\log_3\left(-\frac{3}{2}\right)$ which is not a real number.

64. $\ln(x+1) - \ln(x-2) = \ln 4$

$$\ln \frac{x+1}{x-2} = \ln 4$$
$$\frac{x+1}{x-2} = 4$$
$$x + 1 = 4(x-2)$$
$$x + 1 = 4x - 8$$
$$1 = 3x - 8$$
$$9 = 3x$$
$$x = 3$$

65.

$$40 = 20e^{0.6t}$$
$$2 = e^{0.6t}$$
$$\ln 2 = \ln e^{0.6t}$$
$$\ln 2 = 0.6t$$
$$\frac{\ln 2}{0.6} = t$$
$$1.1552 \approx t$$

66.

$$100 = A_0 e^{-0.42(3)}$$
$$100 = A_0 e^{-1.26}$$
$$\frac{100}{e^{-1.26}} = A_0$$
$$352.54 \approx A_0$$

67.

$$A = A_0 e^{kt}$$
$$\frac{A}{A_0} = e^{kt}$$
$$\ln \frac{A}{A_0} = \ln e^{kt}$$
$$\ln \frac{A}{A_0} = kt$$
$$\frac{\ln \frac{A}{A_0}}{k} = t$$
$$\frac{\ln A - \ln A_0}{k} = t \text{ or } t = \frac{\ln A - \ln A_0}{k}$$

68.
$$150 = 600e^{kt}$$
$$\frac{150}{600} = e^{kt}$$
$$0.25 = e^{kt}$$
$$\ln 0.25 = \ln e^{kt}$$
$$\ln 0.25 = kt$$
$$\frac{\ln 0.25}{t} = k \text{ or } k = \frac{\ln 0.25}{t}$$

69. $\ln y - \ln x = 2$
$$\ln \frac{y}{x} = 2$$
$$e^{\ln \frac{y}{x}} = e^2$$
$$\frac{y}{x} = e^2$$
$$y = xe^2$$

70. $\ln(y + 3) - \ln(x + 1) = \ln 5$
$$\ln \frac{y+3}{x+1} = \ln 5$$
$$\frac{y+3}{x+1} = 5$$
$$y + 3 = 5(x + 1)$$
$$y = 5(x + 1) - 3$$
$$y = 5x + 5 - 3$$
$$y = 5x + 2$$

71. $\ln 450 = \dfrac{\log 450}{\log e} \approx 6.1092$

72. $\log_3 50 = \dfrac{\log 50}{\log 3} \approx 3.5609$

73.
$$A = P(1 + r)^n$$
$$= 12,000(1 + 0.1)^8$$
$$= 12,000(1.1)^8$$
$$\approx 25,723.07$$
The amount is \$25,723.07.

74.
$$R = 1000(0.5)^{0.000041t}$$
$$R(20,000) = 1000(0.5)^{0.000041(20,000)}$$
$$= 1000(0.5^{0.82})$$
$$\approx 1000(0.56644194)$$
$$\approx 566.4$$
There are about 566.4 mg after 20,000.

75. $N(t) = 2000(2)^{0.05t}$

a. Let $N(t) = 50,000$.
$$50,000 = 2000(2)^{0.05t}$$
$$\frac{50,000}{2000} = 2^{0.05t}$$
$$25 = 2^{0.05t}$$
$$\log 25 = \log 2^{0.05t}$$
$$\log 25 = 0.05t \log 2$$
$$\frac{\log 25}{0.05 \log 2} = t$$
$$92.88 \approx t$$
The time is 92.88 minutes.

b. Let $N(t) = 120,000$.
$$120,000 = 2000(2)^{0.05t}$$
$$\frac{120,000}{2000} = 2^{0.05t}$$
$$60 = 2^{0.05t}$$
$$\log 60 = \log 2^{0.05t}$$
$$\log 60 = 0.05t \log 2$$
$$\frac{\log 60}{0.05 \log 2} = t$$
$$118.14 \approx t$$
The time is 118.14 minutes.

76. $A(n) = 72 - 18\log(n + 1)$

a.
$$A(0) = 72 - 18\log(0 + 1)$$
$$= 72 - 18\log(1)$$
$$= 72 - 18(0)$$
$$= 72$$
The original class average was 72.

b. $A(2) = 72 - 18\log(2 + 1)$
$$= 72 - 18\log(3)$$
$$\approx 72 - 8.6$$
$$= 63.4$$
After 2 months, the class average was
63.4.

c. Let $A(n) = 59.4$.
$$59.4 = 72 - 18\log(n + 1)$$
$$-12.6 = -18\log(n + 1)$$
$$\frac{-12.6}{-18} = \log(n + 1)$$
$$0.7 = \log(n + 1)$$
$$10^{0.7} = 10^{\log(n+1)}$$
$$5.01 \approx n + 1$$
$$4.01 \approx n$$
It takes about 4 months.

77. $P = 14.7e^{-0.00004x}$
$$P = 14.7e^{-0.00004(12,000)}$$
$$P = 14.7e^{-0.48}$$
$$P \approx 14.7(0.6187834)$$
$$P \approx 9.10$$
The atmospheric pressure is 9.10 pounds per
square inch at 12,000 feet above sea level.

78. $P = P_0 e^{kt}$
$P_0 = 10,000, \ k = 0.07, \text{ and } P = 20,000$
$$20,000 = 10,000e^{(0.07)t}$$
$$2 = e^{0.07t}$$
$$\ln 2 = 0.07t$$
$$t = \frac{\ln 2}{0.07}$$
$$t \approx 9.90$$
It will take about 9.90 years for the $10,000
to double.

Practice Test

1. a. Yes, $\{(4, 2), (-3, 8), (-1, 3), (5, 7)\}$ is
one-to-one.

b. $\{(2, 4), (8, -3), (3, -1), (7, 5)\}$ is the
inverse function.

2. a. $(f \circ g)(x) = f[g(x)]$
$$= f(x + 3)$$
$$= (x + 3)^2 - 4$$
$$= x^2 + 6x + 9 - 4$$
$$= x^2 + 6x + 5$$

b. $(f \circ g)(3) = 3^2 + 6(3) + 5$
$$= 9 + 18 + 5$$
$$= 32$$

3. a. $(g \circ f)(x) = g[f(x)]$
$$= g(x^2 + 5)$$
$$= \sqrt{x^2 + 5 - 4}$$
$$= \sqrt{x^2 + 1}$$

b. $(g \circ f)(6) = \sqrt{6^2 + 1}$
$$= \sqrt{36 + 1}$$
$$= \sqrt{37}$$

4. a. $y = f(x) = -3x - 5$
$$x = -3y - 5$$
$$x + 5 = -3y$$
$$\frac{x + 5}{-3} = y$$
$$-\frac{1}{3}(x + 5) = y$$
$$f^{-1}(x) = -\frac{1}{3}(x + 5)$$

b.

x	$f(x)$
0	-5
$-\frac{5}{3}$	0

x	$f^{-1}(x)$
0	$-\frac{5}{3}$
-5	0

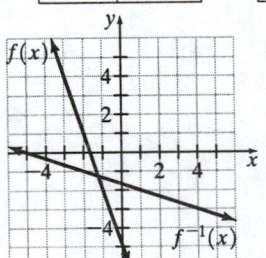

5. a. $y = f(x) = \sqrt{x-1}, \ x \geq 1$

$$x = (y-1)^{1/2}$$
$$x^2 = [(y-1)^{1/2}]^2$$
$$x^2 = y-1$$
$$x^2 + 1 = y$$
$$f^{-1}(x) = x^2 + 1, \ x \geq 0$$

b.

x	$f(x)$
1	0
2	1
5	2

x	$f^{-1}(x)$
0	1
1	2
2	5

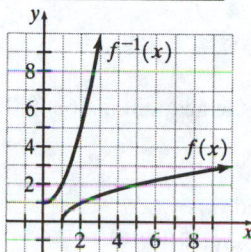

6. The domain of $y = \log_a x$ is $\{x|x \text{ is a real number and } x > 0\}$.

7. $\log_8 \dfrac{1}{64} = \log_8 8^{-2} = -2$

8. $y = 2^x$

x	-2	-1	0	2	3
y	$\frac{1}{4}$	$\frac{1}{2}$	1	4	8

Domain: $\mathbb{R}$
Range: $\{y|y > 0\}$

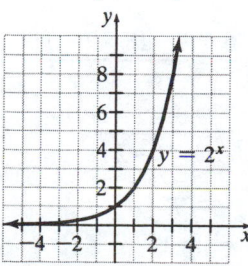

9. $y = \log_2 x$
$x = 2^y$

x	$\frac{1}{4}$	$\frac{1}{2}$	1	2	4
y	-2	-1	0	1	2

Domain: $\{x|x > 0\}$
Range: $\mathbb{R}$

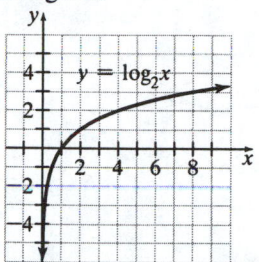

10. $4^{-3} = \dfrac{1}{64}$

$\log_4 \dfrac{1}{64} = -3$

11. $\log_3 243 = 5$
$3^5 = 243$

12. $4 = \log_2 x$
$2^4 = x$
$16 = x$

13. $y = \log_{27} 3$
$27^y = 3$
$3^{3y} = 3^1$
$3y = 1$
$y = \dfrac{1}{3}$

14. $\log_3 \dfrac{x(x-4)}{x^2}$

$= \log_3 x(x-4) - \log_3 x^2$
$= \log_3 x + \log_3(x-4) - 2\log_3 x$

15. $3\log_8(x-4) + 2\log_8(x+1) - \frac{1}{2}\log_8 x$

$= \log_8(x-4)^3 + \log_8(x+1)^2 - \log_8 x^{1/2}$

$= \log_8 \dfrac{(x-4)^3(x+1)^2}{\sqrt{x}}$

16. $2\log\sqrt{9}$

$= \log_9(\sqrt{9})^2$

$= \log_9 9$

$= 1$

17. a. $\log 4620 \approx 3.6646$

 b. $\log 0.000638 \approx -3.1952$

18. $3^x = 123$

$\log 3^x = \log 123$

$x\log 3 = \log 123$

$x = \dfrac{\log 123}{\log 3}$

$x \approx 4.38$

19. $\log 4x = \log(x+3) + \log 2$

$\log 4x = \log 2(x+3)$

$4x = 2(x+3)$

$4x = 2x + 6$

$2x = 6$

$x = 3$

20. $\log(x+5) - \log(x-2) = \log 6$

$\log\dfrac{x+5}{x-2} = \log 6$

$\dfrac{x+5}{x-2} = 6$

$x+5 = 6x - 12$

$17 = 5x$

$\dfrac{17}{5} = x$

21. $\ln N = 3.52$

$e^{3.52} = N$

$33.7844 \approx N$

22. $\log_6 40 = \dfrac{\ln 40}{\ln 6} \approx \dfrac{3.6889}{1.7918} \approx 2.0588$

23. $200 = 500e^{-0.03t}$

$\dfrac{200}{500} = e^{-0.03t}$

$\ln\dfrac{200}{500} = -0.03t$

$\dfrac{\ln\frac{200}{500}}{-0.03} = t$

$t \approx 30.5430$

24. $A = p\left(1 + \dfrac{r}{n}\right)^{nt}$

Use $p = 3500$, $r = 0.06$ and $n = 4$

$t = 10$

$A = 3500\left(1 + \dfrac{0.06}{4}\right)^{4\cdot 10}$

$= 3500(1.015)^{40}$

≈ 6349.06

The amount in the account is \$6349.06.

25. $v = v_0 e^{-0.0001205t}$

Use $v = 40$, and $v_0 = 60$.

$40 = 60e^{-0.0001205t}$

$\dfrac{40}{60} = e^{-0.0001205t}$

$\dfrac{2}{3} = e^{-0.0001205t}$

$\ln\dfrac{2}{3} = \ln e^{-0.0001205t}$

$\ln\dfrac{2}{3} = -0.0001205t$

$\dfrac{\ln\frac{2}{3}}{-0.0001205} = t$

$3364.86 \approx t$

The fossil is approximately 3364.86 years old.

Cumulative Review Test

1. $\dfrac{6 - |-18| \div 3^2 - 6}{4 - |-8| \div 2^2} = \dfrac{6 - 18 \div 9 - 6}{4 - 8 \div 4}$

$= \dfrac{6 - 2 - 6}{4 - 2}$

$= \dfrac{-2}{2}$

$= -1$

2. $\left(\dfrac{3x^4y^{-3}}{6xy^4z^2}\right)^{-3} = \left(\dfrac{x^3}{2y^7z^2}\right)^{-3}$

$= \left(\dfrac{2y^7z^2}{x^3}\right)^{3}$

$= \dfrac{2^3 y^{7\cdot3} z^{2\cdot3}}{x^{3\cdot3}}$

$= \dfrac{8y^{21}z^6}{x^9}$

3. Let x = the time Kendra runs until they meet

	rate	time	distance
Jason	4	$x+\frac{1}{2}$	$4\left(x+\frac{1}{2}\right)$
Kendra	5	x	$5x$

a. Their distances are equal.

$4\left(x+\dfrac{1}{2}\right) = 5x$

$4x + 2 = 5x$

$2 = x$

They will meet in 2 hours.

b. Substitute 2 for x in either distance expression.

$5x = 5(2) = 10$

They will each run 10 miles before meeting.

4. $\dfrac{1}{2}(2x+12) \ge 3x - 6 + x$

$x + 6 \ge 4x - 6$

$12 \ge 3x$

$4 \ge x$

or $x \le 4$

5. Domain: $\{x | x \text{ is a real number, } -1 \le x \le 4\}$
Range: $\{y | y \text{ is a real number, } -3 \le y \le 2\}$

6. a. Slope $= \dfrac{y_2 - y_1}{x_2 - x_1}$

$= \dfrac{4 - (-3)}{1 - (-2)}$

$= \dfrac{4 + 3}{1 + 2}$

$= \dfrac{7}{3}$

b. $y - y_1 = m(x - x_1)$

$y - 4 = \dfrac{7}{3}(x - 1)$

$3y - 12 = 7(x - 1)$

$3y - 12 = 7x - 7$

$-5 = 7x - 3y$

or $7x - 3y = -5$

7. $y < \dfrac{2}{3}x + 3$

Plot a dashed line at $y = \dfrac{2}{3}x + 3$.

Use the point (0, 0) as the check point.

$y < \dfrac{2}{3}x + 3$

$0 < \dfrac{2}{3}(0) + 3$

$0 < 3$ True

Therefore, shade the half-plane containing (0, 0).

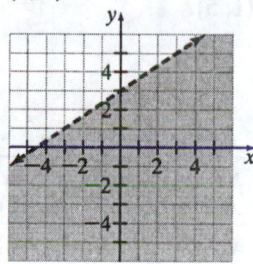

8. $0.4x + 0.6y = 3.2$
$1.4x - 0.3y = 1.6$
or
$4x + 6y = 32$
$14x - 3y = 16$
Multiply the second equation by 2 and add to the first equation.

$$4x + 6y = 32$$
$$2[14x - 3y = 16]$$
gives
$$4x + 6y = 32$$
$$28x - 6y = 32$$
Add: $\overline{32x = 64}$
$$x = 2$$

Substitute 2 for x in the first equation.
$$4(2) + 6y = 32$$
$$6y = 24$$
$$y = 4$$
The solution is (2, 4).

9.
$$x + y = 6$$
$$-2x + y = 3$$

$$\begin{bmatrix} 1 & 1 & | & 6 \\ -2 & 1 & | & 3 \end{bmatrix}$$

$$\begin{bmatrix} 1 & 1 & | & 6 \\ 0 & 3 & | & 15 \end{bmatrix} 2R_1 + R_2$$

$$\begin{bmatrix} 1 & 1 & | & 6 \\ 0 & 1 & | & 5 \end{bmatrix} \frac{1}{3}R_2$$

The system is
$$x + y = 6$$
$$y = 5$$
Substitute 5 for y in the first equation.
$$x + 5 = 6$$
$$x = 1$$
The solution is (1, 5).

10. $\begin{vmatrix} 3 & 0 & -1 \\ 2 & 5 & 3 \\ -1 & 4 & 6 \end{vmatrix}$

$$= 3\begin{vmatrix} 5 & 3 \\ 4 & 6 \end{vmatrix} - 2\begin{vmatrix} 0 & -1 \\ 4 & 6 \end{vmatrix} + (-1)\begin{vmatrix} 0 & -1 \\ 5 & 3 \end{vmatrix}$$
$$= 3(30 - 12) - 2(0 + 4) - (0 + 5)$$
$$= 3(18) - 2(4) - 5$$
$$= 54 - 8 - 5$$
$$= 41$$

11. $12x^2 - 5xy - 3y^2 = (4x - 3y)(3x + y)$

12. $f(x) = 3x^2 - 7x - 11$
$$9 = 3x^2 - 7x - 11$$
$$0 = 3x^2 - 7x - 20$$
$$0 = (3x + 5)(x - 4)$$
$$3x + 5 = 0 \quad \text{or} \quad x - 4 = 0$$
$$x = -\frac{5}{3} \qquad\qquad x = 4$$
The values of x are $-\dfrac{5}{3}$ and 4.

13.

$$4 \overline{\smash{\big|}\ \begin{array}{ccc} 2 & -1 & -26 \\ & 8 & 28 \end{array}}$$
$$\overline{\ \begin{array}{ccc} 2 & 7 & 2 \end{array}}$$

$$\frac{2x^2 - x - 26}{x - 4} = 2x + 7 + \frac{2}{x - 4}$$

14.
$$\frac{x+1}{x+2} + \frac{x-2}{x-3} = \frac{x^2 - 4}{x^2 - x - 6}$$
$$\frac{x+1}{x+2} + \frac{x-2}{x-3} = \frac{(x+2)(x-2)}{(x+2)(x-3)}$$
$$(x+2)(x-3)\left(\frac{x+1}{x+2} + \frac{x-2}{x-3}\right) = (x+2)(x-3)\left[\frac{(x+2)(x-2)}{(x+2)(x-3)}\right]$$
$$(x-3)(x+1) + (x+2)(x-2) = (x+2)(x-2)$$
$$(x-3)(x+1) = 0$$
$$x - 3 = 0 \quad \text{or} \quad x + 1 = 0$$
$$x = 3 \quad \text{or} \qquad x = -1$$

The only solution is $x = -1$ because $x = 3$ yields an undefined fraction $\dfrac{x-2}{x-3} = \dfrac{3-2}{3-3} = \dfrac{1}{0}$.

15.
$$\frac{x^{-3} + x^{-2}}{x^{-1} - x^{-2}} = \frac{\frac{1}{x^3} + \frac{1}{x^2}}{\frac{1}{x} - \frac{1}{x^2}}$$

$$= \frac{x^3 \left(\frac{1}{x^3} + \frac{1}{x^2}\right)}{x^3 \left(\frac{1}{x} - \frac{1}{x^2}\right)}$$

$$= \frac{1 + x}{x^2 - x}$$

16.
$$\sqrt[3]{\frac{27x^4 y^5}{5xy^7}} = \sqrt[3]{\frac{27x^3}{5y^2}}$$

$$= \frac{3x}{\sqrt[3]{5y^2}}$$

$$= \frac{3x}{\sqrt[3]{5y^2}} \cdot \frac{\sqrt[3]{25y}}{\sqrt[3]{25y}}$$

$$= \frac{3x\sqrt[3]{25y}}{5y}$$

17.
$$\sqrt[3]{5x+1} + 4 = 0$$
$$\sqrt[3]{5x+1} = -4$$
$$\left(\sqrt[3]{5x+1}\right)^3 = (-4)^3$$
$$5x + 1 = -64$$
$$5x = -65$$
$$x = -13$$

18. $y = (x-4)^2 + 1$

This is a parabola that opens upward.
vertex: (4, 1)

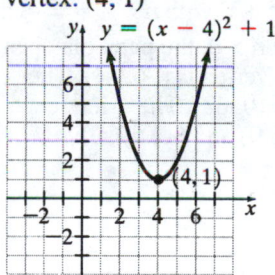

19. $f(x) = x^3 - 6x^2 + 5x$
$$x^3 - 6x^2 + 5x \geq 0$$
$$x(x^2 - 6x + 5) \geq 0$$
$$x(x - 5)(x - 1) \geq 0$$

The solution is $[0, \ 1] \cup [5, \ \infty)$.

20. $P = P_0 e^{-0.028t}$

a. $P = 600e^{-0.028(60)}$
$P \approx 111.82$
There are about 111.82 grams left after 60 years.

b.
$$300 = 600e^{-0.028t}$$
$$\frac{1}{2} = e^{-0.028t}$$
$$\ln \frac{1}{2} = -0.028t$$
$$\frac{\ln \frac{1}{2}}{-0.028} = t$$
$$t \approx 24.8$$

The half-life of strontium 90 is about 24.8 years.

Chapter 10

Exercise Set 10.1

1.

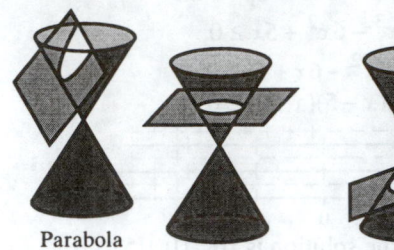

Parabola   Circle   Ellipse   Hyperbola

3. Yes, any parabola in the form
$y = a(x-h)^2 + k$ is a function because each
value of x corresponds to only one value of
y. The domain is $\mathbb{R}$, the set of all real
numbers. Since the vertex is at (h, k) and
$a > 0$, the range is $\{y \mid y \geq k\}$.

5. The graphs have the same vertex, $(3, 4)$. The
first graph opens upward, and the second
one opens downward.

7. The distance is always a positive number
because both distances are squared and we
use the principal square root.

9. A circle is the set of all points in a plane that
are the same distance from a fixed point.

11. $y = (x-2)^2 + 3$
This is a parabola in the form
$y = a(x-h)^2 + k$ with $a = 1$, $h = 2$ and
$k = 3$. Since $a > 0$, the parabola opens
upward. The vertex is $(2, 3)$. The y-intercept
is $(0, 7)$. There are no x-intercepts.

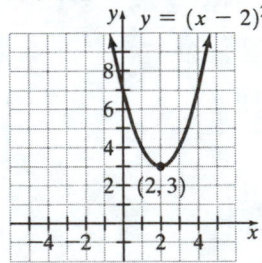

13. $x = (y-4)^2 - 3$
This is a parabola in the form
$x = a(y-k)^2 + h$ with $a = 1$, $h = -3$ and
$k = 4$. Since $a > 0$, the parabola opens to the
right. The vertex is $(-3, 4)$. The y-intercepts
are about $(0, 2.27)$ and $(0, 5.73)$. The
x-intercept is $(13, 0)$.

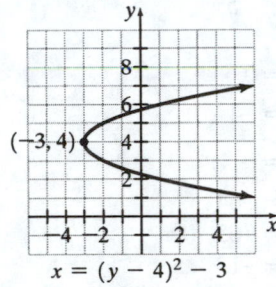

$$x = (y - 4)^2 - 3$$

15. $y = 2(x+6)^2 - 4$
This is a parabola in the form
$y = a(x-h)^2 + k$ with $a = 2$, $h = -6$ and
$k = -4$. Since $a > 0$, the parabola opens
upward. The vertex is $(-6, -4)$. The
y-intercept is $(0, 68)$. The x-intercepts are
about $(-7.41, 0)$ and $(-4.59, 0)$.

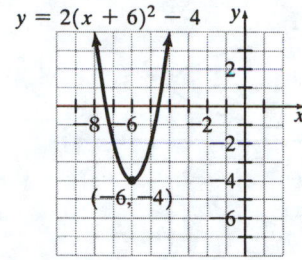

$y = 2(x + 6)^2 - 4$

17. $x = -5(y+3)^2 - 6$

This is a parabola in the form
$x = a(y-k)^2 + h$ with $a = -5$, $h = -6$ and
$k = -3$. Since $a < 0$, the parabola opens to the
left. The vertex is $(-6, -3)$. There are no
y-intercepts. The x-intercept is $(-51, 0)$

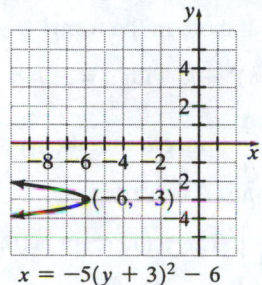

$x = -5(y + 3)^2 - 6$

19. $y = -2\left(x + \dfrac{1}{2}\right)^2 + 6$

This is a parabola in the form

$y = a(x-h)^2 + k$ with $a = -2$, $h = -\dfrac{1}{2}$ and

$k = 6$. Since $a < 0$, the parabola opens

downward. The vertex is $\left(-\dfrac{1}{2},\ 6\right)$. The

y-intercept is $\left(0,\ \dfrac{11}{2}\right)$. The x-intercepts are

about $(-2.23, 0)$ and $(1.23, 0)$.

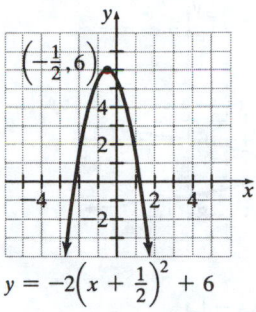

$y = -2\left(x + \dfrac{1}{2}\right)^2 + 6$

21. a. $y = x^2 + 2x$

$y = (x^2 + 2x + 1) - 1$

$y = (x+1)^2 - 1$

b. This is a parabola in the form
$y = a(x-h)^2 + k$ with $a = 1$, $h = -1$ and
$k = -1$. Since $a > 0$, the parabola opens
upward. The vertex is $(-1, -1)$. The
y-intercept is $(0, 0)$. The x-intercepts are
$(-2, 0)$ and $(0, 0)$.

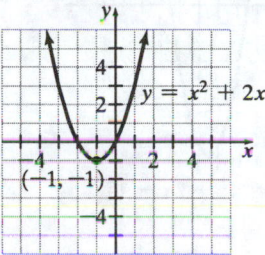

$y = x^2 + 2x$

23. a. $x = y^2 + 6y$

$x = (y^2 + 6y + 9) - 9$

$x = (y+3)^2 - 9$

b. This is a parabola in the form
$x = a(y-k)^2 + h$ with $a = 1$, $h = -9$ and
$k = -3$. Since $a > 0$, the parabola opens
to the right. The vertex is $(-9, -3)$. The
y-intercepts are $(0, -6)$ and $(0, 0)$. The
x-intercept is $(0, 0)$.

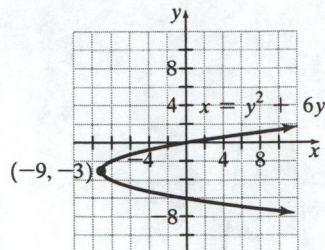

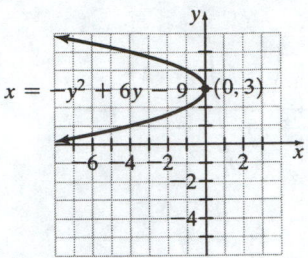

25. a. $y = x^2 + 2x - 15$

$y = (x^2 + 2x + 1) - 1 - 15$

$y = (x + 1)^2 - 16$

b. This is a parabola in the form
$y = a(x - h)^2 + k$ with $a = 1, h = -1$,
and $k = -16$. Since $a > 0$, the parabola
opens upward. The vertex is $(-1, -16)$.
The y-intercept is $(0, -15)$. The
x-intercepts are $(-5, 0)$ and $(3, 0)$.

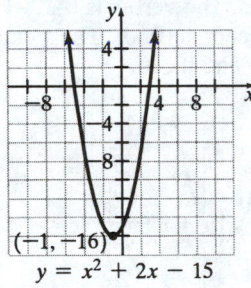

27. a. $x = -y^2 + 6y - 9$

$x = -(y^2 - 6y) - 9$

$x = -(y^2 - 6y + 9) + 9 - 9$

$x = -(y - 3)^2$

b. This is a parabola in the form
$x = a(y - k)^2 + h$ with $a = -1, h = 0$ and
$k = 3$. Since $a < 0$, the parabola opens to
the left. The vertex is $(0, 3)$. The
y-intercept is $(0, 3)$. The x-intercept is
$(-9, 0)$.

29. a. $y = x^2 + 7x + 10$

$y = \left(x^2 + 7x + \dfrac{49}{4} \right) - \dfrac{49}{4} + 10$

$y = \left(x + \dfrac{7}{2} \right)^2 - \dfrac{9}{4}$

b. This is a parabola in the form
$y = a(x - h)^2 + k$ with $a = 1, h = -\dfrac{7}{2}$

and $k = -\dfrac{9}{4}$. Since $a > 0$, the parabola

opens upward. The vertex is

$\left(-\dfrac{7}{2}, -\dfrac{9}{4} \right)$. The y-intercept is $(0, 10)$.

The x-intercepts are $(-5, 0)$ and $(-2, 0)$.

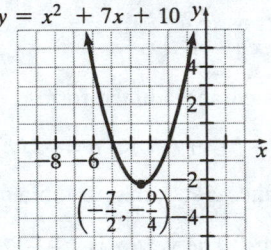

31. a. $y = 2x^2 - 4x - 4$

$y = 2(x^2 - 2x) - 4$

$y = 2(x^2 - 2x + 1) - 2 - 4$

$y = 2(x - 1)^2 - 6$

b. This is a parabola in the form
$y = a(x - h)^2 + k$ with $a = 2, h = 1$ and
$k = -6$. Since $a > 0$, the parabola opens
upward. The vertex is $(1, -6)$. The
y-intercept is $(0, -4)$. The x-intercepts
are about $(-0.73, 0)$ and $(2.73, 0)$.

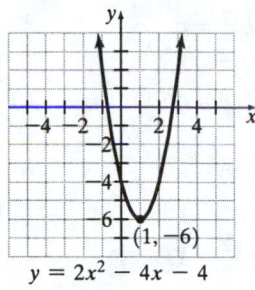

$$y = 2x^2 - 4x - 4$$

33. $d = \sqrt{(x_2 - x_1)^2 + (y_2 - y_1)^2}$

$= \sqrt{(2-2)^2 + [-5-(-2)]^2}$

$= \sqrt{0^2 + (-3)^2}$

$= \sqrt{0+9}$

$= \sqrt{9}$

$= 3$

35. $d = \sqrt{(x_2 - x_1)^2 + (y_2 - y_1)^2}$

$= \sqrt{[5-(-4)]^2 + (3-3)^2}$

$= \sqrt{9^2 + 0^2}$

$= \sqrt{81+0}$

$= \sqrt{81}$

$= 9$

37. $d = \sqrt{(x_2 - x_1)^2 + (y_2 - y_1)^2}$

$= \sqrt{[6-(-3)]^2 + [-2-(-5)]^2}$

$= \sqrt{9^2 + 3^2}$

$= \sqrt{81+9}$

$= \sqrt{90}$

≈ 9.49

39. $d = \sqrt{(x_2 - x_1)^2 + (y_2 - y_1)^2}$

$= \sqrt{(5-0)^2 + (-1-6)^2}$

$= \sqrt{5^2 + (-7)^2}$

$= \sqrt{25+49}$

$= \sqrt{74}$

≈ 8.60

41. $d = \sqrt{(x_2 - x_1)^2 + (y_2 - y_1)^2}$

$= \sqrt{[-4.3-(-1.6)]^2 + (-1.7-3.5)^2}$

$= \sqrt{(-2.7)^2 + (-5.2)^2}$

$= \sqrt{7.29+27.04}$

$= \sqrt{34.33}$

≈ 5.86

43. $d = \sqrt{(x_2 - x_1)^2 + (y_2 - y_1)^2}$

$= \sqrt{\left(-\dfrac{1}{2} - \dfrac{3}{4}\right)^2 + (6-2)^2}$

$= \sqrt{\left(-\dfrac{5}{4}\right)^2 + 4^2}$

$= \sqrt{\dfrac{25}{16} + 16}$

$= \sqrt{\dfrac{281}{16}}$

≈ 4.19

45. $d = \sqrt{(x_2 - x_1)^2 + (y_2 - y_1)^2}$

$= \sqrt{\left(0-\sqrt{5}\right)^2 + \left[0-\left(-\sqrt{2}\right)\right]^2}$

$= \sqrt{\left(-\sqrt{5}\right)^2 + \left(\sqrt{2}\right)^2}$

$= \sqrt{5+2}$

$= \sqrt{7}$

≈ 2.65

47. $d = \sqrt{(x_2 - x_1)^2 + (y_2 - y_1)^2}$

$= \sqrt{\left(\sqrt{2}-\sqrt{3}\right)^2 + \left(\sqrt{3}-\sqrt{7}\right)^2}$

$= \sqrt{\left(2-2\sqrt{6}+3\right) + \left(3-2\sqrt{21}+7\right)}$

$= \sqrt{15-2\sqrt{6}-2\sqrt{21}}$

≈ 0.97

49. Midpoint $= \left(\dfrac{x_1 + x_2}{2},\ \dfrac{y_1 + y_2}{2} \right)$

$\qquad = \left(\dfrac{1+2}{2},\ \dfrac{4+6}{2} \right)$

$\qquad = \left(\dfrac{3}{2},\ 5 \right)$

51. Midpoint $= \left(\dfrac{x_1 + x_2}{2},\ \dfrac{y_1 + y_2}{2} \right)$

$\qquad = \left(\dfrac{0+4}{2},\ \dfrac{8+(-6)}{2} \right)$

$\qquad = (2,\ 1)$

53. Midpoint $= \left(\dfrac{x_1 + x_2}{2},\ \dfrac{y_1 + y_2}{2} \right)$

$\qquad = \left(\dfrac{4+1}{2},\ \dfrac{7+(-3)}{2} \right)$

$\qquad = \left(\dfrac{5}{2},\ 2 \right)$

55. Midpoint $= \left(\dfrac{x_1 + x_2}{2},\ \dfrac{y_1 + y_2}{2} \right)$

$\qquad = \left(\dfrac{-9.62 + 3.52}{2},\ \dfrac{12.58 + 6.57}{2} \right)$

$\qquad = (-3.05,\ 9.575)$

57. Midpoint $= \left(\dfrac{x_1 + x_2}{2},\ \dfrac{y_1 + y_2}{2} \right)$

$\qquad = \left(\dfrac{\frac{5}{2}+2}{2},\ \dfrac{3+\frac{9}{2}}{2} \right)$

$\qquad = \left(\dfrac{9}{4},\ \dfrac{15}{4} \right)$

59. Midpoint $= \left(\dfrac{x_1 + x_2}{2},\ \dfrac{y_1 + y_2}{2} \right)$

$\qquad = \left(\dfrac{\sqrt{5} + \sqrt{3}}{2},\ \dfrac{1+4}{2} \right)$

$\qquad = \left(\dfrac{\sqrt{3} + \sqrt{5}}{2},\ \dfrac{5}{2} \right)$

61. $(x-h)^2 + (y-k)^2 = r^2$

$\qquad (x-0)^2 + (y-0)^2 = 3^2$

$\qquad\qquad x^2 + y^2 = 9$

63. $(x-h)^2 + (y-k)^2 = r^2$

$\qquad (x-3)^2 + (y-0)^2 = 1^2$

$\qquad\quad (x-3)^2 + y^2 = 1$

65. $(x-h)^2 + (y-k)^2 = r^2$

$\qquad [x-(-6)]^2 + (y-5)^2 = 5^2$

$\qquad\quad (x+6)^2 + (y-5)^2 = 25$

67. $(x-h)^2 + (y-k)^2 = r^2$

$\qquad (x-4)^2 + (y-7)^2 = \left(\sqrt{8}\right)^2$

$\qquad (x-4)^2 + (y-7)^2 = 8$

69. The center is $(0, 0)$ and the radius is 4.

$\qquad (x-h)^2 + (y-k)^2 = r^2$

$\qquad (x-0)^2 + (y-0)^2 = 4^2$

$\qquad\qquad x^2 + y^2 = 16$

71. The center is $(3, -2)$ and the radius is 3.

$\qquad (x-h)^2 + (y-k)^2 = r^2$

$\qquad (x-3)^2 + [y-(-2)]^2 = 3^2$

$\qquad\quad (x-3)^2 + (y+2)^2 = 9$

73. $x^2 + y^2 = 16$

$\qquad x^2 + y^2 = 4^2$

The graph is a circle with its center at the origin and radius 4.

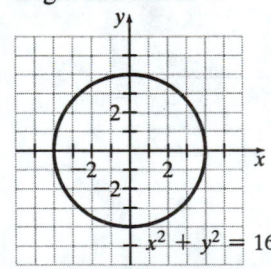

75. $x^2 + y^2 = 10$

$$x^2 + y^2 = \left(\sqrt{10}\right)^2$$

The graph is a circle with its center at the origin and radius $\sqrt{10}$.

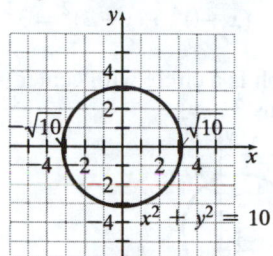

77. $(x+4)^2 + y^2 = 25$

$$(x+4)^2 + (y-0)^2 = 5^2$$

The graph is a circle with its center at (–4, 0) and radius 5.

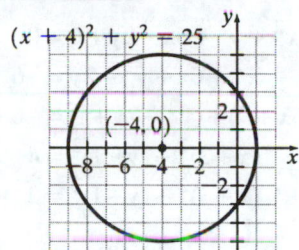

79. $(x+8)^2 + (y+2)^2 = 9$

$$(x+8)^2 + (y+2)^2 = 3^2$$

The graph is a circle with its center at (–8, –2) and radius 3.

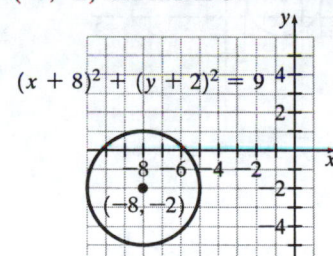

81. $y = \sqrt{16 - x^2}$

If we solve $x^2 + y^2 = 16$ for y, we obtain $y = \pm\sqrt{16 - x^2}$. Therefore, the graph of $y = \sqrt{16 - x^2}$ is the upper half ($y \geq 0$) of a circle with its center at the origin and radius 4.

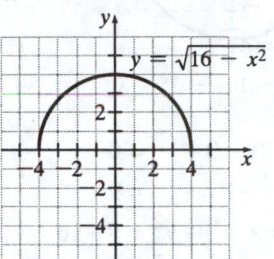

83. $y = -\sqrt{4 - x^2}$

If we solve $x^2 + y^2 = 4$ for y, we obtain $y = \pm\sqrt{4 - x^2}$. Therefore, the graph of $y = -\sqrt{4 - x^2}$ is the lower half ($y \leq 0$) of a circle with its center at the origin and radius 2.

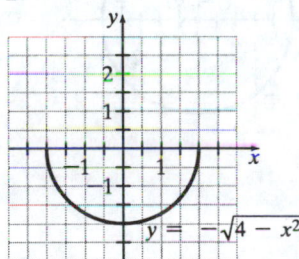

85. a.
$$x^2 + y^2 + 10y - 75 = 0$$
$$x^2 + y^2 + 10y = 75$$
$$x^2 + (y^2 + 10y + 25) = 75 + 25$$
$$x^2 + (y+5)^2 = 100$$
$$x^2 + (y+5)^2 = 10^2$$

b. The graph is a circle with center $(0, -5)$ and radius 10.

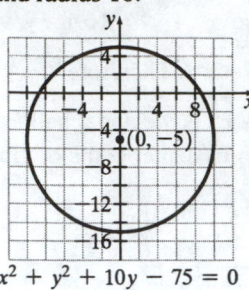

$$x^2 + y^2 + 10y - 75 = 0$$

87. a.
$$x^2 + 8x - 9 + y^2 = 0$$
$$x^2 + 8x + y^2 = 9$$
$$(x^2 + 8x + 16) + y^2 = 9 + 16$$
$$(x + 4)^2 + y^2 = 25$$
$$(x + 4)^2 + y^2 = 5^2$$

b. The graph is a circle with center $(-4, 0)$ and radius 5.

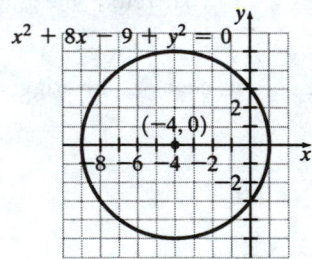

$$x^2 + 8x - 9 + y^2 = 0$$

89. a.
$$x^2 + y^2 + 2x - 4y - 4 = 0$$
$$x^2 + 2x + y^2 - 4y = 4$$
$$(x^2 + 2x + 1) + (y^2 - 4y + 4) = 4 + 1 + 4$$
$$(x + 1)^2 + (y - 2)^2 = 9$$
$$(x + 1)^2 + (y - 2)^2 = 3^2$$

b. The graph is a circle with center $(-1, 2)$ and radius 3.

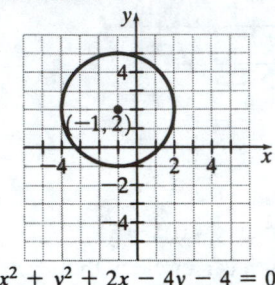

$$x^2 + y^2 + 2x - 4y - 4 = 0$$

91. a.
$$x^2 + y^2 + 6x - 2y + 6 = 0$$
$$x^2 + 6x + y^2 - 2y = -6$$
$$(x^2 + 6x + 9) + (y^2 - 2y + 1) = -6 + 9 + 1$$
$$(x + 3)^2 + (y - 1)^2 = 4$$
$$(x + 3)^2 + (y - 1)^2 = 2^2$$

b. The graph is a circle with center $(-3, 1)$ and radius 2.

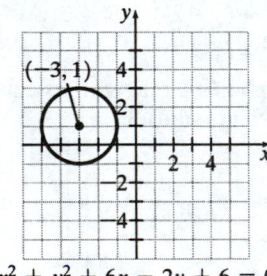

$$x^2 + y^2 + 6x - 2y + 6 = 0$$

93. a.
$$x^2 + y^2 - 8x + 2y + 13 = 0$$
$$x^2 - 8x + y^2 + 2y = -13$$
$$(x^2 - 8x + 16) + (y^2 + 2y + 1) = -13 + 16 + 1$$
$$(x - 4)^2 + (y + 1)^2 = 4$$
$$(x - 4)^2 + (y + 1)^2 = 2^2$$

b. The graph is a circle with center $(4, -1)$ and radius 2.

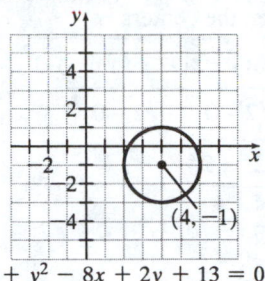

$$x^2 + y^2 - 8x + 2y + 13 = 0$$

95. x-intercept:
$$x = 0^2 - 6(0) - 7$$
$$x = -7$$
The x-intercept is $(-7, 0)$
y-intercepts:
$$0 = y^2 - 6y - 7$$
$$0 = (y+1)(y-7)$$
$$y = -1 \text{ or } y = 7$$
The y-intercepts are $(0, -1)$ and $(0, 7)$.

97. x-intercept:
$$x = 2(0 - 5)^2 + 6$$
$$x = 56$$
The x-intercept is $(56, 0)$.
y-intercepts:
$$0 = 2(y - 5)^2 + 6$$
Since $2(y - 5)^2 + 6 \geq 6$ for all real values of y, this equation has no real solutions.
There are no y-intercepts.

99. No. For example, the origin is the midpoint of both the segment from $(1, 1)$ to $(-1, -1)$ and the segment from $(2, 2)$ to $(-2, -2)$, but these segments have different lengths.

101. The distance from the midpoint $(4, -6)$ to the endpoint $(7, -2)$ is half the length of the line segment.
$$\frac{d}{2} = \sqrt{(7-4)^2 + [-2-(-6)]^2}$$
$$= \sqrt{3^2 + 4^2}$$
$$= \sqrt{25}$$
$$= 5$$
Since $\frac{d}{2} = 5$, $d = 10$. The length is 10 units.

103. Since $(-5, 2)$ is 2 units above the x-axis, the radius is 2.
$$(x - h)^2 + (y - k)^2 = r^2$$
$$(x + 5)^2 + (y - 2)^2 = 2^2$$
$$(x + 5)^2 + (y - 2)^2 = 4$$

105. a. Diameter $= \sqrt{(x_2 - x_1)^2 + (y_2 - y_1)^2}$
$$= \sqrt{(9-5)^2 + (8-4)^2}$$
$$= \sqrt{4^2 + 4^2}$$
$$= \sqrt{16 + 16}$$
$$= \sqrt{32}$$
$$= 4\sqrt{2}$$
Since the diameter is $4\sqrt{2}$ units, the radius is $2\sqrt{2}$ units.

b. Midpoint $= \left(\dfrac{x_1 + x_2}{2}, \dfrac{y_1 + y_2}{2} \right)$
$$= \left(\frac{5+9}{2}, \frac{4+8}{2} \right)$$
$$= (7, 6)$$
The center is $(7, 6)$.

c. $(x - h)^2 + (y - k)^2 = r^2$
$$(x - 7)^2 + (y - 6)^2 = \left(2\sqrt{2}\right)^2$$
$$(x - 7)^2 + (y - 6)^2 = 8$$

107. The minimum number is 0 and the maximum number is 4 as shown in the diagrams.

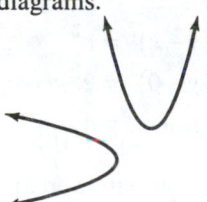

No points of intersection

4 points of intersection

109. a. Since $150 - 2(68.2) = 13.6$, the clearance is 13.6 feet.

b. Since $150 - 68.2 = 81.8$, the center of the wheel is 81.8 feet above the ground.

c.
$$(x-h)^2 + (y-k)^2 = r^2$$
$$(x-0)^2 + (y-81.8)^2 = 68.2^2$$
$$x^2 + (y-81.8)^2 = 68.2^2$$
$$x^2 + (y-81.8)^2 = 4651.24$$

111. a. The center of the blue circle is the origin, and the radius is 4.
$$x^2 + y^2 = r^2$$
$$x^2 + y^2 = 4^2$$
$$x^2 + y^2 = 16$$

b. The center of the red circle is (2, 0), and the radius is 2.
$$(x-h)^2 + (y-k)^2 = r^2$$
$$(x-2)^2 + (y-0)^2 = 2^2$$
$$(x-2)^2 + y^2 = 4$$

c. The center of the green circle is (–2, 0), and the radius is 2.
$$(x-h)^2 + (y-k)^2 = r^2$$
$$[x-(-2)]^2 + (y-0)^2 = 2^2$$
$$(x+2)^2 + y^2 = 4$$

d. Shaded area = (blue circle area) – (red circle area) – (green circle area)
$$= \pi(4^2) - \pi(2^2) - \pi(2^2)$$
$$= 16\pi - 4\pi - 4\pi$$
$$= 8\pi$$

113. Each circle has radius 4, and the centers are (0, 0) and (2, 2), respectively. The distance between the centers is
$$d = \sqrt{(x_2 - x_1)^2 + (y_2 - y_1)^2}$$
$$= \sqrt{(2-0)^2 + (2-0)^2}$$
$$= \sqrt{2^2 + 2^2}$$
$$= \sqrt{4+4}$$
$$= \sqrt{8}$$
$$\approx 2.83$$
Since the circles have the same radius and the distance between the centers is less than twice the radius, there are 2 intersection points.

115.

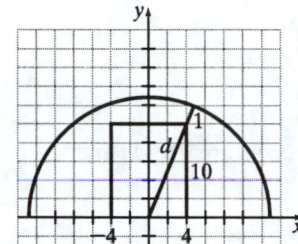

In the illustration,
$$d = \sqrt{4^2 + 10^2}$$
$$= \sqrt{16 + 100}$$
$$= \sqrt{116}$$

In order to have 1 foot to spare, the radius should be $1 + \sqrt{116} \approx 11.77$ feet.

117. First find the slope.
$$m = \frac{y_2 - y_1}{x_2 - x_1} = \frac{2-4}{-2-(-6)} = \frac{-2}{4} = -\frac{1}{2}$$
Use the point-slope form.
$$y - y_1 = m(x - x_1)$$
$$y - 4 = -\frac{1}{2}[x - (-6)]$$
$$y - 4 = -\frac{1}{2}x - 3$$
$$y = -\frac{1}{2}x + 1$$

118. $\begin{vmatrix} 4 & 0 & 3 \\ 5 & 2 & -1 \\ 3 & 6 & 4 \end{vmatrix} = 4\begin{vmatrix} 2 & -1 \\ 6 & 4 \end{vmatrix} - 5\begin{vmatrix} 0 & 3 \\ 6 & 4 \end{vmatrix} + 3\begin{vmatrix} 0 & 3 \\ 2 & -1 \end{vmatrix}$

$$= 4(8+6) - 5(0-18) + 3(0-6)$$
$$= 4(14) - 5(-18) + 3(-6)$$
$$= 56 + 90 - 18$$
$$= 128$$

119. $P\left(\dfrac{1}{3}\right) = -2\left(\dfrac{1}{3}\right)^2 - \dfrac{1}{3} + 2$

$$= -\dfrac{2}{9} - \dfrac{3}{9} + \dfrac{18}{9}$$
$$= \dfrac{13}{9}$$

120. First find k.

$$T = \frac{k m_1 m_2}{R^2}$$
$$\frac{3}{2} = \frac{k(6)(4)}{4^2}$$
$$\frac{3}{2} = \frac{24k}{16}$$
$$\frac{48}{48} = k$$
$$k = 1$$

Now substitute the new values.

$$T = \frac{k m_1 m_2}{R^2}$$
$$T = \frac{1(6)(10)}{2^2}$$
$$T = 15$$

Exercise Set 10.2

1. An ellipse is a set of points in a plane, the sum of whose distances from two fixed points is constant.

3. $\dfrac{(x-h)^2}{a^2} + \dfrac{(y-k)^2}{b^2} = 1$

5. If $a = b$, the formula for a circle is obtained.

7. $\dfrac{x^2}{4} + \dfrac{y^2}{1} = 1$

Since $a^2 = 4$, $a = 2$.
Since $b^2 = 1$, $b = 1$.

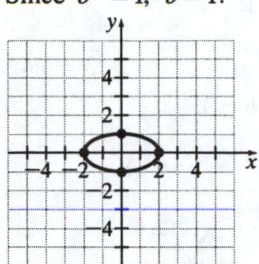

9. $\dfrac{x^2}{4} + \dfrac{y^2}{9} = 1$

Since $a^2 = 4$, $a = 2$.
Since $b^2 = 9$, $b = 3$.

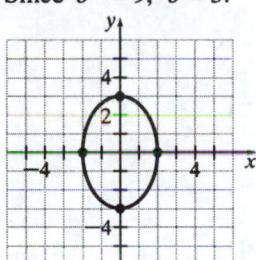

11. $\dfrac{x^2}{16} + \dfrac{y^2}{25} = 1$

Since $a^2 = 16$, $a = 4$.
Since $b^2 = 25$, $b = 5$.

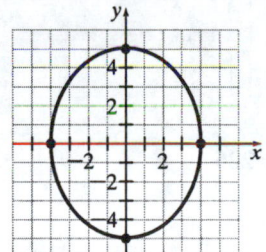

13. $9x^2 + 12y^2 = 108$

$$\frac{9x^2}{108} + \frac{12y^2}{108} = 1$$

$$\frac{y^2}{12} + \frac{y^2}{9} = 1$$

Since $a^2 = 12$, $a = \sqrt{12} = 2\sqrt{3} \approx 3.46$.

Since $b^2 = 9$, $b = 3$.

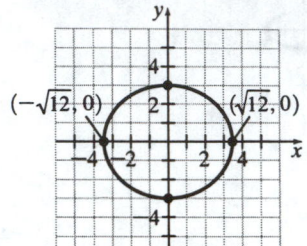

15. $100x^2 + 25y^2 = 400$

$$4x^2 + y^2 = 16$$

$$\frac{4x^2}{16} + \frac{y^2}{16} = 1$$

$$\frac{x^2}{4} + \frac{y^2}{16} = 1$$

Since $a^2 = 4$, $a = 2$.

Since $b^2 = 16$, $b = 4$.

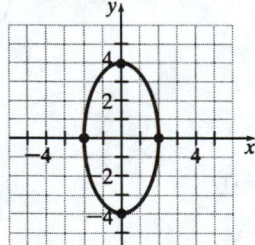

17. $9x^2 + 25y^2 = 225$

$$\frac{9x^2}{225} + \frac{25y^2}{225} = 1$$

$$\frac{x^2}{25} + \frac{y^2}{9} = 1$$

Since $a^2 = 25$, $a = 5$.

Since $b^2 = 9$, $b = 3$.

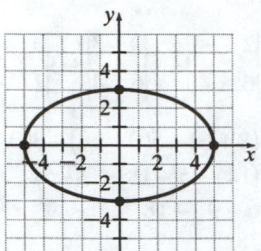

19. $\dfrac{x^2}{16} + \dfrac{(y-2)^2}{9} = 1$

The center is $(0, 2)$.

Since $a^2 = 16$, $a = 4$.

Since $b^2 = 9$, $b = 3$.

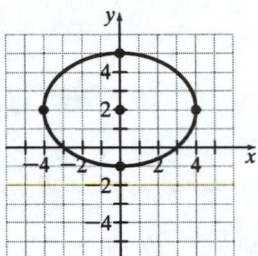

21. $\dfrac{(x+1)^2}{9} + \dfrac{(y-2)^2}{4} = 1$

The center is $(-1, 2)$.

Since $a^2 = 9$, $a = 3$.

Since $b^2 = 4$, $b = 2$.

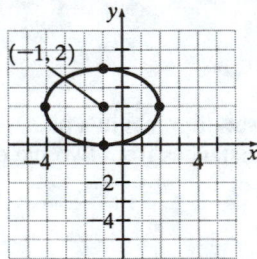

23. $4(x-2)^2 + 9(y+2)^2 = 36$

$$\frac{4(x-2)^2}{36} + \frac{9(y+2)^2}{36} = 1$$

$$\frac{(x-2)^2}{9} + \frac{(y+2)^2}{4} = 1$$

The center is $(2, -2)$.

Since $a^2 = 9$, $a = 3$.

Since $b^2 = 4$, $b = 2$.

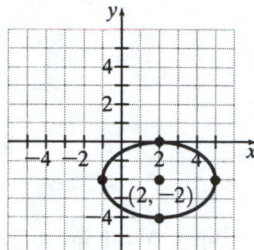

25. $12(x+4)^2 + 3(y-1)^2 = 48$

$$\frac{12(x+4)^2}{48} + \frac{3(y-1)^2}{48} = 1$$

$$\frac{(x+4)^2}{4} + \frac{(y-1)^2}{16} = 1$$

The center is $(-4, 1)$.

Since $a^2 = 4$, $a = 2$.

Since $b^2 = 16$, $b = 4$.

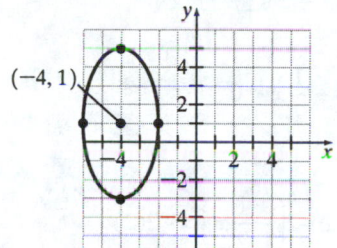

27. There is only one point, at $(0, 0)$. The only way two non-negative numbers can sum to 0 is if they are both 0.

29. The center is the origin, $a = 2$, and $b = 3$.

$$\frac{x^2}{a^2} + \frac{y^2}{b^2} = 1$$

$$\frac{x^2}{2^2} + \frac{y^2}{3^2} = 1$$

$$\frac{x^2}{4} + \frac{y^2}{9} = 1$$

31. There are no points of intersection, because the ellipse with $a = 2$ and $b = 3$ is completely inside the circle of radius 4.

33.
$$x^2 + 4y^2 - 4x - 8y - 92 = 0$$
$$x^2 - 4x + 4y^2 - 8y = 92$$
$$(x^2 - 4x + 4) + 4(y^2 - 2y + 1) = 92 + 4 + 4$$
$$(x-2)^2 + 4(y-1)^2 = 100$$
$$\frac{(x-2)^2}{100} + \frac{(y-1)^2}{25} = 1$$

The center is $(2, 1)$.

35. Since $90.2 - 20.7 = 69.5$, the distance between the foci is 69.5 feet.

37. Using $a = 3$ and $b = 2$, we may assume that the ellipse has the equation $\dfrac{x^2}{9} + \dfrac{y^2}{4} = 0$ and that the foci are located at $(\pm c, \ 0)$. Apply the definition of an ellipse using the points $(3, 0)$ and $(0, 2)$. That is, the distance from $(3, 0)$ to $(-c, 0)$ plus the distance from $(3, 0)$ to $(c, 0)$ is the same as the sum of the distance from $(0, 2)$ to $(-c, 0)$ and the distance from $(0, 2)$ to $(c, 0)$.

$$\sqrt{[3-(-c)]^2 + (0-0)^2} + \sqrt{(3-c)^2 + (0-0)^2} = \sqrt{[0-(-c)]^2 + (2-0)^2} + \sqrt{(0-c)^2 + (2-0)^2}$$

$$|3+c| + |3-c| = \sqrt{c^2 + 4} + \sqrt{(-c)^2 + 4}$$

Note that the foci are inside the ellipse, so $3 + c > 0$ and $3 - c > 0$. So, $|3+c| = 3 + c$ and $|3-c| = 3 - c$.

$$(3+c)+(3-c) = 2\sqrt{c^2+4}$$
$$6 = 2\sqrt{c^2+4}$$
$$3 = \sqrt{c^2+4}$$
$$9 = c^2+4$$
$$5 = c^2$$
$$c = \pm\sqrt{5}$$

The foci are located at $\left(\pm\sqrt{5},\ 0\right)$.

That is, the foci are $\sqrt{5} \approx 2.24$ feet, in both directions, from the center of the ellipse, along the major axis.

39. Answers will vary.

41. Solve for y.

$$\frac{x^2}{4} + \frac{y^2}{1} = 1$$
$$x^2 + 4y^2 = 4$$
$$4y^2 = 4 - x^2$$
$$y^2 = \frac{4-x^2}{4}$$
$$y = \pm\frac{1}{2}\sqrt{4-x^2}$$

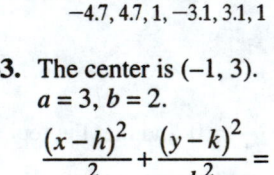

$$-4.7, 4.7, 1, -3.1, 3.1, 1$$

43. The center is $(-1, 3)$.
$a = 3$, $b = 2$.

$$\frac{(x-h)^2}{a^2} + \frac{(y-k)^2}{b^2} = 1$$
$$\frac{(x+1)^2}{3^2} + \frac{(y-3)^2}{2^2} = 1$$
$$\frac{(x+1)^2}{9} + \frac{(y-3)^2}{4} = 1$$

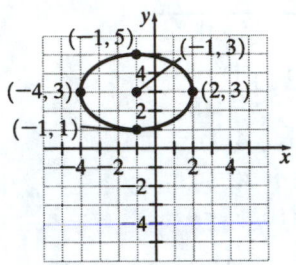

46.
$$-3 \le 4 - \frac{1}{2}x < 6$$
$$-7 \le -\frac{1}{2}x < 2$$
$$-2(-7) \ge -2\left(-\frac{1}{2}x\right) > -2(2)$$
$$14 \ge x > -4$$
$$-4 < x \le 14$$

47. $|2x-4| = 8$
$$2x - 4 = -8 \quad \text{or} \quad 2x - 4 = 8$$
$$2x = -4 \qquad\qquad 2x = 12$$
$$x = -2 \qquad\qquad x = 6$$

48. $|2x-4| \le 8$
$$-8 \le 2x - 4 \le 8$$
$$-4 \le 2x \le 12$$
$$-2 \le x \le 6$$

49. $|2x - 4| > 8$

$2x - 4 < -8$ or $2x - 4 > 8$

$2x < -4$ $2x > 12$

$x < -2$ $x > 6$

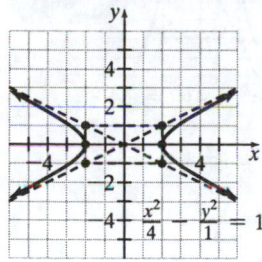

Exercise Set 10.3

1. A hyperbola is the set of points in a plane, the difference of whose distances from two fixed points is a constant.

3. The graph of $\dfrac{x^2}{a^2} - \dfrac{y^2}{b^2} = 1$ is a hyperbola with vertices at $(a, 0)$ and $(-a, 0)$. Its transverse axis lies along the x-axis. The asymptotes are $y = \pm\dfrac{b}{a}x$.

5. a. $\dfrac{x^2}{4} - \dfrac{y^2}{1} = 1$

Since $a^2 = 4$ and $b^2 = 1$, $a = 2$ and $b = 1$. The equations of the asymptotes are $y = \pm\dfrac{b}{a}x$, or $y = \pm\dfrac{1}{2}x$.

b. To graph the asymptotes, plot the points $(2, 1)$, $(-2, 1)$, $(2, -1)$, and $(-2, -1)$. The graph intersects the x-axis at $(-2, 0)$ and $(2, 0)$.

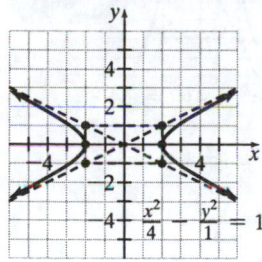

7. a. $\dfrac{y^2}{9} - \dfrac{x^2}{16} = 1$

Since $a^2 = 16$ and $b^2 = 9$, $a = 4$ and $b = 3$. The equations of the asymptotes are $y = \pm\dfrac{b}{a}x$, or $= y \pm\dfrac{3}{4}x$.

b. To graph the asymptotes, plot the points $(4, 3)$, $(-4, 3)$, $(4, -3)$ and $(-4, -3)$. The graph intersects the y-axis at $(0, -3)$ and $(0, 3)$.

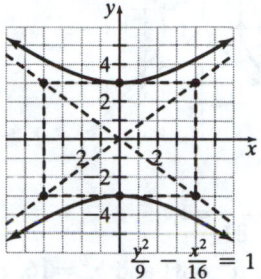

9. a. $\dfrac{y^2}{25} - \dfrac{x^2}{36} = 1$

Since $a^2 = 36$ and $b^2 = 25$, $a = 6$ and $b = 5$. The equations of the asymptotes are $y = \pm\dfrac{b}{a}x$, or $y = \pm\dfrac{5}{6}x$.

b. To graph the asymptotes, plot the points $(6, 5)$, $(-6, 5)$, $(6, -5)$, and $(-6, -5)$. The graph intersects the y-axis at $(0, -5)$ and $(0, 5)$.

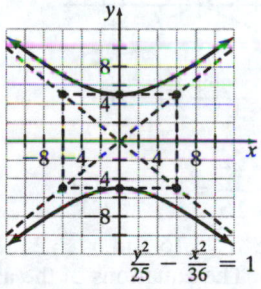

11. a. $\dfrac{x^2}{4} - \dfrac{y^2}{4} = 1$

Since $a^2 = 4$ and $b^2 = 4$, $a = 2$ and $b = 2$. The equations of the asymptotes are $y = \pm\dfrac{b}{a}x$, or $y = \pm x$.

b. To graph the asymptotes, plot the points $(2, 2)$, $(-2, 2)$, $(2, -2)$, and $(-2, -2)$. The graph intersects the x-axis at $(-2, 0)$ and $(2, 0)$.

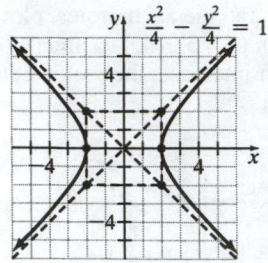

13. a. $\dfrac{y^2}{16} - \dfrac{x^2}{81} = 1$

Since $a^2 = 81$ and $b^2 = 16$, $a = 9$ and $b = 4$. The equations of the asymptotes are $y = \pm\dfrac{b}{a}x$, or $y = \pm\dfrac{4}{9}x$.

b. To graph the asymptotes, plot the points $(9, 4)$, $(-9, 4)$, $(9, -4)$, and $(-9, -4)$. The graph intersects the y-axis at $(0, -4)$ and $(0, 4)$.

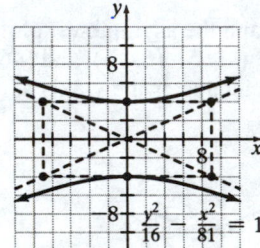

15. a. $\dfrac{y^2}{25} - \dfrac{x^2}{16} = 1$

Since $a^2 = 16$ and $b^2 = 25$, $a = 4$ and $b = 5$. The equations of the asymptotes are $y = \pm\dfrac{b}{a}x$, or $y = \pm\dfrac{5}{4}x$.

b. To graph the asymptotes, plot the points $(4, 5)$, $(-4, 5)$, $(4, -5)$, and $(-4, -5)$. The graph intersects the y-axis at $(0, -5)$ and $(0, 5)$.

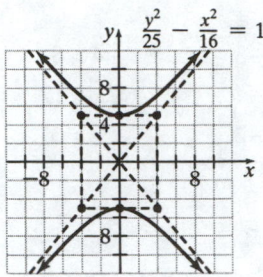

17. a. $16x^2 - 4y^2 = 64$

$\dfrac{16x^2}{64} - \dfrac{4y^2}{64} = 1$

$\dfrac{x^2}{4} - \dfrac{y^2}{16} = 1$

Since $a^2 = 4$ and $b^2 = 16$, $a = 2$ and $b = 4$. The equations of the asymptotes are $y = \pm\dfrac{b}{a}x$, or $y = \pm2x$.

b. To graph the asymptotes, plot the points $(2, 4)$, $(-2, 4)$, $(2 -4)$, and $(-2, -4)$. The graph intersects the x-axis at $(-2, 0)$ and $(2, 0)$.

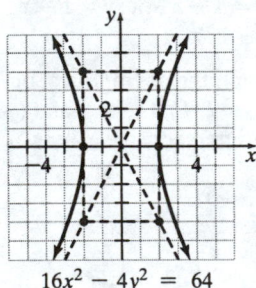

$16x^2 - 4y^2 = 64$

19. a. $9y^2 - x^2 = 9$

$\dfrac{9y^2}{9} - \dfrac{x^2}{9} = 1$

$\dfrac{y^2}{1} - \dfrac{x^2}{9} = 1$

Since $a^2 = 9$ and $b^2 = 1$, $a = 3$ and $b = 1$. The equations of the asymptotes are $y = \pm\dfrac{b}{a}x$, or $\pm\dfrac{1}{3}x$.

b. To graph the asymptotes, plot the points $(3, 1)$, $(-3, 1)$, $(3, -1)$, and $(-3, -1)$. The graph intersects the y-axis at $(0, -1)$ and $(0, 1)$.

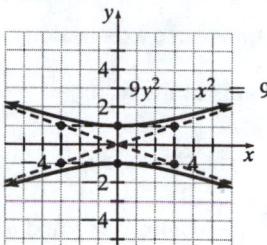

21. a. $4y^2 - 36x^2 = 144$

$$\frac{4y^2}{144} - \frac{36x^2}{144} = 1$$

$$\frac{y^2}{36} - \frac{x^2}{4} = 1$$

Since $a^2 = 4$ and $b^2 = 36$, $a = 2$ and $b = 6$. The equations of the asymptotes are $y = \pm\frac{b}{a}x$, or $y = \pm 3x$.

b. To graph the asymptotes, plot the points $(2, 6)$, $(-2, 6)$, $(2, -6)$, and $(-2, -6)$. The graph intersects the y-axis at $(0, -6)$ and $(0, 6)$.

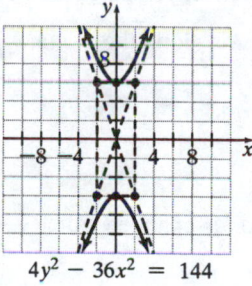

23. a. $25x^2 - 9y^2 = 225$

$$\frac{25x^2}{225} - \frac{9y^2}{225} = 1$$

$$\frac{x^2}{9} - \frac{y^2}{25} = 1$$

Since $a^2 = 9$ and $b^2 = 25$, $a = 3$ and $b = 5$. The equations of the asymptotes are $y = \pm\frac{b}{a}x$, or $y = \pm\frac{5}{3}x$.

b. To graph the asymptotes, plot the points $(3, 5)$, $(-3, 5)$, $(3, -5)$, and $(-3, -5)$. The graph intersects the x-axis at $(-3, 0)$ and $(3, 0)$.

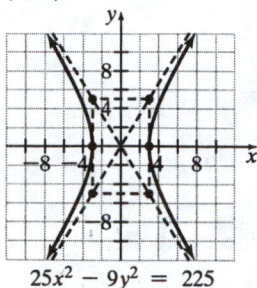

25. $4x = 6x^2 + y + 3$

$$-y = 6x^2 - 4x + 3$$

$$y = -6x^2 - 4x + 3$$

The graph is a parabola.

27. $4x^2 - 4y^2 = 29$

$$\frac{4x^2}{29} - \frac{4y^2}{29} = 1$$

$$\frac{x^2}{\frac{29}{4}} - \frac{y^2}{\frac{29}{4}} = 1$$

The graph is a hyperbola.

29. $x = y^2 + 6y - 7$

The graph is a parabola.

31. $-2x^2 + 4y^2 = 16$

$$\frac{-2x^2}{16} + \frac{4y^2}{16} = 1$$

$$-\frac{x^2}{8} + \frac{y^2}{4} = 1$$

$$\frac{y^2}{4} - \frac{x^2}{8} = 1$$

The graph is a hyperbola.

33. $5x^2 + 10y^2 = 12$

$$\frac{5x^2}{12} + \frac{10y^2}{12} = 1$$

$$\frac{x^2}{\frac{12}{5}} + \frac{y^2}{\frac{6}{5}} = 1$$

The graph is an ellipse.

35. $x + y = 2y^2 + 9$

$$x = 2y^2 - y + 9$$

The graph is a parabola.

37. $6x^2 + 6y^2 = 36$

$$\frac{6x^2}{36} + \frac{6y^2}{36} = 1$$

$$\frac{x^2}{6} + \frac{y^2}{6} = 1$$

$$x^2 + y^2 = 6$$

The graph is a circle.

39. $-3x^2 - 3y^2 = -27$

$$-\frac{3x^2}{-27} - \frac{3y^2}{-27} = 1$$

$$\frac{x^2}{9} + \frac{y^2}{9} = 1$$

$$x^2 + y^2 = 9$$

The graph is a circle.

41. $-6y^2 + x^2 = -9$

$$\frac{-6y^2}{-9} + \frac{x^2}{-9} = 1$$

$$\frac{y^2}{\frac{3}{2}} - \frac{x^2}{9} = 1$$

The graph is a hyperbola.

43. Since the intercepts are $(0, \pm 2)$, the hyperbola is of the form $\dfrac{y^2}{b^2} - \dfrac{x^2}{a^2} = 1$ with

$b = 2$. Since the asymptotes are $y = \pm\dfrac{1}{2}x$,

we have $\dfrac{b}{a} = \dfrac{1}{2}$. Therefore, $\dfrac{2}{a} = \dfrac{1}{2}$, so $a = 4$.

The equation of the hyperbola is

$$\frac{y^2}{2^2} - \frac{x^2}{4^2} = 1, \text{ or } \frac{y^2}{4} - \frac{x^2}{16} = 1.$$

45. Since the transverse axis is along the x-axis, the equation is of the form $\dfrac{x^2}{a^2} - \dfrac{y^2}{b^2} = 1$

Since the asymptotes are $y = \pm\dfrac{5}{3}x$, we

require $\dfrac{b}{a} = \dfrac{5}{3}$. Using $a = 3$ and $b = 5$, the

equation of the hyperbola is $\dfrac{x^2}{3^2} - \dfrac{y^2}{5^2} = 1$, or

$$\frac{x^2}{9} - \frac{y^2}{25} = 1.$$

No, this is not the only possible answer, because a and b are not uniquely

determined. $\dfrac{x^2}{18} - \dfrac{y^2}{50} = 1$ and others will

also work.

47. No, for each value of x with $|x| > a$, there are 2 possible values of y.

49. $\dfrac{x^2}{9} - \dfrac{y^2}{4} = 1$. This hyperbola has its

transverse axis along the x-axis with vertices at $(\pm 3, \ 0)$.

Domain: $(-\infty, \ -3] \cup [3, \ \infty)$

Range: $\mathbb{R}$

51. The equation is changed from $\dfrac{x^2}{a^2} - \dfrac{y^2}{b^2} = 1$

to $\dfrac{x^2}{b^2} - \dfrac{y^2}{a^2} = 1$. Both graphs have a

transverse axis along the x-axis. The vertices of the second graph will be closer to the origin, at $(\pm b, \ 0)$ instead of $(\pm a, \ 0)$. The second graph will open wider.

53. $\dfrac{y^2}{9} - \dfrac{x^2}{16} = 1$

$$\dfrac{y^2}{9} = \dfrac{x^2}{16} + 1$$

$$y^2 = \dfrac{9}{16}\left(x^2 + 16\right)$$

$$y = \pm \dfrac{3}{4}\sqrt{x^2 + 16}$$

$-9.4, 9.4, 1, -6.2, 6.2, 1$

55. $\dfrac{x}{2} + \dfrac{2}{3}(x - 6) = x + 4$

$$6\left[\dfrac{x}{2} + \dfrac{2}{3}(x - 6)\right] = 6(x + 4)$$

$$3x + 4(x - 6) = 6x + 24$$

$$7x - 24 = 6x + 24$$

$$7x = 6x + 48$$

$$x = 48$$

56. The domain of a function is the set of values for the independent variable. The range is the set of values obtained for the dependent variable.

57. $6x - 2y < 12$

$y \geq -2x + 3$

58. $-2x^2 + 6x - 5 = 0$

$$x = \dfrac{-b \pm \sqrt{b^2 - 4ac}}{2a}$$

$$x = \dfrac{-6 \pm \sqrt{6^2 - 4(-2)(-5)}}{2(-2)}$$

$$= \dfrac{-6 \pm \sqrt{-4}}{-4}$$

$$= \dfrac{-6 \pm 2i}{-4}$$

$$= \dfrac{3 \pm i}{2}$$

Exercise Set 10.4

1. A nonlinear system of equations is a system in which at least one equation is nonlinear.

3. $x^2 + y^2 = 9$

$x + 2y = 3$

Solve $x + 2y = 3$ for x: $x = 3 - 2y$.

Substitute $3 - 2y$ for x in $x^2 + y^2 = 9$.

$$x^2 + y^2 = 9$$

$$(3 - 2y)^2 + y^2 = 9$$

$$9 - 12y + 4y^2 + y^2 = 9$$

$$5y^2 - 12y = 0$$

$$y(5y - 12) = 0$$

$y = 0$ or $y = \dfrac{12}{5}$

$x = 3 - 2y$ $x = 3 - 2y$

$x = 3 - 2(0)$ $x = 3 - 2\left(\dfrac{12}{5}\right)$

$x = 3$ $x = -\dfrac{9}{5}$

The solutions are $(3, 0)$ and $\left(-\dfrac{9}{5}, \dfrac{12}{5}\right)$.

5. $y = x^2 - 5$
$3x + 2y = 10$

Substitute $x^2 - 5$ for y in $3x + 2y = 10$
$$3x + 2y = 10$$
$$3x + 2(x^2 - 5) = 10$$
$$3x + 2x^2 - 10 = 10$$
$$2x^2 + 3x - 20 = 0$$
$$(x + 4)(2x - 5) = 0$$

$x = -4$ or $x = \dfrac{5}{2}$

$y = x^2 - 5$ $y = x^2 - 5$
$y = (-4)^2 - 5$ $y = \left(\dfrac{5}{2}\right)^2 - 5$
$y = 11$
$y = \dfrac{5}{4}$

The solutions are $(-4,\ 11)$ and $\left(\dfrac{5}{2},\ \dfrac{5}{4}\right)$.

7. $2x^2 - y^2 = -8$
    $x - y = 6$

Solve the second equation for y: $y = x - 6$.
Substitute $x - 6$ for y in $2x^2 - y^2 = -8$.
$$2x^2 - y^2 = -8$$
$$2x^2 - (x - 6)^2 = -8$$
$$2x^2 - (x^2 - 12x + 36) = -8$$
$$2x^2 - x^2 + 12x - 36 = -8$$
$$x^2 + 12x - 28 = 0$$
$$(x - 2)(x + 14) = 0$$

$x = 2$ or $x = -14$

$y = x - 6$ $y = x - 6$
$y = 2 - 6$ $y = -14 - 6$
$y = -4$ $y = -20$

The solutions are $(2, -4)$ and $(-14, -20)$.

9. $x^2 - 4y^2 = 16$
    $x^2 + y^2 = 1$

Solve $x^2 + y^2 = 1$ for x^2: $x^2 = 1 - y^2$.
Substitute $1 - y^2$ for x^2 in $x^2 - 4y^2 = 16$.

$$x^2 - 4y^2 = 16$$
$$(1 - y^2) - 4y^2 = 16$$
$$1 - 5y^2 = 16$$
$$-5y^2 = 17$$
$$y^2 = -\dfrac{17}{5}$$
$$y = \pm i\sqrt{\dfrac{17}{5}}$$

There is no real solution.

11. $y = x^2 - 3$
$x^2 + y^2 = 9$

Solve $y = x^2 - 3$ for x^2: $x^2 = y + 3$.
Substitute $y + 3$ for x^2 in $x^2 + y^2 = 9$.
$$x^2 + y^2 = 9$$
$$(y + 3) + y^2 = 9$$
$$y^2 + y - 6 = 0$$
$$(y - 2)(y + 3) = 0$$
$y = 2$ or $y = -3$

$x^2 = y + 3$ $x^2 = y + 3$
$x^2 = 2 + 3$ $x^2 = -3 + 3$
$x^2 = 5$ $x^2 = 0$
$x = \pm\sqrt{5}$ $x = 0$

The solutions are $(0, -3)$, $\left(\sqrt{5}, 2\right)$, and $\left(-\sqrt{5}, 2\right)$.

13. $x^2 - y^2 = 4$
$\underline{x^2 + y^2 = 4}$
    $2x^2 = 8$
    $x^2 = 4$
    $x = 2$ or $x = -2$

$x^2 + y^2 = 4$ $x^2 + y^2 = 4$
$2^2 + y^2 = 4$ $(-2)^2 + y^2 = 4$
$y^2 = 0$ $y^2 = 0$
$y = 0$ $y = 0$

The solutions are $(2, 0)$ and $(-2, 0)$.

15.

$$x^2 + y^2 = 13 \quad (1)$$
$$2x^2 + 3y^2 = 30 \quad (2)$$
$$-2x^2 - 2y^2 = -26 \quad (1) \text{ multiplied by } -2$$
$$\underline{2x^2 + 3y^2 = 30 \quad (2)}$$
$$y^2 = 4$$
$$y = 2 \qquad \text{or} \qquad y = -2$$

$$x^2 + y^2 = 13 \qquad\qquad x^2 + y^2 = 13$$
$$x^2 + 2^2 = 13 \qquad\qquad x^2 + (-2)^2 = 13$$
$$x^2 = 9 \qquad\qquad\qquad x^2 = 9$$
$$x = \pm 3 \qquad\qquad\qquad x = \pm 3$$

The solutions are (3, 2), (3, –2), (–3, 2), and (–3, –2).

17.

$$4x^2 + 9y^2 = 36$$
$$\underline{2x^2 - 9y^2 = 18}$$
$$6x^2 = 54$$
$$x^2 = 9$$
$$x = 3 \qquad \text{or} \qquad x = -3$$

$$4x^2 + 9y^2 = 36 \qquad\qquad 4x^2 + 9y^2 = 36$$
$$4(3)^2 + 9y^2 = 36 \qquad\qquad 4(-3)^2 + 9y^2 = 36$$
$$9y^2 = 0 \qquad\qquad\qquad 9y^2 = 0$$
$$y^2 = 0 \qquad\qquad\qquad y^2 = 0$$
$$y = 0 \qquad\qquad\qquad y = 0$$

The solutions are (3, 0) and (–3, 0).

19.

$$2x^2 + 3y^2 = 21 \quad (1)$$
$$x^2 + 2y^2 = 12 \quad (2)$$
$$-2x^2 - 3y^2 = -21 \quad (1) \text{ multiplied by } -1$$
$$\underline{2x^2 + 4y^2 = 24 \quad (2) \text{ multiplied by } 2}$$
$$y^2 = 3$$
$$y = \sqrt{3}$$

$$x^2 + 2y^2 = 12$$
$$x^2 + 2\left(\sqrt{3}\right)^2 = 12$$
$$x^2 + 6 = 12$$
$$x^2 = 6$$
$$x = \pm\sqrt{6}$$

or

$$y = -\sqrt{3}$$

$$x^2 + 2y^2 = 12$$
$$x^2 + 2\left(-\sqrt{3}\right)^2 = 12$$
$$x^2 + 6 = 12$$
$$x^2 = 6$$
$$x = \pm\sqrt{6}$$

The solutions are $\left(\sqrt{6},\ \sqrt{3}\right)$, $\left(\sqrt{6},\ -\sqrt{3}\right)$, $\left(-\sqrt{6},\ \sqrt{3}\right)$, and $\left(-\sqrt{6},\ -\sqrt{3}\right)$.

21.

$$-x^2 - 2y^2 = 6 \quad (1)$$
$$5x^2 + 15y^2 = 20 \quad (2)$$
$$-x^2 - 2y^2 = 6 \quad (1)$$
$$\underline{x^2 + 3y^2 = 4 \quad (2) \text{ divided by } 5}$$
$$y^2 = 10$$
$$y = \sqrt{10}$$

$$x^2 + 3y^2 = 4$$
$$x^2 + 3\left(\sqrt{10}\right)^2 = 4$$
$$x^2 = -26$$
$$x = \pm i\sqrt{26}$$

or

$$y = -\sqrt{10}$$

$$x^2 + 3y^2 = 4$$
$$x^2 + 3\left(-\sqrt{10}\right)^2 = 4$$
$$x^2 = -26$$
$$x = \pm i\sqrt{26}$$

There is no real solution.

23.
$$x^2 + y^2 = 9 \quad (1)$$
$$16x^2 - 4y^2 = 44 \quad (2)$$
$$4x^2 + 4y^2 = 36 \quad (1) \text{ multiplied by 4}$$
$$\underline{16x^2 - 4y^2 = 64 \quad (2)}$$
$$20x^2 = 100$$
$$x^2 = 5$$
$$x = \sqrt{5} \quad \text{or} \qquad x = -\sqrt{5}$$

$$x^2 + y^2 = 9 \qquad x^2 + y^2 = 9$$
$$\left(\sqrt{5}\right)^2 + y^2 = 9 \qquad \left(-\sqrt{5}\right)^2 + y^2 = 9$$
$$y^2 = 4 \qquad\qquad y^2 = 4$$
$$y = \pm 2 \qquad\qquad y = \pm 2$$

The solutions are $\left(\sqrt{5},\ 2\right)$, $\left(\sqrt{5},\ -2\right)$,
$\left(-\sqrt{5},\ 2\right)$, and $\left(-\sqrt{5},\ -2\right)$.

25. Answers will vary.

27. Let x = length
$\quad\quad y$ = width
$$xy = 240$$
$$2x + 2y = 68$$
Solve $2x + 2y = 68$ for y: $y = 34 - x$.
Substitute $34 - x$ for y in $xy = 240$.
$$xy = 240$$
$$x(34 - x) = 240$$
$$34x - x^2 = 240$$
$$x^2 - 34x + 240 = 0$$
$$(x - 10)(x - 24) = 0$$
$$x - 10 = 0 \qquad \text{or} \quad x - 24 = 0$$
$$x = 10 \qquad\qquad\quad x = 24$$

$$y = 34 - x \qquad\qquad y = 34 - x$$
$$y = 34 - 10 \qquad\qquad y = 34 - 24$$
$$y = 24 \qquad\qquad\quad y = 10$$
The solutions are (10, 24) and (24, 10).
The dimensions of the dance floor are 10
feet by 24 feet.

29. Let x = length
$\quad\quad y$ = width
$$xy = 112$$
$$x^2 + y^2 = \left(\sqrt{260}\right)^2$$
Solve $xy = 112$ for y: $y = \dfrac{112}{x}$.
$$x^2 + y^2 = 260$$
$$x^2 + \left(\frac{112}{x}\right)^2 = 260$$
$$x^2 + \frac{12,544}{x^2} = 260$$
$$x^4 + 12,544 = 260x^2$$
$$x^4 - 260x^2 + 12,544 = 0$$
$$(x^2 - 64)(x^2 - 196) = 0$$
$$x^2 - 64 = 0 \qquad \text{or} \quad x^2 - 196 = 0$$
$$x^2 = 64 \qquad\qquad x^2 = 196$$
$$x = \pm 8 \qquad\qquad x = \pm 14$$
Since x must be positive, $x = 8$ or $x = 14$.
If $x = 8$, then $y = \dfrac{112}{8} = 14$.

If $x = 14$, then $y = \dfrac{112}{14} = 8$.
The dimensions of the new bill are 8 cm by
14 cm.

31. Let x = length
$\quad\quad y$ = width
$$x^2 + y^2 = 17^2$$
$$x + y + 17 = 40$$
Solve $x + y + 17 = 40$ for y: $y = 23 - x$.
Substitute $23 - x$ for y in $x^2 + y^2 = 17^2$.
$$x^2 + y^2 = 17^2$$
$$x^2 + (23 - x)^2 = 17^2$$
$$x^2 + (529 - 46x + x^2) = 289$$
$$2x^2 - 46 + 529 = 289$$
$$2x^2 - 46x + 240 = 0$$
$$x^2 - 23x + 120 = 0$$
$$(x - 8)(x - 15) = 0$$

$$x - 8 = 0 \quad \text{or} \quad x - 15 = 0$$
$$x = 8 \qquad\qquad x = 15$$

$$y = 23 - x \qquad\qquad y = 23 - x$$
$$y = 23 - 8 \qquad\qquad y = 23 - 15$$
$$y = 15 \qquad\qquad y = 8$$

The solutions are (8, 15) and (15, 8).
The dimensions of the piece of wood are 8 in. by 15 in.

33.
$$xy = 48$$
$$2x + y = 20$$

Solve $2x + y = 20$ for y: $y = 20 - 2x$.
Substitute $20 - 2x$ for y in $xy = 48$.

$$xy = 48$$
$$x(20 - 2x) = 48$$
$$20x - 2x^2 = 48$$
$$2x^2 - 20x + 48 = 0$$
$$x^2 - 10x + 24 = 0$$
$$(x - 4)(x - 6) = 0$$
$$x - 4 = 0 \quad \text{or} \quad x - 6 = 0$$
$$x = 4 \qquad\qquad x = 6$$

$$y = 20 - 2x \qquad\qquad y = 20 - 2x$$
$$y = 20 - 2(4) \qquad\qquad y = 20 - 2(6)$$
$$y = 12 \qquad\qquad y = 8$$

The solutions are (4, 12) and (6, 8).
The dimensions are 6 ft by 8 ft or 4 ft by 12 ft.

35.
$$d = -16t^2 + 64t$$
$$d = -16t^2 + 16t + 80$$

Substitute $-16t^2 + 64t$ for d in
$d = -16t^2 + 16t + 80$.

$$d = -16t^2 + 16t + 80$$
$$-16t^2 + 64t = -16t^2 + 16t + 80$$
$$64t = 16t + 80$$
$$48t = 80$$
$$t = \frac{80}{48} = \frac{5}{3} \approx 1.67$$

The balls are the same height above the ground at $t \approx 1.67$ sec.

37. Since $t = 1$ year, we may write the formula
as $i = pr$.

$$7.50 = pr$$
$$7.50 = (p + 25)(r - 0.01)$$

Rewrite the second equation by multiplying
the binomials. Then substitute 7.50 for pr
and solve for r.

$$7.50 = (p + 25)(r - 0.01)$$
$$7.50 = pr - 0.01p + 25r - 0.25$$
$$7.50 = 7.50 - 0.01p + 25r - 0.25$$
$$0 = -0.01p + 25r - 0.25$$
$$0.01p + 0.25 = 25r$$
$$\frac{0.01p}{25} + \frac{0.25}{25} = \frac{25r}{25}$$
$$r = 0.0004p + 0.01$$

Substitute $0.0004p + 0.01$ for r in
$7.50 = pr$.

$$7.50 = pr$$
$$7.50 = p(0.0004p + 0.01)$$
$$7.50 = 0.0004p^2 + 0.01p$$
$$0 = 0.0004p^2 + 0.01p - 7.50$$
$$0 = p^2 + 25p - 18{,}750$$
$$0 = (p - 125)(p + 150)$$

$$p - 125 = 0 \quad \text{or} \quad p + 150 = 0$$
$$p = 125 \qquad\qquad p = -150$$

Since the principal must be positive, use
$p = 125$.

$$r = 0.0004p + 0.01$$
$$r = 0.0004(125) + 0.01$$
$$r = 0.06$$

The principal is \$125 and the interest rate is
6%.

39.
$$C = 10x + 300$$
$$R = 30x - 0.1x^2$$
$$C = R$$
$$10x + 300 = 30x - 0.1x^2$$
$$0.1x^2 - 20x + 300 = 0$$
$$x = \frac{-b \pm \sqrt{b^2 - 4ac}}{2a}$$
$$= \frac{-(-20) \pm \sqrt{(-20)^2 - 4(0.1)(300)}}{2(0.1)}$$
$$= \frac{20 \pm \sqrt{280}}{0.2}$$

$$x = \frac{20 + \sqrt{280}}{0.2} \approx 183.7 \text{ or}$$

$$x = \frac{20 - \sqrt{280}}{0.2} \approx 16.3$$

The break-even points are ≈ 16 and ≈ 184.

41. $C = 80x + 900$

$R = 120x - 0.2x^2$

$$C = R$$

$$80x + 900 = 120x - 0.2x^2$$

$$0.2x^2 - 40x + 900 = 0$$

$$x = \frac{-b \pm \sqrt{b^2 - 4ac}}{2a}$$

$$= \frac{-(-40) \pm \sqrt{(-40)^2 - 4(0.2)(900)}}{2(0.2)}$$

$$= \frac{40 \pm \sqrt{880}}{0.4}$$

$$x = \frac{40 + \sqrt{880}}{0.4} \approx 174.2 \text{ or}$$

$$x = \frac{40 - \sqrt{880}}{0.4} \approx 25.8$$

The break-even points are ≈ 26 and ≈ 174.

43. Solve each equation for y.

$3x - 5y = 12$

$$-5y = -3x + 12$$

$$y = \frac{3}{5}x - \frac{12}{5}$$

$x^2 + y^2 = 10$

$$y^2 = 10 - x^2$$

$$y = \pm\sqrt{10 - x^2}$$

Use $y_1 = \frac{3}{5}x - \frac{12}{5}$, $y_2 = \sqrt{10 - x^2}$, and

$y_2 = -\sqrt{10 - x^2}$.

−9.4, 9.4, 1, −6.2, 6.2, 1

Approximate solutions: $(-1, -3)$, $(3.12, -0.53)$

45. Let $x =$ length of one leg

$y =$ length of other leg

$$x^2 + y^2 = 26^2$$

$$\frac{1}{2}xy = 120$$

Solve $\frac{1}{2}xy = 120$ for y: $y = \frac{240}{x}$.

Substitute $\frac{240}{x}$ for y in $x^2 + y^2 = 26^2$

$$x^2 + y^2 = 26^2$$

$$x^2 + \left(\frac{240}{x}\right)^2 = 676$$

$$x^2 + \frac{57,600}{x^2} = 676$$

$$x^4 + 57,600 = 676x^2$$

$$x^4 - 676x^2 + 57,600 = 0$$

$$\left(x^2 - 100\right)\left(x^2 - 576\right) = 0$$

$$x^2 - 100 = 0 \qquad \text{or} \quad x^2 - 576 = 0$$

$$x^2 = 100 \qquad\qquad\quad x^2 = 576$$

$$x = \pm 10 \qquad\qquad\quad x = \pm 24$$

Since x is a length, x must be positive.

If $x = 10$, then $y = \frac{240}{10} = 24$.

If $x = 24$, then $y = \frac{240}{24} = 10$.

The legs have lengths 10 yards and 24 yards.

47. The operations are evaluated in the following order: parentheses, exponents, multiplication or division, addition or subtraction.

48.
$$\frac{3}{5}(2x - y) = \frac{3}{4}(2x - 3y) + 6$$

$$20\left[\frac{3}{5}(2x - y)\right] = 20\left[\frac{3}{4}(2x - 3y) + 6\right]$$

$$12(2x - y) = 15(2x - 3y) + 120$$

$$24x - 12y = 30x - 45y + 120$$

$$-12y = 6x - 45y + 120$$

$$33y = 6x + 120$$

$$y = \frac{6x + 120}{33}$$

$$y = \frac{2x + 40}{11}$$

49. $A = p\left(1 + \dfrac{r}{n}\right)^{nt}$

$= 5000\left(1 + \dfrac{0.08}{2}\right)^{2(2)}$

$= 5000(1.04)^4$

≈ 5849.29

The amount is $5849.29.

50.
$$\dfrac{3-4y}{3} \geq \dfrac{2y-6}{4} - \dfrac{7}{6}$$
$$12\left(\dfrac{3-4y}{3}\right) \geq 12\left(\dfrac{2y-6}{4} - \dfrac{7}{6}\right)$$
$$4(3-4y) \geq 3(2y-6) - 14$$
$$12 - 16y \geq 6y - 18 - 14$$
$$-16y \geq 6y - 44$$
$$-22y \geq -44$$
$$\dfrac{-22y}{-22} \leq \dfrac{-44}{-22}$$
$$y \leq 2$$

51. $f(x) = \sqrt{x+2}$

52.
$$f(x) = g(x)$$
$$\sqrt{x^2 - 3x - 9} = \sqrt{2x-3}$$
$$x^2 - 3x - 9 = 2x - 3$$
$$x^2 - 5x - 6 = 0$$
$$(x+1)(x-6) = 0$$
$$x = -1 \text{ or } x = 6$$
Check: $x = 1$
$$\sqrt{(-1)^2 - 3(-1) - 9} \overset{?}{=} \sqrt{2(-1) - 3}$$
$$\sqrt{-5} \overset{?}{=} \sqrt{-5}$$
not a real number

$$x = 6$$
$$\sqrt{6^2 - 3(6) - 9} \overset{?}{=} \sqrt{2(6) - 3}$$
$$\sqrt{9} \overset{?}{=} \sqrt{9}$$
$$3 = 3 \text{ True}$$
The only solution is $x = 6$.

Review Exercises

1. $d = \sqrt{(x_2 - x_1)^2 + (y_2 - y_1)^2}$

$= \sqrt{(3-0)^2 + (-4-0)^2}$

$= \sqrt{3^2 + (-4)^2}$

$= \sqrt{9 + 16}$

$= \sqrt{25}$

$= 5$

$\text{Midpoint} = \left(\dfrac{x_1 + x_2}{2}, \dfrac{y_1 + y_2}{2}\right)$

$= \left(\dfrac{0+3}{2}, \dfrac{0 + (-4)}{2}\right)$

$= \left(\dfrac{3}{2}, -2\right)$

2. $d = \sqrt{(x_2 - x_1)^2 + (y_2 - y_1)^2}$

$= \sqrt{(2-6)^2 + (-1-2)^2}$

$= \sqrt{(-4)^2 + (-3)^2}$

$= \sqrt{16 + 9}$

$= \sqrt{25}$

$= 5$

$\text{Midpoint} = \left(\dfrac{x_1 + x_2}{2}, \dfrac{y_1 + y_2}{2}\right)$

$= \left(\dfrac{6+2}{2}, \dfrac{2 + (-1)}{2}\right)$

$= \left(4, \dfrac{1}{2}\right)$

3. $d = \sqrt{(x_2 - x_1)^2 + (y_2 - y_1)^2}$

    $= \sqrt{[3 - (-2)]^2 + [9 - (-3)]^2}$

    $= \sqrt{5^2 + 12^2}$

    $= \sqrt{25 + 144}$

    $= \sqrt{169}$

    $= 13$

    Midpoint $= \left(\dfrac{x_1 + x_2}{2}, \dfrac{y_1 + y_2}{2} \right)$

    $= \left(\dfrac{-2 + 3}{2}, \dfrac{-3 + 9}{2} \right)$

    $= \left(\dfrac{1}{2}, 3 \right)$

4. $d = \sqrt{(x_2 - x_1)^2 + (y_2 - y_1)^2}$

    $= \sqrt{[-2 - (-4)]^2 + (5 - 3)^2}$

    $= \sqrt{2^2 + 2^2}$

    $= \sqrt{4 + 4}$

    $= \sqrt{8}$

    ≈ 2.83

    Midpoint $= \left(\dfrac{x_1 + x_2}{2}, \dfrac{y_1 + y_2}{2} \right)$

    $= \left(\dfrac{-4 + (-2)}{2}, \dfrac{3 + 5}{2} \right)$

    $= (-3, 4)$

5. $y = (x - 3)^2 + 4$

    This is a parabola in the form
    $y = a(x - h)^2 + k$ with $a = 1$, $h = 3$, and
    $k = 4$. Since $a > 0$, the parabola opens
    upward. The vertex is (3, 4). The y-intercept
    is (0, 13).
    There are no x-intercepts.

    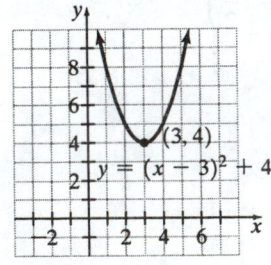

6. $y = (x + 4)^2 - 5$

    This is a parabola in the form
    $y = a(x - h)^2 + k$ with $a = 1$, $h = -4$, and
    $k = -5$. Since $a > 0$, the parabola opens
    upward. The vertex is (–4, –5). The y-
    intercept is (0, 11).
    The x-intercepts are about (–6.24, 0) and
    (–1.76, 0).

    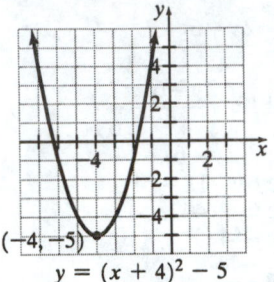

    $y = (x + 4)^2 - 5$

7. $x = (y - 1)^2 + 4$

    This is a parabola in the form
    $x = a(y - k)^2 + h$ with $a = 1$, $h = 4$, and
    $k = 1$. Since $a > 0$, the parabola opens to
    the right. The vertex is (4, 1). There are no
    y-intercepts.
    The x-intercept is (5, 0).

    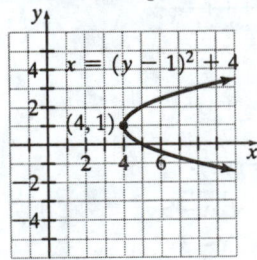

8. $x = -2(y + 4)^2 - 3$

    This is a parabola in the form
    $x = a(y - k)^2 + h$ with $a = -2$, $h = -3$, and
    $k = -4$. Since $a < 0$, the parabola opens to
    the left. The vertex is (–3, –4). There are no
    y-intercepts.
    The x-intercept is (–35, 0).

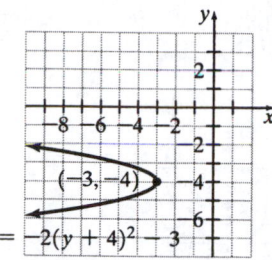

$$y = -2(y+4)^2 - 3$$

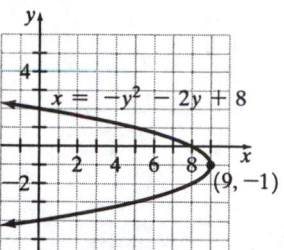

$$x = -y^2 - 2y + 8$$

$(9, -1)$

9. a. $y = x^2 - 6x$

$$y = \left(x^2 - 6x + 9\right) - 9$$

$$y = (x-3)^2 - 9$$

b. This is a parabola in the form
$y = a(x-h)^2 + k$ with $a = 1$, $h = 3$,
and $k = -9$. Since $a > 0$, the parabola
opens upward. The vertex is (3, –9).
The *y*-intercept is (0, 0). The
x-intercepts are (0, 0) and (6, 0).

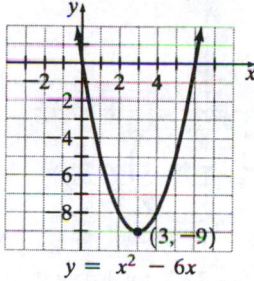

$$y = x^2 - 6x$$

10. a. $x = -y^2 - 2y + 8$

$$x = -(y^2 + 2y) + 8$$

$$x = -(y^2 + 2y + 1) + 1 + 8$$

$$x = -(y+1)^2 + 9$$

b. This is a parabola in the form
$x = a(y-k)^2 + h$ with $a = -1$, $h = 9$,
and $k = -1$. Since $a < 0$, the parabola
opens to the left. The vertex is (9, –1).
The *y*-intercepts are (0, –4) and (0, 2).
The *x*-intercept is (8, 0).

11. a. $x = y^2 + 5y + 4$

$$x = \left(y^2 + 5y + \frac{25}{4}\right) - \frac{25}{4} + 4$$

$$x = \left(y + \frac{5}{2}\right)^2 - \frac{9}{4}$$

b. This is a parabola in the form
$x = a(y-k)^2 + h$ with $a = 1$, $h = -\dfrac{9}{4}$,
and $k = -\dfrac{5}{2}$. Since $a > 0$, the parabola

opens to the right. The vertex is
$\left(-\dfrac{9}{4}, -\dfrac{5}{2}\right)$.
The *y*-intercepts are (0, –4) and (0, –1).
The *x*-intercept is (4, 0).

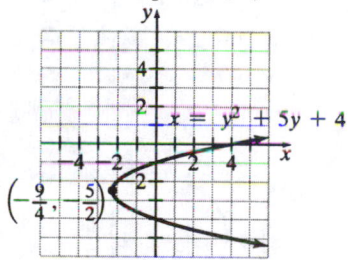

$$x = y^2 + 5y + 4$$

$\left(-\dfrac{9}{4}, -\dfrac{5}{2}\right)$

12. a. $y = 2x^2 - 8x - 24$

$$y = 2(x^2 - 4x) - 24$$

$$y = 2(x^2 - 4x + 4) - 8 - 24$$

$$y = 2(x-2)^2 - 32$$

b. This is a parabola in the form
$y = a(x-h)^2 + k$ with $a = 2$, $h = 2$,
and $k = -32$. Since $a > 0$, the parabola
opens upward. The vertex is (2, –32).
The *y*-intercept is (0, –24). The
x-intercepts are (–2, 0) and (6, 0).

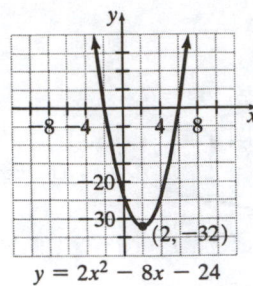

$$y = 2x^2 - 8x - 24$$

13. a.
$$(x-h)^2 + (y-k)^2 = r^2$$
$$(x-0)^2 + (y-0)^2 = 5^2$$
$$x^2 + y^2 = 5^2$$

b.

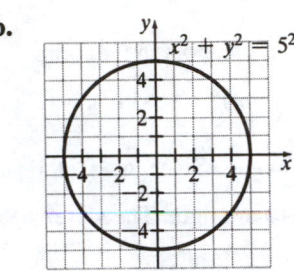

14. a.
$$(x-h)^2 + (y-k)^2 = r^2$$
$$[x-(-3)]^2 + (y-4)^2 = 3^2$$
$$(x+3)^2 + (y-4)^2 = 3^2$$

b.

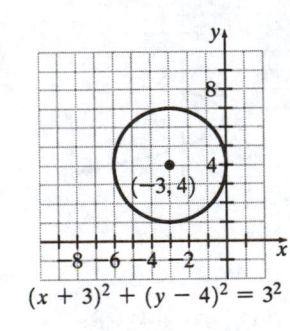

$$(x+3)^2 + (y-4)^2 = 3^2$$

15. a.
$$x^2 + y^2 - 4y = 0$$
$$x^2 + (y^2 - 4y + 4) = 4$$
$$x^2 + (y-2)^2 = 2^2$$

b. The graph is a circle with center $(0, 2)$ and radius 2.

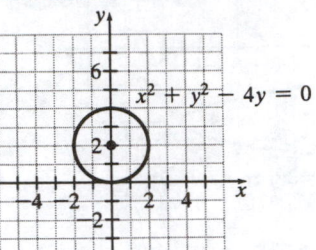

16. a.
$$x^2 + y^2 - 2x + 6y + 1 = 0$$
$$x^2 - 2x + y^2 + 6y = -1$$
$$(x^2 - 2x + 1) + (y^2 + 6y + 9) = -1 + 1 + 9$$
$$(x-1)^2 + (y+3)^2 = 9$$
$$(x-1)^2 + (y+3)^2 = 3^2$$

b. The graph is a circle with center $(1, -3)$ and radius 3.

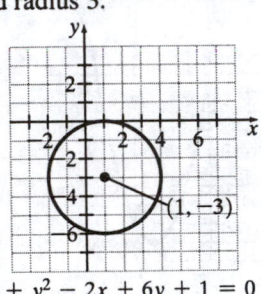

$$x^2 + y^2 - 2x + 6y + 1 = 0$$

17. a.

$$x^2 - 8x + y^2 - 10y + 40 = 0$$
$$(x^2 - 8x + 16) + (y^2 - 10y + 25) = -40 + 16 + 25$$
$$(x-4)^2 + (y-5)^2 = 1$$
$$(x-4)^2 + (y-5)^2 = 1^2$$

b. The graph is a circle with center (4, 5) and radius 1.

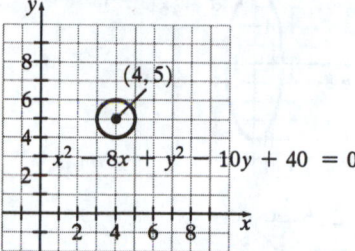

18. a.

$$x^2 + y^2 - 4x + 10y + 17 = 0$$
$$x^2 - 4x + y^2 + 10y = -17$$
$$(x^2 - 4x + 4) + (y^2 + 10y + 25) = -17 + 4 + 25$$
$$(x-2)^2 + (y+5)^2 = 12$$
$$(x-2)^2 + (y+5)^2 = \left(\sqrt{12}\right)^2$$

b. The graph is a circle with center (2, –5) and radius $\sqrt{12} \approx 3.46$.

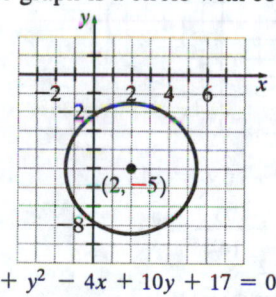

19. $y = \sqrt{16 - x^2}$

If we solve $x^2 + y^2 = 16$ for y, we obtain

$y = \pm\sqrt{16 - x^2}$. Therefore, the graph of

$y = \sqrt{16 - x^2}$ is the upper half $(y \geq 0)$ of a

circle with its center at the origin and radius 4.

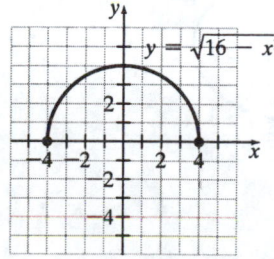

20. $y = -\sqrt{25 - x^2}$

If we solve $x^2 + y^2 = 25$ for y, we obtain $y = \pm\sqrt{25 - x^2}$. Therefore, the graph of $y = -\sqrt{25 - x^2}$ is the lower half $(y \le 0)$ of a circle with its center at the origin and radius 5.

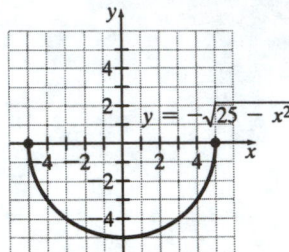

21. The center is (–1, 1) and the radius is 2.
$$(x - h)^2 + (y - k)^2 = r^2$$
$$[x - (-1)]^2 + (y - 1)^2 = 2^2$$
$$(x + 1)^2 + (y - 1)^2 = 4$$

22. The center is (5, –3) and the radius is 3.
$$(x - h)^2 + (y - k)^2 = r^2$$
$$(x - 5)^2 + [y - (-3)]^2 = 3^2$$
$$(x - 5)^2 + (y + 3)^2 = 9$$

23. $\dfrac{x^2}{9} + \dfrac{y^2}{4} = 1$

Since $a^2 = 9$, $a = 3$.
Since $b^2 = 4$, $b = 2$.

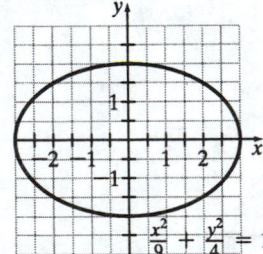

24. $\dfrac{x^2}{9} + \dfrac{y^2}{64} = 1$

Since $a^2 = 9$, $a = 3$.
Since $b^2 = 64$, $b = 8$.

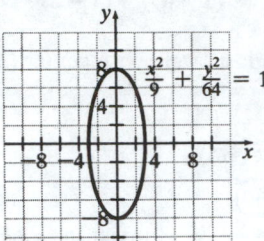

25. $4x^2 + 9y^2 = 36$

$$\dfrac{4x^2}{36} + \dfrac{9y^2}{36} = 1$$

$$\dfrac{x^2}{9} + \dfrac{y^2}{4} = 1$$

Since $a^2 = 9$, $a = 3$.
Since $b^2 = 4$, $b = 2$.

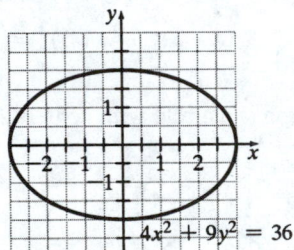

26. $9x^2 + 16y^2 = 144$

$$\dfrac{9x^2}{144} + \dfrac{16y^2}{144} = 1$$

$$\dfrac{x^2}{16} + \dfrac{y^2}{9} = 1$$

Since $a^2 = 16$, $a = 4$.
Since $b^2 = 9$, $b = 3$.

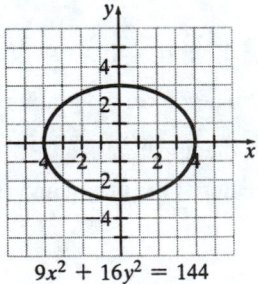

27. $\dfrac{(x-3)^2}{16}+\dfrac{(y+2)^2}{4}=1$

The center is (3, –2).

Since $a^2=16$, $a=4$.

Since $b^2=4$, $b=2$.

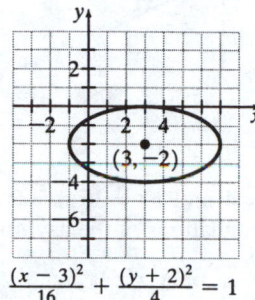

$\dfrac{(x-3)^2}{16}+\dfrac{(y+2)^2}{4}=1$

28. $\dfrac{(x+3)^2}{9}+\dfrac{y^2}{25}=1$

The center is (–3, 0).

Since $a^2=9$, $a=3$.

Since $b^2=25$, $b=5$.

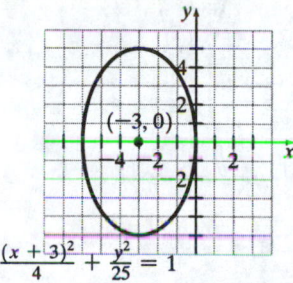

$\dfrac{(x+3)^2}{4}+\dfrac{y^2}{25}=1$

29. $25(x-2)^2+9(y-1)^2=225$

$\dfrac{25(x-2)^2}{225}+\dfrac{9(y-1)^2}{225}=1$

$\dfrac{(x-2)^2}{9}+\dfrac{(y-1)^2}{25}=1$

The center is (2, 1).

Since $a^2=9$, $a=3$.

Since $b^2=25$, $b=5$.

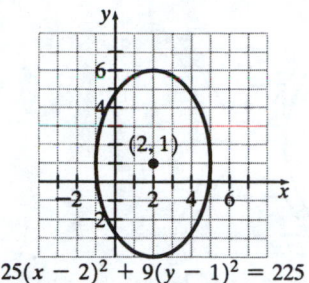

$25(x-2)^2+9(y-1)^2=225$

30. $4(x+3)^2+25(y-2)^2=100$

$\dfrac{4(x+3)^2}{100}+\dfrac{25(y-2)^2}{100}=1$

$\dfrac{(x+3)^2}{25}+\dfrac{(y-2)^2}{4}=1$

The center is (–3, 2).

Since $a^2=25$, $a=5$.

Since $b^2=4$, $b=2$.

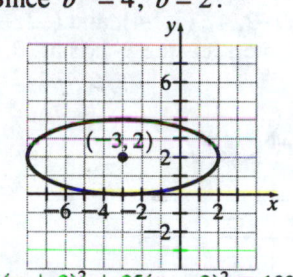

$4(x+3)^2+25(y-2)^2=100$

31. a. $\dfrac{x^2}{4}-\dfrac{y^2}{9}=1$

Since $a^2=4$ and $b^2=9$, $a=2$ and $b=3$. The equations of the asymptotes are $y=\pm\dfrac{b}{a}x$, or $y=\pm\dfrac{3}{2}x$.

b. To graph the asymptotes, plot the points (2, 3), (–2, 3), (2, –3), and (–2, –3). The graph intersects the *x*-axis at (–2, 0) and (2, 0).

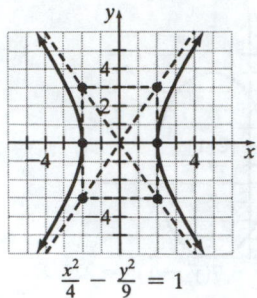

$$\frac{x^2}{4} - \frac{y^2}{9} = 1$$

32. a. $\dfrac{y^2}{16} - \dfrac{x^2}{4} = 1$

Since $a^2 = 4$ and $b^2 = 16$, $a = 2$ and $b = 4$. The equations of the asymptotes are $y = \pm\dfrac{b}{a}x$, or $y = \pm 2x$.

b. To graph the asymptotes, plot the points $(2, 4)$, $(-2, 4)$, $(2, -4)$, and $(-2, -4)$. The graph intersects the y-axis at $(0, -4)$ and $(0, 4)$.

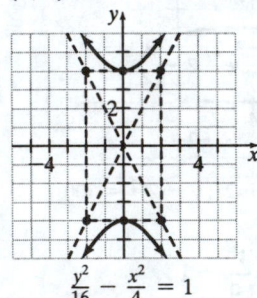

$$\frac{y^2}{16} - \frac{x^2}{4} = 1$$

33. a. $\dfrac{y^2}{9} - \dfrac{x^2}{25} = 1$

Since $a^2 = 25$ and $b^2 = 9$, $a = 5$ and $b = 3$. The equations of the asymptotes are $y = \pm\dfrac{b}{a}x$, or $y = \pm\dfrac{3}{5}x$.

b. To graph the asymptotes, plot the points $(5, 3)$, $(-5, 3)$, $(5, -3)$, and $(-5, -3)$. The graph intersects the y-axis at $(0, -3)$ and $(0, 3)$.

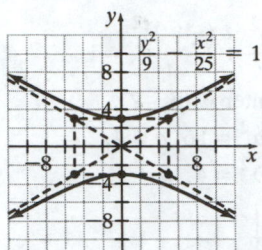

34. a. $\dfrac{x^2}{4} - \dfrac{y^2}{36} = 1$

Since $a^2 = 4$ and $b^2 = 36$, $a = 2$ and $b = 6$. The equations of the asymptotes are $y = \pm\dfrac{b}{a}x$, or $y = \pm 3x$.

b. To graph the asymptotes, plot the points $(2, 6)$, $(-2, 6)$, $(2, -6)$, and $(-2, -6)$. The graph intersects the x-axis at $(-2, 0)$ and $(2, 0)$.

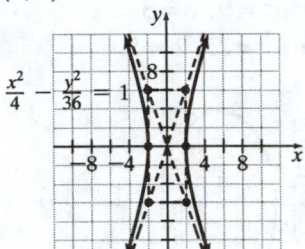

$$\frac{x^2}{4} - \frac{y^2}{36} = 1$$

35. a. $9y^2 - 4x^2 = 36$

$$\frac{9y^2}{36} - \frac{4x^2}{36} = 1$$

$$\frac{y^2}{4} - \frac{x^2}{9} = 1$$

b. Since $a^2 = 9$ and $b^2 = 4$, $a = 3$ and $b = 2$. The equations of the asymptotes are $y = \pm\dfrac{b}{a}x$, or $y = \pm\dfrac{2}{3}x$.

c. To graph the asymptotes, plot the points $(3, 2)$, $(-3, 2)$, $(3, -2)$, and $(-3, -2)$. The graph intersects the y-axis at $(0, -2)$ and $(0, 2)$.

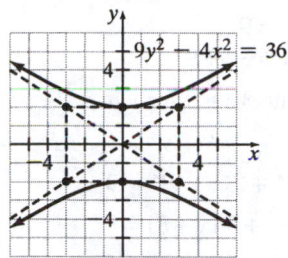

36. a. $x^2 - 16y^2 = 16$

$$\frac{x^2}{16} - \frac{16y^2}{16} = 1$$

$$\frac{x^2}{16} - \frac{y^2}{1} = 1$$

b. Since $a^2 = 16$ and $b^2 = 1$, $a = 4$ and $b = 1$. The equations of the asymptotes are $y = \pm\frac{b}{a}x$, or $y = \pm\frac{1}{4}x$.

c. To graph the asymptotes, plot the points (4, 1), (–4, 1), (4, –1), and (–4, –1). The graph intersects the x-axis at (–4, 0) and (4, 0).

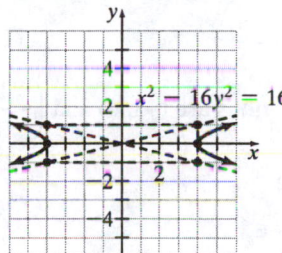

37. a. $25x^2 - 16y^2 = 400$

$$\frac{25x^2}{400} - \frac{16y^2}{400} = 1$$

$$\frac{x^2}{16} - \frac{y^2}{25} = 1$$

b. Since $a^2 = 16$ and $b^2 = 25$, $a = 4$ and $b = 5$. The equations of the asymptotes are $y = \pm\frac{b}{a}x$, or $y = \pm\frac{5}{4}x$.

c. To graph the asymptotes, plot the points (4, 5), (–4, 5), (4, –5), and (–4, –5). The graph intersects the x-axis at (–4, 0) and (4, 0).

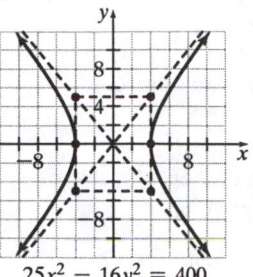

38. a. $49y^2 - 9x^2 = 441$

$$\frac{49y^2}{441} - \frac{9x^2}{441} = 1$$

$$\frac{y^2}{9} - \frac{x^2}{49} = 1$$

b. Since $a^2 = 49$ and $b^2 = 9$, $a = 7$ and $b = 3$. The equations of the asymptotes are $y = \pm\frac{b}{a}x$, or $y = \pm\frac{3}{7}x$.

c. To graph the asymptotes, plot the points (7, 3), (–7, 3), (7, –3), and (–7, –3). The graph intersects the y-axis at (0, –3) and (0, 3).

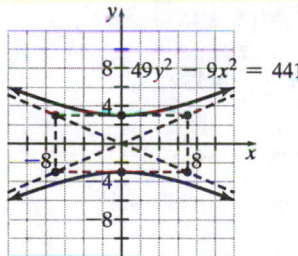

39. $\frac{x^2}{4} - \frac{y^2}{25} = 1$

The graph is a hyperbola.

40. $4x^2 + 9y^2 = 144$

$$\frac{4x^2}{144} + \frac{9y^2}{144} = 1$$

$$\frac{x^2}{36} + \frac{y^2}{16} = 1$$

The graph is an ellipse.

41. $4x^2 + 4y^2 = 16$

$$\frac{4x^2}{4} + \frac{4y^2}{4} = \frac{16}{4}$$

$$x^2 + y^2 = 4$$

The graph is a circle.

42. $x^2 - 25y^2 = 25$

$$\frac{x^2}{25} - \frac{25y^2}{25} = 1$$

$$\frac{x^2}{25} - \frac{y^2}{1} = 1$$

The graph is a hyperbola.

43. $\dfrac{x^2}{16} + \dfrac{y^2}{9} = 1$

The graph is an ellipse.

44. $y = (x-3)^2 + 4$

The graph is a parabola.

45. $4x^2 + 9y^2 = 36$

$$\frac{4x^2}{36} + \frac{9y^2}{36} = 1$$

$$\frac{x^2}{9} + \frac{y^2}{4} = 1$$

The graph is an ellipse.

46. $x = -y^2 - 6y - 7$

The graph is a parabola.

47. $x^2 + y^2 = 9$

$\qquad y = 3x + 9$

Substitute $3x + 9$ for y in $x^2 + y^2 = 9$.

$$x^2 + y^2 = 9$$
$$x^2 + (3x+9)^2 = 9$$
$$x^2 + 9x^2 + 54x + 81 = 9$$
$$10x^2 + 54x + 72 = 0$$
$$5x^2 + 27x + 36 = 0$$
$$(x+3)(5x+12) = 0$$

$x + 3 = 0 \qquad\qquad$ or

$x = -3$

$y = 3x + 9$
$y = 3(-3) + 9$
$y = 0$

$5x + 12 = 0$

$$x = -\frac{12}{5}$$

$y = 3x + 9$

$$y = 3\left(-\frac{12}{5}\right) + 9$$

$$y = \frac{9}{5}$$

The solutions are (–3, 0) and $\left(-\dfrac{12}{5},\ \dfrac{9}{5}\right)$.

48. $x^2 - y^2 = 4$

$2x + 2y = 8$

Solve $2x + 2y = 8$ for y: $y = 4 - x$.

Substitute $4 - x$ for y in $x^2 - y^2 = 4$.

$$x^2 - y^2 = 4$$
$$x^2 - (4 - x)^2 = 4$$
$$x^2 - (16 - 8x + x^2) = 4$$
$$8x - 16 = 4$$
$$8x = 20$$
$$x = \frac{5}{2}$$

$$y = 4 - x$$
$$y = 4 - \frac{5}{2}$$
$$y = \frac{3}{2}$$

The solution is $\left(\dfrac{5}{2},\ \dfrac{3}{2}\right)$.

49. $x^2 + y^2 = 4$
$x^2 - y^2 = 4$

Solve $x^2 - y^2 = 4$ for x^2: $x^2 = y^2 + 4$.

Substitute $y^2 + 4$ for x^2 in $x^2 + y^2 = 4$.

$$x^2 + y^2 = 4$$
$$(y^2 + 4) + y^2 = 4$$
$$2y^2 + 4 = 4$$
$$2y^2 = 0$$
$$y^2 = 0$$
$$y = 0$$

$$x^2 = y^2 + 4$$
$$x^2 = 0^2 + 4$$
$$x^2 = 4$$
$$x = \pm 2$$

The solutions are (2, 0) and (–2, 0).

50. $x^2 + 4y^2 = 4$
$x^2 - 6y^2 = 12$

Solve $x^2 + 4y^2 = 4$ for x^2: $x^2 = 4 - 4y^2$.

Substitute $4 - 4y^2$ for x^2 in $x^2 - 6y^2 = 12$.

$$x^2 - 6y^2 = 12$$
$$(4 - 4y^2) - 6y^2 = 12$$
$$4 - 10y^2 = 12$$
$$-10y^2 = 8$$
$$y^2 = -\frac{4}{5}$$
$$y = \pm i \sqrt{\frac{2}{5}}$$

There is no real solution.

51. $x^2 + y^2 = 16$
$$\underline{x^2 - y^2 = 16}$$
$$2x^2 = 32$$
$$x^2 = 16$$
$$x = 4 \qquad\qquad x = -4$$

$$x^2 + y^2 = 16 \qquad\qquad x^2 + y^2 = 16$$
$$4^2 + y^2 = 16 \qquad\qquad (-4)^2 + y^2 = 16$$
$$y^2 = 0 \qquad\qquad\qquad y^2 = 0$$
$$y = 0 \qquad\qquad\qquad y = 0$$

The solutions are (4, 0) and (–4, 0).

52.
$$x^2 + y^2 = 25 \quad (1)$$
$$x^2 - 2y^2 = -2 \quad (2)$$
$$2x^2 + 2y^2 = 50 \quad (1) \text{ multiplied by } 2$$
$$\underline{x^2 - 2y^2 = -2 \quad (2)}$$
$$3x^2 = 48$$
$$x^2 = 16$$
$$x = 4$$

$$x^2 + y^2 = 25$$
$$4^2 + y^2 = 25$$
$$y^2 = 9$$
$$y = \pm 3$$

or

$$x = -4$$

$$x^2 + y^2 = 25$$
$$(-4)^2 + y^2 = 25$$
$$y^2 = 9$$
$$y = \pm 3$$

The solutions are (4, 3), (4, −3), (−4, 3) and (−4, −3).

53. $-4x^2 + y^2 = -12$ (1)

$8x^2 + 2y^2 = -8$ (2)

$-4x^2 + y^2 = -12$ (1)

$\underline{4x^2 + y^2 = -4}$ (2) divided by 2

$2y^2 = -16$

$y^2 = -8$

$y = \pm i\sqrt{8}$

$= \pm 2i\sqrt{2}$

There is no real solution.

54. $-2x^2 - 3y^2 = -6$ (1)

$5x^2 + 4y^2 = 15$ (2)

$-10x^2 - 15y^2 = -30$ (1) multiplied by 5

$\underline{10x^2 + 8y^2 = 30}$ (2) multiplied by 2

$-7y^2 = 0$

$y^2 = 0$

$y = 0$

$5x^2 + 4y^2 = 15$

$5x^2 + 4(0)^2 = 15$

$5x^2 = 15$

$x^2 = 3$

$x = \pm\sqrt{3}$

The solutions are $\left(\sqrt{3}, 0\right)$ and $\left(-\sqrt{3}, 0\right)$.

55. Let $x = $ length

$y = $ width

$xy = 32$

$2x + 2y = 24$

Solve $2x + 2y = 24$ for y: $y = 12 - x$.

Substitute $12 - x$ for y in $xy = 32$.

$xy = 32$

$x(12 - x) = 32$

$12x - x^2 = 32$

$x^2 - 12x + 32 = 0$

$(x - 4)(x - 8) = 0$

$x - 4 = 0$ or $x - 8 = 0$

$x = 4$ $x = 8$

$y = 12 - x$ $y = 12 - x$

$y = 12 - 4$ $y = 12 - 8$

$y = 8$ $y = 4$

The solutions are (4, 8) and (8, 4).
The dimensions of the table are 4 feet by 8 feet.

56. $C = 20.3x + 120$

$R = 50.2x - 0.2x^2$

$C = R$

$20.3x + 120 = 50.2x - 0.2x^2$

$0.2x^2 - 29.9 + 120 = 0$

$x = \dfrac{-b \pm \sqrt{b^2 - 4ac}}{2a}$

$= \dfrac{-(-29.9) \pm \sqrt{(-29.9)^2 - 4(0.2)(120)}}{2(0.2)}$

$= \dfrac{29.9 \pm \sqrt{798.01}}{0.4}$

$x = \dfrac{29.9 + \sqrt{798.01}}{0.4} \approx 145.4$ or

$x = \dfrac{29.9 - \sqrt{798.01}}{0.4} \approx 4.1$

The break-even points are ≈ 4 and ≈ 145.

57. Let $x = $ length

$y = $ width

$xy = 300$

$x^2 + y^2 = 25^2$

Solve $xy = 300$ for y: $y = \dfrac{300}{x}$.

Substitute $\dfrac{300}{x}$ for y in $x^2 + y^2 = 25^2$.

$$x^2 + y^2 = 25^2$$

$$x^2 + \left(\frac{300}{x}\right)^2 = 625$$

$$x^2 + \frac{90,000}{x^2} = 625$$

$$x^4 + 90,000 = 625x^2$$

$$x^4 - 625x^2 + 90,000 = 0$$

$$(x^2 - 225)(x^2 - 400) = 0$$

$$x^2 - 225 = 0 \qquad \text{or} \quad x^2 - 400 = 0$$

$$x^2 = 225 \qquad\qquad x^2 = 400$$

$$x = \pm 15 \qquad\qquad x = \pm 20$$

Since x must be positive, $x = 15$ or $x = 20$.

If $x = 15$, then $y = \dfrac{300}{15} = 20$.

If $x = 20$, then $y = \dfrac{300}{20} = 15$.

The dimensions of the carpet are 15 feet by 20 feet.

58. Since $t = 1$ year, we may rewrite the formula $i = prt$ as $i = pr$.

$$250 = pr$$

$$250 = (p + 1250)(r - 0.01)$$

Rewrite the second equation by multiplying the binomials. Then substitute 250 for pr and solve for r.

$$250 = pr - 0.01p + 1250r - 12.5$$

$$250 = 250 - 0.01p + 1250r - 12.5$$

$$0 = -0.01p + 1250r - 12.5$$

$$0.01p + 12.5 = 1250r$$

$$\frac{0.01p}{1250} + \frac{12.5}{1250} = r$$

$$r = 0.000008p + 0.01$$

Substitute $0.000008p + 0.01$ for r in

$$250 = pr.$$

$$250 = pr$$

$$250 = p(0.000008p + 0.01)$$

$$250 = 0.000008p^2 + 0.01p$$

$$0 = 0.000008p^2 + 0.01p - 250$$

$$0 = p^2 + 1250p - 31,250,000$$

$$0 = (p - 5000)(p + 6250)$$

$$p - 5000 = 0 \qquad \text{or} \quad p + 6250 = 0$$

$$p = 5000 \qquad\qquad p = -6250$$

The principal must be positive, so use $p = 5000$.

$$r = 0.000008p + 0.01$$

$$r = 0.000008(5000) + 0.01$$

$$r = 0.05$$

The principal is \$5000 and the rate is 5%.

Practice Test

1. They are formed by cutting a cone or pair of cones.

2.
$$d = \sqrt{(x_2 - x_1)^2 + (y_2 - y_1)^2}$$
$$= \sqrt{[3 - (-4)]^2 + (4 - 5)^2}$$
$$= \sqrt{7^2 + (-1)^2}$$
$$= \sqrt{49 + 1}$$
$$= \sqrt{50}$$

The length is $\sqrt{50} \approx 7.07$ units.

3.
$$\text{Midpoint} = \left(\frac{x_1 + x_2}{2}, \frac{y_1 + y_2}{2}\right)$$
$$= \left(\frac{4 + (-6)}{2}, \frac{2 + 5}{2}\right)$$
$$= \left(-1, \frac{7}{2}\right)$$

4. $y = -2(x - 3)^2 + 4$

This is a parabola in the form
$$y = a(x - h)^2 + k$$
with $a = -2$, $h = 3$, and $k = 4$. Since $a < 0$, the parabola opens downward. The vertex is $(3, 4)$. The y-intercept is $(0, -14)$. The x-intercepts are about $(1.59, 0)$ and $(4.41, 0)$.

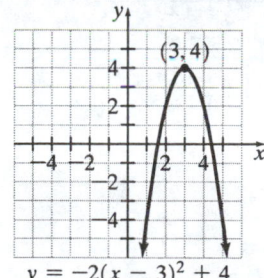

$$y = -2(x - 3)^2 + 4$$

5. $x = y^2 - 2y + 4$

$x = (y^2 - 2y + 1) - 1 + 4$

$x = (y-1)^2 + 3$

This is a parabola in the form

$x = a(y-k)^2 + h$

with $a = 1$, $h = 3$ and $k = 1$. Since $a > 0$, the parabola opens to the right. The vertex is $(3, 1)$.

There is no y-intercept. The x-intercept is $(4, 0)$.

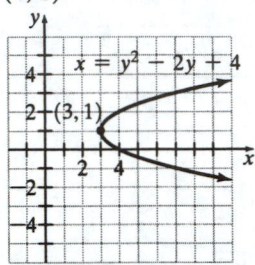

6. $x = -3y^2 + 12y - 8$

$x = -3(y^2 - 4y) - 8$

$x = -3(y^2 - 4y + 4) + 12 - 8$

$x = -3(y-2)^2 + 4$

This is a parabola in the form

$x = a(y-k)^2 + h$

with $a = -3$, $h = 4$, and $k = 2$. Since $a < 0$, the parabola opens to the left. The vertex is $(4, 2)$. The y-intercepts are about $(0, 0.85)$ and $(0, 3.15)$. The x-intercept is $(-8, 0)$.

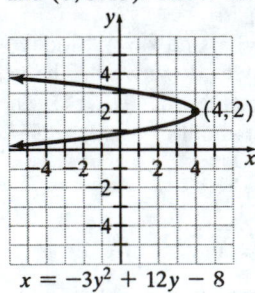

$x = -3y^2 + 12y - 8$

7. $(x-h)^2 + (y-k)^2 = r^2$

$[x-(-3)]^2 + [y-(-1)]^2 = 4^2$

$(x+3)^2 + (y+1)^2 = 16$

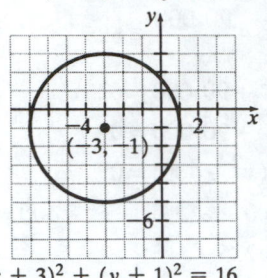

$(x + 3)^2 + (y + 1)^2 = 16$

8. $(x+1)^2 + (y-1)^2 = 9$. The graph of this equation is a circle with center $(-1, 1)$ and radius 3. The possible values of x are from $-1 - 3$ to $-1 + 3$, or $[-4, 2]$. The possible values of y are from $1 - 3$ to $1 + 3$, or $[-2, 4]$.
Domain: $[-4, 2]$, Range: $[-2, 4]$

9. The center is $(3, -1)$ and the radius is 4.

$(x-h)^2 + (y-k)^2 = r^2$

$(x-3)^2 + [y-(-1)]^2 = 4^2$

$(x-3)^2 + (y+1)^2 = 4^2$

10. $y = -\sqrt{16 - x^2}$

If we solve $x^2 + y^2 = 16$ for y, we obtain

$y = \pm\sqrt{16 - x^2}$. Therefore, the graph of

$y = -\sqrt{16 - x^2}$ is the lower half $(y \le 0)$ of a circle with its center at the origin and radius 4.

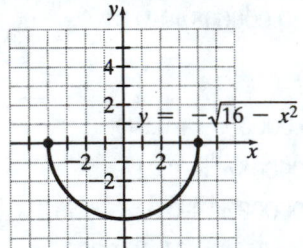

11.
$$x^2 + y^2 - 2x - 6y + 1 = 0$$
$$x^2 - 2x + y^2 - 6y = -1$$
$$(x^2 - 2x + 1) + (y^2 - 6y + 9) = -1 + 1 + 9$$
$$(x-1)^2 + (y-3)^2 = 9$$
The graph is a circle with center (1, 3) and radius 3.

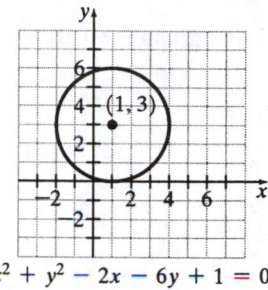

$$x^2 + y^2 - 2x - 6y + 1 = 0$$

12. $9x^2 + 16y^2 = 144$
$$\frac{9x^2}{144} + \frac{16y^2}{144} = 1$$
$$\frac{x^2}{16} + \frac{y^2}{9} = 1$$
Since $a^2 = 16,\ a = 4.$
Since $b^2 = 9,\ b = 3.$

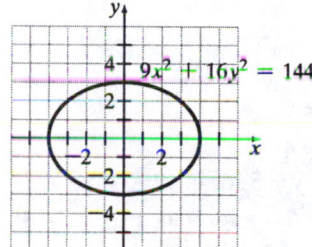

13. The center is (−2, −1), $a = 4$, and $b = 2$.
$$\frac{(x-h)^2}{a^2} + \frac{(y-k)^2}{b^2} = 1$$
$$\frac{[x-(-2)]^2}{4^2} + \frac{[y-(-1)]^2}{2^2} = 1$$
$$\frac{(x+2)^2}{16} + \frac{(y+1)^2}{4} = 1$$
The values of a^2 and b^2 are switched, so this is not the graph of the given equation.

14. $4(x-3)^2 + 16(y+1)^2 = 64$
$$\frac{4(x-3)^2}{64} + \frac{16(y+1)^2}{64} = 1$$
$$\frac{(x-3)^2}{16} + \frac{(y+1)^2}{4} = 1$$
The center is (3, −1). Since $a^2 = 16,\ a = 4.$
Since $b^2 = 4,\ b = 2.$

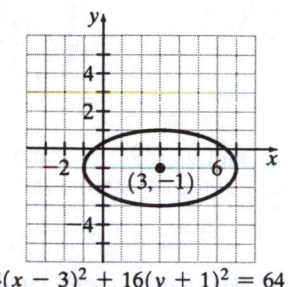

$$4(x - 3)^2 + 16(y + 1)^2 = 64$$

15.
$$x^2 + y^2 + 4x + 8y - 12 = 0$$
$$x^2 + 4x + y^2 + 8y = 12$$
$$(x^2 + 4x + 4) + (y^2 + 8y + 16) = 12 + 4 + 16$$
$$(x+2)^2 + (y+4)^2 = 32$$
The center is (−2, −4).

16. The transverse axis lies along the axis corresponding to the positive term of the equation in standard form.

17. $\dfrac{x^2}{16} - \dfrac{y^2}{36} = 1$
Since $a^2 = 16$ and $b^2 = 36$, $a = 4$ and $b = 6$. The equations of the asymptotes are
$$y = \pm \frac{b}{a}x,\ \text{or}\ y = \pm \frac{3}{2}x.$$

18. $\dfrac{y^2}{25} - \dfrac{x^2}{1} = 1$
Since $a^2 = 1$ and $b^2 = 25$, $a = 1$ and $b = 5$. The equations of the asymptotes are
$$y = \pm \frac{b}{a}x,\ \text{or}\ y = \pm 5x.$$
To graph the asymptotes, plot the points (1, 5), (−1, 5), (1, −5), and (−1, −5). The graph intersects the y-axis at (0, −5) and

$(0, 5)$.

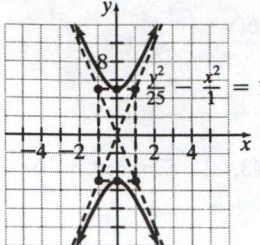

19. $\dfrac{x^2}{4} - \dfrac{y^2}{9} = 1$

Since $a^2 = 4$ and $b^2 = 9$, $a = 2$ and $b = 3$.
The equations of the asymptotes are

$y = \pm\dfrac{b}{a}x$, or $y = \pm\dfrac{3}{2}x$.

To graph the asymptotes, plot the points
$(2, 3)$, $(-2, 3)$, $(2, -3)$, and $(-2, -3)$. The
graph intersects the x-axis at $(-2, 0)$ and
$(2, 0)$.

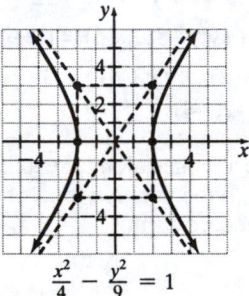

20. $4x^2 - 16y^2 = 48$

$\dfrac{4x^2}{48} - \dfrac{16y^2}{48} = 1$

$\dfrac{x^2}{12} - \dfrac{y^2}{3} = 1$

Since the equation is of the form

$\dfrac{x^2}{a^2} - \dfrac{y^2}{b^2} = 1$, the graph is a hyperbola.

21. $16x^2 + 4y^2 = 64$

$\dfrac{16x^2}{64} + \dfrac{4y^2}{64} = 1$

$\dfrac{x^2}{4} + \dfrac{y^2}{16} = 1$

Since the equation is of the form

$\dfrac{x^2}{a^2} + \dfrac{y^2}{b^2} = 1$, the graph is an ellipse.

22. $\begin{aligned} x^2 + y^2 &= 6 \\ 2x^2 - y^2 &= 3 \\ \hline 3x^2 &= 9 \\ x^2 &= 3 \end{aligned}$

$\begin{array}{ccc} x = \sqrt{3} & \text{or} & x = -\sqrt{3} \\ x^2 + y^2 = 6 & & x^2 + y^2 = 6 \\ \left(\sqrt{3}\right)^2 + y^2 = 6 & & \left(-\sqrt{3}\right)^2 + y^2 = 6 \\ 3 + y^2 = 6 & & 3 + y^2 = 6 \\ y^2 = 3 & & y^2 = 3 \\ y = \pm\sqrt{3} & & y = \pm\sqrt{3} \end{array}$

The solutions are $\left(\sqrt{3},\ \sqrt{3}\right)$, $\left(\sqrt{3},\ -\sqrt{3}\right)$,

$\left(-\sqrt{3},\ \sqrt{3}\right)$, and $\left(-\sqrt{3},\ -\sqrt{3}\right)$.

23. $\begin{aligned} x + y &= 5 \\ x^2 + y^2 &= 4 \end{aligned}$

Solve $x + y = 5$ for y: $y = 5 - x$.
Substitute $5 - x$ for y in $x^2 + y^2 = 4$.

$\begin{aligned} x^2 + y^2 &= 4 \\ x^2 + (5 - x)^2 &= 4 \\ x^2 + 25 - 10x + x^2 &= 4 \\ 2x^2 - 10x + 21 &= 0 \end{aligned}$

$$x = \frac{-b \pm \sqrt{b^2 - 4ac}}{2a}$$

$$= \frac{-(-10) \pm \sqrt{(-10)^2 - 4(2)(21)}}{2(2)}$$

$$= \frac{10 \pm \sqrt{-68}}{4}$$

$$= \frac{10 \pm 2i\sqrt{17}}{4}$$

$$= \frac{5 \pm i\sqrt{17}}{2}$$

There is no real solution.

24. Let $x = $ length,
$y = $ width.
$$xy = 6000$$
$$2x + 2y = 320$$
Solve $2x + 2y = 320$ for y: $y = 160 - x$.
Substitute $160 - x$ for y in $xy = 6000$.
$$xy = 6000$$
$$x(160 - x) = 6000$$
$$160x - x^2 = 6000$$
$$x^2 - 160x + 6000 = 0$$
$$(x - 60)(x - 100) = 0$$
$$x - 60 = 0 \qquad \text{or} \quad x - 100 = 0$$

$x = 60$	$x = 100$
$y = 160 - x$	$y = 160 - x$
$y = 160 - 60$	$y = 160 - 100$
$y = 100$	$y = 60$

The solutions are (60, 100) and (100, 60).
The dimensions are 60 feet by 100 feet.

25. Let $x = $ length
$y = $ width
$$xy = 60$$
$$x^2 + y^2 = 13^2$$

Solve $xy = 60$ for y: $y = \dfrac{60}{x}$.

Substitute $\dfrac{60}{x}$ for y in $x^2 + y^2 = 13^2$.

$$x^2 + y^2 = 13^2$$

$$x^2 + \left(\frac{60}{x}\right)^2 = 169$$

$$x^2 + \frac{3600}{x^2} = 169$$

$$x^4 + 3600 = 169x^2$$

$$x^4 - 169x^2 + 3600 = 0$$

$$(x^2 - 25)(x^2 - 144) = 0$$

$$x^2 - 25 = 0 \quad \text{or} \quad x^2 - 144 = 0$$

$x^2 = 25$	$x^2 = 144$
$x = \pm 5$	$x^2 = \pm 12$

Since x must be positive, $x = 5$ or $x = 12$.

If $x = 5$, then $y = \dfrac{60}{5} = 12$. If $x = 12$, then

$y = \dfrac{60}{12} = 5$. The dimensions of the bed of

the truck are 5 feet by 12 feet.

Cumulative Review Test

1. $\quad 2x - y = 6 \qquad (1)$
$\quad 3x + 2y = 4 \qquad (2)$
Multiply equation (1) by 2 and add.
$2[2x - y = 6]$
$\quad 3x + 2y = 4$
gives
$$4x - 2y = 12$$
$$\underline{3x + 2y = 4}$$
Add: $7x \qquad = 16$
$$x = \frac{16}{7}$$

Substitute $\dfrac{16}{7}$ for x in equation (1).

$$2\left(\frac{16}{7}\right) - y = 6$$

$$\frac{32}{7} - y = \frac{42}{7}$$

$$-y = \frac{10}{7}$$

$$y = -\frac{10}{7}$$

The solution is $\left(\dfrac{16}{7}, \, -\dfrac{10}{7}\right)$.

2. $f(a+3) = (a+3)^2 + 2(a+3) + 5$
$\qquad\qquad = a^2 + 6a + 9 + 2a + 6 + 5$
$\qquad\qquad = a^2 + 8a + 20$

3. $\dfrac{6x^2 + 5x - 4}{2x^2 - 3x + 1} \cdot \dfrac{4x^2 - 1}{8x^3 + 1} = \dfrac{(3x+4)(2x-1)}{(x-1)(2x-1)} \cdot \dfrac{(2x+1)(2x-1)}{(2x+1)(4x^2 - 2x + 1)}$

$\qquad\qquad\qquad\qquad = \dfrac{(3x+4)}{(x-1)} \cdot \dfrac{(2x-1)}{(4x^2 - 2x + 1)}$

$\qquad\qquad\qquad\qquad = \dfrac{(3x+4)(2x-1)}{(x-1)(4x^2 - 2x + 1)}$

4. $\dfrac{x}{x+3} - \dfrac{x+1}{2x^2 - 2x - 24} = \dfrac{x}{x+3} - \dfrac{x+1}{2(x+3)(x-4)}$

$\qquad\qquad\qquad\qquad = \dfrac{2x(x-4)}{2(x+3)(x-4)} - \dfrac{x+1}{2(x+3)(x-4)}$

$\qquad\qquad\qquad\qquad = \dfrac{2x^2 - 8x - (x+1)}{2(x+3)(x-4)}$

$\qquad\qquad\qquad\qquad = \dfrac{2x^2 - 9x - 1}{2(x+3)(x-4)}$

5. $\qquad\qquad\qquad \dfrac{y+1}{y+3} + \dfrac{y-3}{y-2} = \dfrac{2y^2 - 15}{y^2 + y - 6}$

$(y+3)(y-2)\left(\dfrac{y+1}{y+3} + \dfrac{y-3}{y-2}\right) = \left(\dfrac{2y^2 - 15}{(y+3)(y-2)}\right)(y+3)(y-2)$

$(y-2)(y+1) + (y+3)(y-3) = 2y^2 - 15$

$\qquad\qquad y^2 - y - 2 + y^2 - 9 = 2y^2 - 15$

$\qquad\qquad\qquad\quad 2y^2 - y - 11 = 2y^2 - 15$

$\qquad\qquad\qquad\qquad\quad -y - 11 = -15$

$\qquad\qquad\qquad\qquad\qquad\quad -y = -4$

$\qquad\qquad\qquad\qquad\qquad\quad\; y = 4$

6. Assume all variables represent positive numbers.

$$\sqrt{\frac{12x^5y^3}{8z}} = \sqrt{\frac{6x^5y^3z}{4z^2}}$$

$$= \frac{\sqrt{6x^5y^3z}}{\sqrt{4z^2}}$$

$$= \frac{\sqrt{6x^5y^3z}}{2z}$$

$$= \frac{\sqrt{x^4y^2}\sqrt{6xyz}}{2z}$$

$$= \frac{x^2y\sqrt{6xyz}}{2z}$$

7. $\dfrac{6}{\sqrt{3}-\sqrt{5}} = \dfrac{6}{\sqrt{3}-\sqrt{5}} \cdot \dfrac{\sqrt{3}+\sqrt{5}}{\sqrt{3}+\sqrt{5}}$

$$= \frac{6\left(\sqrt{3}+\sqrt{5}\right)}{3-5}$$

$$= \frac{6\left(\sqrt{3}+\sqrt{5}\right)}{-2}$$

$$= -3\left(\sqrt{3}+\sqrt{5}\right)$$

8. $3\sqrt[3]{2x+2} = \sqrt[3]{80x-24}$

$\left(3\sqrt[3]{2x+2}\right)^3 = \left(\sqrt[3]{80x-24}\right)^3$

$27(2x+2) = (80x-24)$

$54x+54 = 80x-24$

$54x = 80x-78$

$-26x = -78$

$x = 3$

Check: $3\sqrt[3]{2(3)+2} \overset{?}{=} \sqrt[3]{80(3)-24}$

$3\sqrt[3]{8} \overset{?}{=} \sqrt[3]{216}$

$3 \cdot 2 = 6$ True

The solution is 3.

9. $(5-4i)(5+4i) = 5^2 - (4i)^2$

$$= 25 - (-16)$$

$$= 41$$

10. $(x-3)^2 = 28$

$x-3 = \pm\sqrt{28}$

$x-3 = \pm\sqrt{4 \cdot 7}$

$x-3 = \pm 2\sqrt{7}$

$x = 3 \pm 2\sqrt{7}$

11. $3x^2 - 4x - 8 = 0$

$$x = \frac{-b \pm \sqrt{b^2 - 4ac}}{2a}$$

$$= \frac{-(-4) \pm \sqrt{(-4)^2 - 4(3)(-8)}}{2(3)}$$

$$= \frac{4 \pm \sqrt{16 + 96}}{6}$$

$$= \frac{4 \pm \sqrt{112}}{6}$$

$$= \frac{4 \pm 4\sqrt{7}}{6}$$

$$= \frac{2 \pm 2\sqrt{7}}{3}$$

12. $3p^{2/3} + 14p^{1/3} - 24 = 0$

Let $x = p^{1/3}$.

$3x^2 + 14x - 24 = 0$

$(3x-4)(x+6) = 0$

$x = \dfrac{4}{3}$ or $x = -6$

$p^{1/3} = \dfrac{4}{3}$ $p^{1/3} = -6$

$p = \dfrac{64}{27}$ $p = -216$

The solutions are $\dfrac{64}{27}$ and -216.

13. $\dfrac{3x-2}{x+4} \geq 0$

$x \neq -4$

$x+4 = 0$ $3x-2 = 0$

$x = -4$ $x = \dfrac{2}{3}$

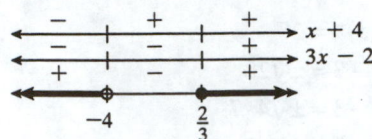

14. a. $f \circ g(x) = f(2x - 3)$

$\qquad = (2x - 3)^2 + 6(2x - 3)$

$\qquad = 4x^2 - 12x + 9 + 12x - 18$

$\qquad = 4x^2 - 9$

b. $g \circ f(x) = g(x^2 + 6x)$

$\qquad = 2(x^2 + 6x) - 3$

$\qquad = 2x^2 + 12x - 3$

15. $9x^2 + 4y^2 = 36$

$\dfrac{9x^2}{36} + \dfrac{4y^2}{36} = 1$

$\dfrac{x^2}{4} + \dfrac{y^2}{9} = 1$

Since $a^2 = 4$, $a = 2$.

Since $b^2 = 9$, $b = 3$

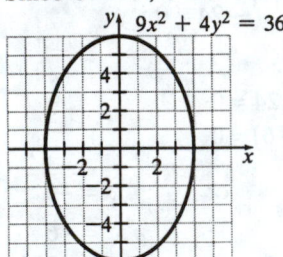

16. $\dfrac{y^2}{25} - \dfrac{x^2}{16} = 1$

Since $a^2 = 25$ and $b^2 = 16$, $a = 5$ and $b = 4$

The equations of the asymptotes are

$y = \pm\dfrac{b}{a}x$, or $y = \pm\dfrac{5}{4}x$. To graph the

asymptotes, plot the points (4, 5), (4, –5), (–4, 5), and (–4, –5).

The graph intersects the y-axis at (0, –5) and (0, 5).

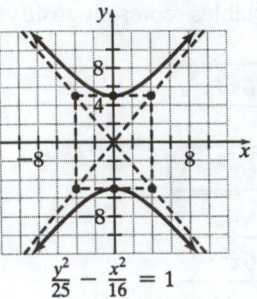

$\dfrac{y^2}{25} - \dfrac{x^2}{16} = 1$

17. $\log(3x - 4) + \log(4) = \log(x + 6)$

$\qquad \log 4(3x - 4) = \log(x + 6)$

$\qquad \log(12x - 16) = \log(x + 6)$

$\qquad 12x - 16 = x + 6$

$\qquad 12x = x + 22$

$\qquad 11x = 22$

$\qquad x = 2$

Check: $\log[3(2) - 4] + \log 4 \stackrel{?}{=} \log(2 + 6)$

$\qquad\qquad \log 2 + \log 4 \stackrel{?}{=} \log 8$

$\qquad\qquad \log(2 \cdot 4) = \log(8)$ True

The solution is 2.

18. $250 = 500e^{-0.3t}$

$\qquad \dfrac{1}{2} = e^{-0.3t}$

$\qquad \ln\dfrac{1}{2} = \ln e^{-0.3t}$

$\qquad \ln\dfrac{1}{2} = -0.3t$

$\qquad -\dfrac{1}{0.3}\ln\dfrac{1}{2} = t$

$\qquad t = -\dfrac{1}{0.3}\ln\dfrac{1}{2} \approx 2.31$

19. Let x be the original cost of the suit.

$\qquad x - 0.2x - 25 = 155$

$\qquad 0.8x - 25 = 155$

$\qquad 0.8x = 180$

$\qquad x = \dfrac{180}{0.8}$

$\qquad x = 225$

The original cost was \$225.

20. Let x be the number of pounds of cashews and let y be the number of pounds of peanuts.

$$x + y = 4$$
$$7x + 5y = 25$$

Solve $x + y = 4$ for y: $y = 4 - x$.
Substitute $4 - x$ for y in the second equation.

$$7x + 5(4 - x) = 25$$
$$2x + 20 = 25$$
$$2x = 5$$
$$x = 2.5$$

Then $y = 4 - x = 4 - 2.5 = 1.5$

The mixture should contain $2\frac{1}{2}$ pounds of cashews and $1\frac{1}{2}$ pounds of peanuts.

Chapter 11

Exercise Set 11.1

1. A sequence is a list of numbers arranged in a specific order.

3. A finite sequence is a function whose domain includes only the first n natural numbers.

5. In a decreasing sequence, the terms decrease.

7. A series is the sum of the terms of a sequence.

9. $\displaystyle\sum_{n=1}^{5}(n+2)$

   The sum as n goes from 1 to 5 of $n+2$.

11. $a_n = 3n$
    $a_1 = 3(1) = 3$
    $a_2 = 3(2) = 6$
    $a_3 = 3(3) = 9$
    $a_4 = 3(4) = 12$
    $a_5 = 3(5) = 15$
    The terms are 3, 6, 9, 12, 15.

13. $a_n = \dfrac{n+4}{n}$
    $a_1 = \dfrac{1+4}{1} = 5$
    $a_2 = \dfrac{2+4}{2} = 3$
    $a_3 = \dfrac{3+4}{3} = \dfrac{7}{3}$
    $a_4 = \dfrac{4+4}{4} = 2$
    $a_5 = \dfrac{5+4}{5} = \dfrac{9}{5}$
    The terms are $5, 3, \dfrac{7}{3}, 2, \dfrac{9}{5}$.

15. $a_n = \dfrac{3}{n^2}$
    $a_1 = \dfrac{3}{1^2} = 3$
    $a_2 = \dfrac{3}{2^2} = \dfrac{3}{4}$
    $a_3 = \dfrac{3}{3^2} = \dfrac{3}{9} = \dfrac{1}{3}$
    $a_4 = \dfrac{3}{4^2} = \dfrac{3}{16}$
    $a_5 = \dfrac{3}{5^2} = \dfrac{3}{25}$
    The terms are $3, \dfrac{3}{4}, \dfrac{1}{3}, \dfrac{3}{16}, \dfrac{3}{25}$.

17. $a_n = \dfrac{n+2}{n+1}$
    $a_1 = \dfrac{1+2}{1+1} = \dfrac{3}{2}$
    $a_2 = \dfrac{2+2}{2+1} = \dfrac{4}{3}$
    $a_3 = \dfrac{3+2}{3+1} = \dfrac{5}{4}$
    $a_4 = \dfrac{4+2}{4+1} = \dfrac{6}{5}$
    $a_5 = \dfrac{5+2}{5+1} = \dfrac{7}{6}$
    The terms are $\dfrac{3}{2}, \dfrac{4}{3}, \dfrac{5}{4}, \dfrac{6}{5}, \dfrac{7}{6}$.

19. $a_n = (-1)^n$
    $a_1 = (-1)^1 = -1$
    $a_2 = (-1)^2 = 1$
    $a_3 = (-1)^3 = -1$
    $a_4 = (-1)^4 = 1$
    $a_5 = (-1)^5 = -1$
    The terms are $-1, 1, -1, 1, -1$.

21. $a_n = (-2)^{n+1}$

$a_1 = (-2)^{1+1} = (-2)^2 = 4$

$a_2 = (-2)^{2+1} = (-2)^3 = -8$

$a_3 = (-2)^{3+1} = (-2)^4 = 16$

$a_4 = (-2)^{4+1} = (-2)^5 = -32$

$a_5 = (-2)^{5+1} = (-2)^6 = 64$

The terms are 4, –8, 16, –32, 64.

23. $a_n = 2n + 7$

$\begin{aligned} a_{12} &= 2(12) + 7 \\ &= 24 + 7 \\ &= 31 \end{aligned}$

25. $a_n = 2n - 4$

$\begin{aligned} a_5 &= 2(5) - 4 \\ &= 10 - 4 \\ &= 6 \end{aligned}$

27. $a_n = (-2)^n$

$\begin{aligned} a_4 &= (-2)^4 \\ &= 16 \end{aligned}$

29. $a_n = \dfrac{n^2}{2n+1}$

$\begin{aligned} a_9 &= \dfrac{9^2}{2(9)+1} \\ &= \dfrac{81}{18+1} \\ &= \dfrac{81}{19} \end{aligned}$

31. $a_n = 2n + 3$

$a_1 = 2(1) + 3 = 2 + 3 = 5$

$a_2 = 2(2) + 3 = 4 + 3 = 7$

$a_3 = 2(3) + 3 = 6 + 3 = 9$

$s_1 = a_1 = 5$

$\begin{aligned} s_3 &= a_1 + a_2 + a_3 \\ &= 5 + 7 + 9 \\ &= 21 \end{aligned}$

33. $a_n = 2^n + 1$

$a_1 = 2^1 + 1 = 3$

$a_2 = 2^2 + 1 = 4 + 1 = 5$

$a_3 = 2^3 + 1 = 8 + 1 = 9$

$s_1 = a_1 = 3$

$\begin{aligned} s_3 &= a_1 + a_2 + a_3 \\ &= 3 + 5 + 9 \\ &= 17 \end{aligned}$

35. $a_n = (-1)^{2n}$

$a_1 = (-1)^{2(1)} = (-1)^2 = 1$

$a_2 = (-1)^{2(2)} = (-1)^4 = 1$

$a_3 = (-1)^{2(3)} = (-1)^6 = 1$

$s_1 = a_1 = 1$

$\begin{aligned} s_3 &= a_1 + a_2 + a_3 \\ &= 1 + 1 + 1 \\ &= 3 \end{aligned}$

37. $a_n = \dfrac{n^2}{2}$

$a_1 = \dfrac{1^2}{2} = \dfrac{1}{2}$

$a_2 = \dfrac{2^2}{2} = \dfrac{4}{2} = 2$

$a_3 = \dfrac{3^2}{2} = \dfrac{9}{2}$

$s_1 = a_1 = \dfrac{1}{2}$

$\begin{aligned} s_3 &= a_1 + a_2 + a_3 \\ &= \dfrac{1}{2} + \dfrac{4}{2} + \dfrac{9}{2} \\ &= \dfrac{14}{2} \\ &= 7 \end{aligned}$

39. Each term is twice the preceding term. The next three terms are 64, 128, 256.

41. Each term is two more than the preceding term. The next three terms are 15, 17, 19.

43. Each denominator is one more than the preceding one while each numerator is one. The next three terms are $\dfrac{1}{6}, \dfrac{1}{7}, \dfrac{1}{8}$.

45. Each term is –1 times the previous term. The next three terms are 1, –1, 1.

47. Each denominator is three times the previous one while each numerator is one. The next three terms are $\dfrac{1}{81}, \dfrac{1}{243}, \dfrac{1}{729}$.

49. Each term is $-\dfrac{1}{2}$ times the preceding term. The next three terms are $\dfrac{1}{16}, -\dfrac{1}{32}, \dfrac{1}{64}$.

51. Each term is eight less than the previous term. The next three terms are –25, –33, –41.

53. $\displaystyle\sum_{n=1}^{5}(3n-1) = \big[3(1)-1\big]+\big[3(2)-1\big]+\big[3(3)-1\big]+\big[3(4)-1\big]+\big[3(5)-1\big]$
$$= 2+5+8+11+14$$
$$= 40$$

55. $\displaystyle\sum_{k=1}^{6}\left(2k^2-3\right) = \big[2(1)^2-3\big]+\big[2(2)^2-3\big]+\big[2(3)^2-3\big]+\big[2(4)^2-3\big]+\big[2(5)^2-3\big]+\big[2(6)^2-3\big]$
$$= -1+5+15+29+47+69$$
$$= 164$$

57. $\displaystyle\sum_{n=2}^{4}\dfrac{n^2+n}{n+1} = \dfrac{2^2+2}{2+1}+\dfrac{3^2+3}{3+1}+\dfrac{4^2+4}{4+1}$
$$= \dfrac{6}{3}+\dfrac{12}{4}+\dfrac{20}{5}$$
$$= 2+3+4$$
$$= 9$$

59. $a_n = n+3$

The fifth partial sum is $\displaystyle\sum_{n=1}^{5}(n+3)$.

61. $a_n = \dfrac{n^2}{4}$

The third partial sum is $\displaystyle\sum_{n=1}^{3}\dfrac{n^2}{4}$.

63. $\displaystyle\sum_{i=1}^{5}x_i = x_1+x_2+x_3+x_4+x_5$
$$= 2+3+5+(-1)+4$$
$$= 13$$

65. $\displaystyle\left(\sum_{i=1}^{5}x_i\right)^2 = \left(x_1+x_2+x_3+x_4+x_5\right)^2$
$$= \big(2+3+5+(-1)+4\big)^2$$
$$= 13^2$$
$$= 169$$

67. $\bar{x} = \dfrac{15+20+25+30+35}{5} = \dfrac{125}{5} = 25$

69. $\bar{x} = \dfrac{72 + 83 + 4 + 60 + 18 + 20}{6}$

$= \dfrac{257}{6}$

≈ 42.83

71. Answers will vary.

73. Answers will vary.

75. $\bar{x} = \dfrac{\sum x}{n}$

$n\bar{x} = n \cdot \dfrac{\sum x}{n}$

$n\bar{x} = \sum x$ or

$\sum x = n\bar{x}$

77. Yes; $\displaystyle\sum_{i=1}^{n} 2x_i = 2\sum_{i=1}^{n} x_i$; Examples will vary.

79. a. $\displaystyle\sum x = x_1 + x_2 + x_3$

$= 3 + 5 + 2$

$= 10$

b. $\displaystyle\sum y = y_1 + y_2 + y_3$

$= 4 + 1 + 6$

$= 11$

c. $\displaystyle\sum x \cdot \sum y = 10 \cdot 11$

$= 110$

d. $\displaystyle\sum xy = x_1 y_1 + x_2 y_2 + x_3 y_3$

$= 3(4) + 5(1) + 2(6)$

$= 12 + 5 + 12$

$= 29$

e. No, $\displaystyle\sum xy \ne \sum x \cdot \sum y$

86. $2x^2 + 15 = 13x$

$2x^2 - 13x + 15 = 0$

$2x^2 - 10x - 3x + 15 = 0$

$2x(x - 5) - 3(x - 5) = 0$

$(2x - 3)(x - 5) = 0$

$2x - 3 = 0 \quad$ or $\quad x - 5 = 0$

$2x = 3 \qquad\qquad x = 5$

$x = \dfrac{3}{2}$

The solutions are $\dfrac{3}{2}$ and 5.

87. $6x^2 - 3x - 4 = 2$

$6x^2 - 3x - 6 = 0 \qquad a = 6,\, b = -3,\, c = -6$

This quadratic has two real solutions since the discriminant, $b^2 - 4ac$, is

$(-3)^2 - 4(6)(-6) = 153$ which is greater than zero.

88. $\dfrac{x^2}{4} + \dfrac{y^2}{1} = 1$ can be written as

$\dfrac{x^2}{2^2} + \dfrac{y^2}{1^2} = 2$. The graph is an ellipse

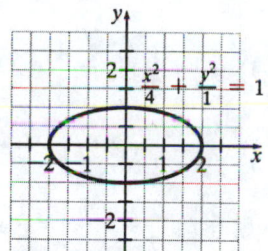

89. $x^2 + y^2 = 5$

$x = 2y$

Substitute $2y$ for x into the first equation.

$x^2 + y^2 = 5$

$(2y)^2 + y^2 = 5$

$4y^2 + y^2 = 5$

$5y^2 = 5$

$y^2 = 1$

$y = \pm 1$

If $y = 1$ then $x = 2(1) = 2$. If $y = -1$, then $x = 2(-1) = -2$. The solutions are $(2, 1)$ and $(-2, -1)$.

Exercise Set 11.2

1. In an arithmetic sequence, each term differs by a constant amount.

3. It is called the common difference.

5. $a_1 = 4$
   $a_2 = 4 + (2-1)(3) = 4 + 3 = 7$
   $a_3 = 4 + (3-1)(3) = 4 + 2(3) = 4 + 6 = 10$
   $a_4 = 4 + (4-1)(3) = 4 + 3(3) = 4 + 9 = 13$
   $a_5 = 4 + (5-1)(3) = 4 + 4(3) = 4 + 12 = 16$
   The terms are 4, 7, 10, 13, 16.
   The general term is $a_n = 4 + (n-1)3$ or $a_n = 3n + 1$.

7. $a_1 = -5$
   $a_2 = -5 + (2-1)(2) = -5 + 2 = -3$
   $a_3 = -5 + (3-1)(2) = -5 + 2(2) = -5 + 4 = -1$
   $a_4 = -5 + (4-1)(2) = -5 + 3(2) = -5 + 6 = 1$
   $a_5 = -5 + (5-1)(2) = -5 + 4(2) = -5 + 8 = 3$
   The terms are $-5, -3, -1, 1, 3$. The general term is $a_n = -5 + (n-1)2$ or $a_n = 2n - 7$.

9. $a_1 = \dfrac{1}{2}$
   $a_2 = \dfrac{1}{2} + (2-1)\left(\dfrac{3}{2}\right) = \dfrac{1}{2} + \dfrac{3}{2} = \dfrac{4}{2} = 2$
   $a_3 = \dfrac{1}{2} + (3-1)\left(\dfrac{3}{2}\right) = \dfrac{1}{2} + 2\left(\dfrac{3}{2}\right) = \dfrac{1}{2} + \dfrac{6}{2} = \dfrac{7}{2}$
   $a_4 = \dfrac{1}{2} + (4-1)\left(\dfrac{3}{2}\right) = \dfrac{1}{2} + 3\left(\dfrac{3}{2}\right) = \dfrac{1}{2} + \dfrac{9}{2} = \dfrac{10}{2} = 5$
   $a_5 = \dfrac{1}{2} + (5-1)\left(\dfrac{3}{2}\right) = \dfrac{1}{2} + 4\left(\dfrac{3}{2}\right) = \dfrac{1}{2} + \dfrac{12}{2} = \dfrac{13}{2}$
   The terms are $\dfrac{1}{2}, 2, \dfrac{7}{2}, 5, \dfrac{13}{2}$. The general term is $a_n = \dfrac{1}{2} + (n-1)\dfrac{3}{2}$ or $a_n = \dfrac{3}{2}n - 1$.

11. $a_1 = 100$
    $a_2 = 100 + (2-1)(-5) = 100 + (-5) = 100 - 5 = 95$
    $a_3 = 100 + (3-1)(-5) = 100 + 2(-5) = 100 - 10 = 90$
    $a_4 = 100 + (4-1)(-5) = 100 + 3(-5) = 100 - 15 = 85$
    $a_5 = 100 + (5-1)(-5) = 100 + 4(-5) = 100 - 20 = 80$
    The terms are 100, 95, 90, 85, 80. The general term is $a_n = 100 + (n-1)(-5)$ or $a_n = -5n + 105$.

13. $a_4 = a_1 + (4-1)d$
$a_4 = 5 + 3(3) = 5 + 9 = 14$

15. $a_{18} = a_1 + (18-1)d$
$a_{18} = -6 + 17(-1) = -6 - 17 = -23$

17. $a_{10} = a_1 + (10-1)d$
$a_{10} = -2 + 9\left(\dfrac{5}{3}\right) = -\dfrac{6}{3} + \dfrac{45}{3} = \dfrac{39}{3} = 13$

19. $a_9 = a_1 + (9-1)d$
$19 = 3 + 8d$
$16 = 8d$
$\dfrac{16}{8} = d$ or $d = 2$

21. $a_n = a_1 + (n-1)d$
$28 = 4 + (n-1)(3)$
$28 = 4 + 3n - 3$
$28 = 1 + 3n$
$27 = 3n$
$\dfrac{27}{3} = n$ or $n = 9$

23. $a_n = a_1 + (n-1)d$
$-\dfrac{17}{3} = -\dfrac{7}{3} + (n-1)\left(-\dfrac{2}{3}\right)$
$-17 = -7 + (n-1)(-2)$
$-17 = -7 - 2n + 2$
$-17 = -5 - 2n$
$-12 = -2n$
$\dfrac{-12}{-2} = n$ or $n = 6$

25. $s_{10} = \dfrac{10(a_1 + a_{10})}{2}$
$\quad = \dfrac{10(1+19)}{2}$
$\quad = 5(20)$
$\quad = 100$
$a_{10} = a_1 + (10-1)d$
$a_{10} = a_1 + 9d$
$19 = 1 + 9d$
$18 = 9d$
$\dfrac{18}{9} = d$ or $d = 2$

27. $s_8 = \dfrac{8(a_1 + a_8)}{2}$
$\quad = \dfrac{8\left(\frac{3}{5}+2\right)}{2}$
$\quad = 4\left(\dfrac{3}{5} + 2\right)$
$\quad = 4\left(\dfrac{3}{5} + \dfrac{10}{5}\right)$
$\quad = 4\left(\dfrac{13}{5}\right)$
$\quad = \dfrac{52}{5}$

$a_8 = a_1 + (8-1)d$
$a_8 = a_1 + 7d$
$2 = \dfrac{3}{5} + 7d$
$\dfrac{7}{5} = 7d$
$\dfrac{1}{5} = d$ or $d = \dfrac{1}{5}$

29. $s_5 = \dfrac{5(a_1 + a_5)}{2}$

$= \dfrac{5\left(\frac{12}{5} + \frac{28}{5}\right)}{2}$

$= \dfrac{5\left(\frac{40}{5}\right)}{2}$

$= \dfrac{40}{2}$

$= 20$

$a_5 = a_1 + (5-1)d$

$a_5 = a_1 + 4d$

$\dfrac{28}{5} = \dfrac{12}{5} + 4d$

$\dfrac{16}{5} = 4d$

$\dfrac{4}{5} = d$ or $d = \dfrac{4}{5}$

31. $s_{11} = \dfrac{11(a_1 + a_{11})}{2}$

$= \dfrac{11(7 + 67)}{2}$

$= \dfrac{11(74)}{2}$

$= 407$

$a_{11} = a_1(11-1)d$

$a_{11} = a_1 + 10d$

$67 = 7 + 10d$

$60 = 10d$

$\dfrac{60}{10} = d$ or $d = 6$

33. $a_1 = 5$

$a_2 = 5 + (2-1)(3) = 5 + 3 = 8$

$a_3 = 5 + (3-1)(3) = 5 + 2(3) = 5 + 6 = 11$

$a_4 = 5 + (4-1)(3) = 5 + 3(3) = 5 + 9 = 14$

The terms are 5, 8, 11, 14.

$a_{10} = 5 + (10-1)(3) = 5 + 9(3) = 5 + 27 = 32$

$s_{10} = \dfrac{10(5+32)}{2} = \dfrac{10(37)}{2} = 185$

35. $a_1 = -8$

$a_2 = -8 + (2-1)(-5)$

$\quad = -8 - 5$

$\quad = -13$

$a_3 = -8 + (3-1)(-5)$

$\quad = -8 + 2(-5)$

$\quad = -8 - 10$

$\quad = -18$

$a_4 = -8 + (4-1)(-5)$

$\quad = -8 + 3(-5)$

$\quad = -8 - 15$

$\quad = -23$

The terms are $-8, -13, -18, -23$.

$a_{10} = -8 + (10-1)(-5)$

$\quad = -8 + 9(-5)$

$\quad = -8 - 45$

$\quad = -53$

$s_{10} = \dfrac{10[-8 + (-53)]}{2} = 5(-61) = -305$

37. $a_1 = 100$

$a_2 = 100 + (2-1)(-7)$

$\quad = 100 - 7$

$\quad = 93$

$a_3 = 100 + (3-1)(-7)$

$\quad = 100 + 2(-7)$

$\quad = 100 - 14$

$\quad = 86$

$a_4 = 100 + (4-1)(-7)$

$\quad = 100 + 3(-7)$

$\quad = 100 - 21$

$\quad = 79$

The terms are 100, 93, 86, 79.

$a_{10} = 100 + (10-1)(-7)$

$\quad = 100 + 9(-7)$

$\quad = 100 - 63$

$\quad = 37$

$s_{10} = \dfrac{10(100 + 37)}{2} = 5(137) = 685$

39. $a_1 = \frac{9}{5}$

$a_2 = \frac{9}{5} + (2-1)\left(\frac{3}{5}\right) = \frac{9}{5} + \frac{3}{5} = \frac{12}{5}$

$a_3 = \frac{9}{5} + (3-1)\left(\frac{3}{5}\right) = \frac{9}{5} + 2\left(\frac{3}{5}\right) = \frac{9}{5} + \frac{6}{5} = \frac{15}{5} = 3$ The terms are $\frac{9}{5}, \frac{12}{5}, 3, \frac{18}{5}$.

$a_4 = \frac{9}{5} + (4-1)\left(\frac{3}{5}\right) = \frac{9}{5} + 3\left(\frac{3}{5}\right) = \frac{9}{5} + \frac{9}{5} = \frac{18}{5}$

$a_{10} = \frac{9}{5} + (10-1)\left(\frac{3}{5}\right) = \frac{9}{5} + 9\left(\frac{3}{5}\right) = \frac{9}{5} + \frac{27}{5} = \frac{36}{5}$

$s_{10} = \frac{10\left(\frac{9}{5} + \frac{36}{5}\right)}{2} = 5\left(\frac{45}{5}\right) = 45$

41. $d = 4 - 1 = 3$

$a_n = a_1 + (n-1)(d)$

$\quad = 1 + (n-1)(3)$

$\quad = 1 + 3n - 3$

$\quad = -2 + 3n$

$43 = -2 + 3n$

$45 = 3n$

$\frac{45}{3} = n$ or $n = 15$

$s_{15} = \frac{15(a_1 + a_{15})}{2}$

$\quad = \frac{15(1 + 43)}{2}$

$\quad = \frac{15(44)}{2}$

$\quad = 330$

43. $d = -5 - (-9) = -5 + 9 = 4$

$a_n = a_1 + (n-1)d$

$\quad = -9 + (n-1)(4)$

$\quad = -9 + 4n - 4$

$\quad = -13 + 4n$

$27 = -13 + 4n$

$40 = 4n$

$\frac{40}{4} = n$ or $n = 10$

$s_{10} = \frac{10(a_1 + a_{10})}{2}$

$\quad = \frac{10(-9 + 27)}{2}$

$\quad = 5(18)$

$\quad = 90$

45. $d = -\frac{7}{6} - \left(-\frac{5}{6}\right) = -\frac{7}{6} + \frac{5}{6} = -\frac{2}{6}$

$a_n = a_1 + (n-1)d$

$\quad = -\frac{5}{6} + (n-1)\left(-\frac{2}{6}\right)$

$\quad = -\frac{5}{6} - \frac{2}{6}n + \frac{2}{6}$

$\quad = -\frac{3}{6} - \frac{2}{6}n$

$-\frac{21}{6} = -\frac{3}{6} - \frac{2}{6}n$

$-\frac{18}{6} = -\frac{2}{6}n$

$-18 = -2n$

$\frac{-18}{-2} = n$ or $n = 9$

$$s_9 = \frac{9(a_1 + a_9)}{2} = \frac{9\left[-\frac{5}{6} + \left(-\frac{21}{6}\right)\right]}{2}$$

$$= \frac{9\left(-\frac{26}{6}\right)}{2} = 9\left(-\frac{26}{6}\right)\left(\frac{1}{2}\right) = -\frac{39}{2}$$

47. $d = -16 - (-12) = -16 + 12 = -4$

$a_n = a_1 + (n - 1)$

$d = -12 + (n - 1)(-4)$

$= -12 - 4n + 4$

$= -8 - 4n$

$-52 = -8 - 4n$

$-44 = -4n$

$\dfrac{-44}{-4} = n$ or $n = 11$

$s_{11} = \dfrac{11(a_1 + a_{11})}{2}$

$= \dfrac{11[-12 + (-52)]}{2}$

$= \dfrac{11(-64)}{2}$

$= -352$

49. $s_n = \dfrac{n(a_1 + a_n)}{2}$

$s_{50} = \dfrac{50(1 + 50)}{2}$

$s_{50} = 1275$

51. $s_n = \dfrac{n(a_1 + a_n)}{2}$

$s_{50} = \dfrac{50(1 + 99)}{2}$

$s_{50} = 2500$

53. $s_n = \dfrac{n(a_1 + a_n)}{2}$

$s_{20} = \dfrac{20(3 + 60)}{2}$

$s_{20} = 630$

55. The smallest number greater than 7 that is divisible by 6 is 12. The largest number less than 1610 that is divisible by 6 is 1608. Now find n in the equation $a_n = a_1 + (n - 1)d$.

$1608 = 12 + (n - 1)6$

$1596 = 6(n - 1)$

$266 = n - 1$

$267 = n$

There are 267 numbers between 7 and 1610 that are divisible by 6.

57. $1 + 2 + 3 + \cdots + 100$

$= (1 + 100) + (2 + 99) + \cdots + (50 + 51)$

$= 101 + 101 + \cdots + 101$

$= 50(101)$

$= 5050$

59. a. $a_1 = 22, d = -\dfrac{1}{2}, n = 7$

$a_n = a_1 + (n - 1)d$

$a_7 = 22 + (7 - 1)\left(-\dfrac{1}{2}\right) = 22 - 3 = 19$

Her seventh swing is 19 feet.

b. $s_n = \dfrac{n(a_1 + a_n)}{2}$

$s_7 = \dfrac{7(22 + 19)}{2} = 143.5$

She travels 143.5 feet during the seven swings.

61. $d = -6$ in. $= -\dfrac{1}{2}$ ft, $a_1 = 6$

$a_n = a_1 + (n - 1)d$

$a_{11} = 6 + (11 - 1)\left(-\dfrac{1}{2}\right) = 6 - 5 = 1$

The ball bounces 1 foot on the eleventh bounce.

63. $20 + 19 + 18 + \cdots + 1$ or $1 + 2 + 3 + \cdots + 20$

$$s_n = \frac{n(a_1 + a_n)}{2}$$

$$s_{20} = \frac{20(1 + 20)}{2} = 10(21) = 210$$

There are 210 logs in the pile.

65. $s_n = \frac{n(a_1 + a_n)}{2}$

$$s_{31} = \frac{31(1 + 31)}{2} = \frac{31(32)}{2} = 496$$

On day 31, Craig will have saved $496.

67. a. $a_{10} = 32,000 + (10 - 1)(400) = 35,600$

She will receive $35,600 in her tenth year of retirement.

b. $s_{10} = \frac{10(32,000 + 35,600)}{2}$

$$= 5(67,600)$$

$$= 338,000$$

In her first 10 years, she will receive a total of $338,000.

69. $360 - 180 = 180$
$540 - 360 = 180$
$720 - 540 = 180$

The terms form an arithmetic sequence with $d = 180$ and $a_3 = 180$.

$$a_n = a_1 + (n - 1)d$$
$$180 = a_1 + (3 - 1)180$$
$$180 = a_1 + 360$$
$$-180 = a_1$$
$$a_n = a_1 + (n - 1)d$$
$$a_n = -180 + (n - 1)(180)$$
$$= -180 + 180n - 180$$
$$= 180n - 360$$
$$= 180(n - 2)$$

77. The system will have only one solution since the slopes of the two lines are different and they must intersect.

78. $2x + 3y = -4$
$-x - y = -1$

To eliminate x, multiply the second equation by 2 and then add.

$2x + 3y = -4$

$2[-x - y = -1]$

gives

$$\begin{aligned} 2x + 3y &= -4 \\ -2x - 2y &= -2 \end{aligned}$$

Add: $y = -6$

Substitute -6 for y in the first equation.

$$2x + 3y = -4$$
$$2x + 3(-6) = -4$$
$$2x - 18 = -4$$
$$2x = 14$$
$$x = 7$$

The solution is $(7, -6)$.

79. $|x - 2| < 4$

$$-4 < x - 2 < 4$$
$$-2 < x < 6$$

The solution is the region between the dashed lines $x = -2$ and $x = 6$.

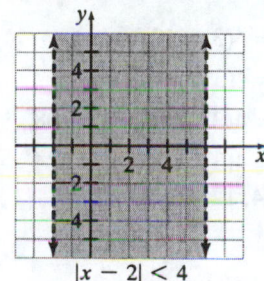

$|x - 2| < 4$

80. $\dfrac{(x + 2)^2}{4} + \dfrac{(y - 3)^2}{9} = 1$

This is an ellipse with center at $(-2, 3)$.

$a = \sqrt{4} = 2$ and $b = \sqrt{9} = 3$

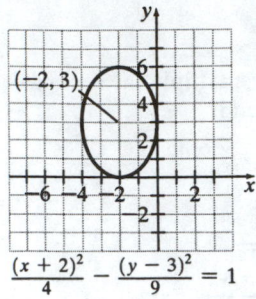

$$\frac{(x+2)^2}{4} - \frac{(y-3)^2}{9} = 1$$

Exercise Set 11.3

1. A geometric sequence is a sequence in which each term after the first is the same multiple of the preceding term.

3. To find the common ratio, take any term except the first and divide by the term that precedes it.

5. r^n approaches 0 as n gets larger and larger when $|r| < 1$.

7. $a_1 = 4$

   $a_2 = 4(3)^{2-1} = 4(3) = 12$

   $a_3 = 4(3)^{3-1} = 4(3)^2 = 4(9) = 36$

   $a_4 = 4(3)^{4-1} = 4(3)^3 = 4(27) = 108$

   $a_5 = 4(3)^{5-1} = 4(3)^4 = 4(81) = 324$

   The terms are 4, 12, 36, 108, 324.

9. $a_1 = 90$

   $a_2 = 90\left(\frac{1}{3}\right)^{2-1} = 90\left(\frac{1}{3}\right) = 30$

   $a_3 = 90\left(\frac{1}{3}\right)^{3-1} = 90\left(\frac{1}{3}\right)^2 = 90\left(\frac{1}{9}\right) = 10$

   $a_4 = 90\left(\frac{1}{3}\right)^{4-1} = 90\left(\frac{1}{3}\right)^3 = 90\left(\frac{1}{27}\right) = \frac{10}{3}$

   $a_5 = 90\left(\frac{1}{3}\right)^{5-1} = 90\left(\frac{1}{3}\right)^4 = 90\left(\frac{1}{81}\right) = \frac{10}{9}$

   The terms are 90, 30, 10, $\frac{10}{3}$, $\frac{10}{9}$.

11. $a_1 = -5$

    $a_2 = -5(-2)^{2-1} = -5(-2)^1 = -5(-2) = 10$

    $a_3 = -5(-2)^{3-1} = -5(-2)^2 = -5(4) = -20$

    $a_4 = -5(-2)^{4-1} = -5(-2)^3 = -5(-8) = 40$

    $a_5 = -5(-2)^{5-1} = -5(-2)^4 = -5(16) = -80$

    The terms are -5, 10, -20, 40, -80.

13. $a_1 = 3$

    $a_2 = 3\left(\frac{3}{2}\right)^{2-1} = 3\left(\frac{3}{2}\right) = \frac{9}{2}$

    $a_3 = 3\left(\frac{3}{2}\right)^{3-1} = 3\left(\frac{3}{2}\right)^2 = 3\left(\frac{9}{4}\right) = \frac{27}{4}$

    $a_4 = 3\left(\frac{3}{2}\right)^{4-1} = 3\left(\frac{3}{2}\right)^3 = 3\left(\frac{27}{8}\right) = \frac{81}{8}$

    $a_5 = 3\left(\frac{3}{2}\right)^{5-1} = 3\left(\frac{3}{2}\right)^4 = 3\left(\frac{81}{16}\right) = \frac{243}{16}$

    The terms are 3, $\frac{9}{2}$, $\frac{27}{4}$, $\frac{81}{8}$, $\frac{243}{16}$.

15. $a_6 = a_1 r^{6-1}$

    $a_6 = 4(2)^{6-1} = 4(2)^5 = 4(32) = 128$

17. $a_7 = a_1 r^{7-1}$

    $a_7 = 15(3)^{7-1} = 15(3)^6 = 15(729) = 10{,}935$

19. $a_8 = a_1 r^{8-1}$

    $a_8 = 2\left(\frac{1}{2}\right)^{8-1} = 2\left(\frac{1}{2}\right)^7 = 2\left(\frac{1}{128}\right) = \frac{1}{64}$

21. $a_{12} = a_1 r^{12-1}$

    $a_{12} = -3(-2)^{12-1}$

    $\quad = -3(-2)^{11}$

    $\quad = -3(-2048)$

    $\quad = 6144$

23. $s_6 = \dfrac{a_1(1-r^6)}{1-r}$

$s_6 = \dfrac{4(1-5^6)}{1-5}$

$= \dfrac{4(1-15,625)}{-4}$

$= \dfrac{4(-15,624)}{-4}$

$= \dfrac{-15,624}{-1}$

$= 15,624$

25. $s_7 = \dfrac{a_1(1-r^7)}{1-r}$

$s_7 = \dfrac{80(1-2^7)}{1-2}$

$= \dfrac{80(1-128)}{-1}$

$= \dfrac{80(-127)}{-1}$

$= \dfrac{-10,160}{-1}$

$= 10,160$

27. $s_9 = \dfrac{a_1(1-r^9)}{1-r}$

$s_9 = \dfrac{-30\left[1-\left(-\frac{1}{2}\right)^9\right]}{1-\left(-\frac{1}{2}\right)}$

$= \dfrac{-30\left[1-\left(-\frac{1}{512}\right)\right]}{\frac{3}{2}}$

$= \dfrac{-30\left(1+\frac{1}{512}\right)}{\frac{3}{2}}$

$= \dfrac{-30\left(\frac{513}{512}\right)}{\frac{3}{2}}$

$= -30\left(\dfrac{513}{512}\right)\left(\dfrac{2}{3}\right)$

$= -\dfrac{2565}{128}$

29. $s_5 = \dfrac{a_1(1-r^5)}{1-r}$

$s_5 = \dfrac{-9\left[1-\left(\frac{2}{5}\right)^5\right]}{1-\frac{2}{5}}$

$= \dfrac{-9\left(1-\frac{32}{3125}\right)}{\frac{3}{5}}$

$= \dfrac{-9\left(\frac{3093}{3125}\right)}{\frac{3}{5}}$

$= -9\left(\dfrac{3093}{3125}\right)\left(\dfrac{5}{3}\right)$

$= -\dfrac{9279}{625}$

31. $r = \dfrac{5}{2} \div 5 = \dfrac{5}{2} \cdot \dfrac{1}{5} = \dfrac{1}{2}$

$a_n = 5\left(\dfrac{1}{2}\right)^{n-1}$

33. $r = -6 \div 2 = -3$

$a_n = 2(-3)^{n-1}$

35. $r = -3 \div -1 = 3$

$a_n = -1(3)^{n-1}$

37. $r = 3 \div 6 = \dfrac{1}{2}$

$s_\infty = \dfrac{6}{1-\frac{1}{2}} = \dfrac{6}{\frac{1}{2}} = 6\left(\dfrac{2}{1}\right) = 12$

39. $r = 2 \div 5 = \dfrac{2}{5}$

$s_\infty = \dfrac{5}{1-\frac{2}{5}} = \dfrac{5}{\frac{3}{5}} = 5\left(\dfrac{5}{3}\right) = \dfrac{25}{3}$

41. $r = \dfrac{4}{15} \div \dfrac{1}{3} = \dfrac{4}{15}\left(\dfrac{3}{1}\right) = \dfrac{4}{5}$

$s_\infty = \dfrac{\frac{1}{3}}{1-\frac{4}{5}} = \dfrac{\frac{1}{3}}{\frac{1}{5}} = \dfrac{1}{3}\left(\dfrac{5}{1}\right) = \dfrac{5}{3}$

43. $r = -1 \div 9 = -\frac{1}{9}$

$$s_\infty = \frac{9}{1 - \left(-\frac{1}{9}\right)}$$

$$= \frac{9}{1 + \frac{1}{9}}$$

$$= \frac{9}{\frac{10}{9}}$$

$$= 9\left(\frac{9}{10}\right)$$

$$= \frac{81}{10}$$

45. $r = \frac{1}{2} \div 1 = \frac{1}{2}$

$$s_\infty = \frac{1}{1 - \frac{1}{2}} = \frac{1}{\frac{1}{2}} = 1\left(\frac{2}{1}\right) = 2$$

47. $r = \frac{16}{3} \div 8 = \frac{16}{3}\left(\frac{1}{8}\right) = \frac{2}{3}$

$$s_\infty = \frac{8}{1 - \frac{2}{3}} = \frac{8}{\frac{1}{3}} = 8\left(\frac{3}{1}\right) = 24$$

49. $r = 20 \div -60 = \frac{20}{-60} = -\frac{1}{3}$

$$s_\infty = \frac{-60}{1 - \left(-\frac{1}{3}\right)} = \frac{-60}{1 + \frac{1}{3}} = \frac{-60}{\frac{4}{3}} = -60\left(\frac{3}{4}\right) = -45$$

51. $r = -\frac{12}{5} \div -12 = -\frac{12}{5}\left(-\frac{1}{12}\right) = \frac{1}{5}$

$$s_\infty = \frac{-12}{1 - \frac{1}{5}} = \frac{-12}{\frac{4}{5}} = -12\left(\frac{5}{4}\right) = -15$$

53. $0.2727\cdots$
$= 0.27 + 0.0027 + 0.000027 + \cdots$
$= 0.27 + 0.27(0.01) + 0.27(0.01)^2 + \cdots$
$r = 0.01$ and $a_1 = 0.27$
$$s_\infty = \frac{0.27}{1 - 0.01} = \frac{0.27}{0.99} = \frac{27}{99} = \frac{3}{11}$$

55. $0.7777\ldots = 0.7 + 0.07 + 0.007 + \cdots$
$= 0.7 + 0.7(0.1) + 0.7(0.1)^2 + \cdots$
$r = 0.1$ and $a_1 = 0.7$
$$s_\infty = \frac{0.7}{1 - 0.1} = \frac{0.7}{0.9} = \frac{7}{9}$$

57. $0.515151\cdots$
$= 0.51 + 0.0051 + 0.000051 + \cdots$
$= 0.51 + 0.51(0.01) + 0.51(0.01)^2 + \cdots$
$r = 0.01$ and $a_1 = 0.51$
$$s_\infty = \frac{0.51}{1 - 0.01} = \frac{0.51}{0.99} = \frac{51}{99} = \frac{17}{33}$$

59. Consider a new series b_1, b_2, b_3, ... where $b_1 = 28$ and $b_3 = 112$. Now $b_3 = b_1 r^{3-1}$ becomes
$112 = 28r^2$
$\frac{112}{28} = r^2$
$4 = r^2$
so that $r = 2$ or $r = -2$. From the original series
$$a_1 = \frac{a_3}{r^2} = \frac{28}{4} = 7.$$

61. Consider a new series b_1, b_2, b_3, ... where $b_1 = 15$ and $b_4 = 405$. Now $b_4 = b_1 r^{4-1}$ becomes
$405 = 15r^3$
$\frac{405}{15} = r^3$
$27 = r^3$
$\sqrt[3]{27} = r$
$3 = r$
From the original series, $a_1 = \frac{a_2}{r} = \frac{15}{3} = 5.$

63. $a_1 = 1.40$, $n = 8$, $r = 1.03$
$a_n = a_1 r^{n-1}$
$a_8 = 1.4(1.03)^{8-1} \approx 1.72$
In 8 years, a loaf of bread would cost \$1.72.

65. $r = \frac{1}{2}$. Let a_n be the amount left after the nth day. After 1 day there are $300\left(\frac{1}{2}\right) = 150$ grams left, so $a_1 = 150$.

a. $37.5 = 150\left(\dfrac{1}{2}\right)^{n-1}$

$\left(\dfrac{1}{2}\right)^{n-1} = \dfrac{37.5}{150} = \dfrac{1}{4} = \left(\dfrac{1}{2}\right)^{2}$

$n - 1 = 2$

$n = 3$

37.5 grams are left after 3 days.

b. $a_8 = 150\left(\dfrac{1}{2}\right)^{8-1} = 150\left(\dfrac{1}{128}\right) \approx 1.172$

After 8 days, about 1.172 grams of the substance remain.

67. Each year the population is 1.024 times the population in the previous year, so $r = 1.024$. Let a_n be the population after the nth year. After the first year, the population is

$a_1 = 271(1.024) = 277.504$ million

a. $a_{12} = a_1 r^{12-1}$

$= 277.504(1.024)^{11}$

$= 360.22$ million

After 12 years, the population is about 360.22 million people.

b. $a_n = 2(271) = 542$

Now, use $a_n = a_1 r^{n-1}$

$542 = 277.504(1.024)^{n-1}$

$\dfrac{542}{277.504} = (1.024)^{n-1}$

Now, use logarithms.

$\log \dfrac{542}{277.504} = \log(1.024)^{n-1}$

$\log \dfrac{542}{277.504} = (n-1)\log 1.024$

$\dfrac{\log \frac{542}{277.504}}{\log 1.024} = n - 1$

$28.23 \approx n - 1$

$29.23 \approx n$

The population will double after about 29.23 years.

69. $r = \dfrac{1}{2}$

a. After 1 meter there is $\dfrac{1}{2}$ of the original light remaining, so $a_1 = \dfrac{1}{2}$.

$a_1 = \dfrac{1}{2}$

$a_2 = \dfrac{1}{2}\left(\dfrac{1}{2}\right)^{2-1} = \left(\dfrac{1}{2}\right)\left(\dfrac{1}{2}\right) = \dfrac{1}{4}$

$a_3 = \dfrac{1}{2}\left(\dfrac{1}{2}\right)^{3-1} = \left(\dfrac{1}{2}\right)\left(\dfrac{1}{4}\right) = \dfrac{1}{8}$

$a_4 = \dfrac{1}{2}\left(\dfrac{1}{2}\right)^{4-1} = \left(\dfrac{1}{2}\right)\left(\dfrac{1}{8}\right) = \dfrac{1}{16}$

$a_5 = \dfrac{1}{2}\left(\dfrac{1}{2}\right)^{5-1} = \left(\dfrac{1}{2}\right)\left(\dfrac{1}{16}\right) = \dfrac{1}{32}$

b. $a_n = \dfrac{1}{2}\left(\dfrac{1}{2}\right)^{n-1} = \left(\dfrac{1}{2}\right)^{n}$

c. $a_7 = \left(\dfrac{1}{2}\right)^{7} = \dfrac{1}{128} \approx 0.0078$ or 0.78%

71. After 1 hour there is $\dfrac{2}{3}$ of the original dye left, so $a_1 = \dfrac{2}{3}$. Also, $r = \dfrac{2}{3}$ and $n = 10$.

$a_{10} = \dfrac{2}{3}\left(\dfrac{2}{3}\right)^{10-1}$

$= \dfrac{2}{3}\left(\dfrac{2}{3}\right)^{9}$

$= \left(\dfrac{2}{3}\right)^{10}$

≈ 0.017

0.017 or 1.7% of the dye remains afer 10 hours.

73. a. $a_1 = 0.6(220) = 132$, $r = 0.6$

$a_n = a_1 r^{n-1}$

$a_3 = 132(0.6)^{3-1}$

$a_3 = 47.52$

The height of the third bounce is 47.52 feet.

b. $s_\infty = \dfrac{a_1}{1-r}$

$s_\infty = \dfrac{220}{1-0.6} = 550$

She travels a total of 550 feet in the downward direction.

75. a. $a_1 = 30(0.7) = 21$, $r = 0.7$

$a_n = a_1 r^{n-1}$

$a_4 = 21(0.7)^{4-1}$

$a_4 = 7.203$

The ball will bounce 7.203 inches on the fourth bounce.

b. $s_\infty = \dfrac{a_1}{1-r}$

$s_\infty = \dfrac{30}{1-0.7}$

$s_\infty = 100$

The ball travels a total of 100 inches in the downward direction.

77. Blue: $a_1 = 1$, $r = 2$
Red: $a_1 = 1$, $r = 3$

$a_6 = a_1 r^{6-1}$

Blue: $a_6 = 1(2)^5 = 32$

Red: $a_6 = 1(3)^5 = 243$

$243 - 32 = 211$

There are 211 more chips in the sixth stack of red chips.

79. Let a_n = value left after the nth year. After the first year there is $9800\left(\dfrac{4}{5}\right) = 7840$ of the value left, so $a_1 = 7840$, $r = \dfrac{4}{5}$.

$a_2 = 7840\left(\dfrac{4}{5}\right)^{2-1} = 7840\left(\dfrac{4}{5}\right) = 6272$

$a_3 = 7840\left(\dfrac{4}{5}\right)^{3-1} = 7840\left(\dfrac{16}{25}\right) = 5017.60$

a. 7840, 6272, $5017.60

b. $a_n = 7840\left(\dfrac{4}{5}\right)^{n-1}$

c. $a_5 = 7840\left(\dfrac{4}{5}\right)^{5-1}$

$= 7840\left(\dfrac{256}{625}\right)$

≈ 3211.26

After 5 years, the value of the car is $3211.26.

81. Each time the ball bounces it goes up and then comes down the same distance. Therefore, the total vertical distance will be twice the height it rises after each bounce plus the initial 10 feet. The heights after each bounce form an infinite geometric sequence with $r = 0.9$ and $a_1 = 9$.

$s_\infty = \dfrac{9}{1-0.9} = \dfrac{9}{0.1} = 90$

Total distance: 6

$2(s_\infty) + 10 = 2(90) + 10 = 190$

The total vertical distance is 190 feet.

83. a. y_2 goes up more steeply.

b.

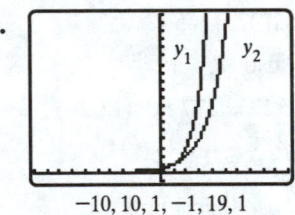

$-10, 10, 1, -1, 19, 1$

y_2 goes up more steeply.

85. This is a geometric sequence with $r = \dfrac{2}{1} = 2$. Also, $a_n = 1,048,576 = 2^{20}$.

Using $a_n = a_1 r^{n-1}$ gives

$2^{20} = 1(2)^{n-1}$

$2^{20} = 2^{n-1}$

$20 = n - 1$

$21 = n$

Thus, there are 21 terms in the sequence.

$$s_{21} = \frac{a_1\left(1 - r^{21}\right)}{1 - r}$$

$$= \frac{1\left(1 - 2^{21}\right)}{1 - 2}$$

$$= \frac{1 - 2{,}097{,}152}{-1}$$

$$= \frac{-2{,}097{,}151}{-1}$$

$$= 2{,}097{,}151$$

87.
$$\begin{array}{r} 4x^2 - 3x - 6 \\ \times \qquad\quad 2x - 3 \\ \hline -12x^2 + 9x + 18 \\ 8x^3 - 6x^2 - 12x \qquad\quad \\ \hline 8x^3 - 18x^2 - 3x + 18 \end{array}$$

88.
$$\begin{array}{r} 8x - 15 \\ 2x+5\overline{\smash{\big)}\,16x^2 + 10x - 18} \\ \underline{16x^2 + 40x} \\ -30x - 18 \\ \underline{-30x - 75} \\ 57 \end{array}$$

$$\left(16x^2 + 10x - 18\right) \div \left(2x + 5\right)$$
$$= 8x - 15 + \frac{57}{2x + 5}$$

89. Let x be the amount of time it takes Mr. Donovan to load the truck. Then $2x$ is the amount of time it takes Mrs. Donovan to load the truck.

$$\frac{8}{x} + \frac{8}{2x} = 1$$

$$2x\left(\frac{8}{x}\right) + 2x\left(\frac{8}{2x}\right) = 2x(1)$$

$$16 + 8 = 2x$$

$$24 = 2x$$

$$12 = x$$

It takes Mr. Donovan 12 hours to load the truck by himself.

90. $\left(\dfrac{9}{100}\right)^{-\frac{1}{2}} = \left(\dfrac{100}{9}\right)^{\frac{1}{2}} = \dfrac{\sqrt{100}}{\sqrt{9}} = \dfrac{10}{3}$

91. $\sqrt[3]{9x^2y}\left(\sqrt[3]{3x^4y^6} - \sqrt[3]{8xy^4}\right)$

$$= \sqrt[3]{9x^2y}\left(\sqrt[3]{3x^4y^6}\right) - \sqrt[3]{9x^2y}\left(\sqrt[3]{8xy^4}\right)$$

$$= \sqrt[3]{27x^6y^7} - \sqrt[3]{72x^3y^5}$$

$$= \sqrt[3]{27x^6y^6 \cdot y} - \sqrt[3]{8x^3y^3 \cdot 9y^2}$$

$$= \sqrt[3]{27x^6y^6}\sqrt[3]{y} - \sqrt[3]{8x^3y^3}\sqrt[3]{9y^2}$$

$$= 3x^2y^2\sqrt[3]{y} - 2xy\sqrt[3]{9y^2}$$

92. $x\sqrt{y} - 2\sqrt{x^2y} + \sqrt{4x^2y}$

$$= x\sqrt{y} - 2\sqrt{x^2 \cdot y} + \sqrt{4x^2 \cdot y}$$

$$= x\sqrt{y} - 2\sqrt{x^2}\sqrt{y} + \sqrt{4x^2}\sqrt{y}$$

$$= x\sqrt{y} - 2x\sqrt{y} + 2x\sqrt{y}$$

$$= x\sqrt{y}$$

93. $\sqrt{a^2 + 9a + 3} = -a$

$$a^2 + 9a + 3 = a^2$$

$$9a + 3 = 0$$

$$9a = -3$$

$$a = \frac{-3}{9}$$

$$= -\frac{1}{3}$$

Exercise Set 11.4

1. The first and last numbers in each row are 1 and the inner numbers are obtained by adding the two numbers in the row above (to the right and left).

$$\begin{array}{ccccccccc} & & & & 1 & & & & \\ & & & 1 & & 1 & & & \\ & & 1 & & 2 & & 1 & & \\ & 1 & & 3 & & 3 & & 1 & \\ 1 & & 4 & & 6 & & 4 & & 1 \end{array}$$

3. $\dbinom{5}{2} = \dfrac{5!}{2! \cdot (5-2)!}$

$= \dfrac{5!}{2! \cdot 3!}$

$= \dfrac{5 \cdot 4 \cdot 3 \cdot 2 \cdot 1}{(2 \cdot 1)(3 \cdot 2 \cdot 1)}$

$= \dfrac{20}{2}$

$= 10$

5. $\dbinom{5}{5} = \dfrac{5!}{5!(5-5)!} = \dfrac{1}{0!} = 1$

7. $\dbinom{7}{0} = \dfrac{7!}{0!(7-0)!} = \dfrac{7!}{7!} = 1$

9. $\dbinom{8}{4} = \dfrac{8!}{4! \cdot (8-4)!}$

$= \dfrac{8!}{4! \cdot 4!}$

$= \dfrac{8 \cdot 7 \cdot 6 \cdot 5 \cdot 4 \cdot 3 \cdot 2 \cdot 1}{(4 \cdot 3 \cdot 2 \cdot 1)(4 \cdot 3 \cdot 2 \cdot 1)}$

$= \dfrac{1680}{24}$

$= 70$

11. $\dbinom{8}{2} = \dfrac{8!}{2! \cdot (8-2)!}$

$= \dfrac{8!}{2! \cdot 6!}$

$= \dfrac{8 \cdot 7 \cdot 6 \cdot 5 \cdot 4 \cdot 3 \cdot 2 \cdot 1}{(2 \cdot 1) \cdot (6 \cdot 5 \cdot 4 \cdot 3 \cdot 2 \cdot 1)}$

$= \dfrac{56}{2}$

$= 28$

13. $(x+4)^3 = \dbinom{3}{0}x^3 4^0 + \dbinom{3}{1}x^2 4^1 + \dbinom{3}{2}x^1 4^2 + \dbinom{3}{3}x^0 4^3$

$= 1x^3(1) + 3x^2(4) + 3x(16) + 1(1)64$

$= x^3 + 12x^2 + 48x + 64$

15. $(a-b)^4 = \dbinom{4}{0}a^4(-b)^0 + \dbinom{4}{1}a^3(-b)^1 + \dbinom{4}{2}a^2(-b)^2 + \dbinom{4}{3}a^1(-b)^3 + \dbinom{4}{4}a^0(-b)^4$

$= 1a^4(1) + 4a^3(-b) + 6a^2 b^2 + 4a(-b^3) + 1(1)b^4$

$= a^4 - 4a^3 b + 6a^2 b^2 - 4ab^3 + b^4$

17. $(3a-b)^5$

$= \dbinom{5}{0}(3a)^5(-b)^0 + \dbinom{5}{1}(3a)^4(-b)^1 + \dbinom{5}{2}(3a)^3(-b)^2 + \dbinom{5}{3}(3a)^2(-b)^3 + \dbinom{5}{4}(3a)^1(-b)^4 + \dbinom{5}{5}(3a)^0(-b)^5$

$= 1(243a^5)(1) + 5(81a^4)(-b) + 10(27a^3)b^2 + 10(9a^2)(-b^3) + 5(3a)b^4 + 1(1)(-b^5)$

$= 243a^5 - 405a^4 b + 270a^3 b^2 - 90a^2 b^3 + 15ab^4 - b^5$

19. $\left(2x + \dfrac{1}{2}\right)^4 = \dbinom{4}{0}(2x)^4\left(\dfrac{1}{2}\right)^0 + \dbinom{4}{1}(2x)^3\left(\dfrac{1}{2}\right)^1 + \dbinom{4}{2}(2x)^2\left(\dfrac{1}{2}\right)^2 + \dbinom{4}{3}(2x)^1\left(\dfrac{1}{2}\right)^3 + \dbinom{4}{4}(2x)^0\left(\dfrac{1}{2}\right)^4$

$= 1(16x^4)(1) + 4(8x^3)\left(\dfrac{1}{2}\right) + 6(4x^2)\left(\dfrac{1}{4}\right) + 4(2x)\left(\dfrac{1}{8}\right) + 1(1)\left(\dfrac{1}{16}\right)$

$= 16x^4 + 16x^3 + 6x^2 + x + \dfrac{1}{16}$

21. $\left(\dfrac{x}{2}-3\right)^4 = \dbinom{4}{0}\left(\dfrac{x}{2}\right)^4(-3)^0 + \dbinom{4}{1}\left(\dfrac{x}{2}\right)^3(-3)^1 + \dbinom{4}{2}\left(\dfrac{x}{2}\right)^2(-3)^2 + \dbinom{4}{3}\left(\dfrac{x}{2}\right)^1(-3)^3 + \dbinom{4}{4}\left(\dfrac{x}{2}\right)^0(-3)^4$

$= 1\left(\dfrac{x^4}{16}\right)(1) + 4\left(\dfrac{x^3}{8}\right)(-3) + 6\left(\dfrac{x^2}{4}\right)(9) + 4\left(\dfrac{x}{2}\right)(-27) + 1(1)(81)$

$= \dfrac{x^4}{16} - \dfrac{3x^3}{2} + \dfrac{27x^2}{2} - 54x + 81$

23. $(x+y)^{10} = \dbinom{10}{0}x^{10}y^0 + \dbinom{10}{1}x^9y^1 + \dbinom{10}{2}x^8y^2 + \dbinom{10}{3}x^7y^3 + \cdots$

$= 1x^{10}(1) + \dfrac{10}{1}x^9y + \dfrac{10 \cdot 9}{2 \cdot 1}x^8y^2 + \dfrac{10 \cdot 9 \cdot 8}{3 \cdot 2 \cdot 1}x^7y^3 + \cdots$

$= x^{10} + 10x^9y + 45x^8y^2 + 120x^7y^3 + \cdots$

25. $(3x-y)^7 = \dbinom{7}{0}(3x)^7(-y)^0 + \dbinom{7}{1}(3x)^6(-y)^1 + \dbinom{7}{2}(3x)^5(-y)^2 + \dbinom{7}{3}(3x)^4(-y)^3 + \cdots$

$= 1(2187x^7)(1) + \dfrac{7}{1}(729x^6)(-y) + \dfrac{7 \cdot 6}{2 \cdot 1}(243x^5)(y^2) + \dfrac{7 \cdot 6 \cdot 5}{3 \cdot 2 \cdot 1}(81x^4)(-y^3) + \cdots$

$= 2187x^7 + 7(729x^6)(-y) + 21(243x^5)(y^2) + 35(81x^4)(-y^3) + \cdots$

$= 2187x^7 - 5103x^6y + 5103x^5y^2 - 2835x^4y^3 + \cdots$

27. $(x^2-3y)^8 = \dbinom{8}{0}(x^2)^8(-3y)^0 + \dbinom{8}{1}(x^2)^7(-3y)^1 + \dbinom{8}{2}(x^2)^6(-3y)^2 + \dbinom{8}{3}(x^2)^5(-3y)^3 + \cdots$

$= 1(x^{16})(1) + \dfrac{8}{1}(x^{14})(-3y) + \dfrac{8 \cdot 7}{2 \cdot 1}(x^{12})(9y^2) + \dfrac{8 \cdot 7 \cdot 6}{3 \cdot 2 \cdot 1}(x^{10})(-27y^3) + \cdots$

$= x^{16} + 8(x^{14})(-3y) + 28(x^{12})(9y^2) + 56(x^{10})(-27y^3) + \cdots$

$= x^{16} - 24x^{14}y + 252x^{12}y^2 - 1512x^{10}y^3 + \cdots$

29. Yes, $n! = n \cdot (n-1)!$
$4! = 4 \cdot 3 \cdot 2 \cdot 1 = 4 \cdot (3 \cdot 2 \cdot 1) = 4 \cdot (3)! = 4 \cdot (4-1)!$

31. Yes, $(n-3)! = (n-3)(n-4)(n-5)!$ for $n \geq 5$.
Let $n = 7$:
$(7-3)! = (7-3)(7-4)(7-5)!$ or
$4! = 4 \cdot 3 \cdot 2! = 4 \cdot 3 \cdot 2 \cdot 1 = 4!$

33. $(x+3)^8$

First term is $\dbinom{8}{0}(x)^8(3)^0 = 1(x^8)(1) = x^8$.

Second term is
$\dbinom{8}{1}(x)^7(3)^1 = 8(x^7)(3) = 24x^7$.

Next to last term is
$\dbinom{8}{7}(x)^1(3)^7 = 8(x)(2187) = 17{,}496x$

Last term is

$$\binom{8}{8}(x)^0(3)^8 = 1(1)(6561) = 6561.$$

35. $(a+b)^n = \sum_{i=0}^{n}\binom{n}{i}a^{n-i}b^i$

39. $s_n - s_n r = a_1 - a_1 r^n$

$$s_n(1-r) = a_1 - a_1 r^n$$

$$\frac{s_n(1-r)}{1-r} = \frac{a_1 - a_1 r^n}{1-r}$$

$$s_n = \frac{a_1 - a_1 r^n}{1-r} = \frac{a_1(1-r^n)}{1-r}$$

40. $s_n - s_n r = a_1 - a_1 r^n$

$$s_n - s_n r = a_1(1-r^n)$$

$$\frac{s_n - s_n r}{1-r^n} = \frac{a_1(1-r^n)}{1-r^n}$$

$$\frac{s_n - s_n r}{1-r^n} = a_1$$

$$a_1 = \frac{s_n - s_n r}{1-r^n} \text{ or } a_1 = \frac{s_n(1-r)}{1-r^n}$$

41. $16x^2 - 8x - 3$

Observe that $16(-3) = -48$

The two numbers whose product is -48 and whose sum is -8 are -12 and 4 since $(-12)(4) = -48$ and $-12 + 4 = -8$. Thus, the middle term, $-8x$, can be written as $-12x + 4x$ and the factorization is

$$16x^2 - 8x - 3 = 16x^2 - 12x + 4x - 3$$

$$= 4x(4x-3) + 1(4x-3)$$

$$= (4x+1)(4x-3)$$

42. $3a^2b^2 - 24ab^2 + 48b^2 = 3b^2(a^2 - 8a + 16)$

$$= 3b^2(a-4)^2$$

Review Exercises

1. $a_1 = 1 + 2 = 3$

$a_2 = 2 + 2 = 4$

$a_3 = 3 + 2 = 5$

$a_4 = 4 + 2 = 6$

$a_5 = 5 + 2 = 7$

The terms are 3, 4, 5, 6, 7.

2. $a_1 = \frac{1}{1} = 1$

$a_2 = \frac{1}{2}$

$a_3 = \frac{1}{3}$

$a_4 = \frac{1}{4}$

$a_5 = \frac{1}{5}$

The terms are $1, \frac{1}{2}, \frac{1}{3}, \frac{1}{4}, \frac{1}{5}$.

3. $a_1 = 1(1+1) = 1(2) = 2$

$a_2 = 2(2+1) = 2(3) = 6$

$a_3 = 3(3+1) = 3(4) = 12$

$a_4 = 4(4+1) + 4(5) = 20$

$a_5 = 5(5+1) = 5(6) = 30$

The terms are 2, 6, 12, 20, 30.

4. $a_1 = \frac{1^2}{1+4} = \frac{1}{5}$

$a_2 = \frac{2^2}{2+4} = \frac{4}{6} = \frac{2}{3}$

$a_3 = \frac{3^2}{3+4} = \frac{9}{7}$

$a_4 = \frac{4^2}{4+4} = \frac{16}{8} = 2$

$a_5 = \frac{5^2}{5+4} = \frac{25}{9}$

The terms are $\frac{1}{5}, \frac{2}{3}, \frac{9}{7}, 2, \frac{25}{9}$.

5. $a_7 = 3(7) + 4 = 21 + 4 = 25$

6. $a_7 = (-1)^7 + 3 = -1 + 3 = 2$

7. $a_9 = \dfrac{9+7}{9^2} = \dfrac{16}{81}$

8. $a_{11} = 11(11-3) = 11(8) = 88$

9. $a_1 = 3(1) + 2 = 3 + 2 = 5$
$a_2 = 3(2) + 2 = 6 + 2 = 8$
$a_3 = 3(3) + 2 = 9 + 2 = 11$
$s_1 = a_1 = 5$
$s_3 = a_1 + a_2 + a_3 = 5 + 8 + 11 = 24$

10. $a_1 = 2(1)^2 = 2$
$a_2 = 2(2)^2 = 2(4) = 8$
$a_3 = 2(3)^2 = 2(9) = 18$
$s_1 = a_1 = 2$
$s_3 = a_1 + a_2 + a_3 = 2 + 8 + 18 = 28$

11. $a_1 = \dfrac{1+3}{1+2} = \dfrac{4}{3}$
$a_2 = \dfrac{2+3}{2+2} = \dfrac{5}{4}$
$a_3 = \dfrac{3+3}{3+2} = \dfrac{6}{5}$
$s_1 = a_1 = \dfrac{4}{3}$
$s_3 = a_1 + a_2 + a_3$
$\quad = \dfrac{4}{3} + \dfrac{5}{4} + \dfrac{6}{5}$
$\quad = \dfrac{80}{60} + \dfrac{75}{60} + \dfrac{72}{60}$
$\quad = \dfrac{227}{60}$

12. $a_1 = (-1)^1(1+2) = -1(3) = -3$
$a_2 = (-1)^2(2+2) = 1(4) = 4$
$a_3 = (-1)^3(3+2) = -1(5) = -5$
$s_1 = a_1 = -3$
$s_3 = a_1 + a_2 + a_3 = -3 + 4 - 5 = -4$

13. This is a geometric sequence. $r = 2 \div 1 = 2$
$a_1 = 1$
$a_5 = 1(2)^{5-1} = 2^4 = 16$
$a_6 = 1(2)^{6-1} = 2^5 = 32$
$a_7 = 1(2)^{7-1} = 2^6 = 64$

The terms are 16, 32, 64.
$a_n = 1(2)^{n-1} = 2^{n-1}$

14. This is a geometric sequence.
$r = 4 \div (-8) = \dfrac{4}{-8} = -\dfrac{1}{2}$
$a_1 = -8$
$a_5 = -8\left(-\dfrac{1}{2}\right)^4 = -8\left(\dfrac{1}{16}\right) = -\dfrac{1}{2}$
$a_6 = -8\left(-\dfrac{1}{2}\right)^5 = -8\left(-\dfrac{1}{32}\right) = \dfrac{1}{4}$
$a_7 = -8\left(-\dfrac{1}{2}\right)^6 = -8\left(\dfrac{1}{64}\right) = -\dfrac{1}{8}$
The terms are $-\dfrac{1}{2}, \ \dfrac{1}{4}, \ -\dfrac{1}{8}$.
$a_n = -8\left(-\dfrac{1}{2}\right)^{n-1}$

15. This is a geometric sequence.
$r = \dfrac{4}{3} \div \dfrac{2}{3} = \dfrac{4}{3}\left(\dfrac{3}{2}\right) = 2$
$a_1 = \dfrac{2}{3}$
$a_5 = \dfrac{2}{3}(2)^{5-1} = \dfrac{2^5}{3} = \dfrac{32}{3}$
$a_6 = \dfrac{2}{3}(2)^{6-1} = \dfrac{2^6}{3} = \dfrac{64}{3}$
$a_7 = \dfrac{2}{3}(2)^{7-1} = \dfrac{2^7}{3} = \dfrac{128}{3}$
The terms are $\dfrac{32}{3}, \ \dfrac{64}{3}, \ \dfrac{128}{3}$.
$a_n = \dfrac{2}{3}(2)^{n-1} = \dfrac{2^n}{3}$

16. This is an arithmetic sequence.
$d = 6 - 9 = -3$
$a_1 = 9$
$a_5 = 9 + (5-1)(-3) = 9 - 12 = -3$
$a_6 = 9 + (6-1)(-3) = 9 - 15 = -6$
$a_7 = 9 + (7-1)(-3) = 9 - 18 = -9$
The terms are $-3, -6, -9$.
$a_n = a_1 + (n-1)d$
$\quad = 9 + (n-1)(-3)$
$\quad = 9 - 3n + 3$
$\quad = 12 - 3n$

17. $\displaystyle\sum_{n=1}^{3}(n^2+2) = (1^2+2)+(2^2+2)+(3^2+2)$

$$= 3+6+11$$
$$= 20$$

18. $\displaystyle\sum_{k=1}^{4}k(k+2)$
$$= 1(1+2)+2(2+2)+3(3+2)+4(4+2)$$
$$= 1(3)+2(4)+3(5)+4(6)$$
$$= 3+8+15+24$$
$$= 50$$

19. $\displaystyle\sum_{k=1}^{5}\frac{k^2}{3} = \frac{1^2}{3}+\frac{2^2}{3}+\frac{3^2}{3}+\frac{4^2}{3}+\frac{5^2}{3}$

$$= \frac{1}{3}+\frac{4}{3}+\frac{9}{3}+\frac{16}{3}+\frac{25}{3}$$
$$= \frac{55}{3}$$

20. $\displaystyle\sum_{n=1}^{4}\frac{n}{n+1} = \frac{1}{1+1}+\frac{2}{2+1}+\frac{3}{3+1}+\frac{4}{4+1}$

$$= \frac{1}{2}+\frac{2}{3}+\frac{3}{4}+\frac{4}{5}$$
$$= \frac{163}{60}$$

21. $\displaystyle\sum_{i=1}^{4}x_i = x_1+x_2+x_3+x_4$

$$= 3+9+5+10$$
$$= 27$$

22. $\displaystyle\sum_{i=1}^{4}x_i^2 = x_1^2+x_2^2+x_3^2+x_4^2$

$$= 3^2+9^2+5^2+10^2$$
$$= 9+81+25+100$$
$$= 215$$

23. $\displaystyle\sum_{i=2}^{3}(x_i^2+1) = (x_2^2+1)+(x_3^2+1)$

$$= (9^2+1)+(5^2+1)$$
$$= (81+1)+(25+1)$$
$$= 82+26$$
$$= 108$$

24. $\displaystyle\left(\sum_{i=1}^{4}x_i\right)^2 = (x_1+x_2+x_3+x_4)^2$

$$= (3+9+5+10)^2$$
$$= 27^2$$
$$= 729$$

25. $a_1 = 5$
$a_2 = 5+(2-1)(2) = 5+2 = 7$
$a_3 = 5+(3-1)(2) = 5+2(2) = 9$
$a_4 = 5+(4-1)(2) = 5+3(2) = 11$
$a_5 = 5+(5-1)(2) = 5+4(2) = 13$
The terms are 5, 7, 9, 11, 13.

26. $a_1 = \dfrac{1}{2}$
$a_2 = \dfrac{1}{2}+(2-1)(-2) = \dfrac{1}{2}-2 = -\dfrac{3}{2}$
$a_3 = \dfrac{1}{2}+(3-1)(-2) = \dfrac{1}{2}-4 = -\dfrac{7}{2}$
$a_4 = \dfrac{1}{2}+(4-1)(-2) = \dfrac{1}{2}-6 = -\dfrac{11}{2}$
$a_5 = \dfrac{1}{2}+(5-1)(-2) = \dfrac{1}{2}-8 = -\dfrac{15}{2}$
The terms are $\dfrac{1}{2},\ -\dfrac{3}{2},\ -\dfrac{7}{2},\ -\dfrac{11}{2},\ -\dfrac{15}{2}.$

27. $a_1 = -12$
$a_2 = -12+(2-1)\left(-\dfrac{1}{2}\right) = -12-\dfrac{1}{2} = -\dfrac{25}{2}$
$a_3 = -12+(3-1)\left(-\dfrac{1}{2}\right)$
$ = -12+2\left(-\dfrac{1}{2}\right)$
$ = -12-1$
$ = -13$

$$a_4 = -12 + (4-1)\left(-\frac{1}{2}\right)$$

$$= -12 + 3\left(-\frac{1}{2}\right)$$

$$= -12 - \frac{3}{2}$$

$$= -\frac{27}{2}$$

$$a_5 = -12 + (5-1)\left(-\frac{1}{2}\right)$$

$$= -12 + 4\left(-\frac{1}{2}\right)$$

$$= -12 - 2$$

$$= -14$$

The terms are $-12,\ -\frac{25}{2},\ -13,\ -\frac{27}{2},\ -14$.

28. $a_1 = -100$

$$a_2 = -100 + (2-1)\left(\frac{1}{5}\right) = -100 + \frac{1}{5} = -\frac{499}{5}$$

$$a_3 = -100 + (3-1)\left(\frac{1}{5}\right) = -100 + \frac{2}{5} = -\frac{498}{5}$$

$$a_4 = -100 + (4-1)\left(\frac{1}{5}\right) = -100 + \frac{3}{5} = -\frac{497}{5}$$

$$a_5 = -100 + (5-1)\left(\frac{1}{5}\right) = -100 + \frac{4}{5} = -\frac{496}{5}$$

The terms are $-100,\ -\frac{499}{5},\ -\frac{498}{5},\ -\frac{497}{5},$

$-\frac{496}{5}$.

29. $a_9 = a_1 + (9-1)d$
$a_9 = 2 + (9-1)(3) = 2 + 8(3) = 26$

30. $a_5 = a_1 + (5-1)d$
$34 = 50 + 4d$
$-16 = 4d$

$-\dfrac{16}{4} = d$ or $d = -4$

31. $a_7 = a_1 + (7-1)d$
$0 = -3 + 6d$
$3 = 6d$
$\dfrac{3}{6} = d$
$\dfrac{1}{2} = d$

32. $a_n = a_1 + (n-1)d$
$-13 = 12 + (n-1)(-5)$
$-13 = 12 - 5n + 5$
$-13 = 17 - 05n$
$-30 = -5n$
$\dfrac{-30}{-5} = n$ or $n = 6$

33. $a_8 = a_1 + (8-1)d$
$21 = 7 + 7d$
$14 = 7d$
$2 = d$ or $d = 2$

$$s_8 = \frac{8(a_1 + a_8)}{2}$$

$$= \frac{8(7 + 21)}{2}$$

$$= \frac{8(28)}{2}$$

$$= 4(28)$$

$$= 112$$

34. $a_7 = a_1 + (7-1)d$
$-48 = -12 + 6d$
$-36 = 6d$
$\dfrac{-36}{6} = d$
$-6 = d$ or $d = -6$

$$s_7 = \frac{7(a_1 + a_7)}{2}$$

$$= \frac{7(-12 - 48)}{2}$$

$$= \frac{7(-60)}{2}$$

$$= -210$$

35. $a_7 = a_1 + (7-1)d$

$$3 = \frac{3}{5} + 6d$$

$$3 - \frac{3}{5} = 6d$$

$$\frac{12}{5} = 6d$$

$$\frac{1}{6}\left(\frac{12}{5}\right) = \frac{1}{6}(6d)$$

$$\frac{2}{5} = d \text{ or } d = \frac{2}{5}$$

$$s_7 = \frac{7(a_1 + a_7)}{2}$$

$$= \frac{7\left(\frac{3}{5} + 3\right)}{2}$$

$$= \frac{7\left(\frac{18}{5}\right)}{2}$$

$$= 7\left(\frac{18}{5}\right)\left(\frac{1}{2}\right)$$

$$= \frac{63}{5}$$

36. $a_9 = a_1 + (9-1)d$

$$-6 = -\frac{10}{3} + 8d$$

$$-6 + \frac{10}{3} = 8d$$

$$-\frac{8}{3} = 8d$$

$$\frac{1}{8}\left(-\frac{8}{3}\right) = \frac{1}{8}(8d)$$

$$-\frac{1}{3} = d \text{ or } d = -\frac{1}{3}$$

$$s_9 = \frac{9(a_1 + a_n)}{2}$$

$$= \frac{9\left(-\frac{10}{3} - 6\right)}{2}$$

$$= \frac{9\left(-\frac{28}{3}\right)}{2}$$

$$= 9\left(-\frac{28}{3}\right)\left(\frac{1}{2}\right)$$

$$= -42$$

37. $a_1 = 2$

$a_2 = 2 + (2-1)(4) = 2 + 4 = 6$

$a_3 = 2 + (3-1)(4) = 2 + 2(4) = 2 + 8 = 10$

$a_4 = 2 + (4-1)(4) = 2 + 3(4) = 2 + 12 = 14$

The terms are 2, 6, 10, 14.

$a_{10} = 2 + (10-1)(4)$

$\phantom{a_{10}} = 2 + 9(4)$

$\phantom{a_{10}} = 2 + 36$

$\phantom{a_{10}} = 38$

$s_{10} = \dfrac{10(2+38)}{2} = 5(40) = 200$

38. $a_1 = -8$

$a_2 = -8 + (2-1)(-3) = -8 - 3 = -11$

$a_3 = -8 + (3-1)(-3) = -8 - 6 = -14$

$a_4 = -8 + (4-1)(-3) = -8 - 9 = -17$

The terms are $-8, -11, -14, -17$.

$a_{10} = -8 + (10-1)(-3) = -8 - 27 = -35$

$s_{10} = \dfrac{10(-8-35)}{2} = 5(-43) = -215$

39. $a_1 = \dfrac{5}{6}$

$a_2 = \dfrac{5}{6} + (2-1)\left(\dfrac{2}{3}\right) = \dfrac{5}{6} + \dfrac{2}{3} = \dfrac{9}{6} = \dfrac{3}{2}$

$a_3 = \dfrac{5}{6} + (3-1)\left(\dfrac{2}{3}\right)$

$ = \dfrac{5}{6} + 2\left(\dfrac{2}{3}\right)$

$ = \dfrac{5}{6} + \dfrac{4}{3}$

$ = \dfrac{13}{6}$

$$a_4 = \frac{5}{6} + (4-1)\left(\frac{2}{3}\right)$$
$$= \frac{5}{6} + 3\left(\frac{2}{3}\right)$$
$$= \frac{5}{6} + \frac{6}{3}$$
$$= \frac{17}{6}$$

The terms are $\frac{5}{6}, \frac{3}{2}, \frac{13}{6}, \frac{17}{6}$.

$$a_{10} = \frac{5}{6} + (10-1)\left(\frac{2}{3}\right)$$
$$= \frac{5}{6} + 9\left(\frac{2}{3}\right)$$
$$= \frac{5}{6} + \frac{18}{3}$$
$$= \frac{41}{6}$$

$$s_{10} = \frac{10\left(\frac{5}{6} + \frac{41}{6}\right)}{2}$$
$$= \frac{10\left(\frac{46}{6}\right)}{2}$$
$$= 5\left(\frac{46}{6}\right)$$
$$= 5\left(\frac{23}{3}\right)$$
$$= \frac{115}{3}$$

40. $a_1 = -80$
$a_2 = -80 + (2-1)(4) = -80 + 4 = -76$
$a_3 = -80 + (3-1)(4) = -80 + 8 = -72$
$a_4 = -80 + (4-1)(4) = -80 + 12 = -68$
The terms are $-80, -76, -72, -68$.
$a_{10} = -80 + (10-1)(4) = -80 + 36 = -44$
$s_{10} = \frac{10(-80-44)}{2} = 5(-124) = -620$

41. $d = 8 - 3 = 5$
$a_n = a_1 + (n-1)d$
$53 = 3 + (n-1)5$
$53 = 3 + 5n - 5$
$53 = -2 + 5n$
$55 = 5n$
$11 = n$ or $n = 11$
$$s_{11} = \frac{11(a_1 + a_{11})}{2}$$
$$= \frac{11(3+53)}{2}$$
$$= \frac{11(56)}{2}$$
$$= 11(28)$$
$$= 308$$

42. $d = -11 - (-16) = -11 + 16 = 5$
$a_n = a_1 + (n-1)d$
$24 = -16 + (n-1)5$
$24 = -16 + 5n - 5$
$24 = -21 + 5n$
$45 = 5n$
$9 = n$ or $n = 9$
$$s_9 = \frac{9(a_1 + a_9)}{2} = \frac{9(-16+24)}{2} = \frac{9(8)}{2} = 36$$

43. $d = \frac{9}{10} - \frac{6}{10} = \frac{3}{10}$
$a_n = a_1 + (n-1)d$
$$\frac{36}{10} = \frac{6}{10} + (n-1)\frac{3}{10}$$
$$\frac{36}{10} = \frac{6}{10} + \frac{3}{10}n - \frac{3}{10}$$
$$\frac{36}{10} = \frac{3}{10} + \frac{3}{10}n$$
$$\frac{33}{10} = \frac{3}{10}n$$
$$\frac{10}{3}\left(\frac{33}{10}\right) = \frac{10}{3}\left(\frac{3}{10}n\right)$$
$$11 = n \text{ or } n = 11$$

$$s_{11} = \frac{11(a_1 + a_{11})}{2}$$

$$= \frac{11\left(\frac{6}{10} + \frac{36}{10}\right)}{2}$$

$$= \frac{11\left(\frac{42}{10}\right)}{2}$$

$$= 11\left(\frac{42}{10}\right)\left(\frac{1}{2}\right)$$

$$= \frac{231}{10}$$

44. $d = 0 - (-5) = 5$

$a_n = a_1 + (n-1)d$

$85 = -5 + (n-1)5$

$85 = -5 + 5n - 5$

$85 = -10 + 5n$

$95 = 5n$

$19 = n$ or $n = 19$

$$s_{19} = \frac{19(a_1 + a_{19})}{2}$$

$$= \frac{19(-5 + 85)}{2}$$

$$= \frac{19(80)}{2}$$

$$= 19(40)$$

$$= 760$$

45. $a_1 = 5$

$a_2 = 5(2)^{2-1} = 5(2) = 10$

$a_3 = 5(2)^{3-1} = 5(2)^2 = 5(4) = 20$

$a_4 = 5(2)^{4-1} = 5(2)^3 = 5(8) = 40$

$a_5 = 5(2)^{5-1} = 5(2)^4 = 5(16) = 80$

The terms are 5, 10, 20, 40, 80.

46. $a_1 = -12$

$a_2 = -12\left(\frac{1}{2}\right)^{2-1} = -12\left(\frac{1}{2}\right) = -6$

$a_3 = -12\left(\frac{1}{2}\right)^{3-1} = -12\left(\frac{1}{4}\right) = -3$

$a_4 = -12\left(\frac{1}{2}\right)^{4-1} = -12\left(\frac{1}{8}\right) = -\frac{3}{2}$

$a_5 = -12\left(\frac{1}{2}\right)^{5-1} = -12\left(\frac{1}{16}\right) = -\frac{3}{4}$

The terms are $-12, -6, -3, -\frac{3}{2}, -\frac{3}{4}$.

47. $a_1 = 20$

$a_2 = 20\left(-\frac{2}{3}\right)^{2-1} = 20\left(-\frac{2}{3}\right) = -\frac{40}{3}$

$a_3 = 20\left(-\frac{2}{3}\right)^{3-1} = 20\left(-\frac{2}{3}\right)^2 = 20\left(\frac{4}{9}\right) = \frac{80}{9}$

$a_4 = 20\left(-\frac{2}{3}\right)^{4-1} = 20\left(-\frac{2}{3}\right)^3 = 20\left(-\frac{8}{27}\right) = -\frac{160}{27}$

$a_5 = 20\left(-\frac{2}{3}\right)^{5-1} = 20\left(-\frac{2}{3}\right)^4 = 20\left(\frac{16}{81}\right) = \frac{320}{81}$

The terms are $20, -\frac{40}{3}, \frac{80}{9}, -\frac{160}{27}, \frac{320}{81}$.

48. $a_1 = -100$

$a_2 = -100\left(\frac{1}{5}\right)^{2-1} = -100\left(\frac{1}{5}\right) = -20$

$a_3 = -100\left(\frac{1}{5}\right)^{3-1} = -100\left(\frac{1}{25}\right) = -4$

$a_4 = -100\left(\frac{1}{5}\right)^{4-1} = -100\left(\frac{1}{125}\right) = -\frac{4}{5}$

$a_5 = -100\left(\frac{1}{5}\right)^{5-1} = -100\left(\frac{1}{625}\right) = -\frac{4}{25}$

The terms are $-100, -20, -4, -\frac{4}{5}, -\frac{4}{25}$.

49. $a_7 = 12\left(\frac{1}{3}\right)^{7-1} = 12\left(\frac{1}{729}\right) = \frac{4}{243}$

50. $a_9 = 25(2)^{9-1} = 25(256) = 6400$

51. $a_9 = -8(-2)^{9-1} = -8(256) = -2048$

52. $a_8 = \frac{5}{12}\left(\frac{2}{3}\right)^{8-1} = \frac{5}{12}\left(\frac{128}{2187}\right) = \frac{160}{6561}$

53. $s_8 = \frac{12(1-2^8)}{1-2}$

$= \frac{12(1-256)}{-1}$

$= \frac{12(-255)}{-1}$

$= 3060$

54.

$s_7 = \frac{\frac{3}{5}\left[1-\left(\frac{5}{3}\right)^7\right]}{1-\frac{5}{3}}$

$= \frac{\frac{3}{5}\left(1-\frac{78,125}{2187}\right)}{-\frac{2}{3}}$

$= \frac{3}{5}\left(-\frac{75,938}{2187}\right)\left(-\frac{3}{2}\right)$

$= \frac{37,969}{1215}$

55. $s_5 = \frac{-84\left[1-\left(-\frac{1}{4}\right)^5\right]}{1-\left(-\frac{1}{4}\right)}$

$= \frac{-84\left[1-\left(-\frac{1}{1024}\right)\right]}{1+\frac{1}{4}}$

$= \frac{-84\left(1+\frac{1}{1024}\right)}{\frac{5}{4}}$

$= -84 \cdot \left(\frac{1025}{1024}\right) \cdot \frac{4}{5}$

$= -\frac{4305}{64}$

56. $s_9 = \frac{9\left[1-\left(\frac{3}{2}\right)^9\right]}{1-\frac{3}{2}}$

$= \frac{9\left(1-\frac{19,863}{512}\right)}{-\frac{1}{2}}$

$= 9\left(-\frac{19,171}{512}\right)\left(-\frac{2}{1}\right)$

$= \frac{172,539}{256}$

57. $r = 12 \div 6 = 2$

$a_n = 6(2)^{n-1}$

58. $r = \frac{8}{3} \div 8 = \frac{8}{3} \cdot \frac{1}{8} = \frac{1}{3}$

$a_n = 8\left(\frac{1}{3}\right)^{n-1}$

59. $r = -20 \div -4 = \frac{-20}{-4} = 5$

$a_n = -4(5)^{n-1}$

60. $r = \frac{18}{15} \div \frac{9}{5} = \frac{18}{15} \cdot \frac{5}{9} = \frac{2}{3}$

$a_n = \frac{9}{5}\left(\frac{2}{3}\right)^{n-1}$

61. $r = \frac{7}{2} \div 7 = \frac{7}{2} \cdot \frac{1}{7} = \frac{1}{2}$

$s_\infty = \frac{7}{1 - \frac{1}{2}} = \frac{7}{\frac{1}{2}} = 7\left(\frac{2}{1}\right) = 14$

62. $r = \frac{8}{3} \div (-8) = \frac{8}{3}\left(-\frac{1}{8}\right) = -\frac{1}{3}$

$s_\infty = \frac{-8}{1 - \left(-\frac{1}{3}\right)} = \frac{-8}{\frac{4}{3}} = -8\left(\frac{3}{4}\right) = -6$

63. $r = -\frac{10}{3} \div -5 = \left(-\frac{10}{3}\right)\left(-\frac{1}{5}\right) = \frac{2}{3}$

$s_\infty = \frac{-5}{1 - \frac{2}{3}} = \frac{-5}{\frac{1}{3}} = -5\left(\frac{3}{1}\right) = -15$

64. $r = 1 \div \frac{7}{2} = 1\left(\frac{2}{7}\right) = \frac{2}{7}$

$s_\infty = \frac{\frac{7}{2}}{1 - \frac{2}{7}} = \frac{\frac{7}{2}}{\frac{5}{7}} = \frac{7}{2}\left(\frac{7}{5}\right) = \frac{49}{10}$

65. $r = 1 \div 2 = \frac{1}{2}$

$s_\infty = \frac{2}{1 - \frac{1}{2}} = \frac{2}{\frac{1}{2}} = 2\left(\frac{2}{1}\right) = 4$

66. $r = \frac{7}{3} \div 7 = \frac{7}{3} \cdot \frac{1}{7} = \frac{1}{3}$

$s_\infty = \frac{7}{1 - \frac{1}{3}} = \frac{7}{\frac{2}{3}} = 7\left(\frac{3}{2}\right) = \frac{21}{2}$

67. $r = -\frac{24}{3} \div -12 = -\frac{24}{3}\left(-\frac{1}{12}\right) = \frac{2}{3}$

$s_\infty = \frac{-12}{1 - \frac{2}{3}} = \frac{-12}{\frac{1}{3}} = -12\left(\frac{3}{1}\right) = -36$

68. $r = -1 \div 5 = -\frac{1}{5}$

$s_\infty = \frac{5}{1 - \left(-\frac{1}{5}\right)} = \frac{5}{\frac{6}{5}} = 5\left(\frac{5}{6}\right) = \frac{25}{6}$

69. $0.5252\cdots$

$= 0.52 + 0.0052 + 0.000052 + \cdots$

$= 0.52 + 0.52(0.01) + 0.52(0.01)^2 + \cdots$

$a_1 = 0.52$ and $r = 0.01$

$s_\infty = \frac{0.52}{1 - 0.01} = \frac{0.52}{0.99} = \frac{52}{99}$

70. $0.531531\ldots = 0.531 + 0.000531 + 0.000000531 + \cdots = 0.531 + 0.531(0.001) + 0.531(0.001)^2 + \cdots$

$a_1 = 0.531$ and $r = 0.001$

$s_\infty = \frac{0.531}{1 - 0.001} = \frac{0.531}{0.999} = \frac{531}{999} = \frac{59}{111}$

71. $(3x + y)^4 = \binom{4}{0}(3x)^4(y)^0 + \binom{4}{1}(3x)^3(y)^1 + \binom{4}{2}(3x)^2(y)^2 + \binom{4}{3}(3x)^1(y)^3 + \binom{4}{4}(3x)^0(y)^4$

$= 1(81x^4)(1) + 4(27x^3)(y) + 6(9x^2)(y^2) + 4(3x)(y^3) + 1(1)(y^4)$

$= 81x^4 + 108x^3y + 54x^2y^2 + 12xy^3 + y^4$

72. $(2x - 3y^2)^3 = \binom{3}{0}(2x)^3(-3y^2)^0 + \binom{3}{1}(2x)^2(-3y^2)^1 + \binom{3}{2}(2x)^1(-3y^2)^2 + \binom{3}{3}(2x)^0(-3y^2)^3$

$= 1(8x^3)(1) + 3(4x^2)(-3y^2) + 3(2x)(9y^4) + 1(1)(-27y^6)$

$= 8x^3 - 36x^2y^2 + 54xy^4 - 27y^6$

73. $(x-2y)^9 = \binom{9}{0}(x)^9(-2y)^0 + \binom{9}{1}(x)^8(-2y)^1 + \binom{9}{2}(x)^7(-2y)^2 + \binom{9}{3}(x)^6(-2y)^3 + \cdots$

$\qquad = 1(x^9)(1) + 9(x^8)(-2y) + 36(x^7)(4y^2) + 84(x^6)(-8y^3) + \cdots$

$\qquad = x^9 - 18x^8y + 144x^7y^2 - 672x^6y^3 + \cdots$

74. $(2a^2 + 3b)^8 = \binom{8}{0}(2a^2)^8(3b)^0 + \binom{8}{1}(2a^2)^7(3b)^1 + \binom{8}{2}(2a^2)^6(3b)^2 + \binom{8}{3}(2a^2)^5(3b)^3 + \cdots$

$\qquad = 1(256a^{16})(1) + 8(128a^{14})(3b) + 28(64a^{12})(9b^2) + 56(32a^{10})(27b^3) + \cdots$

$\qquad = 256a^{16} + 3072a^{14}b + 16,128a^{12}b^2 + 48,384a^{10}b^3 + \cdots$

75. This is an arithmetic series with $d = 1$, $a_1 = 100$, and $a_n = 200$.

$\quad a_n = a_1 + (n-1)d$

$200 = 100 + (n-1)(1)$

$200 = 99 + n$

$101 = n$

The sum is

$s_{101} = \dfrac{101(100 + 200)}{2}$

$\qquad = \dfrac{101(300)}{2}$

$\qquad = 101(150)$

$\qquad = 15,150$

76. This is an arithmetic sequence with $d = 1000$

a. $a_1 = 30,000$

$a_2 = 30,000 + (2-1)(1000)$

$\quad = 30,000 + 1000$

$\quad = 31,000$

$a_3 = 30,000 + (3-1)(1000)$

$\quad = 30,000 + 2000$

$\quad = 32,000$

$a_4 = 30,000 + (4-1)(1000)$

$\quad = 30,000 + 3000$

$\quad = 33,000$

$a_5 = 30,000 + (5-1)(1000)$

$\quad = 30,000 + 4000$

$\quad = 34,000$

His salaries for the first five years are $30,000, $31,000, $32,000, $33,000, and $34,000.

b. $a_n = 30,000 + (n-1)(1000)$

$\quad = 30,000 + 1000(n-1)$

$\quad = 29,000 + 1000n$

c. After 9 years is the 10th year.

$a_{10} = 29,000 + 1000(10)$

$\quad = 29,000 + 10,000$

$\quad = 39,000$

His salary would be $39,000.

77. This is a geometric series with $r = 2$, $a_1 = 100$ and $n = 11$.

(There are 11 terms here since 200 represents the first doubling.)

$a_{11} = a_1 r^{10}$

$\quad = 100(2)^{10}$

$\quad = 100(1024)$

$\quad = 102,400$

You would have $102,400.

78. Each year, the cost of the object will be 1.08 times greater than the previous year. After 12 years will be the 13th year. Therefore,

$a_{13} = 200(1.08)^{13-1}$

$\quad = 200(1.08)^{12}$

$\quad \approx 200(2.51817)$

$\quad \approx 503.63$

The item would cost $503.63.

79. This is an infinite geometric series with $r = 0.92$ and $a_1 = 8$.

$$s_\infty = \frac{8}{1 - 0.92} = \frac{8}{0.08} = 100$$

The pendulum travels a total distance of 100 feet.

Practice Test

1. a. An infinite sequence is a function whose domain is the set of natural numbers.

   b. A finite sequence is a function whose domain includes only the first n natural numbers.

2. A series is the sum of the terms of a sequence

3. a. An arithmetic sequence is one whose terms differ by a constant amount.

   b. A geometric sequence is one whose terms differ by a common multiple.

4. a. This sequence is neither arithmetic or geometric because the terms do not differ by a constant amount nor by a common multiple.

   b. This sequence is arithmetic because the terms differ by -3.

c. This sequence is geometric because the terms differ by the multiple $-\dfrac{1}{2}$.

5. $a_1 = \dfrac{1+2}{1^2} = \dfrac{3}{1} = 3$

$a_2 = \dfrac{2+2}{2^2} = \dfrac{4}{4} = 1$

$a_3 = \dfrac{3+2}{3^2} = \dfrac{5}{9}$

$a_4 = \dfrac{4+2}{4^2} = \dfrac{6}{16} = \dfrac{3}{8}$

$a_5 = \dfrac{5+2}{5^2} = \dfrac{7}{25}$

The terms are $3, 1, \dfrac{5}{9}, \dfrac{3}{8}, \dfrac{7}{25}$.

6. $a_1 = \dfrac{2 \cdot 1 + 3}{1} = \dfrac{2+3}{1} = \dfrac{5}{1} = 5$

$a_2 = \dfrac{2 \cdot 2 + 3}{2} = \dfrac{4+3}{2} = \dfrac{7}{2}$

$a_3 = \dfrac{2 \cdot 3 + 3}{3} = \dfrac{6+3}{3} = \dfrac{9}{3} = 3$

$s_1 = a_1 = 5$

$s_3 = a_1 + a_2 + a_3$

$\quad = 5 + \dfrac{7}{2} + 3$

$\quad = \dfrac{10}{2} + \dfrac{7}{2} + \dfrac{6}{2}$

$\quad = \dfrac{23}{2}$

7. $\displaystyle\sum_{n=1}^{5} (2n^2 + 3) = [2(1^2) + 3] + [2(2^2) + 3] + [2(3^2) + 3] + [2(4^2) + 3] + [2(5^2) + 3]$

$\qquad = (2 + 3) + (8 + 3) + (18 + 3) + (32 + 3) + (50 + 3)$

$\qquad = 5 + 11 + 21 + 35 + 53$

$\qquad = 125$

8. $\displaystyle\sum_{i=1}^{4} x_i^2 = x_1^2 + x_2^2 + x_3^2 + x_4^2$

$\qquad = 4^2 + 2^2 + 8^2 + 12^2$

$\qquad = 16 + 4 + 64 + 144$

$\qquad = 228$

9. $d = \frac{2}{3} - \frac{1}{3} = \frac{1}{3}$

$a_n = a_1 + (n-1)d$

$\quad = \frac{1}{3} + (n-1)\left(\frac{1}{3}\right)$

$\quad = \frac{1}{3} + \frac{1}{3}n - \frac{1}{3}$

$\quad = \frac{1}{3}n$

10. $r = 10 \div 5 = \frac{10}{5} = 2$

$a_n = a_1 r^{n-1} = 5(2)^{n-1}$

11. $a_1 = 12$

$a_2 = a_1 + (2-1)d = 12 + 1(-3) = 12 - 3 = 9$

$a_3 = a_1 + (3-1)d = 12 + 2(-3) = 12 - 6 = 6$

$a_4 = a_1 + (4-1)d = 12 + 3(-3) = 12 - 9 = 3$

The terms are 12, 9, 6, 3.

12. $a_1 = \frac{5}{8}$

$a_2 = a_1 r^1 = \frac{5}{8}\left(\frac{2}{3}\right) = \frac{10}{24} = \frac{5}{12}$

$a_3 = a_1 r^2 = \frac{5}{8}\left(\frac{2}{3}\right)^2 = \frac{5}{8}\left(\frac{4}{9}\right) = \frac{20}{72} = \frac{5}{18}$

$a_4 = a_1 r^3 = \frac{5}{8}\left(\frac{2}{3}\right)^3 = \frac{5}{8}\left(\frac{8}{27}\right) = \frac{40}{216} = \frac{5}{27}$

The terms are $\frac{5}{8}, \frac{5}{12}, \frac{5}{18}, \frac{5}{27}$.

13. $a_8 = a_1 + (8-1)d$

$\quad = 100 + (7)(-12)$

$\quad = 100 - 84$

$\quad = 16$

14. $s_8 = \frac{8(a_1 + a_8)}{2}$

$\quad = \frac{8[3 + (-11)]}{2}$

$\quad = \frac{8(-8)}{2}$

$\quad = -32$

15. $d = -16 - (-4) = -16 + 4 = -12$

$a_n = a_1 + (n-1)d$

$-148 = -4 + (n-1)(-12)$

$-148 = -4 - 12n + 12$

$-148 = 8 - 12n$

$-156 = -12n$

$\frac{-156}{-12} = n$

$13 = n$

16. $a_7 = a_1 r^6 = 8\left(\frac{2}{3}\right)^6 = 8\left(\frac{64}{729}\right) = \frac{512}{729}$

17. $s_7 = \frac{a_1(1 - r^7)}{1 - r}$

$\quad = \frac{\frac{3}{5}\left[1 - (-5)^7\right]}{1 - (-5)}$

$\quad = \frac{\frac{3}{5}\left[1 - (-78,125)\right]}{1 - (-5)}$

$\quad = \frac{\frac{3}{5}(1 + 78,125)}{1 + 5}$

$\quad = \frac{\frac{3}{5}(78,126)}{6}$

$\quad = \frac{3(78,126)}{5 \cdot 6}$

$\quad = \frac{78,126}{5 \cdot 2}$

$\quad = \frac{39,063}{5}$

18. $r = 6 \div 12 = \frac{6}{12} = \frac{1}{2}$

$a_n = a_1 r^{n-1} = 12\left(\frac{1}{2}\right)^{n-1}$

19. $r = \dfrac{2}{3}$

$$s_\infty = \dfrac{3}{1 - \frac{2}{3}} = \dfrac{3}{\frac{1}{3}} = 3 \cdot \dfrac{3}{1} = 9$$

20. $0.6262 \cdots = 0.62 + 0.0062 + 0.000062 + \cdots$

$\qquad\qquad = 0.62 + 0.62(0.01) + 0.62(0.01)^2 + \cdots$

$\quad r = 0.01$ and $a_1 = 0.62$

$$s_\infty = \dfrac{0.62}{1 - 0.01} = \dfrac{0.62}{0.99} = \dfrac{62}{99}$$

21. $\dbinom{7}{2} = \dfrac{7!}{2!(7-2)!}$

$\qquad = \dfrac{7!}{2!\,5!}$

$\qquad = \dfrac{7 \cdot 6 \cdot 5 \cdot 4 \cdot 3 \cdot 2 \cdot 1}{2 \cdot 1 \cdot 5 \cdot 4 \cdot 3 \cdot 2 \cdot 1}$

$\qquad = \dfrac{42}{2}$

$\qquad = 21$

22. $(x + 2y)^4 = \dbinom{4}{0}(x)^4(2y)^0 + \dbinom{4}{1}(x)^3(2y)^1 + \dbinom{4}{2}(x)^2(2y)^2 + \dbinom{4}{3}(x)^1(2y)^3 + \dbinom{4}{4}(x)^0(2y)^4$

$\qquad\qquad\quad = 1(x^4)(1) + 4(x^3)(2y) + 6(x^2)(4y^2) + 4(x)(8y^3) + 1(1)(16y^4)$

$\qquad\qquad\quad = x^4 + 8x^3y + 24x^2y^2 + 32xy^3 + 16y^4$

23. $\bar{x} = \dfrac{\sum x}{n} = \dfrac{74 + 93 + 83 + 87 + 68}{5} = \dfrac{405}{5} = 81$

24. $a_1 = 1000$, $n = 20$

$\quad a_n = a_1 + (n-1)d$

$\quad a_{20} = 1000 + (20 - 1)(1000) = 20,000$

$\quad s_{20} = \dfrac{20(1000 + 20,000)}{2} = 210,000$

After 20 years, she will have saved \$210,000.

25. $r = 3$, $a_1 = 500(3) = 1500$

$\quad a_8 = a_1 r^7$

$\qquad = 1500(3)^7$

$\qquad = 1500(2187)$

$\qquad = 3,280,500$

Cumulative Review Test

1. $\dfrac{1}{2}x + \dfrac{1}{3}(x - 2) = \dfrac{3}{4}(x - 5)$

$\quad 12\left(\dfrac{1}{2}x\right) + 12\left[\dfrac{1}{3}(x - 2)\right] = 12\left[\dfrac{3}{4}(x - 5)\right]$

$\qquad\qquad 6x + 4(x - 2) = 9(x - 5)$

$\qquad\qquad 6x + 4x - 8 = 9x - 45$

$\qquad\qquad\quad 10x - 8 = 9x - 45$

$\qquad\qquad\qquad 10x = 9x - 37$

$\qquad\qquad\qquad\quad x = -37$

2. $|2x-3|-4>10$

$|2x-3|>14$

$2x-3<-14 \quad \text{or} \quad 2x-3>14$

$\quad 2x<-11 \qquad\qquad 2x>17$

$\quad\quad x<-\dfrac{11}{2} \qquad\qquad x>\dfrac{17}{2}$

The solution is $x<-\dfrac{11}{2}$ or $x>\dfrac{17}{2}$.

3. $3x-5y>10$

Graph the line $3x-5y=10$ using a dashed line. Use (0, 0) for the checkpoint.

$3x-5y>10$

$3(0)-5(0)>10$

$\qquad 0>10 \quad$ False

Since this is a false statement, shade the region which does not contain the point (0, 0).

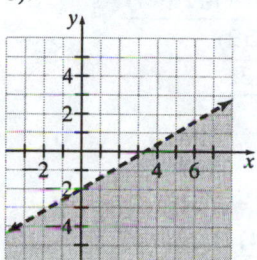

4. $5x-2y=8$

$x-y=4$

Solve the second equation for x.

$x-y=4$

$x=y+4$

Substitute $y+4$ for x in the first equation.

$\qquad 5x-2y=8$

$5(y+4)-2y=8$

$5y+20-2y=8$

$\qquad 3y+20=8$

$\qquad\quad 3y=-12$

$\qquad\quad y=-\dfrac{12}{3}=-4$

Now substitute -4 for y in the equation $x=y+4$.

$x=y+4$

$x=-4+4=0$

The solution is (0, –4).

5. $A=\dfrac{pt}{p+t}$

$A(p+t)=pt$

$Ap+At=pt$

$\qquad At=pt-Ap$

$\qquad At=p(t-A)$

$\qquad \dfrac{At}{t-A}=\dfrac{p(t-A)}{t-A}$

$\qquad \dfrac{At}{t-A}=p \quad \text{or} \quad p=\dfrac{At}{t-A}$

6. $\dfrac{x}{x^2+x-12}+\dfrac{x+2}{3x^2+16x+16}$

$=\dfrac{x}{(x+4)(x-3)}+\dfrac{x+2}{(x+4)(3x+4)}$

$=\dfrac{x}{(x+4)(x-3)}\cdot\dfrac{(3x+4)}{(3x+4)}+\dfrac{x+2}{(x+4)(3x+4)}\cdot\dfrac{(x-3)}{(x-3)}$

$=\dfrac{x(3x+4)}{(x+4)(x-3)(3x+4)}+\dfrac{(x+2)(x-3)}{(x+4)(x-3)(3x+4)}$

$=\dfrac{3x^2+4x}{(x+4)(x-3)(3x+4)}+\dfrac{x^2-x-6}{(x+4)(x-3)(3x+4)}$

$=\dfrac{4x^2+3x-6}{(x+4)(x-3)(3x+4)}$

7. $\dfrac{\sqrt[3]{24x^6y^3}}{\sqrt[3]{2x^2y^5}} = \sqrt[3]{\dfrac{24x^6y^3}{2x^2y^5}}$

$= \sqrt[3]{\dfrac{12x^4}{y^2}}$

$= \dfrac{\sqrt[3]{12x^4}}{\sqrt[3]{y^2}}$

$= \dfrac{\sqrt[3]{x^3}\sqrt[3]{12x}}{\sqrt[3]{y^2}}$

$= \dfrac{x\sqrt[3]{12x}}{\sqrt[3]{y^2}} \cdot \dfrac{\sqrt[3]{y}}{\sqrt[3]{y}}$

$= \dfrac{x\sqrt[3]{12xy}}{y}$

8. $\sqrt{28} - 3\sqrt{7} + \sqrt{63} = \sqrt{4 \cdot 7} - 3\sqrt{7} + \sqrt{9 \cdot 7}$

$= \sqrt{4}\sqrt{7} - 3\sqrt{7} + \sqrt{9}\sqrt{7}$

$= 2\sqrt{7} - 3\sqrt{7} + 3\sqrt{7}$

$= 2\sqrt{7}$

9. $\sqrt{5x+1} - \sqrt{2x-2} = 2$

$\sqrt{5x+1} = 2 + \sqrt{2x-2}$

$\left(\sqrt{5x+1}\right)^2 = \left(2 + \sqrt{2x-2}\right)^2$

$5x+1 = 4 + 4\sqrt{2x-2} + 2x - 2$

$5x+1 = 2x + 2 + 4\sqrt{2x-2}$

$3x-1 = 4\sqrt{2x-2}$

$(3x-1)^2 = \left(4\sqrt{2x-2}\right)^2$

$9x^2 - 6x + 1 = 16(2x-2)$

$9x^2 - 6x + 1 = 32x - 32$

$9x^2 - 38x + 33 = 0$

$(x-3)(9x-11) = 0$

$x - 3 = 0$ or $9x - 11 = 0$

$\qquad x = 3 \qquad\qquad x = \dfrac{11}{9}$

The solutions are 3 and $\dfrac{11}{9}$.

10. $\dfrac{5-2i}{3+4i} = \dfrac{5-2i}{3+4i} \cdot \dfrac{3-4i}{3-4i}$

$= \dfrac{15 - 20i - 6i + 8i^2}{(3)^2 - (4i)^2}$

$= \dfrac{15 - 26i - 8}{9 - (-16)}$

$= \dfrac{7 - 26i}{25}$

11. $-4x^2 - 2x + 8 = 0$

$\dfrac{-4x^2}{-2} - \dfrac{2x}{-2} + \dfrac{8}{-2} = \dfrac{0}{-2}$

$2x^2 + x - 4 = 0,\, a = 2,\, b = 1,\, c = -4$

$x = \dfrac{-b \pm \sqrt{b^2 - 4ac}}{2a}$

$= \dfrac{-1 \pm \sqrt{(1)^2 - 4(2)(-4)}}{2(2)}$

$= \dfrac{-1 \pm \sqrt{1 + 32}}{4}$

$= \dfrac{-1 \pm \sqrt{33}}{4}$

12. $\dfrac{(2x-3)(x-4)}{x+1} < 0$

$2x - 3 = 0$ for $x = \dfrac{3}{2}$, $x - 4 = 0$ for $x = 4$,

and $x + 1 = 0$ for $x = -1$.

Use number lines to determine the intervals

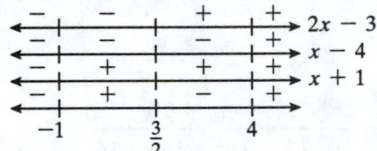

True   False   True   False

The solution is $x < -1$ or $\dfrac{3}{2} < x < 4$.

13. $4x^2 + 4y^2 = 36$

$\dfrac{4x^2}{4} + \dfrac{4y^2}{4} = \dfrac{36}{4}$

$x^2 + y^2 = 9$

The graph is a circle with a center at $(0, 0)$ and radius 3.

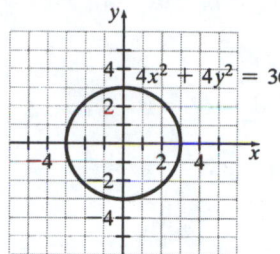

14. a.
$$y = x^2 + 2x - 3$$
$$y = x^2 + 2x + 1 - 1 - 3$$
$$y = x^2 + 2x + 1 - 4$$
$$y = (x+1)^2 - 4$$

b. The graph is a parabola with vertex at (−1, −4).

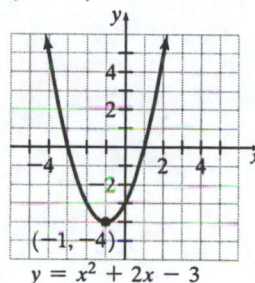

$$y = x^2 + 2x - 3$$

15.

16. $\log(4x - 1) + \log 3 = \log(8x + 13)$
$$\log 3(4x - 1) = \log(8x + 13)$$
$$3(4x - 1) = 8x + 13$$
$$12x - 3 = 8x + 13$$
$$4x - 3 = 13$$
$$4x = 16$$
$$x = 4$$

17. $a_1 = 8$ and $a_6 = 28$
$$a_6 = a_1 + (6-1)d$$
$$28 = 8 + (6-1)d$$
$$28 = 8 + 5d$$
$$20 = 5d$$
$$4 = d$$

18. $a_1 = 5$ and $r = 3$
$$s_3 = \frac{a_1(1 - r^3)}{1 - r}$$
$$s_3 = \frac{5(1 - 3^3)}{1 - 3}$$
$$= \frac{5(1 - 27)}{-2}$$
$$= \frac{5(-26)}{-2}$$
$$= 5(13)$$
$$= 65$$

19. Let d be the distance between the two towns.

	d	r	$t = \frac{d}{r}$
Car	d	40	$\frac{d}{40}$
Train	d	60	$\frac{d}{60}$

The difference in time is 2 hours.
$$\frac{d}{40} - \frac{d}{60} = 2$$
$$120\left(\frac{d}{40}\right) - 120\left(\frac{d}{60}\right) = 120(2)$$
$$3d - 2d = 240$$
$$d = 240$$
The distance is 240 miles.

20. Let x be the width. Then $x + 20$ is the length.
$$x(x + 20) = 300$$
$$x^2 + 20x = 300$$
$$x^2 + 20x - 300 = 0$$
$$(x - 10)(x + 30) = 0$$
$$x - 10 = 0 \quad \text{or} \quad x + 30 = 0$$
$$x = 10 \qquad x = -30$$
Reject $x = -30$, since width cannot be negative. Thus, the width is 10 feet and the length is 10 + 20 = 30 feet.